ANNUAL REVIEW OF PHYTOPATHOLOGY

ANNUAL REVIEW OF PHYTOPATHOLOGY

VOLUME 33, 1995

ROBERT K. WEBSTER, *Editor*
The University of California

GEORGE A. ZENTMYER, *Associate Editor*
The University of California

GREGORY SHANER, *Associate Editor*
Purdue University

ANNUAL REVIEWS INC. 4139 EL CAMINO WAY P.O. BOX 10139 PALO ALTO, CALIFORNIA 94303-0897

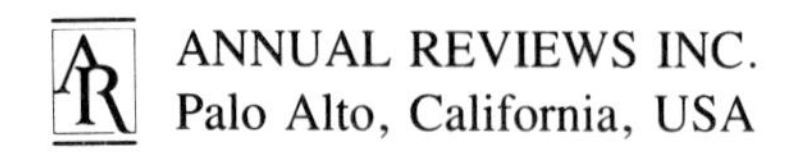

International Standard Serial Number: 0066–4286
International Standard Book Number: 0–8243–1333-x
Library of Congress Catalog Card Number: 63-8847

♾ The paper used in this publication meets the minimum requirements of American National Standard for Information Sciences—Permanence of Paper for Printed Library Materials, ANSI Z39.48-1984.

Typesetting by Kachina Typesetting Inc., Tempe, Arizona; John Olson, President; Marty Mullins, Typesetting Coordinator; and by the Annual Reviews Inc. Editorial Staff

PRINTED AND BOUND IN THE UNITED STATES OF AMERICA

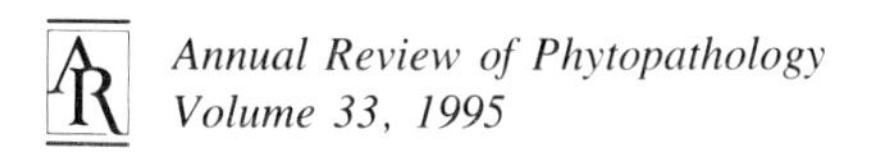

CONTENTS

SOME RELATED ARTICLES IN OTHER ANNUAL REVIEWS

From the *Annual Review of Entomology*, Volume 40, 1995:

The Sweetpotato or Silverleaf Whiteflies: Biotypes of Bemisia tabaci *or a Species Complex?* JK Brown, DR Frohlich, and RC Rosell

From the *Annual Review of Genetics*, Volume 29, 1995:

Genetics of Ustilago maydis, *A Fungal Pathogen that Induces Tumors in Maize,* F Banuett

The Plant Response in Pathogenesis, Symbiosis, and Wounding: Variations on a Common Theme, PC Zambryski and C Baron

From the *Annual Review of Microbiology*, Volume 49, 1995:

New Mechanisms of Drug Resistance in Parasitic Protozoa, P Borst and M Ouellette

From the *Annual Review of Plant Physiology and Plant Molecular Biology*, Volume 46, 1995:

Cellular Mechanisms of Aluminum Toxicity and Resistance in Plants, LV Kochian

Regulation of Metabolism in Transgenic Plants, M Stitt and U Sonnewald

Heterologous Expression of Genes in Bacterial, Fungal, Animal, and Plant Cells, WB Frommer and O Ninnemann

ANNUAL REVIEWS INC. is a nonprofit scientific publisher established to promote the advancement of the sciences. Beginning in 1932 with the *Annual Review of Biochemistry,* the Company has pursued as its principal function the publication of high-quality, reasonably priced *Annual Review* volumes. The volumes are organized by Editors and Editorial Committees who invite qualified authors to contribute critical articles reviewing significant developments within each major discipline. The Editor-in-Chief invites those interested in serving as future Editorial Committee members to communicate directly with him. Annual Reviews Inc. is administered by a Board of Directors, whose members serve without compensation.

ANNUAL REVIEWS OF
Anthropology
Astronomy and Astrophysics
Biochemistry
Biophysics and Biomolecular Structure
Cell Biology
Computer Science
Earth and Planetary Sciences
Ecology and Systematics
Energy and the Environment
Entomology
Fluid Mechanics
Genetics
Immunology
Materials Science
Medicine
Microbiology
Neuroscience
Nuclear and Particle Science
Nutrition
Pharmacology and Toxicology
Physical Chemistry
Physiology
Phytopathology
Plant Physiology and Plant Molecular Biology
Psychology
Public Health
Sociology

SPECIAL PUBLICATIONS

Excitement and Fascination of Science, Vols. 1, 2, and 3

Intelligence and Affectivity, by Jean Piaget

For the convenience of readers, a detachable order form/envelope is bound into the back of this volume.

Arthur Kelman

Annu. Rev. Phytopathol. 1995. 33:1–21

CONTRIBUTIONS OF PLANT PATHOLOGY TO THE BIOLOGICAL SCIENCES AND INDUSTRY[1]

Arthur Kelman
Department of Plant Pathology, North Carolina State University, Raleigh, North Carolina 27695

KEY WORDS: viruses as plant pathogens, spiroplasma, ice-nucleation positive bacteria, effect of plant diseases on breeding strategies, Wisconsin fast plants

ABSTRACT

Research in plant pathology has made major contributions to knowledge of basic biology and genetics of plants and microorganisms as well as to development of new products for industry. Among examples cited are first evidence of the nature of viruses as agents of disease; development of density gradient ultracentrifugation, a powerful tool for research in virology and cell biology; studies on the unique properties of xanthan gum, the extracellular polysaccharide of a plant pathogenic bacterium; development of a method for introduction of beneficial genes into plants via a tumorigenic bacterium minus its tumor-inducing capability; and discovery that epiphytic ice-nucleating bacteria can trigger frost damage in plants.

INTRODUCTION

This is a period when the dark clouds of decreased funding create pressures for reassessment of priorities in both basic and mission-oriented research in the agricultural sciences and land grant universities (22). Members of Con-

[1]Adapted from Chapter 4, "Contributions of Plant Pathology to the Biological Sciences," in *Historical Perspectives in Plant Science,* ed. Kenneth J Fry. 1994. Ames, IA: Iowa State Univ. Press

0066-4286/95/0901-0001$05.00

gress, growers, and representatives of commodity groups as well as public interest groups with special agendas in the area of sustainable agriculture are all questioning the benefits of current research programs and are seeking rapid solutions, not long-term basic studies. Numerous regional and national conferences and workshops have been convened to discuss whether current research objectives are designed to meet the long-term needs of our society. Our professional society (APS) is also seeking to establish priorities for future research in plant pathology and to assess the advances in increasing crop productivity and effective control of plant diseases (14, 40, 59). The specific contributions of plant pathology have also been well documented in recent symposia and reports. Emphasis has been placed on the potential for future progress in effective control of disease that will contribute not only to increased productivity, but also to sustainable agricultural systems and an increased concern for environmental quality (14, 15, 40, 59). However, the specific advances in related basic and applied biological and agricultural sciences as well as certain industries that have received their impetus from studies of plant diseases have not been enumerated or described in detail. Rarely has in-depth consideration been given to the positive contributions of plant pathology to other fields of science and to some industries that have been established directly or indirectly from the studies on plant diseases and their causal agents.

Furthermore, when we introduce students to the study of plant pathology, we usually emphasize how dependent our science is on mathematics, the physical sciences, the basic biological sciences, and to a lesser degree on other agricultural sciences (1). The manner in which these areas impinge on the science and art of plant pathology is reasonably well recognized and documented. The general historical development of the field of plant pathology has been thoroughly reviewed in a large number of texts and articles (37, 43, 61, 68) and in numerous articles in the *Annual Review of Phytopathology*. Thus, this article aims to examine the significant contributions to other disciplines and, to a limited extent, to new industries that have developed from studies on mechanisms of pathogenesis and biology of plant pathogens. Selected examples are cited to indicate how studies on the stresses that diseases impose on plants have provided new insights on plant growth and development. It is proposed that these findings may not have been discovered in investigations solely concerned with normal processes in plants. In evaluating the benefits arising from research in plant pathology, it is important to consider not only those areas viewed to be the normal responsibility of our science, but also to expand our perspectives and boundaries. Thus, the examples selected will serve to illustrate how plant pathology serves as a key contributor to basic understanding of biological systems and the nature of disease in all organisms, and how it contributes to other fields in agriculture and industry in ways not fully recognized.

THE GERM THEORY OF DISEASE

The earliest written records provide vivid evidence of the devastating impact of plant diseases on mankind (11, 43, 61, 68). In the absence of either any explanation for destructive epidemics of plant diseases or adequate means of developing control measures, it was only natural that these outbreaks would be attributed to supernatural forces. In numerous biblical references, outbreaks of crop diseases such as blight, mildews, and rusts were interpreted as manifestations of the wrath of a supreme being and as a punishment for immoral behavior. The Romans offered sacrifices to Robigus, a god who had to be propitiated to prevent the rust that periodically ravaged the wheat fields of Rome and its territories. In the Middle Ages the dread effects of eating bread made from rye infected by the ergot fungus came to be called the Holy Fire and then Saint Anthony's Fire. For several centuries thousands of people in central and western Europe died because of this plant disease. They suffered extreme agony from the hallucinogenic effects and gangrene that resulted from restriction of movement of blood to extremities after ingestion of the various alkaloids present in bread made with ergotized grain (11, 61).

Another pervasive belief was that disease in plants and animals resulted from the machinations of an evil spirit. Thus, in King Lear Shakespeare mentioned the foul fiend, Flibbertigibbet, who "mildewed the white wheat" among other evil deeds (61). As more knowledge was obtained, the concept evolved that diseases resulted mainly from adverse environmental factors; numerous observations indicated close correlations between disease incidence and specific extremes of temperature and moisture as well as adverse soil conditions.

All students of microbiology and general biology have learned about the classic contributions of Robert Koch and Louis Pasteur in establishing the germ theory of disease (9). However, many decades before these discoveries, other scientists interested in the nature of plant diseases had clearly demonstrated that specific microorganisms caused disease in plants. Thus, early investigators are truly the unrecognized heroes of microbiology and medicine. The leaders in science and medicine early in the last century did not accept the fact that the microorganisms present in the diseased tissue of plants and animals actually could be the causal agents. Finally, acceptance of these revolutionary new ideas brought within the grasp of humans the potential ability to control disease in man, animals, and plants and to enhance the quality of life for mankind.

Many scientists contributed to the establishment of the concept that disease in plants could be caused by parasitic microorganisms (37, 61, 68). The first critical experimental evidence that fungi can cause disease in plants was presented in 1803 by IB Prevost in his study of covered smut of wheat, and after a lapse of half a century by HA de Bary. In 1853, at the age of 22, he published a comprehensive paper on the rusts and smuts as causes of disease in plants; this paper has become a classic in the field of plant pathology (33). Note that this publication was issued about two decades before Koch published his classical studies on the anthrax disease of cattle. De Bary was not afraid to take a stand; he characterized as "inexact and based upon illusion" the reports of the two leading distinguished proponents of the concept that fungi in diseased plants arose by spontaneous generation. With the publication of his initial paper and subsequent books and journal articles, de Bary became recognized as a world leader in mycology, and students from many countries came to work with him. Many of his 68 students established centers for research in the new field of plant pathology in the United States and other countries.

One of de Bary's major contributions resulted from his investigations on late blight disease of potato and presentation of evidence that the cause was a fungus. Few other diseases have had as devastating an effect as the major epidemics of this disease in Ireland, western Europe, and the USA in 1845 and in years thereafter (11, 43). It was a source of great concern both politically and scientifically that a plant disease of this magnitude could not be controlled and that no one really knew what the cause was although many theories were proposed. The initial effects in Ireland were death by starvation and disease of over 1 million people and the mass migration to Canada and the United States of another 1.5 million refugees. This was unequivocal evidence of the helplessness of mankind in coping with a disease affecting a major food crop, and it served as a major incentive for research on this and related diseases (11, 43, 61).

In a survey of available major texts on the history of biological sciences, particularly microbiology (9), it is disconcerting that the real pioneers in the establishment of the germ theory of disease are rarely even mentioned. However, it is difficult to believe that the leaders in the effort to establish the germ theory of disease in man were totally unaware of the publications on the nature of disease in plants. In addition, these early studies probably aided in creating the intellectual climate in which the landmark studies of Robert Koch and Louis Pasteur would be acceptable.

Paradoxically, many scientists in Germany, the very country in which the germ theory of disease for plants(fungi), animals, and humans (bacteria) was given solid grounding, were unwilling to accept the evidence that bacteria could cause disease in plants. From this circumstance there arose a very famous controversy (1897–1901) between the leading researcher on bacterial disease of plants in the United States, Erwin F Smith, and a senior German scientist, Alfred Fischer (10). The publicity given to this acrimonious debate undoubtedly hastened acceptance of the evidence that plant disease could be caused by bacteria as well as fungi. It undoubtedly also gave impetus to the developing field of microbiology as a science that would lead to control of the diseases of mankind.

VIRUSES AS AGENTS OF DISEASE

The demonstration that the tobacco mosaic disease was caused by a virus probably had greater impact on other sciences, particularly molecular biology and human medicine, than any other finding in studies on plant diseases (36, 53). In 1882, Adolf Mayer completed the first experiments that conclusively demonstrated that a disease of tobacco, which he named mosaic (TMV), was caused by a transmissible biologic agent. The next key experiments were completed by Dimitri Iwanowski in 1892. In addition to confirming Mayer's work, he completed a critical experiment in which he demonstrated that the causal agent could pass through a filter that prevented the passage of bacteria. Six years later it was Martinus Beijerinck who realized the remarkable nature of TMV and was perhaps the first to suggest that it should be called a "virus." There has been some controversy over which of these early investigators should be designated the first to recognize the novel nature of the causal agent of the tobacco mosaic disease. Recently, Bos (5) proposed that the credit should go to Mayer rather than Iwanowski. Lustig & Levine (50) selected 1892, the date of publication of Iwanowski's paper on TMV, as the starting point for 100 years of virology, but Bos provided valid reasons for setting this date ten years earlier, the publication in 1882 of Mayer's research. This was based on the fact that, two decades after Mayer's work was published, Iwanowski still insisted that TMV was caused by a microorganism small enough (in some resting or spore stage) to pass through filters that prevented the passage of typical bacterial cells.

Three decades were to pass, however, before the true nature of the plant viruses was recognized (5, 50, 53). In 1935, Wendell Stanley, a biochemist working with a group of plant pathologists, reported in *Science* that he could obtain protein crystals from juice from infected plants and that he could reproduce symptoms of TMV with these preparations. In Judson's (36) fascinating text, *The Eight Days of Creation*, he presents a historical account of the seminal discoveries of molecular biology. He evaluated Stanley's contribution as "the most portentous and publicized biological discovery of the decade." The idea that a living, self-reproducing entity could be a crystallizable protein captured the imagination of scientists and laymen alike.

Major advances in research often follow and build on the discovery of new simple techniques. Thus, Stanley's work as well as subsequent studies by many other virologists were greatly dependent on the finding of Francis Holmes, a plant pathologist, that TMV caused hypersensitive reactions, "local lesions" in certain resistant plants. This finding made it possible in each of the purification steps or treatment procedures to determine the concentration of TMV in a given preparation. Although Stanley received great acclaim for his discovery, including the Nobel Prize in chemistry, there was a major flaw in his conclusion that the crystal of TMV was a large protein. Two years after Stanley's paper was published, Bawden & Pirie, in the plant pathology group at the Rothamsted Agricultural Experiment Station in England, published the evidence that TMV was a nucleoprotein (3). Harrison (27) notes with respect to this important paper by Bawden & Pirie that "not only were their observations and conclusions correct, but, more remarkably, the accuracy of their quantitative estimates have scarcely been bettered by more modern techniques after more than fifty years." The ease with which large quantities of TMV could be obtained, its stability, high degree of infectiousness, and related properties made it the model of choice for basic studies in virology. These initial studies of a plant virus opened the door for research in the field of general virology and virus diseases of humans and animals.

The true significance of the presence of RNA as a component of viruses was demonstrated in the elegant experiments reported by Frankel-Conrat & Williams in 1955 and Gierer & Schramm in 1956. They demonstrated independently that the naked RNA of TMV, free of its protein coat, was the infectious agent that entered the cell and initiated replication of new virus. The protein served as a protective shell around the RNA (53).

When James Watson decided to learn about the techniques of crystallography in his work with Francis Crick at the University of Cambridge, he selected TMV as the model structure to be examined initially in X-ray diffraction studies. At the end of about six months on the project, he obtained some first-rate results and observed the helical arrangement of the viral subunits. Thus, his own early observations on a plant virus may have enhanced progress in developing the concept of the helical structure of DNA (36).

Density Gradient Ultracentrifugation

Among other techniques that emerged from studies on plant viruses and that contributed to advancing the field of general virology was the procedure developed by Myron Brakke while he worked with Lindsay Black at the Brooklyn Botanic Garden (6–8). This technique, density gradient centrifugation, was described by Matthews (53) as "one of the most influential developments in virology and molecular biology." Although the method had a high potential for wide application, it did not become widely used until almost a decade after the paper describing the technique was published in 1951. Brakke (7) attributed this delay to the fact that few biochemists read the literature on plant viruses and that initial studies on fractionation of subcellular components were still in their infancy. However, after the value of the technique was recognized by several prominent animal virologists, it became a standard procedure in hundreds of research laboratories (7). This is evident from the numerous citations in the literature that appeared after 1960. In 1970, Beckman Instruments published a bibliography covering the period from 1960–1970 of all the papers in which this technique was used; 100 pages of citations were listed.

Although few other contributions in plant virology compare in importance with these early contributions, studies on plant viruses continue to contribute to the advancement of the fields of general virology and molecular biology (8). Recent studies on the movement of viruses in plant tissues have revealed a remarkable interaction between viruses and host cells that results in modifications in how plasmodesmata function (49). These studies are another example of how continuing research on plant viruses expands our knowledge of normal structure and function in plants.

MYCOPLASMA-LIKE ORGANISMS (MLOs) AND SPIROPLASMAS AS DISEASE AGENTS

For many decades plant pathologists were frustrated in their studies on a large group of "yellows" diseases of plants because they were unable to characterize the nature and structure of viruses presumed to be the causal agents (51, 71). The mystery associated with these diseases was in part resolved when Doi and associates, working in Japan, recognized that wall-less prokaryotes were present in the phloem of plants considered to be affected by yellows viruses (19, 51). Doi's discovery was in part accidental; he was working in an electron microscope facility that was also being used by Kaoru Koshimuzu, a veterinarian. Koshimuzu saw Doi's electron photomicrographs and noted a remarkable similarity of the MLOs in these preparations to the mycoplasma in cells of diseased birds that he was studying. Although it was shown that these organisms in plants were highly sensitive to certain antibiotics, they could not be cultured and to this day, defy the best efforts of all who have tried to do so (46). Thus they continue to be described as mycoplasma-like organisms (MLOs). Although over 300 diseases are now considered to be caused by MLOs, some investigators think even after three decades of study that only a small percentage of the total number of these organisms present in nature has been isolated, identified, and described.

Closely linked to the studies on MLOs was the discovery of a previously undescribed group of organisms, the spiroplasmas, by Robert Davis (16). Davis was attempting to determine the nature of the causal agent of the corn stunt disease and sought other means than the electron microscope to examine the structure of the causal agent. When phase contrast and dark field microscopy were used, he found, to his surprise, that crude extracts from corn stunt plants contained the remarkable tiny spiral-shaped motile cells that he named spiroplasmas (16). Subsequent examination of cultures of the pathogen of citrus stubborn disease revealed that this organism was also a spiroplasma. Since that time a large number of spiroplasmas have been identified; these include pathogens of insects and animals and many different saprophytic forms that appear to be widely distributed on plants. Thus, a new area of scientific specialization has evolved as the study of these strange prokaryotes has been expanding in recent years.

DISCOVERY OF OTHER PREVIOUSLY UNDESCRIBED ORGANISMS OR AGENTS OF DISEASE

In addition to the recent developments in studies on mycoplasma-like organisms, other organisms and pathogens new to science have been discovered in connection with investigations of several diseases in which specific causal agents had not been identified. A full discussion of the research that has followed recognition of these pathogens is beyond the scope of this paper; however, a few examples can be cited

The potato spindle tuber disease is one example of a disease that was presumed to be caused by a virus, but no specific virus had been found in infected plants. The causal agent, a viroid, was discovered by Diener (17, 18). Viroids are unique in that they are the smallest of all known agents of disease; they lack the protein coat that characterizes plant viruses and exist as tiny naked single-stranded RNA molecules. As yet, only a few other diseases of plants have been found to be caused by viroids, and their potential importance as agents of disease in man remains to be determined. However, the discovery of viroids has opened a new area of research on mechanisms of disease induction since viroids do not behave in plant cells in the same manner as viruses and apparently are able to induce changes in metabolism of host cells by interference with gene regulation.

For many years a serious disease of grapevines known as Pierce's disease was thought to be caused by a virus, but, as was the case with potato stunt disease, the specific causal agent could not be isolated or characterized. The causal agent is a fastidious xylem-limited prokaryote now classified in the genus *Xylella* (31). Many different woody plants are susceptible to strains of this pathogen, including some forest trees that show symptoms of decline. Diseases of this type will provide a fertile new area for future study on mechanisms of wilt induction and the relationship of stress factors to disease development in woody plants.

In efforts to explain the appearance of plaques in cultures of a plant pathogenic pseudomonad in the absence of a phage, Stolp (63) discovered a hitherto unknown and unusual bacterium, which he named *Bdellovibrio bacteriovorus*.

The parasitic strains of these remarkable gram-negative bacteria have the unique ability to attach themselves to other bacteria, to bore through their cell walls, multiply, and cause the attacked cells to lyse and thereby release more parasite cells. Here, too, the discovery of previously unreported organisms has indirectly resulted from studies of plant pathogenic organisms. The roles of this pathogen of bacteria in the microbial ecology of soils and as a potential biological control agent are still to be explored fully.

MODIFICATION OF GENERAL MANAGEMENT PRACTICES

Modifications in management practices of both crops and forest trees (11a) have received their impetus from studies on the biology of a number of plant pathogens. Numerous examples can be cited in which standard crop management practices in horticulture and agronomy are governed primarily by an understanding of factors influencing disease development in a given crop. These factors include the sequence in which crops are rotated, methods for weed control, regulation of environmental factors in particular irrigation practices, and control of insects as well as postharvest handling and storage practices. Most of these approaches have been well documented (1, 68).

Under forest conditions direct intervention for control of specific diseases may be very difficult (11a, 29). Thus, application of standard practices that may be effective in the absence of disease or pathogens can result in disaster if imposed on situations where aggressive pathogens may be present. This is well illustrated by the shift in the prevalence and importance of dwarf mistletoe that occurred in certain stands of ponderosa pine in the southwestern USA because the biology of this tree pathogen was not well understood (28). In national forests under the management of the Forest Service, a specific number of seed trees had to be left to insure a more uniform and rapid regeneration of stands after cutting operations. The requirement for seed trees, ostensibly a sound practice in other areas, had a very unfortunate impact on the new stands on thousands of acres. Scattered in the virgin forests of the region was a parasitic seed plant, the dwarf mistletoe (*Arceuthobium*). This parasite has a destructive impact on parasitized trees and results in witches brooms and severe stunting (62). The female plant produces seed that are discharged at a high velocity and can be projected for distances of 30–50 feet from parasitized trees. In many large areas, seed trees that were left were often of no commercial value because of heavy dwarf mistletoe infestations. As a result, the new stands became so severely affected that it will be extremely difficult to salvage them. In sharp contrast, the new stands after clear-cutting by commercial companies in which no seed trees were left were usually free of mistletoe infestations.

Detailed studies on the biology of the mistletoe have provided the background needed for the development of a new management program to minimize mistletoe infections in the new stands. This experience also emphasized the importance of selecting high-quality pathogen-free seed trees for the regeneration of forest stands and the importance that knowledge of forest pathogens has in establishing acceptable forestry management practices in stands in which diseases are endemic (11a, 29).

PLANT PATHOGENS AS MYCOHERBICIDES

Discovery of a biological control of northern jointvetch (*Aeschynomene virginica* [L.] B.S.P.), a weed in rice fields, by a strain of *Colletotrichum gloeosporiodes* (66) evolved from studies to determine the causal agent of a destructive disease of this weed. The damage from the pathogen was so severe that adequate data in the field experiment to evaluate impact of the weed on rice plantings could not be obtained. Basic studies on the biology of the pathogen indicated that the mechanism for dissemination of spores was not effective. If an effective means of obtaining uniform infection could be developed the possibility existed for use of the fungus as a mycoherbicide. The initial research findings were promising, and a collaborative effort was initiated at the University of Arkansas headed by George E Templeton to develop effective means of inoculation of weed plants under field conditions (65, 66). Subsequently, a commercial product, *Collego®*, was developed that is now used effectively to control this weed. A second product, *DeVine*, has been used effectively in Florida to control stranglervine (*Morrenia odorata* Lindl.) in citrus groves. In this instance, *Phytophthora palmivora* was the pathogen found to be highly effective for use as a mycoherbicide. A number of research and development programs are now exploring other pathogens with weed-host specificity (65).

NOVEL CHEMICAL COMPOUNDS PRODUCED BY PLANT PATHOGENS

Gibberellins

Research on the alteration in physiological processes in diseased plants and the chemicals involved has resulted in the identification and characterization of some very unusual chemicals, many with remarkable physiological effects at incredibly low concentrations. Among these compounds are the gibberellins, which were discovered by a Japanese scientist, E Kurosawa (42), in a study on the disease of rice known as the bakanae or "foolish seedling" disease caused by *Gibberella fujikori* (24, 54). Seedlings affected by the bakanae disease grow faster and much taller than healthy plants. Kurosawa's paper was overlooked until after World War II. Subsequent research resulted in the discovery that gibberellins are formed normally in higher plants as well as by a number of microorganisms (54). At present more than 65 gibberellins have been characterized and an entire field of research developed that is providing new insights into basic aspects of plant growth and development (24). A literature survey for the period from January 1, 1992, to December 30, 1994, revealed a listing of 285 publications on the gibberellins and related compounds. Practical applications involve the use of gibberellic acid and related gibberellins to promote growth, flowering, and germination of seed.

Xanthan Gums

Investigators who studied the complex extracellular polysaccharide of *Xanthomonas campestris,* causal agent of black rot of cabbage and diseases of many other crops, were always impressed with the high viscosity of this capsular material (45, 64). Based in part on the recommendation of Mortimer Starr at the University of California at Davis, the Kelco Company, then engaged in the production of agar from seaweed, explored the possible commercial production and use of the polysaccharide as a viscosifier and suspending agent in oil well–drilling fluids. Extensive studies at the USDA Northern Regional Laboratory in Peoria, Illinois, in cooperation with the Kelco Company, elucidated the structure and promising properties of xanthan gum. Xanthan gums have unusual properties including the ability to maintain their structure under relatively high temperatures (38). The polysaccharide has a high viscosity at low concentrations, is relatively stable over a range in pH levels, and is resistant to various acids and bases (35). Commercialization began in 1964 and has increased rapidly; annual production is now over 100,000 tons. Currently, xanthan gum is incorporated in many different food products such as batters, baked goods, beverages, candy, frozen dairy products, low-calorie foods as a replacement for starch, salad dressing, etc; pharmaceuticals such as cough medicines; and personal care products such as toothpaste, liquid soaps, and shampoos. In addition, it is used extensively in a wide range of industrial applications and in oil fields in drilling fluids and oil recovery processes (38). No investigator who was engaged in the original studies on the extracellular slime of *Xanthomonas campestris* could have possibly envisioned the current industrial and potential future uses of this unique material (45, 64). Few other major industries can point to an origin as unusual as this one involving development of multiple uses of the extracellular slime of a bacterial pathogen of cabbage.

Secondary Metabolites Formed by Infected Plants

Plants respond to infection by synthesizing a number of different secondary metabolites associated with possible resistant reactions. Considerable effort has been invested in the identification and characterization of these compounds. Many of these metabolites also have novel and rare structures that have attracted the interest of chemists (56). These include capsidiol (41), a sesquiterpenoid; isoflavonoids such as pisatin and glyceollin (34); stilbenes such as viniferin (44), casbene (60) and many other unusual compounds with remarkable biological effects at extremely low concentrations. Pimentel (56), in his survey of opportunities for research in chemistry, emphasized the unexploited opportunities to increase knowledge of the chemistry of these secondary metabolites. A few specific examples are described below.

Nematode Egg-Hatching Factor

The soybean cyst nematode is one of the more difficult of plant parasitic nematodes to control. Eggs persist in the cysts formed by the body of the female and can survive for long periods of time in the soil. Roots of host plants were found to release a substance that stimulates hatching of the eggs. The chemical structure of this compound, glycinoeclepin A, has been elucidated. The physiological activity of this compound is truly remarkable. It can induce hatching at concentrations as low as a few parts per trillion (21, 52, 69). Its potential as a control agent or in physiological studies has not yet been exploited, although the compound was recently synthesized (69).

Witchweed Seed Germination Factor

One of the most destructive of all parasitic seed plants is witchweed (*Striga asiatica*). The tiny seeds of witchweed remain dormant in the soil until a compound released by roots of certain plants triggers the germination process. The chemical structure of the compound (strigol) and closely related compounds have now been determined (56, 62). Strigol and its synthetic analogs have a potential for application as possible chemical control compounds and provide a means of gaining an improved understanding of seed germination processes.

Tentoxin

Studies on the nature of several compounds produced by fungi, in particular host-specific toxins such as tentoxin, have provided new knowledge of basic cellular and enzymatic processes in plants that had not been defined previously (20, 25). A full assessment of recent research on tentoxin is beyond the scope of this report, but the scope of this research and new knowledge obtained is very extensive. This is further demonstration of the value of studies on extracellular products of plant pathogenic fungi.

ADVANCES IN PLANT BREEDING AND GENETICS

The evolution of cultivated plants from wild species occurred through a continuous selection in which a broad range of useful characteristics was sought, resulting in plants adapted to regional environments and with desirable characteristics. Throughout this process, plants with relatively high levels of resistance to local diseases and insects were selected either consciously or by chance (26). However, the important discovery by Biffen in 1905 that resistance to stripe rust in wheat was governed by the then recently rediscovered laws of Mendelian inheritance provided a scientific basis for breeding for disease resistance in general (68). Biffen's discovery also called attention to the importance of incorporating disease-resistance factors in programs that had as their primary objectives increasing quality and productivity of crops (25, 68).

Nature of Resistance to a Host-Specific Toxin

No other disease of a food crop ever received publicity in the United States as extensive as that given to the outbreak of the southern corn leaf blight epidemic in 1970 (67). The impact of the disease on the hybrids that carried the Texas cytoplasmic male sterility (Tcms) factor has been well documented. Fortunately, the resources were available for a rapid shift in the breeding lines, the use of Tcms for large-scale production of hybrids was stopped, and the problem that was so threatening initially has been resolved. However, as a result of this outbreak, major advances have been made in our understanding of the relationship between Tcms and susceptibility to *Bipolaris* (*Helminthosporium*) *maydis* Race T (47). This was the first major widespread example of non-Mendelian maternal inheritance of disease susceptibility in higher plants (57). The characteristics of male sterility and disease susceptibility are closely linked and are carried in the mitochondria on a 13-kDa polypeptide (URF13). In addition to pathotoxin sensitivity and male sterility, it has been found that plants containing the URF 13 protein are also sensitive to the carbamate insecticide methomyl. Tcms-maize lines are also highly susceptible to another fungus disease, yellow leaf blight caused by *Phyllosticta maydis*. This fungus produces a host-specific toxin with a chemical structure similar to the toxin of the southern corn leaf blight pathogen (47). Chaumont et al (12) recently reported on a study to determine whether introduction of the T-urf13 gene into a species other than corn would also confer sensitivity to the pathotoxin and methomyl, as well as conferring male sterility. Transgenic tobacco plants in which high concentrations of the polypeptide were present also were sensitive to methomyl; however, in these transgenic plants the polypeptide was not solely associated with the mitochondria, but with other organelles as well. Also in transgenic plants expression of URF13 was not correlated with male sterility. Either male sterility is not conferred by the presence of the protein in transgenic tobacco or the polypeptide is not formed in sufficient amounts in the anther cells to affect pollen formation. Additional studies are required to gain a full understanding of the close association between male sterility and sensitivity to toxic moieties such as a fungal pathotoxin and a carbamate insecticide. These and related studies have reemphasized the need for constant vigilance in those breeding programs that result in widespread introduction into any major crops of specific genetic factors.

Wisconsin Fast Plants

In the search for improved procedures to facilitate the breeding program for resistance to pathogens of cabbage and related species of brassicas, Paul Williams at the University of Wisconsin-Madison conceived a novel approach (72). To develop a model plant he decided to capitalize on the ability of certain brassicas to flower quickly and complete their life cycles in a relatively short time. He initiated an ambitious screening program and began to examine a world collection of over 2000 brassicas. Fortunately, he found a few plants with relatively short life cycles. Rapid-cycling strains of *Brassica rapa* and five other related species were developed in an intensive effort that extended over a decade. The product of selection with *B. rapa* was a model plant that flowered 14 days after planting the seed, was about 6 inches tall at maturity, and could complete 10 generations in 1 year. It was soon recognized that these plants developed for the breeding program in disease resistance had qualities that make them very attractive for basic investigations on cell and molecular biology, physiology, genetics, and plant breeding. The demand for seed from these unique plants was so great that in 1982 Williams established a Crucifer Genetics Cooperative that now has on its roster over 1600 scientists in 56 countries examining over 100 specific genetic traits.

Because these plants can be grown in large numbers under fluorescent light (up to 2500 plants per m^2), are easily pollinated, have a large number of variant types, and are self-incompatible, they are very adaptable for classroom projects. With the support of the Educational Materials Development Program of the National Science Foundation, the Wisconsin Fast Plants Program was established; it is now actively designing materials for instruction in various aspects of plant biology for use in classrooms from kindergarten through college. The influence of the complete range of environmental and nutritional factors on the development of plants can be examined readily. The possibility of making crosses and following several generations of plants offers students unparalleled opportunities to design simple experiments and obtain results in one semester. The educational materials developed in connection with the rapid cycling plants now known as Wisconsin Fast Plants are available to elementary, high school, and college teachers. Over 35,000 biology teachers and their students are now participating in the Fast Plants Program (P Williams, personal communication).

Declining enrollments in graduate programs in agriculture, and the plant sciences in particular, are a source of growing concern. The decline in funding and number of available positions are obvious factors to be considered in this concomitant decrease in enrollments. The long-term benefits of Fast Plants and related programs will not only foster learning and demystify subjects such as genetics as an essential component of the education of all students, but may also attract students to careers in the plant sciences.

Genetic Engineering Plants with a Modified Tumor-Inducing Factor

The rewards of basic research on a plant disease often arrive in unanticipated ways. One striking example is illustrated by results of the search for the basic mechanism of tumor induction by the soilborne pathogen, *Agrobacterium tumefaciens.* Hundreds of papers record the efforts of a large number of scientists who struggled for many years to unravel the remarkable process that enables the bacterium, *A. tumefaciens,* to transform plant cells so that they grow in an autonomous fashion (55). However, it is now apparent that the salient experiments to decipher the enigma could only have been completed after the powerful techniques of recombinant DNA were discovered. Much of the credit for these studies on tumor induction goes to Eugene Nester and his coworkers at the University of Washington (55, 58). In particular, the experiments of Mary Dell Chilton and associates (13) provided the definitive data indicating that the tumor-inducing segment of the DNA of the Ti plasmid introduced into the plant cell by the bacterium actually is integrated into the chromosome of the cell. Since the DNA from the bacterium could be engineered in such a way that the genetic factors for tumor induction could be eliminated, a remarkable system became available for introduction of other desirable genes into plants. With the development of new techniques for regeneration of plants that could flower and produce seed carrying introduced genes, the prospect for producing transgenic plants became a reality.

The development of transgenic plants with virus resistance (4, 23, 39) constitutes a milestone in disease control. This resistance is conferred by expression of viral coat protein genes in transgenic crop plants (4). This evolution of basic research findings to an unexpected practical application illustrates how basic research on a bacterial disease has aided in establishment of a new technology. It has already shown its value to plant breeders seeking higher precision in introducing genetic traits not available via standard breeding procedures. The technology can also be used in the development of mutations that will make it possible to analyze the functions of growth hormones and fundamental processes of plant biology (39).

ROLE OF EPIPHYTIC BACTERIA IN FROST INJURY

Abiotic injuries and diseases are generally caused by a broad range of environmental extremes with adverse temperatures and moisture relationships as the major factors. Plant pathologists have developed innovative approaches to minimize adverse effects caused by environmental extremes, misuse of chemicals, mineral deficiency diseases, and injuries resulting from harvesting, shipment, and storage procedures. However, reducing losses from frost damage has been extremely difficult to implement. The discovery of the relationship of epiphytic bacteria with frost injury is another excellent example of how studies in plant pathology lead to unexpected discoveries in the fields of microbiology and to the establishment of new industries.

Many frost-sensitive fruit crops periodically suffer severe damage to blossoms and subsequent loss of fruit that may be measured in terms of millions of dollars annually. If a grant proposal had been prepared to study populations of bacteria on fruit blossoms and leaf surfaces as a means of increasing the understanding of frost damage, it is unlikely that such a grant proposal would have received approval for funding. That epiphytic bacteria influence frost damage to plants was an unexpected outcome from a study of the presumed effect of a plant pathogen on a crop plant under frost conditions. Paul Hoppe, a USDA corn pathologist working at the University of Wisconsin-Madison in the mid-1960s, was evaluating corn lines for resistance to northern leaf blight (*Helminthosporium turcicum*). In one experiment in 1964 his inoculation procedure involved spraying young plants in his field plots with a spore suspension of the causal fungus (32). A water suspension had been prepared by grinding heavily infected leaves on which lesions with abundant sporulation of the fungus were present. A period of low temperatures occurred several days after the inoculations were made (32). An unforeseen result was the fact that severe symptoms of frost injury appeared on the leaves of corn lines that had been sprayed with the spore suspension; damage was markedly less severe on the plants in unsprayed plots. The assumption was made that incipient infections had in some manner predisposed leaves to frost injury. For several years after these initial observations, Hoppe conducted experiments in growth chambers in which he reproduced the same pattern of increased frost injury on plants treated with water suspensions containing powdered infected corn leaves obtained from field plants. However, the specific causal factors in these interactions were not determined. After his retirement the project was continued by Deane Arny and Chris Upper in the Department of Plant Pathology at the University of Wisconsin-Madison. After additional tests with appropriate controls, it became evident that the ice-nucleation factor was an entity present on field-grown corn leaves and not related to the action or products of the fungus (2).

In 1973, Steven Lindow initiated a doctoral program with Upper. Soon he determined that ice-nucleation active (INA^+) bacteria in the nonsterile powdered corn-leaf suspensions were involved in this unique effect (2). He modified an ingenious technique developed by Gabor Vali in Wyoming for use with bacteria to test the ice-nucleating ability of test strains. This involved floating a sheet of aluminum foil on the surface of a refrigerated constant-temperature bath held at -5°C. Droplets of water placed on the sheet at this temperature do not freeze unless ice nuclei are present, while those containing INA^+ bacteria froze quickly. One group of the INA^+ cultures were found to be similar in all of their key reactions with *Pseudomonas syringae* (48). A modification of procedures developed by Lindow was subsequently applied in elegant epidemiological studies on brown spot of bean (30). Insertion of leaves with epiphytic populations of *P. syringae* into test tubes with water at –5°C results in rapid freezing and provides a means of following shifts in populations under different environmental conditions. New understanding of the epidemiology of epiphytic bacterial pathogens has been gained from these studies.

Upon completion of his degree at the University of Wisconsin, Lindow was appointed to the faculty at the University of California, Berkeley, and continued his studies on the ice-nucleating bacteria. In one phase of his research he and coworkers genetically engineered one strain of *P. syringae* from potato so that the gene for ice nucleation was deleted (48). This provided the model strain for testing the hypothesis that INA^- bacteria could be used to prevent frost damage. Since the INA^- strain was a good colonizer of leaf surfaces, the possibility existed that it would prevent the populations of INA^+ strains from increasing on leaves of frost-sensitive plants. On the basis of inoculation tests on a large number of possible host plants, it had been found that the epiphytic INA^+ strain from potato also was not pathogenic on any of a broad range of the crop species tested (48).

At that time Jeremy Rifkin, head of an environmental group, was leading a campaign to prevent the introduction of any genetically modified organism into the environment. As a result of various lawsuits and associated publicity, the experiments to test the genetically modified INA^- strain under field conditions became the most highly publicized field tests in the history of plant pathology, or indeed any of the biological or agricultural sciences. The experiments were delayed almost five years until all the requirements of the US Environmental Protection Agency were satisfied. In cooperation with Lindow, one of the first field experiments was completed by a biotechnology company, Advanced Genetic Sciences. A photograph of Julie Lindeman of Advanced Genetic Sciences in a protective space suit spraying strawberry plants with a suspension of the INA^- bacterium appeared on the front page of the *New York Times* and was shown on many television news reports. In the background of the photograph reporters were visible drinking coffee in apparent unconcern about the presumed dangers of the experiment. When Lindow finally established his field plots, attempts were made to block progress in the experiments by vandalism that involved destruction of test plants and damage to equipment.

Large-scale commercial application of INA^- bacteria as a means of reducing frost damage on a commercial basis is still under development. A naturally occurring ice^- strain of *Pseudomonas fluorescence* is now registered and will be sold commercially as "Blight Ban A506" for frost and fire blight control in 1995 (S Lindow, personal communication). One commercial application has been the use of INA^+ bacteria in a patented product, *Snowmax,* which is now used to improve the efficiency of snow production on ski slopes. An additional application now under study is the use of INA^+ bacteria as an aid in freezing foods more rapidly at higher temperatures than normal (72a).

SUMMARY AND CONCLUSIONS

The above examples illustrate some of the ways in which the study of plant diseases has led to significant advances in other biological sciences and to the establishment of new industries. Many other examples can be cited, and the prospects are high that many similar discoveries will be made in the near future. The powerful tools of molecular biology can be applied to gain an understanding of a host of compounds still not fully characterized that are produced by plant pathogens and by plants in the act of defending themselves against pathogens. The prospects are high that our understanding of the nature of tumorigenesis and differentiation of self and nonself may best be resolved in studies on pathogen-host reactions.

Plant pathologists have played a leading role in the debate about release of genetically engineered organisms (GEMS). Wilson & Lindow (70) have recently summarized the evidence for low risk in properly designed and managed releases and emphasized the potential benefits of research in this area, including the prospects for use of GEMS in bioremediation, basic studies in soil ecology, and biocontrol.

In several instances, long delays occurred in wide application of an innovative technique because of the high degree of insularity that, unfortunately, still separates many agricultural scientists from basic biological scientists. The long delay before the density gradient centrifugation technique of Brakke (7) was widely adopted is a case in point. The conclusive evidence that a microorganism can be the cause of disease in a living organism was ignored for over 50 years by scientists investigating human disease. Medical mycologists initially overlooked the close taxonomic relationship between fungi that cause disease in humans and those that infect plants.

Under our current system of funding and with intense competition for the available funds, there is a natural tendency for young scientists to avoid the research projects that explore new areas in innovative or unconventional ways. A number of the advances cited here exemplify instances in which individuals had the courage to venture into unexplored territory. Thus, the prospects are bright for additional major basic contributions to other biological sciences as well as in the development of new technology and beneficial products of value to agriculture and industry resulting from research in plant pathology.

References have been made to discoveries that resulted from the cooperative efforts of researchers with many different scientific backgrounds and scientists who did not have in-depth training in plant pathology. However, they were intrigued by the advantages of using a plant disease as a model system to gain an understanding of normal plant growth, development, and reproduction.

Thus, in a period of change and stress, plant pathologists have reason to point with pride not only to the significant contributions that our field has made to insure the availability of food and fiber and the beauty of ornamental plants to the world but to other fields of science and industry. We also need to recognize the fundamental importance of investigations into the nature of disease and biology of pathogens. These studies are an aid to development of improved controls as well as a means of advancing our understanding of normal physiology and genetics of plants and the microorganisms that either attack them or exist in nature as helpful saprophytes. Major benefits can continue to accrue to science and humanity by these investigations.

Literature Cited

1. Agrios G. 1988. *Plant Pathology*. San Diego, CA: Academic. 803 pp. 3rd ed.
2. Arny DC, Lindow SE, Upper CD. 1976. Frost sensitivity of *Zea mays* by application of *Pseudomonas syringae*. *Nature* 262:282–84
3. Bawden F C. 1970. Musings of an erstwhile plant pathologist. *Annu. Rev. Phytopathol.* 8:1–12
4. Beachy RN, Loesch-Fries S, Tumer NE. 1990. Coat protein mediated resistance against virus infection. *Annu. Rev. Phytopathol.* 28:451–74
5. Bos L. 1995. One hundred years of virology? *ASM News* 61:53–54
6. Brakke MK. 1951. Density gradient centrifugation, a new separation technique. *J. Am. Chem. Soc.* 73:1847–48
7. Brakke MK. 1979. The origins of density gradient centrifugation. *Fractions* 1:1–9
8. Brakke MK. 1988. Perspectives on progress in plant virology. *Annu. Rev. Phytopathol.* 26:331–50
9. Bulloch W. 1938. *The History of Bacteriology*. London: Oxford Univ. Press. 422 pp.
10. Campbell CL. 1982. Erwin Frink Smith, pioneer plant pathologist. *Annu. Rev. Phytopathol.* 21:21–27
11. Carefoot GL, Sprott, CK. 1969. *Famine on the Wind; Plant Diseases and Human History*. London: Angus & Robertson. 222 pp.
11a. Castello JD, Leopold DJ, Smallidge PJ. 1995. Pathogens, patterns, and processes in forest ecosystems. *BioScience* 45:16–24
12. Chaumont F, Bernier B, Buxant R, Williams ME, Levings CS III, Boutry M. 1995. Targeting the maize T-urf13 product into tobacco mitochondria confers methomyl sensitivity to mitochondrial

respiration. *Proc. Natl. Acad. Sci. USA* 92:1167–71
13. Chilton MD, Drummond MH, Merlo DJ, Sciaky D, Montoya AL. 1977. Stable incorporation of plasmid DNA into higher plant cells: the molecular basis of crown gall tumorigenesis. *Cell* 11: 263–71
14. Cook RJ. 1994. The future of plant pathology: a senior scientist's perspective. *Phytopathol. News* 28:47–48
15. Cook RJ, Gabriel CJ, Kelman A, Tolin S, Vidaver AK. 1995. Research on plant disease and pest management is essential to sustainable agriculture. *BioScience* 45:354–57
16. Davis RE. 1979. Spiroplasmas, newly recognized arthropod borne pathogens. In *Leafhopper Vectors and Plant Disease Agents,* ed. KF Harris, K Maramorosch, pp. 451–88. New York: Academic. 654 pp.
17. Diener TO. 1972. Viroids, the smallest known agents of infectious disease. *Annu. Rev. Microbiol.* 28:23–39
18. Diener TO. 1982. Viroids and their interactions with host cells. *Annu. Rev. Microbiol.* 36:239–58
19. Doi Y, Terenaka M, Yora K, Asuyama H. 1967. Mycoplasma or PLT group-like microorganisms found in phloem elements of plants infected with mulberry dwarf, potato witches' broom, aster yellows, or paulownia witches' broom. *Nippon Shokubutsu Byori Gokkaiho* 33: 259–66
20. Durbin RD. 1981. Applications. In *Toxins in Plant Disease,* ed. RD Durbin, pp. 495–505. New York: Academic. 515 pp.
21. Fukuzawa A, Furusaki A, Ikura M, Masamune T. 1985. Glycinoeclepin as a natural hatching stimulus for the soybean cyst nematode. *J. Chem. Soc.* 4: 222–24
22. Fischer JR, Zuiches JJ. 1994. *Challenges confronting agricultural research at land grant universities.* Issue Pap. 5. Ames, IA: Counc. Agric. Sci. Technol. 12 pp.
23. Fitchen JH, Beachy RN. 1993. Genetically engineered protection against viruses in transgenic plants. *Annu. Rev. Microbiol.* 47:913–44
24. Graebe JE. 1987. Gibberellin biosynthesis and control. *Annu. Rev. Plant Physiol.* 38:419–65
25. Graniti A, Durbin RD, Ballio A. 1988. *Phytotoxins and Plant Pathogenesis.* Berlin: Springer-Verlag. 508 pp.
26. Harlan JR. 1976. Diseases as a factor in plant evolution. *Annu. Rev. Phytopathol.* 14:31–51
27. Harrison BD. 1994. Frederick Charles Bawden: plant pathologist and pioneer in plant virus research. *Annu. Rev. Phytopathol.* 32:39–48
28. Hawksworth FG, Wiens D. 1970. Biology and taxonomy of the dwarf mistletoes. *Annu. Rev. Phytopathol.* 8:187–208
29. Hepting GH, Cowling EB. 1977. Forest pathology: unique features and prospects. *Annu. Rev. Phytopathol.* 15:431–50
30. Hirano SS, Upper CD. 1983. Ecology and epidemiology of foliar bacterial plant pathogens. *Annu. Rev. Phytopathol.* 21:243–69
31. Hopkins DL. 1989. *Xylella fastidiosa:* xylem-limited bacterial pathogen of plants. *Annu. Rev. Phytopathol.* 27:271–90
32. Hoppe PE, Arny DC, Martens JW. 1964. Frost susceptibility in corn increased by leaf blight infections. *Plant Dis. Rep.* 48:815–16
33. Horsfall JG, Wilhelm S. 1982. Heinrich Anton de Bary: Nach Einhundertfunfzig Jahren. *Annu. Rev. Phytopathol.* 20:27–32
34. Ingham JL. 1982. Phytoalexins from the Leguminosae. In *Phytoalexins,* ed. JA Bailey, JW Mansfield, pp. 21–80. New York: Wiley. 334 pp.
35. Jeanes A, Pittsley JE, Senti FR. 1961. Polysaccharide B-1459: a new hydrocolloid polyelectrolyte produced from glucose by bacterial fermentation. *J. Appl. Polym. Sci.* 5:519–26
36. Judson HF. 1979. *The Eighth Day of Creation.* New York: Simon & Schuster. 686 pp.
37. Keitt GW. 1959. History of plant pathology. In *Plant Pathology,* ed. JG Horsfall, AE Dimond, pp. 61–97. New York: Academic
38. Kelco Div. 1988. *Xanthan Gum: Natural Biogum for Scientific Water Control.* Rahway, NJ: Merck & Co. 3rd ed.
39. Klee H, Horsch R, Rogers S. 1987. *Agrobacterium*-mediated plant transformation and its further applications to plant biology. *Annu. Rev. Plant Physiol.* 38:467–86
40. Kommedahl T, Williams PH. 1983. *Challenging Problems in Plant Health.* St. Paul, MN: Am. Phytopathol. Soc. 538 pp.
41. Kuć J. 1982. Phytoalexins from the Solanaceae. In *Phytoalexins,* ed. JA Bailey, JW Mansfield, pp. 81–105. New York: Wiley. 334 pp.
42. Kurosawa E. 1926. Experimental studies on the secretion of *Fusarium heterosporum* on rice plants. *J. Nat. Hist. Soc. Formosa* 16:213–27 (In Japanese)

43. Large EC. 1962. *The Advance of the Fungi.* New York: Dover. 488 pp.
44. Langcake P, Pryce JC. 1977. A new class of phytoalexins from grapevines. *Experientia* 33:151–52
45. Leach JG, Lilly UG, Wilson HA, Purvis MR Jr. 1957. Bacterial polysaccharides: the nature and function of the exudate produced by *Xanthomonas phaseoli. Phytopathology* 47:113–20
46. Lee IM, Davis RE. 1986. Prospects for *in vitro* culture of plant pathogenic mycoplasma-like organisms. *Annu. Rev. Phytopathol.* 24:339–54
47. Levings CS. 1990. The Texas cytoplasm of maize: cytoplasmic male sterility and disease susceptibility. *Science* 250:942–49
48. Lindow SE. 1983. The role of bacterial ice nucleation in frost injury to plants. *Annu. Rev. Phytopathol.* 21:363–84
49. Lucas WJ, Gilbertson RL. 1994. Plasmodesmata in relation to viral movement within leaf tissues. *Annu. Rev. Phytopathol.* 32:387–411
50. Lustig A, Levine AJ. 1992. One hundred years of virology. *J. Virol.* 66:4629–31
51. Maramorosch K. 1979. How mycoplasmas and rickettsias induce plant disease. In *Plant Disease,* ed. JG Horsfall, EB Cowling, 4:203–17. New York: Academic
52. Masamune T, Anetai M, Takasuoi M, Katsui M. 1982. Isolation of a natural hatching stimulus, glycineoclepin A, for the soybean cyst nematode. *Nature* 297: 495–96
53. Matthews REF. 1981. *Plant Virology.* London: Academic. 897 pp.
54. Moore TC. 1989. *Biochemistry and Physiology of Plant Hormones.* New York: Springer-Verlag. 855 pp. 3rd ed.
55. Nester EW, Gordon MP, Amasino RM, Yanofsky MF. 1984. Crown gall: a molecular and physiological analysis. *Annu. Rev. Plant Physiol.* 37:387–413
56. Pimentel GC. 1985. *Opportunities in Chemistry.* Washington, DC: Natl. Res. Counc./Natl. Acad. Press. 244 pp.
57. Pring DR, Lonsdale DM. 1989. Cytoplasmic male sterility and maternal inheritance of disease susceptibility in maize. *Annu. Rev. Phytopathol.* 27:483–502
58. Ream W. 1989. *Agrobacterium tumefaciens* and interkingdom genetic exchange. *Annu. Rev. Phytopathol.* 27: 583–618
59. Rowe, RC. 1994. Plant pathology: where are we headed beyond 2000. *Phytopathol. News* 28:39–40
60. Sitton D, West CA. 1975. Casbene: an antifungal diterpene produced in cell-free extracts of *Ricinus communis* seedlings. *Phytochemistry* 14:1921–25
61. Stakman EC. 1959. The role of plant pathology in the scientific and social development of the world. In *Plant Pathology: Problems and Progress. 1908–1950,* ed. CS Holton, pp. 3–13. Madison, WI: Univ. Wisc. Press. 588 pp.
62. Stewart GR, Press MC. 1990. The physiology and biochemistry of parasitic angiosperms. *Annu. Rev. Plant Physiol. Plant Mol. Biol.* 41:127–51
63. Stolp H. 1973. The Bdellovibrios: bacterial parasites of bacteria. *Annu. Rev. Phytopathol.* 11:53–76
64. Sutton JC, Williams PH. 1970. Comparison of extracellular polysaccharide of *Xanthomonas campestris* from culture and from infected cabbage leaves. *Can. J. Bot.* 48:645–51
65. TeBeest DO, Yang XB, Cisar CR. 1992. The status of biocontrol of weeds with fungal pathogens. *Annu. Rev. Phytopathol.* 30:637–38
66. Templeton GE, TeBeest DO, Smith RJ Jr. 1984. Biological weed control in rice with a strain of *Colletotrichum gloesporiodes* (Penz.) Sacc. used as a mycoherbicide. *Crop Prot.* 3:409–22
67. Ullstrup AJ. 1972. The impacts of the southern corn leaf blight epidemics of 1970–1971. *Annu. Rev. Phytopathol.* 10: 37–47
68. Walker JC. 1969. *Plant Pathology.* New York: McGraw-Hill. 819 pp. 3rd ed.
69. Watanabe H, Mori K. 1991. Triterpenoid total synthesis. Part 2. Synthesis of glycinoeclepin A, a potent hatching stimulus for the soybean cyst nematode. *J. Chem. Soc. Perkin. Trans.* 1:2919–32
70. Wilson M, Lindow SE. 1993. Release of recombinant microorganisms. *Annu. Rev. Microbiol.* 47:913–44
71. Whitcomb RF, Tully RG. 1979. *The Mycoplasmas: Plant and Insect Mycoplasmas.* New York: Academic. Vol. 3
72. Williams PH, Hill CB. 1986. Rapid cycling populations of *Brassica. Science* 232:1385–89

ADDED IN PROOF

72a. Lee RE Jr., Warren GJ, Gusta LV. 1995. *Biological Ice Nucleation and Its Applications.* St. Paul, MN: Am. Phytopathol. Soc. Press

ES LUTTRELL

Annu. Rev. Phytopathol. 1995. 33:23–35

PIONEER LEADERS IN PLANT PATHOLOGY: ES Luttrell

Richard T. Hanlin
Department of Plant Pathology, University of Georgia, Athens, Georgia 30602-7274

KEY WORDS: ascomycetes, ascomal ontogeny, biography, deuteromycetes, morphology

---

ABSTRACT

ES Luttrell was a mycologist who spent 42 of his 44 professional years at the University of Georgia, first at the Georgia Experiment Station in Griffin and later at the main campus in Athens. He is best known for his innovative classification scheme for the perithecial ascomycetes, in which orders were based on patterns of ascomal ontogeny and mode of ascus dehiscence. This work established him as an authority on ascocarp development. His later studies on the classification of the *Helminthosporium* complex, in which he showed that small differences in conidium germination and structure could be correlated with distinct teleomorphs, brought him recognition from plant pathologists as well. These studies were notable for his attention to detail and the quality of his work. His contributions to mycology and plant pathology were numerous and varied and brought him various awards later in his career. To his colleagues he served as administrator, teacher, mentor and friend.

ES LUTTRELL

When ES Luttrell joined the staff of the Georgia Experiment Station on October 1, 1942, as Associate Botanist in the Department of Plant Pathology, there was little in his record to suggest that when he retired 44 years later he would be widely recognized for his outstanding research both in mycology and in plant pathology. Before long, however, his scholarship and intellect became evident and set him apart from his colleagues. To those of us who worked with him, his publications represented only a portion of his contributions to mycology

0066-4286/95/0901-0023$05.00

and plant pathology. Anyone who had personal contact with him came away enhanced both personally and professionally.

Lutt was a man totally immersed in and dedicated to his research. His brilliant mind, insatiable intellectual curiosity, and professional dedication were the source of his outstanding achievements. Others may have had longer publication records, but none exceeded him in the quality and intellectual insight of his work. And although he published a number of coauthored papers, most of these came late in his career. Significantly, in all of his major contributions he was sole author. He published only three papers on ascomycete morphology jointly, and these were with postdoctoral fellows and students.

EARLY YEARS

Everett Stanley Luttrell was born January 10, 1916, in Richmond, Virginia, the son of Ralph Edgar and Mae Flippen Luttrell. He contracted polio as an infant, which left him with difficulty in walking and with pain a common companion. As a child he frequented the rivers and bays of the Richmond area, and he developed an interest in the natural history of the region. He also became familiar with the local folklore and in later years he would often relate stories about his experiences there. Lutt attended the Richmond Public Schools and in due course graduated from high school, but then was faced with the problem of what to do. These were depression years and jobs of any kind were scarce. His mother wanted him to attend college, but as the time approached for classes to begin, he still had not enrolled.

Mae Luttrell was a strong-willed woman (a quality her son came to share), insistent that her close associates should not easily avoid following her wishes. When she discovered that her son had not registered for college, she went to the Registrar's Office at the University of Richmond and enrolled him herself. Lutt dutifully began to attend classes. This was certainly a pivotal time in Lutt's life. His hesitation in enrolling in college may well have reflected the fact that he found classes intellectually boring and unchallenging, or that he harbored a lifelong dislike for fetters of any kind. Yet he had a strong sense of duty and responsibility to fulfill obligations. His mother undoubtedly felt that academic life, such as a teaching position, was a career at which her son could succeed without undue physical hardship. Besides the value of the college education, however, Mae Luttrell pushed her son to be independent, and to view his physical shortcomings as inconveniences rather than as a handicap. By the time he had graduated, these qualities were ingrained in his behavior and they were a prominent feature of his character thereafter.

At the University of Richmond Lutt majored in biology, continuing his earlier interests in natural history. Although his undergraduate career was not outstanding, one of his professors, a Dr. Tucker, recognized his latent talent and encouraged him to continue his studies at Duke University. Here he came under the tutelage of Dr. Frederick A Wolf, mycologist in the Botany Department, who was widely known for his studies on the biology of plant pathogenic ascomycetes. Under Dr. Wolf's direction Lutt began a study of foliar pathogens of trees in Duke Forest, adjacent to the university campus. In 1939 he presented his M.S. thesis entitled "A tar spot disease of American holly and the life history of the causal organism." One year later he completed his doctoral dissertation, "The morphology and development of some fungi parasitic on trees within Duke Forest." This included studies of two ascomycetes and a new genus of conidial fungi. All three studies were published in 1940. The clarity of his prose and high quality of his research are already evident in these papers, and they include the first of the detailed line drawings that would be a prominent feature of his publications.

Unable to find professional employment after graduation, Lutt took temporary work as a surveyor's assistant with the USDA Soil Conservation Service (SCS) in Virginia. During this time he took additional graduate courses at the University of North Carolina and the University of Virginia. Although he could not conduct morphological research, Lutt took advantage of his travels around the State of Virginia with the SCS to assemble a large (ca 5000 specimens) collection of lichens. These later formed the basis for his monograph of the Cladoniaceae of Virginia (4). Interestingly, this was the only paper he published on lichens, and the only organized herbarium collection that he made during his long career.

PROFESSIONAL YEARS

Research

Finally, in 1942, Lutt received his first professional (albeit temporary) appointment with the Department of Plant Pathology at Georgia Experiment Station, located at (Experiment) Griffin. When the faculty member he had replaced returned from military service in 1945, nothing was said about Lutt's position being temporary, so he stayed on.

Once in his new position, Lutt immediately resumed his morphological and taxonomic studies of ascomycetes. As he studied additional genera he broadened his review of the literature on developmental morphology, but he was handicapped by the limited resources of the Station library in that pre-computer era. Then, in 1947, Lutt accepted a position as Assistant Professor of Botany at the University of Missouri, where he had access to a much better library. With these resources available, he undertook a thorough review of the literature on ascomycete morphology. He perceived distinct patterns of development of the ascocarp that were characteristic of certain groups of ascomycetes. This led to the 1951 publication, "Taxonomy of the Pyrenomycetes" (3), in which he revised the classification of the perithecial ascomycetes, basing the new scheme on the ontogenetic pattern of the ascocarp. He also separated those ascomycetes with bitunicate asci from the unitunicate forms, and noted the correlation of these modes of ascus dehiscence and the basic pattern of ascocarp development. Later he proposed that the bitunicate forms be segregated in a new subclass (6, 9). This book firmly established him as an authority on ascomycete morphology, both nationally and internationally. It is somewhat ironic that the work for which he is best known was published not in Georgia, but by the University of Missouri, although by the time he had completed the manuscript he had returned to his old position at Griffin. He was happy to exchange the rigorous teaching schedule of bacteriology, mycology, plant pathology and forest pathology for the slower-paced life in Georgia.

Throughout his career Lutt worked on a wide range of ascomycetes and conidial fungi, but the developmental morphology of the ascocarp remained his dominant interest. Regardless of what else he was studying, he was also working on the morphology of some ascomycete. The majority of his morphological studies were conducted on bitunicate species, but he studied unitunicate forms as well. All of these studies were characterized by detailed observations of the various stages of ascocarp development, and illustrated by photographs and line drawings. In studying the development of a fungus, he used a variety of techniques including whole mounts, squash mounts, slide culture, and microtome sections. Later, in his work on host-parasite relations, he included electron microscopy. As he observed the fungus under the microscope, he made notes and rough sketches. Once satisfied that he had worked out the details of the ontogeny, he made the illustrations for the manuscript. Although he used photographs, line drawings were his preferred method of illustrating his work, as he could emphasize features that were difficult to photograph. Many hours were spent in preparing these drawings. Pencil sketches were first made on standard sheets of paper with the aid of a camera lucida, and these were then arranged into plates. Each entire plate was then traced on a single sheet of drawing paper and inked. Because his plates were usually large (some nearly a meter high) and therefore had to be reduced for publication, the density of the stippling was a problem. To help with this he used a reducing glass to determine the optimal spacing, then he meticulously added each dot. Only rarely did he have to change a plate once it was finished. The illustrations were then merged with the text to produce the final manuscript. Besides his own contributions, his work stimulated many additional studies on ascomycete morphology. Despite the wealth of new data, he resisted suggestions that he revise his book. In addition to his morphological studies, Lutt produced three monographs, on *Stomiopeltis*, *Leptosphaerulina*, and the Cladoniaceae of Virginia.

Although not trained as a plant pathologist, Lutt was confronted by plant pathology problems soon after he arrived in Georgia. The day of the extension specialist had not yet arrived and farmers were accustomed to taking their problems directly to the Experiment Station, especially on rainy Saturday mornings. This led Lutt to initiate research on various local disease problems, especially those associated with beans and grapes, and he published a series of papers on diseases of these and other crop plants. He also conducted surveys of the diseases associated with several crops in Georgia. Because most of the plant diseases he studied were caused by conidial fungi, Lutt began to look at them in more detail. Shortly after his return to Georgia he began a series of studies that were mycological in nature, but which were of direct interest to plant pathologists; his studies on the taxonomy of *Helminthosporium* species continued for several years, during which he published a key to the graminicolous species and a revision of the *Helminthosporium sativum* complex, and culminated in a pair of papers on the taxonomic criteria (7) and systematics (8) of the *Helminthosporium* complex. In this classic study he demonstrated that subtle differences in conidium morphology and pattern of germination were correlated with clearly distinct teleomorphs, and supported the segregation of *Helminthosporium* into four anamorphic genera. He also introduced new terms to describe modes of conidiogenesis and conidium germination. These papers underscored the importance and value of a detailed understanding of the biology of plant pathogenic fungi. He liked to cite these studies to emphasize what he referred to as the "predictive value" of taxonomic data (12, 13).

As he worked with these conidial fungi, he discovered the ascigerous states for several of them, and he described a number of new species of ascomycetes and conidial states, many of which were sent to him by other workers.

In 1979 Lutt published a new system of classification for the conidial fungi (Deuteromycetes) (14) that was a parallel of his earlier treatment of the ascomycetes: Orders were based on the fundamental mode of conidiogenesis, with families being based on such characters as conidium arrangement and type of fructification. While it is now generally recognized that conidium ontogeny can be too variable to serve as the basis for a taxonomic system, this paper points up again his innovative approach to solving practical problems.

In later years he became more interested in host-parasite interactions, and turned his attention to what he termed "replacement" diseases (16), in which host structures are replaced by fungal tissues. He published studies on the infection of *Sporobolus* ovaries by *Bipolaris* and of the development of the sclerotium in dallisgrass and in *Claviceps purpurea* (15). He was actively pursuing the study of the smut fungi and their effects on the morphology of their hosts (1, 18) at the time of his death.

Lutt felt strongly that mycologists and plant pathologists should study fungi of economic importance, to justify their expenditure of public funds. In this sense, he felt that the efforts directed toward such genera as *Aspergillus, Blastocladiella,* and *Neurospora* in the development of model systems were misdirected, and that more would be gained by studying plant pathogens instead. He also lamented the separation between mycology and plant pathology, as interfering with their common goal of developing a better understanding of the fungi. As an active member of both the Mycological Society of America (MSA) and the American Phytopathological Society (APS), he strove to bring the two groups together. To MSA members he stressed the value of working on the biology of plant pathogenic fungi (11), whereas at APS meetings he emphasized the value of morphology and taxonomy to plant pathology (9). He also believed that one should work with living systems whenever possible. He regarded it as remarkable that with all of the research on the fungi, there still are very few species whose biology is well understood.

Although about one third of his papers were jointly authored, Lutt preferred to work alone, especially in the area of ascomycete morphology. Even though he worked only 145 km from Julian H Miller, himself a distinguished and recognized authority on pyrenomycete taxonomy, they never collaborated, and so far as I am aware, never discussed ascomycete taxonomy. Likewise, he never collaborated with BB Higgins, his department head, whose early work involved developmental studies of plant pathogenic ascomycetes, and though Lutt and I frequently discussed our research, we never published together.

Teaching

Faculty positions at Griffin entailed full-time research, hence little or no formal teaching was involved. During his two years at Missouri (1947–49) Lutt labored under a heavy teaching load of 11 classes in 4 subjects (bacteriology, mycology, plant pathology, and forest pathology). He did not teach again until he moved to the main campus of the University of Georgia in Athens in 1966. Athens had been without a mycologist since the retirement of Julian Miller in 1958, so Lutt took over the introductory mycology class that had been taught by John Owen, the former department head. When I moved to Athens in January 1967, he turned both the introductory and advanced mycology courses over to me, and later he developed a new graduate course, "Phytopathology: Principles and Theory." This great course was soon required of all students. Lutt prepared his own manual, drawing upon examples from all areas of plant pathology and many from medicine as well. The course, which included considerable epidemiology, proved challenging to most students, who were unaccustomed to thinking in such broad terms, and often lacked the requisite background to profit from it.

Although Lutt enjoyed teaching, he also found it frustrating. He tried to be innovative, but was often stymied by the students who tended to study only for exams. In the second year he announced that everyone would receive an 'A', so they could forget exams and concentrate on learning the material. Much to his dismay, word spread of his grading policy, and many students with little interest in plant pathology came for an easy 'A'. Reluctantly, he returned to giving exams.

As a way of evaluating the variable background of his students, Lutt instituted an exam on the first day of class. This, of course, caused great consternation, even though he made it clear that the quiz was for his information only. Students seemed to be so indoctrinated with the importance of exams that they were unbelieving when told otherwise. Lutt finally abandoned this practice as well.

With his complete dedication to his profession, Lutt had difficulty accepting that others had lesser devotion. He often complained that students seemed uninterested in their studies and were distracted by too many other activities.

After moving to Athens, Lutt had only a few graduate students late in his career. He never recruited students; he preferred to assist whenever needed without being involved in the paperwork that was a part of their programs.

Administration

In 1955 Lutt succeeded BB Higgins as Head of the Department of Plant Pathology at the Griffin Station. His attention to detail and sense of responsibility made him a good administrator. As the department was small and administration less complicated than now, he had ample time for research. He continued in this position until July 1966, when he became head of the department at Athens, replacing John Owen, who became Director of Experiment Stations. The position at Athens was actually a dual position: both Head of the Department of Plant Pathology and Plant Genetics, as well as Chairman of the statewide Division of Plant Pathology. The Division Chairman was responsible for overseeing the activities of all three departments at the Athens, Griffin, and Tifton stations, along with the extension personnel at Athens and Tifton. It thus required considerably more time than did the headship at Griffin. Lutt's move to Athens was a fortuitous and timely one. The State Legislature of Georgia had decided in the early 1960s to improve and expand the facilities in the sciences at the University, and allocated funds for new buildings and faculty positions. The size of the Athens department doubled; a new Plant Sciences Building was built; the plant pathology curriculum was expanded; and the graduate program was upgraded, as were the research programs. Lutt brought vision and foresight to the expansion process, and stimulated others to greater achievement. With his broad experience, he could discuss research with anyone and offer useful insights. He never hesitated to share his knowledge, especially with younger faculty. One of his greatest strengths as department head, however, came from the high degree of respect and confidence shown him by his faculty. With his unquestioned integrity one could always be certain that he acted in the best interests of the department and of the individual. This lessened tensions among the faculty and made difficult decisions easier.

Lutt was left with little time for the research that he loved and he tired of the frustrations of dealing with the higher administration, so in 1970 he resigned his administrative duties to return to full-time research and teaching. He never considered himself an administrator, and he could be quite critical of administrative activities.

Scholarship

Relatively few individuals can be regarded as truly brilliant, but Lutt was one of those. His intellectual curiosity was broad and he was a prodigious reader. More importantly, he understood, assimilated, and retained what he had read. This ability enabled him to look at things in a broad perspective and to discern patterns or correlations not evident to others. From this quality of his mind came the innovative ideas that he applied to the fungi he studied. His reading extended far beyond mycology and plant pathology. A predictable feature of our weekly departmental seminar was that, regardless of the topic, Lutt would ask pertinent and relevant questions. He could seek out the essential facts of a paper and relate these to other things he had read or observed. He enjoyed intellectual challenges. When he discovered a key to *Fusarium* species in Russian and was unable to obtain a translation, with the help of a Russian-speaking faculty member at the Griffin Station, he learned sufficient Russian to be able to translate the article. He also learned Latin so he could write his own descriptions, and he could read French, German, and Spanish. Learning was a continuous process for him. He constantly sought ways to improve his drawing techniques, by studying the works of earlier, skilled artists, and by experimenting with new pens and inks. When it became apparent that electron microscopy (EM) techniques could add valuable information to studies on host-parasite relations, he enrolled in the EM course at Georgia so he could incorporate this into his research program.

Lutt was a master of the English language, and he was frequently asked to review manuscripts from a broad range of disciplines. He would conscientiously critique the paper, writing his comments in longhand on canary-colored paper, then give the philosophical basis for his comments. These sometimes exceeded the manuscript in length; they were not always easy reading for the author, but they invariably improved the manuscript.

He was often distressed by what he regarded as improper or imprecise usage of technical terms in publications. This caused him to write two papers discussing terminology (5, 10). Perhaps the term that distressed him the most was the use of "incite" with reference to the initiation of plant disease infection.

Despite his lucid discussion of the reasons this term is inappropriate to plant pathology, it unfortunately continues to appear in the literature.

Rewards

Lutt was a member of several professional societies including the American Phytopathological Society (APS), Botanical Society of America, British Mycological Society, and the Mycological Society of America (MSA). He served the MSA as counselor (1967–69), vice president (1969–70), president-elect (1971–72), and as president (1972–73). In 1981 he was honored by his selection as the MSA Annual Lecturer, and in 1983 he received the society's Distinguished Mycologist Award. In 1964 he served as president of the Southern Division of APS, and in 1972 was elected an APS Fellow. The genus *Luttrellia* was named in his honor (21), as were several species. Then, in 1978, the University of Georgia appointed him as DW Brooks Distinguished Professor of Plant Pathology, a position he held until his retirement in 1986. With donations from family and friends, the Department of Plant Pathology at the University of Georgia established the ES Luttrell Lecture Series. Each spring a noted mycologist or plant pathologist is invited to campus to present a lecture and meet with students and faculty.

FAMILY

The one concession Lutt made to his research was to his family. He was a devoted husband and father, and later grandfather. In April 1944, Lutt married Margaret Muse, a public health nurse from Albany, Georgia, who often visited relatives who worked at the Griffin Station. It was there that she met Lutt. Margaret and the three children shared Lutt with his beloved fungi. In later years, when the children were grown, Margaret began to accompany Lutt to meetings, where they formed close friendships with other pathologists and mycologists whose wives also regularly attended the meetings. They continued this practice until Lutt retired.

Besides his family, Lutt relaxed by gardening. In this, as in other aspects of his life, he was competitive. A major challenge for a gardener in the South is to grow English peas. They will not mature in the heat of summer, and so must be planted in late winter, but not so early that a late freeze will catch them. Each spring, Lutt would proudly announce the date his peas emerged, certain that he had beat other gardening faculty members. Even in his garden, though, he was never far from his fungi, as it also contained numerous diseased plants on which he could observe the effects and progress of the pathogen. He grew these from transplants or seeds so they would be close at hand to study.

RETIREMENT

Research had been too great a part of Lutt's life to think of abandoning it through mandatory retirement. So, with space available to him in the department, he decided to continue his studies. Unfortunately, the illness that was to claim him began soon after; he was able to complete only two more manuscripts (19, 20). Characteristically, one paper dealt with plant pathology, the other with ascomycete morphology.

REMINISCENCES

I first met Lutt at the railroad station in Griffin, Georgia, in March, 1960. My presence resulted from a rather curious series of events. My doctoral dissertation involved studies on ascomycete morphology, so I had read Lutt's book and other morphology papers. In December, 1959, I wrote him to ask some questions about his studies, and closed the letter with the comment that I would complete my PhD the following summer. He promptly answered my questions, and said that he did not know of any jobs available. This was not a major concern because university and college positions in the early 1960s were not difficult to find. I was greatly surprised, therefore, to receive a second letter from him a few weeks later asking if I was interested in working with him. He explained that he had just received a National Science Foundation grant for research on ascomycete morphology and it contained funds for a technician. With the assistance of the Experiment Station Director, he had converted this into a temporary faculty position and he invited me to interview for it. So a few weeks later we met in the small textile mill town of Griffin, site of the Georgia Experiment Station, a branch of the University of Georgia College of Agriculture, located some 65 km south of Atlanta. Thus began my introduction to Lutt and to the South. Little did I dream that we would spend our professional careers together.

Lutt had an engaging personality, with a ready smile and friendly handshake. It was difficult not to like him. We collected my suitcase and he led me to an early 1950s vintage Ford sedan, which his colleagues had affectionately dubbed the "Blue Bullet," referring, of course, to his habit of driving slowly. After meeting the faculty I was shown around the large laboratory where he worked. It was not long before I encountered his subtle sense of humor. At one point he asked what equipment I would need, and I mentioned an incubator for growing cultures at room temperature. Lutt looked at me with a slight smile and said, "You might want to consider a refrigerator." After I spent my first summer there I learned the full impact of his comment. The laboratory was located on the top floor of a three-story building with no air conditioning. The temperature was such that paraffin blocks softened and sectioning was impossible. We purchased a refrigerated microtome but it would not function in midsummer because of the heat. Finally, during my third year there, a central air conditioning unit was installed for the laboratory and our adjacent offices. But Lutt had worked like that for 20 years.

Lutt cared little for personal comfort. As long as he had what he needed to do research, he was satisfied. Early in his career funds for research had been inadequate and spending money for comfort or convenience could not be considered. Basically, however, his focus on his research was so intense that he did not want to deal with inessentials. He also disliked anything that kept him from his research. In Venezuela, for example, he became very frustrated with the traditional 2 hour lunch break. To be kept from his studies for so long was unacceptable to him, especially when working on something new and interesting.

Lutt was an avid collector. If left alone outdoors for more than a few minutes, he would invariably head for the nearest plants and begin examining them for fungi with his hand lens. However, accumulating things detracted from time better spent on research, so he kept only what he had immediate use for. He received numerous reprints from all over the world, which he stacked in small piles according to subject around his office. As the piles became larger, it was difficult to remember what was in them. When I moved to Griffin I was assigned the office formerly occupied by BB Higgins, and accepted his offer to take over his collection of reprints assembled during his nearly half-century long career. These I added to my small collection and arranged alphabetically in document boxes. Soon Lutt got into the habit of coming to my office to look for articles he could not locate, but often I lacked what he needed. Then one day he suggested, "Why don't I just give you all of my reprints. You can organize them and I'll know where to find them. And you can use them too." He kept in his office only what he was working on.

Lutt attended his first Mycological Society of America (MSA) meeting in 1962, held at the University of Massachusetts in Amherst. Shortly after we arrived, I introduced Lutt to the local representative, Margaret Barr Bigelow, a former classmate of mine at Michigan. During their conversation Lutt was explaining where he worked and Margaret said, "Oh, that's where Dick Hanlin works." Thereafter, when asked where he worked, Lutt responded, "I work where Hanlin works." It was some time before he let me forget that.

After that first meeting, Lutt regularly attended the MSA, and when possible, APS meetings. He also attended the First International Mycological Congress in England and the precongress foray in Switzerland, and later traveled to Mexico and Venezuela to collect fungi. Meetings were very tiring for him, but he enjoyed the interchange of ideas with professional colleagues, as well as the social contacts. The MSA/AIBS meetings were usually held on university campuses, and this frequently required walking long distances between residence halls and meeting sites. Although walking was difficult for Lutt, he steadfastly refused any assistance. Consequently, several of us who usually drove to the meetings devised a scheme whereby one of us would "happen" to drive close by the meeting room when Lutt was coming out; he would stop and offer a ride, which Lutt readily accepted. Had Lutt known that this had all been arranged, we could not have pushed him into the car.

Lutt's presence at meetings was eagerly anticipated each year. He was always surrounded by colleagues engaged in a stimulating discussion of some technical topic, or enjoying the humor of his stories. Students found him easy to approach and talk to, as he always took an interest in their studies. He had a knack of making you feel at ease, regardless of your status in your profession.

One cannot think of Lutt without recalling his great sense of humor and his prowess as a storyteller. Although his research papers were written in a serious vein, he often injected humor into his talks. He once presented a paper entitled "Significant studies of insignificant diseases," and when invited to give a talk on single spore isolations he began with "The first principle to remember in making single ascospore isolations is that it is usually unnecessary" (2). His history of the Georgia Experiment Station (17) is filled with his subtle humor. Because of it he was often the target of jokes by other faculty members, who could be certain that he would retaliate when an appropriate moment arose. For some reason he had a knack of having humorous things not of his own doing happen to him, and these became the source of numerous anecdotes among his friends. He took all of this in good spirits.

EPILOGUE

Working with Lutt was a rare privilege. Professionally he is remembered for the quality and innovativeness of his research, but those associated closely with him will recall his personal qualities. He was a man of integrity and sincerity, caring and sensitive, with a deep sense of fairness and responsibility. His convictions were strongly held, but he was willing to listen to opposing views. Despite his recognition, he remained modest and dedicated to his fungi, never deviating from his chosen path. Perhaps the best indication of how much he meant to his former colleagues is the frequency with which they still mention him.

Literature Cited

1. Cashion NL, Luttrell ES. 1988. Host-parasite relationships in karnal bunt of wheat. *Phytopathology* 78:75–84
2. Kendrick B, Samuels GJ, Webster J, Luttrell ES. 1979. Techniques for establishing connections between anamorph and teleomorph. In *The Whole Fungus*, ed. B Kendrick, 2:635–51. Ottawa: Natl. Mus. Can.
3. Luttrell ES. 1951. Taxonomy of the pyrenomycetes. *Univ. Missouri Stud.* 24: 1–120
4. Luttrell ES. 1954. The Cladoniaceae of Virginia. *Lloydia* 17:275–306
5. Luttrell ES. 1954. Incite, incitant, and incitement. *Plant Dis. Reptr.* 38:321–22
6. Luttrell ES. 1955. The ascostromatic ascomycetes. *Mycologia* 47:511–32
7. Luttrell ES. 1963. Taxonomic criteria in *Helminthosporium. Mycologia* 55: 643–74
8. Luttrell ES. 1964. Systematics of *Helminthosporium* and related genera. *Mycologia* 56:119–32
9. Luttrell ES. 1965. Classification of the Loculoascomycetes. *Phytopathology* 55: 828–33
10. Luttrell ES. 1965. Paraphysoids, pseudoparaphyses, and apical paraphyses. *Trans. Br. Mycol. Soc.* 48:135–44
11. Luttrell ES. 1974. Parasitism of fungi on vascular plants. *Mycologia* 66:1–15
12. Luttrell ES. 1977. Correlations between conidial and ascigerous state characters in *Pyrenophora, Cochliobolus,* and *Setosphaeria. Rev. Mycol.* 41:271–79
13. Luttrell ES. 1978. Biosystematics of *Helminthosporium:* impact on agriculture. In *Beltsville Symp. Agric. Res. 2. Biosystemat. Agriculture,* pp. 193–209. Montclair, NJ: Allanheld, Osmon & Co.
14. Luttrell ES. 1979. Deuteromycetes and their relationships. See Ref. 2, 1:241–64
15. Luttrell ES. 1980. Host-parasite relationships and development of the ergot sclerotium in *Claviceps purpurea. Can. J. Bot.* 58:942–58
16. Luttrell ES. 1981. Tissue replacement diseases caused by fungi. *Annu. Rev. Phytopathol.* 19:373–89
17. Luttrell ES. 1985. An account of the origins and development of plant pathology in the University of Georgia and the several Experiment Stations. *Univ. Georgia Agric. Exp. Stn. Spec. Publ.* 35:1–56
18. Luttrell ES. 1987. Relations of hyphae to host cells in smut galls caused by species of *Tilletia, Tolyposporium,* and *Ustilago* (Ustilaginales). *Can. J. Bot.* 65:2581–91
19. Luttrell ES. 1989. The package approach to growing peanuts. *Annu. Rev. Phytopathol.* 27:1–10
20. Luttrell ES. 1989. Morphology of *Meliola floridensis. Mycologia* 81:192–204
21. Shearer CA. 1978. Fungi of the Chesapeake Bay and its tributaries VII. *Luttrellia estuarina* gen. et sp. nov. (Ascomycetes). *Mycologia* 70:692–97

Annu. Rev. Phytopathol. 1995. 33:37–67

THE IMPACT OF MOLECULAR CHARACTERS ON SYSTEMATICS OF FILAMENTOUS ASCOMYCETES[1,2]

G. J. Samuels
United States Department of Agriculture, Agriculture Research Service, Systematic Botany and Mycology Laboratory, Room 304, B-011A, BARC-W, Beltsville, Maryland 20705-2350

K. A. Seifert
Centre for Land and Biological Resources Research, Agriculture and Agrifood Canada, Research Branch, Ottawa, Ontario K1A 0C6, Canada

KEY WORDS: systematics, Ascomycetes, Deuteromycetes, phylogeny

ABSTRACT

Information derived from nucleic acid analyses either has complemented phylogenetic arguments based on phenetic characters or facilitated choice among competing hypotheses. Despite limited taxon sampling, a picture of the interrelationships of filamentous Ascomycetes at higher taxonomic levels is developing. Intergeneric relationships within groups that include economically important fungi (e.g. Eurotiales, Hypocreales) are being clarified, and generic circumscriptions redefined. Molecular analyses have supported predictions of links between individual asexual species of groups or asexual species of the Fungi Imperfecti, and groups of Ascomycete genera and species. However,

[1]This review is dedicated to the memory of FA "Bud" Uecker, a friend and colleague who bridged the gap between phenetic and molecular phylogenetics.

individual asexual species have not been linked unequivocally to individual Ascomycete species. Anamorph names are necessary and should be retained because teleomorphs may not be recognized in vivo nor formed in vitro. In the few cases where phenetic and molecular phylogenies seem irreconcilable, the ribosomal genes may not give the most parsimonious explanation. The taxon name "Plectomycetes" is confused and should be dropped.

INTRODUCTION

The primary aim of systematic mycology is to arrange fungi in a taxonomy that facilitates identification of species and enables prediction of biological activity. The ability to provide such a taxonomy is directly related to the number of characters that can be used to delimit species and other taxa. A succession of tools that included increasingly powerful lenses, pure culture techniques, and the ability to compare enzymes and secondary metabolites have led to increasingly useful taxonomies. The ability to look directly at nucleic acid sequences, through polymerase chain reaction (PCR) technology, tantalizes with the possibility that identification and phylogenetic reconstruction can be undertaken without even seeing the fungi in question. The respective roles of nucleic acid and phenetic characters in systematic biology are rapidly evolving, and the process is a topic of great debate. The long-term interaction of these two types of characters on fungal diagnostics and understanding of fungal phylogeny is unclear.

The methods employed for gathering molecular data have been outlined in other publications (21, 66). Molecular data are often analyzed using maximum parsimony within cladistic analysis, or less frequently by phenetic or numerical taxonomy methods. The process and interpretation of the analysis are beyond the scope of this review. Here, we do not examine the methods of gathering data nor of their analysis, but rather accept at face value results presented by the respective authors. However, it is important to realize that the traditional Linnaean, hierarchical classification system (classes, orders, family, genera, etc) is not always easily compared with the results of cladistic analysis, which identify only monophyletic groups. Although in phenetics, for example, it might be possible to designate a similarity index as the equivalent of a genus, in cladistics this is not done. Certain clades might be considered equivalent to a Linnaean genus, but other clades will have no obvious Linnaean equivalents (156). The monophyly of certain Linnaean taxa may be confirmed, but to date, a satisfactory means of handling taxa that are not monophyletic has yet to be developed. Examples of reviews of data analysis and congruence between phenetic and genetic analyses include Swofford (142) and Patterson et al (97).

A major impediment to achieving natural and usable taxonomies for filamentous Ascomycetes and the related Deuteromycetes is the difficulty in recognizing the difference between homology and convergence in morphology. Many of the fungi that the phytopathologist encounters are strictly mitotic, lacking a sexual reproductive phase. These are classified in the Fungi Imperfecti, a system that was based from the outset on convergence in form. For this reason, it is widely recognized that species included in one anamorph genus sometimes may only be distantly related. Morphology by itself may not permit prediction of biological relatedness among anamorph genera or even among species within a genus. Moreover, the phenetic characters used in classification and phylogenetic reconstruction of the sexually reproducing Ascomycetes can be interpreted differently by different individuals, a tendency that frustrates students and users of the taxonomic system. As a consequence, there is as yet no generally acceptable higher level (class, order, family) classification of filamentous Ascomycetes. Characters derived from DNA offer data that are determined independently of morphology and can therefore be useful in evaluating homology versus convergence in form.

The modern era of Ascomycete systematics began in 1932 when Nannfeldt (91) divided the filamentous Ascomycetes into the large groups Discomycetes, Pyrenomycetes, Loculoascomycetes, and Plectomycetes, largely on the nature of their ascomata and asci. The phylogenetic significance of ascomatal characters has been evaluated using 18S and 28S rDNA sequences. Is the perithecium (Pyrenomycetes), apothecium (Discomycetes), or ascostroma (Loculoascomycetes) indicative of a monophyletic group? Are all cleistothecial Ascomycetes—Nannfeldt's Plectascales—more closely related among themselves, or are groups of them more closely related to Pyrenomycetes, Discomycetes, or Loculoascomycetes than to other cleistothecial fungi? What is the phylogenetic significance of sterile tissue in the ascoma; or ascal type, whether it is unitunicate, bitunicate (= fissitunicate), operculate, or inoperculate? In this paper we review the interaction as it stands today in relation to filamentous Ascomycetes and the associated mitotic or anamorphic phases, the Deuteromycetes, which together include many of the plant pathogenic fungi.

PYRENOMYCETES

Pyrenomycetes are Ascomycetes that have unitunicate asci and perithecial ascomata that open through a narrow apical pore. The arrangement of the pyrenomycetous fungi into orders and families is far from settled because of differing interpretations of relatively few studies of perithecial ontogeny, and other morphological characters. One of the most important advances in the systematics of the pyrenomycetous fungi was the recognition of several different types of perithecial organization (76, 76a). However, there is a tendency for large groups of species to lack salient characters such as characteristic ascal apices, sterile filaments among asci, or anamorphs that give clues as to their proper placement. The taxonomic placement of important plant pathogenic fungi such as *Ceratocystis* and *Ophiostoma,* and the consequent recognition of their nearest relatives, for example, was only partially resolved through phenetic methods. Molecular methods have enabled the development of trees that indicate interrelationships of the Pyrenomycetes and suggest placement for morphologically ambiguous taxa and resolution for persisting conflicts resulting from differing interpretations of phenetic characters.

Homology and convergence in centrum ontogeny was discussed in the context of 18S rDNA sequences in the Pyrenomycetes (135). Two major lines of evolution were defined, one that included the Hypocreales (*Nectria, Fusarium,* etc), Clavicipitales (*Claviceps, Epichloë*), *Ceratocystis*, and *Glomerella*; and one that included *Diaporthe, Ophiostoma, Sordaria,* and *Xylaria.* Centrum types, as defined by many authors, were congruent with the DNA gene tree. Given limited taxon sampling, the topology of the gene tree mapped onto the strict consensus tree of centrum types supports strong phylogenetic significance of centrum type at the ordinal level.

The Hypocreales and Clavicipitales and Their Anamorphs: Acremonium, Fusarium, Cylindrocladium, Cylindrocarpon, Gliocladium, Trichoderma, Verticillium

Pyrenomycetes with brightly or lightly pigmented ascomata have, since at least the mid-nineteenth century, been placed together as one or more orders or families. These fungi have been collectively referred to as the "hypocrealean" fungi (115) and include perhaps a larger concentration of plant pathogenic and biologically useful fungi than any other group. Well-known members include the teleomorphic genera *Gibberella, Nectria, Epichloë,* and the anamorphic genera *Acremonium, Cylindrocarpon, Cylindrocladium, Fusarium, Gliocladium, Trichoderma,* and *Verticillium.* The Hypocreales *s. str.* has been recognized to comprise one family, the Hypocreaceae, or three families, the Hypocreaceae for *Hypomyces, Hypocrea* and including *Trichoderma*, Nectriaceae for *Nectria, Gibberella,* and relatives and including *Fusarium, Cylindrocarpon,* etc, and Hypomycetaceae for *Hypomyces*, a genus of species parasitic on agarics and polypores (see review in 115).

Molecular systematics of the order, studied directly or indirectly (48, 93, 110–112, 133, 135), indicate that the Hypocreales is paraphyletic. The hypocrealean fungi form a single clade that also includes the Clavicipitales. The perceived relationship between the Hypocreales, in the traditional sense, and the Clavicipitales has varied from time to time and individual to individual. In recent years, the groups have been given ordinal status in part because of centrum ontogeny (76) but also because of differences in biology (5, 115). The seemingly different ontogenetic types of ascomata have been proposed to be homologous (135). As discussed below, sequences of 18S and 28S rDNA indicate that the Clavicipitales are a sister group to non-nectriaceous members of the Hypocreales (112, 133, 134). Derivation of the Clavicipitales from within the Hypocreales is also suggested on the basis of sequences of the orotodine-5′-monophosphate decarboxylase gene (110).

Cladistic analysis of 18S rDNA (133, 134) and 28S rDNA sequences (112) suggests the existence of three major clades that may be interpreted as correlating with families. These include a basal group of nectriaceous fungi with red or purple perithecia, including among others *Nectria haematococca, Gibberella, Neocosmospora* and anamorphs in *Fusarium* and *Cylindrocarpon,* and a separate clade including species of *Nectria* with orange perithecia (*N. ochroleuca* and *Gliocladium roseum* anamorph), which together can be considered the Nectriaceae. The Hypocreaceae clade includes *Hypocrea* and *Trichoderma,* as well as *Hypomyces,* and the Clavicipitaceae includes *Epichloë, Balansia,* and the *Acremonium* grass endophytes (111, 112, 135). The relative positions of the clades differ according to which gene is analyzed. Sequences of the 18S rDNA gene indicated that the Clavicipitales is a sister group to a clade that includes *Hypocrea* and *Hypomyces,* both on a clade that has the "red" *Nectria* (including *Fusarium*) at its base (133). Nearly the opposite arrangement was found when the 18S gene was sequenced (112): *Hypocrea* and *Hypomyces* were on a clade that was basal to a clade that included *Nectria* and the Clavicipitales.

The clade that includes *Nectria* species with red perithecia and *Gibberella,* and their anamorphs, includes the largest number of species in the order and the most plant pathogens. Differences in perithecial pigmentation and anatomy, anamorph, and habitat indicate where new generic lines could be drawn (Table 1). Species now known as *Nectria* have either yellow/orange or red perithecia, and these differences correlate to types of anamorphs (124). The yellow/orange perithecial forms often have anamorphs similar to *Gliocladium roseum,* a common soil fungus that is useful in biological control; the red perithecial forms include, among others, *Fusarium, Cylindrocarpon,* and *Cylindrocladium* anamorphs. In more detailed analysis of the red clade, using 18S rDNA data, *N. cinnabarina,* the cause of coral spot on a wide variety of temperate woody hosts and the type species of the genus, was shown to be basal to all other members of that clade (112). Many saprobic species of the Hypocreales have not been included in molecular-phylogenetic studies, so the final organization of the order is still unresolved.

The taxonomy of *Fusarium* has been perhaps the most controversial of all fungi. Much of the debate has centered on species concepts, with the acceptance that the genus itself was well defined by the distinctive falcate phragmoconidia with notched foot cell. But the existence of so-called *Fusarium* anamorphs in different Ascomycete orders has led the generic concept to be

Table 1 Natural groups of *Nectria, Gibberella,* and their anamorphs, and *Fusarium* and *Fusarium*-like anamorphs and their teleomorphs supported by DNA sequence data

Teleomorph*	Anamorph*	Reference
Nectria cinnabarina (type)	*Tubercularia*	93, 111, 112
N. subg. *Dialonectria*	*Fusarium* sect. *Eupionnotes*	48, 93
N. rigidiuscula-Group	*Fusarium decemcellulare, Fusarium* spp.	48, 93
N. haematococca/Neocosmospora	*Fusarium* sect. *Martiella "Acremonium"*	93, 112
N. atrofusca/N. desmazieresii	*Fusarium staphyleae, F. buxicola*	93
Gibberella	*Fusarium s. str.* (sects.) *Elegans [F. oxysporum], Liseola*	48, 93
Hypocreales (unknown)	*Fusarium dimerum*	93
*Monographella nivalis**	*Microdochium (F. nivale)**	42, 48, 122
N. radicicola, N. coccinea/galligena	*Cylindrocarpon*	93, 112
N. ochroleuca (Bionectria)	*Gliocladium roseum (= Dendrodochium)*	93, 111
Calonectria	*Cylindrocladium*	93, 112
*Plectosphaerella**	*Plectosporium (F. tabacinum)**	95, 112, 148

*Nonhypocreaceous taxa

questioned, and some species to be removed (42, 95, 122). Guadet et al (48) sequenced two highly variable stretches of the 5′ end of the 28S rRNA gene to evaluate divergence within *Fusarium.* Their results indicated two monophyletic groups: One comprised *F. oxysporum,* which is a strictly asexual anamorph, *Gibberella,* and *N. rigidiuscula* (*F. decemcellulare,* as *Calonectria*); and the second group comprised *N. haematococca* and *Fusarium* sect. *Martiella* (= *F. solani sensu* Snyder & Hansen), in which the anamorphs of *N. haematococca* and its relatives can be classified. O'Donnell (93) sampled a broader range of *Fusarium* species and nectriaceous members of the Hypocreales and concluded that the sections that are currently recognized in *Fusarium* are artificial and should be modified or abandoned. The fusaria were widely distributed among the teleomorph groups. Most fusaria fell into a single group that included two monophyletic subgroups: (*a*) *Gibberella-N. haematococca-N. rigidiuscula*; and its sister group (*b*) *N. episphaeria* (anamorph = *F.* sect. *Eupionnotes*)-*N. flammea* (anamorph = *F.* sect. *Coccophilum*)-*N. ventricosa.* Group (*a*) was a sister group to *Calonectria* (anamorph = *Cylindrocladium*) and *N. penicillioides* (anamorph = *Flagellospora*). Two additional species, *F.*

buxicola (*N. desmazieresii*) and *F. staphyleae* (*N. atrofusca*), were much more closely related to *Tubercularia* (*N. cinnabarina*-Group) and *Cylindrocarpon* (*N. coccinea/galligena*-Group) than to the rest of the fusaria. Guadet et al (48) found that *F. nivale* grouped as far away from the rest of the *Fusarium* isolates as did *Neurospora crassa*. Their results thus supported the previous removal of *F. nivale* to *Microdochium* (42, 122) as *Microdochium nivale* (teleomorph = *Monographella nivalis*, Xylariales: Hyponectriaceae, 5). Most strains identified as *F. nivale* that produce the trichothecene mycotoxin nivalenol are now known to have been misidentified and are actually other *Fusarium* species, although one strain (NRRL 3289) that produces the toxin at low levels may be true *Microdochium nivale* (79, 100). The groupings of teleomorphs that have *Fusarium* anamorphs generally conform to perithecial anatomy, habitat (116), and mycotoxin profiles.

The taxonomy of particular species complexes within *Fusarium* is certain to remain unsettled for some time. In two groups of species, the first including *Fusarium moniliforme* and its relatives (grouped together as varieties of *Gibberella fujikuroi*, 71) and the second including *Fusarium solani* (*Nectria haematococca*, 153), several intersterile mating populations have been identified using classical genetic mating techniques. These so-called mating populations, which are equivalent to biological species, are being studied intensively, but unfortunately morphological taxonomy is lagging behind. In *G. fujikuroi*, which presently has six recognized mating populations, characters of the anamorph correlate with some of the mating populations, but in each case, there is more than one mating population with a particular "species" of anamorph. Sequences of the 28S rRNA did not exhibit enough variation to allow the biological species within *G. fujikuroi* to be distinguished with any certainty (102). The situation in the *Nectria haematococca* group may be even more complex, but there has been little published on its molecular genetics. RAPD analysis was used to characterize mating populations I and VI within *F. solani* f.sp. *cucurbitae* (28). *Gibberella fujikuroi* and *N. haematococca* are possible examples of fungi that have been undergoing fairly rapid speciation in response to agricultural activities. If speciation is a relatively recent event, detecting phenotypic characters (or even genotypic characters) that reflect the speciation may prove challenging.

Gliocladium was originally described as an analogue for *Penicillium*-like hyphomycetes with slimy conidia. As teleomorph affinities for described species were discovered, it became clear that these anamorphic characters were found in several different groups of Ascomycetes, or in some cases, that the concept of "penicillate conidiophore branching" had been too broadly interpreted. Focusing on the type species of the genus, *G. penicillioides*, and two biocontrol species *G. virens* and *G. roseum*, Rehner & Samuels (111) studied the 28S rDNA from conidial and ascospore isolates of anamorphs that could be determined as *Gliocladium*, and other hypocrealean taxa that do not have *Gliocladium* anamorphs. Soil isolates of *Gliocladium* that had not been directly derived from teleomorphs grouped with their respective teleomorph groups. As was suspected from the diverse teleomorph relations, *G. virens, G. roseum*, and *G. penicillioides* were shown to represent phylogenetically distinct lineages within the Hypocreales. Because *G. virens* was derived from within *Hypocrea*, near *H. gelatinosa*, the anamorph of which is morphologically very similar to *G. virens* (160), there is little doubt that *G. virens* is phylogenetically an asexual species of *Hypocrea*. These results indicate that von Arx (159) and Bissett (14) were correct in placing the species in *Trichoderma* as *T. virens*. Conidial isolates of *G. roseum*, a commonly encountered soil fungus that is the anamorph of *Nectria ochroleuca* (34), formed a group with *N. ochroleuca* and similar *Nectria* species that have pallid perithecia (= *Bionectria*). The so-called *Gliocladium* anamorphs in this group are morphologically distinct from all other *Gliocladium* species and should probably be classified in the anamorph genus *Clonostachys* (34).

Two additional anamorph genera that have species with teleomorphs in the Hypocreaceae and Clavicipitaceae are *Verticillium* and *Acremonium* (synonym *Cephalosporium*). The insecticolous *Verticillium lecanii*, and the cephalosporin antibiotic-producing species *Acremonium chrysogenum* were shown by sequences of the 28S rDNA gene to have been derived from within the Hypocreales (including the Clavicipitales) (112), but could not be linked to any specific teleomorph genus. *Acremonium chrysogenum* was closely allied to the group that includes *N. ochroleuca*. Surprisingly, the insect parasite *Verticillium lecanii* did not cluster with insecticolous species of *Cordyceps*, but was basal to a clade that included *Nectria* and its segregates. *Verticillium dahliae*, an asexual plant pathogenic species grouped with *Plectosphaerella cucumerina* in a clade that included *Glomerella cingulata* and this *V. dahliae* species, is therefore not hypocreaceous (112). The nonhypocreaceous nature of *V. dahliae* could have been predicted by its production of dark microsclerotia.

Acremonium is a common anamorph in the Hypocreales and Clavicipitales, as well as in at least 40 other unrelated teleomorph genera (G Murase & B Kendrick, personal communication). This is an example of the morphologically simplest phialidic anamorph with light-colored, slimy unicellular conidia ("Moniliaceae"). *Acremonium* species are known to cause disease in plants and animals, including humans, to produce antibiotics and proteases, and to be endophytes of grasses. Because *Acremonium* anamorphs are part of the life cycles of such diverse Ascomycetes, the *Acremonium* form has little phylogenetic significance.

Species of *Acremonium* sect. *Albolanosum* are characterized by subtle morphological characters, occur as endophytes of grasses, and are characterized by the anamorph of *Epichloë* typhina, the cause of choke disease in the grass family Poaceae, subfamily Pooideae (88, 164). Although *Epichloë* has included several species at one time or another, White (162, 163) restricted the genus to *E. typhina* and a few morphologically similar but reproductively isolated, pathogenic species. He regarded their anamorphs to be varieties of *A. typhinum* (teleomorph = *E. clarkii*), which can be distinguished on subtle cultural characters. Additional nonpathogenic species of *Acremonium* sect. *Albolanosum* occur as endophytes in grasses but induce no symptoms. These nonpathogenic fungi are biologically and morphologically similar to *A. typhinum* and have been considered to be *Epichloë* species despite lacking a sexual phase. Isozyme banding patterns were not sufficient to separate *Epichloë typhina* isolates from the nonpathogenic, seed-borne *Acremonium* endophytes, and some *Acremonium* isolates are frequently more similar to *Epichloë* isolates in isozyme characters than to other *Acremonium* isolates, and vice versa (72). Sequences of the nuclear ITS1 and ITS2 rDNA could not distinguish some strains of *E. typhina* from some strains of the nonpathogenic endophytes *A. coenophialum, A. lolii*, and *A. starrii* (125). Schardl et al (125) and Tsai et al (146) concluded that nonpathogenic *Acremonium* endophytes, which are apparently strictly asexual, have derived from *E. typhina* on multiple occasions by hybridization with *Epichloë* species. The various studies, including isozyme results (26, 72), have demonstrated a complex population structure in these endophytes.

Trichoderma species are known only to be linked to *Hypocrea* and the closely related genera *Podostroma* and *Sarawakus*. Species of *Trichoderma* are well known in biological control of plant disease, as well as in the production of cellulase and as models in the study of genetic regulation. The most widely used taxonomic system of *Trichoderma* recognizes nine aggregate species, within which the limits of individual species are not clear (114). Bissett (12–14) recognized morphological sections and discrete species, but these sections and species have not been examined by phylogenetic analysis. Molecular studies in *Trichoderma* have often not included detailed attempts to correlate morphology with molecular characters. Furthermore, because the *Hypocrea* sexual states of *Trichoderma* species have only rarely been observed in culture, it has not been possible to calibrate any morphological or molecular features against any kind of biological species concept. Nucleic acids have been used with varying degrees of success to recognize strains and a few species in *Trichoderma*. Zimand et al (167) found that ten strains of *T. harzianum* had similar RAPD patterns, whereas few strains determined as *T. viride* or *T. hamatum* showed any similarities. Muthumeenakshi et al (89) found three discrete ITS types within the morphological species *T. harzianum*, of which one was highly pathogenic to commercially grown mushrooms. A RAPD technique was also used to eliminate duplicated strains and to identify genomic diversity of *Trichoderma* isolates in microbial screening (41). In none of these studies was there a detailed attempt to correlate morphology with molecular characters. The correlation between conidial ornamentation and two types of mtDNA fragment patterns suggested the possibility of at least two species within the morphological species *T. viride* (80). RFLPs (81) and a combined morphometric and isozyme study (123) indicated that the industrial species *T. reesei* is distinct from the morphologically similar *T. longibrachiatum*. ITS1 sequences of rDNA of *T. reesei* and the Ascomycete *Hypocrea jecorina* are identical (K Kuhls, personal communication), and both exhibit equally high production of cellulase enzymes (D Turner & C Kubicek, personal communication), leading to the conclusion that *T. reesei* is the anamorph of *H. jecorina*. However, isozyme profiles and cultural characters (123) indicate a genetic difference that was not detected by sequencing of the ITS1 region. Kuhls (personal communication) concluded that *T. reesei* was probably recently derived from *H. jecorina* and has maintained overall genomic similarity.

Ceratocystis and Ophiostoma

The taxonomy of fungi included in *Ceratocystis* and *Ophiostoma* has been controversial at all taxonomic levels and represents one of the best examples of morphological convergence that has been examined by molecular techniques. Species of these genera are well known pathogens of trees and some vegetable crops. They produce globose, usually black, ascomata with thin walls, a centrum filled with deliquescing asci that do not discharge, and variously shaped ascospores that are usually extruded through a long ascomatal neck. The appropriate ordinal classification of these fungi was uncertain because they included a mixture of characteristics considered typical of Pyrenomycetes (ascomata with preformed openings) and Plectomycetes (centrum development, deliquescing asci). Nannfeldt (91) included them in the Plectascales, in the order Ophiostomatales, while recognizing that they have both plectomycetous and pyrenomycetous characters. A relationship of *Ophiostoma* and *Ceratocystis* to the Endomycetaceae (yeasts) was proposed under the assumption that hat-shaped ascospores, formed in both groups, are unlikely to have been evolved more than once (109). However, differences in ascospore ontogeny and 18S rDNA sequences have demonstrated the multiple development of this distinctive form (51).

In recent years, the correct ordinal placement of these fungi was entangled in disagreements over generic concepts. Samuels (121) reviewed the history of disagreement over the generic concepts. Because of the morphological similarities of the ascocarps, *Ophiostoma* was considered a synonym of *Ceratocystis* by Upadhyay (151), a conclusion generally not accepted in Europe. Because of differences in anamorphs, cycloheximide sensitivity, cell wall chemistry (see reviews in 33, 134), pathogenicity (50, 58, 132), and dispersal biology (78), the two genera were considered distinct by an increasingly large number of scientists. Furthermore, not only were the genera sometimes considered distinct, but they were not even considered closely related by some workers. For example, Barr (5), utilizing phenetic and biological characters, placed *Ceratocystis* and *Ophiostoma* in separate orders, the Sordariales and the Microascales, respectively.

Two independent sets of molecular studies have examined the taxonomic affinities of these fungi and the question of whether *Ceratocystis, Ophiostoma,* and other segregate genera such as *Europhium* and *Ceratocystiopsis* should be considered synonyms. Hausner et al (52) obtained sequences of the 18S and 28S rDNA of several species within the complex, and found that *Ceratocystis* and *Ophiostoma* formed separate monophyletic lines, a finding confirmed by Spatafora & Blackwell (134) from sequences of 18S rDNA of a broad range of pyrenomycetous genera. The synonymy with *Ophiostoma* of *Europhium,* proposed (96) for species lacking ascomatal necks, and *Ceratocystiopsis,* accepted by Upadhyay (151) for species with falcate ascopores, was suggested by sequences of both the 18S and 28S rRNA genes (51, 53).

The molecular studies have also confirmed that *Ceratocystis* and *Ophiostoma* are not closely related. In the analysis by Spatafora & Blackwell (134), *Ceratocystis* grouped on a major clade with the Microascales, Clavicipitales, and Hypocreales, whereas *Ophiostoma* grouped on a second major clade with the Diaporthales, Sordariales, and Xylariales. In the study by Hausner et al (52), *Ceratocystis* also grouped with a member of the Microascales. The sampling of outgroup taxa was too small to determine the affinities of *Ophiostoma,* but Hausner et al (54) did not identify particularly close relatives and suggested that the genus should occupy its own order. No close relationship was found between either *Ceratocystis* or *Ophiostoma* and the Eurotiales.

Species concepts in economically important species of *Ophiostoma* have also been the focus of many molecular studies. The study by Hausner et al (54) represents one of the most detailed examinations of species within a genus by sequencing. The relationships between 55 species were inferred from 28S rDNA sequences. The resulting cladogram failed to support the natural subdivision of the genus based on characters of the ascospore or the anamorph.

Brasier (19) formally proposed a new species, *Ophiostoma novoulmi,* as the cause of the current pandemic of aggressive Dutch Elm disease. Isolates formerly considered an aggressive subgroup within *O. ulmi* were found to be reproductively isolated from the nonaggressive subgroup. Although the species were thus considered distinct for classical genetic reasons, they could also be differentiated by morphological and cultural characters, as well as by RFLPs of mitochondrial DNA.

One anamorph genus linked to *Ophiostoma, Sporothrix,* has been used extensively in the arguments over abandoning the Deuteromycetes, as discussed below. *Sporothrix* species have a simple morphology, consisting of sparingly branched, hyaline conidiophores producing single or chained conidia from clusters of denticles. Similar anamorphs, also referred to *Sporothrix*, are also found in the hyphal yeasts and phragmobasidiomycetes (Dacrymycetales) (32). Despite the simple morphology, however, the species can readily be distributed among the main taxonomic groups based on colony characteristics, coenzyme Q systems, cell wall carbohydrate composition, and ultrastructure of cell wall and septal pores. Using some of these characters, de Hoog (31) intercalated one group of *Sporothrix* species, including *Sporothrix schenkii*, the cause of sporotrichosis in humans, with *Ophiostoma*. This was supported by results obtained by Berbee & Taylor (8) using 18S ribosomal RNA gene sequence characters.

Glomerella and *Colletotrichum*

Colletotrichum gloeosporioides is the anamorph of the Pyrenomycete *Glomerella cingulata*. The two stages of this holomorph species occur on species of at least 175 genera of plants in the United States (37a). Despite the distinctive morphological characters of *Glomerella*, its taxonomic position has been unsettled (149). The genus is usually included in the Phyllachoraceae, which has been included in the Phyllachorales (3, 37) and in the Xylariales (5). Uecker (149) found the centrum of *G. cingulata* to be of the *Sordaria* type and very similar in development to *Plectosphaerella cucumerina* (149; anamorph = *Fusarium tabacinum*, which is *Plectosporium tabacinum*, 95) and *Triangularia backusii*. Sequences 18S rDNA of *Glomerella cingulata* and *Colletotrichum gloeosporioides* placed them together on a clade with the Microascales, Sordariales, Hypocreales, and Clavicipitales (135). This argues that at least *Glomerella* is a representative of a distinct order, but the molecular phylogeny proposed by Spatafora & Blackwell (135) did not consider possible close relatives. *Glomerella* and *Plectosphaerella* clustered together when the 18S rDNA gene was sequenced (112).

Colletotrichum species have received some attention from molecular biologists. The confusion over species concepts arose from the work of von Arx (157), who synonymized a vast number of species under the name *C. gloeosporioides*. Sutton (140, 141) has shown that many species in this complex can be distinguished easily from each other by features of conidial shape and dimensions, colony characters, and appressorial shape. The relationships of some of the species in the genus were inferred from studies of the ITS2 and 28S rDNA domains (127). The distinctions between some of the species segregated from *C. gloeosporioides* were confirmed, as were the concepts of some of the species with falcate conidia. However, Sherriff et al (127) concluded that *C. gloeosporioides*, as it is presently defined, is still somewhat heterogeneous. Braithwaite et al (17) distinguished between two of the forms of *C. gloeosporioides* occurring on *Stylosanthes* by using RFLPs obtained from the hybridization of probes with digested genomic DNA. Many molecular studies have concentrated on distinguishing the restricted number of species found on particular crops, or the population biology of particular species. For example, species on strawberry, which could be distinguished by conidial shape supplemented by colony characters (49), were also distinguished using ITS1 sequences (136) and isozymes (15). Vaillancourt & Hanau (152) used RAPD and RFLPs of mitochondrial DNA to distinguish between *C. graminicola* isolates from corn and sorghum, which were subsequently shown to have distinct *Glomerella* teleomorphs and were hence considered separate species. The ITS1 sequence of the coffee pathogen *C. kahawae* was identical among ten strains originating from several African countries (137).

Populations of *Colletotrichum gloeosporioides* are thought to have limited host ranges and to be genetically distinct, and only isolates found within the narrow host range have been thought to be sexually compatible. However, Cisar et al (27) found sexual compatibility between *C. gloeosporioides* from distantly related hosts, although the pattern of inheritance of RFLP markers was complex.

THE CLEISTOTHECIUM AND THE PLECTOMYCETES

The Plectomycetes, based roughly in Nannfeldt's (91) supraordinal group Plectascales, is characterized by fungi that have cleistothecial ascomata (i.e. asci enclosed in an ascoma that does not have a preformed opening, where the asci tend to be globose, lack an apical discharge mechanism, and deliquesce early in development, and that generally have unicellular and often globose ascospores). Morphologists have recognized that cleistothecia have been derived many times from hymenial Ascomycetes, including Pyrenomycetes, Loculoascomycetes, and Discomycetes (see e.g. 3, 7, 77).

Berbee & Taylor (9, 10) adopted a more restrictive view of the Plectomycetes. This view included fungi with *Aspergillus, Penicillium*, and *Paecilomyces* anamorphs, as well as the phylogenetically ambiguous *Ajellomyces capsulatus* (cause of human histoplasmosis), *Ascosphaera apis* (cause of chalk brood of bees), and *Monascus ruber* (cause of spoilage of starchy foods and a source of red food dye in some Asian countries). Ascomata of all of these fungi are closed, thus cleistothecial and plectomycetes in Nannfeldt's sense. Sequences of the 18S rDNA gene (9) indicated that this group of fungi clustered independently of a group of more or less typical pyrenomycetous fungi. Because the cleistothecial and perithecial fungi clustered independently, based on their narrow sampling, Berbee & Taylor (9) suggested that their molecular phylogeny could be interpreted as corresponding to ascomal morphology. However, limiting Ascomycetes with cleistothecia to the same class (9) is simplistic. Sequences of 18S and 28S rDNA (10, 112, 134) provided molecular evidence for morphologically based arguments (77, 158) that forcible and passive discharge of ascospores is interspersed among the pyrenomycetous fungi; the cleistothecial condition per se is not therefore necessarily an indicator of relationships.

The core of the Plectomycetes (sensu Berbee & Taylor, 9), the Eurotiales and Onygenales (including *Ajellomyces*), have been placed in their own subclass on morphological characters (5). Additional molecular evidence, based on sequences of the 18S rDNA, supports a supraordinal position for these fungi, possibly as a sister group to the Pyrenomycetes (45, 134, 150). In a study that emphasized Discomycetes, the Eurotiales and Onygenales clustered on a clade with the Pyrenomycetes that had the inoperculate Discomycete order Helotiales at its base (45). Two other fungi included by Berbee & Taylor (9) in the class Plectomycetes, *Monascus ruber* and *Ascosphaera,* have not previously been linked to the Eurotiales or Onygenales, and their phylogenetic affinities have always been obscure. Malloch (77) and Barr (3) placed *Monascus* in the operculate Discomycete order Pezizales. Skou (131) considered *Monascus* to belong to the Plectomycetes, which was not defined but was viewed broadly, and retained the order Ascosphaeriales for *Ascosphaera,* which he considered related to the Plectomycetes. Barr (3) and Currah (30) considered the Ascosphaeriales and Onygenales to be at best distantly related to each other, and Barr (3) proposed the class Ascosphaeromycetes for *Ascosphaera.*

Powdery mildews also could be interpreted as Plectomycetes, in the broad sense, because their ascomata are closed. However, because their asci are basal, whether solitary or multiple, they have been included among the Pyrenomycetes (e.g. 1), or thought to have been derived from some group of Discomycetes (e.g. 20, 22). On the basis of sequence analysis of 18S rDNA, the closest neighbor for the powdery mildew *Blumeria graminis* f.sp. *hordei* was the inoperculate Discomycete *Sclerotinia sclerotiorum* (45, 118), which supports derivation of the powdery mildews from inoperculate Discomycetes.

Penicillium, Aspergillus, and the Eurotiales

Most microbiologists are familiar with species of *Penicillium* and *Aspergillus.* They are common colonizers of most organic matter, produce harmful mycotoxins, useful organic acids, and cause diseases of plants and animals. Because of their biological, economic, and social impact, their taxonomy has been studied in great detail. Despite this close scrutiny, species remain difficult to identify, especially in *Penicillium.* Generic concepts can be questioned because intermediate forms between the two genera occur, and because extreme morphological forms have sometimes been accorded generic rank. Anamorph names are far more familiar than the corresponding teleomorph names, but the existence of different teleomorph genera within both *Penicillium* and *Asper-*

gillus has led to questions about the monophyly of both the anamorphic and teleomorphic genera. Furthermore, the anamorphic genus *Paecilomyces* shares some teleomorph genera with *Penicillium*, and the distinction between the two anamorph genera has been problematic. Molecular techniques have contributed to solving some of these persistent questions. For example, partial sequences of the 18S rRNA have shown that *Aspergillus* and *Penicillium* seem to be distinct monophyletic lines, and the morphological characters that have traditionally been used to distinguish them appear to be phylogenetically accurate (25). For a review of recent trends in *Aspergillus* and *Penicillium* systematics see (105).

The classical morphological, culture-based taxonomy of *Penicillium* was established by Thom (144) and Raper & Thom (108), who described teleomorphs but continued to use only anamorphic names. Pitt (104) refined *Penicillium* taxonomy with the addition of physiological characters, and clarified the nomenclature of the affiliated teleomorph genera *Eupenicillium* and *Talaromyces*. Species concepts were further refined by chemotaxonomic application of mycotoxin profiles pioneered by Frisvad & Filtenborg (38).

Many *Penicillium* species classified in the subgenera *Penicillium, Furcatum,* and *Aspergillioides* are linked to teleomorphs in *Eupenicillium* (104, 138). Frisvad et al (39) proposed links between individual *Eupenicillium* species and species in subg. *Penicillium* (the *Asymmetrica* of Raper & Thom, 108) on the basis of mycotoxin profiles. Peterson (101), using sequences from the ITS and D1-D2 regions of the 28S rDNA gene, showed a close relationship between three anamorphic species of subg. *Penicillium* (*P. commune, P. roqueforti,* and *P. digitatum*), and *Eupenicillium crustaceum.* According to Peterson (101), anamorphs in subg. *Penicillium* and some anamorphs of *Eupenicillium* species are closely related in morphology, mycotoxin production, and growth and temperature requirements.

The other teleomorph genus linked to *Penicillium, Talaromyces,* also has species with *Paecilomyces* anamorphs. Anamorphic species classified in *Penicillium* subg. *Biverticillium* (the *Biverticillata-Symmetrica* of Raper & Thom, 108) are morphologically similar to *Penicillium* anamorphs of *Talaromyces* (104) and have similar mycotoxin profiles (40). Sequence analysis of the small mitochondrial rDNA, and the ITS1, 5.8S and ITS2 rDNA genes that intercalated species from subg. *Biverticillium* among various *Talaromyces* species, indicates that the derivation of anamorphic species by loss of the teleomorph has occurred on multiple occasions (73, 74). This finding contradicted earlier results using RFLPs of rDNA, which suggested that subg. *Biverticillium* and *Talaromyces* were separate lines and that the teleomorph had been lost only once (143). *Talaromyces/Penicillium* subg. *Biverticillium* form a separate lineage from *Eupenicillium/Penicillium* subg. *Penicillium* (75).

Morphologically, *Paecilomyces* is differentiated from *Penicillium* by the "nongreen" conidia and also by the shapes of the phialides, which have a long neck in *Paecilomyces* and are variously shaped in *Penicillium*. In addition to teleomorphs in the genus *Talaromyces, Paecilomyces* is linked to the eurotiaceous genera *Byssochlamys* and *Thermoascus* (119) as well as to the noneurotiaceous Clavicipitales (*Torrubiella*, 56) and the Hypocreales (GJ Samuels, personal observation). No comprehensive molecular phylogenetic study of *Paecilomyces* has been published, but it is clear from the teleomorph data that the genus is polyphyletic. Also, the one *Byssochlamys* species with a *Paecilomyces* anamorph (the group that probably includes the type species, *P. variotii*) was intermediate between the *Talaromyces* and *Eupenicillium* clades in the analysis by LoBuglio et al (75), indicating that *Penicillium*, as it is currently defined, is paraphyletic. Revision of the generic concept of *Paecilomyces* need not await molecular data. For example, it is possible that several of the species allied with *P. farinosus* in section *Isarioidea* (119) may form a monophyletic group, perhaps allied with species now classified in *Nomuraea*.

The occurrence of species with *Penicillium*- or *Aspergillus*-like conidiophores on synnemata has caused some taxonomic problems. In the early years of *Penicillium* taxonomy, the segregate genus *Coremium* was widely adopted for synnematous species. The abandonment of *Coremium* as a separate genus was proposed as long ago as 1910 (144), but the classification of synnematous species within the genus has been problematic. Interpretations of the conidiophore branching and phialide shape led some workers to include *P. vulpinum* and *P. clavigerum* in subg. *Penicillium*, and the synnematous *P. duclauxii* in subg. *Biverticillium* (e.g. 108, 139), and other workers to include all three in subg. *Biverticillium* (104). LoBuglio et al (75) reviewed the literature surrounding the confusion and noted that ubiquinones, mycotoxin profiles, and secondary metabolites favored the separation of *P. clavigerum* and *P. duclauxii* into different subgenera. None of these three species is known to have a teleomorph. Using sequences from the mitochondrial small rDNA and the ITS1, 5.8S, ITS2 rDNA region of these synnematous species, they found that *P. vulpinum* and *P. clavigerum* grouped with subg. *Penicillium* and *Eupenicillium*, and *P. duclauxii* grouped with subg. *Biverticillium* and *Talaromyces*.

Synnematous species with *Aspergillus*-like conidiophores also occur, although they are not as well known as synnematous species of *Penicillium*. Samson & Seifert (120) monographed a group of species with luxuriant, yellowish-green synnemata, often up to 6 cm tall, that occur on the seeds of tropical plants, especially persimmons. The teleomorphic core of these anamorphs was considered to be *Penicilliopsis*, characterized by the production of sclerotic, stipitate cleistothecia at the base of the synnemata. *Penicilliopsis* was positively linked to anamorphs in the genera *Sarophorum* and *Stilbodendron*, which have somewhat penicillate conidiophores. Another similar anamorph genus, *Stilbothamnium*, was reduced in rank as a subgenus of *Asper-*

gillus because of the obvious morphological similarity of the conidiophores. Because of the production of apparently sterile sclerotia similar to ascomata of *Penicilliopsis* spp., these species were assumed also to be related to this teleomorph genus. After analysis of the D2 domain of the 28S rDNA gene (35), *Stilbothamnium togoense* (as *S. nudipes*) was found to be closely related to *A. flavus* in section *Circumdati*. This result confirmed the somewhat controversial inclusion of *Stilbothamnium* in *Aspergillus*, but rejected its recognition as a distinct subgenus within *Aspergillus*. Surprisingly, the proposed connection of the synnematous aspergilli with *Penicilliopsis* was not confirmed. Species of *Penicilliopsis* and *Stilbodendron* formed a monophyletic group within the aspergilli, but on a separate clade from the synnematous species formerly assigned to *Stilbothamnium*.

Mononematous *Aspergillus* species are linked with several other teleomorph genera, the best known of which are *Eurotium*, which contains xerophilic species that often cause food spoilage, and *Emericella*, which includes the commonly used experimental organism *Aspergillus nidulans*. In general, the teleomorph genera correspond to subgenera of *Aspergillus*. Sequence analysis of part of the 18S rDNA gene of representatives of the teleomorph genera and selected *Aspergillus* species showed the expected intercalation of anamorphs and teleomorphs (25). The existence of morphologically distinct teleomorphs that share an apparently monophyletic anamorph morphology led Gams (43) to question whether there was any practical reason to create segregate anamorph genera to correspond with the distinct teleomorph genera.

The taxonomy of other plant parasitic species of *Aspergillus* also has been clarified by molecular data. Species concepts in the aflatoxin-producing species *A. flavus* and *A. parasiticus* and their nontoxigenic, presumably domesticated derivatives, *A. sojae* and *A. oryzae*, have been especially controversial. Kurtzman (62) considered the four taxa conspecific based on the high degree of nuclear DNA complementarity. A group of *A. flavus*-like isolates with unusual, conical sclerotia had a low DNA complementarity with the *A. flavus* complex and was later described as the new species *A. nomius* (63). Klich & Pitt (60, 61) argued that this methodology was too insensitive, that *A. flavus* and its sister species could be distinguished readily morphologically, and that the differentiation was necessary because of differences in aflatoxin-producing abilities. Consistent differences between *A. flavus* and its assumed domesticated form *A. oryzae* were found using RFLPs of total DNA (59). Similarly, *A. flavus* and *A. parasiticus* could be distinguished by RFLPs of total DNA using probes (86), RFLPs of mitochondrial DNA (85), and isozymes (29). Egel et al (36) examined the relationships between many strains of these species by studying RFLPs of PCR-amplified Taka-amylase A genes. Their methods could not detect any differences between *A. parasiticus* and *A. sojae*; they split *A. flavus* into two groups, one of which included *A. oryzae*, and suggested that

A. nomius might be heterogeneous. The number of biological and genetic differences between the aflatoxin-producing aspergilli and their domesticated relatives indicates that there is at least a practical value to recognizing these as distinct species. Whether these differences can be detected in the ribosomal RNA gene will be an acid test of the utility of the methodology for resolving species complexes.

Species concepts in the black aspergilli, centered around *Aspergillus niger*, the cause of onion and date rots, have also been controversial. The morphological concepts of Raper & Fennell (107), which included 13 species, were revised (2) to include 6 species with numerous infraspecific taxa. Questions about the phylogenetic reliability of morphological characters, especially conidial dimensions and surface ornamentation, led to further studies. Kusters-van Someren et al (65) studied pectin lyase isozymes and RFLPs of nuclear DNA of type and representative isolates from the complex and recognized four species plus an aggregate around *A. niger*. Using RFLPs of total nuclear DNA (64), total mitochondrial DNA (154), and the ribosomal DNA (155), the *A. niger* aggregate has been shown to consist of two species, *A. niger* and *A. tubingensis*. Although the resulting number of species of black aspergilli matches the treatment by Al-Mussalam (2), no correlation of the taxa recognized by RFLPs with morphologically based taxa has been published.

THE DISCOMYCETES

The Discomycetes are characterized by an ascoma that opens widely at maturity to expose a hymenium of asci. The group has traditionally been divided into two major orders primarily on the basis of ascal type: the Helotiales (= Leotiales, inoperculate Discomycetes including many plant pathogens such as *Sclerotinia*) and the Pezizales (operculate Discomycetes including primarily saprotrophic fungi in genera such as *Peziza*). Sequences of the 18S rDNA demonstrate monophyly of the traditional Discomycete orders Pezizales and Helotiales as well as the lichenized inoperculate Discomycetes of the Lecanorales (45). Gargas & Taylor (45) found that the Pezizales are basal to the rest of the filamentous Ascomycetes, including the Helotiales, Pyrenomycetes, and Eurotiales. This study did not include any Loculoascomycetes, including members of the Patellariales that have apothecial ascomata. As with the Plectomycetes, the Discomycetes are characterized in part by the apothecium, but ascal characters are indispensable in delimiting groups of Discomycetes. It is far more likely that apothecial fungi that have bitunicate asci, such as *Rhytidhysteron*, are related to other nonapothecial Loculoascomycetes (e.g. *Botytryosphaeria*) than to either the Pezizales or Helotiales.

As was discussed above, the closest relative for the powdery mildew *Blumeria graminis* f.sp. *hordei* is in the Sclerotiniaceae (Helotiales), a relationship that was predicted by morphology (20, 22) as well as by nucleic acid sequences (45, 118).

The Sclerotiniaceae

The Sclerotiniaceae is the only family of Discomycetes that has been the subject of molecular-phylogenetic analysis. Carbone & Kohn (23) sequenced the ITS1 of several members of the inoperculate Discomycete family Sclerotiniaceae (Helotiales). They reported a separate and recently evolved lineage for plant pathogenic genera that have sclerotial stromata. This *Sclerotinia* cluster includes *Ciborina, Monilinia, Myriosclerotinia*, and *Verpatinia* as well as anamorphs in *Botrytis, Cristulariella*, and *Sclerotium*. Sequences of the asexual *Sclerotium cepivorum* and *Cristulariella moricola* were 98% similar to those of *Sclerotinia sclerotiorum*, as was predicted by other biochemical and morphological characters. Species of the asexual genus *Botrytis* formed a subcluster within the sclerotial cluster. Species of the asexual *Monilinia* were widely distributed among the sclerotial and substratal members of the family. Apothecia of *Sclerotinia homoeocarpa*, cause of dollar spot of turf, have not been seen since the fungus was originally described and thus there is no way to assess its taxonomy based on apothecial microanatomy. On the basis of other characters, though, it was predicted to be related to saprobic members of the Sclerotiniaceae (e.g. *Rutstroemia*) that have substratal stromata. While the species with substratal stromata were not shown to represent a distinct line within the Sclerotiniaceae, *S. homoeocarpa* clustered with the saprophytic genus *Rutstroemia* and not with the sclerotial *Sclerotinia* species.

Scleroderris cankers

Three species of inoperculate Discomycetes (Helotiales) that cause Scleroderris cankers of conifers in Europe and North America have been taxonomically confusing. Characters of the teleomorph as well as serological and electrophoretic soluble proteins lead to their inclusion in *Ascocalyx* along with saprobic species (87). In contrast, differences in biology, anamorphs, and isozyme patterns indicated that *Gremmeniella* and *Ascocalyx* are distinct (103). RFLP analysis of 18S and ITS rDNA supports the retention of two genera (11).

LOCULOASCOMYCETES

A wide variety of plant pathogens are Loculoascomycetes, including *Botryosphaeria, Cochliobolus, Leptosphaeria, Mycosphaerella, Phaeosphaeria*, and anamorphs in genera such as *Alternaria, Cercospora, Drechslera*, and *Diplodia*. Despite their economic importance, these fungi have attracted less attention from molecular geneticists than have any of the major groups of filamentous Ascomycetes. Interrelationships among the Loculoascomycetes have tended to be viewed differently on opposite sides of the Atlantic. This difference may reflect opposing views of the importance of ascomatal ontogeny and sterile filaments in the ascomatal locule in systematics. There is a tendency for European authors (e.g. 37) to accept few orders, whereas North American authors (e.g. 4) have accepted more orders. Taxon sampling, although small, in the only major molecular phylogenetic study of Loculoascomycetes presents a picture of paraphyly within the group. Major clades include the Pleosporales (including *Leptosphaeria* and *Cochliobolus*), its sister group the Dothideales (including *Mycosphaerella*), and the Herpotrichiellaceae (including the black yeasts that are animal pathogens), which is on a clade that is basal to the other Loculoascomycetes and to the Pyrenomycetes (150). These results argue that anatomical characters of ascomata, including the type and origin of sterile tissue within the developing locule, are phylogenetically significant.

Mycosphaerella

RFLP (24) and RAPD (57) techniques distinguished species of *Mycosphaerella* (Dothideales) causing Sigatoka leaf disease of banana in the Pacific region and in tropical America. Despite reportedly indistinguishable teleomorphs, *M. fijiensis* and *M. musicola* can be separated through their anamorphs as well as on the symptoms that they induce in banana leaves.

Alternaria

Alternaria species, the cause of leaf spots and blights in many crop plants, are the anamorphs of genera in the Pleosporales, including *Pleospora* and *Leptosphaeria*. Most species, however, have not been linked to a sexual phase, which makes it difficult to objectively assess intraspecific variation. Several plant diseases are caused by host-specific toxins produced by *Alternaria* strains (see review in 117). The taxonomy of those strains is unsettled, and they have been referred to as *formae speciales* or pathotypes of *A. alternata*, or in some cases as species (see 69, 117, 129, 130). Simmons (129) and Simmons & Roberts (130) found close correlation between morphology and toxin production in characterizing short-conidial species of *Alternaria* associated with Black Spot Disease of Japanese Pear. They (130) emphasized that it was essential to examine strains under defined conditions for particular features, as is commonly done for other fungi (e.g. *Penicillium*, 104), and that conidial measurements alone could not distinguish species. RFLP analysis of total DNA was unable to resolve short-conidial *Alternaria* strains to species distinct from *A. alternata* (6). However, because strains that had been identified by Simmons & Roberts (130) as typical of *A. gaisen* were not included, there is no basis for comparison of the RFLP work; nor did the authors follow the cultivation protocol offered (130) for morphological identification of strains.

DISCUSSION

Most molecular studies with fungi have addressed problems of recognizing plant pathogenic species and of documenting genetic diversity among populations of plant pathogens (e.g. 6, 17, 44, 47, 67, 68, 70, 82–84, 92, 94, 98, 128, 136, 165, 166). The use of molecular techniques in the study of fungal populations could itself easily be the subject of an entire review.

Given that no more than 5% of the estimated 1.5 million species of all groups of fungi have been described (55), taxon sampling for any phylogenetic analysis is certain to be deficient. Most sampling in the filamentous Ascomycetes has been done in the Eurotiales and only a few orders of Pyrenomycetes, especially the Hypocreales, and very little among the Discomycetes and Loculoascomycetes, both of which include many important plant pathogens. One feature of any kind of taxonomic study is that as more specimens are studied for each species, and as more species are added, the analysis becomes more complex and the resulting patterns less clearly defined.

Taxonomic hypotheses originally based in morphology that have been tested by molecular methods have generally been corroborated. It has been difficult to obtain a picture of Ascomycete phylogeny because no single study has included all the major elements (Pyrenomycetes, Discomycetes, Loculoascomycetes, and Eurotiales). Ascomatal form per se is not necessarily a good indicator of phylogenetic relationships, but when considered with centrum ontogeny and ascal type, it is predictive of monophyletic supraordinal taxa. Thus the Pyrenomycetes appear to be monophyletic. Among the apothecial fungi, the Pezizales, Helotiales, and Lecanorales are monophyletic, but the Pezizales are basal to a clade that includes most of the rest of the filamentous ascomycetes (exclusive of the Loculoascomycetes, which were not included in the study, 45). Loculoascomycetes in the Pleosporales and Dothideales share a clade with the Pyrenomycetes, and that clade has at its base another group of Loculoascomycetes that have black yeasts as their anamorphs (Herpotrichiellaceae; 150). Cleistothecial fungi are polyphyletic but the cleistothecial Eurotiales and Onygenales represent a monophyletic group that warrants supraordinal status. The same may be said of *Ascosphaera*. We recommend that the name "Plectomycete" not be used for these fungi because it misleadingly implies that all cleistothecial fungi are Plectomycetes. The term plectomycete has had different meanings, from very broad to very narrow, and its continued use as a formal taxon would only continue to cause confusion.

In some cases the resolution of conflicting, morphologically based hypotheses has been achieved from molecular studies. In this regard, the powdery mildews are not Pyrenomycetes (e.g. 5) but are derived from some group of Discomycetes (e.g. 20, 22, 45, 118); the Eurotiales are not derived from the

Hypocreales (77) but may represent a sister group to the Pyrenomycetes and possibly a subclass (5, 9).

In a few cases, molecular studies have resulted in the proposal of new hypotheses that either conflict with morphologically based hypotheses or were not anticipated by morphologists. For example, it is now clear that *Ceratocystis* and *Ophiostoma* are distinct pyrenomycetous genera and not members of the same order. Their respective anamorphs, however, suggest that *Ceratocystis* is most closely related to the Sordariales and *Ophiostoma* to the Diatrypales (5, 121). 18S rDNA sequences indicate that *Ceratocystis* is closest to the Microascales and *Ophiostoma* to the Diaporthales (133). When molecular phylogenies do not correspond with established morphological phylogenies, it is often the case that morphology is neutral and can be rationalized within either hypothesis (e.g. the type of anamorphs formed by species of *Ceratocystis* and *Ophiostoma*, respectively, is also found in a diversity of fungi and so may have limited phylogenetic significance).

There are only a few cases in the filamentous Ascomycetes studied so-far in which the molecular and morphological phylogenies seem unreconcilable. An example is the Ascomycete genus *Melanospora*, which produces black ascospores with polar germination pores of a type common only in the Sordariales. Sequencing studies of 18S and 28S rDNA have suggested that the genus is linked to (134), or derived from within (112), the Hypocreales. For this to be true, these ascospores must have evolved de novo within an order of species that have either colorless or golden-brown, nonporous ascospores. From a morphological perspective, this is not a parsimonious explanation and is one example where ribosomal genes and morphological genes seem to be telling different stories.

Whether there remains any formal need for the Deuteromycetes or for separate generic names for anamorphs has been discussed for decades, but the debate has taken on a new intensity with the advent of DNA-based classifications. Bruns et al (21) wrote, "If all fungi can be compared through their nucleic acids and placed on a single phylogenetic tree, do we need to maintain the Deuteromycota?" Berbee & Taylor (9) argued that the asexual fungus *Sporothrix schenkii* should be formally incorporated into the Ascomycete genus *Ophiostoma*. There is no doubt that individual anamorphs can be placed with corresponding groups of holomorph genera or species. There is no philosophical reason why the Deuteromycetes cannot be integrated into the teleomorph taxonomy, but we consider it more important to be aware of the phylogenetic relationships of groups of anamorph species than to abandon anamorph names. We agree with Gams (43) and Seifert et al (126) that there are serious practical and logical considerations to overcome before wholesale integration can be attempted. One objection springs from the lack of scientists and requisite funding to undertake the study of more than random links between individual anamorphs and teleomorphs. A second objection is the real problem of the diagnostician's need to identify species within polyphyletic genera because convergent morphology among unrelated species and a tendency for simplification within a natural series make the practical task of identifying natural genera of Deuteromycetes difficult. A third major problem in amalgamating the Deuteromycetes into the Ascomycetes is the difficulty of linking apparently strictly asexual fungi to individual holomorph species. This has rarely been achieved. The difficulty may relate to the nature of the anamorph species and how it developed from the holomorph.

Seifert et al (126) argued that taxonomy is based on the classification of species, and that it would be difficult to formally classify anamorphic fungi in the teleomorph taxonomic system in the absence of any kind of equivalence in species concepts. General mechanisms of speciation were discussed by Brasier (18) and Perkins (99) wherein a new population becomes reproductively isolated from an outcrossing basal population, resulting in the formation of biological species that may or may not be distinguishable morphologically, and may or may not be sexually competent. Given the apparent lability of the genes controlling sexual reproduction (over 200 mutants affect sexual development in *Neurospora crassa*, 106), it is easy to envision the development of sexually incompetent, clonally reproducing strains from sexually reproducing basal populations, as proposed for *Talaromyces* (74) and *Epichloë* (125). Further, because many genes regulate the ultimate form and physiology of the conidium-producing apparatus (145), there are many opportunities for mutations that could lead to the recognition of new "species" of Deuteromycetes. Perkins (99) questioned whether the term species can even be applied to anamorphic fungi because they are not only reproductively isolated, but apparently exist as reproductively incompetent clonal lines. As Seifert et al (126) have suggested, this may satisfy some taxonomists who would rather not deal with anamorphs, but it hardly solves the problem of how to classify mitosporic organisms.

It may be difficult or impossible to know whether a Deuteromycete is sexually incompetent because ascomata rarely form in nature or in agar culture. There is, however, evidence that fungi thought to be primarily asexual are capable of recombination events, either through meiosis or through vegetative fusion with phylogenetically closely related sexual populations (125). *Aspergillus nidullelus* in Britain exists as primarily clonal individuals, but because of infrequent meiotic recombination they do not exist as distinct, diverging lineages (46). Genetic diversity among the nonpathogenic, mitotic *Epichloë* (*Acremonium* spp.) endophytes of grasses may be maintained without meiosis through interspecific hybridization with sexual *Epichloe* species through fusion of vegetative hyphae and subsequent nuclear fusion (146).

Host specificity has long been regarded to be a powerful force in fungal speciation. Indeed, there is ample pathogenic and molecular evidence for host-specificity at the species and *forma specialis* level (e.g. 16, 90). However, not all molecular evidence supports the idea of coevolution of host and pathogen (e.g. 27, 125). Species in mitotic genera such as *Phomopsis*, where there are few morphological characters with which to distinguish species, are particularly rich in host-specific taxa, if only as reflected in the number of host names that appear as fungal-species epithets (147). In an examination of the importance of host in taxonomy of *Phomopsis*, Rehner & Uecker (113) used rDNA ITS1 and ITS2 sequences in an examination of isolates of unidentified strains that originated from a variety of woody and herbaceous, dicotyledonous plants; they found that isolates that had identical or similar ITS sequences occur on different, and often taxonomically unrelated, host species. They posed two alternative hypotheses to explain their observations: If ITS similar sequences indicate conspecificity, then many *Phomopsis* species may have a broad host range; or if different *Phomopsis* species have similar ITS sequences, then the diversity of host taxa at the terminal clades could indicate host switching during speciation.

Data derived from DNA have become a useful tool in recognizing and characterizing species in groups where identification is difficult. Great care must be taken to select isolates for study that have been rigorously identified (and can be reidentified if necessary) and to ensure that intraspecific variation is considered. The data are only as accurate as the names applied to the strains, and it is common for cultures to be misidentified in studies of morphologically difficult or unmonographed genera. In this regard, O'Donnell (93) found that each of three isolates determined as *Nectria lugdunensis* in culture collections occupied phylogenetically different positions. The problem of infraspecific variation is every bit as serious for molecular data as it is for morphological data. O'Donnell (92) found three ITS types in the morphological species *Gibberella pulicaris* (anamorph = *Fusarium sambucinum*) and cautioned that the phylogenetic value of ITS sequences and other molecular and genetic data in *Fusarium* might be limited until the source of intraspecific variability is understood.

Contemporary taxonomic research should, wherever possible, include concurrent collaborative reviews of morphology, isozymes, DNA, and toxin or other secondary metabolite production. Experienced, morphologically oriented taxonomists are becoming increasingly rare, and it is impossible to produce meaningful phylogenies without them. Morphologically difficult genera of plant pathogens such as *Alternaria, Phoma*, and *Colletotrichum* are taxonomically complex because they are biologically complex and evolutionarily inventive. Significant morphological traits or differences are rarely obvious to the untrained eye. Taxonomic programs focused on molecular genetics require active, classically oriented mycologists who can propose hypotheses for testing, and can interpret or reinterpret classical criteria. The results of such collaborations will certainly be useful systems of classification of economically important fungi.

ACKNOWLEDGMENTS

We are indebted to A Gargas, S Rehner and FA Uecker, and W Untereiner for allowing us use of manuscripts in press. We are also grateful to K Kuhls, D Turner, and C Kubicek for allowing us access to manuscripts in preparation. Gracia Murase and Bryce Kendrick kindly let us use their unpublished list of anamorph/teleomorph connections. Drs J Bissett, T Ouellet, ME Palm, AY Rossman, and S Rehner offered critical comments on various drafts of this manuscript.

Literature Cited

1. Alexopoulos CJ. 1962. *Introductory Mycology*. New York: Wiley. 2nd ed.
2. Al-Mussalam Z. 1980. *Revision of the black Aspergillus species*. PhD thesis. Rijksuniversiteit, Utrecht, Netherlands. 91 pp.
3. Barr ME. 1983. The Ascomycete connection. *Mycologia* 75:1–13
4. Barr ME. 1987. *Prodromus to Class Loculo Ascomycetes*. Amherst: published by the author
5. Barr ME. 1990. Prodromus to nonlichenized, pyrenomycetous members of class Hymenoascomycetes. *Mycotaxon* 39:43–184
6. Bates MR, Buck KW, Brasier CM. 1993. Molecular relationships of the mitochondrial DNA of *Ophiostoma ulmi* and the NAN and EAN races of *O. novo-ulmi* determined by restriction fragment length polymorphisms. *Mycol. Res.* 97:1093–100
7. Benny GL, Kimbrough JW. 1980. A synopsis of the orders and families of Plectomycetes with keys to genera. *Mycotaxon* 12:1–91
8. Berbee ML, Taylor JW. 1992. 18S ribosomal RNA gene sequence characters place the human pathogen *Sporothrix schenkii* in the genus *Ophiostoma*. *Exp. Mycol.* 16:87–91
9. Berbee ML, Taylor JW. 1992. Two Ascomycete classes based on fruiting-body characters and ribosomal DNA sequence. *Mol. Biol. Evol.* 9:278–84
10. Berbee ML, Taylor JW. 1992. Convergence in ascospore discharge mechanism among Pyrenomycete fungi based on 18S ribosomal RNA gene sequence. *Mol. Phylogenet. Evol.* 1:59–71
11. Bernier L, Hamelin RC, Ouellette GB. 1994. Comparison of ribosomal DNA length and restriction site polymorphisms in *Gremmeniella* and *Ascocalyx* isolates. *Appl. Environ. Microbiol.* 60: 1279–86
12. Bissett J. 1984. A revision of the genus *Trichoderma*. I. Section *Longibrachiatum* sect. nov. *Can. J. Bot.* 62: 924–31
13. Bissett J. 1991. A revision of the genus *Trichoderma*. II. Infrageneric classification. *Can. J. Bot.* 69:2357–72
14. Bissett J. 1991. A revision of the genus *Trichoderma*. III. Section *Pachybasium*. *Can. J. Bot.* 69:2372–417
15. Bonde MR, Peterson GL, Maas JL. 1991. Isozyme comparisons for identification of *Colletotrichum* species pathogenic to strawberry. *Phytopathology* 81:1523–28
16. Borromeo ES, Nelson RJ, Bonman JM, Leung H. 1993. Genetic differentiation among isolates of *Pyricularia* infecting rice and weed hosts. *Phytopathology* 83:393–99
17. Braithwaite KS, Irwin JAG, Manners JM. 1990. Restriction fragment length

polymorphisms in *Colletotrichum gloeosporioides* infecting *Stylosanthes* spp. in Australia. *Mycol. Res.* 94:1129–37

18. Brasier CM. 1987. The dynamics of fungal speciation. In *Evolutionary Biology of the Fungi*, ed. ADM Rayner, CM Brasier, D Moore, 16:231–60. Cambridge: Cambridge Univ. Press
19. Brasier CM. 1991. *Ophiostoma novo-ulmi* sp. nov., causative agent of current Dutch elm disease pandemics. *Mycopathologia* 115:151–61
20. Braun U. 1987. A monograph of the Erysiphales (powdery mildews). *Nova Hedwigia Z. Kryptogamenk.* 89:1–700
21. Bruns TD, White TJ, Taylor JW. 1991. Fungal molecular systematics. *Annu. Rev. Ecol. Syst.* 22:525–64
22. Cain RF. 1972. Evolution of the fungi. *Mycologia* 64:1–14
23. Carbone I, Kohn LM. 1993. Ribosomal DNA sequence divergence within internal transcribed spacer 1 of the Sclerotiniaceae. *Mycologia* 85:415–27
24. Carlier J, Mourichon X, González-de-Léon D, Zapater MF, Lebrun MH. 1994. DNA restriction fragment length polymorphisms in *Mycosphaerella* species that cause banana leaf spot diseases. *Phytopathology* 84:751–56
25. Chang J-Y, Oyaizu H, Sugiyama J. 1991. Phylogenetic relationships among eleven selected species of *Aspergillus* and associated teleomorphic genera estimated from 18S ribosomal RNA partial sequences. *J. Gen. Appl. Microbiol.* 37:289–308
26. Christensen MJ, Leuchtmann A, Rowan DD, Tapper BA. 1993. Taxonomy of *Acremonium* endophytes of tall fescue (*Festuca arundinacea*), meadow fescue (*F. pratenis*) and perennial ryegrass (*Lolium perenne*). *Mycol. Res.* 97:1083–92
27. Cisar CR, TeBeest DO, Spiegel FW. 1994. Sequence similarity of mating type idiomorphs: a method which detects similarity among the Sordariaceae fails to detect similar sequences in other filamentous Ascomycetes. *Mycologia* 86:540–46
28. Crowhurst RN, Hawthorne BT, Rikkerink EHA, Templeton MD. 1991. Differentiation of *Fusarium solani* f.sp. *cucurbitae* races 1 and 2 by random amplification of polymoprhic DNA. *Curr. Genet.* 20:391–96
29. Cruikshank RH, Pitt JI. 1990. Isoenzyme patterns in *Aspergillus flavus* and closely related species. See Ref. 119b, pp. 259–65
30. Currah RS. 1985. Taxonomy of the Onygenales: Arthrodermataceae, Gymnoascaceae, Myxotrichaceae and Onygenaceae. *Mycotaxon* 24:1–216
31. de Hoog GS. 1974. The genera *Blastobotrys*, *Sporothrix*, *Calcarisporium* and *Calcarisporiella* gen. nov. *Stud. Mycol.* 7:1–84
32. de Hoog GS. 1993. *Sporothrix*-like anamorphs of *Ophiostoma* species and other fungi. See Ref. 161, 6:53–60
33. de Hoog GS, Scheffer RJ. 1984. *Ceratocytis* versus *Ophiostoma*: a reappraisal. *Mycologia* 76:292–99
34. Domsch KH, Gams W, Anderson T-H. 1980.*Compendium of Soil Fungi*. New York: Academic. Vol. 1
35. Dupont J, Dutertre M, Lafay J-F, Roquebert M-F, Brygoo Y. 1990. A molecular assessment of the position of *Stilbothamnium* in the genus *Aspergillus*. See Ref. 119b, pp. 335–42
36. Egel DS, Cotty PJ, Elias KS. 1994. Relationships among isolates of *Aspergillus* sect. *flavi* that vary in aflatoxin production. *Phytopathology* 84:906–12
37. Eriksson E, Hawksworth DL. 1993. Outline of the Ascomycetes—1993. *Syst. Ascom.* 12:51–257

37a. Farr DF, Bills GF, Chamuris GP, Rossman AY. 1989. *Fungi on Plants and Plant Products in the United States*. St. Paul, MN: APS

38. Frisvad JC, Filtenborg O. 1983. Classification of terverticillate penicillia based on profiles of mycotoxins and other secondary metabolites. *Appl. Environ. Microbiol.* 46:1301–10
39. Frisvad JC, Samson RA, Stolk AC. 1990. Chemotaxonomy of *Eupenicillium javanicum* and related species. See Ref. 119b, pp. 445–54
40. Frisvad JC, Filtenborg O, Samson RA, Stolk AC. 1990. Chemotaxonomy of the genus *Talaromyces*. *Antonie van Leeuwenhoek* 57:178–89
41. Fujimori F, Okuda T. 1994. Application of the random amplified polymorphic DNA using the polymerase chain reaction for efficient elimination of duplicate strains in microbial screening. *J. Antibiot.* 47:173–82
42. Gams W, Müller E. 1980. Conidiogenesis of *Fusarium nivale* and *Rhynchosporium oryzae* and its taxonomic implications. *Neth. J. Plant Pathol.* 86: 45–53
43. Gams W. 1995. How natural should anamorph genera be? *Can. J. Bot.* In press
44. Garber RC, Yoder OC. 1984. Mitochondrial DNA of the filamentous Ascomycete *Cochliobolus heterostrophus*. *Curr. Genet.* 8:621–28

45. Gargas A, Taylor JW. 1995. Phylogeny of Discomycetes and early radiations of the apothecial Ascomycotina inferred from SSU rDNA sequence data. *Exp. Mycol.* 18:7–15
46. Geiser DM, Arnold ML, Timberlake WE. 1994. Sexual origins of British *Aspergillus nidulans* isolates. *Proc. Natl. Acad. Sci. USA* 91:2349–52
47. Gowan SP, Vilgalys R. 1991. Ribosomal DNA length polymorphisms within populations of *Xylaria magnoliae* (Ascomycotina). *Am. J. Bot.* 78:1603–7
48. Guadet J, Julien J, Lafay JF, Brygoo Y. 1989. Phylogeny of some *Fusarium* species as determined by large-subunit rRNA sequence comparison. *Mol. Biol. Evol.* 6:227–42
49. Gunnell P, Gubler W. 1992. Taxonomy and morphology of *Colletotrichum* species pathogenic to strawberry. *Mycologia* 84:157–65
50. Harrington TC. 1993. Diseases of conifers caused by species of *Ophiostoma* and *Leptographium*. See Ref. 161, 18: 161–72
51. Hausner G, Reid J, Klassen GR. 1992. Do galeate-ascospore members of the Cephaloascaceae, Endomycetaceae and Ophiostomataceae share a common phylogeny? *Mycologia* 84:870–81
52. Hausner G, Reid J, Klassen GR. 1993. On the subdivision of *Ceratocystis* s.l., based on partial ribosomal DNA sequences. *Can. J. Bot.* 71:52–63
53. Hausner G, Reid J, Klassen GR. 1993. *Ceratocystiopsis*: a reappraisal based on molecular criteria. *Mycol. Res.* 97:625–33
54. Hausner G, Reid J, Klassen GR. 1993. On the phylogeny of *Ophiostoma, Ceratocystis* s.s. and *Microascus*, and relationships within *Ophiostoma* based on partial ribosomal DNA sequences. *Can. J. Bot.* 71:1249–65
55. Hawksworth DL. 1991. The fungal dimension of biodiversity: magnitude, significance, and conservation. *Mycol. Res.* 95:641–55
56. Hywell-Jones N. 1993. *Torrubiella luteorostrata*: a pathogen of scale insects and its association with *Paecilomyces cinnamomeus* with a note on *Torrubiella tenuis*. *Mycol. Res.* 97: 1126–30
57. Johanson A, Crowhurst RN, Rikkerink EHA, Fullerton RA, Templeton MD. 1994. The use of species-specific DNA probes for the identification of *Mycospharella fijiensis* and *M. musicola*, the causal agents of Sigatoka disease of banana. *Plant Pathol.* 43:701–7
58. Kile GA. 1993. Plant diseases caused by species of *Ceratocystis sensu stricto* and *Chalara*. See Ref. 161, 19:173–83
59. Klich MA, Mullaney EJ. 1987. DNA restriction enzyme fragment polymorphisms as a tool for rapid differentiation of *Aspergillus flavus* from *Aspergillus oryzae*. *Exp. Mycol.* 11:170–75
60. Klich MA, Pitt JI. 1985. The theory and practice of distinguishing species of the *Aspergillus flavus* group. See Ref. 119a, pp. 211–20
61. Klich MA, Pitt JI. 1988. Differentiation of *Aspergillus flavus* from *A. parasiticus* and other closely related species. *Trans. Br. Mycol. Soc.* 91:99–108
62. Kurtzman CP. 1985. Classification of fungi through nucleic acid relatedness. See Ref. 119a, pp. 233–54
63. Kurtzman CP, Horn BW, Hesseltine CW. 1987. *Aspergillus nomius*, a new aflatoxin-producing species related to *Aspergillus flavus* and *Aspergillus tamarii*. *Antonie van Leeuwenhoek* 53: 147–58
64. Kusters-van Someren MA, Samson RA, Visser J. 1990. Variation in pectinolytic enzymes of the black Aspergilli: a biochemical and genetic approach. See Ref. 119b, pp. 321–34
65. Kusters-van Someren MA, Samson RA, Visser J. 1991. The use of RFLP analysis in classification of the black Aspergilli: reinterpretation of the *Aspergillus niger* aggregate. *Curr. Genet.* 19: 21–26
66. Kohn LM. 1992. Developing new characters for fungal systematics: an experimental approach for determining the rank of resolution. *Mycologia* 84:139–53
67. Kohn LM, Petsche DM, Bailey SR, Novak LA, Anderson JB. 1988. Restriction fragment length polymorphisms in nuclear and mitochondrial DNA of *Sclerotinia* species. *Phytopathology* 78: 1047–51
68. Kohn LM, Stasovski E, Carbone I, Royer J, Anderson JB. 1991. Mycelial incompatibility and molecular markers identify genetic variability in field populations of *Sclerotinia sclerotiorum*. *Phytopathology* 81:480–85
69. Kusaba M, Tsuge T. 1994. Nuclear ribosomal DNA variation and pathogenic specialization in *Alternaria* fungi known to produce host-specific toxins. *Appl. Environ. Microbiol.* 60:3055–62
70. Leonard KJ, Leath S. 1990. Genetic diversity in field populations of *Cochliobolus carbonum* on corn in North Carolina. *Phytopathology* 80: 1154–59
71. Leslie JF. 1991. Mating populations

in *Gibberella fujikuroi* (*Fusarium* Section *Liseola*). *Phytopathology* 81: 1058–60
72. Leuchtmann A, Clay K. 1990. Isozyme variation in the *Acremonium/ Epichloë* fungal endophyte complex. *Phytopathology* 80:1133–39
73. LoBuglio KF, Taylor JW. 1993. Molecular phylogeny of *Talaromyces* and *Penicillium* species in subgenus *Biverticillium*. In *The Fungal Holomorph: Mitotic, Meiotic and Pleomorphic Speciation in Fungal Systematics*, ed. DR Reynolds, JW Taylor, 11:115–19. Wallingford: CAB Int.
74. LoBuglio KF, Pitt JI, Taylor JW. 1993. Phylogenetic analysis of two ribosomal DNA regions indicates multiple independent losses of a sexual *Talaromyces* state among asexual *Penicillium* species in subgenus *Biverticillium*. *Mycologia* 85:592–604
75. LoBuglio KF, Pitt JI, Taylor JW. 1994. Independent origins of the synnematous *Penicillium* species, *P. duclauxii*, *P. clavigerum*, and *P. vulpinum*, as assessed by two ribosomal DNA regions. *Mycol. Res.* 98:250–56
76. Luttrell ES. 1951. Taxonomy of the Pyrenomycetes. *Univ. Missouri Stud.* 24:1–120
76a. Luttrell ES. 1955. The ascostromatic ascomycetes. *Mycologia* 47:511–32
77. Malloch D. 1981. The plectomycete centrum. In *Ascomycete Systematics. The Luttrellian Concept*, ed. DR Reynolds, 6:73–91. New York: Springer-Verlag
78. Malloch D, Blackwell M. 1993. Dispersal biology of the ophiostomatoid fungi. See Ref. 161, 21:195–206
79. Marasas WFO, Nelson PE, Toussoun TA. 1984. *Toxigenic* Fusarium *Species*. University Park: Pennsylvania State Univ. Press
80. Meyer RJ. 1991. Mitochondrial DNAs and plasmids as taxonomic characteristics in *Trichoderma viride*. *Appl. Environ. Microbiol.* 57:2269–76
81. Meyer W, Morawetz R, Börner T, Kubicek CP. 1992. The use of DNA-fingerprint analysis in the classification of some species of the *Trichoderma* aggregate. *Curr. Genet.* 21:27–30
82. McDonald BA, Martinez JP. 1990. DNA restriction fragment length polymorphisms among *Mycosphaerella graminicola* (anamorph *Septoria tritici*) isolates collected from a single wheat field. *Phytopathology* 80:1368–73
83. McDonald BA, McDermott JM. 1993. Population genetics of plant pathogenic fungi. *BioScience* 43:311–19
84. Mes JJ, Van Doorn J, Roebroeck EJA, Van Egmond E, Van Aartrijk J, Boonekamp PM. 1994. Restriction fragment length polymorphisms, races and vegetative compatibility groups within a worldwide collection of *Fusarium oxysporum* f.sp. *gladioli*. *Plant Pathol.* 43:362–70
85. Moody SF, Tyler BM. 1990. Restriction enzyme analysis of mitochondrial DNA of the *Aspergillus flavus* group: *A. flavus*, *A. parasiticus*, and *A. nomius*. *Appl. Environ. Microbiol.* 56:2441–52
86. Moody SF, Tyler BM. 1990. Use of nuclear DNA restriction fragment length polymorphisms to analyze the diversity of the *Aspergillus flavus* group: *A. flavus*, *A. parasiticus*, and *A. nomius*. *Appl. Environ. Microbiol.* 56:2453–61
87. Müller E, Dorworth CE. 1983. On the discomycetous genera *Ascocalyx* Naumov and *Gremmeniella* Morelet. *Sydowia* 36:193–203
88. Morgan-Jones G, Gams W. 1982. Notes on hyphomycetes. XLI. An endophyte of *Festuca arundinacea* and the anamorph of *Epichloë typhina*, new taxa in one of two new sections of *Acremonium*. *Mycotaxon* 15:311–18
89. Muthumeenakshi S, Mills PR, Brown AE, Seaby DA. 1994. Intraspecific molecular variation among *Trichoderma harzianum* isolates colonizing mushroom compost in the British Isles. *Microbiology* 140:769–77
90. Namiki F, Shiomi T, Kayamura T, Tsuge T. 1994. Characterization of the formae speciales of *Fusarium oxysporum* causing wilts of cucurbits by DNA fingerprinting with nuclear repetitive DNA sequences. *Appl. Environ. Microbiol.* 60:2684–91
91. Nannfeldt JA. 1932. Studien über die Morphologie und Systematik der nichtlichenisierten inoperculaten Discomyceten. *Nova Acta Reg. Soc. Sci. Upsal. Ser. IV*, 8(2):1–368. (In German)
92. O'Donnell K. 1992. Ribosomal DNA internal transcribed spacers are highly divergent in the phytopathogenic Ascomycete *Fusarium sambucinum* (*Gibberella pulicaris*). *Curr. Genet.* 22: 213–20
93. O'Donnell K. 1993. *Fusarium* and its near relatives. See Ref. 113a, 24:225–33
94. Ouellet T, Seifert KA. 1993. Genetic characterization of *Fusarium graminearum* strains using RAPD and PCR amplification. *Phytopathology* 83:1001–7
95. Palm ME, Gams W, Nirenberg H. 1995. *Plectosporium*, a new genus for *Fusarium tabacinum*, the anamorph of *Plec-*

tosphaerella cucumerina. *Mycologia* 87:In press
96. Parker AK. 1957. *Europhium*, a new genus of the Ascomycetes with a *Leptographium* imperfect state. *Can. J. Bot.* 35:173–79
97. Patterson C, Williams DM, Humphries CJ. 1993. Congruence between molecular and morphological phylogenies. *Annu. Rev. Ecol. Syst.* 24:153–88
98. Peever TL, Milgroom MG. 1994. Genetic structure of *Pyrenophora teres* populations determined with random amplified polymorphic DNA markers. *Can. J. Bot.* 72:915–23
99. Perkins DD. 1991. In praise of diversity. In *More Gene Manipulations in Fungi*, ed. JW Bennett, LL Lasure, 1:3–26. New York: Academic
100. Peterson SW. 1991. Phylogenetic analysis of *Fusarium* species using ribosomal RNA sequence comparisons. *Phytopathology* 81:1051–54
101. Peterson SW. 1993. Molecular genetic assessment of relatedness of *Penicillium* subgenus *Penicillium*. See Ref. 113a, 12:121–28
102. Peterson SW, Logrieco A. 1991. Ribosomal RNA sequence variation among interfertile strains of some *Gibberella* species. *Mycologia* 83:397–2
103. Petrini O, Petrini LE, Laflamme G, Ouellette GB. 1989. Taxonomic position of *Gremmeniella abietina* and related species: a reappraisal. *Can. J. Bot.* 67:2805–14
104. Pitt JI. 1979. *The Genus* Penicillium. London: Academic
105. Pitt JI, Samson RA. 1990. Approaches to *Penicillium* and *Aspergillus* systematics. *Stud. Mycol.* 32:77–90
106. Raju NB. 1992. Genetic control of the sexual cycle in *Neurospora*. *Mycol. Res.* 96:241–62
107. Raper KB, Fennell DI. 1965. *The Genus* Aspergillus. Baltimore: Williams & Wilkins
108. Raper KB, Thom C. 1949. *A Manual of the Penicillia*. Baltimore: Williams & Wilkins
109. Redhead SA, Malloch DW. 1977. The Endomycetaceae: new concepts, new taxa. *Can. J. Bot.* 55:1701–11
110. Rehner SA. 1993. Orotodine-5′-monophosphate decarboxylase: a single copy nuclear gene for molecular phylogenetic analyses of Pyrenomycetes. *Inoculum: Newsl. Mycol. Soc. Am.* 44:55 (Abstr.)
111. Rehner SA, Samuels GJ. 1994. Taxonomy and phylogeny of *Gliocladium* analyzed from nuclear large subunit ribosomal DNA sequences. *Mycol. Res.* 98:625–34
112. Rehner SA, Samuels GJ. 1995. Molecular systematics of the Hypocreales: a teleomorph gene phylogeny and the status of their anamorphs. *Can. J. Bot.* 73:In press
113. Rehner SA, Uecker FA. 1995. Nuclear ribosomal internal transcribed spacer phylogeny and host diversity in the coelomycete *Phomopsis*. *Can. J. Bot.*72:1666–74
113a. Reynolds DR, Taylor JW, eds. 1993. *The Fungal Holomorph: Mitotic, Meiotic, and Pleomorphic Speciation in Fungal Systematics*. Wallingford, UK: CAB Int.
114. Rifai MA. 1969. A revision of the genus *Trichoderma*. *Mycol. Pap.* 116:1–56
115. Rogerson CT. 1970. The hypocrealean fungi (Ascomycetes, Hypocreales). *Mycologia* 73:865–910
116. Rossman AY, Samuels GJ, Rogerson CT, Lowen R. 1993. Genera of the Hypocreales. *Int. Works. Ascomycete Systematics, 1st, Paris*. (Abstr.)
117. Rotem J. 1994. *The Genus* Alternaria. St. Paul, MN: APS
118. Saenz GS, Taylor JW, Gargas A. 1994. 18S rRNA gene sequences and supraordinal classification of the Erysiphales. *Mycologia* 86:212–16
119. Samson RA. 1974. *Paecilomyces* and some allied hyphomycetes. *Stud. Mycol.* 6:1–117
119a. Samson RA, Pitt JI, eds. 1985. *Advances in* Penicillium *and* Aspergillus *Systematics*. New York/London: Plenum
119b. Samson RA, Pitt JI, eds. 1990. *Modern Concepts in* Penicillium *and* Aspergillus *Classification*. New York: Plenum
120. Samson RA, Seifert KA. 1985. The Ascomycete genus *Penicilliopsis* and its anamorphs. See Ref. 119a, pp. 397–428
121. Samuels GJ. 1993. The case for distinguishing *Ceratocystis* and *Ophiostoma*. See Ref. 161, 2:15–20
122. Samuels GJ, Hallett IC. 1983. *Microdochium stoveri* and *Monographella stoveri*, new combinations for *Fusarium stoveri* and *Micronectriella stoveri*. *Trans. Br. Mycol. Soc.* 81:473–83
123. Samuels GJ, Petrini O, Manguin S. 1994. Morphological and macromolecular characterization of *Hypocrea schweinitzii* and its *Trichoderma* anamorph. *Mycologia* 86:421–35
124. Samuels GJ, Seifert KA. 1987. Kinds of Pleoanamorphy in the Hypocreales. In *Pleomorphic Fungi: The Diversity and its Taxonomic Implications*, ed. J Sugiyama, 3:29–56. Tokyo: Kodansha
125. Schardl CL, Liu J-S, White JF Jr, Finkel RA, An Z, Siegel MR. 1991. Molecular phylogenetic relationships of nonpatho-

genic grass mycosymbionts and clavicipitaceous plant pathogens. *Plant Syst. Evol.* 178:27–41
126. Seifert KA, Wingfield BD, Wingfield MJ. 1995. A critique of DNA sequence analysis in the taxonomy of filamentous Ascomycetes and ascomycetous anamorphs. *Can. J. Bot.* In press
127. Sherriff C, Whelan MJ, Arnold GM, Lafay J-F, Brygoo Y, Bailey JA. 1994. Ribosomal DNA sequence analysis reveals new species groupings in the genus *Colletotrichum*. *Exp. Mycol.* 18:121–38
128. Simcox KD, Nickrent D, Pedersen WL. 1992. Comparison of isozyme polymorphism in races of *Cochliobolus carbonum*. *Phytopathology* 82:621–24
129. Simmons EG. 1993. *Alternaria* themes and variations (63–72). *Mycotaxon* 48: 91–107
130. Simmons EG, Roberts RG. 1993. *Alternaria* themes and variations (73). *Mycotaxon* 48:109–40
131. Skou JP. 1982. Ascosphaeriales and their unique ascomata. *Mycotaxon* 15: 487–99
132. Smalley EB, Raffa KF, Proctor RH, Klepzig KD. 1993. Tree responses to infection by species of *Ophiostoma* and *Ceratocystis*. See Ref. 161, 22:207–17
133. Spatafora JW, Blackwell M. 1993. Molecular systematics of unitunicate perithecial Ascomycetes: the Clavicipitales-Hypocreales connection. *Mycologia* 85:912–22
134. Spatafora JW, Blackwell M. 1994. The polyphyletic origins of the ophiostomatoid fungi. *Mycol. Res.* 98:1–9
135. Spatafora JW, Blackwell M. 1994. Cladistic analysis of partial ssrDNA sequences among unitunicate perithecial Ascomycetes and its implications on the evolution of centrum development. In *Ascomycete Systematics. Problems and Perspectives in the Nineties*, ed. DL Hawksworth, pp. 233–42. New York/ London: Plenum
136. Sreenivasaprasad S, Brown AE, Mills PR. 1992. DNA sequence variation and interelationships among *Colletotrichum* species causing strawberry anthracnose. *Physiol. Mol. Plant Pathol.* 41:265–81
137. Sreenivasaprasad S, Brown AE, Mills PR. 1993. Coffee berry disease pathogen in Africa: genetic structure and relationship to the group species *Colletotrichum gloeosporioides*. *Mycol. Res.* 97:995–1000
138. Stolk AC, Samson RA. 1983. The Ascomycete genus *Eupenicillium* and related *Penicillium* anamorphs. *Stud. Mycol.* 23:1–149
139. Stolk AC, Samson RA, Frisvad JC, Filtenborg O. 1990. The systematics of the terverticillate penicillia. See Ref. 119b, pp. 121–37
140. Sutton BC. 1980. *The Coelomycetes*. Kew: Commonw. Mycol. Inst.
141. Sutton BC. 1992. The genus *Glomerella* and its anamorph *Colletotrichum*. In Colletotrichum: *Biology, Pathology and Control*, ed. JA Bailey, MJ Jeger, pp. 1–26. Wallingford, UK: CAB Int.
142. Swofford DL. 1991. When are phylogeny estimates from molecular and morphologcal data incongruent? In *Phylogenetic Analysis of DNA Sequences*, ed. MM Miyamoto, J Cracraft, pp. 295–333. New York: Oxford Univ. Press
143. Taylor JW, Pitt JI, Hocking AD. 1990. Ribosomal DNA restriction studies of *Talaromyces* species with *Paecilomyces* and *Penicillium* anamorphs. See Ref. 119b, pp. 357–70
144. Thom C. 1910. *The Penicillia*. Baltimore: Williams & Wilkins
145. Timberlake WE. 1990. Molecular genetics of *Aspergillus* development. *Annu. Rev. Genet.* 24:5–36
146. Tsai H-F, Liu J-S, Staben C, Christensen MJ, Latch GCM, et al. 1994. Evolutionary diversification of fungal endophytes of tall fescue grass by hybridization with *Epichloë* species. *Proc. Natl. Acad. Sci. USA* 91:2542–46
147. Uecker FA. 1988. A world list of *Phomopsis* names with notes on nomenclature, morphology and biology. *Mycol. Mem.* 13:1–231
148. Uecker FA. 1993. Development and cytology of *Plectosphaerella cucumerina*. *Mycologia* 85:470–79
149. Uecker FA. 1994. Ontogeny of the ascoma of *Glomerella cingulata*. *Mycologia* 86:82–88
150. Untereiner WA, Straus NA, Malloch D. 1995. A molecular-morphometric approach to the systematics of the Herpotrichiellaceae and allied black yeasts. *Mycol. Res.* 99: In press
151. Upadhyay HP. 1981. *A Monograph of* Ceratocystis *and* Ceratocystiopsis. Athens, GA: Univ. Georgia Press
152. Vaillancourt LJ, Hanau RM. 1992. Genetic and morphological comparisons of *Glomerella* (*Colletotrichum*) isolates from maize and from sorghum. *Exp. Mycol.* 16:219–29
153. Van Etten HD, Kistler HC. 1988. *Nectria haematococca*, mating populations I and VI. *Adv. Phytopathol.* 6:189–206
154. Varga J, Kevei F, Fekete C, Coenen, A, Kozakiewicz Z, Croft JH. 1993. Restriction fragment length polymorphisms in the mitochondrial DNAs of

the *Aspergillus niger* aggregate. *Mycol. Res.* 97:1207–12
155. Varga J, Kevei F, Vriesema A, Debets F, Kozakiewicz Z, Croft JH. 1994. Mitochondrial DNA restriction fragment length polymorphisms in field isolates of the *Aspergillus niger* aggregate. *Can. J. Microbiol.* 40:612–21
156. Vilgalys R, Hibbett DS. 1993. Phylogenetic classification of fungi and our Linnaean heritage. See Ref. 113a, pp. 255–60
157. von Arx JA. 1957, Die Arten der Gattung *Colletotrichum* Cda. *Phytopathol. Lab. "Willie Commelin Scholten."* 17: 413–68
158. von Arx JA. 1973. Ostiolate and nonostiolate Pyrenomycetes. *Konikl. Nederl. Akad. Wetensch. Amsterdam. Proc. Ser. C* 76:289–96
159. von Arx JA. 1987. Plant pathogenic fungi. *Beih. Nova Hedwigia* 87:288
160. Webster J. 1964. Culture studies on *Hypocrea* and *Trichoderma* I. Comparison of perfect and imperfect states of *H. gelatinosa*, *H. rufa* and *Hypocrea* sp. I. *Trans. Br. Mycol. Soc.* 47:75–96
161. Wingfield MJ, Seifert KA, Webber JF, eds. 1993. Ceratocystis *and* Ophiostoma. *Taxonomy, Ecology and Pathogenicity.* St. Paul: APS
162. White JF Jr. 1993. Endophyte-host associations in grasses. XIX. A systematic study of some sympatric species of *Epichloë* in England. *Mycologia* 85: 444–55
163. White JF Jr. 1994. Endophyte-host associations in grasses. XX. Structural and reproductive studies of *Epichloë armillarians* sp. nov. and comparisons to *E. typhina*. *Mycologia* 86:571–80
164. White JF Jr, Morgan-Jones G, Morrow AC. 1993. Taxonomy, life cycle, reproduction and detection of *Acremonium* endophytes. *Agric. Ecosyst. Environ.* 44: 13–37
165. Wilson R, Wheatcroft R, Miller JD, Whitney NJ. 1994. Genetic diversity among natural populations of endophytic *Lophodermium pinastri* from *Pinus resinosa*. *Mycol. Res.* 98:740–44
166. Xia JQ, Correll JC, Lee FN, Marchetti MA, Rhoads DD. 1993. Fingerprinting to examine microgeographic variation in the *Magnaporthe grisea* (*Pyricularia grisea*) population in two rice fields in Arkansas. *Phytopathology* 83:1929–35
167. Zimand G, Valinsky L, Elad Y, Chet I, Manulis S. 1994. The use of the RAPD procedure for the identification of *Trichoderma* strains. *Mycol. Res.* 98: 531–34

Annu. Rev. Phytopathol. 1995. 33:269–102

CONCEPTS AND TERMINOLOGY ON PLANT/PEST RELATIONSHIPS: Toward Consensus in Plant Pathology and Crop Protection

L. Bos
DLO Research Institute for Plant Protection, IPO-DLO, P.O. Box 9060, 6700 GW Wageningen, The Netherlands

J.E. Parlevliet
Department of Plant Breeding, Agricultural University, P.O. Box 386, 6700 AJ Wageningen, The Netherlands

KEY WORDS: crop protection science, damage, disease, pest resistance, vulnerability

> Logical organization and precision of expression are the very essence of good teaching whether in classroom or field. Even in the realm of research orderly thinking and clear expression must sooner or later prevail if our science is to attain to that position of authority, confidence, and dignity which it deserves.
>
> Whetzel (1929)

ABSTRACT

In plant pathology, terminological confusion still reigns despite national attempts at standardization. Terminological agreements reached within the crop protection community in The Netherlands are elaborated here and presented as an endeavor toward international consensus. Much of the on-going terminological disconcert derives from differences in outlook between academically oriented biologists (including biologically trained pathologists) and pathologists working in and for agricultural institutions where disease and harm have anthropocentric connotations. The name crop protection science more realistically covers and marks the field dealt with by most plant pathologists, and

0066-4286/95/0901-0069$05.00

adoption of the FAO-defined term pest to encompass all biotic factors that are harmful to plants and their products is advocated.

The effect of pests on plants and the interrelationships between pests and plants in dependence upon the environment, topical in resistance breeding, are especially dealt with. A diagrammatic model is used to better describe these relationships and to define the terms that denote the phenomena and mechanisms involved.

INTRODUCTION

In plant pathology, accurate characterization of the relationships between plants and their harmful or hostile agents is essential for understanding plant disease and injury. The state of our knowledge is reflected in the terminology we use and in how we define concepts. As the discipline of plant pathology and its subspecialties has developed, so too has confusion over terminology—to the detriment of study and communication within the discipline. Whetzel first pointed out this trend in 1926 (64). In 1935, Wilbrink (65) called for uniform international terminology on resistance, but controversy continues, particularly regarding terminology on plant/pathogen relationships, as is evident from editorial comments (16) or letters to the editors of scientific journals (3, 5, 14, 23, 35). A recent treatise on the nomenclature and concepts of pathogenicity and virulence by Shaner et al (55) addressed the unresolved terminological confusion and inconsistency, and appealed for international consultation.

The landmark *Advanced Treatise on Plant Pathology*, later *Plant Disease*, published between 1959–1980 (39, 40), made valuable contributions to our understanding of plant/pathogen relationships, although individual authors offered conflicting interpretations of terms. The CMI's *Plant Pathologist's Pocketbook* (22), the handbook on *Plant Pathology* by Agrios (1), and English and German dictionaries of plant pathology (31, 36) and microbiology (45) all helped standardize scientific language. However, consensus and consistency in phytopathological terminology can only be achieved through scientific societies, as first advocated in 1935 by Whetzel (65) and more recently by Shaner et al (55). Terminology committees in various countries have undertaken national initiatives, for example the American Phytopathological Society (APS) in 1940 (2), the British Mycological Society (MBS) in 1950 (12), the Federation of British Plant Pathologists (FBPP) in 1973 (28), and the (now Royal) Netherlands Society of Plant Pathology (NPV, Nederlandse Plantenziektenkundige Vereniging) in 1968 and 1985 (20, 21). The Canadian Phytopathological Society (CPS) (63) compiled a list of terms concerning plant viral diseases, and the Canadian Ministry of Agriculture (4) prepared a phytopathological glossary with definitions in French.

However, these semiauthorized lists are now partially outdated, and more recent comments discuss single terms out of their overall phytopathological context. The "Commissie voor de Terminologie" of the Netherlands Society of Plant Pathology reached agreement within the nation's crop protection community on the whole complex of plant/parasite relationships (21). This list deserves wider attention and could be a starting point for future international discussion. As members of the NPV terminology committee, we now discuss these proposals within the broad framework of plant/pest relationships and the context of crop protection. We aim for conciseness and clarity rather than full reference to the literature.

TERMINOLOGY AND COMMON LANGUAGE

Science aims at describing reality and communicating about it; both require unambiguous terminology. The more precise the definition of concepts behind terms, the better the tools to describe reality, and the less the chance of confusion. Language and terminology are constantly evolving to accommodate scientific progress. Scientific terminology therefore must be reevaluated, redefined, and restandardized on a regular basis.

Not only are there problems over interpretation of terminology within the discipline of plant pathology, but there has also been a drift away from usage common both to biology and to standard speech. Wilbrink (65) indicated that "Our lack of appropriate terms as well as our indiscriminate use of the available ones may be misleading to those who have to apply our results in agriculture and horticulture." Shaner et al (55) also pointed out that "our literature is extremely difficult for persons outside our discipline to understand ... it is wise to comply with standard linguistic, and medical and scientific usage wherever possible." For instance, a babel of tongues has arisen among plant pathologists and breeders when discussing virulence, avirulence, and aggressiveness (3, 55); and phytopathological interpretation has drifted completely away from original linguistic meaning. Following NPV (21), we consulted Henderson's *Dictionary of Biological Terms* (37), Webster's *Third New International Dictionary of the English Language* (33), and the *Oxford Advanced Learner's Dictionary of Current English* (38).

From the outset, it should be made clear that *plant pathology*, though often studied academically by biologists, or by pathologists trained as biologists, for all intents and purposes is a branch of agricultural sciences. Plant pathogens and parasites first drew attention and are studied primarily for the damage they do to crops. Research into these organisms has always been geared to *crop protection*, and most of it continues to be performed in agricultural research institutions. The different perspectives of biologists, plant pathologists, resistance breeders, and growers have produced discrepancies in the interpretation of terms, confused terminology, and led to miscommunication. *Biologists* in particular bring their own definitions. Their main focus is on how all organisms (including plants) live together, whereas *plant pathologists* look to the harmful effects of biotic and abiotic agents on plants and crops. *Growers*, and in their wake *resistance breeders*, are principally interested in the economic implications of agents that reduce plant yield. These divergent interests may explain the controversy over terminology in plant pathology (or crop protection).

DISEASE, INJURY, DAMAGE, AND LOSS

To start with, clarification is needed on the meaning of basic terms such as disease, injury, damage, and loss as they relate to the effect of elements in the environment hostile to plants and crops. Their definition bears on the scope and limitations of the plant pathology discipline.

Disease

Disease or "illness" literally implies that the organism concerned is "ill at ease." The terms *pathology* and *pathogen* are derived from the Greek word "pathos" for emotion, passion, or suffering. However, nature does not discriminate between good or bad, normal or abnormal, and healthy or diseased; biological dictionaries (37, 45) do not even list the term disease. Judgment as to what is good or bad is a human construct. The characterization of disease in plants usually carries an implication of economic loss to the grower. Philosophically, disease-inducing agents, like any other organism, exercise their own "rights" in seeking ecological niches, irrespective of economic consequence. In that perspective, *abnormality, aberration,* or *disorder* (all implying deviation from normal), and disease can be defined only in terms of statistics. Whether the mosaic of virus-infected *Abutilon* plants, long ascribed to genetic variation (*Abutilon striatum* var. *Thompsonii*), should be considered (*a*) a disease, (*b*) an anatomic or physiological aberration (variegation), or (*c*) a phenomenon beneficial to the species for having caused its large-scale vegetative propagation and wide distribution by humans, may seem a matter of mere speculation, but the terms merit specification.

Within plant pathology, *disease* is widely accepted to be any "deviation from normal functioning of physiological processes, of sufficient duration to cause disturbance or cessation of vital activity" (2, 12, 28). The BMS definition (12) starts with the adjective "harmful," but the harmfulness of the deviation is already clearly indicated in the second part of the definition and thus seems superfluous. Horsfall & Dimond (41) particularly stressed that disease implies the prolonged action of the causal agent in contrast to injury, which comes and goes suddenly (see below). Cowling & Horsfall (24) emphasized that disease results when plants are subject to sustained impairment or disruption [or continuous irritation (41)] by a pathogen. The pathogen drives the disease; when the pathogen dies, the plant may recover. Disease is a *dynamic process* that encompasses a series of events, each often leading autonomously to the next, once pathogenesis has been triggered by the pathogen. Examples are yellows diseases, where yellowing, premature senescence and decline result from degeneration in the phloem caused primarily by a phloem-inhabiting virus or mycoplasma. NPV (21) therefore added to the above definition that, in contrast to deviation or abnormality and injury, "disease mostly is a dynamic process involving a series of *symptoms,* occurring simultaneously or in sequence and constituting the *syndrome.*"

The corpuscular presence of the biotic agent may be considered a *sign* of disease consisting of "visible structures [of the pathogen] produced in, or on, diseased tissues" (28). Signs are "the pathogen or its parts or products seen on a host plant" (1). They are thus objective tags of the pathogen itself, giving a label to the disease and helping in its etiological diagnosis. The syndrome may be characteristic of the interaction between the victim and causative agent (harmful factor or disease incitant). Proficiency in symptomatology is obviously an asset in field and clinical diagnosis.

The diseased condition, especially reduction in vitality, undermines the plant's competitiveness. Nonspecific growth reduction, early senescence, premature death, as well as increased (or sometimes decreased) susceptibility and sensitivity to other pathogens often result as well. Finally, the quality and quantity of yield may also be reduced. Quality may be impaired by external abnormalities (e.g. cosmetic defects associated with discolorations of, or necrosis in, leaves, flowers, and fruits, or mechanical injury or feeding damage) or by internal deviations (e.g. chemical composition that reduces nutritive value or taste). Thus, the underlying abnormalities, i.e. the deviations that determine final economic loss, if observed and recorded, are part of the syndrome rather than the result. Their being earmarked as symptoms extends the usual definition of disease. Disease is the sum total of all measurable deviations, and it is more than the mere visible abnormalities. In disease, there is a continuum of abnormalities eventually culminating in loss in terms of quantity and quality at harvest time, and this will prove to bear on our definitions of other terms.

Whether and to what extent a plant is considered diseased depends not only on the precision of observation but also on the definition of disease. In practice, this determination depends largely on the outlook and interests of the observer (biologist, pathologist, breeder, or grower), hence the persistence of controversy over the use of terms such as *latency,* for symptomless infection, tolerance, and vulnerability (defined below). For example, growth reduction is an important symptom to be taken into account in deciding whether a plant is diseased. With many so-called latent viruses, e.g. in potatoes, careful comparison of infected and virus-free plants revealed slight, though perceptible, external abnormalities and/or yield reduction of up to 12% (for References see 8). Such infected plants are thus diseased and, in a strict sense, the infection is not latent. However, growth reduction and loss with no further indication of disease, as in genotypes with heritably lower yielding capacity, do not necessarily imply the presence of disease. Disease is a complex phenomenon, and final yield loss is determined not only by external symptoms (see Figure 5*A*) but also by invisible physiological aberrations (9). Visible symptoms are not therefore always reliable parameters of final loss (9, 61, 62), an important fact for breeders in selecting for resistance.

The above definition of disease does not specify causes. Webster's (33) states that in disease, impairment of the normal state of the living body and its malfunctioning is in "response to environmental factors (as malnutrition, industrial hazards, or climate), to specific infective agents (as worms, bacteria, or viruses), to inherent defects of the organism (as various genetic anomalies), or to combinations of these factors." This definition states that the incitant or inducer of disease (24) may be either biotic or abiotic. FBPP (28) therefore concluded that definitions proposed by the APS (2) and BMS (12) are broad enough to include *infectious* and *noninfectious* diseases. Confusingly, FBPP then recommends "that the term "disease" should apply only to malfunctions caused by pathogenic organisms or viruses, and that those caused by other factors should be termed *disorders*," a distinction supported by Holliday (36). However, disorder means simply "derangement" or "malfunctioning"; it refers to disturbances of physiological processes, and disorder underlies any type of disease. Nutritional disorder (e.g. mineral deficiency or toxicity) may lead to disease, particularly if a continuing stimulus is involved. If no obvious disease results, then specific deficiencies, e.g. magnesium, may be mentioned irrespective of their physiological consequence. In common parlance, another synonym of "disorder" is "confusion" (e.g. 38). Adoption of the FBPP definition of disorder for "a harmful deviation from normal functioning of physiological processes arising from causes other than pathogenic organisms or viruses, e.g. mineral deficiency or toxicity, genetic anomaly, low-temperature injury, etc" (28, 36), or of CMI's more concise definition (22) for "a harmful nonpathogenic deviation from normal growth," would cause confusion and should be discouraged. Agrios (1) ignores the term disorder, and we do not know of any acceptable equivalent, for example, in German and Dutch.

Injury

Disease presupposes malfunction, and thus reaction by the victim or suscept in an ongoing process that differs from injury, which as a term is not defined by APS (2), BMS (12), FBPP (28), CMI (22), and Holliday (36). *Injury* is the simple mechanical effect of a parasite or other harmful agent upon the *victim*, which elicits no other reaction than some wound-healing response. Mechanical damage may be caused by single events such as invertebrate feeding or the burning effect of chemicals or heat. Plants consumed by herbivores are *food plants* rather than hosts. However, in closer and more prolonged relationships between invertebrates (e.g. small insects, mites, and nematodes) and plants, the victim, which acts as a host, may also be "irritated" into a pathological reaction and consequently become diseased (see section on Pathogenism). Likewise, air pollutants may affect plants mechanically (or chemically) and cause injury, or may cause disease through continuous irritation. Horsfall & Dimond (41) wrote: "Disease is not the same as injury ... Disease results from continuous irritation." Cowling & Horsfall (24) were more specific: "injury results when plants are disrupted momentarily, as by a mowing machine or grasshopper. In contrast, disease results when plants are subject to sustained impairment (or, as some authors prefer, "continuous irritation") by a pathogen." Therefore, conceptually (and thus terminologically), disease and injury are distinct, although both terms indicate that the plant is affected. Determining whether disease or injury has occurred is dependent on the tools and precision of observation, as is the determination of whether a potentially harmful agent induces symptoms or its presence remains symptomless (*inapparency* or *latency*).

Damage and Loss

Disease and injury both imply, and eventually lead to, physiological and structural *damage* and finally to *loss* in quantity and quality of harvested products (*yield loss* and *economic loss*). Because of the losses sustained, the hostile agents are called *harmful factors*. Their effects may be denoted as *harm*, a term that also covers damage, injury, and loss, but linguistic confusion lingers here. Webster's (33) treats injury, hurt, damage, harm, and mischief as synonymous, having "in common the act or result of inflicting on a person or thing something that causes loss, pain, distress, or impairment." Injury "is the most comprehensive, applying to the act or result involving in impairment or destruction of right, health, freedom, soundness, or loss of something of value." Hurt "applies chiefly to physical injury but in any application it stresses pain or suffering whether injury is involved or not." Damage "applies to injury involving loss, as of property, value or usefulness." Harm "applies to any evil that injures or may injure." Hornby (38) defined harm to cover both damage and injury, injury as "the act that hurts, or the place in the body that is hurt or wounded," and damage as "harm or injury that causes loss of value." NPV (21) summarized damage as "the deleterious effect(s) of attack, injury, disease, or weed infestation" and added that "the term is mostly used in an economic sense, but may also hold for any biologically deleterious effect."

The economic and anthropocentric connotation of damage and of the related terms harm, loss, and disease bears on our definition of other terms such as pest, parasite, and parasitism. Strictly speaking, damage and loss are the final stages in a continuum of symptoms, and yield loss may even occur without any visible symptom. Phytopathologists, particularly those trained as biologists, tend to concentrate on biological phenomena that result from attack or infection. Breeders and growers, with their goal of improved crop yield, are concerned with any impairment of productive capacity; other symptoms are only of interest to them for any possible influence on yield and for their potential as parameters of final yield loss.

PEST AND RELATED TERMS

The above definitions now permit clarification of the terms that denote harmful factors, such as pests, parasites and pathogens, that inflict damage and loss upon plants. These terms reflect the relationships with their victims or suscepts to be discussed in more detail in the next sections.

All Harmful Biotic Factors are Pests

Any organism that is harmful, troublesome, or destructive could be, and often is, described as a *pest*, although the term is often reserved for harmful insects and other invertebrates (36). Like harm, the term pest is not in the biologist's lexicon (37), nor is it defined by APS (2), BMS (12), FBPP (28), or CMI (22). The Oxford Dictionary (38) defines pest in common parlance to indicate insects, rodents, or snails. Webster's (33) is broader: A pest is "(*a*) an epidemic disease associated with high mortality; specifically: plague, (*b*) something resembling a pest especially in destructiveness or noxiousness; especially: a plant or animal detrimental to man or to his interests, (*c*) one that pesters or annoys: nuisance." This definition confuses effect (disease or harm) with cause (the pestering organism). In agriculture we need an unambiguous definition. The term pest has already found wide application (in the term pesticides) for all sorts of chemical and biological controls of harmful agents (which should then be called pests). In 1951, FAO defined the term pest with respect to plants as "any form of plant or animal life, or any pathogenic agent, injurious or potentially injurious to plants or plant products" (30). NPV (21) urged the adoption of the FAO definition of pest in plant pathology and crop protection. The definition encompasses all biotic factors that are harmful to plants or their products and could also include abiotic factors, although the verb "pester" suggests biological activity. Since the Dutch equivalent "plaag" has the epidemic connotation of occurrence in high incidence, NPV (21) defined the term as "population of an organism at such a density that hindrance ensues." The concept then approaches that of epidemic, which is indicated by the English word plague. Thus, the term pest should not be limited to harmful insects or

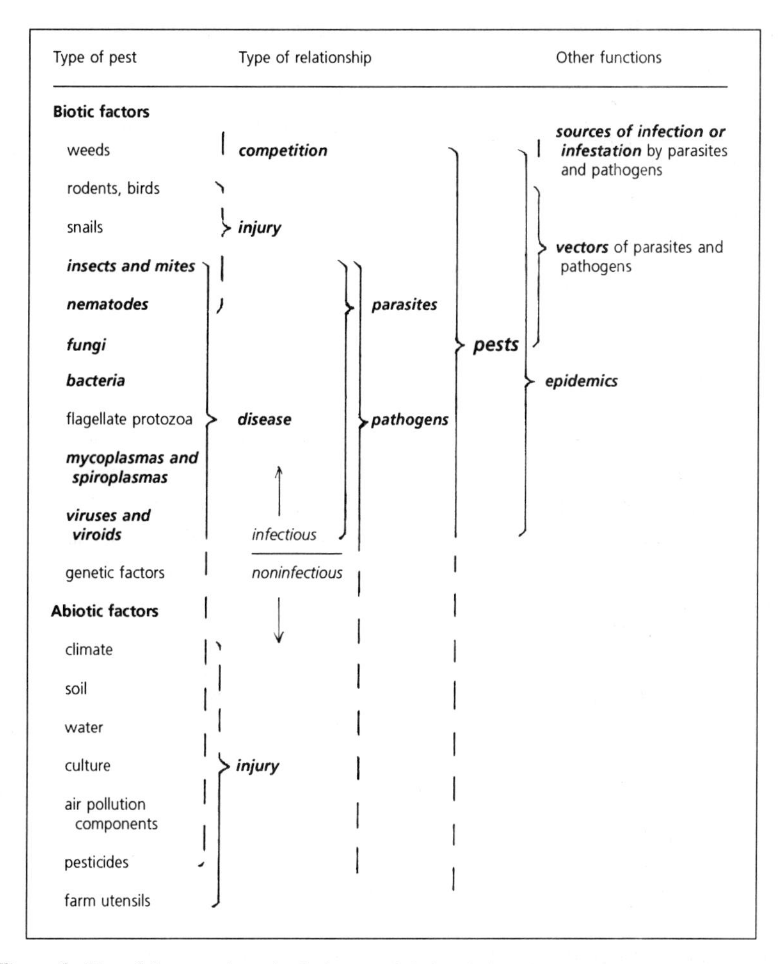

Figure 1 Harmful agents (pests) of plants and their relationships with plants and crops. Major groups of pests that are treated in handbooks on plant pathology are printed in bold italics.

animals, but should include weeds and pathogens such as fungi and viruses and their epidemic potential.

The terms "disease" and "pest," often bracketed together, are of totally different categories, whereas "disease" and "injury" are related. Both are caused by "pests" (possibly including abiotic factors). The causes of the disorders of plants, rather than their effects, need to be addressed for control.

Pest, like disease, has an ethical and economic connotation. The same holds true for *weed*, a type of pest that is "a plant (group of plants or species) growing where it is not wanted" (21). This definition also includes volunteer plants of crops growing in other crops. The term comes from crop husbandry, and biologists disagree with the above definition in that "to the forum of nature, weeds are equally entitled to grow as what is considered useful" (60). The most important types of plant pests and their relationships with plants are listed in Figure 1.

Types of Relationships with Plants

A range of terms is used to describe the various types of plant/pest relationships and the differing degrees of interaction involved. Plant/pest interactions are mainly, though not exclusively, nutritional.

COMPETITION In the first category of relationships, that is in *competition*, interaction is not by contact but occurs via the environment. Most weeds, for example, have no direct contact with crop plants, but are in competition for space, light, water, and nutrition. Weeds often harbor other crop pests and in turn act as sources of infection or infestation for crop plants. Some weeds harbor predators of harmful organisms or release chemical compounds that repel crop pests, whereas others release chemicals toxic to neighboring crop plants and thus are pathogens (allelopathy) (29, 51, 66).

EPHEMERAL FOOD DEPRIVATION In the second category, food is directly derived by one type of free-living organism from another, usually in brief contact, and no continuous irritation occurs. An animal that kills and feeds upon other animals is called a *predator* (21, 33, 37) or *carnivore,* if it eats flesh. A predator is usually larger than its prey and does not develop on or in its prey like a parasite or a parasitoid (see below). An animal that feeds for a brief period on plants (and in the process injures them mechanically) is a *herbivore* (37). The plant merely provides food and is partially, largely, or entirely consumed. If it survives, the plant is injured but no disease results. Interaction and feeding by animals like nematodes and small insects often is of longer duration, which then causes irritation in the plant and the possible onset of disease. These animals then are pathogenic herbivores or just pathogens.

PROLONGED NUTRITIVE RELATIONSHIPS Symbiosis, parasitism, and other related terms are relevant to the third category, which comprises closer and more prolonged contacts.

Symbiosis is an ambiguous term because it literally suggests only the (permanent) living together of different types of organisms irrespective of the type of relationship. It is often ignored in plant pathology (2, 12, 28) or is discussed "because some biologists consider that its definition should cover parasitism" (36). Webster's (33) defines symbiosis as "(*a*) the living together in more or less intimate association or even close union of two dissimilar organisms (as in parasitism, mutualism, or commensalism), (*b*) the intimate living together of two dissimilar organisms in any of various mutually beneficial relationships; often: mutualism." Some plant pathologists use the term in a broad sense to include parasitism (6, 7), but others use it more narrowly for the "harmonic association of different organisms" (38) or for "mutual dependence" (31). Still others remain ambiguous, as in the biological dictionary's (37) definition of symbiosis as the "living together of different species not necessarily only those of mutual benefit; the term is often used for an association of mutual benefit which is more properly called mutualism." Confusion started with De Bary (25), the father of plant pathology; he first defined symbiosis in 1879 as "the living together of two phylogenetically unrelated species of organisms" when referring to the "symbiosis" of *Phytophthora infestans* with potato, thus including a harmful partnership. In his discussion of interorganismal genetics in host/pathogen relationships, Loegering (43) stressed similarities in these relationships with those in lichens and mycorrhiza and strongly favored denoting all as forms of symbiosis.

Symbiosis comprises commensalism and mutualism. There is general accord on the meaning of *commensalism,* a partnership benefical to one and not harmful to the other partner, and of *mutualism,* beneficial to both. But commensalism is sometimes extended to include "both species as a rule benefiting from the association" (37) or "mutual convenience" (36). Potentially pathogenic organisms or viruses (especially latent viruses) may also live in harmony with their hosts without causing any apparent disease. This relationship typically represents commensalism. However, there is no clear borderline between commensalism and parasitism when one partner is weakened and consequently suffers.

To lessen confusion, NPV (21) proposed to restrict the meaning of symbiosis to "the more or less intimate living together of two dissimilar organisms to the benefit of one (commensalism) or of both partners (mutualism) ... it is uncommon in plant pathology to classify parasitism with symbiosis although parasite and host are closely living together." The definition of symbiosis should therefore be restricted to forms of harmonious coexistence. A *symbiont* is characterized as "one of the partners in symbiosis" (37), or "an organism living in symbiosis" (33), but also as "usually the smaller member of a symbiotic pair of dissimilar size" (33). The organism with which it coexists is a *host,* especially organisms that harbor endosymbionts. Examples are nitrogen-binding bacteria in root nodules of leguminous plant species.

Parasitism is a widely used term in plant pathology, entomology and nematology, but its interpretation varies. Food deprivation dominates in most definitions: parasitism thus is the "partial or complete nutritional dependence of one organism or virus on the tissues of another living organism" (2, 12, 22,

28). However, there is often an additional modification that the parasite is "conferring no benefit in return" (12, 28) or that it is to the detriment of the host (31, 37). FBPP (28) claimed that "the BMS addendum, that the parasite confers "no benefit in return" is difficult to establish in practice, and was presumably added to enable a distinction to be made between parasitic and symbiotic relationships ... However, there seems no reason why, for example, a mycorrhizal fungus should not be regarded as a parasite having a symbiotic relationship with its host." The addition ambiguously extends the definition of parasitism to all types of symbiotic relationships, using withdrawal of nutrients as a basic criterion, irrespective of the effects of withdrawal on one or both partners. Bateman (6) used the term parasitism for the whole complex of simultaneous actions and thus as synonymous with symbiosis, but also concurred with the condition that in parasitism "one party, the parasite, benefits from the other," and then added that "it is often assumed, sometimes incorrectly, that the parasitic relationship is necessarily harmful to the host." He divided parasitism into mutualism, commensalism, and pathogenism. The same ambiguity shows up in Webster's (33) first reference to parasitism as "the willingness to live luxuriously at the expense of the poor," followed by "a relationship that involves intimate association of organisms of two kinds including commensalism, symbiosis, parasitism, parasitoidism." Other authors (1, 36) try to remain on the safe side by recording nutritional dependence as the sole criterion for parasitism, which then equals symbiosis.

These citations on parasitism and symbiosis reflect the reasoning among biologists not to interpret phenomena from an ethical or economic standpoint. In our view, this ambiguity should be discouraged. In everyday speech, parasitism is associated with one partner, the *parasite*, living at the expense of the other. Webster's (33) is explicit when defining parasite as "(*a*) one frequenting the tables of the rich or living at another's expense ... (*b*) an organism living in or on another living organism, obtaining from it part or all of its organic nutriment, and commonly exhibiting some degree of adaptive structural modification ... (*c*) such an organism that causes some degree of real damage to its host; compare commensal ... symbiont." Hornby (38) also specified a parasite as a "person supported by another and giving him nothing in return." NPV (21) accordingly defined parasite as "an organism or virus, more or less permanently or for part of its life cycle living in more or less close association with a living organism (host), from the tissues and at the detriment of which it derives part or all of its food."

Parasitism involves development of the parasite on or in its host. *Ectoparasites* feed on plants from the outside; examples include powdery mildews on leaves and free-living plant-parasitic nematodes on roots. Most fungi and all viruses are *endoparasites*. A *parasitoid* is an insect that during its larval stage lives off, and in, another insect, which is eventually killed.

Pathogenism, a qualitative term for the phenomenon of pathogenicity of agents, is often the outcome of parasitism, but not always. The denotion parasitism for a plant/pest relationship does not specify the type of damage the pest (parasite) inflicts upon its host as a consequence of interaction with the plant. It does not indicate whether the result is only injury to the plant, or whether there is continuous irritation evoking reaction in the plant, followed by disease—in which case the parasite acts as a *pathogen*. Whether parasite or pathogen is the appropriate term also depends on the tools and the accuracy with which the host is examined.

Usually, arthropods are thought only to injure plants, but in large numbers they also cause disease, particularly when establishing a parasitic connection that may cause continuous irritation. For example, phytotoxic or phytohormonal salivary compounds can induce a range of toxemias such as chlorosis and necrosis of foliage, disturbed food and water translocation, leaf curling, tissue swelling, galls, and other growth abnormalities (17, 47). When feeding on plant roots, ectoparasitic nematodes inject saliva into plant cells, thereby reducing growth of cells, tissues, and organs. Parasitic weeds live in such close and continuing contact with crop plants that they not only directly derive food from their host, but may also cause disease that culminates in premature plant death (e.g. *Striga* and *Orobanche* spp.), or abnormal growth (e.g. *Arceutobium* spp.) (42). Thus parasitism often leads to disease, but not all parasites are pathogens. Parasites of all sorts are mainly characterized as biotic entities deriving food from and at the expense of other living organisms; they become pathogens when inducing harmful reaction in their hosts (24).

Not all pathogens are parasites, e.g. plants secreting allelopathic substances into the soil affect other organisms through the environment. Such pathogens could be considered *ectopathogens* (66) and are not parasitic. Pathogens living on the surface of plants fall into the same category. Abiotic pathogenic agents do not have a parasitic relationship with their victims, and these cannot be considered hosts. These agents, although pathogenic, cannot be considered to be parasites inasmuch as parasites are, by definition, biological entities (organisms or viruses). Viruses and viroids are an intermediate class of pathogens between biotic and abiotic agents, with genetic features in common with organisms. Although not considered to be organisms, they are parasites in their relation with their host.

Other Functions

Other pests affect plant and crop health in ways other than or in addition to competition and food deprivation. Weeds are often important *sources of infection* or *infestation* of crops with parasites and pathogens and may assist in their perennation when susceptible crops are temporarily absent. Various pests act as vehicles of spread (*vectors*) of other pests. In nature, viruses, for example, move mainly from plant to plant through insects, nematodes, and fungi, and long-distance dissemination of pests, particularly viruses, often is in wind-blown or water-floating seeds as of weeds.

It is hard to distinguish between disease and injury (Figure 1), and between parasitism and pathogenism to delimit *plant pathology* as a discipline. It is the discourse or science of "suffering" of plants (41). It includes the pathogenic effects on plants of insects, mites, nematodes, weeds; of abiotic factors; and fungi, bacteria and viruses, but excludes their nonpathogenic injurious effects. *Crop or plant protection science* is more descriptive of the field of interest. It also covers the ecology of the harmful agents and emphasizes the practical relevance better.

ESTABLISHMENT OF HOST/PARASITE RELATIONSHIP

The onset and/or establishment of host/parasite relationship, i.e. the aggression by the parasite, is called *attack*. This is a highly descriptive, though never scientifically defined, term. It may entail local and systemic ingress or invasion and colonization of the host by the parasite (e.g. 27).

Infestation refers to attack by an ectoparasite, which causes injury but also establishes an external relationship with the plant or crop. Webster's (33) defines infest broadly: "(*a*) to attack or harass persistently: worry, annoy, (*b*) to visit persistently or in large numbers: overrun, haunt, (*c*) to live in or on as a parasite." In biology, infestation is used specifically for "invasion by exterior organisms, as by ectoparasites" (37). There is consensus within the plant pathology community on its use "especially ... for insect and other animal pests ... that overrun the surface of a plant" (28, 36), and on its applying to their "dispersion through soil or other substrate" (28), or "the occurrence of great numbers of insects, mites, nematodes, etc, in an area or field, on a plant surface or in soil" (1). Attack by an endoparasite, usually a pathogen, is called *infection*. Again, modern interpretation has narrowed from the Latin root, inficere, meaning to put or dip into, to taint. Infection is generally accepted as the "process or state of establishment of a pathogenic microorganism or virus in a living organism" (2), or "the entry of an organism or virus into a host and the establishment of a permanent or temporary parasitic relationship" (28, 36). NPV (21) clearly defined infection as "the landing of a pathogen on a host and the initiation of any type of parasitic activity." Infection is thus closely associated with pathogenic activity. In nature it is often preceded by accidental, superficial contact or *contamination*. Infection includes establishment (invasion and colonization of the parasite) and may involve several mechanisms in a continuous process that affects one cell, tissue, or organ after the other. The term covers the entire compatible relationship between parasite and host.

Infestation and contamination may also be used to characterize the soil containing soil-borne plant parasites (e.g. nematodes) and propagules of pathogens. *Inoculation* is the "application of microorganisms or virus particles to a host or into a culture medium" (28, 36), or, better, "the transfer of material containing a parasite or its propagules (*a*) into or onto tissues of an organism to initiate infection, (*b*) into or onto a culture medium for propagation" (21). It is a human activity, even when vectors are used for transfer.

Attack includes nonparasitic plant-pest relationships leading to injury or gross damage and also covers for short-duration activity by insects or nematodes on food plants. It infers aggression and implies hostility and also applies to the activity of predators.

The term attack has good equivalents in German (Befall) and Dutch (aantasting). NPV (21) defines attack as "the act of attacking or the state of being attacked" ... a "conseqence of aggressiveness" (see below) "of the parasite or phytophagous organism and of the susceptibility of the host or victim, and it does not necessarily result in disease." Several host-parasite relationships are especially characterized by attack, i.e. the visible physical presence of the parasite rather than the resultant disease. Examples are mildews and rusts. These attacks are named after the presence of the parasites (signs) rather than the reactions of their hosts (symptoms).

A plant that is attacked by pests in their widest sense is a *suscept,* i.e. "an organism affected or capable of being affected," and "by a given disease," (2, 28), or is "a host affected by, or prone to, disease" (36). This would, however, imply host reaction and thus sensitivity, which we distinguish from susceptibility (see below). We therefore prefer to describe a suscept as "an organism attacked or capable of being attacked by a pest," irrespective of the type and duration of interaction. This goes slightly beyond Webster's definition (33) of a suscept as "an organism upon or in which another organism is or may become parasitic." When a parasitic relationship is establised, the organism under attack serves as a *host,* which is "an organism harboring a parasite" (12, 36) or "a living organism harboring another organism or virus dependent on it for existence" (2); both definitions are used by FBPP (28). The attacking organism or virus acts as a parasite or, if no damage results (i.e. infection is symptomless or latent), the attacking organism or virus is a commensal and thus is a symbiont rather than a parasite.

The type of relationship is determined by the genotype of both the parasite and the host (see below). An organism that is pathogenic on one genotype (cultivar) of host may be a commensal on another, which then acts as a symptomless source of infection. In nature, pathogenic relationships are exceptions rather than the rule, such as with many viruses that occur without symptoms in wild vegetation. These viruses remain potentially pathogenic to other genotypes, as of genetically homogeneous crops. Most higher organisms, including plants, are immune from attack or infection by most microorganisms and viruses in their habitat (i.e. "the place where an organism grows or lives naturally"; 28). They are *nonsuscept* or *nonhost* to those microorganisms and viruses, and most combinations of plant and potentially hostile agent are incompatible. *Incompatibility* does not exclude attempts at aggression by the potential parasite and defensive reactions by the potential host (34, 35), but the terms nonsuscept and nonhost have so far not been adopted by terminology committees. The phenomenon of incompatibility between plant and pest is generally referred to as *immunity,* which ranges in definition from "freedom from disease" to "freedom from attack by a pathogenic organism or virus" (2) to "exemption from infection" (12). FBPP (28) added to these definitions that "immunity implies total exclusion of the potential pathogen." NPV (21) called for the term to be used only in the sense of insusceptibility, "for instances where after inoculation the parasite is no more detectable and no macroscopically visible symptoms develop," but added that "microscopical symptoms may occur; then extreme *hypersensitivity* is at hand."

ROLE OF THE ENVIRONMENT

Whether and to what extent there is interaction between a living potential pest (or parasite or pathogen) and a plant (potential victim or host) at and after confrontation depends on the environment as well as on the genotype of both. For host/pathogen relationships this interaction is often expressed by the disease triangle of host, pathogen, and environment (Figure 2). Such a triangular interaction also holds for plant/pest (or suscept/pest) relationships in general.

Important environmental factors are biotic—temperature, light, and cultural conditions such as nutrition and water supply. They all influence the physiology of both plant and pest. Thus plant/pest interactions, even if genotypically controlled, are extremely variable and are phenotypically conditioned. Liability to disease usually is referred to as *predisposition* (19, 67) or disease proneness (36, 67); it has been defined as "the tendency of nongenetic conditions, acting

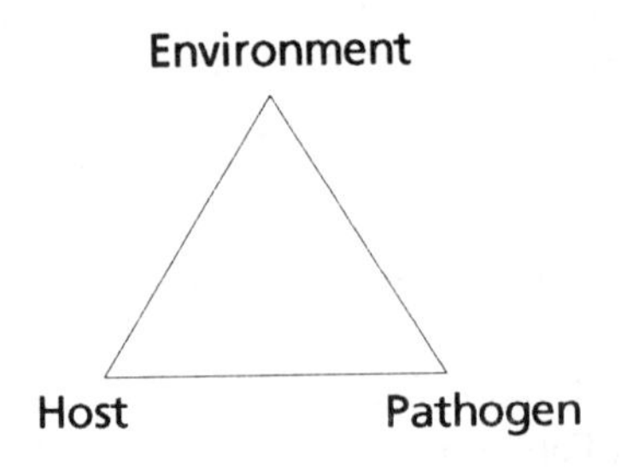

Figure 2 Disease triangle.

before infection, to affect the susceptibility of plants to disease" (67) or to pathogens (19). NPV (21) defined predisposition more precisely as "the pre-infection state of an organism or organ, depending upon external conditions and stage of development, that determines the severity of disease." If the definition were broadened to cover all plant/pest relationships, it would read: "Predisposition is the tendency of nongenetic conditions, acting before plant/pest confrontation, to affect the liability (susceptibility) to attack and damage."

One environmental factor physiologically predisposing the potential suscept is previous or simultaneous attack by other pests, particularly infection by other pathogens. Apart from a wide range of direct interactions, parasites or pathogens mostly interact mutually via their host's physiology, so that susceptibility and sensitivity (see below) to one parasite or pathogen is being increased or decreased by the other.

*Pre*disposition linguistically implies only the effect of preattack or preinfection conditions, but other environmental factors that emerge during attack or infection may alter the course of a disease by aggravating symptoms, *masking* them temporarily, or causing them to disappear permanently (*recovery* from disease). The effects of plant and pest characteristics must be clearly distinguished from environmental effects in any assessment of plant/pest interactions, a difficult proposition unless environmental conditions are standardized.

A SCHEMATIC ANALYSIS OF HOST/PARASITE INTERACTIONS

The relationships between host and parasite, especially those between plant host and pathogen, have been extensively studied for their genetics with a view to possible interference through plant breeding. This requires knowledge of variations in both parasite or pathogen and host and their genotypes. The natural coevolution of parasite or pathogen and host has led to narrow specialization. Further mutual adaptation, down to gene-for-gene interaction, has resulted from continued breeding for resistance.

Description and quantification of host/parasite or host/pathogen relationships require that the part played by each of the two organisms be differentiated and analyzed. The goal is to detect and identify underlying mechanisms, possibly in terms of molecular biology and gene activity. The Terminology Committee of the Royal Netherlands Society of Plant Pathology (21) found the scheme of Figure 3 (now broadened to comprise environmentally dependent host/parasite relationships) helpful in identifying components of host/pathogen relationships. It has already been used for the interaction between viruses and plants (10). Each aspect or component phenomenon may represent a separate mechanism that involves specific genes.

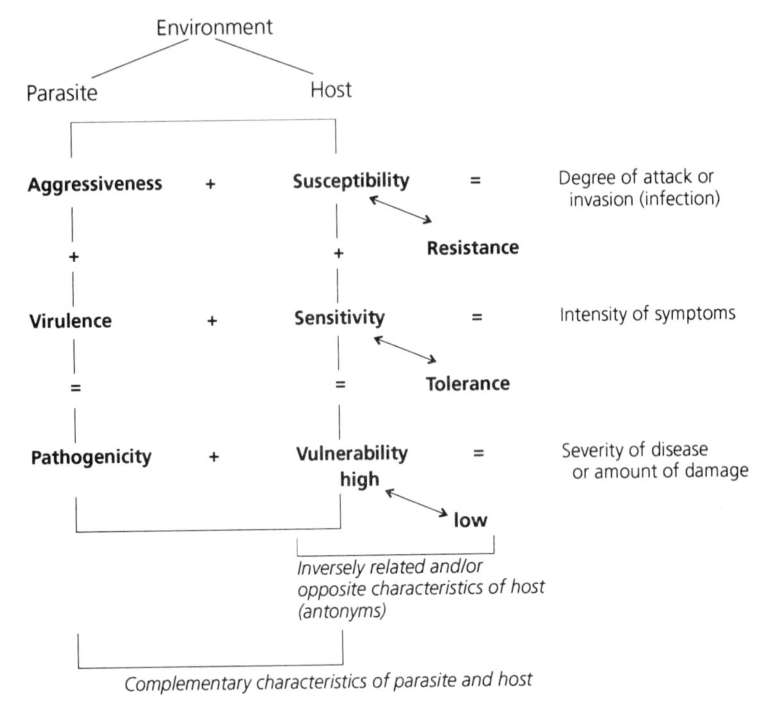

Figure 3 Diagrammatic representation of environmentally dependent host/parasite relationships. (Completed after NPV, 1985 (21)).

Combined Effects

The behavior of a parasite cannot be studied without a host, and the capability of a plant genotype to act as a host cannot be described without a parasite. The characteristics of parasite and host listed in Figure 3 are therefore *complementary characteristics.* It is their combined effect that is observed, and the impact of each in the host/parasite interaction cannot be measured independently, but only in relation to, and with the help of, a range of the other's genotypes. In single host/parasite interactions, the part played by one is confounded with that of the other. The effects of each cannot then be singled out and measured individually, e.g. a low population density of an insect, or poor colonization of a fungus on a given plant may be due to low aggressiveness of the pest, to low susceptibility of the plant, or to both (for definitions of the terms, see below).

Characteristics of the Parasite

The literature on plant pathology is extremely controversial with respect to the terms pathogenicity, aggressiveness, and virulence when used to describe the role of the parasites (and pathogens) in attacking their hosts. Shaner et al (55) recently reviewed the literature with respect to pathogenicity and virulence, and included standard English usage and medical and scientific terminology. They concluded that "obviously, confusion reigns," and hoped by their paper to stimulate discussion.

AGGRESSIVENESS The term *aggressiveness* or aggressivity denotes whether and to what extent a parasite (organism or virus) can attack other organisms. Webster's (33) describes aggressive as "marked by combative readiness or bold determination: not conciliatory, militant; promoting or accessory to aggression in predacious animals (as insects), spreading with vigor (aggressive weeds); chemically active; tending or able to utilize a variety of habitats, able to encroach on occupied areas." Terminology committees of APS (2) and BMS (12) and the Plant Pathologist's Pocketbook (22) did not include aggressiveness as a term, and FBPP (28) rejected it because it is often used incorrectly for "horizontal pathogenicity" (see Host and Parasite Specificity). However, the term aggressiveness is linguistically highly descriptive, and NPV (21) defined it as the "ability of a parasite to attack," to establish a parasitic relationship and develop on a given host or invade it in space and time. Shaner et al (55) preferred the term "parasitic fitness." Fitness, however, encompasses all characteristics of the organism to survive, of which the ability to parasitize is only a part, albeit an important one. For pathogens, it also includes infectivity. Aggressiveness alone is not sufficient for pathogenicity. A nonpathogenic parasite may be very aggressive and even injurious, but may lack virulence (see below) and hence pathogenicity.

NPV (21) added that the term aggressiveness is quantitative and assessed by the degree of host attack, but that this is equally influenced by the host's susceptibility and/or resistance (see below). Aggressiveness must therefore be assessed on a range of host genotypes by measuring the amount of pathogen or parasite developing on or in the given host species, the extent of host plant and of host population that is attacked or invaded, and the population density of the pathogen/parasite on plants and crops, or the number of plants infected or diseased. For example, the recurrence of Dutch elm disease in Europe during the 1970s, which involved hitherto resistant cultivars of elm, has been ascribed to the emergence of a more aggressive "strain" of the fungus *Ceratocystis (Ophiostoma) ulmi.* The new type not only attacks plants faster and more intensively, but it also develops more rapidly in vitro and produces more aerial and fluffy mycelium than the conventional waxy and yeast-like type (32). Scanning by electron microscope of vascular tissue of diseased trees indicated that the new type developed faster in xylem vessels and penetrated the vessel pits more rapidly (44). It was therefore more vigorous and more disposed to attack than the regular strain; it was more aggressive.

VIRULENCE The term *virulence* is often used imprecisely for the "relative capacity to cause disease; degree or measure of pathogenicity of a parasitic organism or virus" (2, 12). FBPP (28) and Holliday (36) are not clear, but recommend against its use for vertical pathogenicity (see Host and Parasite Specificity). NPV (21) defined virulence as the "capacity of a pathogen to cause symptoms"; it "indicates the severity of the symptoms a pathogen causes in its host per unit of attack (e.g. quantity of pathogen)." The term is highly descriptive deriving as it does from the Latin word, virus, for poison or venom, and thus it connotes toxicity, noxiousness, or malignancy. Shaner et al (55) agree that consistency of the meaning of this term with its meaning in standard English and medicine should have considerable weight. The term refers to the ability of a parasite to incite reaction and cause disease. Qualitatively, it makes a parasite into a pathogen; without virulence a parasite is not a pathogen.

One parasite, species, subspecies, or biotype may be highly virulent on one host genotype and of low or no virulence on another. The mechanism of virulence is then easily confused with differences in sensitivity between the different genotypes of the host. To separate the virulence of the pathogen from the effect of host sensitivity, the severity of symptoms must be measured on a range of host genotypes in relation to the density of pathogen.

PATHOGENICITY *Pathogenicity* is the quantitative capacity to cause disease; the "overall disease-inducing capacity of a biotic or abiotic factor" (21), and it is determined jointly by the aggressiveness and virulence of the parasite. It is not an absolute capacity, with virulence being an indication of the degree of pathogenicity, as is often claimed (2, 14), nor is it synonymous with virulence when used quantitatively (28). Its degree can be assessed by measuring the amount of parasite that develops (degree of infection as a parameter of aggressiveness) and by the severity of the symptoms produced on a series of host genotypes in relation to the amount of parasite. Without virulence, the parasite does not act as a pathogen but only causes mechanical injury or is a mere commensal. Pathogenicity is not restricted to host-parasite interactions but also results from the activity of ectopathogens, even when not in direct contact with plants (see allelopathy).

Characteristics of the Host

Events that occur in the host during and after the parasitic relationship is established are also subject to terminological inconsistency. Much of what happens depends on the host's genotype, either allowing, supporting, or counteracting the parasite's activities. For instance, poor parasite development may be greatly affected by the host's poor nutritive support and/or passive or active defence.

The three groups of terms relevant to host behavior—susceptibilty and resistance, sensitivity and tolerance, and high and low vulnerability (Figure 3)—may be *inversely related characteristics* denoting the same mechanism (e.g. the higher the susceptibility, the lower the resistance); the two terms are antonyms. They may also be counterparts or *opposite characteristics*; in some instances, susceptibility and resistance mechanisms are likely to act simultaneously and independently, together determining the host's response to attack by the parasite. In discussing the characteristics of the host in determining what is going on in plants that are under attack and how they react, it must be kept in mind that characteristics of both the plant and of the parasite complement each other in the final reaction of plants. The individual effect of each must then be deduced from what is observable on or in the host, i.e. the extent of attack and the severity of reaction and outcome of disease and/or damage.

SUSCEPTIBILITY AND RESISTANCE The *degree of attack or invasion* (infection) of the host by a given parasite depends on the host's susceptibility and/or resistance as well as on the parasite's aggressiveness. The extent and rapidity of multiplication of the parasite also determine relationships with plants at the population level—information crucial to our understanding of epidemiology (48, 53, 58, 59). Susceptibility and resistance may both include a range of mechanisms, each controlled by one or more different genes, which must be identified. These genes may be switched on before or during ingress of the parasite, or during access by the vector (e.g. of a virus), and/or during establishment of attack or infection, i.e. multiplication, colonization, and internal spread or invasion. Prevailing definitions differ widely in their emphasis.

Susceptibility has been defined as the "inability of an organism to oppose the operation or to overcome the effects of an injurious or pathogenic factor, or the inability to defend itself against or to overcome the effects of invasion by a pathogenic organism or virus" (2), or as "nonimmunity" (12, 22). According to FBPP (28), "susceptible should only be used in the sense of nonimmune" even though it is often used to indicate the presence of severe symptoms. Cooper & Jones (23) noted that "susceptible has been used in contrasting senses: as the opposite of immune, as a descriptor of the ease of infection in individual plants, as a descriptor of the prevalence of disease in a plant population, or as an indicator of disease severity in individual plants." They suggested the term infectible, but this would only apply if the parasite acts as a pathogen. NPV (21) defined susceptibility as the "inability of an organism to impede the growth/development of a parasite; the complex of characteristics making an organism suitable for hosting a parasite."

Most of these definitions characterize susceptibility as an antonym of resistance, even though susceptibility and resistance may act simultaneously as counterparts involving different mechanisms. We propose rewording the NPV definition by changing the order of the two components. Suceptibility then is "the complex of characteristics making an organism suitable for hosting a parasite, and/or its relative inability to impede the attack (growth/development) of the parasite (and thus the infection by a pathogen)," or briefly in line with the definition of suscept, "the capacity of an organism of being attacked by a pest." Any reference to the effects of the invasion (2, 28, 23) should be omitted. Susceptibility should not be considered "an indicator of disease severity in individual plants," as Cooper & Jones proposed (23), since this implies sensitivity (see below). Plants may be susceptible without showing symptoms (see also Tolerance).

Resistance refers to the ability to "oppose; use force against in order to prevent the advance of: resistance to the enemy/an attack/authority/the police" (38). As such it is primarily directed toward the incitant of damage or injury rather than toward its effect. Control of the pathogen itself is a more effective strategy to counter the damage it causes than control of the resultant disease. As with suceptibility, cause and effect (parasite or pathogen and injury or disease) must be distinguished. The definition of resistance proposed by APS (2) and FBPP (28) is inadequate—"(*a*) the ability of an organism to withstand or oppose the operation of or to lessen or overcome the effects of an injurious or pathogenic factor, (*b*) the ability of the host to suppress or retard the activity of a pathogenic organism or virus"—because it covers both resistance to the pathogen and tolerance (see below). CMI (22) yields to the latter by defining as: "the power of an organism to overcome, completely or in some degree, the effect of a pathogen or other damaging factor." Cooper & Jones (23) justly advocated "to reserve the terms resistant and susceptible to denote the opposite ends of a scale covering the effects of an infectible individual on ... infection, multiplication, and invasion." NPV (21) further defined resistance as "(*a*) the ability of a host to hinder growth and activity of a parasite or phytophagous organism and the multiplication of a virus, (*b*) the ability of an organism to neutralize an injurious chemical factor." Resistance, like susceptibility, may incorporate a range of mechanisms including resistance to the vector of a pathogen, especially a virus, and resistance to the ingress, establishment, and spread of the parasite itself.

SENSITIVITY AND TOLERANCE The *intensity of symptoms* produced by the host is determined by two or three factors: the host's sensitivity and/or tolerance and the virulence of the parasite.

Sensitivity is generally used to denote the severity of reaction by the host (2, 12, 36), and is sometimes used nonquantitatively as an alternative to susceptible (28). NPV (21) more precisely defines sensitivity as the "characteristic of an organism to react with relatively severe symptoms (including yield reduction) to a parasite, phytophagous organism or abiotic factor." It is the host's ability to respond to attack or infection. If the organism does not react, then the attack leads to latent infection or causes mechanical injury, and the host cannot be considered sensitive. *Hypersensitivity* of plants (a resistance mechanism) refers to "the violent reaction of an organism to attack by a pathogenic organism or a virus resulting in prompt death of invaded tissue, thus preventing further spread of infection" (2, 12, 28, 36). However, this also may result from nonnecrotic reactions involving other obstructions caused by infection, as with many viruses.

Tolerance is usually defined as "the ability of the affected organism to endure the operation of a pathogenic factor or invasion by a pathogenic organism or virus with little or no reaction, as shown by more or less complete absence of symptom expression and damage" (2) or "the ability to endure infection by a particular pathogen, without showing severe disease" (12). The definition by FBPP (28) is in accord, and CMI (22) adds the qualification "or giving little reaction to the effect of other factors." However, these definitions, as illustrated in Figure 4, reflect the biologist's or pathologist's outlook. They do not address the recurrent problem of how to define disease. Disease severity is judged by severity and extent of visible symptoms (symptom score), and reduced vitality and yield loss are usually not included in assessing the syndrome. Growers and resistance breeders must look beyond overt symptoms to effects on yield (21), and they should discriminate between yield loss (qualitatively and quantitatively) and overall disease severity (extent and intensity

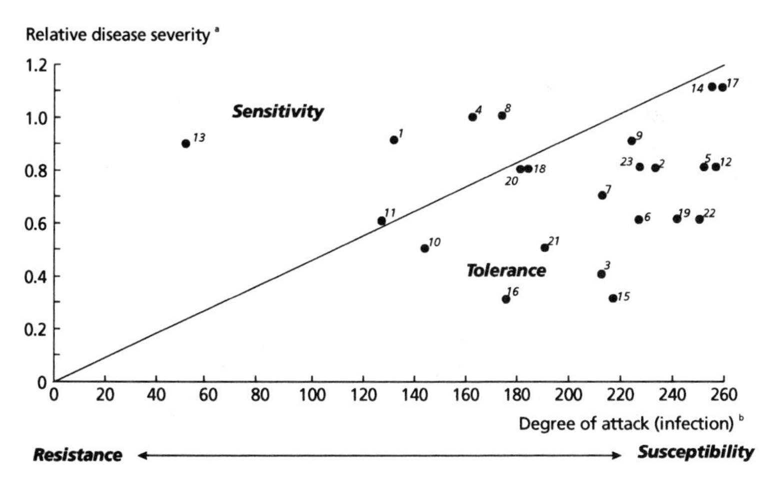

Figure 4 Inverse relationship between tolerance and sensitivity of crop genotypes (cultivars), as viewed by biologists and pathologists when not discriminating between overall damage and final yield loss. Data are from observations on the reaction of butterhead lettuce cultivars to cucumber mosaic virus (62). *a* Symptom score at harvest based on a 0 (no symptoms) to 5 (severe symptoms) scale. *b* Relative virus concentration determined by ELISA serology. Mild disease may be due to resistance to the pathogen and/or to tolerance to infection.

of symptoms). Schafer (54), following Caldwell et al (15), emphasized the relatively minor effect on yield when defining tolerance as "the capacity of a cultivar resulting in less yield or quality loss relative to disease severity or pathogen development when compared with other cultivars of the crop."

However, this definition does not answer the question of what is tolerated, the disease or the pathogen. Only the first is referred to in the title of Schafer's paper (54). For host/parasite relationships, Clarke (18)) distinguished between *tolerance of (or to) the parasite* (for tolerance with respect to the physical presence of the parasite) and *tolerance of (or to) disease* (for tolerance to disease resulting from invasion by a pathogen as measured by plant growth or yield) (Figure 5*A* and *B*). Clarke further distinguished "overall tolerance" of a plant's ability to "endure parasitic infection and disease" with relatively little "impairment of growth or yield," and concluded that Schafer's definition encompasses overall tolerance, but some objections are valid. NPV (21) unambiguously defined tolerance as the "ability of a host to limit the harmful effects of a parasite or phytophagous organism or of an abiotic factor." Extreme tolerance allows normal yield and symptomless infection (tolerance to both attack and disease).

Tolerance is the opposite of or inversely related to sensitivity (Figures 4, 5). Cooper & Jones (23) stated that "the terms tolerant and sensitive are the opposite ends of a scale covering the disease reaction of the plant to ... infection and establishment." However, their definition only pertains to tolerance to attack (Figure 5*B*). The terms tolerant and sensitive are often used in an absolute sense, as of a tolerant or sensitive cultivar, but they are gradual, and the regression lines drawn in Figures 4 and 5 are arbitrary.

Tolerance was often used incorrectly in the past in the sense of partial or incomplete resistance. FBPP issued a legitimate warning (28) that "absence of symptoms because colonization has been restricted in extent or intensity by host-defensive responses implies resistance, not tolerance." Breeders should specify whether absence or poor expression of symptoms in their breeding materials is due to resistance or to tolerance. Resistance of a host to a parasite or pathogen and tolerance are independent mechanisms, as are aggressiveness and virulence. If a parasite is not pathogenic, then the host does not react with symptoms; it is insensitive. The effect of the relationship between host and nonpathogenic parasite may be described and quantified in terms and degree of intensity of injury or damage.

VULNERABILITY The ultimate *amount of damage and the severity of disease* on or in a host that is caused by attack by a parasite, given its injuriousness or pathogenicity, depend on the host's vulnerability: the combined effects of the host's susceptibility and sensitivity on the one hand, and of the host's resistance and tolerance on the other. Here also the inputs by the parasite must be distinguished from those of the host.

Vulnerability has never been defined in the international literature on plant pathology. It is not a scientific term, but is nevertheless increasingly used in

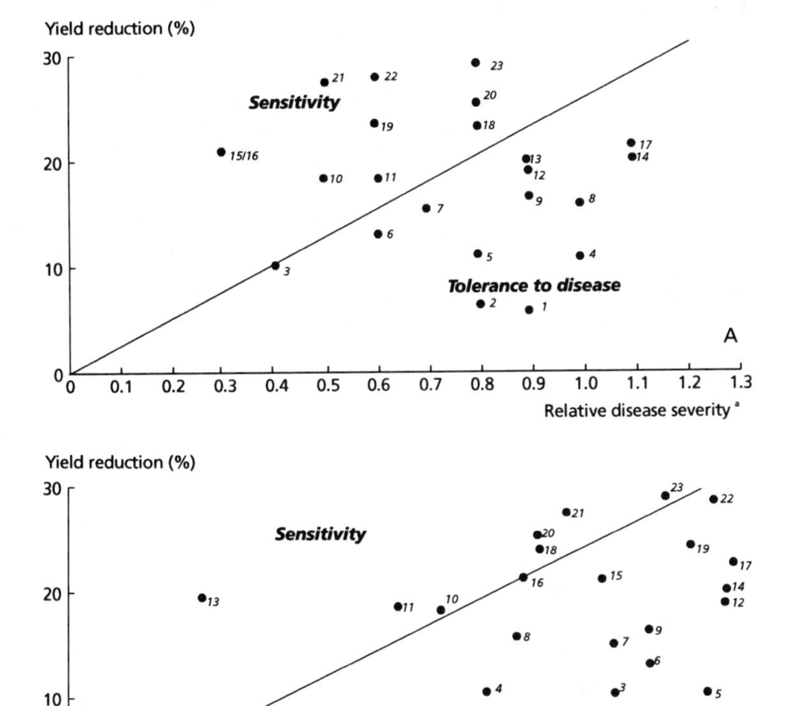

Figure 5 Graphical illustration of tolerance to disease (*A*) and tolerance to attack (*B*) (in inverse relationships to sensitivity) of a range of crop genotypes as viewed by growers and breeders emphasizing yield reduction in relation to amount of disease (*A*) and amount of attack (*B*). [Graphs drawn from data also used for Figure 4 on the reaction of lettuce cultivars to cucumber mosaic virus (62); [a] Symptom score at harvest based on a 0 (no symptoms) to 5 (severe symptoms) scale. [b] Relative virus concentration determined by ELISA serology. For data on tobacco etch virus in flue-cured tobacco cultivars, similar to those of Figure 5*A*, see (9), and for comparable data on powdery mildew in barley cultivars, see (50). Note that cultivars 14 and 15 in Figure 5*A* sustain the same yield reduction although 14 shows considerably more visible symptoms of disease than 15; cultivar 14 is more tolerant to disease than 15. Furthermore, cultivar 2 suffers less from yield loss than 23, although both display a similar degree of disease; cultivar 2 is more tolerant to disease and also to yield reduction.

the titles of publications (13, 46). Vulnerable means "(*a*) capable of being wounded: defenseless against injury, (*b*) open to attack or damage" (33). In the current context, it expresses that a plant subject to attack or infection may be damaged or undergo an adverse reaction. NPV (21) has adequately defined vulnerability as the "inability of a plant (or organism) to resist attack by a parasite or a phytophagous organism and to counteract the effect of attack." The adoption of vulnerability as a phytopathological term has already been advocated (11) because it encompasses both susceptibility and sensitivity (Figure 6). Pathologists and biologists (Figure 6*A*) view vulnerability differently than do growers and breeders (Figure 6*B*). *Natural vulnerability*, which expresses a relatively low capacity of genotypes in wild vegetation to survive natural selection pressure, is distinct from *agronomic vulnerability*, which expresses relatively low capacity of genotypes, breeding lines, or new cultivars to pass final critical agronomic evaluation.

There is no good antonym of vulnerability. Using high and low vulnerability

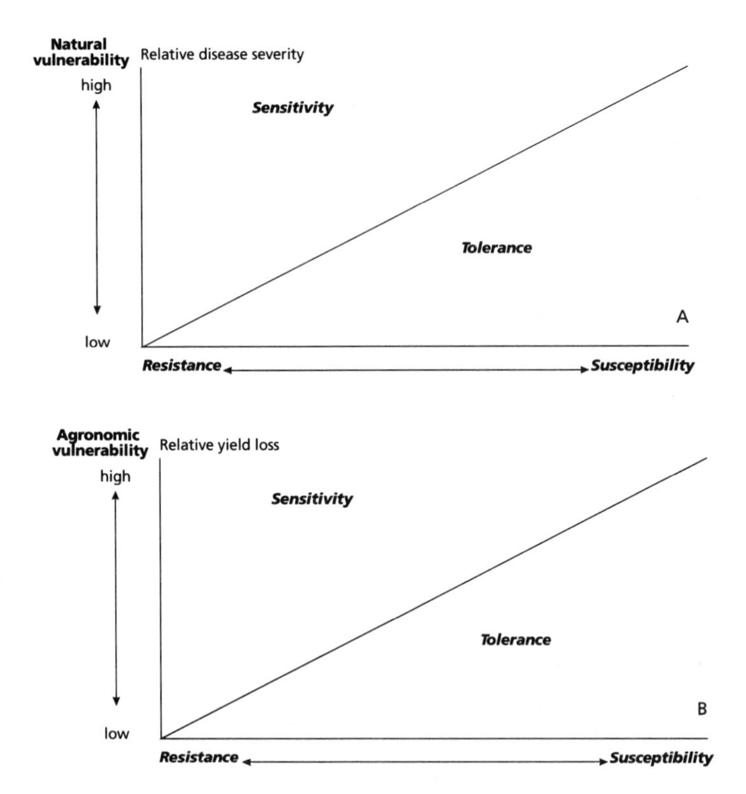

Figure 6 Graphical illustration of two types of vulnerability; *A*, natural vulnerability; *B*, agronomic vulnerability.

as two opposites in a range of degrees in which plants may "suffer" would be less confusing than the often used term "disease resistance." This term lacks precision with respect to underlying mechanisms and has never been adequately defined—except as part of the definition of resistance (see above). It does not indicate whether mild symptom expression and/or minimal damage result from resistance to the parasite, and thus from a low rate of growth or multiplication of the parasite, or from tolerance to attack. Disease is a phenomenon; an organism cannot resist a phenomenon but only its causative agent. We propose the use of *high vulnerability* and *low vulnerability* as antonyms, with the latter to describe "the capability of an organism to prevent the development of damage as judged by the relatively low severity of disease (irrespective of the degree of attack by the pathogen)." Use of the term "disease resistance" as an antonym of vulnerability without specifying whether resistance to the pathogen (resistance per se or sensu stricto) or mere tolerance (so-called "disease resistance") is involved should be discouraged. The fate of the parasite or pathogen must be assessed when screening genotypes for resistance per se by measuring the number or amount of parasite or pathogen per unit of host tissue. This is easiest for insects of which population densities can be determined in relation to area of leaf surface and by their increase or decrease as a function of time. Fungi are best assessed by counting leaf spots or pustules and their change over time. Sophisticated tests, such as ELISA, are now routinely applied to gain quantitative insight into the fate of viruses in different genotypes (Figures 4, 5). Resistance to individual pests thus assessed must then be specified, e.g. as resistance to potato virus Y. or resistance to *Phytophthora infestans.*

HOST AND PARASITE SPECIFICITY

In nature, all plant species employ resistance or defence mechanisms against parasites. Mechanisms such as those using phytoalexins, tannins, leaf cuticle thickness, and conditions of the stem bark are directed against broad groups of parasites or may make plants into nonhosts. Such *broad resistance* has been overcome by some parasites that specialized in plants with such resistance mechanisms. The phaseolin produced by beans, for instance, is circumvented by the bean rust *Uromyces phaseoli* and degraded or tolerated by *Colletotrichum lindemuthianum.* In agriculture, deployment of an endlessly diverging array of *cultivars,* often bred for *parasite-specific resistance* (49) and the inevitable adaptation by the parasites, has led to a wide and evolving array of *specific host/parasite interactions.* Crops and their parasites are often both highly specialized, and the resistances used by breeders against them are also highly specific.

Specific Host/Parasite Interaction

Parasite-specific resistance induced by breeding is often overcome by mutation and genetic adaptation of the parasite. The genotype of the parasite that neutralizes the resistance is indicated as a new *biotype*, race, pathotype, or strain (see below). Resistance is often described as *race-specific* or *vertical resistance*; it holds only for specific types of the parasite and is mostly monogenic. When two host genotypes, H1 and H2, are exposed to two genotypes of the parasite, P1 and P2, one can observe the differential interaction when H1 is resistant to P1 and susceptible to P2, whereas H2 is resistant to P2 and susceptible to P1. With a set of known genotypes of the one actor, the genotypes of the other actor are identifiable. If there is no such specific interaction, *horizontal or race-nonspecific resistance* may be involved.

Breeding for pest resistance is now a major goal of plant breeding to improve crop productivity. Cultivars are frequently chosen for their resistance to specific parasites (as recorded in official lists of approved crop cultivars). Large-scale introduction of such crop genotypes exerts selection pressure on the parasite population, allowing existing biotypes of the parasite or new mutants able to overcome the resistance to predominate. Frequently, resistance breaks down, and the search for new resistance goes on. Within the current context of impetus toward sustainable agriculture, the feasibility of reducing specificity to ensure more durable resistance (e.g. 53, 56) has been hotly debated. As resistance breeding has progressed, so too has the terminology in respect to both host and parasite in their mutual relationships. Confusion is rampant, particularly with respect to virulence and aggressiveness.

Van der Plank (59) used virulence quite differently from the original, linguistic meaning of the word and interpreted it in the sense of *vertical or cultivar-specific pathogenicity* (for the ability to overcome vertical resistance) and aggressiveness for *horizontal or cultivar-nonspecific pathogenicity*. Avoidance of the terms virulence and aggressiveness has therefore been advocated (53), at least in this restricted sense (28). Differential interaction is caused by gene-for-gene interaction; each resistance gene in the host interacts with a specific gene in the parasite. The product of the resistance gene interacts with the product of the parasite's gene, which leads to *incompatibility*. Van der Plank (59) described compatibility as the result of the interaction between the suceptibility gene in the host and the "virulence" gene in the pathogen. In his view, overcoming resistance was caused by a change in virulence, but the "avirulence" gene is the active gene, and the so-called "virulence" is the absence of avirulence (the resistance gene no longer recognizes the product of the "avirulence" gene) and is thus a return to the original aggressiveness. Virulence/avirulence are therefore inappropriate to describe what actually happens. A better description is contained in the term induction. Resistance is induced by the elicitor produced by the gene in the pathogen (the inductor gene). Absence of this gene, or a mutation making it malfunctional, results in the resistance reaction not being induced and, hence, normal aggressiveness.

Parasite Specificity

The study of variation relies on segregation of parasite offspring in pure culture from host or substrate, and this yields *isolates*. The term suggests purity, as of single-spore isolates of fungi and single-lesion isolates of viruses (1, 21, 28). Microorganisms may be grown in pure culture via selective media, and many viruses can be separated from host material and other viruses by selective physicochemical techniques. For several viruses isolation is still only by transfer from an infected plant by inoculation of a selective host or with a special vector. The name of a particular isolate usually indicates its origin in terms of host or location (21), without information on its identity.

Taxonomically recognized variants within a parasite species are *subspecies*, *variety*, and *forma specialis*. The variants do not differ sufficiently morphologically to warrant description as a species. According to FBPP (28), *forma specialis* is "a subdivision of a species of a parasitic or symbiotic microorganism distinguished primarily by adaptation to a particular host (Bacterial Code, Rec. 8a(5))" or "a taxon characterized from a physiological [especially host adaptation] point of view but scarcely or not at all from a morphological standpoint (Botanical Code, Art. 4 Note)". Examples are *Puccinia graminis* f.sp. *tritici* and *P. graminis* f.sp. *avenae*, for types of cereal rust occurring on wheat and oats, respectively. Similar subdivisions of bacterial species have more recently been called *pathovar* (1, 21, 27a), e.g. *Xanthomonas campestris* pv. *oryzae* for *X. campestris* pathogenic on rice.

Further merely biological variants are biotypes. The definition of *biotype* does not clearly demarcate it from *forma specialis* and *pathovar*. This definition varies from "a group of individuals of identical" genotype (37) or "largely identical genotype" (45), or such "a group differing from similar groups only physiologically (as in aggressiveness and pathogenicity to their host)" (31), then including a physiologic race, up to a "subdivision of a physiologic race." FBPP (28) listed these definitions indiscriminately. It does not seem wise to use the term biotype in a sense other than its linguistic meaning for variants only characterized biologically and not covered by taxonomic rules. Differentiation between subconcepts has also proved risky and there is much overlap, as, for example, in plant nematology (5, 26, 57).

Consensus exists with respect to *physiologic race* (usually race). Such races are groups [not taxons, as indicated by FBPP (28)] "of parasites (particularly fungi) characterized by specialization to certain cultivars of one host species" (28). An example is *Fusarium oxysporum* f.sp. *pisi* r.2.

The term *pathotype* has been defined as "a subdivision of a species, distinguished by common characters of pathogenicity, particularly in host range" (28), then merging with physiologic race (21). Like strain for many viruses, it is used more specifically for biotypes that differ in the host-varietal interaction encountered in breeding programs and determined by their reaction on a set of different cultivars. One might agree with Trudgill (57) "that pathotype should be used only in relation to gene-for-gene interactions."

The wider term *strain* denotes all sorts of characterized and thus recognizable variants. The best of Webster's (33) biological definitions of strain are (*a*) "a line descended or derived from a particular ancestral individual: progeny, descendants; also lineage, ancestry," (*b*) "a selected group of organisms sharing or presumed to share a common ancestry and usually lacking clear-cut morphological distinctions from related forms but having distinguishing physiological qualities," and (*c*) "broadly: a specified intraspecific group (as a stock, line or ecotype)." FBPP (28) is also not much help for "biologic strain" referring to physiologic or biologic race. For viruses, where the term is widely used for all sorts of intraspecific variants, FBPP refers exclusively to serological or immunological characters and it considers strain to be nearly synonymous to *serotype*. Bos et al (11) recently stressed that no taxonomic definition of strain exists in virology: "For the sake of reproducibility and for recognition purposes, major criteria usually are biological criteria that can be standardized, particularly varietal interaction with a major natural host (pathotype) or serological specificity (serotype)." They further emphasized that specific host interaction and serological specificity often do not coincide because they may reside in completely different parts of the viral genome. The fact that "serotype and pathotype are not synonymous" undermines the increasing reliance on serological methods for the rapid detection of viruses and their strains. This objection also addresses the application of monoclonal antisera employing single epitopes (amino-acid configurations) on the surface of viral proteins only. Reliance on the structural characterization of viruses by molecular biology, and on their detection with representative probes of complementary nucleic-acid sequences (cDNA), which merely distinguish *chemotypes*, is also problematic, as earlier discussed for pathotypes and strains of nematodes (26). Can such minor serological and molecular-biological characteristics be considered representative parameters of biological identity? Probably not.

NONPATHOGENIC PARASITISM

The terminology that was originally developed for host-pathogen relationships (21) has now been adapted to accommodate nonpathogenic host/parasite relationships as well.

In nonpathogenic/parasite relationships, there is no virulence and pathogenicity, and no disease or characteristic host reactions (symptoms) result, but typical patterns of injury may result nevertheless. This type of interaction is therefore usually described in terms of attack or infestation rather than disease, particularly when no host invasion is involved. The "disorder" is then named after the parasite instead of after the symptom or syndrome, sometimes in combination with the type of injury. Examples are many rusts, mildews, and leaf-miners, where there is often no characteristic host abnormality, except the detectable presence of the parasite or of its pattern of injury. In infection by *Puccinia graminis,* for example, rust pustules are observed emerging through the cuticle, and in powdery mildews the mycelial pathogen is observed developing on the host. These are signs of infection rather than symptoms reflecting host reaction or damage.

Brief or prolonged feeding activity by phytophagous insects on plant parts or on whole plants may also be described in terms of aggressiveness. The insects attack plants as a source of food. The plants do not act as hosts, since no parasitic relationship is established. Phytophagous invertebrates inflict (mechanical) injury but not disease. The plants they feed upon are not infected and these plants do not react physiologically by developing disease; the plants are not sensitive, although they are liable to injury and ensuing damage. They may tolerate a high degree of damage and recover, may die from severe loss of biomass and/or essential organs, or they may be partly or entirely consumed.

ABIOTIC FACTORS

Harmful abiotic factors such as air pollutants, drought, soil pH, or nutritive imbalance often lead to tissue injury, plant malfunction, or suboptimal growth and development leading to visible reaction and thus to disease. Specific symptoms may develop, as with mineral deficiencies (deficiency diseases). These factors are then pathogenic and thus aggressive and even virulent. Plants subjected to abiotic factors are not susceptible because there is no infection or multiplication, but they may be more or less sensitive or tolerant to attack. The term resistance is inappropriate here, except where plants can neutralize a harmful chemical such as a herbicide. Plants or plant genotypes that survive with no apparent damage or loss are tolerant.

CONCLUSION

We agree with Cowling & Horsfall (24) that nature is a "continuum which we try to divide into components and divisions within which the entities are sufficiently similar and between which the entities are sufficiently different that comprehension of the whole is easier to achieve." In classifying information for storage and future retrieval "we divide the continuum into arbitrary segments." That is why the demarcation between healthy and diseased, normal and abnormal, is not precise but rather is a matter of statistics. Whether a host is considered diseased or normal further depends upon the tools and the precision with which the host has been examined. This methodology also determines whether the parasite in question is considered a pathogen, a harmful agent, or a mere commensal. The concepts of disease and abnormality, parasitism and pathogenicity, and related terms are man-made constructs. Only through precise definition of the underlying concepts and by consensus within the discipline can the concepts themselves be made comprehensible. Definitions must also be continuously revised to accommodate our ever-increasing knowledge. Any problems that may arise in describing plant/pest and host/parasite relationships should not be attributable to terminology.

It should be re-emphasized that our plant pathology terminology is oriented to crops and agriculture. Terms like pest, disease, damage, and abnormal all have a strong economic connotation. Disease comprises a continuum of abnormalities from reduction in growth and vitality to yield loss. Such loss may be expressed in terms of quantity, quality, and capital. Quantity and quality can to a greater or lesser extent be described biologically or chemically; they may alter the capacity of the affected organism to survive and then are of pure biologic interest. If expressed in market terms, biologists and pathologists often lose interest. This dichotomy is the source of confusion over terms like latency, resistance, tolerance, vulnerability, pest, disease, and loss. We need terms to discuss plant disease problems within the agro-economic or crop protection context in which most of us operate. However, this condition does not preclude precise definition of terms for the sake of more accurate description of nature, that is, to help us better to see, comprehend, and describe actual events in plant/pest and host/parasite relationships. Consensus, or at least improved communication, on the terminology to describe plant/pest relationships is long overdue.

ACKNOWLEDGMENTS

The authors (virologist and pathologist-geneticist, respectively) are deeply indebted to past and present members of the "Commissie voor de Terminologie" of the Royal Netherlands Society for Plant Pathology (20, 21) for intensive discussions within the committee, and for valuable criticism during the preparation of this manuscript. The Committee represented various subdisciplines in the field of crop protection research: GW Ankersmit (entomology), HH Evenhuis (entomology), T Limonard (mycology), PC Scheepens (weed science), AF van der Wal (nematology), and W van der Zweep (weed science). Critical review by colleagues, notably PWT Maas, is deeply appreciated.

Literature Cited

1. Agrios GN. 1988. *Plant Pathology*. San Diego, CA: Academic. 803 pp. 3rd ed.
2. Am. Phytopathol. Soc. 1940. Report of the Committee on Technical Words. *Phytopathology* 30:361–68
3. Andrivon D. 1993. Nomenclature for pathogenicity and virulence: the need for precision. *Plant Dis.* 83:889–90
4. Aubé C. 1971. Glossaire des termes utilisés en pathologie végétale. *Phytoprotection* 52:1–23
5. Barker KR. 1993. Resistance/tolerance and related concepts/terminology in plant nematology. *Plant Dis.* 77:111–13
6. Bateman DF. 1978. The dynamic nature of disease. See Ref. 39, 3:53–58
7. Bird AF. 1975. Symbiotic relationships between nematodes and plants. *Symp. Soc. Exp. Biol.* 29:351–71
8. Bos L. 1978. *Symptoms of Virus Diseases in Plants; With Indexes of Names of Symptoms in English, Dutch, German, French, Italian, and Spanish.* Wageningen, Netherlands: PUDOC. 225 pp. 3rd ed.
9. Bos L. 1982. Crop losses caused by viruses. *Crop Prot.* 1:263–82
10. Bos L. 1983. *Introduction to Plant Virology.* Wageningen, Netherlands: PUDOC/London/New York: Longman. 160 pp.
11. Bos L, Huijberts N, Cuperus C. 1994. Further observations on variation of lettuce mosaic virus in relation to lettuce (*Lactuca sativa*), and a discussion of resistance terminology. *Eur. J. Plant Pathol.* 100:293–314
12. Br. Mycol. Soc. Plant Pathol. Comm. 1950. Definitions of some terms used in plant pathology. *Trans. Br. Mycol. Soc.* 33:154–60
13. Buddenhagen IW. 1977. Resistance and vulnerability of tropical crops in relation to their evolution and breeding. *Ann. NY Acad. Sci.* 287:309–26
14. Burpee L. 1983. Pathogenicity: coming to terms with another term. *Plant Dis.* 67:129
15. Caldwell RM, Schafer JF, Compton LE, Patterson FL. 1958. Tolerance to cereal leaf rusts. *Science* 128:714–15
16. Calvert OH. 1982. Coming to terms with terms. *Plant Dis.* 66:760
17. Carter W. 1973. *Insects in Relation to Plant Disease.* New York: Wiley. 759 pp. 2nd ed.
18. Clarke DD. 1986. Tolerance of parasites and disease in plants and its significance in host-parasite interactions. *Adv. Plant Pathol.* 5:161–97
19. Colhoun J. 1979. Predisposition by the environment. See Ref. 39, 4:75–96
20. Comm. Terminol. Ned. Planteziektenkd. Ver. 1968. List of phytopathological terms. *Neth. J. Plant Pathol.* 74:65–84 (In Dutch)
21. Comm. Terminol. Ned. Planteziektenkd. Ver. 1985. List of crop protection terms. *Gewasbescherming* 16(Suppl. 1) 64 pp. (In Dutch)
22. Commonw. Mycol. Inst. 1983. *Plant Pathologist's Pocketbook.* Slough, UK: CAB. 439 pp. 2nd ed.
23. Cooper JI, Jones AT. 1983. Responses of plants to viruses: proposals for the use of terms. *Phytopathology* 73:127–28
24. Cowling EB, Horsfall JG. 1979. Prologue: How pathogens induce disease. See Ref. 39, 4:1–21
25. De Bary A. 1979. *Die Erscheinung der Symbiose.* Strassburg: Trübner
26. Dropkin VH. 1988. The concept of race in phytonematology. *Annu. Rev. Phytopathol.* 26:145–61
27. Durbin RD. 1979. How the beachhead is widened. See Ref. 39, 4:155–62

27a. Dye DW, Bradbury JF, Goto M, Hayward AC, Lelliott RA, Schroth MN. 1980. International standards for naming pathovars of phytopathogenic bacteria and a list of pathovar names and pathotype strains. *Rev. Plant Pathol.* 59:153–68

28. Fed. Br. Plant Pathol. Terminol. Subcomm. 1973. A guide to the use of terms in plant pathology. *Commonw. Mycol. Inst., Phytopathol. Pap.* 17. 54 pp.
29. Fischer RF. 1979. Allelopathy. See Ref. 39, 4:313–30
30. Food Agric. Organ. 1991. *International Plant Protection Convention.* Rome: FAO UN. 17 pp.
31. Fröhlich G, ed. 1979. *Wörterbücher der Biologie: Phytopathologie und Pflanzenschutz.* Stuttgart: Fischer Verlag. 291 pp.
32. Gibbs JN, Brasier CM. 1973. Correlation between cultural characteristics and pathogenicity of *Ceratocystis ulmi* from Britain, Europe and America. *Nature* 241:381–83
33. Gove PB, ed. 1976. *Webster's Third New International Dictionary of the English Language, Unabridged.* Springfield, MA: Merriam. 2662 pp.
34. Heath MC. 1980. Reactions of nonsuscepts to fungal pathogens. *Annu. Rev. Phytopathol.* 18:211–36
35. Heath MC. 1981. A generalized concept

of host-parasite specificity. *Phytopathology* 71:1121–23
36. Holliday P. 1989. *A Dictionary of Plant Pathology.* Cambridge: Cambridge Univ. Press. 369 pp.
37. Holmes S. 1979. *Henderson's Dictionary of Biological Terms.* London/New York: Longman. 510 pp. 9th ed.
38. Hornby AS, ed. 1981. *Oxford Advanced Learner's Dictionary of Current English.* Oxford: Oxford Univ. Press. 1037 pp.
39. Horsfall JG, Cowling EB, eds. 1977–1980. *Plant Disease; An Advanced Treatise.* Vol. 1. *How Disease Is Managed.* 1977. 488 pp.; Vol. 2. *How Disease Develops in Populations.* 1978. 464 pp.; Vol. 3. *How Plants Suffer from Disease.* 1978. 488 pp.; Vol. 4. *How Pathogens Induce Disease.* 1979. 466 pp.; Vol. 5. *How Plants Defend Themselves.* 1980. 554 pp. New York: Academic
40. Horsfall JG, Dimond AE, eds. 1959–1960. *Plant Pathology; An Advanced Treatise.* Vol. 1. *The Diseased Plant.* 1959. 674 pp; Vol. 2. *The Pathogen.* 1960. 715 pp.; Vol. 3. *The Diseased Population; Epidemics and Control.* 1960. 675 pp. New York: Academic
41. Horsfall JG, Dimond AE. 1959. The diseased plant. See Ref. 40, 1:1–17
42. Knutson DM. 1979. How parasitic seed plants induce disease. See Ref. 39, 4:293–312
43. Loegering WQ. 1978. Current concepts in interorganismal genetics. *Annu. Rev. Phytopathol.* 16:309–20
44. Miller HJ, Elgersma DM. 1976. The growth of aggressive and non-aggressive strains of *Ophiostoma ulmi* in susceptible and resistant elms, a scanning electron microscopical study. *Neth. J. Plant Pathol.* 82:51–65
45. Müller G, ed. 1980. *Wörterbücher der Biologie: Mikrobiologie.* Jena: Fischer Verlag. 401 pp.
46. Natl. Res. Counc. 1972. *Genetic Vulnerability of Major Crops.* Washington, DC: Natl. Acad. Sci. 307 pp.
47. Norris DM. 1979. How insects induce disease. See Ref. 39, 4:239–55
48. Parlevliet JE. 1979. Components of resistance that reduce the rate of epidemic development. *Annu. Rev. Phytopathol.* 17:203–22
49. Parlevliet JE. 1981. Race-non-specific disease resistance. In *Strategies for the Control of Cereal Diseases,* ed. FJ Jenkyn, RT Plumb, pp. 47–54. Oxford: Blackwell Sci.
50. Parlevliet JE. 1981. Crop loss assessment as an aid in the screening for resistance and tolerance. In *Crop Loss Assessment Methods,* ed. L Chiarappa, Suppl. 3:111–14. Slough, UK: CAB
51. Putnam AR, Duke WB. 1978. Allelopathy in agroecosystems. *Annu. Rev. Phytopathol.* 16:431–51
52. Robinson RA. 1969. Disease resistance terminology. *Rev. Appl. Mycol.* 48:593–606
53. Robinson RA. 1976. *Plant Pathosystems.* Berlin/Heidelberg/New York: Springer-Verlag. 184 pp.
54. Schafer JF. 1971. Tolerance to plant disease. *Annu. Rev. Phytopathol.* 9:235–52
55. Shaner G, Stromberg EL, Lacy GH, Barker KR, Pirone TP. 1992. Nomenclature and concepts of pathogenicity and virulence. *Annu. Rev. Phytopathol.* 30:47–66
56. Simmonds NW. 1985. A plant breeder's perspective of durable resistance. *FAO Plant Prot. Bull.* 33:13–17
57. Trudgill DL. 1991. Resistance to and tolerance of plant parasitic nematodes in plants. *Annu. Rev. Phytopathol.* 29:167–92
58. Van der Plank JE. 1963. *Plant Diseases: Epidemics and Control.* New York: Academic. 349 pp.
59. Van der Plank JE. 1968. *Disease Resistance in Plants.* New York: Academic. 206 pp.
60. Van der Zweep W. 1979. Het begrip onkruid. *Gewasbescherming* 10:168–73
61. Walkey DGA, Pink DAC. 1990. Studies on resistance to beet western yellows virus in lettuce (*Lactuca sativa*) and the occurrence of field sources of the virus. *Plant Pathol.* 39:141–55
62. Walkey DGA, Ward CM, Phelps K. 1985. The reaction of lettuce (*Lactuca sativa* L.) cultivars to cucumber mosaic virus. *J. Agric. Sci.* 105:291–97
63. Welsh MF. 1961. Terminology of plant virus diseases. *Can. J. Bot.* 39:1773–80
64. Whetzel HH. 1929. The terminology of phytopathology. *Proc. Int. Congr. Plant Sci., 2nd, Ithaca, NY,* pp. 1204–15
65. Wilbrink G. 1935. The need of uniform terms in plant immunology. *Proc. Int. Bot. Congr., 6th, Amsterdam,* pp. 204–7. Brill/Leiden
66. Woltz SS. 1978. Nonparasitic plant pathogens. *Annu. Rev. Phytopathol.* 16:403–30
67. Yarwood CE. 1959. Predisposition. See Ref. 40, 1:521–62

Annu. Rev. Phytopathol. 1995. 33:103–18

THE OAK WILT ENIGMA: Perspectives from the Texas Epidemic

D. N. Appel

Department of Plant Pathology and Microbiology, Texas A&M University, College Station, Texas 77843-2132

KEY WORDS: *Ceratocystis fagacearum*, oak wilt, forest pathology, epidemiology, disease control

---

ABSTRACT

Ceratocystis fagacearum (Bretz) Hunt, the oak wilt pathogen, is currently causing massive losses of semievergreen live oaks (*Quercus fusiformis* Small and *Q. virginiana* Mill.) in central Texas. Given the relatively limited oak mortality caused by *C. fagacearum* in the deciduous forests of the North Central, Midwestern, and Mid-Atlantic United States, this Texas epidemic was not anticipated. The intensity of oak wilt in Texas is attributed to a number of factors related to host characteristics and the ability of the pathogen to adapt to limiting environmental conditions. Oak wilt management in semievergreen oaks requires considerable revision of the control techniques previously designed for deciduous oaks. The Texas oak wilt epidemic provides a new perspective from which to evaluate questions concerning oak wilt, including the origins of the pathogen as well as the potential for future losses in unaffected oak forests.

INTRODUCTION

The vascular pathogen *Ceratocystis fagacearum* Bretz Hunt, causal agent of oak wilt, is potentially the most destructive of all tree pathogens (34, 94). This potential is expressed most dramatically when infected red oaks (genus *Quercus*, subgenus *Erythrobalanus*), common constituents of midwestern forests, invariably die within a few weeks after symptoms appear. The destructive

0066-4286/95/0901-0103$05.00

nature of oak wilt in the U.S. throughout the Upper Mississippi River Valley during the decade following World War II raised considerable alarm for the threat to the nation's valuable oak resources (17, 25, 41). The concern at the time was particularly acute due to the relatively recent devastation left in the wake of the epidemics of chestnut blight and Dutch elm disease. Numerous states subsequently devoted considerable resources to prevent another catastrophe. However, despite the widespread existence of susceptible oaks in the U.S., *C. fagacearum* has still caused only limited losses when compared to those more widely recognized pathogens, *Cryphonectria parasitica* (Murr.) Barr and *Ophiostoma ulmi* (Buisman) Nannf., respectively.

Texas is the most recent state to undertake extensive oak wilt research and control programs. These efforts were stimulated by massive losses of live oaks (*Q. fusiformis* Small and *Q. virginiana* Mill.) throughout a 40–50 county region over the past 30 years (10, 52). The losses are largely focused on the geographically unique Balcones Escarpment of central Texas, where semievergreen live oaks are the most valuable woodland and urban tree species (57). Oak wilt research and the implementation of statewide control programs in Texas have followed a pattern common to other states previously affected by the disease. Following confirmation of the causal agent, systematic detection and survey of affected areas has been the usual first response. In the case of oak wilt, strong public education programs and widespread demonstration projects for disease control were often implemented prematurely, with questionable results. Fortunately, throughout most of the oak wilt range in the U.S., localized epidemics have abated (56). In many cases they never developed at all and interest in the disease declined. This is not the case in Texas, where annual losses are occurring at immeasurable rates and significant resources are spent on control (7, 9, 11).

There exist some excellent reviews of the extensive research on oak wilt (34, 56). Each is written under the premise that the damage that *C. fagacearum* could potentially cause has not been realized because of some limitation in the life history of the pathogen. These previous reviews also were prepared from the perspective of oak wilt as it occurs on deciduous oaks in the forests of the Mid-Atlantic, Midwestern, and North Central U.S. Observations of oak wilt in the semideciduous live oaks of Texas indicate that the concerns about the destructive potential of oak wilt were warranted. Many natural resource managers and landowners in Texas are convinced that *C. fagacearum* is causing a detrimental, permanent impact on the fragile, oak savannah ecosystem of central Texas. These observations should elicit concern for the well-being of our extremely valuable resources of *Quercus* wordwide. Although some details of past research on the biology of oak wilt are pertinent and are discussed here (64), this review covers the more recent literature focusing on comparative epidemiology of oak wilt in Texas and the occurrence of the disease elsewhere. This analysis allows some of the more puzzling aspects of oak wilt to be addressed: Why has the pathogen incited an epidemic in a forest ecosystem previously considered as safe from the disease; what is the capacity for *C. fagacearum* to cause losses in presently unaffected oak forests; and what is the origin of *C. fagacearum?*

INFLUENCE OF HOST COMPOSITION ON PATHOGEN BEHAVIOR

Subgeneric Classification and Host Response

RED VS WHITE OAKS One approach to describing the biology of oak wilt is to analyze how various oaks respond to infection by *C. fagacearum*. The genus *Quercus* is an important, complex group of trees and shrubs representing at least 500 species worldwide. In the U.S., there are approximately 50 oak species (45). The majority are evenly distributed between two subgenera: *Quercus* subg. *Erythrobalanus*, the red or black oaks, and *Quercus* subg. *Quercus* (synonym *Lepidobalanus*), the white oaks. Members of these two subgenera are reproductively isolated, display distinctively different anatomical characteristics, but usually will be found growing in close association with one another (65). Differences between the groups, combined with the propensity to cohabitate, have an important influence on the potential for a developing oak wilt epidemic.

No *Quercus* species is immune to infection by *C. fagacearum*. However, species within the two subgenera are not equally susceptible (31, 41, 42). As mentioned above, red oaks are extremely susceptible, whereas white oaks typically respond with a tolerant reaction by limiting symptom development to a few branches. White oaks rarely die from oak wilt (56). This contrast in response to infection between the two subgenera was noted in the definitive description of oak wilt (41). In that report, the species most seriously affected were *Quercus velutina* Lam. (black oak), *Q. borealis* Michx. f. (northern red oak), and *Q. coccinea* Muench. (scarlet oak). The white oaks originally noted as tolerant were *Q. alba* L. (white oak) and *Q. macrocarpa* Michx. (bur oak). Since then, the outcomes of natural and artificial inoculations of red and white oaks have proven to be consistent for all deciduous oaks tested (74).

There is another, important response to infection by *C. fagacearum* that contrasts the two oak subgenera. Under conducive environmental conditions, the only known inoculum source for pathogen transmission by insects is produced on diseased red oaks. Reproductive structures of *C. fagacearum*, called fungal mats, form below the inner bark on the surface of the sapwood (24, 66). These structures appear to represent a brief proliferation of saprophytic growth that occurs just as the tree undergoes the last stages of dying

(34). No similar structures are found on diseased white oaks. As a result of a two-allele, heterothallic mating system, each mat may be one of two mating types (designated *a* or *b*). Depending on the state of maturity and potential for crossing with an opposite mating type, mats will be covered with conidia and/or perithecia that contain ascospores. The relative importance of the two spore types or the influence of sexual recombination on a developing oak wilt epidemic have not been determined (27, 56). The teleomorph is formed when insects contaminated with spermatia, in the form of endoconidia, fly from a mat growing on one tree to a mat of the opposite mating type on another tree. Naturally infected trees are rarely found to be colonized by both mating types (6, 14). Insects that serve to fertilize fungal mats are not necessarily the same as those implicated in long-distance spread of the pathogen (34). Any types of insects inclined to repeatedly visit different mats will aid in fertilization. Fruit flies, for example, are ideal spermitizing agents because they are abundantly attracted to the sweet-smelling mats. Fungal mats form on only a small proportion of diseased red oaks and last only a few weeks before they deteriorate (24). Mat formation occurs mainly on moribund trees during the spring following the season of infection and is the only known saprophytic phase of the pathogen. The reliance of *C. fagacearum* on mat formation for long-distance transmission is given as one factor involved in the limited impact of oak wilt (8, 56).

An additional limiting factor for pathogen transmission is an inefficient vector relationship. Those types of insects involved in recombination of opposite mating types may not have the ability to transmit *C. fagacearum* to new, healthy trees. Contaminated fruit flies are not attracted to healthy oaks, nor do they have the means to inoculate the fungus into vascular systems. Of the dozens of insects known to visit fungal mats, only sap-feeding nitidulid beetles (Coleoptera: nitidulidae) fulfill the requirements to act as long-distance vectors for *C. fagacearum*. Nitidulids are attracted to the sweet-smelling mats to feed and breed (86). As the mats deteriorate, the contaminated nitidulids emerge and may then be attracted to fresh wounds for further feeding (48, 49). Wounds remain receptive to infection for only a few days, and are considered to be one of the most important factors in determining whether successful transmission will occur (77). During these activities, the vector occasionally introduces the pathogen into new trees (46, 69). Because of these requirements, very few trees probably die from insect transmission of *C. fagacearum* (34). This contrasts with the more efficient vector system of Dutch elm disease (DED), where every diseased elm is a potential source of inoculum for elm bark beetles (*Scolytus multistriatus* and *Hylurgopinus rufipes*) (51). Elm bark beetles are also better vectors because they make their own infection courts on healthy trees during feeding, whereas the nitidulids only feed on fresh wounds. An oak bark beetle (*Pseudopityophthorus* spp.) with some similarities to elm bark beetles has been implicated in spread of *C. fagacearum* (73). As with the nitidulids, the oak bark beetles have limiting characteristics that decrease the efficiency with which they may transmit the pathogen and are considered inconsequential in some regions where oak wilt occurs (34).

LIVE OAKS Live oaks do not conform to the distinct pattern of resistance and susceptibility exhibited by white and red oaks. Although most live oaks die within 3–6 months following infection, a small proportion (5–20%) survive indefinitely in various stages of crown loss (11, 19). The cause for this intermediate response between the total susceptibility of red oaks and tolerance of white oaks is unknown. Semievergreen live oaks are difficult to classify according to the two traditional subgenera and the reasons for these difficulties may relate to the variable reaction to infection. Resistance to oak wilt is a poorly understood phenomenon in white oaks, but is suspected to relate to anatomical characteristics and host response that serve to localize colonization within the vascular system and allow the host to produce healthy, unaffected xylem (56). The deciduous red and white oaks are ring porous species, meaning there is an abrupt decrease in size from large, "springwood" vessels formed at the initiation of the growing season to the smaller "summerwood" pores formed as the season progresses. Deciduous, white oak summerwood vessels contrast with those of red oaks by being smaller, thinner-walled, and more angular in shape (85, 91, 92). Also, tyloses, the bubble-like structures believed to be useful in limiting the spread of vascular parasites, form more readily in the vessels of white oak xylem. Although live oaks have numerous similarities to white oaks with regard to acorn maturity, floral anatomy, and leaf characters, the vessels resemble those of the red oaks. Live oaks are further distinguished as being semidiffuse porous, an intermediate pattern between the two subgenera. Due to these characteristics, live oaks are classified by some as subgenus Quercus (68) and by others as Erythrobalanus (92). These anatomical distinctions may also be responsible for the variable response exhibited by live oaks to infection by *C. fagacearum*.

Stand Composition

In Texas, fungal mats have been found on two deciduous red oak species, Spanish oak (*Q. texana* Small) and blackjack oak (*Q. marilandica*) (12). These species comprise only a minor component of the central Texas oak savannahs when compared to live oak. Even though live oak is clearly the most seriously affected species, no fungal mat has ever been found on a diseased live oak. This disparity in the availability of inoculum for insect transmission has a distinct influence on the spatial pattern of disease incidence. As in other states, the majority of losses result from transmission of the pathogen through root connections between diseased and healthy trees rather than through insect transmission (34). Root graft transmission is a well-documented phenomenon in deciduous oaks and is recognized as an important factor in spread of *C. fagacearum*. Root grafts are particularly evident among live oaks with crowded roots growing in the thin, rocky soils of central Texas. In addition to root grafts, live oaks form root connections through the ability to propagate vegetatively by means of root suckering (67). This habit is considered advantageous to the trees during long periods of deficient soil moisture and contributes to their ability to rapidly colonize disturbed sites. Rhizomatous live oaks appear to maintain many of their juvenile root connections through maturity. As a result, live oak stands become highly interconnected systems of clonal mixtures with root:shoot ratios approaching 10:1 (16). Introduction of a vascular parasite into this massive system initiates rapidly expanding infection foci that have steep disease gradients, typical of a disease caused by a pathogen with a poor long-distance dispersal mechanism (22). In live oak, however, a typical focus is several hectares in size, expands at rates of up to 45 m/year, and results in the death of hundreds of trees annually (11). This contrasts with the slower growth and smaller sizes of foci in deciduous oaks (1, 34).

The most severe oak wilt epidemics occur in those areas where the *Quercus* composition and stand densities have been altered by land-management practices. The relatively high incidence of oak wilt in Wisconsin and Minnesota has been attributed to the availability of a recently developed highly susceptible host population (34). In these regions, dense stands of a red oak species, *Q. ellipsoidalis* E.J. Hill (northern pin oak), have become established owing to a reduction in forest diversity from logging operations and fire. Northern pin oak has a vigorous habit for coppicing, or reproduction through stump sprouts following disturbance. A similar pattern has been repeated in central Texas, where widespread, homogenous stands of the rhizomatous live oak have developed as a result of fire control, overgrazing, and selective thinning practices (10, 21). The ability to produce root sprouts is probably responsible, in part, for the widespread homogenous stands of live oak in the region. Prior to widespread settlement in the early 1830s, central Texas was part of the southernmost extension of the Great Plains where live oaks grew on isolated, selected sites. In both Minnesota and Texas, widespread site disturbance and subsequent vegetative reproductive habits of the oak hosts have predisposed generations of individuals to destruction by *C. fagacearum*.

Large numbers of red oaks will provide abundant primary inoculum for random dispersal of *C. fagacearum* over long distances. This type of transmission has been termed overland spread and has been viewed to be any infection occurring at distances greater than 15 m from a known inoculum source (34). Presumably, any species of oak is liable to become infected by feeding of contaminated nitidulids, as long as it has suitable infection courts. However, only the infection of more red oaks will result in further production of primary infections over long distances. White oak infections result in little or no additional disease development. Live oak infections, in contrast, will allow for rapid development of secondary infections through root connections to adjacent trees, but will not support long-distance transmission. The stand density and species composition of oaks will have a strong impact on the character of a localized epidemic.

ENVIRONMENTAL CONSIDERATIONS

Temperature Constraints

The oak wilt epidemic in Texas was unexpected in that it contradicted many assumptions concerning the influence of environmental conditions on pathogen behavior. In natural forest ecosystems, various factors are believed to effectively suppress the development of pathogen populations and epidemic losses. This attribute is called functional diversity (76). In the case of oak wilt, high temperatures and biological controls, combined with inefficient vectors, were believed to be largely responsible for a lack of spread into southern forest ecosystems (34, 76). Temperatures in the range of 30–35°C significantly limit growth of the fungus in vitro. Higher temperatures are lethal (54, 80). Similar temperatures also retard symptom development in artificially inoculated, containerized trees (43). In apparent support of these conclusions, high summertime temperatures are believed to be partly responsible for survival of susceptible turkey oaks (*Q. laevis* Walt. subg. *Erythrobalanus*) in South Carolina (82). Inoculum production, as well as disease development, was also thought to be suppressed by conditions found in the southwestern portion of the oak wilt range. Mat formation occurs in moribund trees with a moisture content of 14% or greater (78). High air temperatures and low relative humidities were assumed to be instrumental in accelerating drying in dying trees and account for the low incidence of mat production in Missouri (34). A canker-causing facultative saprophyte, *Hypoxylon atropunctatum*, is common in dying oaks in the southwestern U.S., and was believed to directly compete with the saprophytic phase of *C. fagacearum* and prevent mat formation (79, 81). In an experiment to demonstrate the consequence of those factors on mat formation, colonized logs of diseased red oaks were transported from Missouri to Pennsylvania and a similar load of trees was brought from Pennsylvania to Missouri (34). Mats formed on the Missouri trees in Pennsylvania, but none formed in Missouri. The environment was considered to be too extreme for pathogen growth and reproduction, contributing to the apparent southern limit to expansion by the oak wilt pathogen.

Survival in Texas

Temperatures throughout the range of the disease in Texas often exceed 35°C, yet the epidemic seems to be little affected. The pathogen is apparently able to survive in the boles of live oaks, where temperatures do not exceed those considered to be limiting to the growth of the pathogen (53). It is also reasonable to assume that the pathogen can survive below ground in the extensive root systems of live oaks during periods of temperature extremes. Even though high temperatures are not sufficient to prevent oak wilt epidemics, heat may contribute to the variable survival rates observed in live oak. Heat is believed to eliminate *C. fagacearum* from the smaller limbs and branches, explaining why the pathogen is difficult to isolate from those tissues during high summertime temperatures (10, 18, 52). At 32°C, hyphal growth, conidial formation, and production of mucilage are inhibited in vitro (81). If heat operates in a similar manner in vivo, high temperatures would eradicate the pathogen from smaller limbs and branches and serve to limit the extent of colonization in live oaks.

C. fagacearum obviously overcomes the detrimental effects of high temperatures on fungal growth and survival to cause widespread, epidemic losses. The temperatures proven to be limiting to northern isolates of *C. fagacearum* in vitro are also limiting to Texas isolates (53). Selection in the pathogen population toward environmental races with heat tolerance is not responsible for the extent of oak wilt in Texas. Another explanation is provided by the Hypotheses of Compensation, advanced to explain the occurrence of epidemics of some pathogens in a variety of climatic regions (75). According to one of those hypotheses, a pathogen may compensate for marginal environmental conditions in a phase of its life cycle by vigorous growth and reproduction in another phase where conditions are favorable. Although high temperatures, low humidity, and *H. atropuctatum* may suppress oak wilt development during the summer, the pathogen can survive in boles and roots of infected trees. Climatic conditions during the remainder of the year in Texas are conducive to *C. fagacearum*. Limitations in inoculum production and inefficient insect vectors are overcome by means of the highly interconnected live oak root systems and large, dense host stands.

IMPLICATIONS FOR OAK WILT MANAGEMENT

Traditional Control Principles

A reasonably complete understanding of the oak wilt disease cycle led to various strategies to prevent losses. The development of recommendations for oak wilt management was believed to be a successful chapter in the history of North American forest pathology (30). The management system has traditionally relied on a mixture of preventive measures aimed at reducing potential sources of inoculum for long-distance spread, disrupting root connections for local spread, and eliminating potential infection courts (31). Throughout much of the range of oak wilt, inoculum has been reduced through some means of preventing fungal mat formation. Trees may be felled and burned or buried. In some cases, spraying felled trees with diesel oil has proven effective in discouraging nitidulid beetles (86). A laborious, but simple, technique to prevent fungal mats is to deep-girdle the tree at two feet above the soil surface and then strip the bark from the trunk below the girdle (87). This promotes drying of the tree and invasion of saprophytes that prevent the mats from forming.

Felling of red oaks has also been proposed to prevent transmission of *C. fagacearum* through root grafts among deciduous trees. Establishment of a 15-m zone of dead trees around the perimeter of a disease focus is intended to create a barrier of dead and dying roots. Trunk injection of poisons, such as cacodylic acid, was also recommended to kill roots and further contribute to reduction of inoculum formation on red oaks (72). When these disease-management programs were implemented statewide in Pennsylvania and West Virginia, the results proved to be only marginally successful (47). Subsequent analyses of oak wilt epidemiology indicated that rates in disease progress, as measured by losses of trees, did not justify major expenditures in disease management (56). Many states ceased to implement region-wide disease control efforts, and since the late 1960s few advances have been made in the technology for oak wilt control in the deciduous forests of the eastern and midwestern U.S. One exception has been in devising models to predict rates of local spread of the pathogen through root grafts (20, 59, 60). From these studies have developed decision keys for establishing trenches around the perimeters of disease centers to prevent root transmission. The keys are based on the probability of annual pathogen spread between healthy and diseased trees of certain sizes and at specific distances from one another.

Modifications for Oak Wilt Control in Live Oak

Upon recognition of oak wilt as the major cause of widespread live oak mortality in central Texas, large-scale control and demonstration programs for disease management were implemented, based on the technology developed for oak wilt control in deciduous trees. Although their basic principles were sound, the control programs often failed to contain the pathogen. For example, placement of a trench at 15 m beyond the perimeter of the focus—based on the distance considered to be typical for root graft transmission among deciduous oaks—was insufficient to prevent root transmission of *C. fagacearum* (9). The extensive, complex root connections and relatively longer latent period for symptom development obscure the limits of colonization by *C. fagacearum* in a stand of live oaks. The 15-m barrier was thus inadequate for judging the potential distance for spread of the pathogen through the common root systems of live oaks (11). The decision keys developed outside of Texas (20, 59, 60) are not effective and have been replaced by acceptance of a 30-m barrier for trenching as a suitable convention for stopping underground pathogen spread. Trenches supplemented with roguing of trees within the 30-m barrier are the most reliable technique for disrupting root systems and subsequent pathogen spread (11).

In Texas, the patterns of pathogen spread observed on rangelands are also found in towns and cities. Because live oaks are very tolerant to site disturbances, urban expansion has encroached on the oak savannah while maintaining the stand structure of the original woodlands (3, 5). As a result, focal expansion in the urban forests of central Texas is rapid and extremely destructive. Conventional tools for trenching and roguing trees, such as bulldozers, backhoes, and ditching machines, are often impractical for disease control in parks and lawns. Therefore, direct control of individual high-value trees became a higher priority than previously experienced with oak wilt in deciduous forests (3, 5). Injection of a relatively new fungicide, propiconazole, was tested and found to be efficacious for preventing losses, although the fungicide does not prevent spread of the pathogen through the root systems of treated trees (7). Propiconazole is an ergosterol-biosynthesis-inhibitor (EBI) and has many desirable characteristics for successfully inhibiting colonization of a vascular pathogen in infected trees. This fungicide in combination with existing techniques now provides a successful strategy to reduce losses from oak wilt under most situations where the disease is found.

THE ORIGINS OF *CERATOCYSTIS FAGACEARUM*

A better understanding of the origins of *C. fagacearum* would greatly improve our ability to determine the probability of severe oak wilt epidemics in areas that are currently unaffected. In comparison to *O. ulmi* and *C. parasitica*, which have clearly demonstrated the capacity for intercontinental spread (32), the known geographic range of *C. fagacearum* is extremely limited. Although first described in 1944, there were earlier descriptions of extensive oak mortality in Wisconsin and Minnesota resembling oak wilt (41). Oak wilt was subsequently reported in adjacent states and eventually a range covering 20 states was established over a 20-year period (29, 74). This apparent spread, in all likelihood, consisted of verifying a long-established range rather than documenting a rapidly spreading pathogen. Although occurrences of oak wilt are occasionally reported outside the 22-state region (93), the fungus has never been positively identified in those areas.

By 1965, *C. fagacearum* was thought to be at the limit of its potential range in the U.S. (74). This range included only one county in Texas. Ten years later a survey in the southwestern range of the disease found no further infections in Texas (70). However, this survey extended no further than 18 counties into the northeastern corner of the state, under the presumption that southward expansion would result from contiguous spread of the pathogen out of southern Arkansas. Meanwhile, widespread live oak mortality in central Texas was reported and attributed to *Cephalosporium diospyri*, the causal agent of persimmon wilt and live oak decline (89, 90).

Historical records of oak mortality in Texas are ambiguous, obscuring the origins of the disease in the state. Foresters from the Bureau of Plant Industry conducted the earliest recorded systematic disease survey in Kerrville during 1909 and 1910 (40), the area currently with the highest incidence and severity of oak wilt. No unusual oak mortality was reported. In 1934 and 1935, widespread live oak mortality with features similar to oak wilt was reported from Austin (83, 84). Over the next 35 years, live oak mortality received much attention and was attributed to a variety of causal agents other than *C. fagacearum* (26, 38). By the time the causal agent of oak wilt, *C. fagacearum*, was verified (52) and the extent of the epidemic in Texas fully appreciated (10), the disease had probably been long established within that range. Both mating types of the pathogen were distributed throughout the range in equal proportions, similar to mating type distributions in other states (6, 95). Likewise, most new reports of oak wilt in Texas are the result of locating well-established disease centers rather than recent introductions. One exception is identification of the pathogen in Houston, TX (DN Appel, unpublished); this represents the eastern-most expansion of the pathogen into the valuable live oak population that extends into the Gulf Coast States. There is also evidence that the pathogen is extending into planted live oak populations in communities located in the north Texas high plains.

Ceratocystis fagacearum may have been introduced into North America relatively recently, and the limited range is a result of insufficient time for the poorly vectored pathogen to distribute throughout the available host population. Alternatively, the pathogenic *C. fagacearum* may represent a recent speciation event from a fungus that previously was not a pathogen of oaks. In either case, some authors assume that the disease originated in the Upper Mississippi River Valley of Wisconsin and Minnesota and extended southward (1, 70, 74). However, the history and current status of the oak wilt epidemic in Texas cast doubt on this assumption.

The population of *C. fagacearum* has proven to be fairly homogeneous. Although a few studies have noted variability in some cultural characters and pathogenicity (13, 23, 39), these traits are insufficient to complete the analyses needed to elucidate basic questions concerning population biology of the fungus. A series of experiments utilizing RFLPs (restriction fragment length polymorphisms) as genetic markers for *C. fagacearum* was initiated to better understand oak wilt epidemiology. Depending on the sample of the pathogen population analyzed, these markers can also be used to study diversity in mitochondrial and nuclear genomes that will help determine the origin of the fungus (58). For example, RFLPs in nuDNA of US isolates of the chestnut blight pathogen, *C. parasitica*, are lower in diversity than isolates from China (62, 63). This finding supports the hypothesis that the North American epidemic originated from China. A remarkably low level of variation in mtDNA and nuDNA has been detected in isolates within Texas and among isolates from throughout the United States (50). Because oak wilt, as caused by *C. fagacearum*, is as yet unknown outside the United States, the possibility of an introduction as the source of the disease becomes increasingly unlikely. However, oak wilt may easily be overlooked in an area where a host population has coevolved. Even in Texas, where the epidemic was intense, foliar symptoms on live oak were sufficiently unusual when compared to those on deciduous trees to make the disease unrecognizable (2, 90). The search for *C. fagacearum* outside of the current range should continue, especially in Mexico and Central America, where there is a high diversity of deciduous and evergreen *Quercus* spp. (44). Alternatively, several close relatives of *C. fagacearum* that are saprophytes or weak parasites of trees could be considered as predecessors in evolution of the pathogen (50). A more exhaustive survey of genetic variation in the genus may provide further insights into the origins of the pathogen.

CONSIDERATIONS FOR THE FUTURE OF OAK WILT

Oak wilt was one of the most important pathogens affecting North American forests in the mid-20th century (28). The impact on developments in the forest pathology community probably far exceeded actual losses of trees. When expected losses did not occur, interest in oak wilt diminished (56, 61). However, the Texas epidemic has shown that the potential expansion of *C. fagacearum* into the oak forests of unaffected regions should be taken seriously.

The impact of oak wilt on currently infected deciduous oak forests is not entirely negative. Surveys of stand regeneration in small openings caused by *C. fagacearum* in West Virginia present promising results for the projected compositions of future stands (88). Disease-free regeneration in the form of seedlings and stump sprouts was similar to that expected of selection cuttings in unaffected stands and normal development was anticipated. Similar, disease-free regeneration from root sprouts is regularly observed for live oaks in Texas (DN Appel, unpublished). In addition, attempts have been made to determine the possibility of selection for disease-resistant trees in Texas during the course of fungal transmission through the highly interconnected live oak stands. Isozyme analyses of pre-and post-epidemic live oak populations revealed significant differences in the genetic compositions of the two populations (15). The low proportion of surviving live oaks may therefore have potential as disease-resistant selections for improved live oak propagation. These survivors may be the progenitors of native live oak populations with increased resistance to the disease. Further evidence for these conclusions is provided by artificial inoculation of seedlings derived from acorns produced by live oak "survivors" (36). The responses of the "survivor" seedlings differed significantly from those observed in progeny from trees growing in pre-epidemic, unaffected populations. Unexpectedly, the "survivor" seedlings died at faster rates, but resprouted with healthy shoots more vigorously than did seedlings from the general population. Resistance and survival to *C. fagacearum* in live oak may relate to pathogen recognition and rapid host response in the form of growth initiation to aid in compartmentalizing the pathogen (36).

The impact of *C. fagacearum* on unaffected oak forests is difficult to predict. In California, there are several valuable oak populations consisting of varying proportions of red and white oaks and deciduous and evergreen oaks (37, 71). Seedlings or young trees of many of those species were susceptible to artificial inoculation by the oak wilt pathogen (4). The structures of California oak woodlands have similarities to those of Texas. For example, a deciduous red oak, *Q. kelloggii* Newb., is intermingled with coast live oak (*Q. agrifolia* Nee) in the Coast Ranges and interior live oak (*Q. wizlesnii*A. DC.) in the Sierra Nevada-Cascades (37). These and similar regions could thus be extremely vulnerable to an oak wilt epidemic (71).

The likelihood of transport of *C. fagacearum* over longer distances must also be considered. The risk of an oak wilt epidemic has been assessed by the European forest pathology community (33, 35) and precautions are being taken to insure the fungus is not introduced through importation of contaminated timber from the United States (55). Given the versatility exhibited by *C. fagacearum* in North American oak forests and the lessons learned specifically from the Texas epidemic, it is important that assumptions concerning the behavior of *C. fagacearum* be carefully analyzed and due recognition be given to the possibility that the fungus may yet exhibit unexpected abilities to incite further epidemic losses.

Literature Cited

1. Anderson GW, Anderson RL. 1963. The rate of spread of oak wilt in the Lake States. *J. For.* 61:823–25
2. Appel DN. 1986. Recognition of oak wilt in live oak. *J. Arboric.* 12:213–18
3. Appel DN. 1989. Tree disease in the urban environment—spatial effects and consequences of human disturbance. In *Spatial Components of Plant Disease Epidemics,* ed. MJ Jeger, Chap. 11. Englewood Cliffs, NJ: Prentice Hall. 243 pp.
4. Appel DN. 1994. The potential for a California oak wilt epidemic. *J. Arboric.* 20:79–86
5. Appel DN. 1994. Identification and control of oak wilt in Texas urban forests. *J. Arboric.* 20:250–58
6. Appel DN, Dress CF, Johnson J. 1985. An extended range for oak wilt and *Ceratocystis fagacearum* compatibility types in the United States. *Can. J. Bot.* 63:1325–28
7. Appel DN, Kurdyla T. 1992. Intravascular injection with propiconazole in live oak for oak wilt control. *Plant Dis.* 76:1120–24
8. Appel DN, Kurdyla T, Lewis R. 1990. Nitidulids as vectors of the oak wilt fungus and other *Ceratocystis* spp. in Texas. *Eur. J. For. Pathol.* 20:412–17
9. Appel DN, Lewis R. 1985. Prospects for oak wilt control in Texas. In *Insects and Diseases of Southern Forests, 34th Annu. Symp.,* ed. RA Goyer, JP Jones, pp. 60–68. Baton Rouge, LA: Agric. Exp. Stn., LA State Univ. 135 pp.
10. Appel DN, Maggio RC. 1984. Aerial survey for oak wilt incidence at three locations in central Texas. *Plant Dis.* 68:661–64
11. Appel DN, Maggio RC, Nelson EL, Jeger MJ. 1989. Measurement of expanding oak wilt centers in live oak. *Phytopathology* 79:1318–22
12. Appel DN, Peters R, Lewis R. 1987. Tree susceptibility, inoculum availability, and potential vectors in a Texas oak wilt center. *J. Arboric.* 13:169–73
13. Barnett HL. 1953. A unisexual culture of *Chalara quercina. Mycologia* 45: 450–57
14. Barnett HL, Jewell FF. 1954. Recovery of isolates of *Endoconidiophora fagacearum* from oak trees following mixed culture inoculations. *Plant Dis. Rep.* 38:359–61
15. Bellamy BK. 1992. *Genetic variation in post-epidemic live oak populations subject to oak wilt.* MS thesis. Texas A&M Univ., College Station. 117 pp.
16. Boo RM, Pettit RD. 1975. Carbohydrate reserves in roots of sand shin oak in west Texas. *J. Range Manage.* 23:15–19
17. Bretz TW. 1953. Oak wilt, a new threat. In *Plant Diseases: The Yearbook of Agriculture, 1953.* Washington, DC: US Dep. Agric. 940 pp.
18. Bretz TW, Morison DW. 1953. Effect of time and temperature on isolation of the oak wilt fungus from infected twig samples. *Plant Dis. Rep.* 37:162–63
19. Brooks D, Gonzalez CF, Appel DN, Filer TH. 1995. Evaluation of endophytic bacteria as potential biological control agents for oak wilt. *Biol. Control.* 4:373–81
20. Bruhn JN, Pickens JB, Stanfield DB. 1991. Probit analysis of oak wilt transmission through root grafts in red oak stands. *For. Sci.* 37:28–44
21. Buechner HK. 1944. The range vegetation of Kerr County, Texas, in relation to livestock and white tailed deer. *Am. Midl. Nat.* 31:697–743
22. Burdon J. 1987. *Disease and Plant Population Biology.* London/New York: Cambridge Univ. Press. 208 pp.
23. Cobb FW, Fergus CL. 1964. Pathogenicity, host specificity, and mat production of seven isolates of the oak wilt fungus. *Phytopathology* 54:865–66
24. Curl EA. 1955. Natural availability of oak wilt inocula. *Ill. Nat. Hist. Surv. Bull.* 26:277–323
25. Dietz SM, Young RA. 1948. Oak wilt—a serious disease in Iowa. *Iowa Agric. Exp. Stn. Bull.* P91:1–20
26. Dunlap AA, Harrison AL. 1949. Dying of live oaks in Texas. *Phytopathology* 39:715–17
27. Engelhard AW. 1956. Influence of time of year and type of inoculum on infection of oak trees inoculated with the oak wilt fungus. *Plant Dis. Rep.* 40:1010–14
28. Fowler ME. 1951. Surveys for oak wilt. *Plant Dis. Rep.* 35:112–18
29. Fowler ME. 1952. Oak wilt surveys in 1951. *Plant Dis. Rep.* 36:162–65
30. French DW. 1983. Successes and failures in forest pathology. *Phytopathology* 73:774 (Abstr.)
31. French DW, Stienstra WC. 1980. Oak wilt. *University of Minnesota Agricultural Extension Service Extension Folder 310.* 6 pp.

32. Gibbs JN. 1981. European forestry and *Ceratocystis* species. *EPPO Bull.* 11: 193–97
33. Gibbs JN. 1978. Intercontinental epidemiology of Dutch elm disease. *Annu. Rev. Phytopathol.* 16:287–307
34. Gibbs JN, French DW. 1980. The transmission of oak wilt. *USDA For. Serv. Res. Pap. NC-185.* 17 pp.
35. Gibbs JN, Liese W, Pinon J. 1984. Oak wilt for Europe? *Outlook Agric.* 13:203–7
36. Greene TA, Appel DN. 1994. Response of live oak selections to inoculation with *Ceratocystis fagacearum*. *Can. J. For. Res.* 24:603–8
37. Griffin JR. 1977. Oak woodland. In *Terrestrial Vegetation of California*, ed. MG Barbour, J. Major, Chap. 11. New York: Wiley. 1002 pp.
38. Halliwell RS. 1966. Association of *Cephalosporium* with a decline of oak in Texas. *Plant Dis. Rep.* 50:75–78
39. Haynes SC. 1976. *Variation in pathogenicity of Ceratocystis fagacearum isolates.* MS thesis. West VA Univ., Morgantown. 61 pp.
40. Heald FD, Wolf FA. 1910. A plant-disease survey in the vicinity of San Antonio, Texas. In *USDA Bur. Plant Ind. Bull.* 685, pp. 11–127
41. Henry BW, Moses CS, Richards CA, Riker AJ. 1944. Oak wilt, its significance, symptoms and cause. *Phytopathology* 34:636–47
42. Henry BW, Riker AJ. 1947. Wound infection of oak trees with *Chalara quercina* and its distribution within the host. *Phytopathology* 37:735–43
43. Houston DR, Drake CR, Kuntz JE. 1965. Effects of environment on oak wilt development. *Phytopathology* 55: 1114–21
44. Irgens-Moller H. 1955. Forest-tree genetics research: *Quercus* L. *Econ. Bot.* 9:53–71
45. Jensen RJ, Hokanson SC, Isebrands JG, Hancock JF. 1993. Morphometric variation in oaks of the Apostle Islands in Wisconsin: evidence of hybridization between *Quercus rubra* and *Q. ellipsoidalis* (Fagaceae). *Am. J. Bot.* 80:1358–66
46. Jewell FF. 1956. Insect transmission of oak wilt. *Phytopathology* 46:244–57
47. Jones TW. 1971. An appraisal of oak wilt control programs in Pennsylvania and West Virginia. *USDA For. Serv. Res. Pap. NE-240.* 15 pp.
48. Juzwick J, French DW. 1983. *Ceratocystis fagacearum* and *C. picea* on the surfaces of free-flying and fungus-mat-inhabiting nitidulids. *Phytopathology* 73:1164–68
49. Juzwick J, French DW, Jeresek J. 1985. Overland spread of the oak wilt fungus in Minnesota. *J. Arboric.* 11:323–27
50. Kurdyla TM, Guthrie PAI, McDonald BA, Appel DN. 1995. RFLPs in mitochondrial and nuclear DNA indicate low levels of genetic diversity in the oak wilt pathogen *Ceratocystis fagacearum*. *Curr. Genet.* In press
51. Lanier GN. 1978. Vectors. In *Dutch Elm Disease, Perspectives After 60 Years*, ed. WA Sinclair, RJ Campana. Cornell Univ. State Agric. Exp. Stn. *SEARCH Agric.* 8:13–17
52. Lewis R. 1985. Temperature tolerance and survival of *Ceratocystis fagacearum* in Texas. *Plant Dis.* 69:443–44
53. Lewis R, Brook AR. 1985. An evaluation of Arbotect and Lignasan trunk injections as potential treatments for oak wilt in live oaks. *J. Arboric.* 11: 125–28
54. Lewis R, Oliveria FL. 1979. Live oak decline in Texas. *J. Arboric.* 5:241–44
55. Liese W, Ruetze M. 1987. On the risk of introducing oak wilt on white oak logs from North America. *Arboric. J.* 11:237–44
56. MacDonald WL, Hindal DF. 1981. Life cycle and epidemiology of *Ceratocystis*. In *Fungal Wilt Diseases in Plants*, ed. ME Mace, AA Bell, pp. 113–44. New York: Academic. 640 pp.
57. Martin CW, Maggio RC, Appel DN. 1989. The contributory value of trees to residential property in the Austin, Texas metropolitan area. *J. Arboric.* 15: 72–76
58. McDermott JM, McDonald BA. 1993. Gene flow in plant pathosystems. *Annu. Rev. Phytopathol.* 31:353–73
59. Menges ES, Kuntz JE. 1985. Predictive equations for local spread of oak wilt in Southern Wisconsin. *For. Sci.* 31:43–51
60. Menges ES, Loucks OL. 1984. Modeling a disease-caused patch disturbance: oak wilt in the midwestern United States. *Ecology* 65:487–98
61. Merrill W. 1967. The oak wilt epidemics in Pennsylvania and West Virginia: an analysis. *Phytopathology* 57:1206–10
62. Milgroom MG, Lipari SE. 1993. Maternal inheritance and diversity of mitochondrial DNA in the chestnut blight fungus *Cryphonectria parasitica*. *Phytopathology* 83:563–67
63. Milgroom MG, Lipari SE, Powell WA. 1992. DNA fingerprinting and analysis of population structure in the chestnut

blight fungus, *Cryphonectria parasitica. Genetics* 131:297–306
64. Mistretta AP, Anderson RL, MacDonald WL, Lewis R. 1984. Annotated bibliography of oak wilt 1943–80. *USDA For. Serv.* GTR WO-45. 132 pp.
65. Mohler CL. 1990. Co-occurrence of oak subgenera: implications for niche differentiation. *Bull. Torr. Bot. Club* 117: 247–55
66. Morris CL, Fergus CL. 1952. Observations on the production of mycelial mats of the oak wilt fungus in Pennsylvania. *Phytopathology* 42:681–82
67. Muller CH. 1951. The significance of vegetative reproduction in *Quercus. Madrono* 11:129–37
68. Muller CH. 1961. The live oaks of the series Virentes. *Am. Mid. Nat.* 65:17–39
69. Norris DM. 1953. Insect transmission of oak wilt in Iowa. *Plant Dis. Rep.* 37:417–18
70. Peacher PH, Weiss MJ, Wolf JF. 1975. Southward spread of oak wilt remains static. *Plant Dis. Rep.* 59:303–4
71. Plumb TR, McDonald PM. 1981. Oak management in California. *USDA For. Serv. Gen. Tech. Rep. PSW-54.* 22 pp.
72. Rexrode CO. 1977. Cacodylic acid reduces the spread of oak wilt. *Plant Dis. Rep.* 61:972–75
73. Rexrode CO, Jones TW. 1970. Oak bark beetles, important vectors of oak wilt. *J. For.* 68:294–97
74. Rexrode CO, Lincoln AC. 1965. Distribution of oak wilt. *Plant Dis. Rep.* 49: 1007–10
75. Rotem J. 1978. Climatic and weather influences on epidemics. In *Plant Disease, An Advanced Treatise,* ed. JG Horsfall, EB Cowling, 2:317–37. New York: Academic. 436 pp.
76. Schmidt RA. 1978. Disease in forest ecosystems: the importance of functional diversity. In *Plant Disease, An Advanced Treatise,* ed. JG Horsfall and EB Cowling, 2:287–315. New York: Academic. 436 pp.
77. Shelstad D, Queen L, French D, Fitzpatrick D. 1991. Describing the spread of oak wilt using a geographic information system. *J. Arboric.* 17:192–93
78. Spilker OW, Young HC. 1955. Longevity of *Endoconidiophora fagacearum* in lumber. *Plant Dis. Rep.* 39:429–32
79. Tainter FH. 1982. Nonstructural carbohydrate contents of trees affected with Texas live oak decline. *Plant Dis.* 66: 120–22
80. Tainter FH. 1986. Growth, sporulation, and mucilage production by *Ceratocystis fagacearum* at high temperatures. *Plant Dis.* 70:339–42
81. Tainter FH, Gubler WD. 1973. Natural biological control of oak wilt in Arkansas. *Phytopathology* 63:1027–34
82. Tainter FH, Ham DI. 1983. The survival of *Ceratocystis fagacearum* in South Carolina. *Eur. J. For. Pathol.* 13:102–9
83. Taubenhaus JJ. 1934. Live oak disease in Austin. *Tex. Agric. Exp. Stn. Annu. Rep.* 47:97–98
84. Taubenhaus JJ. 1935. Live oak disease in Austin. *Tex. Agric. Exp. Stn. Annu. Rep.* 48:99–100
85. Tillson AH, Muller CH. 1942. Anatomical and taxonomic approaches to subgeneric segregation in American *Quercus. Am. J. Bot.* 29:523–29
86. True RP, Barnett HL, Dorsey CK, Leach JG. 1960. Oak wilt in West Virginia. *W. Va. Agric. Exp. Stn. Bull.* 448T. 119 pp.
87. True RP, Gillespie WH. 1961. Oak wilt and its control in West Virginia. *W. Va. Agric. Exp. Stn.* Circ. 112. 18 pp.
88. Tryon EH, Martin JP, MacDonald WL. 1983. Natural regeneration in oak wilt centers. *For. Ecol. Manage.* 7:149–55
89. Van Arsdel EP. 1972. Some cankers on oaks in Texas. *Plant Dis. Rep.* 56:300–4
90. Van Arsdel EP, Bush DL, Kaufman H. 1975. A comparison of *Cephalosporium diospyri* from Texas oaks with *Ceratocystis fagacearum. Proc. Am. Phytopathol. Soc.* 2:142 (Abstr.)
91. Williams S. 1939. Secondary vascular tissues of the oaks indigenous to the United States-I. The importance of secondary xylem in delimiting Erythrobalanus and Leucobalanus. *Bull. Torrey Club* 66:353–65
92. Williams S. 1942. Secondary vascular tissues of the oaks indigenous to the United States-II. A comparative anatomical study of the wood of Leucobalanus and Erythrobalanus. *Bull. Torry Club* 69:115–29
93. Wingfield MJ, Seifert KA, Webber JF. 1993. *Ceratocystis and Ophiostoma—Taxonomy, Ecology, and Pathogenicity,* pp. 1–4. St. Paul, MN: APS. 293 pp.
94. Young RA. 1949. Studies on oak wilt, caused by *Chalara quercina. Phytopathology* 39:425–41
95. Yount WL. 1954. Identification of oak wilt isolates as related to kind of inoculum and pattern of disease spread. *Plant Dis. Rep.* 38:293–96

Annu. Rev. Phytopathol. 1995. 33:119–44

THE RELATIONSHIP BETWEEN PLANT DISEASE SEVERITY AND YIELD

R. E. Gaunt
Department of Plant Science, Lincoln University, Post Office Box 84, Lincoln, Canterbury, New Zealand

KEY WORDS: disease assessment, potential yield targets, yield sensitivity, disease models

ABSTRACT

The relationship between disease and yield is most often summarized as a simple empirical model that describes average crop performances in the presence of a pathogen. Such models may be robust and useful for surveys but their use is usually constrained to the specific conditions under which the model was developed. Changes in production system usually invalidate the relationship. The alternative is to base the relationship on an epidemiological analysis of the pathogen population and a physiological concept of host growth and development. This review provides the knowledge and conceptual basis and discusses the limitations to progress in the development of such models. It is shown that a host-based assessment of disease is well suited to yield investigations and to multiple pest constraints, and that disease is logically related to yield via radiation interceptions and radiation use efficiency.

INTRODUCTION

Infection by plant pathogens and yield are linked by epidemiological and physiological processes that may be considered as three major functional relationships. Disease severity is determined by a function of the degree of infection, colonization, and damage of host tissues. The amount of host development and growth is a function of disease severity, and yield realization is a function of host development and growth. In this review I examine functions

0066-4286/95/0901-0119$05.00

that define the relationship, using examples mostly from staple crops and above-ground fungal diseases. Other diseases and crops that have been less studied are referred to where information is available and to widen the scope of the review. Readers are referred to other related reviews (58, 74, 83, 113) for more detailed information on specific aspects.

THE AMOUNT OF DISEASE

Many reviews have been published on the measurement of disease (44, 45, 61, 112, 120), so only aspects of disease measurement relevant to the main topic are discussed here. The amount of disease, referred to generally as disease intensity (112), is commonly assessed either on the presence of disease, as disease incidence or prevalence, or on symptoms, as disease severity. Disease severity implies that the host is damaged, even though that is not necessarily measured, especially when based on the amount of pathogen development rather than the amount of host reaction. The lack of precision and accuracy in measuring disease is a major constraint in relating disease to yield (74, 84). The relevance and appropriateness of disease measurements to yield production are also important considerations. Rouse (95) pointed out that disease, as conventionally defined, is not directly related to disease as usually measured. He distinguished between actual disease (departure from normal plant function) and visible disease. If measurement of the "departure from normal function" were easier many of the difficulties experienced in relating disease to yield, other than descriptively, would be overcome.

A dilemma faced by plant pathologists is whether to measure disease by methods that allow comparison of epidemics, such as the number of lesions, or that are related to host productivity and yield. Intuitively, and practically, these are opposed objectives and usually no single method can achieve both satisfactorily. For example, measurements of severity based on lesion number or lesion area may be less related to yield, but more to disease progress, than green leaf area duration, even though both are descriptors of disease. Pathogen population perspectives bias measurement toward more epidemiologically meaningful measures such as pustule days per leaf (7, 26, 27), whereas the yield perspective leads toward measurement of green leaf area duration, intercepted photosynthetically active radiation, reflectance, and other host-based variables. The latter are addressed below and the physiological basis of such measurements is explained in terms of host productivity.

Most progress toward physiologically meaningful assessments of disease severity has been made with above-ground diseases, for reasons of accessibility, perceived importance, and the amount of underlying knowledge of their effects. The physiological consequences of shoot diseases, complex though they may be, are more amenable to investigation and measurement than diseases of the root system, stem bases, and underground organs. However, some progress has been made with the below-ground diseases, as follows. Physiological assessment of root diseases has been based on indirect host reactions such as water status and turgor. Correlating these effects with the pathogen population and degree of infection is complex and not often attempted (108). Alternatively, root disease may be quantified indirectly by yield response. Controlling disease by fumigation of wheat fields resulted in significant yield response (up to 70%) in the US Pacific Northwest (24) and in Australia (96). Relating these responses to root volume as a measure of disease is more useful than relating the yield responses to root infection (23). Root infections may be assessed by destructive sampling and washing of the root system, or by inspection methods using sight tubes and appropriate camera technology. At best the estimates are proportional to the amount of disease, except perhaps in the case of nematodes, which are in some respects more straightforward (36). Many subchronic and chronic diseases may not be recognized at all. For example, root pruning by *Pythium* species is difficult to detect by conventional washing techniques. The development of three-dimensional root models (12) provides the opportunity for assessing sampling and measurement methods on computer-generated, realistic root volumes, which avoids the problems of destructive sampling and also allows validation in the field. Stem base diseases such as eye spot in wheat represent a disease type with special assessment problems. Lesion penetration into tissues is more important than lesion area, and lodging may account for greater effects on yield than lesion area alone (40). Synergistic development of root diseases has been studied in some detail, especially on potatoes (97, 104).

Above-ground disease assessment has focused on fungal pathogens, although significant work has also been reported with virus-infected plants (73), nematodes (111), and bacterial diseases. Leaf diseases may have a complex effect on the ability of plants to grow and develop. Yield response was explained by leaf necrosis in wheat infected by *Septoria tritici* (41). Such direct effects on leaf tissue destruction as defoliation and premature senescence are relatively easy to measure but may be satisfactory only for short-term epidemics caused by necrotrophic pathogens. Effects on radiation use efficiency, such as photosynthetic rate, are more difficult to measure. Baastians (5) suggested the use of virtual lesion size, which describes the effect on photosynthetic rate in host area equivalents. Longer-term effects on the plant, including future leaf development, are more difficult to measure. Epidemics that develop while plants are still developing vegetatively may cause delays in leaf emergence, leaf extension, and final leaf size (67–69). Estimating these effects requires coupling of epidemic and plant growth models, but few investigators have addressed this problem (8, 65). There are two possible approaches, one mechanistic and the other more empirical. In the first, the effect of a leaf disease

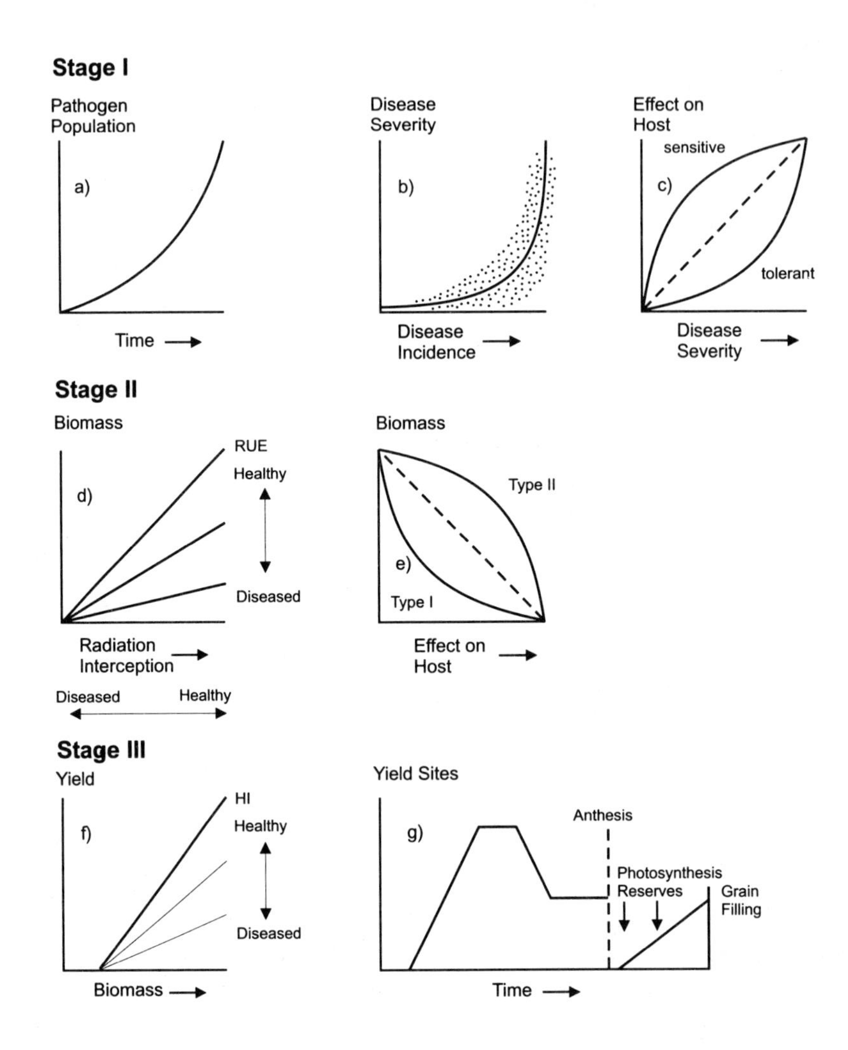

Figure 1 Interactions between processes relevant to severity/yield relationships. Stage I: pathogen population dynamics and effects on host. (*a*) Exponential growth of pathogen. (*b*) Relationship between disease incidence and disease severity. (*c*) Relation between severity and effect on host, showing likely curves for sensitive and tolerant crops. Stage II: production of biomass relative to radiation interruption (RI) and radiation use efficiency (RUE) effects of disease. (*d*) Proportionality of biomass to RI and RUE. (*e*) Type I curve typical of dominant effect on RUE, and Type II curve typical of dominant effect on RI (59, 82). Stage III: Production of yield. (*f*) Yield as a function of disease effects on harvest index. (*g*) Dynamics of development of number of yield sites and of grain yield by the retention of sites and storage as in determinate crops.

may be accommodated in leaf-emergence, extension, and development models as submodels of crop growth and development models. Alternatively, the effects on plant growth may be incorporated into integrated expressions of disease such as the area under the disease progress curve or, more appropriately, as the integrated healthy area duration (59, 120), but there has been little incorporation of these methods of measurement into physiological models. Assessment of disease severity is the linkage between an understanding of epidemic progress and the host processes related to growth and development. This may be represented diagrammatically (Figure 1). Madden & Nutter (74) suggested a framework for advances in this area and readers are referred to their review for further comment.

THE PRODUCTION OF BIOMASS

Plant Growth and Development

Biomass production is affected by many interacting factors but, assuming water, nutrients, and similar inputs are optimal, it is often related directly to solar radiation. Monteith (80) suggested that crop productivity was equal to interception (megajoules per m^2) of photosynthetically active radiation and radiation use efficiency (grams dry matter per megajoule). Many authors have reported relationships, often linear, between biomass increase and radiation interception, especially if canopy interception of photsynthetically active radiation is considered rather than single-leaf interception (38, 42, 48). The relationship is often independent of genotype and season, as in potatoes (106). Equally clear is the relationship between leaf area index and radiation interception; Beer's Law defines light interception in a complex canopy. From the viewpoint of a pathologist interested in the potential of disease to reduce productivity, leaf area index is often an adequate method of assessment especially if leaf area index is not much greater than the optimum. The duration of green area and the wasteful interception of photsynthetically active radiation by damaged photosynthetic tissues are key factors associated with reduced yield.

Solar radiation interception can be measured as that amount of solar radiation intercepted by green tissues [Healthy Area Absorption (120)]. The relation between biomass production and healthy area absorption is asymptotic and constraints will have less effect in crops with high green leaf area index than those with low green leaf area index. There have been many reports of the factors affecting the production of leaves in cereals and in other crops, but fewer on the decline in green area. There has been little interest in understanding the regulation of this decline or the effect on production. This may be considered an oversight, given the importance of green area duration for yield and the effect of adversity on leaf senescence. Reduction in radiation use efficiency can be caused by several factors, none of which is measured by assessment of green area. Solar use efficiencies in nonconstrained crops of genotypes selected by decades of plant breeding are of the order of 3%, the product of photosynthetic and respiratory efficiencies related to total biomass. Radiation use efficiency is more stable in many situations than radiation interception, partly because plants compensate in suboptimal conditions. Compensation occurs in many ways in individual plants and in crops, but the initial reaction to constraints is often associated with metabolic responses of tissues. Compensation may also occur by the reallocation of metabolism to nonaffected tissues. Thus, for example, a noninfected leaf may photosynthesize at a greater rate in compensation for reduced photosynthesis by an infected leaf. In many cases, carbon dioxide efficiency ratios, a measure of photosynthetic efficiency, decrease when plants are manipulated to reduce demand for assimilates. This response is common provided that alternative sinks do not exist. It has been demonstrated that carbon dioxide efficiency ratios may increase in response to increased demand, though such manipulations are more difficult to achieve and more likely to introduce artifacts. Disease, as discussed later, may cause increased output by plant tissues.

Many computer-based models predict growth and development well, at least in the environment for which they were developed. For example, the MACROS model simulates the effect of meteorological factors on physiological processes in rice (54). Porter et al (87) compared the advantages and disadvantages of three wheat models: AFRCWHEAT2 modeled intercepted photsynthetically active radiation well but overestimated the rate of decline in green area after anthesis; in contrast, CERES-WHEAT overestimated pre-anthesis photsynthetically active radiation interception but was better than AFRCWHEAT2 after anthesis; and SWHEAT was inferior to the others for photosynthetically active radiation interception prediction but predicted phenology very well. Development of the models, despite these limitations, has increased our understanding of these processes, and the models incorporate sufficient knowledge to be useful to pathologists.

The Effect of Disease on Plant Growth and Development

Plant diseases may be categorized according to the type of causal agent, the tissues infected, the epidemic characteristics, tissue utilization, resistance reactions, and others, all reflecting specific perspectives of disease. Here I consider disease in the context of disruption of host tissues, as originally suggested by McNew (79), and expanded by Boote et al (9) and Teng et al

(115). The effects of stand reducers, photosynthetic rate reducers, senescence accelerators, light stealers, assimilate sappers, tissue consumers, and turgor reducers can be related to major effects on radiation interception and radiation use efficiency. Simple assumptions, such as the effect on green area duration, may account for much of the observed effects through light interception but would not include, for example, effects on radiation use efficiency. Charles-Edwards (19) listed major effects on respiration and partitioning of assimilates, which are not included above. Pathogens that cause root pruning and other damage to root systems are likely to affect the sensitivity of crops to drought. Variation in root volume and distribution interacts with water availability and disease severity, as do foliar diseases—by depriving roots of assimilates and thus reducing root growth and affecting water-use efficiency of plants (4). Water is commonly the second most yield-constraining factor after radiation interception, and the period and duration of water deficit is commonly linearly related to yield.

Radiation use efficiency is not affected in many pathosystems (93, 94), and the major effect of disease is on radiation interception (52, 120). Photosynthetic rates in potato infected by *Phytophthora infestans* were not affected (119), but leaf blast affected photosynthesis and dark respiration in rice (6). Photosynthetic rate was reduced indirectly by *Verticillum dahliae* on potatoes, via an effect on stomatal conductance and transpiration, although the total effect was less than on individual leaves (53). Photosynthetic rates decreased in parts of barley leaves infected by leaf rust (103), but overall photosynthetic rates increased until major tissue disruption occurred. Farrar & Lewis (34) reported examples of responses in a range of pathosystems; the outcome is thus dependent on plant genotype, pathogen, and environment but the net effects on radiation use efficiency are often negligible. Rabbinge et al (89) related effects on photosynthesis and related processes to disease location and damage thresholds in winter wheat infected with powdery mildew. They concluded that disease assessment should take account of canopy positioning of disease, that the effects on net assimilation may often be small, and that thresholds are likely to be dependent on mildew distribution, leaf area index, and crop development stage. Johnson (59) summarized the outcomes in relation to the cause of Type I and Type II damage responses (74; see Figure 1). Pests affecting radiation use efficiency induce Type I responses, typified by leaf hopper (59). Type II responses are more typical of pests that affect radiation interception. Some pests may affect both radiation use efficiency and radiation interception and thus the response may not fit neatly either Type I or Type II pattern. The observed responses are influenced by the relative importance of radiation interception and radiation use efficiency and by yield strategies, as discussed in the next section.

THE PRODUCTION OF YIELD

Biomass and Yield

In many domesticated species, reproduction and the potential yield are controlled by regulation of development rather than by assimilate and biomass production. For example, Bush & Evans (15) and Evans (31) reported that, in rice, day length influenced both the duration to flower initiation and the duration from initiation to anthesis. This regulation influenced the balance between production and storage of photoassimilates, and therefore the yield potential. Recent research on the potential effects of predicted climate change emphasizes this point, especially in relation to anticipated changes in temperature.

HARVEST INDEX Crop species may be categorized into two main types based on their growth and storage strategy. Cereals, many legumes, and some other crops determine the potential achievable yield early in crop growth and a proportion of that potential is subsequently realized. Determinate cereals are more rigid in this regard than many of the legume species, where growth is often indeterminate and the harvest index is more flexible. In contrast, crops such as sugar beet, potatoes, and cassava reach a genetically and environmentally determined harvest index ratio early in crop growth and, in the absence of major constraints, maintain this ratio till maturity (21, 37). There may be time variations in the development of harvest index, even though these crops do not follow the sequential determinate development typified by cereals. For example, the final harvest index was similar in five potato cultivars and in two seasons, but during crop growth the proportion and amount of harvest index achieved was variable (106). Some cultivars were consistently faster than others in approaching the final harvest index, which implies different growth strategies, and possibly different sensitivity to adversity, even in such nondeterminate growth types.

YIELD POTENTIAL Plant breeders have achieved major increases in yield relative to older cultivars, as illustrated by comparisons of old and new cultivars grown under the same conditions at a single location (2, 17). Comparison of different rice types grown in the same field location and in glasshouses illustrated the changes that have occurred in many breeding programs (33). Lowland rice types became shorter, with smaller and more upright leaves and with reduced day length sensitivity. Panicle weight fluctuated and there were correlated responses in panicle number that appeared to be the more dominant character. No associated changes were reported in photosynthesis rate, crop growth rate, or kernel weight, but there was a marked increase in harvest index and grain dry weight production per day. Increased harvest index ratios are largely the result of partitioning biomass from stem dry weight to grain or other yield units, at least in cereals. Dwarfing genes in spring wheat were positively correlated with yield (39), at least under optimal conditions. Some cereals, and especially C4 species, seem not to follow this pattern. Many legume cultivars with increased yield have associated increased total shoot biomass rather than increased harvest index. Whatever the mechanism, potential yields have increased markedly and it is often assumed that this implies greater sensitivity to constraints, including disease. The removal of genetic, and maybe environmental, yield constraints may be associated with less flexibility. The sensitivity of crop species and cultivars in specific locations and seasons requires further research before our understanding of these relationships is sufficient to predict reactions to disease. Some studies have compared the relative loss of different genotypes under constrained versus optimal conditions. Study of physiological constraints such as drought and shading are more instructive about plant sensitivity than disease constraints, because disease is often confounded by different infection levels. The results suggest that yield reductions may be greater in the higher-yielding than in the lower-yielding cultivars. Crossover of performance, or interaction with production environment, is not uncommon and is especially interesting to pathologists concerned with the reaction of crops to disease. The low-yielding cultivars may exhibit "disease tolerance" in specific locations (43, 83, 102). However, the degree of constraint may need to be large before there is a major change in ranked performance of different cultivars (29). Ceccarelli & Grando (18) concluded that crop failure is more likely to occur under adverse conditions in barley cultivars with high-yield potentials than those with low-yield potentials. Despite this knowledge, healthy yield targets are not often considered in models between disease and yield.

ROLE OF RESERVES Plant yield may be accumulated both from current photosynthesis and by the utilization of carbon reserves. It is useful to consider two types of reserves, pathway (or short-term) storage and organ (or midterm) storage. Pathway storage is often diurnal and may occur in the source itself or in tissues closely associated with transport. It is often in the form of simple sugars and it provides buffering for short-term variations in productivity, as in legumes with a buffered N-fixation rate during each 24 h period. Organ storage, on the other hand, supplements shortfalls in current photosynthesis, as, for example, in stem reserves in cereals and legumes and in woody storage tissues in perennial crops. The contribution of stem reserves, or other reserves, to yield by relocation has long been controversial (121). The interpretation of results is dependent on the methods of measurement and on the assumptions made in defining the contribution (see 47). Methods range from assessment of only the relocation of stored material to yield units, or inclusion of all contributions to growth, respiration, and other processes associated with yield production. Many calculations of contributions to cereal yield are based on the common assumption that most stem reserves are stored pre-anthesis in cereals. However, maximum reserves are often not accumulated until 6–10 days or longer after anthesis (11), especially in those crops with a slow rate of initial grain filling. Much of the variation in the estimated contribution is associated with the amount of stress in crops that have been studied. In some circumstances, both storage and relocation will be high, but either component may be negligible, depending on supply and demand. In legumes the position is less ambiguous than in cereals and some other crops. The demand for organic nitrogen compounds imposes a much greater reliance on stored reserves than in non-nitrogen-fixing crop species. In some crops, such as tomatoes (55), there may be a sink-limited system and stem reserves, though accumulated, may have no role in fruit development. Such imbalanced systems are probably rare, but the issue is still relevant to the disease severity relationship with yield of such crops.

Whether reserve utilization is relevant to yield differences between genotypes remains largely unresolved. In some circumstances dwarfing genes were correlated with less deposition of stem reserves. For example, deposition and use of stem reserves was positively correlated with height in near-isogenic wheat lines (10). The advantage of extra reserves in tall plants may be realized only under conditions adverse for photosynthesis during grain filling. A disadvantage may occur when the greater stem storage competes with, and limits, development of the grain sink size. The range of possibilities was demonstrated in different palm species (118). In the sago palm and in other species with a seasonal or single-flowering habit, stem reserves may be very substantial and relocation during fruiting is the dominant source of carbohydrate for the fruit. On the other hand, the coconut palm, which flowers continuously, has little or no temporary storage. Fruit are totally dependent on current photosynthesis, and are produced at a rate directly correlated with photosynthesis. One can conclude that reserve materials are involved to at least some extent in yield production of modern cultivars grown with moderate to high inputs. Contributions may be especially important if photosynthesis is constrained during the realization phase of yield production. From the pathologist's viewpoint, the absolute amount contributed is perhaps less relevant than the reduced contribution in diseased plants, either because less material is stored or because there is greater demand during grain filling because of the pathogen's nutritional requirements.

COMPETITION BETWEEN SITES Individual plants, and crops, are often regarded as assemblages of sources and sinks. The primary sources are those photosynthetic tissues with a net independence for carbon compounds, and the feeder roots for nutrients and water. Temporary storage sites in stems and other tissues may be regarded as secondary sources during periods of relocation. Sinks include new and expanding tissues, short-term, mid-term, and permanent storage sites, the last commonly associated with yield. The potential size of sinks is more important to yield potential than the rate of filling. Knowledge of the pattern and methods of starch deposition in grain is increasing (116) and there is little evidence to suggest that this is a rate-limiting step to yield production. The relative association of sources and sinks varies both temporally and spatially, affecting the relationship between biomass production and yield. The determinate cereals, and forage and many vegetable crops, represent extremes in the development of sources and sinks. The cereals establish most of the assimilatory tissues before major storage in yield-related sinks occurs, whereas at the other extreme, source and sink development are either coincident or the sinks provide for further source development. Although not usually as simple as sometimes portrayed, whether a crop is source or sink limited at specific growth stages is relevant to understanding yield development and realization, and is therefore relevant to the reaction to disease. Crops rarely are predominantly either sink or source limited for extended periods, except in very low yielding situations. Source capacity may be the ultimate limitation in a well balanced system; a sink limitation during, say, grain filling may obscure the fact that at earlier growth stages the same crop was source limited. For pathologists, an understanding of the change in source/sink balances during crop growth and development is of greater interest than the issue of which is most limiting overall.

YIELD BALANCE Increased yields have been achieved by breeding for lower investment in fixed resources, accompanied by more efficient function of those resources. Thus an increased duration of individual leaves, and maybe roots, associated with continued efficient function allows more resources to be allocated to growth and yield. Greater yields may result, but only if the fixed resources are protected adequately from pests and diseases. The implication is that damage threshold values will be reduced proportional to the success of such breeding. In most crops the duration of growth to harvest is closely related to both total biomass produced and yield (109, 110). Short-duration crops set lower yield targets, especially in determinate species, even though the amount of yield per day is usually higher than in long-duration crops. The assimilative capacity in short-duration crops is often in excess of demand, because of reduced sink development. When an extra crop is grown each year greater sink limitation in each crop is balanced by the greater total sink production per year. Longer crop growth durations result in greater yield only if there is sustained photosynthetic production or adequate reserve materials. Exceptions occur, especially under adverse conditions (71). In comparing several cereal species growing in rain-fed environments where water was a substantial constraint, barley yielded better than other species. Rapid early growth, early anthesis, and short duration of grain filling conserved water, but also allowed development of larger sink sizes and greater accumulation of stem reserves. The latter were used to supplement current photosynthesis for grain filling.

YIELD COMPONENTS The value of dividing yield into components for analysis of factors affecting yield has been considered axiomatic by many physiologists, breeders, and pathologists. The analysis is simple and the explanation of yield variation can be useful, especially if based on knowledge of when the components are determined. The sequences of component determination in cereals and other crops are well documented. The subcomponents, such as spikelet initiation and abortion, offer further precision and insight into both yield development and the analysis of specific constraints. Individual plants may compensate fully for the effect of reduced sowing rate, or seedling establishment, even over wide ranges. On the other hand, compensation is often incomplete for events such as frost damage, defoliation, or other factors during crop development that have a direct effect on sink capacity. Compensation for constraints to components developed late in crop growth may be less complete than when the constraint occurs at crop establishment or in early growth stages. A reliance on analysis of components, especially primary components, can be as confusing as it is illuminating. High or low values for individual components have little meaning except in the context of other component values and the production environment. Evans (32) concluded that modern cultivars growing in conditions for which they were selected are likely to have a balance between source and sink activity and that this balance is reflected in the yield components. The balance may be disrupted by seasonal factors and by disease.

It is usual to express yield as the amount of dry matter per hectare, or the equivalent. Increasingly, it is relevant to express yield by taking account of quality or the amounts of protein produced. Yield will likely also be considered in relation to resource materials in limited supply, such as water, fossil-energy, phosphorus, and other nutrients. For the pathologist it is of interest to consider yield in relation to disease, for example, to estimate the "cost" of resistance (1), or to the amount of fungicide in relation to pollution and residues in food.

The Relation Between Disease and Yield

Just as plant growth and development must be understood to explain yield, plant pathogen growth is relevant to considerations of yield response. There has been much emphasis on the outcome of growth, for example, in terms of spore number per day and lesion development, but less attention has been paid to the cost and the consequences of pathogen growth, such as substrate utilization and damage to host function, despite the importance of these effects for plant yield.

It is moot whether diseases cause yield loss (22), except perhaps in the strict sense of postharvest losses, or whether yield potential relative to investment in nutrients and water is not attained. The yield that has not yet been gained can hardly be lost, unless viewed as the loss of potential yield based on the establishment of sinks. This view does not cover all situations commonly referred to as crop or yield loss, given that disease can also reduce sink size and number and thus reduce the potential yield. This is more than a semantic issue, and it usefully focuses attention on the ability of diseases to disrupt plant function and balance. The use of potential, or attainable, yield as the reference yield also encourages a consideration of actual yield in the diseased state. Savary & Zadoks (99, 100) considered adopting diseased yield as a reference in their studies of peanuts, but rejected it in favor of yield of healthy crops receiving average inputs. They considered the variability of the diseased yield too great to be an acceptable reference yield. In other systems with less variation this may not be the case. Root diseases are often not explicitly controlled in experimental determinations of "healthy" crop yield, yet there would be few situations without at least some constraints from this source. There has recently been more focus on increasing yield by disease control, rather than focusing on reducing the losses. This focus is currently more politically acceptable, but is also more relevant to the actual effects of disease.

Analyses of the disease/yield relationship are either empirical or explanatory. Well-parameterized empirical relationships of disease and yield are useful for surveys and general indications of the impact of disease. Most models currently available are limited in their usefulness by the range of conditions to which they can be applied. Explanatory models require physiological inputs but do not necessarily need to be complex. The challenge is to identify the key issues and incorporate these in simple relationship statements, maybe in the form of computer-based models, without being simplistic. The following sections address approaches to derive explanatory and descriptive relationships of disease and yield based on physiological and empirical models.

PHYSIOLOGICAL APPROACHES Physiological approaches to understanding severity/yield relationships include the following major objectives:

1. to identify key physiological processes related to growth and development affected by disease;
2. to understand the interacting effects of such processes;
3. to predict likely outcomes of disease, with less unexplained variability than current models;
4. to provide the basis for improved empirical descriptions of loss;
5. to identify new, more suitable, techniques for both research and management.

Causality and correlation between disease and yield require careful interpretation. The use of relationships based on physiologically based measurements of disease severity and yield do not necessarily define causality. For example, although carbon flow to corn ears grown under different nitrogen availabilities was highly correlated with leaf area per plant, the yield was dependent on the size of the grain sink (66). Pathologists have two main options for the development and use of physiological models. They may base models on environmental inputs to physiological processes related to yield. These models do not cope well with physiological processes other than photosynthesis and plant development. In particular, partitioning is difficult to model other than empirically. Wardlaw (121) pointed out the need for greater knowledge of factors controlling carbon partitioning, organ initiation, and development, the degree of source/sink limitation at any one time, and the role of storage in buffering fluctuations in plant balance within the cropping environment. Any increased knowledge of these factors will increase our understanding of the relationship between yield and disease. The second option is to base models on a reduced set of physiological assumptions, without succumbing to the trap of apparent causality. Tests for this are often based on intuitive reasoning, which may be proved flawed as new knowledge becomes available.

The sensitivity of crop yield varies depending on the production environment and therefore the damage threshold (124) is not the same at all crop-growth stages or in different crops. This was well illustrated by the effect of short-duration water deficits (85) and by shading in field crops. Grain yield was most correlated with radiation during the reproductive phase, ear number with the early reproductive phase, grain number per ear with the late reproductive phase, and grain weight with radiation interception during grain filling (30). Yield reductions associated with foliar diseases have similar correlations and the same has been shown in some tuber crops (120) and in peanuts (50). The frequent reduction of yield potential by premature flower senescence, fertilization failure, and early abortion of young seed suggests that assimilate supply is critical at these stages. The actual requirement for assimilates may be low but if the supply is interrupted for a brief period either by low production or competition, the outcome may be larger than expected. Control processes during these critical periods may provide the key to understanding harvest index and yields, and the effect of disease on yield. The effects of disease may be understood more clearly if partitioned into effects on plant development and plant growth (35, 44, 117, 123). This approach allows the slope of the yield response to be partitioned into host effects and may provide the necessary linkage to plant growth models.

Green area and green area duration have been related both physiologically and by correlation to yield in both healthy and diseased crops of many species. The prediction of green area duration is a useful approach to yield prediction and the effect of disease. Advances in this area are illustrated well by work on sorghum (16, 81), which emphasized the importance of the sustained maintenance of green leaf area index. The simulation of senescence processes has received less attention than desirable relative to leaf area development processes. If crop models are deficient in this regard, they may be less useful for disease severity/yield studies since disease and senescence are similar in effect and are sometimes causally related. Biotrophic disease may be more complex in this regard than necrotrophic diseases because of the demand for assimilates for pathogen growth and maintenance respiration. Nonetheless, in many diseases it has been shown that green area duration is an adequate estimate of the effect on the host. Maas (72) reported using remotely sensed satellite data to reinitialize predicted green leaf area index in a model of crop growth and yield. Average yield predictions were 30% overestimated without this process, but only 2% with the use of the satellite data. Ground-based measurements could be used similarly; presumably this process could include reinitialization to take account of disease just as well as inaccuracies in the model.

The amount of stem reserves stored and utilized in healthy plants was referred to above. Disease in two cultivars of barley, sown at different times and in two markedly different seasons, reduced stem dry weight and the amount of stored carbohydrates, but increased the total amount of stored carbohydrates used for grain filling in most situations when compared with healthy crops (47). The effect of disease on the storage and utilization of stem reserves depended on the timing and duration of the epidemics and the healthy yield potential of the crop. McGrath & Pennypacker (77), working with wheat and rust diseases, demonstrated a relatively simple effect of post-anthesis epidemics on yield and grain growth. In diseased crops, grain filling occurred at a lower rate and for a shorter duration than in healthy crops. These authors associated some differences in response to rust to differing amounts of stem reserve utilization, though the loss in culm weight was not directly associated with disease severity. This effect is not surprising, given that the use of stem reserves is secondary to current photosynthesis and is dependent on several other factors. Current evidence suggests that crops with low stem reserves or high yield potentials would characteristically be sensitive to disease during grain filling, whereas crops with more stem reserves or lower yield potentials would be less sensitive. The challenge for pathologists is to develop methods to identify such crop types and modify management of both crop and disease to reduce risk while maintaining yield.

If seen as a three-stage process [pathogen population to disease effects, disease effects to crop growth, and crop growth and development to yield (see Figure 1)], progress toward greater understanding of the diseased yield relationship may be more rapid. Crops are rarely exposed to single pests, but the basis for experiments is generally single exposure. Even in single-pest situations, it is unusual for there to be single constraints to yield as abiotic factors are also limiting. Crops are usually multi-constraint systems, and it is useful to consider disease as another, arguably more complex, constraint. New ways of dealing with multiple pests focus inevitably on physiological descriptions of the pests (60, 62, 100) to allow valid comparison. Coupling of disease and crop growth has frequently been based on approaches used with insect pests (91), weeds (107), and pollutants (64). This analysis is useful when the effect is simple, as is often the case with insects. Some diseases can be considered similar to leaf-chewing, root-pruning, phloem-feeding, and cell-feeding insects. However, many pathogens have more complex effects. For example, a powdery mildew may be regarded as a cell feeder, but it is also a light stealer and a senescence inducer. Simple functional or analytical models, which include factors such as time and duration of stress, the role of stem reserves, the balance of source and sink, and the radiation use efficiency, all have some degree of empiricism. The role of such models is either for management, in which case the empiricism represents an acceptable averaging of complex effects, or for identifying areas where more understanding is required and where more information is necessary. Physiological assumptions, or knowledge, of sensitivity to disease of crops at different growth stages may be accommodated in response surface models of disease intensity, physiological age, and yield (114). Similar suggestions were made for the effects of insect pests on plant yield (49). Although conceptually useful, these models have proved difficult to translate into practice. Madden & Nutter (74) review further conceptual modeling approaches to this problem. Some progress was reported recently with constraints on cotton insect pests (90). Using previously determined generalized yield responses (86) and their own data for two Lepidopteran pests, Ring et al (90) developed response surfaces for normalized yield. Data for pest levels at key physiological ages were available from field experiments and they fitted the response surfaces well. Although incomplete and marred by some anomalies, their model based on these response surfaces is a useful example of such approaches. It clearly identified pest levels and physiological ages at which more data were required and indicated those parts of the response surfaces that could be improved by prior knowledge and intuitive reasoning.

Aspects of disease/yield interactions that are difficult to analyze are the spatial distributions of individual plant performance in the crop, and disease in the crop and in the canopy. Stand reducers, for example, are likely to widen the variation in plant performance in the population. Compensation by adjacent plants may be more important to crop yield in these plants than the average. Disease on these plants may be of greater impact than on plants with average or reduced performance. Some progress has been made in identifying variation in plant and crop yield performance (46), and in disease distribution. The variability in wheat yields ranged from standard deviations of 4–25% as scale increased from plot to regional values. On the same farm the value may be as high as 15%, and within-field variation (4–9%) is sufficient to consider linkage with disease spatial variation (74). Conceptual approaches to relating pathogen spatial variation to yield have been suggested (56, 57). Spatial pathogen variation is related to the size of the spore source and to canopy accessibility (28), to dispersal systems (76, 125), to plant genotype (13), to age (98, 100), and to dispersal factors (3). The best pathogen spatial information is available for the dynamics of epidemic spread and yield in crops infected by viruses (73), but others have been investigated. Relatively few attempts have been made to relate the pattern in yield with the pattern in disease.

DESCRIPTIVE APPROACHES The major objectives of descriptive empirical models of disease and yield include the following: defining simple relationships between disease and yield for general surveys; estimating yield in specific fields with limited epidemics; and increasing knowledge by testing hypotheses built in as functions or parameters.

A common feature of descriptive models of disease and yield is that they are different! Derived by correlation and regression techniques reviewed elsewhere (74, 114), the origins of these differences warrant scrutiny. Because empirical relationships are not necessarily causal, the difference could arise for several reasons, for instance, the derivation of such models. Whether models are single or multiple point, those reported are commonly selected from other possible models, each of which describes the relationship to varying degrees of accuracy and precision. Models, assuming they fit the data significantly, are frequently selected on the R^2 values. Other criteria are available (25, 114) but are not generally used. The emphasis on selection for best fit, based on an F statistic and the R^2 value, may divert researchers from the selection of models common to other models, based perhaps on the same measure of disease at the same growth stage. Thus similar models to others reported may be hidden or rejected by the selection process. A survey of the similarity of models in cereals (RE Gaunt, unpublished) suggested that more similar models would be reported if some degree of standardization were accepted in the selection process. Standardization of model form (e.g. Richards function rather than quadratic) might also be useful for future models. A further reason for difference is the choice of single-point, multiple-point, or integral models. Single-point models, in which disease measurement at a single time or growth stage is related to yield, are limited in usefulness in variable environments because pest populations vary between seasons and locations. Methods to accommodate different epidemics with similar AUDPC values, by weighting the disease levels by growth stage, have not been widely adopted but would be useful (105). Few direct analyses have been made of the repeatability of empirically derived relationships; rather, greater emphasis has been placed on the development of robust, averaged models with high significance of fit but often with only moderate R^2 values. A model with an R^2 of 70% still leaves 30% variation unexplained. At best such models are useful for gross estimates of loss, and at worst they may provide a poor basis for subsequent analyses and may encourage unwarranted confidence that the relationship has been quantified and explained.

Yield or yield loss is usually expressed either in absolute units or as a proportion of healthy yield. The latter value is frequently not available in surveys, although an estimate or measurement is usually included in disease/yield experiments. The assumptions implicit in these relationships may often be violated: Yield does not vary either in absolute units or proportionally with disease in many situations (78). It may be argued intuitively that empirical models, especially those based on pathogen-oriented assessments of severity, will differ when the healthy yield target varies widely. There have been few specific investigations of this issue. The response of high- and low-yield target crops, and possibly crops with similar yield but different growth and development strategies, are likely to produce different models. Daamen (27) defined a damage function (kg/100m^2 per pustule-day of powdery mildew) in nine wheat crops. The function was not dependent on yield levels, but these varied only between 6 and 7.5t dry matter/ha; it was suggested that the function was only relevant to similar yield and disease levels. Lipps & Madden (70) found greater losses in low- compared to high-yielding wheat crops. Savary & Zadoks (100, 101) demonstrated different thresholds and yield responses in peanut crops with different yield potential (1.08–3.39t dry pods/ha) when exposed to rust and late leaf spot. Studies have limited usefulness in this regard if there are low yield ranges, low disease severities, or low proportions of yield variation attributable to disease. Rossing (91) used a mechanistic model to predict the outcomes of aphid infestation in winter wheat at yield ranges of 5-10 t/ha, finding a linear response up to 9t/ha. Field data were consistent with the model's findings. The greatest contrast in yield response is seen in comparisons between crop species and weed species. Direct comparisons are difficult, since canopy structure and populations are very different. However, several weed species have little or no reduction in yield at disease levels at which crop species would be expected to respond. This result has usually been attributed to lower harvest index, since biomass production is reduced significantly (51). The evidence suggests that some improvement of precision over a wide yield range may be achievable if crop physiological processes are accounted for, or at least considered, in empirical models. For example, absolute yield units and a host-oriented measure of severity have been used in several models. When yield target is unknown, some use can be made of the fact that yield potential is correlated with duration of crop growth. The specific nature and limitations of an empirical loss model may be reduced by including a predictor of crop growth duration such as sowing date (75, 122). Inclusion of crop growth duration significantly improved the ability of empirically derived models to predict loss in crops with widely different yield targets associated with different sowing dates. This approach may be useful in areas that are relatively uniform in soil type, rainfall distribution, and other physical factors. Shtienberg et al (105) developed an empirical model for yield response in potatoes, based on physiological assumptions. Disease severity was adjusted for the seasonal fluctuation in host growth and tuber bulking rates. When validated with independent data for two diseases under a range of environments, the predictions of loss were within 5% of observed loss in 80% of the crops examined.

Seasonal flux may invalidate empirical models, because other factors override the estimated relationship. Such data are contained in synoptic analysis of production over several seasons (88). Burleigh et al (14) reported seasonal flux as the most important variables in empirical prediction of wheat yield in Rwanda, grown under suboptimal conditions. Overall yield was associated with irradiance, field azimuth, and slope in different years and seasons. Yield components were associated with these variables, with aluminum toxicity and nitrogen status, and with disease to differing degrees. Empirical models, other than those based on physiological assumptions, are unlikely to be reliable between seasons and locations in variable environments. On the other hand, such models may be acceptable for crops grown with high inputs in stable environments, especially if the crops are exposed only to short-duration late epidemics.

In summary, simple models based on mathematical descriptions may have physiologically meaningful variables and parameters incorporated by careful consideration of the important yield factors. These models, although still partly descriptive, may be more useful for surveys than at present; they also may be useful for the development of concepts and thus warrant further development (56, 57, 74).

CONCLUSIONS

Crop species have spread from their center of origin (32). The degree of adaptation naturally, and by plant breeding, has influenced the balance of physiological processes in crops. This balance is relevant to the reaction of crops to disease constraints, especially in relation to the ability of the crop to produce energy-rich compounds for both host and pathogen respiration and storage processes. These issues are fundamental to an understanding of the relationship between disease severity and yield. Adaptability in species and cultivars is considered a desirable trait by plant breeders. It ensures that cultivars perform well at different locations, thus increasing the marketability of the cultivars, and ensuring the stability of yield at the same location. Although yield stability, a desirable socioeconomic as well as biologic trait, may often be associated with durable resistance to pests and diseases, it may also be associated with physiological responses to climate variation. Analysis of cultivar performance, relative to mean cultivar performance over a period of time and/or location, may provide information as relevant to disease response as to disease susceptibility.

This review postulates that a greater understanding of the relation between severity and yield is useful for management of diseases. If this postulate is accepted, then a greater understanding of the outcome of genotype and environment interactions for growth, and the relative contributions of processes and organs to yield, is imperative. Knowledge of the mechanisms contributing to yield, and to yield reduction by constraints, inevitably lags behind the development of new genotypes and management systems. Research in both breeding and agronomy has progressed successfully based on empirical tests and observations, largely without the benefit of a detailed knowledge of the processes involved. Similarly, the management of disease epidemics and the maintenance of yield in the presence of disease has progressed empirically rather than mechanistically. This characterization does not devalue the relevance of investigations of the processes involved. As with many complex situations, progress may be empirical but it will be guided better by retrospective analysis of previous advances. Knowledge of the relationships between severity and yield may guide future breeding and research as well as explain current observations.

Experience suggests that plant genotypes and plant management will continue to change, and that potential yields will increase. It is tempting to conclude that crop sensitivity to disease will increase concomitantly. However, this may be a prediction based more on what we know now rather than on what will be discovered in the future. Plants may be developed with greater flexibility as we learn more about the genetic, biochemical, and physiological processes of both plant growth and the effect of pathogens on those processes. It is not inconceivable that plants could support not only more yield but also more yield in the presence of disease. The challenge is to understand these processes better and to gain a better basis for disease management and the maintenance of yield. Pessimists may suggest that, at least in some cases, yields will not continue to increase to any significant degree. The contention holds that nonrecurring innovations have led to yield increases in the past and that there are few, if any, further innovations to be made. By contrast, optimists point out that yield plateaus are often associated with factors such as the reduction of inputs, or shifts to greater use of more marginal land. They may also identify potential for increase, such as the current enthusiasm for the perceived and real opportunities available through molecular technologies in plant breeding. It therefore behooves us to separate the physiologic and genotypic trends from the social and economic.

Caution is required when assuming that severity and yield are independent other than for observed direct effects. Plant growth dynamics influence epidemic progress in a plant tissue feedback loop (8, 61), as well as affecting those yield factors reviewed in this paper. Thus the severity yield relationship should account for these secondary effects if we wish to translate observed relationships into more general ones that are not genotype and environment specific. Rouse (95) predicted significant improvements in better crop growth–based epidemic and yield models. Some progress has been made over the past decade, largely in the slow and laborious generation of additional data to explain the nuances of the relationship. Wide-ranging responses to diseases of many crops have been documented and this information has been useful in defining possible risks to production and establishing priorities. However, continued description of such relationships can be regarded as only fine-tuning our knowledge of the response range, without increasing our understanding of the actual response. New approaches to physiological modeling may offer greater prospects for progress than the repetitive definition of specific responses to disease. New methods to detect and predict the likely sensitivity of crops at different growth stages, and at different locations, will be the most rewarding. Such approaches should widen the scope of current models of yield in the presence of disease, or reduce the variation associated with response estimates. Recently, the models themselves have been subjected to uncertainty analysis (63, 92), in which the sources of uncertainty in models are assessed in relation to attitudes to risk. Areas of the models likely to benefit most from additional research are thus identified.

Chester (20) produced one of the earliest systematic analyses of the appraisal and interpretation of plant disease–related "losses" some forty-five years ago. In the interim, description and documentation of yield responses have progressed, but progress in interpretation and increased understanding have been slower. Steady advances in understanding abiotic constraints to physiological processes have assisted this progress, but key issues remain unresolved. This lack of understanding may well deter many pathologists from undertaking analyses of loss except at the functional/analytical level.

Literature Cited

1. Allard RW. 1990. Future directions in plant population genetics, evolution and breeding. In *Plant Population Genetics, Breeding and Genetic Resources*, ed. AHD Brown, MT Clegg, AL Kahler, BS Weir, pp. 1–19. Sunderland, MA: Sinauer
2. Austin RB, Ford MA, Morgan CL. 1989. Genetic improvement in the yield of winter wheat: a further evaluation. *J. Agric. Sci.* 112:295–301
3. Aylor DE, Wang Y, Miller DR. 1993. Intermittent wind close to the ground within a grass canopy. *Boundary-Layer-Meteorol.* 66:427–48
4. Balasubramaniam R, Gaunt RE. 1990. The effect of fungicide sprays on root development, yield and yield components of wheat in the absence of disease. *Plant Prot. Q.* 4:95–97
5. Bastiaans L. 1991. Ratio between virtual and visual lesion size as a measure to describe reduction in leaf photosynthesis of rice due to leaf blast. *Phytopathology* 81:611–15
6. Bastiaans L. 1993. Effects of leaf blast on photosynthesis of rice. I. Leaf Photosynthesis.*Neth. J. Plant Pathol.* 99: 197–203
7. Beresford RM, Royle DJ. 1991. The assessment of infectious disease for brown rust (*Puccinia hordei*) of barley. *Plant Pathol.* 40:374–81
8. Berger RD, Jones JW. 1985. A general model for disease progress with functions for variable latency and lesion expansion on growing host plants. *Phytopathology* 75:792–97
9. Boote KJ, Jones JW, Mishoe JW, Berger RD. 1983. Coupling pests to crop growth simulators to predict yield reductions. *Phytopathology* 73:1581–87
10. Borrell AK, Incoll LD, Dalling MJ. 1993. The influence of the Rht_1 and Rht_2 alleles on the deposition and use of stem reserves in wheat. *Ann. Bot.* 71:317–26
11. Borrell AK, Incoll LD, Simpson RJ, Dalling MJ. 1989. Partitioning of dry matter and the deposition and use of stem reserves in a semi-dwarf wheat crop. *Ann. Bot.* 63:527–39
12. Brown T. 1995. *Simulation of the development of the root system and associated microbial community of Pinus radiata*. PhD thesis. Lincoln Univ., Canterbury, NZ. 196 pp.
13. Buiel AAM, Verhaar MA, van Den Bosch F, Hoog-Kamer W, Zadoks JC. 1989. Effect of cultivar mixtures on the wave velocity of expanding yellow stripe rust foci in winter wheat. *Neth. J. Agric. Sci.* 37:75–78
14. Burleigh JR, Yamoah CF, Regas JL, Eylands VJ. 1991. Agroclimatology and modelling. *Agron. J.* 83:625–31
15. Bush MG, Evans LT. 1988. Growth and development in tall and dwarf isogenic lines of spring wheat. *Field Crops Res.* 18:243–70
16. Carberry PS, Hammer GL, Muchow RC. 1993. Modelling genotypic and environmental control of leaf area dynamics in grain sorghum. III. Senescence and prediction of green leaf area. *Field Crops Res.* 33:329–51
17. Ceccarelli S. 1989. Wide adaptation: how wide? *Euphytica* 40:197–205
18. Ceccarelli S, Grando S. 1991. Selection environment and environmental sensitivity in barley. *Euphytica* 57:157–67
19. Charles-Edwards DA. 1982. In *Physiological Determinants of Crop Growth.* Sydney: Academic. 161 pp.
20. Chester KS. 1950. Plant diseases losses: their appraisal and interpretation. *Plant Dis. Rep. Suppl.* 193:189–362
21. Cock JH. 1983. Cassava. *Potential Productivity of Field Crops Under Different Environments*, pp. 341–59. Los Baños: IRRI
22. Cook RJ. 1985. Use of the term "crop loss". *Plant Dis.* 69:95
23. Cook RJ, Haglund WA. 1991. Wheat yield depression associated with conservation tillage caused by root pathogens in the soil, not phytotoxins from the straw. *Soil Biol. Biochem.* 23:1125–32
24. Cook RJ, Sitton JW, Haglund WA. 1987. Increased growth and yield responses of wheat to reduction in the *Pythium* populations by soil treatments. *Phytopathology* 77:1192–98
25. Cornell JA, Berger RD. 1987. Factors that influence the value of the coefficient of determination in simple linear and nonlinear regression models. *Phytopathology* 77:63–70
26. Daamen RA. 1988. Effects of nitrogen fertilisation and cultivar on the damage relation of powdery mildew (*Erysiphe graminis*) in winter wheat. *Neth. J. Plant Pathol.* 94:69–80
27. Daamen RA. 1989. Assessment of the profile of powdery mildew and its damage function at low disease intensities in field experiments with winter wheat. *Neth. J. Plant Pathol.* 95:85–105
28. de Jong MD, Wagenmakers PS, Goudriaan J. 1991. Modelling the es-

cape of *Chondrostereum purpureum* spores from a larch forest with biological control of *Prunus serotina*. *Neth J. Plant Pathol.* 97:55–61

29. Edmeades G, Tollenaar M. Genetic and cultural improvements in maize production. *Proc. Intl. Congr. Plant Physiol., New Delhi, India* 1:164–80. New Delhi: Indian Agric. Res. Inst.
30. Evans LT. 1978. The influence of irradiance before and after anthesis on grain yield and its components in microcrops of wheat grown in a constant daylength and temperature regime. *Field Crops Res.* 1:5–19
31. Evans LT. 1987. Short day induction of inflorescence initiation in some winter wheat varieties. *Aust. J. Plant Physiol.* 14:277–86
32. Evans LT. 1993. *Crop Evolution, Adaptation and Yield.* Cambridge, UK: Cambridge Univ. Press. 500 pp.
33. Evans LT, Visperas RM, Vergara BS. 1984. Morphological and physiological changes among rice varieties used in the Philippines over the last seventy years. *Field Crops Res.* 8:105–24
34. Farrar JF, Lewis DH. 1987. Nutrient relations in biotrophic infections. In *Fungal Infection of Plants*, ed. GF Pegg, PG Ayres, pp. 92–132. Cambridge, UK: Cambridge Univ. Press
35. Ferrandino FJ. 1989. A distribution-free method for estimating the effect of aggregated plant damage on crop yield. *Phytopathology* 79:1229–32
36. Ferris H, Schneider SM, Semenoff MC. 1984. Distributed egg production functions for *Meloidogyne arenaria Meloidogyne* in grape varieties, and consideration of the mechanistic relationship between plant and parasite. *J. Nematol.* 16:178–83
37. Fick GW, Loomis RS, Williams WA. 1975. Sugar beet. In *Crop Physiology: Some Case Histories*, ed. LT Evans, pp. 259–95. Cambridge, UK: Cambridge Univ. Press
38. Fischer RA. 1983. Wheat. In *Potential Productivity of Field Crops under Different Environments*, pp. 129–54. Los Baños: IRRI
39. Fischer RA, Quail KJ. 1990. The effect of major dwarfing genes on yield potential in spring wheats. *Euphytica* 46: 51–56
40. Fitt BDL, Goulds A, Polley RW. 1988. Eyespot (*Pseudocercosporella herpotrichoides*) epidemiology in relation to prediction of disease severity and yield loss in winter wheat—a review. *Plant Pathol.* 37:311–28
41. Forrer HR, Zadoks JC. 1983. Yield reduction in wheat in relation to leaf necrosis caused by *Septoria tritici*. *Neth. J. Plant Pathol.* 89:87–98
42. Gallagher JN, Biscoe PV. 1978. Radiation absorption, growth and yield of cereals. *J. Agric. Sci.* 91:47–60
43. Gaunt RE. 1981. Disease tolerance—an indicator of thresholds. *Phytopathology* 71:915–16
44. Gaunt RE. 1987. Measurement of disease and pathogens. In *Epidemiology and Crop Loss Assessment*, ed. PS Teng, pp. 6–18. Minnesota, MN: APS Press
45. Gaunt RE. 1991. Practical methods for quantifying diseases and pathogen populations. In *Crop Loss Assessment in Rice*, ed. PS Teng, pp. 67–74. IRRI: Manila, Philippines.
46. Gaunt RE, Robertson MJ. 1989. Yield variability and the impact on local and regional estimates of yield reductions caused by disease. In *Spatial Components of Plant Disease Epidemics*, ed. MJ Jeger, pp. 204–22. Englewood, NJ: Prentice Hall. 243 pp.
47. Gaunt RE, Wright AC. 1992. Disease-yield relationship in barley. II. Contribution of stored stem reserves to grain filling. *Plant Pathol.* 41:688–701
48. Goudriaan J, Monteith JL. 1990. A mathematical function for crop growth based on light interception and leaf area expansion. *Ann. Bot.* 66:695–701
49. Gutierrez AP, Curry GL. 1989. Conceptual framework for studying crop-pest systems. In *Integrated Pest Management Systems and Cotton Production*, eds. RE Frisbie, KM El-Zik, LT Wilson, pp. 37–64. New York: Wiley
50. Hang AN, McCloud DE, Boote KJ, Duncan WG. 1984. Shade effects on growth, partitioning and yield components of peanuts. *Crop Sci.* 24:109–15
51. Harry IB, Clarke DD. 1992. The effects of powdery mildew (*Erysiphe fischeri*) infection on the development and function of leaf tissue by *Senecio vulgaris*. *Physiol. Mol. Plant Pathol.* 40:211–24
52. Haverkort AJ, Bicamumpaka M. 1986. Correlation between intercepted radiation and yield of potato crops infested by *Phytophthora infestans* in central Africa. *Neth. J. Plant Pathol.* 92:239–47
53. Haverkort AJ, Rouse DI, Turkensteen LJ. 1990. The influence of *Verticillium dahliae* and drought on potato crop growth. 1. Effects on gas exchange and stomatal behaviour of individual leaves and crop canopies. *Neth. J. Plant Pathol.* 96:273–89
54. Herrera-Reyes CG, Penning de Vries FWT. 1990. Computer simulation of the

potential production of rice. *Int. Rice Res. N.* 15:11–12

55. Hocking PJ, Steer BT. 1994. The distribution and identity of assimilates in tomato with special reference to stem reserves. *Ann. Bot.* 73:315–25
56. Hughes G. 1988. Spatial heterogeneity in crop loss assessment models. *Phytopathology* 78:883–84
57. Hughes G. 1990. Characterising crop responses to patchy pathogen attack. *Plant Pathol.* 39:2–4
58. James WC, Teng PS, Nutter FW Jr. 1991. Estimated losses of crops from plant pathogens. In *CRC Handbook of Pest Management*, 1:15–51. 2nd ed.
59. Johnson KB. 1987. Defoliation, disease, and growth: a reply. *Phytopathology* 77:1495–97
60. Johnson KB. 1992. Evaluation of a mechanistic model that describes potato crop losses caused by multiple pests. *Phytopathology* 82:363–69
61. Johnson KB, Teng PS. 1987. Analysis of potato foliage losses caused by interacting infestations of early blight, *Verticillium wilt*, and potato leafhopper; and the relationship to yield. *J. Plant Dis. Prot.* 94:22–33
62. Johnson KB, Teng PS. 1990. Coupling a disease progress model for early blight to a model of potato growth. *Phytopathology* 80:416–24
63. Klepper O, Rouse DI. 1991. A procedure to reduce parameter uncertainty for complex models by comparison with real system output illustrated on a potato growth model. *Agric. Syst.* 36:375–95
64. Kropff MJ, Mooi J, Goudriaan J, Smeets W, Leemans A, et al. 1989. The effects of long-term open-air fumigation with SO2 on a field crop of broad bean (*Vicia faba* L.). I. Depression of growth and yield. *New Phytol.* 113:337–44
65. Kushalappa AC, Ludwig A. 1982. Calculation of apparent infection rate in plant diseases: development of a method to correct for host growth. *Phytopathology* 72:1373–77
66. Lemcoff JH, Looomis RS. 1986. Crop ecology, production and management. Nitrogen influences on yield determination in maize. *Crop Sci.* 26:1017–22
67. Lim LG, Gaunt RE. 1981. Leaf area as a factor in disease assessment. *J. Agric. Sci.* 97:481–83
68. Lim LG, Gaunt RE. 1986. The effect of powdery mildew (*Erysiphe graminis* f. sp. *hordei*) and leaf rust (*Puccinia hordei*) on spring barley in New Zealand. I. Epidemic development, green leaf area and yield. *Plant Pathol.* 35:44–53
69. Lim LG, Gaunt RE. 1986. The effect of powdery mildew (*Erysiphe graminis* f.sp. *hordei*) and leaf rust (*Puccinia hordei*) on spring barley in New Zealand. II. Apical development and yield potential. *Plant Pathol.* 35:54–60
70. Lipps PE, Madden LV. 1989. Effect of fungicide application timing on control of powdery mildew and grain yield of winter wheat. *Plant Dis.* 73:991–94
71. López-Castañeda C, Richards RA. 1994. Variation in temperate cereals in rainfed environments. II. Phasic development and growth. *Field Crops Res.* 37:63–75
72. Maas SJ. 1988. Using satellite data to improve model estimates of crop yield. *Agron. J.* 80:655–62
73. Madden LV. 1990. Understanding and predicting losses due to virus diseases. In *Virus en Plantas. Rev. Invest. Agropecuar.* 22. 328 pp.
74. Madden LV, Nutter FW Jr. 1995. Modeling crop losses at the field scale. *Can. J. Plant Pathol.* In press
75. Manupeerapan T, Pearson CJ. 1993. Apex size, flowering and grain yield of wheat as affected by sowing date. *Field Crops Res.* 32:41–57
76. McCartney HA, Lacey ME. 1992. Release and dispersal of *Sclerotinia* ascospores in relation to infection. *Brighton Crop Prot. Conf. Pests Dis.* 14:109–16
77. McGrath MT, Pennypacker SP. 1991. Reduction in the rate and duration of grain growth in wheat due to stem rust and leaf rust. *Phytopathology* 81:778–87
78. MacKenzie DR, King E. 1980. Developing realistic crop loss models for plant diseases. In *Crop Loss Assessment. Misc. Publ., Univ. Minn. Agric. Exp. Stn.* ed. PS Teng, SV Krupa, 7:85–89. St. Paul
79. McNew GL. 1960. The nature, origin and evolution of parasitism. In *Plant pathology: An Advanced Treatise*, ed. JG Horsfall, AE Dimond, 2:19–69. Madison, WI: Univ. Wisconsin Press
80. Monteith JL. 1977. Climate and the efficiency of crop production in Britain. *Phil. Trans. R. Soc. London Ser. B* 281: 277–94
81. Muchow RC, Carberry PS. 1990. Phenology and leaf-area development in a tropical grain sorghum. *Field Crops Res.* 23:221–37
82. Mumford JD, Norton GA. 1987. Economics of integrated pest control. In *Crop Loss Assessment and Pest Management*, ed. PS Teng, pp. 191–200. St Paul, Minnesota: APS Press
83. Nutter FW Jr, Teng PS, Royer MH. 1993. Terms and concepts for yield,

crop loss and disease thresholds. *Plant Dis.* 77:211–15
84. O'Brien RD, van Bruggen AHC. 1992. Accuracy, precision and correlation to yield loss of disease severity scales for corky root of lettuce. *Phytopathology* 82:91–96
85. O'Toole JC. 1982. Adaptation of rice to drought-prone environments. In *Drought Resistance in Crops, with Emphasis on Rice*, pp. 195–213. Los Baños: IRRI
86. Pedigo LP, Hutchins SH, Higley LG. 1986. Economic injury levels in theory and practice. *Annu. Rev. Entomol.* 31: 341–68
87. Porter JR, Jamieson PD, Wilson DR. 1993. Comparison of the wheat simulation models AFRCWHEAT2, CERES-Wheat and SWHEAT for non-limiting conditions of crop growth. *Field Crops Res.* 33:131–57
88. Prew RD, Church BM, Dewar AM, Lacey J, Magan N, et al. 1985. Some factors limiting the growth and yield of winter wheat and their variation in two seasons. *J. Agric. Sci. Camb.* 104:135–62
89. Rabbinge R, Jorritsma ITM, Schans J. 1985. Damage components of powdery mildew in winter wheat. *Neth. J. Plant Pathol.* 91:235–47
90. Ring DR, Benedict JH, Landivar JA, Eddleman BR. 1993. Economic injury levels, and development and application of response surfaces relating insect injury, normalised yield, and physiological age. *Env. Entomol.* 22:273–82
91. Rossing WAH. 1991. Simulation of damage in winter wheat caused by the grain aphid *Sitobion avenae*. 2. Construction and evaluation of a simulation model. *Neth. J. Plant Pathol.* 97:25–54
92. Rossing WAH, Daamen RA, Jansen MJW. 1994. Uncertainty analysis applied to supervised control of aphids and brown rust in winter wheat. Pt. 2. Relative importance of different components of uncertainty. *Agric. Sys.* 44: 449–60
93. Rotem J, Bashi E, Kranz J. 1983. Studies of crop loss in potato blight caused by *Phytophthora infestans*. *Plant Dis.* 32: 117–22
94. Rotem J, Kranz J, Bashi E. 1983. Measurement of healthy and diseased haulm area for assessing late blight epidemics in potatoes. *Plant Pathol.* 32:109–15
95. Rouse DI. 1988. Use of crop growth-models to predict the effects of disease. *Annu. Rev. Phytopathol.* 26:183–201
96. Rovira AD, Simon A. 1985. Growth, nutrition and yield of wheat in calcareous sandy loams of South Australia: effect of soil fumigation, fungicide, nematicide, and nitrogen fertilisers. *Soil Biol. Biochem.* 17:279–84
97. Rowe RC, Riedel RM, Martin MJ. 1985. Synergistic interactions between *Verticillium dahliae* and *Pratylenchus penetrans* in potato early dying disease. *Phytopathology*. 75:412–18
98. Savary S. 1987. The effect of age of the groundnut crop on the development of primary gradients of *Puccinia arachidis* foci. *Neth. J. Plant Pathol.* 93:15–24
99. Savary S. 1992. Effect of crop age on primary gradients of late leaf spot (*Cercosporidium personatum*) on groundnut. *Plant Pathol.* 41:265–73
100. Savary S. Zadoks JC. 1992. Analysis of crop loss in the multiple pathosystem groundnut-rust-late leaf spot. II. Study of the interactions between diseases and crop intensification in factorial experiments. *Crop Prot.* 11:110–20
101. Savary S, Zadoks JC. 1992. Analysis of crop loss in the multiple pathosystem groundnut-rust-late leaf spot. III. Correspondence analyses. *Crop Prot.* 11: 229–39
102. Schafer JF. 1971. Tolerance to plant disease. *Annu. Rev. Phytopathol.* 9:235–52
103. Scholes JD, Farrar JF. 1986. Increased rates of photosynthesis in localised regions of a barley leaf infected with brown rust. *New Phytol.* 104:601–12
104. Scholte K, s'Jacob JJ. 1989. Synergistic interactions between *Rhizoctonia solani* Kühn, *Verticillium dahliae* Kleb., *Meloidogyne* spp. and *Pratylenchus neglectus* (Rensch) Chitwood & Oteifa, in potato. *Potato Res.* 32:387–95
105. Shtienberg D, Bergeron SN, Nicholson AG, Fry WE, Ewing EE. 1990. Development and evaluation of a general model for yield loss assessment in potatoes. *Phytopathology* 80:466–72
106. Spitters CJT. 1987. An analysis of variation in yield among potato cultivars in terms of light absorption, light utilization and dry matter partitioning. *Acta Hortic.* 214:71–84
107. Spitters CJT. 1989. Weeds: population dynamics, germination and competition. In *Simulation and systems management in crop protection*, eds. R Rabbinge, SA Ward, HH van Laar, pp. 182–216. Pudoc: Wageningen
108. Stanghellini ME, Stowell LJ, Kronland WC, von Bretzel P. 1983. Distribution of *Pythium aphanidermatum* in rhizosphere soil and factors affecting expres-

sion of the absolute inoculum potential. *Phytopathology* 73:1463–66
109. Stapper M, Fischer RA. 1990. Genotype, sowing date and plant spacing influence on high-yielding irrigated wheat in Southern New South Wales. III. Potential yields and optimum flowering dates. *Aust. J. Agric. Res.* 41:1043–56
110. Tanaka A. 1983. Physiological aspects of productivity in field crops. In *Potential Productivity of Field Crops under Different Environments*, pp. 61–80. Los Baños: IRRI
111. Taheri A, Hollamby GJ, Vanstone VA. 1994. Interaction between root lesion nematode, *Pratylenchus neglectus* (Rensch 1924) Chitwood and Oteifa 1952, and root rotting fungi of wheat. *NZ J. Crop Hortic. Sci.* 22:181–85
112. Teng PS. 1983. Estimating and interpreting disease intensity and loss in commercial fields. *Phytopathology* 73:1587–90
113. Teng PS. 1988. Pests and pest-loss models. *Agrotech. Transfer.* 8:4–10
114. Teng PS, Gaunt RE. 1980. Modelling systems of disease and yield loss in cereals. *Agric. Syst.* 6:131–54
115. Teng PS, Johnson KB. 1988. Analysis of epidemiological components in yield loss assessment. In *Experimental Techniques in Plant Disease Epidemiology*, ed. J Kranz, J Rotem, pp. 179–90
116. Ugalde TD, Jenner CF. 1990. Substrate gradients and regional patterns of dry matter deposition within developing wheat endosperm. I. Carbohydrates. *Aust. J. Plant Physiol.* 17:377–94
117. Van Bruggen AHC, Arneson PA. 1986. Path coefficient analysis of effects of *Rhizoctania solani* on growth and development of dry beans. *Phytopathology* 76:874–78
118. Van Kraalingen DWG, Breure CJ, Spitters CJT. 1989. Simulation of oil palm growth and yield. *Agric. For. Meteorol.* 46:227–44
119. Van Oijen M. 1990. Photosynthesis is not impaired in healthy tissue of blighted potato plants. *Neth. J. Plant Pathol.* 96:55–63
120. Waggoner PE, Berger RD. 1987. Defoliation, disease and growth. *Phytopathology*. 77:393–98
121. Wardlaw IF. 1990. The control of carbon partitioning in plants. *New Phytol.* 116:341–81
122. Wright AC, Gaunt RE. 1992. Disease-yield relationship in barley. I. Yield, dry matter accumulation and yield-loss models. *Plant Pathol.* 41:688–98
123. Yang XB, Dowler WM, Tschanz AT, Wang TC. 1992. Comparing the effects of rust on plot yield, plant yield, yield components and vegetative parts of soybean. *J. Phytopathol.* 136:46–56
124. Zadoks JC. 1985. On the conceptual basis of crop loss assessment: the threshold theory. *Annu. Rev. Phytopathol.* 23:455–73
125. Zawolek MW, Zadoks JC. 1992. Studies in focus development: an optimum for the dual dispersal of plant pathogens. *Phytopathology* 82:1288–96

Annu. Rev. Phytopathol. 1995. 33:145–72

THE SECRET LIFE OF FOLIAR BACTERIAL PATHOGENS ON LEAVES

Gwyn A. Beattie and Steven E. Lindow
Department of Environmental Science, Policy and Management, 108 Hilgard Hall, University of California, Berkeley, California 94720-3110

KEY WORDS: adaptation, epiphytic bacteria, epidemiology, phyllosphere, resident phase

---

ABSTRACT

This review focuses on the role of two distinct fitness strategies in the growth, survival, and epidemiology of foliar bacterial pathogens. A *tolerance* strategy requires the ability to tolerate direct exposure to environmental stresses on leaf surfaces, including UV radiation and low water availability. An *avoidance* strategy requires the ability to seek and/or exploit sites that are protected from these stresses, including endophytic sites. The ability to employ an avoidance strategy and grow endophytically may directly influence the potential for pathogenesis, since endophytic populations, not epiphytic populations, are likely responsible for disease induction. Furthermore, exchange between these two populations is probably crucial to the epidemiology of foliar pathogens. While foliar pathogens can grow and survive in both exposed and internal sites, indicating that they can employ both fitness strategies, the poor internal growth of most saprophytes suggests that saprophytes depend primarily on a strategy of tolerance. This difference between pathogens and saprophytes has important implications for predicting the population dynamics of leaf-associated bacterial species and for selecting effective biological control agents.

INTRODUCTION

It has long been recognized that bacteria are common residents on leaves. Although many of these bacteria can influence plant health under suitable

0066-4286/95/0901-0145$05.00

conditions, others appear to have a strictly commensal relationship with their plant hosts. Because of the economic importance of plant health, most research on leaf-associated bacteria has focused on pathogens, as well as incitants of frost injury, and little attention has been given to the strict commensals. The seminal work of Crosse (35) demonstrated that phytopathogenic bacteria can develop large epiphytic populations on healthy plants. Although the role of these populations in the epidemiology of foliar diseases has been examined (75), the traits that influence the ability of bacteria, including strict commensals, to establish and maintain large populations on leaf surfaces are poorly understood. As early as 1903, it was recognized that bacteria on leaves were distinct from those in the soil (28, 48, 83, 156, 157). This distinction suggests that leaf-associated bacteria possess particular adaptations that allow them to exploit the leaf environment. Knowledge of the nature of these adaptations and their role in the growth and survival of bacteria on and in leaves is critical to an understanding of the ecology of leaf-associated bacteria and the epidemiology of foliar pathogens, as well as to strategies for disease control.

Epiphytic bacteria have been defined as bacteria that are capable of living (i.e. multiplying) on plant surfaces (75, 94). From a functional perspective, epiphytic bacteria are generally considered to be those that can be removed from leaves by washing (75) or killed by UV irradiation or chemical surface disinfection (70), although these methods may underestimate the size of the epiphytic populations (75). In contrast, endophytic bacteria have been defined as bacteria living in the leaf intercellular spaces, substomatal cavities, or vascular tissues (70), and have been functionally defined as those bacteria that remain after removal of the epiphytic bacteria. Endophytic populations are usually enumerated by counting the bacteria in the homogenates of washed or surface-sterilized leaves (70, 75). Unfortunately, the interdependency of these functional definitions precludes independent estimations of epiphytic and endophytic population sizes.

The terms "epiphytic" and "endophytic" may be more accurately viewed as two ends of a spectrum reflecting the growth patterns of leaf-associated bacteria than as two distinct groups of organisms. Foliar pathogens can colonize both the surfaces and the internal regions of leaves (70, 77, 94, 119), and active exchange occurs between the internal and external populations, especially through stomata (12, 60, 111, 120, 140). Leben (93) proposed that when bacteria multiply on the surfaces of apparently healthy plants, they are in the resident, or epiphytic, phase. He subsequently modified this definition to include microflora associated with both the surface and the interior of healthy plants (76). He further proposed that when bacteria are associated with diseased plants, they are in the pathogenic phase (97). Since the phases in a bacterium's life cycle should be associated with distinct physiological states, and since the physiological traits required for bacterial growth and survival in the interior spaces of plants are probably different from those required on the surface, we consider bacteria to be in the *resident*, or *epiphytic, phase* when they are on the surface of host and nonhost plants, as first proposed by Leben (97), and to be in an *invasive phase* when they grow endophytically in host or nonhost plants. As is discussed in this review, both phytopathogenic and saprophytic bacteria exhibit growth on leaf surfaces, indicating that both groups of bacteria can have an epiphytic phase. In contrast, the current evidence suggests that only phytopathogenic bacteria are capable of endophytic growth, indicating that only phytopathogens have an invasive phase in their life cycle. An interesting exception may be saprophytes that can colonize the vascular tissues of leaves (72).

Strategies for epiphytic and endophytic growth are probably both important to the fitness of foliar pathogens associated with leaves. Adaptations that promote tolerance to environmental stresses, such as water stress and exposure to UV radiation, may be critical to the survival and growth of cells in exposed sites on the leaf surface. For example, pigment production, which is common among leaf-surface bacteria, may increase tolerance to UV and visible radiation. Bacteria may also cope with these stresses by seeking and/or exploiting sites in which they are protected from exposure; these sites may be on leaf surfaces, such as in depressions and crevices, and in interior regions, such as substomatal chambers and intercellular spaces. This review focuses on the role of these two strategies, *tolerance* and *avoidance* of environmental stress, in the growth, survival, and epidemiology of foliar pathogens.

LOCATION OF BACTERIA ON LEAF SURFACES

Leaf imprint studies demonstrate that bacteria do not occur in a uniform pattern across leaf surfaces, but are localized in particular sites (98, 100, 171). The nature of these sites has been examined by various microscopic techniques. Although the images provided in such studies are extremely useful, they should be evaluated with caution because bacterial cells may be removed or repositioned, or the microbial habitat altered, during sample preparation. In studies using scanning electron microscopy, the most common sites at which bacteria were observed were at the base of the trichomes (12, 95, 110, 113, 119, 163), at stomata (110, 111, 113, 119, 120, 139, 163), and at the epidermal cell wall junctions (23, 39, 95), especially in the grooves along the veins (98, 110, 113). Although bacteria have been observed on the surface of trichomes (40, 95) and even in collapsed areas of deteriorating trichomes (110), trichome-associated bacteria were usually located solely at the bases of the trichomes. This may be due to an abundance of nutrients exuded from cracks in the cuticular layer or from the large number of ectodesmata located around the base of the trichomes (94). Bacteria were also observed in depressions in the cuticle (110), beneath the cuticle (33), near hydathodes (118), and in structures specific to particular plants, such as the stomatal pits in oleander and the shields of pectate hairs in olive (159). Bacteria were observed in the substomatal chambers in the few studies that examined these chambers. Although bacterial colonization usually occurred under wet conditions in these few studies (12, 60, 159), bacterial colonization of substomatal chambers was observed under relatively dry conditions in at least one study (111). Larger numbers of bacteria were generally found on lower than on upper leaf surfaces (30, 98, 120, 159, 171); this was possibly due to a higher density of stomates and/or trichomes on the lower surface (38, 110, 120, 159), to a thinner cuticular layer on the lower surface, or to reduced exposure to UV radiation.

An organism that enters a leaf through a stomate may have access not only to the intercellular space lying below the stomate, i.e. the substomatal chamber, but to a vast array of interconnected spaces within the mesophyll (e.g. 149). Similarly, bacterial entry through a hydathode leads to a network of intercellular spaces between the loosely packed parenchyma cells under the pore (63). In one study of *Phaseolus vulgaris* leaves, the mesophyll air space was found to increase with leaf maturity, and was approximately 7μl per cm^2 in mature leaves (125). This volume comprised as much as 41% of the total leaf volume, when the total leaf volume was calculated based on an estimated thickness of 15–20μm, 65–70μm, and 75–80μm for the epidermal, palisade parenchyma, and spongy parenchyma layers, respectively, of *P. vulgaris* leaves (134). Other studies have found that the mesophyll air space comprised as little as 23% (179) to as much as 50% (122) of the total leaf volume in various plant species. Thus organisms that successfully invade leaves may have access to a vast array of habitats that offer potentially favorable conditions for growth.

ADAPTATIONS TO TOLERATE ENVIRONMENTAL STRESSES ON LEAF SURFACES

A majority of the bacteria located in surface sites may be exposed to one or more stresses that are common to the leaf surface environment. Changes in water availability may present the most difficult challenge to bacterial survival. Because the microclimate in the boundary layer surrounding a leaf is different from that of the ambient air, it is difficult to predict the amount of water available to the leaf residents based on the water content of the ambient air (27). However, numerous studies have found that epiphytic populations tend to increase in the presence of a water film but tend to decrease under dry conditions when a water film was probably absent (75, 77). As a leaf surface dries, the loss of free water may change the physical environment surrounding the resident bacteria. Specifically, this reduction in free water may cause increases in both the concentration of solutes in the remaining water, resulting in osmolarities that may be sufficiently high to damage the bacteria, and in the concentration of plant-derived antimicrobial compounds. Tolerance to such antimicrobial compounds was discussed in a recent review (17). Reduced water availability, resulting from reductions in both the osmotic potential and the matric potential of the water, can eventually lead to bacterial desiccation. Adaptations conferring tolerance to these stresses may be critical to epiphytic survival.

There is considerable evidence that bacteria are surrounded by a layer of extracellular polysaccharides (EPS) on leaf surfaces. Scanning electron micrographs reveal strands of amorphous material that emanate from and between bacterial cells on leaves (39, 111, 113, 120, 139, 159, 163). These strands probably represent the dehydrated remnants of a more complete matrix that originally surrounded the cells. EPS may anchor cells to the leaf surface (161) and prevent cells from desiccation (175), as well as modify the environment around the cell to one more favorable for growth and survival. This matrix may be analogous to that of biofilms, in which many aquatic microorganisms are found. Biofilms can concentrate nutrients from dilute sources, provide protection from predators, and shield cells from lytic enzymes, antibiotics, and other inhibitory compounds (34). Thus bacterial production of such a matrix on leaves could be highly advantageous for growth and survival.

Tolerance to UV and Visible Radiation

Although wavelengths in the far-UV range (< 300 nm) are known to be quite lethal to bacteria, very little far-UV radiation actually reaches the earth's surface. Thus, tolerance to radiation primarily in the near-UV (300–400 nm) and visible range is likely to be important to bacteria on leaf surfaces. Unfortunately, laboratory studies with leaf-associated bacteria have primarily examined tolerance to far-UV radiation (usually 254 nm). The *recA* gene, which is involved in the repair of DNA damage, has been found to contribute to UV tolerance in this range. Inactivating *recA* in *Pseudomonas syringae*, a common leaf resident, decreased its UV tolerance in culture by several orders of magnitude (174). Similarly, *P. syringae* mutants deficient in the production of a siderophore, a compound known to function as a UV chromophore (164), exhibited decreased UV tolerance in culture (107). On bean leaves, the siderophore-deficient mutant and its parent were equally sensitive to UV radiation; however, the role of the siderophore in UV tolerance remains unclear since the strains may simply have not produced the siderophore under the conditions tested. EPS may also play a role in UV tolerance since crude exudate from *Xanthomonas campestris* pv. *phaseoli* cultures exhibits UV absorptive properties (92).

Another protective mechanism in bacteria is the production of carotenoids and other pigments; these neutralize the highly reactive oxygen derivatives that are generated by visible and possibly near-UV radiation. A majority of the bacteria isolated from leaf surfaces produce pigments in culture (6, 45, 156, 157), which suggests that pigment production may be a common adaptation in these bacteria. When genes encoding pigment production were transferred from the common epiphytic resident *Erwinia herbicola* (*Pantoea agglomerans*) into *Escherichia coli*, the recipient strain showed an increased tolerance to near-UV radiation in culture (167), demonstrating that pigments can confer enhanced UV tolerance. However, pigments may not be required for UV tolerance since a pigment-deficient *E. herbicola* mutant was similar to the parental strain in tolerance to near-UV radiation in culture (59). The proposal that melanin production confers protection from UV and visible radiation in fungi (44) suggests that the production of melanoid-type pigments by some leaf surface bacteria (13) may serve a similar role in bacteria.

Osmotolerance

If solutes are abundant in the free water on a leaf surface, they could become sufficiently concentrated upon drying to stress the resident bacteria. An extensive list of organic and inorganic substances have been identified in leaf leachates (165); however, useful quantitative data of their concentrations on leaf surfaces are not available. Skowland & Lindow (unpublished data) found that osmosensitive mutants of a *P. syringae* strain grew similarly to the parental strain on moist leaf surfaces, but when colonized plants were transferred to dry conditions, the mutants experienced larger population decreases than the parental strain in half of the experiments and similar population decreases in the other half. These results suggest that osmotic conditions on leaf surfaces may be highly variable. In other studies, osmosensitive mutants of *P. syringae* also exhibited reduced survival on drying leaf surfaces (15, 16, 104). Although several of the mutants in these studies showed additional phenotypic alterations, the behavior of the remaining mutants suggests that osmotolerance may be required for bacterial survival on leaf surfaces. While osmotolerance may contribute to epiphytic survival, the fact that pre-exposure of cells to a high osmolarity medium did not increase epiphytic fitness (176) suggests that induction of traits conferring osmotolerance on leaves requires more complex signals than simply high osmoticum.

Matric Stress Tolerance/Desiccation Tolerance

The ability to survive in association with dry leaves is probably one of the defining characteristics of leaf-associated bacteria. Several bacterial genera that are not commonly found on leaf surfaces, including *Aeromonas*, *Escherichia*, and *Salmonella* (123), and two subspecies of *Erwinia carotovora* (24, 130), all grew well on wet leaves but did not survive on dry leaves. The water stress encountered by bacteria on drying leaf surfaces can be quantified as the sum of two components: the osmotic potential, which is due to the interaction of water with molecules that can penetrate a membrane, and the matric potential, which is due to the interaction of water with molecules that cannot penetrate a membrane. Matric potential has been studied primarily in soils, where the availability of water to resident organisms is strongly influenced by the soil texture and structure (127). Similarly, availability of water to leaf residents is probably strongly influenced by the structure and surface characteristics of leaves. Some studies have demonstrated that the stress imposed by a low matric potential has a stronger influence on bacteria than an equivalent osmotic potential (116, 136). Whereas osmoprotection mechanisms include both accumulating solutes, via an influx through the membrane, and synthesizing solutes (36), bacteria can tolerate matric stress only by synthesizing solutes or highly hygroscopic polysaccharides (138, 153); thus, tolerating matric stress requires a greater input of energy. Beattie & Lindow (unpublished data) identified a transposon mutant of *P. syringae* that showed an altered response to matric stress in culture, when assayed by growth in the presence of polyethylene glycol (116), and was reduced in its ability to survive on dry leaf surfaces. However, this mutant also showed an altered response to osmotic stress, indicating that a common mechanistic basis of matric and osmotic stress tolerance may complicate the identification of the quantitative contribution of each to the epiphytic survival of leaf-associated bacteria.

ADAPTATIONS TO AVOID ENVIRONMENTAL STRESSES ON LEAF SURFACES

In 1974, Leben (96) proposed that bacteria may survive on dry leaf surfaces in sites that are protected from exposure to harsh physical conditions. Several lines of evidence support his proposal. First, often only a fraction of the populations that develop on leaves in the absence of free water can be washed off, indicating that a large number of cells either are in sites that prevent removal or are firmly attached to the leaves. For example, several days after inoculation of rifampin-resistant *X. campestris* pv. *vesicatoria* or *P. syringae* pv. *phaseolicola* strains onto bean leaves under dry field conditions, only 1 to 35% of the rifampin-resistant bacteria associated with those leaves were recoverable in the leaf washings (29, 155, 171). Similarly, several days after inoculation of various rifampin-resistant *P. syringae* pathovars onto plants under controlled conditions (low relative humidity), only 20 to 24% of the rifampin-resistant bacteria associated with the leaves were recoverable in leaf washings (123). In these studies, the total number of rifampin-resistant bacteria associated with the leaves was considered to be the sum of those in the leaf washings and those in the homogenates of washed leaves. Second, the proportion of the leaf-associated populations that resist removal by washing is inversely correlated with the availability of water on the leaf surface, which suggests that cells in protected refuges may preferentially survive. Specifically, as leaf surfaces dry, the number of bacteria in the leaf washings decrease (e.g. 15, 68, 104, 163, 176, 177), the proportion of the total leaf-associated population recovered in leaf washings decreases (99, 123, 178), and the populations in the leaf homogenates remains constant or increases (163). These studies indicate not only that a large number of leaf-associated bacteria can resist removal by washing, but also that those cells that can resist removal exhibit superior survival on dry leaf surfaces to those cells that can not.

Bacteria may resist removal by attaching firmly to leaf surfaces or by exploiting and/or seeking refuge in protected sites. In one study, large aggregates of *X. campestris* pv. *translucens* were observed on leaves and few bacteria were recovered in leaf washings (163), suggesting that adhesion prevented release during washing. Most of these bacteria were killed by surface sterilants in this study, which confirmed their presence in exposed sites. However, in most studies surface sterilants did not kill large numbers of leaf-associated bacteria. For example, after various phytopathogens were sprayed onto plants and the plants were incubated under conditions in which free water was absent, between 10^2 and 10^7 cells/g of leaf tissue survived exposure of the leaves to sodium hypochlorite (11, 12, 71, 144, 146). Similar results were observed with topical applications of ethanol (154, 163) and hydrogen peroxide (15; M Wilson, SE Lindow & SS Hirano, unpublished data), and with exposure of leaf surfaces to UV radiation (8, 68, 129, 160; M Wilson, SE Lindow & SS Hirano, unpublished data). Since bacteria that adhere to leaf surfaces should be exposed to these treatments and thus should be killed, the large number of survivors provides evidence that bacteria are able to localize in sites that are not exposed directly to the leaf surface.

The exact nature of these "protected sites" is not known, but they probably consist of both surface locations such as depressions and discontinuities in the cuticle and internal locations such as substomatal cavities and intercellular spaces. Trichomes may be one site favoring bacterial survival, since *P. syringae* pv. *tomato* survived much better on dry leaf surfaces of wild-type plants than of tomato mutants that were deficient in trichomes (144), and *P. syringae* pv. *lachrymans* survived better on the pubescent leaves of potato than on the glabrous leaves of pear (66). For the colonized leaf samples used in the majority of scanning electron micrographs described above, bacterial colonization occurred in the absence of a film of water (exceptions: 12, 159); thus, the sites in which bacterial cells were observed, such as the cell wall junctions and depressions in the cuticle, may have been those sites that serve as refuges on dry leaf surfaces. The recessed nature of these sites could allow them to exclude surface sterilants, to protect resident bacteria by blocking exposure to UV radiation, and to retain water as the rest of the leaf surface dries. Unfortunately, because bacterial quantification using microscopy is very difficult (75), the proportional representation of the total leaf-associated population in these sites is not known.

Bacteria must access internal sites to derive protection from them. Foliar pathogens can invade internal leaf tissues from the surface, since the inoculum for foliar diseases originates on leaf surfaces, not in the internal tissues, and since most, if not all, foliar pathogens induce disease symptoms only when they are in the parenchyma or vascular tissue (21). Numerous reports have implicated surface structures as entry sites for foliar pathogens (3, 81), including open and closed stomata (60, 111, 140), hydathodes (25), cracks in the cuticle (110), and wounds such as broken trichomes (133, 144). These sites may be quite abundant. For example, stomata may occur on abaxial leaf surfaces at frequencies of 40 to 300 per mm^2, and the pores of the open stomata may occupy 0.4 to 2% of the total leaf surface area (122). Several microscopy studies have demonstrated that bacteria that were sprayed onto undamaged leaf surfaces, or were inoculated into guttation droplets from the hydathodes, were observed within 3 to 48 h after inoculation in the leaf intercellular spaces (12, 25, 60, 147, 149, 159), in the substomatal chambers (111), emerging from the stomata (139), or in the xylem vessels (25), confirming that pathogens on the surface have access to internal sites. The relative importance of external sites, i.e. outside the leaf epidermis, and internal sites to bacterial survival is unknown; however, the advantages offered by the intercellular environment for bacterial growth, as described below, suggest that cells that localize there may flourish and become a large proportion of the total leaf-associated population.

Motility and Chemotaxis

Motility contributes to bacterial movement to internal sites, since it contributes to invasion by foliar pathogens (14, 69, 125, 137). Haefele & Lindow (68) demonstrated that motility contributes to bacterial movement to sites that are protected from environmental stresses by comparing the behavior of nonmotile mutants of *P. syringae* on leaves to that of the parental strain. After large populations of each strain had developed on wet leaves, the population size of the nonmotile strain decreased more than that of the parental strain when the plants were transferred to dry conditions. Furthermore, after growth on wet leaves, a larger fraction of nonmotile than motile cells were killed by UV irradiation of leaves, although the two strains were identical in sensitivity to UV irradiation in vitro. Kennedy & Ercolani (88) demonstrated a similar phenomenon with a nonmotile mutant of *P. syringae* pv. *glycinea*. Active acquisition of protected sites is the most likely reason that motile cells survived better than nonmotile cells on plant surfaces. Interestingly, motility may not be important within the intercellular spaces, since nonmotile mutants of *P.*

syringae pv. *phaseolicola* and *X. campestris* pv. *malvacearum* were similar to the parental strains in their ability to systemically invade leaves and in virulence after vacuum-infiltration into leaves (55, 125).

Although motility can contribute to the fitness of leaf-associated bacteria, the contribution of chemotaxis remains unclear. Chemotaxis toward plant extracts has been demonstrated in vitro (31, 37, 90), but to our knowledge only one study has examined it in planta. In this study, Mulrean & Schroth (121) observed that *P. syringae* pv. *phaseolicola* exhibited chemotaxis toward wounds on bean leaves. Although it may be advantageous for a bacterium to move in a directed manner toward sites that favor survival, it may also be advantageous for it to move in a nondirected manner, since it may explore a larger fraction of the leaf surface and increase its chance of encountering a favorable site.

Modification of the Internal Environment

Once a bacterial cell has reached the spaces in the leaf mesophyll, it may require particular adaptations to grow and survive there because the leaf mesophyll spaces provide a habitat for microorganisms that is distinct from that in sites outside of the leaf epidermis. The intercellular environment offers several advantages over the leaf surface for bacterial growth: (*a*) The availability of water is probably more constant than that on the surface (122), (*b*) the concentration of nutrients available to the bacteria may be higher than on the surface, (*c*) the intensity of UV radiation is significantly less than on the surface, and (*d*) the competition among organisms may be relatively low due to the presence of fewer types and numbers of organisms. The intercellular spaces may also present several challenges to bacterial growth and survival, most notably a low pH (4, 64) and exposure to plant defense responses. The increased nutrient concentrations and perhaps increased water availability in the intercellular spaces may result partly from the closer proximity of bacterial cells to plant cells, since the mesophyll cell walls lack a waxy, cuticular layer. These differences between the environments may also result from physiological differences between epidermal cells, which primarily influence the surface environment, and mesophyll cells, which primarily influence the intercellular environment, such as in metabolic processes (19, 80), nutrient content (46, 58, 101), and response to stress (9, 115, 168). These differences suggest that although the intercellular spaces provide refuge from the environmental stresses on the leaf surface, bacteria may require adaptations to survive in and exploit this environment that are distinct from those required on the leaf surface.

Foliar pathogens appear to modify the environment in the leaf intercellular spaces, which may make it more favorable for bacterial multiplication. Pathogens have been commonly observed to increase host cell-membrane permeability (1, 26, 173). This leakage may increase the amount of water and nutrients available to bacteria in the intercellular spaces. Plant growth hormones such as auxins, which have been demonstrated to induce plant-cell-wall loosening and membrane leakiness (102), can be produced by many plant-associated bacteria, including *P. syringae* pv. *savastanoi* and *E. herbicola* pv. *gypsophilae* (e.g. 57, 112). Several studies indicate that the production of at least one auxin, 3-indoleacetic acid, can contribute to large leaf-associated populations; these studies have been discussed in a recent review (17). The phytotoxin syringomycin may also induce leakiness in plant cell membranes (65). Studies with mutants deficient in the production of two other phytotoxins, coronatine and tabtoxin, showed that these phytotoxins were not important for establishing large endophytic populations after bacterial infiltration into leaves, but were important for maintaining them (20, 166). While studies with syringomycin-deficient mutants of *P. syringae* pv. *syringae* showed that syringomycin was not required for establishing large endophytic populations after bacterial infiltration into leaves (128, 181), the role of syringomycin on population maintenance was not evaluated. Lastly, phytopathogens may disrupt the H^+gradient across the plant cell membrane, which results in increases in the pH and the nutrient concentration in the intercellular fluid. Atkinson & Baker (4, 5) demonstrated that the presence of pathogens in both host and nonhost plants is associated with a K^+ efflux/H^+ influx exchange across the plant cell membrane, an increase in the intercellular fluid pH, and an increase in the nutrient levels in the intercellular spaces. In support of this model, they found that mutants unable to induce a K^+ efflux and an H^+ influx were reduced in their ability to multiply in host tissue, and the strength of the exchange response induced by various mutants correlated well with their growth rate in leaves. They further suggest that differences in the rate and degree of the exchange response are responsible for differences in bacterial growth in host versus in nonhost plants.

Another adaptation that may alter the intercellular environment is bacterial production of highly hygroscopic EPS. In the mesophyll spaces, as well as in surface sites, water is probably the most important factor influencing microbial growth. The continuous presence of water on the leaf surface often results in larger epiphytic microbial populations (e.g. 68); similarly, the continued water-soaking of leaves after infiltration with bacteria results in larger endophytic populations than in leaves that were allowed to dry (183). The retention of water in the highly hygroscopic polysaccharide matrix that typically envelopes bacteria in the intercellular spaces increases the water available to the bacteria. For example, infiltration of purified EPS from several phytopathogens into leaves resulted in persistent water-soaking (52). Suleman & Steiner (158) demonstrated that the water potential of a bacterial ooze, containing *Erwinia amylovora* cells and their associated EPS, was much lower than the water potential of plant cells in young apple leaves; this indicates that the presence of a bacterial EPS matrix in the mesophyll spaces could cause water to flow from the plant cells to the mesophyll spaces and thus increase the water available to the bacteria. Similar to an EPS matrix on the leaf surface, the EPS matrix in the leaf intercellular spaces may also serve many other functions (42).

Ability to Suppress, Evade, or Resist Plant Defense Responses

For pathogens to grow in the leaf intercellular spaces they must be able to suppress, evade, or resist host defense responses. Although the interactions between pathogens and their hosts are the subject of much research, relatively little is known of the mechanistic bases for these abilities in foliar pathogens. It is known that the presence of avirulence genes restricts bacterial growth in susceptible host tissues (e.g. 126, 172), although the functions of these genes are largely unknown. Bacterial suppression of plant defense responses has been implicated in the interaction of *P. syringae* pv. *phaseolicola* with its host (61, 82). Bacteria may evade plant defenses by producing EPS, which masks bacterial features critical for recognition, and bacterial resistance may result from the insensitivity or detoxification of plant defense compounds, or resistance to immobilization by the plant, although evidence for these is equivocal. These bacterial-plant interactions have been discussed in several reviews (3, 21, 47, 87); they are not discussed further here.

SURVIVAL STRATEGIES OF PATHOGENS VERSUS SAPROPHYTES

We have described two distinct strategies for surviving the environmental stresses present on leaf surfaces: tolerance and avoidance. Leaf-associated populations probably consist of both cells employing tolerance and cells employing avoidance, and the proportions employing each strategy likely depend on the bacterial genotype and the environmental conditions. For example, if the proportion employing a strategy of tolerance is inferred from the proportion of cells that were washed off leaves or that were killed by surface sterilants, then 1 to 35% of the leaf-associated populations frequently adopted a strategy of tolerance under dry conditions (< 60% relative humidity) (29, 123, 144, 155, 171), and > 99% of the population frequently adopted a strategy of tolerance under wet conditions (15, 147). In these studies, phytopathogenic bacteria were a major component of the phyllosphere community. The observation that pathogens and saprophytes respond differently to changes in environmental conditions (76, 78, 171) suggests that pathogens and saprophytes may employ different strategies for survival.

Pathogens

Pathogens may employ both tolerance and avoidance of environmental stress to grow and survive in association with leaves, whereas saprophytes may rely primarily upon tolerance. Foliar pathogens are clearly capable of a strategy of tolerance, since they can establish populations on exposed leaf surfaces under dry conditions, as evidenced by their sensitivity to surface sterilants (15, 68, 88, 144, 163), their removal by washing (29, 68, 88, 99, 155, 171), and direct observation by microscopy (147, 163). In recent studies, Wilson and coworkers (M Wilson, SE Lindow & SS Hirano, unpublished data; 178) found that pathogenic *P. syringae* strains were superior to nonpathogenic *P. syringae* strains and to strains of the saprophytic species *E. herbicola*, *Methylobacterium organophilum*, and *Stenotrophomonas maltophilia* (previously *Xanthomonas maltophilia*), at maintaining populations on dry leaf surfaces for long periods, at localizing in sites protected from surface sterilization with peroxide, and at growing in the leaf intercellular spaces. These correlations suggest that pathogens may also survive on leaves exposed to dry conditions by localizing in protected sites, whereas saprophytes may be limited in this ability. Furthermore, these refuges probably include internal as well as surface locations, since these researchers (M Wilson, SE Lindow & SS Hirano, unpublished data; 178) found that the population sizes established in sites protected from peroxide were significantly correlated with the population sizes established after infiltration into leaves, and since foliar pathogens applied to leaf surfaces have access to internal sites, described above, and have the ability to establish populations within leaves of susceptible and resistant host plants and nonhost plant species (e.g. 2, 11, 12, 50, 70). In support of these conclusions, several studies have identified mutants of foliar pathogens that were reduced in their abilities to maintain populations on dry leaf surfaces, to seek refuge in sites that were protected from surface sterilants, and to establish populations in leaf intercellular spaces (15, 16, 68).

An interesting implication of our proposal that endophytic growth and survival can contribute to epiphytic survival is that pathogenic ability should be associated with strong epiphytic fitness. Specifically, the ability to enter and multiply inside leaf tissue should contribute to large epiphytic population sizes both because a large proportion of the population can survive changing environmental conditions and because the internal population can egress onto the surface. Because pathogens usually establish larger endophytic populations in susceptible hosts than in resistant hosts (e.g. 12, 29, 38, 43, 143, 155, 180) and nonhosts (2, 11, 54), egress by these larger internal populations onto the surface may explain, at least in part, why epiphytic populations of pathogens are usually larger on susceptible hosts than on resistant hosts or nonhosts (70, 75, 77, 119). Furthermore, the observation that *P. syringae* pv. *syringae* exhibits a broader host range than other *P. syringae* pathovars (145) probably reflects an ability to enter and grow inside a broader range of plant species, and may be causally related to its ability to establish large epiphytic populations on a broad range of plant species (22, 77).

Saprophytes

Saprophytic species are commonly isolated from aerial leaf surfaces. Saprophytic bacteria that have been found in large numbers include *E. herbicola, Pseudomonas fluorescens, Corynebacterium* or *Curtobacterium* spp., *Flavobacterium* spp., and *Methylobacterium* spp., which are often referred to as pink-pigmented, facultative methylotrophs (PPFMs) (6, 33, 45, 49, 62, 156, 157). Unfortunately, because of the economic importance of phytopathogenic bacteria, relatively little research has focused on saprophytic bacteria.

Henis & Bashan (70) stated that "epiphytic bacteria can be either pathogenic or saprophytic, whereas endophytic bacteria......are usually pathogenic;" this statement suggests that saprophytes usually do not grow endophytically. Sharon et al (148) demonstrated that after bacterial inoculation onto leaf surfaces, the saprophyte *P. fluorescens* multiplied only on the surface, whereas the pathogens *X. campestris* pv. *vesicatoria* and *P. syringae* pv. *tomato* multiplied both on the surface and inside the leaves of their host plants. This inability of saprophytes to invade and/or multiply in internal tissues has been exploited as a way to eliminate saprophytes during pathogen enrichment from a mixed population (11, 71, 84, 147). Although the ability of saprophytes to invade internal sites has not been examined, numerous studies have demonstrated that saprophytes do not multiply in the leaf intercellular spaces after infiltration. This has been shown for *P. fluorescens* (73, 89, 150, 182), *E. herbicola* (182; M Wilson, SE Lindow & SS Hirano, unpublished data), *P. putida* (182), *Bacillus cereus* (56), and *M. organophilum* and *S. maltophilia* (M Wilson, SE Lindow & SS Hirano, unpublished data). Thus, although saprophytes that are capable of growth in the vascular tissues of leaves may be an exception (72), in general, epiphytic survival strategies of saprophytes probably do not include localization in internal sites.

Saprophytes may, however, survive by occupying protected sites on the leaf surface. Wilson and coworkers (M Wilson, SE Lindow & SS Hirano, unpublished data; 178) found that small populations of each of several saprophytes survived on leaves treated with a topical application of peroxide. In general, the quantitative contribution of populations in surface sites to the total leaf-associated populations is probably much smaller than that of endophytic populations, since only 0.4 to 2% of the leaf-associated population of various saprophytes survived surface sterilization, whereas as much as 22% of the leaf-associated population of various pathogenic *P. syringae* strains survived this treatment (178). In conclusion, although saprophytes may survive on leaf surfaces by occupying surface sites, they probably rely primarily upon a strategy of tolerance for epiphytic survival.

Comparison of Saprophytes to Pathogens

This difference between pathogens and saprophytes in their strategies for survival and growth has several implications for the two groups of bacteria. First, because of the increased dependence of saprophytes on adaptations conferring tolerance to environmental stress, it is inviting to hypothesize that saprophytes have a greater tolerance to environmental stresses than pathogens. In one of two experiments, Wilson and coworkers (M Wilson, SE Lindow & SS Hirano, unpublished data) observed that the population sizes of several saprophytes decreased slower than did those of several pathogens immediately following exposure of colonized leaves to dry conditions; this slower decrease suggests that saprophytes indeed may be more tolerant to the stresses associated with dry leaf surfaces. To our knowledge, the tolerance of saprophytes and pathogens to various stresses in culture has not been directly compared. Second, pathogens should be better than saprophytes at maintaining consistently large leaf-associated populations during periods of rapidly changing environmental conditions. This is because a larger proportion of the pathogen population than the saprophyte population is probably in refuges, and because cells in exposed sites, even highly stress-tolerant cells, are probably more susceptible to death during changing environmental conditions than are cells in protected sites. Only a few studies have compared the population dynamics of phytopathogens and saprophytes under changing environmental conditions. In two such studies, Manceau et al (109) observed that populations of the saprophytes *E. herbicola* and *P. fluorescens* on pear leaves fluctuated greatly in size during a two-month period, while the population size of *P. syringae* remained constant at around 10^4 cells/g. Similarly, Hirano & Upper (76, 78) observed that while population sizes of the saprophytic PPFMs decreased during the day and increased at night on bean leaves under field conditions, populations of the phytopathogenic species *P. syringae* steadily increased in the time interval studied. More recent studies, however, indicate that the PPFMs may be more consistent than *P. syringae* at maintaining populations throughout the growing season (SS Hirano, personal communication). Third, saprophytes may not be ideal choices for use as biological control agents, since their restricted ability to invade and/or multiply in internal sites could drastically reduce their effectiveness in excluding or displacing populations of a pathogenic organism. For example, Lindow et al (105) found that applications of a non-ice-nucleating *E. herbicola* strain were effective at reducing populations of an ice-nucleating *E. herbicola* strain and frost injury caused by that ice-nucleating strain, but were much less effective at reducing frost injury caused by ice-nucleating *P.*

syringae strains. This reduced effectiveness may have been caused by an inability of the saprophytic *E. herbicola* strain to colonize all of the sites inhabited by phytopathogenic *P. syringae* strains.

There is probably substantial variability among pathogenic and saprophytic strains in their reliance on various strategies for survival and growth. Such variability may result from differing abilities to grow endophytically. For example, Wilson and coworkers (M Wilson, SE Lindow & SS Hirano, unpublished data) found that while saprophytes such as *E. herbicola* and *M. organophilum* did not grow inside leaves, several nonpathogenic *P. syringae* strains exhibited a small, but measurable, amount of growth. Furthermore, the relatively poor epiphytic growth of particular pathogens such as *P. syringae* pv. *phaseolicola* suggests that strong endophytic fitness may, in some cases, occur at the expense of epiphytic fitness.

EPIDEMIOLOGY OF FOLIAR PATHOGENS

Epiphytic Population Sizes in Relation to Disease

Large epiphytic populations have been associated with time of disease onset, and with increased amounts of disease, for foliar diseases caused by several pathogens, including *P. syringae* pv. *coronafaciens* (74), *P. syringae* pv. *glycinea* (117), *P. syringae* pv. *papulans* (18), *P. syringae* pv. *syringae* (103, 141), *P. syringae* pv. *tomato* (152), *E. amylovora* (162), and *X. campestris* pv. *phaseoli* (171), and with frost injury caused by ice-nucleating bacteria (106). Furthermore, a clear quantitative relationship between epiphytic population size and probability of disease occurrence has been established for brown spot disease of beans (103, 141), halo blight of oats (74), and frost injury by ice-nucleating bacteria (106). These quantitative relationships have been the subject of several recent reviews (75, 77, 79).

Although the infection process is poorly understood for most foliar pathogens, as early as 1949 it was "generally accepted that disease symptoms due to bacterial leaf infection are correlated rather closely with bacterial multiplication in the intercellular spaces" (2). Using dose-response curves, Ercolani (53) demonstrated the requirement for large endophytic populations in disease induction, and further showed that infiltration of small numbers of bacteria did not always result in disease. It is still generally accepted that bacteria must reach internal tissues and establish large populations there for infection to be successful. The endophytic populations, not the epiphytic populations, are therefore directly responsible for disease induction. This raises the question of why large epiphytic population sizes are correlated with a high probability of disease occurrence, at least in some foliar diseases. The simplest explanation is that endophytic population sizes increase with increasing epiphytic population sizes. Unfortunately, very few reports have examined both epiphytic and endophytic populations in a given set of leaf samples. In two such studies, the observation that epiphytic and endophytic populations of two phytopathogens exhibited similar dynamics over a 20- to 30-day period on bean leaves under field conditions (29, 155) suggests that the sizes of the two populations can be related. Active exchange between epiphytic and endophytic populations, described above, suggests a mechanistic basis for this relationship. In a model developed by Rouse et al (141) that relates brown spot disease incidence to the size and distribution of epiphytic populations on individual leaves, the ED_{50} value for brown spot disease, i.e. the mean epiphytic population size at which 50% of the leaves exhibit disease, may actually be a measure of the efficiency of ingress of the epiphytic populations. The variability in the ED_{50} values among various fields, although not significant (141), may reflect different rates of ingress under various field conditions.

It is generally accepted that large populations of phytopathogens can develop on leaf surfaces in the absence of disease (22, 70, 75, 77, 94). Thus, large epiphytic populations of a phytopathogen may increase the probability of large endophytic populations, but their presence does not ensure the development of endophytic populations that are sufficiently large to induce disease. A major factor influencing disease development in the presence of large epiphytic populations is probably the amount of ingress, which may depend on the number of entry sites available (114, 135) and the environmental conditions (38). The number of natural entry sites is influenced by host genotype (135), leaf age (89), and position on the leaf surface (38, 120, 135, 159). For example, a high stomatal frequency (135) and a wide stomatal aperture (114) were correlated with host susceptibility. Wounds can also be important entry sites for foliar pathogens, since wounded leaves often exhibit greater incidence and/or severity of disease than do unwounded leaves (e.g. 66, 91, 99, 133, 144, 169). Common wounds include epidermal abrasions caused by the impact of sand or broken trichomes caused by the scraping of leaves against each other. Their occurrence probably depends strongly on environmental conditions (133). For example, disease outbreaks following wind-driven rain (75) may result, at least in part, from increased ingress due to (*a*) the presence of an increased number of entry sites for the pathogen because of wounding, (*b*) greater access to those sites because of the presence of a water film, and (*c*) greater penetration into those sites because of water infiltration. Environmental conditions may influence disease development in other ways, such as influencing the size of the epiphytic populations (75, 77, 79) and the endophytic populations, and the susceptibility of the host (see below).

In conclusion, the major role in disease development of epiphytic pathogen populations is probably as the inocula for endophytic populations and for spread to the surface of other plants and plant parts. As such, they play a critical role in the epidemiology of foliar plant diseases. Recent reviews discuss further the epidemiology and ecology of these populations, including their dispersal to other plants (75, 77, 131, 170).

Endophytic Population Sizes in Relation to Disease

In laboratory studies, disease symptoms are often induced in susceptible hosts when endophytic populations achieve a "threshold" level, usually around 10^6 to 10^7 per cm^2 (30, 38, 43, 150, 155, 171, 180). If the initial endophytic population is sufficiently small, substantial endophytic populations may develop but may not achieve this threshold level (53). If the initial endophytic population is sufficiently large, substantial endophytic populations may develop but may not induce visible symptoms during the period before the threshold population is reached. Although this period is short in most laboratory experiments, due to rapid pathogen growth under the conditions provided, the length of the period is influenced by inoculum concentration (54, 86, 99, 143, 180), environmental conditions (29, 155) and the host genotype (29, 142, 150). Theoretically, this period could be lengthened dramatically if the environmental conditions are not favorable for internal growth. Several studies have found large endophytic populations in symptomless leaves of susceptible hosts under greenhouse and field conditions (10, 29, 146, 147). In a striking case, Bashan et al (11) found that large endophytic populations (10^5 to 10^6 cells/g) of *X. campestris* pv. *vesicatoria* were present in bean leaves for at least two months under field conditions without the appearance of visible symptoms. These results provide evidence that susceptible plants may harbor large endophytic pathogenic populations for long periods without the expression of disease symptoms.

Disease induction probably occurs when either the pathogen populations reach a threshold size or the virulence of the pathogen or the susceptibility of the host changes. Environmental conditions are probably the major influences effecting these changes. Temperature, light intensity, day length, leaf age, and water status of the plant have been shown to influence pathogen multiplication and/or virulence (2, 38, 67, 70, 124). Similarly, light intensity (151) and developmental stage of the host (51) have been shown to influence host susceptibility. Although changes in virulence during pathogen growth in planta have not been demonstrated, it is possible that virulence traits could be induced when the pathogen population achieves particularly high densities, since density-dependent expression of traits such as antibiotic (7) and exoenzyme production (85, 132), and plasmid transfer (184) has been demonstrated in several phytopathogens.

Pathogens can develop endophytic populations in resistant cultivars and in nonhost species, but the size of the populations established in resistant plants is usually, if not always, smaller than those in susceptible cultivars (e.g. 2, 29,

38, 41, 43, 54, 142, 150, 155, 180). In fact, it is tempting to speculate that this limitation to the size and/or spread of endophytic pathogen populations in resistant cultivars is causal to their resistance. These populations, as well as those in susceptible cultivars both before and after symptom development, may serve as inoculum for spread to the leaf surface as well as inoculum for the next season by surviving in dried leaves. Since the size of endophytic populations usually peaks at about the time that symptoms are visible (29, 38, 41, 43, 86, 99, 124, 142, 147, 155, 171, 180, 182), symptom expression, per se, probably does not contribute significantly to increases in the endophytic population size. However, the formation of lesions may enhance the amount of egress that occurs, and by doing so, may increase the spread of the pathogen to the leaf surface. Furthermore, since the dried plant tissue surrounding bacteria in lesions may enhance bacterial tolerance to desiccation, Leben et al (99) proposed that the development of lesions may contribute to the survival of pathogens during airborne dispersal as well as from season to season.

CONCLUSIONS AND FUTURE DIRECTIONS

Plant pathogenic and saprophytic bacteria clearly live in commensal relationships with plants and exhibit different levels of intimacy with the leaves that harbor them. The degree to which a bacterial species can penetrate a leaf probably has a strong influence on both the survival and growth strategies of that species and the response of the plant to that species. In evaluating the behavior of various leaf-associated populations, we have considered epiphytic populations to be those populations that are outside of the leaf epidermis, i.e. those on exposed surfaces and in the cracks and crevices of the cuticle, and endophytic populations to be those in the substomatal chambers, intercellular spaces, and vascular tissues. The terms "epiphyte" and "endophyte" as labels for either groups of bacterial species or bacteria recovered from healthy leaves seems inappropriate since these labels may make misleading inferences as to the location of the organisms on or in leaves, and especially since at least phytopathogens may establish both epiphytic and endophytic populations. Furthermore, since phytopathogens commonly occupy leaf sites that are not strictly on the surface of the plant, we feel the term "phylloplane," which has been used to describe the habitat of such strains, should be replaced with the term "phyllosphere"; however, the term "phylloplane" may be used to describe the surface habitats occupied by saprophytic bacteria.

A major role of epiphytic bacterial populations is as reservoirs for the dispersal of bacteria among plants, from which potentially phytopathogenic bacteria gain entrance into plants and to which cells produced within plants escape. Thus these populations function to increase the likelihood of cells entering plants and initiating interactions with the plant that result in either disease or incompatible interactions. The outcome of the plant-bacterial interaction probably often depends on the endophytic bacterial population size; that is, either cells are sufficiently numerous to cause disease or they are not and they simply maintain populations in asymptomatic plants.

The conceptual model of leaf colonization developed here would be strengthened by more experimental data in several areas relating to bacterial biology and ecology. Specific areas where further study should be fruitful include the following:

1. Better information on the location of bacteria within plants, especially under field conditions. Much could probably be learned from quantitative microscopic studies with inoculated leaves in real-world environmental conditions such as in the presence of rainfall. Such studies may better define the sites within leaves that are accessible to bacteria of epiphytic origin, the extent of the mobility of bacteria within plants after their introduction, and the population sizes that develop within plants before plant responses become obvious. In addition, we need to know more about the quantitative dynamics of population fluxes between the surface and the interior of the plant; for example, is invasion of the leaf interior a rare and environmentally specific phenomenon or is it common?
2. Better information on the population dynamics and behavior of saprophytes that are associated with leaves. Because most studies of leaf-associated bacteria have focused on phytopathogenic species, we have a limited perspective of the range of plant-bacterial interactions possible with bacteria that are living in and on asymptomatic leaves. Upon further study we may find that the interactions of plant pathogens with plants are fundamentally different from those of saprophytes, even on asymptomatic plants.
3. If phytopathogenic bacteria actually exist *within* symptomless plants to the extent suggested from surface sterilization studies, we clearly need a better understanding of how these cells can remain inside plants without inducing overt host responses. What factors limit the multiplication of cells inside plants? Is the host response dependent on sensing a threshold bacterial population? Do cells commonly escape detection by the plant or inhibit plant responses? While such questions are normally considered within the realm of disease physiology, the answers to such questions, especially in the context of the small numbers of cells that would typically be introduced into the plant from the plant surface, are important to our understanding of epiphytic populations as inocula for disease.
4. While the physical environment surrounding leaves can be readily measured, the conditions in the microhabitats in which bacteria reside remain the focus largely of conjecture. It will be difficult to speculate further on the potential behavior of leaf-associated bacteria without a better assess-

ment of the resources available to them. New technologies such as the use of biological sensors based on substrate- or environment-responsive promoters linked to reporter genes (108) should enable us to identify some of the features of bacterial habitats on leaves.

5. The behavior of bacteria on and in leaves is currently limited primarily to projections based on their behavior in culture. It seems likely that bacteria express different genes and thus have different and possibly novel phenotypes while associated with leaves (32, 176). Studies using molecular genetic techniques could unambiguously evaluate the contribution of particular traits to the in planta behavior of bacteria (15, 16, 68). Knowledge of these habitat-specific traits should improve our understanding of the behavior of bacteria living on and in asymptomatic plants.

ACKNOWLEDGMENTS

We are grateful to V Elliot, SS Hirano, and M Wilson for their helpful comments on the manuscript, and to SS Hirano, T Skowland, and M Wilson for sharing unpublished information.

Literature Cited

1. Addy SK. 1976. Leakage of electrolytes and phenols from apple leaves caused by virulent and avirulent strains of *Erwinia amylovora*. *Phytopathology* 66: 1403–5
2. Allington WB, Chamberlain DW. 1949. Trends in the population of pathogenic bacteria within leaf tissues of susceptible and immune plant species. *Phytopathology* 39:656–60
3. Anderson AJ. 1982. Preformed resistance mechanisms. See Ref. 120a, 2: 119–36
4. Atkinson MM, Baker CJ. 1987. Association of host plasma membrane K^+/H^+ exchange with multiplication of *Pseudomonas syringae* pv. *syringae* in *Phaseolus vulgaris*. *Phytopathology* 77: 1273–79
5. Atkinson MM, Baker CJ. 1987. Alteration of plasmalemma sucrose transport in *Phaseolus vulgaris* by *Pseudomonas syringae* pv. *syringae* and its association with K^+/H^+ exchange. *Phytopathology* 77:1573–78
6. Austin B, Goodfellow M, Dickinson CH. 1978. Numerical taxonomy of phylloplane bacteria isolated from *Lolium perenne*. *J. Gen. Microbiol.* 104:139–55
7. Bainton NJ, Bycroft BW, Chhabra SR, Stead P, Gledhill L, et al. 1992. A general role for the *lux* autoinducer in bacterial cell signalling: control of antibiotic biosynthesis in *Erwinia*. *Gene* 116:87–91
8. Barnes EH. 1965. Bacteria on leaf surfaces and in intercellular leaf spaces. *Science* 147:1151–52
9. Barratt DHP, Clark JA. 1993. A stress-induced, developmentally regulated, highly polymorphic protein family in *Pisum sativum* L. *Planta* 191:7–17
10. Bashan Y, Azaizeh M, Diab S, Yunis H, Okon Y. 1985. Crop loss of pepper plants artificially infected with *Xanthomonas campestris* pv. *vesicatoria* in relation to symptom expression. *Crop Prot.* 4:77–84
11. Bashan Y, Diab S, Okon Y. 1982. Survival of *Xanthomonas campestris* pv. *vesicatoria* in pepper seeds and roots in symptomless and dry leaves in non-host plants and in the soil. *Plant Soil* 68:161–70

12. Bashan Y, Sharon E, Okon Y, Henis Y. 1981. Scanning electron and light microscopy of infection and symptom development in tomato leaves infected with *Pseudomonas tomato*. *Physiol. Plant Pathol.* 19:139–44
13. Basu PK. 1974. Glucose inhibition of the characteristic melanoid pigment of *Xanthomonas phaseoli* var. *fuscans*. *Can. J. Bot.* 52:2203–6
14. Bayot RG, Ries SM. 1986. Role of motility in apple blossom infection by *Erwinia amylovora* and studies of fire blight control with attractant and repellent compounds. *Phytopathology* 76: 441–45
15. Beattie GA, Lindow SE. 1994. Survival, growth and localization of epiphytic fitness mutants of *Pseudomonas syringae* on leaves. *Appl. Environ. Microbiol.* 60:3790–98
16. Beattie GA, Lindow SE. 1994. Comparison of the behavior of epiphytic fitness mutants of *Pseudomonas syringae* under controlled and field conditions. *Appl. Environ. Microbiol.* 60: 3799–808
17. Beattie GA, Lindow SE. 1994. Epiphytic fitness of phytopathogenic bacteria: physiological adaptations for growth and survival. In *Bacterial Pathogenesis of Plants and Animals: Molecular and Cellular Mechanisms*, ed. JL Dangl, pp. 1–27. New York: Springer-Verlag
18. Bedford KE, MacNeill BH, Bonn WG, Dirks VA. 1988. Population dynamics of *Pseudomonas syringae* pv. *papulans* on Mutsu apple. *Can. J. Plant Pathol.* 10:23–29
19. Beerhues L, Robenek H, Wiermann R. 1988. Chalcone synthases from spinach (*Spinacia oleracea* L.) II. Immunofluorescence and immunogold localization. *Planta* 173:544–53
20. Bender CL, Stone HE, Sims JJ, Cooksey DA. 1987. Reduced pathogen fitness of *Pseudomonas syringae* pv. *tomato* Tn*5* mutants defective in coronatine production. *Physiol. Mol. Plant Pathol.* 30: 273–83
21. Billing E. 1987. *Bacteria as Plant Pathogens*. Washington, DC: Am. Soc. Microbiol. 79 pp.
22. Blakeman JP. 1982. Phylloplane interactions. See Ref. 120a, 1:307–33
23. Blakeman JP. 1985. Ecological succession of leaf surface microorganisms in relation to biological control. In *Biological Control on the Phylloplane*, ed. CE Windels, SE Lindow, pp. 6–30. St. Paul, MN: Am. Phytopathol. Soc.
24. Blakeman JP. 1991. Foliar bacterial pathogens: epiphytic growth and interactions on leaves. *J. Appl. Bacteriol. Symp. Suppl.* 70:49S-59S
25. Bretschneider KE, Gonella MP, Robeson DJ. 1989. A comparative light and electron microscopical study of compatible and incompatible interactions between *Xanthomonas campestris* pv. *campestris* and cabbage (*Brassica oleracea*). *Physiol. Mol. Plant Pathol.* 34: 285–97
26. Burkowicz A, Goodman RN. 1969. Permeability alterations induced in apple leaves by virulent and avirulent strains of *Erwinia amylovora*. *Phytopathology* 59:314–18
27. Burrage SW. 1976. Aerial microclimate around plant surfaces. In *Microbiology of Aerial Plant Surfaces*, ed. CH Dickinson, TF Preece, pp. 173–84. New York: Academic
28. Burri R. 1903. Die Bacterienvegetation auf Oberfläche normal entwickelter Pflanzen. *Zentralbl. Bakteriol. Parasitenkd. Infektionskr. Hyg. Abt.* 2 10:756–63
29. Cafati CR, Saettler AW. 1980. Effect of host on multiplication and distribution of bean common blight bacteria. *Phytopathology* 70:675–79
30. Cafati CR, Saettler AW. 1980. Role of nonhost species as alternate inoculum sources of *Xanthomonas phaseoli*. *Plant Dis.* 64:194–96
31. Chet I, Zilberstein Y, Henis Y. 1973. Chemotaxis of *Pseudomonas lachrymans* to plant extracts and to water droplets collected from the leaf surfaces of resistant and susceptible plants. *Physiol. Plant Pathol.* 3:473–79
32. Cirvilleri G, Lindow SE. 1994. Differential expression of genes of *Pseudomonas syringae* on leaves and in culture evaluated with random genomic *lux* fusions. *Mol. Ecol.* 3:249–57
33. Corpe WA, Rheem S. 1989. Ecology of the methylotrophic bacteria on living leaf surfaces. *FEMS Microbiol. Ecol.* 62:243–50
34. Costerton JW, Cheng KJ, Geesey GG, Ladd TI, Nickel JC, et al. 1987. Bacterial biofilms in nature and disease. *Annu. Rev. Microbiol.* 41:435–64
35. Crosse JE. 1959. Bacterial canker of stonefruits. IV. Investigation of a method for measuring the inoculum potential of cherry trees. *Ann. Appl. Biol.* 47:306–17
36. Csonka LN, Hanson AD. 1991. Prokaryotic osmoregulation: genetics and physiology. *Annu. Rev. Microbiol.* 45: 569–606
37. Cuppels DA. 1988. Chemotaxis by

Pseudomonas syringae pv. *tomato*. *Appl. Environ. Microbiol.* 54:629–32

38. Daub ME, Hagedorn DJ. 1979. Resistance of *Phaseolus* line WBR 133 to *Pseudomonas syringae*. *Phytopathology* 69:946–51
39. Davis CL, Brlansky RH. 1991. Use of immunogold labelling with scanning electron microscopy to identify phytopathogenic bacteria on leaf surfaces. *Appl. Environ. Microbiol.* 57: 3052–55
40. Deasey MC, Matthysee AG. 1988. Characterization, growth, and scanning electron microscopy of mutants of *Pseudomonas syringae* pv. *phaseolicola* which fail to elicit a hypersensitive response in host and non-host plants. *Physiol. Mol. Plant Pathol.* 33:443–57
41. De Cleene M. 1989. Scanning electron microscopy of the establishment of compatible and incompatible *Xanthomonas campestris* pathovars on the leaf surface of Italian ryegrass and maize. *Bull. OEPP/EPPO Bull.* 19:81–88
42. Denny TP. 1995. Bacterial polysaccharides as determinants of pathogenicity and virulence. *Annu. Rev. Phytopathol.* 33:173–97
43. Diachun S, Troutman J. 1954. Multiplication of *Pseudomonas tabaci* in leaves of burley tobacco, *Nicotiana longiflora*, and hybrids. *Phytopathology* 44:186–87
44. Dickinson CH. 1986. Adaptations of micro-organisms to climatic conditions affecting aerial plant surfaces. In *Microbiology of the Phyllosphere*, ed. NJ Fokkema, J van den Heuvel, pp. 77–100. New York: Cambridge Univ. Press
45. Dickinson CH, Austin B, Goodfellow M. 1975. Quantitative and qualitative studies of phylloplane bacteria from *Lolium perenne*. *J. Gen. Microbiol.* 91: 157–66
46. Dietz KJ, Schramm M, Lang B, Lanzl-Schramm A, Dürr C, Martinoia E. 1992. Characterization of the epidermis from barley primary leaves. II. The role of the epidermis in ion compartmentation. *Planta* 187:431–37
47. Dow JM, Daniels MJ. 1994. Pathogenicity determinants and global regulation of pathogenicity of *Xanthomonas campestris* pv. *campestris*. In *Bacterial Pathogenesis of Plants and Animals: Molecular and Cellular Mechanisms*, ed. JL Dangl pp. 29–41. New York: Springer-Verlag
48. Düggeli M. 1904. Die Bakterienflora gesunder Samen und daraus gezogener Keimpflanzchen. *Zentralbl. Bakteriol. Parasitenk. Infektionskr. Hyg. Abt. 2* 13:198–207
49. Dunleavy JM. 1989. *Curtobacterium plantarum* sp. nov. is ubiquitous in plant leaves and is seed transmitted in soybean and corn. *Int. J. Syst. Bacteriol.* 39:240–49
50. Egel DS, Graham JH, Riley TD. 1991. Population dynamics of strains of *Xanthomonas campestris* differing in aggresiveness on Swingle citrumelo and grapefruit. *Phytopathology* 81:666–71
51. Ekpo EJA, Saettler AW. 1976. Pathogenic variation in *Xanthomonas phaseoli* and *X. phaseoli* var. *fuscans*. *Plant Dis. Rep.* 60:80–83
52. El-Banoby FE, Rudolph K. 1979. Induction of water-soaking in plant leaves by extracellular polysaccharides from phytopathogenic pseudomonads and xanthomonads. *Physiol. Plant Pathol.* 15:341–49
53. Ercolani GL. 1973. Two hypotheses on the aetiology of response of plants to phytopathogenic bacteria. *J. Gen. Microbiol.* 74:83–95
54. Ercolani GL, Crosse JE. 1966. The growth of *Pseudomonas phaseolicola* and related plant pathogens in vivo. *J. Gen. Microbiol.* 45:429–39
55. Essenberg M, Cason ET Jr, Hamilton B, Brinkerhoff LA, Gholson RK, Richardson PE. 1979. Single cell colonies of *Xanthomonas malvacearum* in susceptible and immune cotton leaves and the local resistant response to colonies in immune leaves. *Physiol. Plant Pathol.* 15:53–68
56. Fett WF, Jones SB. 1984. Stress metabolite accumulation, bacterial growth and bacterial immobilization during host and nonhost responses of soybean to bacteria. *Physiol. Plant Pathol.* 25:277–96
57. Fett WF, Osman SF, Dunn MF. 1987. Auxin production by plant-pathogenic pseudomonads and xanthomonads. *Appl. Environ. Microbiol.* 53:1839–45
58. Fricke W, Leigh RA, Tomos AD. 1994. Concentrations of inorganic and organic solutes in extracts from individual epidermal, mesophyll and bundle-sheath cells of barley leaves. *Planta* 192:310–16
59. Gibbins LN, Peterson DL. 1978. Responses of *Erwinia herbicola* Y46 and a non-pigmented mutant to ultraviolet radiation and visible light: survival curves and photoreactivation. In *Proc. 4th Int. Conf. Plant Pathog. Bact.*, 2: 443–50. Angers: Inst. Natl. Rech. Agron.
60. Gitaitis RD, Samuelson DA, Strandberg JO. 1981. Scanning electron microscopy of the ingress and establishment of

Pseudomonas albopreciptans in sweet corn leaves. *Phytopathology* 71:171–75
61. Gnanamanickam SS, Patil SS. 1977. Phaseotoxin suppresses bacterially induced hypersensitive reaction and phytoalexin synthesis in bean cultivars. *Physiol. Plant Pathol.* 10:169–79
62. Goodfellow M, Austin B, Dickinson CH. 1976. Numerical taxonomy of some yellow-pigmented bacteria isolated from plants. *J. Gen. Microbiol.* 97:219–33
63. Goodman RN. 1982. The infection process. See Ref. 120a, 1:31–62
64. Grignon C, Sentenac H. 1991. pH and ionic conditions in the apoplast. *Annu. Rev. Plant Physiol. Plant Mol. Biol.* 42:103–28
65. Gross DC. 1991. Molecular and genetic analysis of toxin production by pathovars of *Pseudomonas syringae*. *Annu. Rev. Phytopathol.* 29:247–78
66. Haas JH, Rotem J. 1976. *Pseudomonas lachrymans* adsorption, survival, and infectivity following precision inoculation of leaves. *Phytopathology* 66:992–97
67. Haas JH, Rotem J. 1976. *Pseudomonas lachrymans* inoculum on infected cucumber leaves subjected to dew- and rain-type wetting. *Phytopathology* 66: 1219–23
68. Haefele DM, Lindow SE. 1987. Flagellar motility confers epiphytic fitness advantages upon *Pseudomonas syringae*. *Appl. Environ. Microbiol.* 53:2528–33
69. Hattermann DR, Ries SM. 1989. Motility of *Pseudomonas syringae* pv. *glycinea* and its role in infection. *Phytopathology* 79:284–89
70. Henis Y, Bashan Y. 1986. Epiphytic survival of bacterial leaf pathogens. In *Microbiology of the Phyllosphere*, ed. NJ Fokkema, J van den Heuvel, pp. 252–68. New York: Cambridge Univ. Press
71. Henis Y, Okon Y, Sharon E, Bashan Y. 1980. Detection of small numbers of phytopathogenic bacteria using the host as an enrichment medium. *J. Appl. Bacteriol.* 49:vi
72. Higley PM, Vidaver A. 1994. Isolation and characterization of bacterial endophytes from prairie plants. *Phytopathology* 84:1134
73. Hildebrand DC, Alosi MC, Schroth MN. 1980. Physical entrapment of pseudomonads in bean leaves by films formed at air-water interfaces. *Phytopathology* 70:98–109
74. Hirano SS, Rouse DI, Arny DC, Nordheim EV, Upper CD. 1981. Epiphytic ice nucleation active (INA) bacterial populations in relation to halo blight incidence in oats. *Phytopathology* 71: 881
75. Hirano SS, Upper CD. 1983. Ecology and epidemiology of foliar bacterial plant pathogens. *Annu. Rev. Phytopathol.* 21:243–69
76. Hirano SS, Upper CD. 1989. Diel variation in population size and ice nucleation activity of *Pseudomonas syringae* on snap bean leaflets. *Appl. Environ. Microbiol.* 55:623–30
77. Hirano SS, Upper CD. 1990. Population biology and epidemiology of *Pseudomonas syringae*. *Annu. Rev. Phytopathol.* 28:155–77
78. Hirano SS, Upper CD. 1991. Bacterial community dynamics. In *Microbial Ecology of Leaves*, ed. JH Andrews, SS Hirano, pp. 271–94. New York: Springer-Verlag
79. Hirano SS, Upper CD. 1994. Autecology of foliar pseudomonads. In *Ecology of Plant Pathogens*, ed. JP Blakeman, B Williamson, pp. 227–43. UK: CAB Int.
80. Hite DRC, Outlaw WH Jr, Tarczynski MC. 1993. Elevated levels of both sucrose-phosphate synthase and sucrose synthase in *Vicia* guard cells indicate cell-specific carbohydrate interconversions. *Plant Physiol.* 101:1217–21
81. Huang JS. 1986. Ultrastructure of bacterial penetration in plants. *Annu. Rev. Phytopathol.* 24:141–57
82. Jakobek JL, Smith JA, Lindgren PB. 1993. Suppression of bean defense responses by *Pseudomonas syringae*. *Plant Cell* 5:57–63
83. Jensen V. 1971. The bacterial flora of beech leaves. In *Ecology of Leaf Surface Micro-organisms*, ed. TF Preece, CH Dickinson, pp. 463–69. New York: Academic
84. Jones JB, Pohonrezny KL, Stall RE, Jones JP. 1986. Survival of *Xanthomonas campestris* pv. *vesicatoria* in Florida on tomato crop residue, weeds, seeds, and volunteer tomato plants. *Phytopathology* 76:430–34
85. Jones S, Yu B, Bainton NJ, Birdsall M, Bycroft BW, et al. 1993. The *lux* autoinducer regulates the production of exoenzyme virulence determinants in *Erwinia carotovora* and *Pseudomonas aeruginosa*. *EMBO J.* 12:2477–82
86. Kawamoto SO, Lorbeer JW. 1972. Multiplication of *Pseudomonas cepacia* in onion leaves. *Phytopathology* 62:1263–65
87. Keen NT, Holliday MJ. 1982. Recognition of bacterial pathogens by plants. See Ref. 120a, 2:179–217
88. Kennedy BW, Ercolani GL. 1978. Soy-

bean primary leaves as a site for epiphytic multiplication of *Pseudomonas glycinea*. *Phytopathology* 68:1196–201
89. Klement Z, Farkas GL, Lovrekovich L. 1964. Hypersensitive reaction induced by phytopathogenic bacteria in the tobacco leaf. *Phytopathology* 54:474–77
90. Klopmeyer MJ, Ries SM. 1987. Motility and chemotaxis of *Erwinia herbicola* and its effect on *Erwinia amylovora*. *Phytopathology* 77:909–14
91. Layne REC. 1967. Foliar trichomes and their importance as infection sites for *Corynebacterium michiganense* on tomato. *Phytopathology* 57:981–85
92. Leach JG, Lilly VG, Wilson HA, Purvis MR Jr. 1957. Bacterial polysaccharides: the nature and function of the exudate produced by *Xanthomonas phaseoli*. *Phytopathology* 47:113–20
93. Leben C. 1961. Microorganisms on cucumber seedlings. *Phytopathology* 51: 553–57
94. Leben C. 1965. Epiphytic microorganisms in relation to plant disease. *Annu. Rev. Phytopathol.* 3:209–30
95. Leben C. 1969. Colonization of soybean buds by bacteria: observations with the scanning electron microscope. *Can. J. Microbiol.* 15:319–20
96. Leben C. 1974. Survival of plant pathogenic bacteria. *Ohio Agric. Res. Dev. Ctr. Spec. Circ. 100 Wooster*. 21 pp.
97. Leben C. 1981. How plant-pathogenic bacteria survive. *Plant Dis.* 65:633–37
98. Leben C. 1988. Relative humidity and the survival of epiphytic bacteria with buds and leaves of cucumber plants. *Phytopathology* 78:179–85
99. Leben C, Daft GC, Schmitthenner AF. 1968. Bacterial blight of soybeans: population levels of *Pseudomonas glycinea* in relation to symptom development. *Phytopathology* 58:1143–46
100. Leben C, Schroth MN, Hildebrand DC. 1970. Colonization and movement of *Pseudomonas syringae* on healthy bean seedlings. *Phytopathology* 60: 677–80
101. Leigh RA, Storey R. 1993. Intercellular compartmentation of ions in barley leaves in relation to potassium nutrition and salinity. *J. Exp. Bot.* 44:755–62
102. Leopold AC, Kriedemann PE. 1975. *Plant growth and development*. New York: McGraw-Hill. 545 pp.
103. Lindemann J, Arny DC, Upper CD. 1984. Epiphytic populations of *Pseudomonas syringae* pv. *syringae* on snap bean and nonhost plants and the incidence of bacterial brown spot disease in relation to cropping patterns. *Phytopathology* 74:1329–33
104. Lindow SE, Andersen G, Beattie GA. 1993. Characteristics of insertional mutants of *Pseudomonas syringae* with reduced epiphytic fitness. *Appl. Environ. Microbiol.* 59:1593–601
105. Lindow SE, Arny DC, Upper CD. 1983. Biological control of frost injury: an isolate of *Erwinia herbicola* antagonistic to ice nucleation active bacteria. *Phytopathology* 73:1097–102
106. Lindow SE, Arny DC, Upper CD, Barchet WR. 1978. The role of bacterial ice nuclei in frost injury to sensitive plants. In *Plant Cold Hardiness and Freezing Stress: Mechanisms and Crop Implications*, ed. PH Li, A Sakai, pp. 249–63. New York: Academic
107. Loper JE, Lindow SE. 1987. Lack of evidence for in situ fluorescent pigment production by *Pseudomonas syringae* pv. *syringae* on bean leaf surfaces. *Phytopathology* 77:1449–54
108. Loper JE, Lindow SE. 1994. A biological sensor for iron available to bacteria in their habitats on plant surfaces. *Appl. Environ. Microbiol.* 60:1934–41
109. Manceau C, Lalande JC, Lachaud G, Chartier R, Paulin JP. 1990. Bacterial colonization of flowers and leaf surface of pear trees. *Acta Hort.* 273:73–81
110. Mansvelt EL, Hattingh MJ. 1987. Scanning electron microscopy of colonization of pear leaves by *Pseudomonas syringae* pv. *syringae*. *Can. J. Bot.* 65: 2517–22
111. Mansvelt EL, Hattingh MJ. 1989. Scanning electron microscopy of invasion of apple leaves and blossoms by *Pseudomonas syringae* pv. *syringae*. *Appl. Environ. Microbiol.* 55:533–38
112. Manulis S, Gafni Y, Clark E, Zutra D, Ophir Y, Barash I. 1991. Identification of a plasmid DNA probe for detection of strains of *Erwinia herbicola* pathogenic on *Gypsophila paniculata*. *Phytopathology* 81:54–57
113. Mariano RLR, McCarter SM. 1993. Epiphytic survival of *Pseudomonas viridiflava* on tomato and selected weed species. *Microb. Ecol.* 26:47–58
114. Matthee FN, Daines RH. 1969. The influence of nutrition on susceptibility of peach foliage to water congestion and infection by *Xanthomonas pruni*. *Phytopathology* 59:285–87
115. Mauch F, Meehl JB, Staehelin LA. 1992. Ethylene-induced chitinase and A-1,3-glucanase accumulate specifically in the lower epidermis and along vascular strands of bean leaves. *Planta* 186:367–75
116. McAneney KJ, Harris RF, Gardner WR. 1982. Bacterial water relations using

polyethylene glycol 4000. *Soil Sci. Soc. Am. J.* 46:542–47

117. Mew TW, Kennedy BW. 1982. Seasonal variation in populations of pathogenic pseudomonads on soybean leaves. *Phytopathology* 72:103–5
118. Mew TW, Mew IC, Huang JS. 1984. Scanning electron microscopy of virulent and avirulent strains of *Xanthomonas campestris* pv. *oryzae* on rice leaves. *Phytopathology* 74:635–41
119. Mew TW, Vera Cruz CM. 1986. Epiphytic colonization of host and non-host plants by phytopathogenic bacteria. In *Microbiology of the Phyllosphere*, ed. NJ Fokkema, J van den Heuvel, pp. 269–82. New York: Cambridge University Press
120. Miles WG, Daines RH, Rue JW. 1977. Presymptomatic egress of *Xanthomonas pruni* from infected peach leaves. *Phytopathology* 67:895–97

120a. Mount MS, Lacy GH, eds. 1982. *Phytopathogenic Prokaryotes,* Vols. 1, 2. New York: Academic

121. Mulrean EN, Schroth MN. 1979. In vitro and in vivo chemotaxis by *Pseudomonas phaseolicola. Phytopathology* 69:1039
122. Noble PS. 1974. *Introduction to Biophysical Plant Physiology.* San Francisco: Freeman. 488 pp.
123. O'Brien RD, Lindow SE. 1989. Effect of plant species and environmental conditions on epiphytic population sizes of *Pseudomonas syringae* and other bacteria. *Phytopathology* 79:619–27
124. Oliveira JR, Romeiro RS, Muchovej JJ. 1991. Population tendencies of *Pseudomonas cichorii* and *P. syringae* pv. *garcae* in young and mature coffee leaves. *J. Phytopathol.* 131:210–14
125. Panopoulos NJ, Schroth MN. 1974. Role of flagellar motility in the invasion of bean leaves by *Pseudomonas phaseolicola. Phytopathology* 64:1389–97
126. Parker JE, Barber CE, Fan M-J, Daniels MJ. 1993. Interaction of *Xanthomonas campestris* with *Arabidopsis thaliana*: characterization of a gene from *X. c.* pv. *raphani* that confers avirulence to most *A. thaliana* accessions. *Mol. Plant-Microbe Interact.* 6:216–24
127. Parr JF, Gardner MR, Elliott LF, eds. 1981. *Water Potential Relations in Soil Microbiology*. Madison, WI: Soil Sci. Soc. Am. 151 pp.
128. Patil SS, Hayward AC, Emmons R. 1974. An ultraviolet-induced nontoxigenic mutant of *Pseudomonas phaseolicola* of altered pathogenicity. *Phytopathology* 64:590–95
129. Pennycook SR, Newhook FJ. 1982. Ultraviolet sterilization in phylloplane studies. *Trans. Br. Mycol. Soc.* 78:360–61
130. Pérombelon MCM. 1978. Contamination of potato crops by air-borne *Erwinia.* In *Proc. 4th Int. Conf. Plant Pathog. Bact.*, 2:563–65. Angers: Inst. Natl. Rech. Agron.
131. Pérombelon MCM. 1981. The ecology of erwinias on aerial plant surfaces. In *Microbial Ecology of the Phylloplane,* ed. JP Blakeman, pp. 411–31. New York: Academic
132. Pirhonen M, Flego D, Heikinheimo R, Palva ET. 1993. A small diffusible signal molecule is responsible for the global control of virulence and exoenzyme production in the plant pathogen *Erwinia carotovora. EMBO J.* 12:2467–76
133. Pohronezny K, Hewitt M, Infante J, Datnoff L. 1992. Wind and wind-generated sand injury as factors in infection of pepper by *Xanthomonas campestris* pv. *vesicatoria. Plant Dis.* 76:1036–39
134. Radoglou KM, Jarvis PG. 1992. The effects of CO_2 enrichment and nutrient supply on growth morphology and anatomy of *Phaseolus vulgaris* L. seedlings. *Ann. Bot.* 70:245–56
135. Ramos LJ, Volin RB. 1987. Role of stomatal opening and frequency on infection of *Lycopersicon* spp. by *Xanthomonas campestris* pv. *vesicatoria. Phytopathology* 77:1311–17
136. Rattray EAS, Prosser JI, Glover LA, Killham K. 1992. Matric potential in relation to survival and activity of a genetically modified microbial inoculum in soil. *Soil Biol. Biochem.* 24:421–25
137. Raymundo AK, Ries SM. 1981. Motility of *Erwinia amylovora. Phytopathology* 71:45–49
138. Roberson EB, Firestone MK. 1992. Relationship between desiccation and exopolysaccharide production in a soil *Pseudomonas* sp. *Appl. Environ. Microbiol.* 58:1284–91
139. Roos IMM, Hattingh MJ. 1983. Scanning electron microscopy of *Pseudomonas syringae* pv. *morsprunorum* on sweet cherry leaves. *Phytopathol. Z.* 108:18–25
140. Roos IMM, Hattingh MJ. 1987. Systemic invasion of plum leaves and shoots by *Pseudomonas syringae* pv. *syringae* introduced into petioles. *Phytopathology* 77:1253–57
141. Rouse DI, Nordheim EV, Hirano SS, Upper CD. 1985. A model relating the probability of foliar disease incidence to the population frequencies of bacte-

rial plant pathogens. *Phytopathology* 75: 505–9

142. Rudolph K. 1984. Multiplication of *Pseudomonas syringae* pv. *phaseolicola* "in planta" I. Relation between bacterial concentration and water-congestion in different bean cultivars and plant species. *Phytopathol. Z.* 111:349–62
143. Scharen AL. 1959. Comparative population trends of *Xanthomonas phaseoli* in susceptible, field tolerant and resistant hosts. *Phytopathology* 49:425–28
144. Schneider RW, Grogan RG. 1977. Tomato leaf trichomes, a habitat for resident populations of *Pseudomonas tomato*. *Phytopathology* 67:898–902
145. Schroth MN, Hildebrand DC, Starr MP. 1981. Phytopathogenic members of the genus *Pseudomonas*. In *The Prokaryotes: A Handbook on Habitats, Isolation, and Identification of Bacteria*, ed. MP Starr, H Stolp, HG Trüper, A Balows, HG Schlegel, 1:701–18. New York: Springer-Verlag
146. Schultz T, Gabrielson RL. 1986. *Xanthomonas campestris* pv. *campestris* in western Washington crucifer seed fields: occurrence and survival. *Phytopathology* 76:1306–9
147. Sharon E, Bashan Y, Okon Y, Henis Y. 1982. Presymptomatic multiplication of *Xanthomonas campestris* pv. *vesicatoria* on the surface of pepper leaves. *Can. J. Bot.* 60:1041–45
148. Sharon E, Okon Y, Bashan Y, Henis Y. 1982. Detached leaf enrichment: a method for detecting small numbers of *Pseudomonas syringae* pv. *tomato* and *Xanthomonas campestris* pv. *vesicatoria* in seed and symptomless leaves of tomato and pepper. *J. Appl. Bacteriol.* 53:371–77
149. Shekhawat GS, Patel PN. 1978. Histology of barley plant and rice leaf infected with *Xanthomonas translucens* f. sp. *hordei*. *Phytopathol. Z.* 93:105–12
150. Smith JJ, Mansfield JW. 1981. Interactions between pseudomonads and leaves of oats, wheat and barley. *Physiol. Plant Pathol.* 18:345–56
151. Smith MA, Kennedy BW. 1970. Effect of light on reactions of soybean to *Pseudomonas glycinea*. *Phytopathology* 60:723–25
152. Smitley DR, McCarter SM. 1982. Spread of *Pseudomonas syringae* pv. *tomato* and role of epiphytic populations and environmental conditions in disease development. *Plant Dis.* 66: 713–17
153. Soroker EF. 1990. *Low water content and low water potential as determinants of microbial fate in soil*. PhD thesis. Univ. Calif., Berkeley. 146 pp.
154. Spurr HW Jr. 1979. Ethanol treatment - a valuable technique for foliar biocontrol studies of plant disease. *Phytopathology* 69:773–76
155. Stadt SJ, Saettler AW. 1981. Effect of host genotype on multiplication of *Pseudomonas phaseolicola*. *Phytopathology* 71:1307–10
156. Stout JD. 1960. Bacteria of soil and pasture leaves at Claudelands Showgrounds. *NZ J. Agric. Res.* 3:413–30
157. Stout JD. 1960. Biological studies of some Tussock-Grassland soils. XV. Bacteria of two cultivated soils. *NZ J. Agric. Res.* 3:214–23
158. Suleman P, Steiner PW. 1994. Relationship between sorbitol and solute potential in apple shoots relative to fire blight symptom development after infection by *Erwinia amylovora*. *Phytopathology* 84:1244–50
159. Surico G. 1993. Scanning electron microscopy of olive and oleander leaves colonized by *Pseudomonas syringae* subsp. *savastanoi*. *J. Phytopathol.* 138: 31–40
160. Sztejnberg A, Blakeman JP. 1973. Ultraviolet-induced changes in populations of epiphytic bacteria on beetroot leaves and their effect on germination of *Botrytis cinerea* spores. *Physiol. Plant Pathol.* 3:443–51
161. Takahashi T, Doke N. 1984. A role of extracellular polysaccharides of *Xanthomonas campestris* pv. *citri* in bacterial adhesion to citrus leaf tissues in preinfectious stage. *Annu. Phytopathol. Soc. Jpn.* 50:565–73
162. Thomson SV, Schroth MN, Moller WJ, Reil WO. 1976. Efficacy of bactericides and saprophytic bacteria in reducing colonization and infection of pear flowers by *Erwinia amylovora*. *Phytopathology* 66:1457–59
163. Timmer LW, Marois JJ, Achor D. 1987. Growth and survival of xanthomonads under conditions nonconducive to disease development. *Phytopathology* 77: 1341–45
164. Torres L, Péréz-Ortiacn JE, Tordera V, Beltrán JP. 1986. Isolation and characterization of an Fe(III)-chelating compound produced by *Pseudomonas syringae*. *Appl. Environ. Microbiol.* 52: 157–60
165. Tukey HB Jr. 1970. The leaching of substances from plants. *Annu. Rev. Plant Physiol.* 21:305–24
166. Turner JG, Taha RR. 1984. Contribution of tabtoxin to the pathogenicity of

Pseudomonas syringae pv. *tabaci*. *Physiol. Plant Pathol.* 25:55–69
167. Tuveson RW, Larson RA, Kagan J. 1988. Role of cloned carotenoid genes expressed in *Escherichia coli* in protecting against inactivation by near-UV light and specific phototoxic molecules. *J. Bacteriol.* 170:4675–80
168. Uknes S, Dincher S, Friedrich L, Negrotto D, Williams S, et al. 1993. Regulation of pathogenesis-related protein-1a gene expression in tobacco. *Plant Cell* 5:159–69
169. Vakili NG. 1967. Importance of wounds in bacterial spot (*Xanthomonas vesicatoria*) of tomatoes in the field. *Phytopathology* 57:1099–103
170. Venette JR. 1982. How bacteria find their hosts. See Ref. 120a, 2:3–30
171. Weller DM, Saettler AW. 1980. Colonization and distribution of *Xanthomonas phaseoli* and *Xanthomonas phaseoli* var. *fuscans* in field-grown navy beans. *Phytopathology* 70:500–6
172. Whalen MC, Innes RW, Bent AF, Staskawicz BJ. 1991. Identification of *Pseudomonas syringae* pathogens of *Arabidopsis* and a bacterial locus determining avirulence on both *Arabidopsis* and soybean. *Plant Cell* 3:49–59
173. Wheeler H, Hanchey P. 1968. Permeability phenomena in plant disease. *Annu. Rev. Phytopathol.* 6:331–50
174. Willis DK, Hrabak EM, Lindow SE, Panopoulos NJ. 1988. Construction and characterization of *Pseudomonas syringae recA* mutant strains. *Mol. Plant-Microbe Interact.* 1:80–86
175. Wilson HA, Lilly VG, Leach JG. 1965. Bacterial polysaccharides IV. Longevity of *Xanthomonas phaseoli* and *Serratia marcescens* in bacterial exudates. *Phytopathology* 55:1135–38
176. Wilson M, Lindow SE. 1993. Effect of phenotypic plasticity on epiphytic survival and colonization by *Pseudomonas syringae*. *Appl. Environ. Microbiol.* 59: 410–16
177. Wilson M, Lindow SE. 1994. Inoculum density-dependent mortality and colonization of the phyllosphere by *Pseudomonas syringae*. *Appl. Environ. Microbiol.* 60:2232–37
178. Wilson M, Lindow SE, Hirano SS. 1991. The proportion of different phyllosphere bacteria in sites on or within bean leaves protected from surface sterilization. *Phytopathology* 81:1222
179. Winter H, Robinson DG, Heldt HW. 1993. Subcellular volumes and metabolite concentrations in barley leaves. *Planta* 191:180–90
180. Wyman JG, VanEtten HD. 1982. Isoflavonoid phytoalexins and nonhypersensitive resistance of beans to *Xanthomonas campestris* pv. *phaseoli*. *Phytopathology* 72:1419–24
181. Xu GW, Gross DC. 1988. Evaluation of the role of syringomycin in plant pathogenesis by using Tn*5* mutants of *Pseudomonas syringae* pv. *syringae* defective in syringomycin production. *Appl. Environ. Microbiol.* 54: 1345–53
182. Young JM. 1974. Development of bacterial populations in vivo in relation to plant pathogenicity. *NZ J. Agric. Res.* 17:105–13
183. Young JM. 1974. Effect of water on bacterial multiplication in plant tissue. *NZ J. Agric. Res.* 17:115–19
184. Zhang L, Murphy PJ, Kerr A, Tate ME. 1993. Agrobacterium conjugation and gene regulation by N-acyl-L-homoserine lactones. *Nature* 362:446–48

Annu. Rev. Phytopathol. 1995. 33:173–97

INVOLVEMENT OF BACTERIAL POLYSACCHARIDES IN PLANT PATHOGENESIS

T. P. Denny

Department of Plant Pathology, University of Georgia, Athens, Georgia 30602-7274

KEY WORDS: extracellular polysaccharide, lipopolysaccharide, cyclic β-(1,2)-glucan, membrane-derived oligosaccharide, virulence

---

ABSTRACT

Virulence of phytopathogenic bacteria is often correlated with their ability to produce extracellular polysaccharides (EPSs). The composition and amount of lipopolysaccharide O-antigens and low molecular weight, cell-associated β-linked glucans may also affect virulence of Gram-negative pathogens. For a few species of *Agrobacterium, Clavibacter, Erwinia, Pseudomonas,* and *Xanthomonas* sufficient biochemical and genetical data has accumulated to permit critical evaluation of possible functions of polysaccharides during pathogenesis. It is clear that EPSs are necessary for several pathogens to cause normal disease symptoms such as water-soaking and wilting. Evidence is accumulating that EPSs and cell-associated polysaccharides also promote colonization and enhance survival of some bacteria within host tissues. Further progress will require the use of thoroughly characterized polysaccharide-minus mutants, "natural" inoculation procedures, and careful monitoring of the fate of bacteria within plants.

INTRODUCTION

Many phytopathogenic bacteria produce large amounts of extracellular polysaccharide (EPS) in culture and in host plants during pathogenesis. The EPS may remain closely associated with the cell as a capsule or be shed as a fluidal slime. EPSs may be composed of either a single sugar (homopolysaccharides)

0066-4286/95/0901-0173$05.00

or complex mixtures of sugars in precise, repeating subunits (heteropolysaccharides). In addition, bacteria make other polysaccharides, like the lipopolysaccharide (LPS) O-antigen and small β-glucans, that contribute to cell envelope structure and function. Bacterial polysaccharides have attracted the attention of phytopathologists because their production in culture is often positively correlated with virulence. This review examines efforts to (*a*) demonstrate causal connections between polysaccharide production and virulence and (*b*) determine the likely roles of polysaccharides during pathogenesis. Discussion of the biochemistry and genetics of polysaccharide biosynthesis, which is available elsewhere (18, 58, 102), is kept to a minimum.

This review focuses on selected species of *Pseudomonas, Clavibacter, Erwinia, Xanthomonas*, and *Agrobacterium*. The chosen pathogens cause either wilts, necroses, or hyperplasias, and are likely to be representative of many other species and pathovars in these genera. This is not to imply that polysaccharides may not be important for other phytopathogenic prokaryotes, just that there is little or no information. The genus *Rhizobium* was excluded because the participation of polysaccharides in symbiotic relations is a topic unto itself, and one that is reviewed more frequently (e.g. Reference 59).

EPSs may provide a selective advantage for bacteria in a variety of settings and, mostly based on their generally hydrophilic and anionic properties, they have been suggested to have multiple functions (102). During saprophytic or epiphytic existence, which most phytopathogens routinely experience, EPSs may protect bacteria from desiccation, concentrate minerals and nutrients, reduce contact with hydrophobic or charged macromolecules, and enhance attachment to surfaces. During pathogenesis EPSs might benefit the pathogen by prolonging water-soaking of host tissues, reducing contact with toxic molecules, and minimizing interaction with plant cells so as to reduce host responses and promote colonization. The cell-associated polysaccharides may also affect how bacteria respond to and interact with plant cells. EPSs also likely have a role in causing disease symptoms (especially wilt), which enhances the perceived virulence of the pathogen but may not alter its multiplication or survival in the host.

Despite their probable importance, there are few cases where conclusive tests have been performed to examine the possible roles that polysaccharides may play for phytopathogenic bacteria. Experimentation has been hampered by the difficulties of working with these large, complex polymers; they are difficult to purify, characterize (in terms of composition, structure, size, etc), or visualize microscopically. Colony morphology is not always a reliable indicator of polysaccharide production (58), and precise quantification often requires special procedures. Because polysaccharides are the product of intricate, often overlapping, biosynthetic pathways that require many genes, mutations may have unknown or unexpected side effects, especially if precursor synthesis or cell-associated polysaccharides are affected. Significant progress has only occurred where the work by multiple research programs has collectively provided sufficient information regarding the biochemical, genetic, and pathogenic processes.

PSEUDOMONAS SOLANACEARUM

Pseudomonas solanacearum causes severe, often lethal, wilting diseases on diverse plants, many of which are economically important (14). Strains of this widely distributed and heterogeneous species vary in host range, physiological traits, and genetic lineage (25). However, essentially all our detailed knowledge about how *P. solanacearum* interacts with plants during pathogenesis comes from three strains GMI1000, K60, and AW (25). *P. solanacearum* is treated apart from *Pseudomonas syringae* because of its distinctly different pathology and its taxonomic separation from the fluorescent pseudomonads. Although closely related to species in the new genus *Burkholderia* (120), this review will continue using *P. solanacearum* because its generic assignment is still uncertain (AC Hayward, personal communication).

P. solanacearum is notable for the copious slime that it produces, which is primarily a high-molecular-weight acidic EPS containing three amino sugars (Table 1); minor, low-molecular-weight fractions (< 15,000 Daltons) contain mostly rhamnose or glucose (76). The same acidic EPS is made by strains GMI1000, K60, and AW when grown on either rich or minimal media, and by GMI1000 in tomato plants (M Schell & A Trigalet, personal communications). The 18-kb *eps* operon encodes at least nine structural genes for the biosynthesis and export of the acidic EPS (24, 48), and the *ops* gene cluster contains at least seven genes, some of which appear to be necessary for nucleotide sugars essential for both LPS and EPS biosynthesis (54). Expression of the *eps* operon is controlled by a complex, environmentally sensitive regulatory network (47).

Role of EPS

In early research, a positive correlation between EPS production in culture (mucoid colonies) and virulence was shown, and solutions of crude cell-free EPS wilted tomato stem cuttings (14). However, spontaneous EPS^- mutants of *P. solanacearum* often are pleiotropic (25), making it uncertain whether loss of EPS is solely responsible for reduced virulence. Transposon inactivation of the *eps* operon to create specific EPS^- mutants demonstrated that the acidic EPS is a necessary wilt-inducing factor of *P. solanacearum*. There is now general agreement that, despite multiplying normally in stems near the site of wound inoculation, *eps* mutants produce little if any acidic EPS in planta and are greatly reduced in virulence on tomato and eggplant (24, 25, 53). However,

Table 1 Extracellular polysaccharides (EPS) produced by selected phytopathogenic bacteria[a]

EPS common name	Components	Relevant characteristics	References
Agrobacterium tumefaciens			
succinoglycan	Glc, Gal (7 : 1)	high MW, structure determined, anionic	15, 43
cellulose	Glc	β-(1,4) linked, crystalline microfibrils, production increased by plant compounds	62, 63
cyclic β-(1,2)-glucan	Glc	3–4 kDa (18–24 GLC), β-(1,2) linked, unbranched, some molecules anionic, production stimulated at low osmolarity, also in periplasm (25–50% of total)	12, 16, 44, 69, 83, 122
Clavibacter michiganensis			
subsp. *insidiosus*			
type A	Fuc, Gal, Glc (2 : 1 : 1)	~5 MDa (aggregates >20 mDa), structure determined, anionic, ≥9 polypeptides closely associated, made in planta	35, 112
type B	Gal, Fuc, Rha, Man, Glc	22 kDa, variable composition, neutral, 15% of total EPS	112
subsp. *michiganensis*			
type A	Fuc, Gal, Glc (2 : 1 : 1)	1–10 MDa, structure determined (same as subsp. *insidiosus*), anionic, 4 peptides closely associated, made in planta	115
type B?	uncertain	~35 kDa, poorly characterized	84
subsp. *sepidonicus*			
type A	Man, Fuc, Gal, Glc	2 kDa and ≥1 MDa (dissociates to 2 kDa), variable composition, anionic	40, 119
type B	Gal, Glc, Rha, Man, Rib, Fuc	1–10 kDa, variable composition, neutral	40, 119

Erwinia amylovora			
levan	Fru	high MW, β-(2,6) linked, low viscosity, synthesized extracellularly from sucrose	37
amylovoran	Gal, GlcA (4 : 1?)	>50 MDa, structure proposed, anionic, viscous, made in planta	95, 99
Erwinia stewartii			
stewartan	Glc, Gal, GlcA (3 : 3 : 1?)	>50 MDa; anionic, viscous	20
Pseudomonas solanacearum			
acidic EPS	GalNAc, GalNAcA, BacNAc[30H But] (1 : 1 : 1)	>0.6 MDa, structure determined, linear, anionic, viscous, nitrogen rich, made in planta [minor low MW (<15 kDa) components]	76
Pseudomonas syringae			
levan	Fru	>6 MDa, β-(2,6) linked, highly branched compact molecule, low viscosity, synthesized extracellularly	30, 55
alginate	ManA, GulA	3-50 kDa, β-1(1,4) linked, linear copolymer, anionic, viscous, variable acetylation (1–18%) and percentage GulA (1–30%), made in planta	28, 30, 77, 89
Xanthomonas campestris			
xanthan	Glc, Man, GlcA (2 : 2 : 1)	high MW, structure determined, rod-like conformation, anionic, viscous, made in planta	102, 121

[a] Abbreviations: Glc, glucose; Gal, galactose; GlcA, glucuronic acid; Fuc, fucose; Fru, fructose; Man, mannose; Rib, ribose; Rha, rhamnose; ManA, mannuronic acid; GulA, guluronic acid; GalNAc, *N*-acetylgalactosamine; GalNAcA, 2-*N*-acetyl-2-deoxy-L-galacturonic acid; BacNAc[30HBut], 2-*N*-acetyl-4-*N*(3-hydroxybutanoyl)-2,4,6-trideoxy-D-glucose; MW, molecular weight; Da, daltons.

besides being difficult to quantify and prone to misinterpretation (110, 111), wilt bioassays indicate only whether a given EPS component is sufficient to account for the wilt symptoms. Preliminary results suggest that acidic EPS might also be important for *P. solanacearum* to infect tomato seedlings via undisturbed roots (24) and both Trigalet (personal communication) and my laboratory are investigating the possibility that the acidic EPS aids multiplication of the pathogen in tomato plants. *ops* mutants are deficient in both LPS and EPS, and they are both less virulent and multiply less well in planta than the wild type (19). Agglutination and cell binding assays suggest that acidic EPS and LPS participate in the attachment (or lack thereof) of *P. solanacearum* to plant cell walls (93), but the importance of such phenomena is not known.

CLAVIBACTER MICHIGANENSIS

Most phytopathogenic *Clavibacter* species systemically infect only a few plant species (116). *C. michiganensis* subsp. *insidiosus* (*Cmi*) causes bacterial wilt of alfalfa and other pasture legumes, *C. michiganensis* subsp. *michiganensis* (*Cmm*) causes bacterial canker of tomato, and *C. michiganensis* subsp. *sepidonicus* (*Cms*) causes ring rot of potato. Symptoms on these various hosts, which are often slow to develop, include decreased vigor, stunting, leaf yellowing, stem or tuber necrosis and wilt. Although wilt is commonly seen as an early symptom only with *Cmm* infections, plants infected by all of these pathogens are probably under constant water stress (9, 110).

In culture, these *C. michiganensis* subsp. make two distinct EPSs: a generally high-molecular-weight, acidic polymer, which I call type A EPS, and a low-molecular-weight, neutral type B EPS (Table 1). *Cmi* and *Cmm* appear to make type A EPSs that have the same tetrasaccharide repeating unit, whereas *Cms* makes a distinctly different type A EPS. Composition of the type B EPSs can vary greatly with the strain and the culture conditions. Four naturally occurring, nonmucoid variants of *Cms* produce a low-molecular-weight form of type A EPS and a type B EPS (40, 119). There are no publications describing the genetics of EPS biosynthesis.

Role of EPS

Strobel and associates initially claimed to have isolated glycopeptide "toxins" from *Cms* and *Cmi* that rapidly damaged plant membranes and, in some cases, could be used to differentiate susceptible from resistant cultivars (see Reference 116). Although this work was influential, much of it could not be repeated by others. It is now clear that these EPSs are not glycopeptides, but are closely associated with polypeptides that copurify with type A EPS. In addition, there is no reliable physiological or histological evidence that adequately purified EPS, or the pathogens themselves, rapidly affect plant membranes during the early stages of pathogenesis (5, 8, 110). Nevertheless, the peptides in impure EPS preparations might have toxic or enzymatic activities that contribute to pathogenesis. This possibility is especially intriguing in the case of *Cmm*, where histological and genetic studies suggest that cell wall–degrading enzymes contribute to disease symptoms (5, 66).

Data from both alfalfa and potato indicate that wilt is due to vascular dysfunction that increases resistance to water flow, especially in petiole and leaf traces (8, 9, 110). The ensuing water stress can damage plant cell membranes in potato leaves within 24 h (8). However, evidence that EPS is responsible is still circumstantial, being based on the findings that type A EPS is produced by *Cmi* and *Cmm* in plants (113, 115), and that both type A and B EPSs produced in culture can wilt plant cuttings (113, 114, 119). The current model is that different sizes of polysaccharides obstruct water flow through the pores of corresponding sizes that occur in pit membranes at various locations in the vascular system (110, 113). The observations that in culture *Cmi* mostly makes type A EPS during stationary phase and that the composition of the EPS does not appear to affect growth of *Cmi* in alfalfa suggest that EPS does not participate in specific recognition or colonization in this system (79, 113).

Recent work with genetic variants, however, raises more questions regarding the involvement of type A and B EPSs in vascular dysfunction than it answers. U.v.-induced mutants of *Cmi* that make less or altered EPSs in culture colonized alfalfa normally (79). Similarly, naturally nonmucoid *Cms* strains colonized and wilted potato plants as fast as mucoid strains (1, 40, 119), suggesting that if any EPS is involved, it is the low-molecular-weight types A and B. This result is particularly interesting, since these EPSs are less than half the size of the smallest polymer shown empirically to induce wilt (111, 113). Unfortunately, the EPS produced in plants was not characterized to determine if high-molecular-weight EPS (aggregates of type A or B?) might be present. In contrast, an avirulent mutant of *Cmm* makes normal type A EPS in culture and colonizes tomato plants (66). Although indicating that neither high numbers of bacteria nor EPS (assuming it is made by the mutant in planta) are by themselves sufficient to induce wilt, it does not mean that EPS is unimportant. Indeed, both chemically induced and naturally nonmucoid strains of *Cmm* were less virulent than mucoid strains, and the naturally nonmucoid strains did not colonize tomato plants (R Eichenlaub, personal communication). Satisfying explanations of these uncertainties will require use of mutants specifically lacking the ability to synthesize one or both of the type A and B EPSs.

ERWINIA AMYLOVORA AND *ERWINIA STEWARTII*

Both *Erwinia amylovora* and *Erwinia stewartii*, causal agents of fire blight of apple and pear (and other rosaceous plants) and Stewart's wilt of corn, respectively, are necrogenic pathogens that can also wilt young stems or plants. The bacteria enter via wounds or, in the case of *E. amylovora*, flowers. *E. amylovora* grows primarily in the cortex, except in young shoots where it colonizes all tissues, whereas *E. stewartii* can move systemically in corn through the xylem. Ultrastructural studies (34) indicating that *E. amylovora* (but not *E. stewartii*) rapidly causes plasmolysis and deterioration of parenchyma cells is supported by the electrolyte leakage observed for wilted *Cotoneaster* shoots (96). Neither pathogen is known to make cell wall–degrading enzymes (11, 92), but both absolutely require the functions encoded by the *hrp* (or *wts*) genes for pathogenicity and eliciting the hypersensitive response (HR) on nonhosts (21, 90, 117).

E. amylovora can make both levan and amylovoran, whereas *E. stewartii* makes only stewartan (Table 1). Levan is synthesized from sucrose by an extracellular levansucrase enzyme (37), which does not require additional energy inputs and releases glucose as a byproduct. Despite its high molecular weight, the levan made by *E. amylovora* has a low viscosity (37), which suggests that it may be highly branched and have other properties like the levan made by *Pseudomonas syringae* (see below). Both amylovoran and stewartan are produced as a capsule, but also appear as slime when an easily utilizable sugar is available. Related gene clusters (*ams* and *cps*) for biosynthesis have been identified, and completely sequenced from *E. amylovora* and partially sequenced from *E. stewartii* (6, 22; K Geider & D Coplin, personal communications). The genes for charging the lipid carrier with galactose, polymerization and export of the polymer, and some of the glycosyltransferases appear to be allelic (6, 20).

Role of EPS

Levan-deficient mutants of *E. amylovora* were created by transposon mutagenesis of the levansucrase gene (which was cloned and sequenced), and several of these mutants were found to make normal amounts of amylovoran (33). Although fully virulent on immature pear slices, when injected into pear seedlings the levansucrase-negative mutants caused necrosis several days slower than the wild-type parent (33). The delay in symptoms might reflect slower spread of bacteria in stem tissue. However, sucrose is most prevalent not in stems but in the flower nectaries (37), an important infection court for *E. amylovora*. Thus, production of levan by *E. amylovora* has been suggested to play a more important role in infection of flowers by retaining moisture to protect bacteria from desiccation, reducing the osmolarity of the extracellular fluids in the nectaries by converting sucrose into levan that has low osmotic activity, and by releasing easily metabolizable glucose (37). Testing these possibilities will require careful studies with a levansucrase-negative mutant under natural conditions.

Amylovoran and stewartan were long thought to be responsible for the wilt symptoms caused by *E. amylovora* and *E. stewartii*. Ultrastructural observations of wound-inoculated apple shoots and corn leaves show that within 48 h many xylem vessels are completely filled with bacteria embedded in a fibrillar matrix, which is probably amylovoran (11, 34, 101). Some of the early work with *E. amylovora* showed that crude amylovoran isolated from oozing apple fruit resulted in rapid death of xylem parenchyma cells when taken up by shoot cuttings (49). However, other preparations of amylovoran did not have these toxic effects (96, 101), but instead significantly reduced the water potential and water conductance in stem cuttings (96). Similar to the example of *C. michiganensis*, these results suggest that toxic compounds may be found associated with wilt-inducing EPS in planta. That amylovoran is not in itself toxic was clearly shown by Sijam & Goodman (94), who found that amylovoran solutions containing 100 mM NaCl were much less viscous and did not cause wilt despite spreading throughout the cutting. The salt did not dissociate the amylovoran into lower molecular weight fractions, suggesting that the capacity of amylovoran to cause wilt is due to its high viscosity and not just its large size.

A correlation has been noted between loss of amylovoran or stewartan production in culture and loss of virulence for both spontaneous and uncharacterized transposon mutants of *E. amylovora* and *E. stewartii* (58). The significance of some of this early work is enhanced by the finding that the spontaneous EPS^- strains used, S and E8, are *galE* and *rcsB* mutants, respectively (3, 68). GalE provides UDP-sugar precursors for EPS and LPS biosynthesis and RcsB is part of a system that regulates expression of the *cps* operon (58, 68). Research with strains having transposon insertions in the *ams* or *cps* biosynthetic clusters showed more clearly that production of amylovoran and stewartan are essential for *E. amylovora* and *E. stewartii* to cause normal disease symptoms (4, 6, 22). Amylovoran-negative mutants are commonly described as avirulent, causing no ooze and little or no necrosis on immature pear fruit and no disease symptoms after inoculation of seedlings (4, 6, 100, 108). Multiplication of amylovoran-negative mutants in fruit tissue was reduced 10 to 100 fold (4, 100), and there appears to be less spreading into areas surrounding the site of inoculation (61). Hence, in the case of these mutants, both symptom expression and bacterial multiplication are reduced in planta.

Amylovoran may play an important role in the colonization of stem and leaf tissues by *E. amylovora*. Ultrastructural studies and recovery of viable cells showed that the RcsB mutant E8 remained localized near the site of inoculation and appeared to grow poorly when inoculated into cut petioles or wounded stems (50, 101). In one case, E8 cells within xylem vessels agglutinated within 24 h and deteriorated within 3 days (50). Enumeration of bacteria after inoculation of leaf midribs of apple or pear with several uncharacterized amylovoran-minus transposon mutants showed that viable cells could not be recovered from the leaves or petioles after 3–6 days (74, 108). These results suggest that during the early stages of pathogenesis amylovoran is essential for survival of *E. amylovora*. One possible mechanism is that the negatively charged amylovoran slime "sequesters" positively charged polymers (87) or toxic plant compounds that could immobilize or kill bacteria. Another hypothesis is that amylovoran might prevent close association between bacteria and plant cells (11, 61), thereby reducing possible recognition and response by the host (42, 68). Notably, amylovoran-minus/LPS$^-$ galE mutants cause an HR-like necrosis when infiltrated into pear leaves and fruit slices, but the wild type and an LPS$^-$ mutant did not (42, 68). Perhaps the absence of amylovoran increases exposure of plant cells to the HR-inducing extracellular and cell-associated harpin$_{Ea}$ (117) produced by *E. amylovora*. At later stages of infection, pressure exerted by hydration of amylovoran and levan may help rupture cell walls and enhance movement of bacteria through cortical tissues (91).

Most *E. stewartii* cps mutants were wilt-negative but grew normally during the first day after infiltration into corn leaf mesophyll; some caused fewer water-soaked lesions that turned necrotic sooner than those induced by the wild-type parent, whereas others were almost avirulent (22). Spontaneous EPS$^-$ mutants also grew almost normally in the xylem near the site of inoculation, but systemic colonization was greatly reduced (11). Thus, although stewartan does not appear to be required for initial multiplication of the pathogen within corn leaves, it may aid movement within the xylem and, due to its hygroscopic nature, enhance water-soaking by retaining water released by damaged plant cells. No ultrastructural evidence was found to suggest that wild-type or EPS$^-$ mutants of *E. stewartii* are immobilized or agglutinated in the mesophyll or xylem vessels of corn (11). How stewartan aids bacterial movement is unclear, but one possibility is that it helps to rupture pit membranes (D Coplin, personal communication).

The discovery that the *cps* and *ams* gene clusters are partially allelic has enabled interspecific complementation tests using the cloned gene clusters to restore EPS synthesis (20). In some cases, *E. stewartii cps/ams* merodiploids produced an amylovoran-like EPS and *E. amylovora* merodiploids produced a stewartan-like EPS. Changes in EPS production were measured by sensitivity to stewartan-dependent bacteriophage, agglutination by an amylovoran-specific lectin, and compositional analysis of depolymerase degradation products (20; D Coplin, personal communication). Surprisingly, amylovoran production restored virulence to *E. stewartii cps* mutants, but stewartan-producing *E. amylovora ams* mutants remained avirulent. These results suggest that in Stewart's wilt the physical properties of the EPS are important, whereas in fire blight a structural aspect of amylovoran is critical.

PSEUDOMONAS SYRINGAE AND *XANTHOMONAS CAMPESTRIS*

Pseudomonas syringae and *Xanthomonas campestris* are necrogenic pathogens that cause leaf spots, blights, and canker diseases on a wide variety of plants; they are subdivided into numerous pathovars, which differ in host and tissue specificity. Most enter through stomates or wounds, multiply in the intercellular spaces within mesophyllic and parenchymatous tissues of leaves and fruit, and cause water-soaked spots that later become limited necrotic lesions. Some xanthomonads, such as *X. campestris* pv. *campestris,* enter via hydathodes and primarily colonize the xylem vessels, breaking out into surrounding tissues later in pathogenesis.

Many *P. syringae* pathovars produce levan in culture when grown on media with excess sucrose (30) (Table 1). The levan made by *P. syringae* pv. *phaseolicola* in culture has an exceptionally low intrinsic viscosity due to a highly branched, compact structure (55, 89), which along with other unusual properties might make it particularly effective in coating bacteria (55). Unexpectedly, only one *P. syringae* strain has been observed to make an appreciable amount of levan in infected leaves (28, 89), suggesting that levan may not have a significant role in pathogenesis. Instead, during growth in plants or in culture with sugars other than sucrose *P. syringae* pathovars usually make alginate slime (30, 89) (Table 1). Alginate can form a gel when mixed with plant pectin (although neither pure compound gels), but the process requires nonphysiological conditions (e.g. pH < 3.5) (70). Many of the alginate biosynthetic genes (*alg*) of *Pseudomonas aeruginosa,* a related bacterium, have been cloned and characterized (58, 65), and DNA homologous to some essential *alg* genes has been found in some strains of *P. syringae* (30; C Bender, personal communication). Surprisingly, there are as yet no well characterized alginate-negative mutants of *P. syringae.*

All *X. campestris* pathovars appear to make xanthan (Table 1), which has a cellulose backbone with trisaccharide side chains that give the polymer a rod-like conformation (58, 102). Xanthan can synergistically gel when mixed with galactomannans from plants (102, 121), but the biological significance of this ability is unknown. At least seven genetic loci are required for production of xanthan (39). The 16-kb *xpsI* region contains 12 or more genes (also called *gumB–H*) specific for biosynthesis of the pentasaccharide repeating unit (39, 58). Three of the other regions are required for production of nucleotide sugars necessary for both EPS and LPS.

Role of EPS in Leaf-Spot Diseases

P. syringae pv. *phaseolicola* and *X. campestris* pv. *vesicatoria* cause typical leaf-spot diseases of bean and tomato/pepper, respectively. When infiltrated into leaves, the bacteria grow equally well in susceptible and resistant host cultivars during the 1–2 days required for differential resistance to be expressed (41, 89), and ultrastructural studies showed that colony development and EPS production are comparable for 12–24 h (13, 61). EPS production began within 1–2 h after infiltration into leaves, and microcolonies soon developed a layer of EPS. Specific attachment or envelopment of bacteria in the resistant cultivars was not observed. The cotton leaf-blight pathogen, *X. campestris* pv. *malvacearum*, produced equal amounts of xanthan per bacterium during the first 48 h in isogenic susceptible and resistant cultivars (81). Thus, during the initial stages of pathogenesis EPS production is thought to (*a*) rapidly modify the microenvironment near these bacteria and reduce their continued close contact with plant cells, but (*b*) neither affect nor be affected by plant processes that eventually result in the HR (61).

Several days after inoculation with leaf-spot pathogens the intercellular spaces of infected tissues commonly become filled with water, resulting in the typical water-soaked symptom. There have been reports, mostly by Rudolph and his students (89), that *P. syringae* EPSs by themselves can induce persistent water-soaking symptoms, in some cases with host-specificity. However, these findings generally could not be repeated (88), and this hypothesis is at odds with current thinking regarding the mechanisms of bacterial pathogenicity and host-specificity (7, 52, 72). Instead, it seems likely that leaves appear water-soaked because the hygroscopic alginate retards evaporation of the water leaked from plant cells. In fact, up to one third of the fresh weight of fully water-soaked leaf tissues can be attributed to hydrated alginate (28, 89). In contrast, a lack of water-soaking in resistant leaves is associated with a virtual absence of alginate (77, 89), and only traces of alginate were recovered from brown spot lesions on bean (caused by compatible strains of *P. syringae* pv. *syringae*) that do not have a water-soaked phase (28).

It is not clear what effect loss of EPS production might have on typical leaf-spot pathogens or the symptoms they cause. Preliminary results suggest a role for alginate in the virulence of *P. syringae* pv. *syringae* strain FF5 on pear (CL Bender & LW Moore, personal communication). When alginate-deficient mutants generated by transposon or chemical mutagenesis were used to inoculate sterile pear plantlets in tissue culture, onset of symptoms was delayed and severity of symptoms was reduced for some of the mutants. Whether altered virulence of these mutants can be attributed solely to the lack of alginate remains to be determined. Extending the use of defined alginate-minus mutants to other *P. syringae* pathovars will be essential for determining whether water-soaking is important for maximal multiplication of bacteria in leaves or for increasing lesion size (88). Kingsley et al (57) observed that *opsX* mutants of *X. campestris* pv. *citrumelo,* which make an altered LPS and 75% less EPS than wild type, died within 3–5 days after injection into host citrus leaves. Because the *opsX* mutants multiplied almost normally and were pathogenic on bean, an alternative host, it appears that these polysaccharides are not essential for pathogenicity but may protect this pathogen from an inhospitable environment within citrus leaves. However, alterations in LPS can affect the composition, structure, and integrity of the outer membrane, which could indirectly affect a bacterium's pathogenic ability in unpredictable ways.

Role of EPS in Vascular Diseases

In black rot of crucifers, caused by *X. campestris* pv. *campestris*, bacteria first multiply and spread within the xylem vessels of leaves. Sutton & Williams (103) found that vascular failure occurs at the point where bacterial numbers are greatest and the vessels completely filled with xanthan and plant-produced compounds. Subsequent disorganization of the vessels is followed by vein blackening, the first visible symptom. The ensuing water stress in distal portions of the leaf apparently results in the secondary symptoms of chlorosis, loss of turgor, and electrolyte leakage. Spontaneous and chemically induced mutants that produce less and/or altered xanthan in culture were generally reduced in virulence, although the intrinsic viscosity of the EPS was more important than the quantity made (85, 103). These results must be interpreted cautiously, however, since in *X. campestris* pv. *campestris* EPS production and the multiple extracellular enzymes are coregulated, and xanthan-deficient mutants are often pleiotropic (23).

Newman et al (73) recently reported that one transposon mutant of *X. campestris* pv. *campestris*, which is likely to be specifically xanthan-minus due to an insertion in the *xpsI* operon (MJ Daniels, personal communication), was much less virulent than the wild type when 10^7 bacteria/ml were infiltrated into susceptible turnip leaves. The mutant grew very little in mesophyll tissue but still elicited almost normal production of β-1,3-glucanase, a plant defense-related gene. Very similar results were seen with another transposon mutant deficient in biosynthesis of the outer core of the LPS but not altered in xanthan production (73; MJ Daniels, personal communication). It may be that xanthan and LPS aid bacterial survival and growth in turnip leaves by masking the presence of the bacterium in the intercellular spaces. However, immediate insertion of large numbers of bacteria into healthy mesophyll tissue is not typical of *X. campestris* pv. *campestris* infections and might lead to abnormal plant reactions. To evaluate better the role of xanthan in black rot disease, multiple xanthan and LPS mutants must be tested using a more natural infection method that leads primarily to xylem colonization.

Role of Cell-Associated Polysaccharides

Although there has been substantial interest in the LPS of *P. syringae* with regard to serotyping this pathogen (e.g. Reference 78), its role in pathogenesis is virtually unstudied. In the case of *P. syringae* pv. *morsprunorum*, loss or alteration of the O-polysaccharide is correlated with reduced virulence on cherry and plum (98). On the other hand, differences in LPS were not detected among races of *P. syringae* pv. *glycinea* with different cultivar specificity or between wild-type *P. syringae* pv. *phaseolicola* and its nonpathogenic *hrp* mutants (2, 29).

The role of the *hrpM* locus in *P. syringae* pv. *syringae* (71) is an example of how altering the structure or function of the bacterial cell envelope, in this case by eliminating the membrane-derived oligosaccharides (MDO), can indirectly affect pathogenesis. The MDO produced by *E. coli* are small (6–12 sugar residues), branched glucans (β-(1,2)-and β-(1,6)-linkages) that carry anionic substituents (56). MDO are likely to be made by many gram-negative bacteria. Growth of *E. coli* in low-osmotic-strength medium (e.g. 65–80 mosM) stimulates synthesis and accumulation of MDO in the periplasmic space, which should increase the osmotic strength of this compartment and help protect the cytoplasmic membrane (56). Loss of MDO in *E. coli* results in reduced flagellation, enhanced EPS production, and altered cell envelope structure or function (31, 46). Although multiplication is not affected at osmotic strengths as low as 65 mosM, other research suggests that growth might be reduced at < 30 mosM (17, 56, 122).

Originally selected because it is nonpathogenic, the *hrpM* mutant PS9021 of *P. syringae* pv. *syringae* multiplies poorly in bean leaves and does not elicit the HR on nonhost tobacco (71). The *hrpM* locus is not part of the conserved *hrp* cluster in this strain, and the fluidal colony morphology of PS9021 suggested alterations in surface-associated or extracellular products (71). Sequence analysis of *hrpM* originally did not reveal likely functions for the two putative genes. However, recent analysis of the *mdoA* locus in *E. coli*, which is essential for MDO biosynthesis, revealed two genes (*mdoG* and *mdoH*) that are about 70% identical to those at *hrpM* (60). The wild-type parent of PS9021 makes increased amounts of periplasmic MDO-like glucans in response to reduced osmolarity (107). Complementation studies indicate that the cloned *hrpM* locus restores MDO synthesis to *mdoA* mutants of *E. coli*, and the *mdoA* locus restores virulence to MDO-minus *hrpM* mutants of *P. syringae* pv. *syringae* (D Mills, personal communication), implying that the phenotype of *hrpM* mutants results from loss of MDO biosynthesis.

If MDOs function in *P. syringae* as in *E. coli*, then they are likely to be very important in hypo-osmotic environments, such as might be present in plant intercellular spaces during the early stages of pathogenesis (36). Thus, loss of pathogenicity and HR-induction by *hrpM* (*mdo*) mutants may be the indirect consequence of poor osmoadaptation in planta that results in cell envelope dysfunction. Of the possible consequences, disruption of the essential membrane-associated export machinery encoded by the conserved *hrp* region

(90) would have the greatest impact on pathogen-plant interactions. Further research along these lines may provide new insights into processes that underlie pathogenicity and the HR.

AGROBACTERIUM TUMEFACIENS

Agrobacterium tumefaciens biovars 1 and 2 cause crown gall disease on a variety of dicotyledonous plants. Tumorigenesis begins with invasion of fresh wounds where, as an essential early step, the bacteria attach to plant cells. In response to flavanoids and sugars released by wounded plant cells, the bacteria transfer the T-DNA portion of the tumorigenic Ti plasmid (pTi) into the plant genome using a process similar to bacterial conjugation. Transformed cells autonomously overproduce plant hormones, resulting in tumorous overgrowths. The numerous molecular details of this process can be found in one of the many review articles (e.g. Reference 45). The role of attachment in tumorigenesis has been studied using both tumor-inhibition assays and direct-binding assays (see Reference 62 for more discussion). Direct assessment of binding, whether microscopically or by enumeration of bound bacteria recovered from plant cells, should more accurately determine the mechanisms and cell constituents necessary for attachment. However, many binding assays use suspension culture cells, which generally are not transformed by *A. tumefaciens*, so that they may not reliably report on attachment processes that lead to DNA transfer and tumorigenesis (82).

A. tumefaciens produces a variety of EPSs that could affect its interaction with host plants (Table 1). Potentially important cell-associated polysaccharides are the LPSs, which in biovar 1 and 2 strains generally has up to 20 repeating units in the O-chains (smooth LPS) (67, 118), and periplasmic cyclic β-(1,2)-glucans (16, 69). Many aspects of cyclic β-(1,2)-glucans, which are produced almost exclusively by *Agrobacterium* and *Rhizobium* species, were recently reviewed (12). Under certain conditions, they can comprise up to 10% of the total EPS and 5 to 20% of total cellular dry weight (12, 83). Cyclic β-(1,2)-glucans are structurally similar to the MDO produced by *E. coli* (see above), and since their production likewise is stimulated by low osmotic conditions (69, 122), they also could function in osmoprotection.

Role of EPS

Virulent strains of *A. tumefaciens* normally produce large amounts of succinoglycan slime, but this EPS does not seem to be essential for pathogenesis. Transposon mutants lacking the succinoglycan, but still making cyclic β-(1,2)-glucans, remained fully virulent (15). Although not well characterized at that time, some of the mutants likely were deficient only in succinoglycan production because they were complemented by cloned genes from *R. meliloti* that are now known to encode enzymes specifically for EPS biosynthesis (59). *exoC* mutants, which do not attach to plant cells and are avirulent, lack both succinoglycan and cyclic β-(1,2)-glucans due to inactivation of a phosphoglucomutase necessary for synthesis of UDP-glucose, an essential precursor for a variety of carbohydrate polymers (109). It is the absence of the cyclic β-(1,2)-glucans (see below) that likely explains why *exoC* mutants are avirulent.

All *A. tumefaciens* strains studied to date produce crystalline cellulose fibrils in culture and during infection. Following their initially weak binding, the bacteria appear to be firmly anchored to plant cell surfaces by the fibrils, which also create large bacterial aggregates by trapping other *A. tumefaciens* cells. Nevertheless, cellulose production is not essential or sufficient for pathogenesis, since cellulose-minus mutants still bind to plant cells and cause tumors (64), and attachment-defective *chvA/B* mutants still produce cellulose fibrils (26). However, the firm attachment afforded by cellulose production may increase infection efficiency in dynamic natural settings (62). Aggregation of the *A. tumefaciens* cells may also facilitate cell density–responsive (quorum-sensing) activities (32).

Role of Cell-Associated Polysaccharides

Early work by the Lippincotts (reviewed in Refs. 62, 82) suggested that LPS is involved in attachment of *A. tumefaciens* to plant cells. However, subsequent work in several laboratories has generally not supported the hypothesis that LPS plays a direct role in attachment (64, 82). Mutants specifically defective in production of LPS would be very helpful, but they are not yet available. However, *A. tumefaciens* mutants that produce reduced amounts of core LPS and overproduce O-chain polysaccharides attach normally to carrot suspension culture cells (67), suggesting that LPS structure is not critical. In addition, the observation that, unlike *A. tumefaciens* strains, the few *A. vitis* and *A. rubi* strains examined make LPS essentially lacking the O-chain polysaccharide (rough LPS) (118), suggests that LPS is not essential for tumorigenesis. It is interesting to speculate that production of rough LPS might contribute to either the limited host range of *A. vitis* and *A. rubi* or to their ability to move systemically within their hosts.

The recent discovery that *A. tumefaciens* makes an additional cell-associated polysaccharide, which contains 3-deoxy-D-*manno*-2-octulosonic acid (Kdo) (BL Ruehs, personal communication), complicates interpretation of the LPS literature. This Kdo-rich polysaccharide was first found in hot phenol-water extracts of cells of *Rhizobium fredii* and *R. meliloti* along with lesser amounts of LPS; it is, however, distinct from the LPS (86). The Kdo-rich polysaccharide consists of linear repeating subunits of galactose and Kdo, which makes it structurally analogous to some *E. coli* capsular K antigens (EPS group II) (58,

86), and is thought to be located on the cell surface (80). Current evidence suggests that the Kdo-rich polysaccharide can be important for nodulation of one *R. meliloti* strain (80). Thus this new polysaccharide may also have a role during tumorigenesis by *A. tumefaciens*.

Most research into the mechanism of attachment by *A. tumefaciens* has focused on the role of the β-(1,2)-glucans. Mutants in which the chromosomal *chvA* and *chvB* genes are inactivated do not bind to plant cells, lack extracellular β-(1,2)-glucans, and have altered cell surface properties (12, 27, 83). Extensive work with the *chvA* and *chvB* genes of *A. tumefaciens* has demonstrated that they are required for biosynthesis and export of cyclic β-(1,2)-glucans (12). *chvA* mutants synthesize neutral cyclic β-(1,2)-glucans that remain almost entirely within the cytoplasm (16, 51, 75). ChvB is essential for biosynthesis of cyclic β-(1,2)-glucans from UDP-glucose (123). *chvB* mutants are pleiotropic when grown in low-osmotic-strength media (Table 2), and most, if not all, of the phenotype can be attributed to changes in the cell envelope. That growth in high-osmotic-strength media suppresses most of the affected traits supports the hypothesis that cyclic β-(1,2)-glucans provide osmotic protection when bacteria are subjected to hypo-osmotic conditions.

The role of cyclic β-(1,2)-glucans in attachment is less obvious. The simple experiment of adding purified β-(1,2)-glucans to *chvA* or *chvB* mutants does

Table 2 Phenotype of *chvB* mutants grown at low or high osmolarity

Phenotype at low osmolarity[a]	Suppressed by high osmolarity[b]	References
Cyclic β-glucans not synthesized	no	83, 104
Loss of tumorigenicity	no[e]	17, 26, 83, 106
Reduced attachment to plant cells	no[e]	26, 105, 106
Reduced flagellation	yes	10, 104
Reduced motility/chemotaxis	yes	10, 17, 104
Increased phage resistance[c]	NT	10, 17
Reduced growth rate[d]	yes	17, 122
Altered membrane protein composition	yes	17, 104, 122
Reduced tolerance of Ca^{2+} (>2 mM)	yes	104
Active rhicadhesin not produced	no[e]	106

[a] The mutant phenotype observed in cells grown in low-osmotic-strength media (e.g. about 80 mosM).

[b] Growth in media with osmotic strength of 150 mosM or higher (e.g. >0.1 M sucrose). NT, not tested.

[c] Sensitivity to the typing phage GS2 and GS6 is greatly reduced because they bind to flagella (10). Resistance to other typing phage was unaffected (26).

[d] Observed only in extremely low-osmotic-strength media (e.g. <30 mosM).

[e] Restored if 7 mM $CaCl_2$ is also added (105).

not restore attachment or virulence, suggesting that the extracellular β-(1,2)-glucans do not directly mediate these processes (16, 75, 106). It therefore seems as though a cyclic β-(1,2)-glucan deficiency indirectly reduces binding, either because of improper osmoadaptation or other changes in structure or function of the cell envelope (12). However, neither attachment nor virulence of a *chvB* mutant was restored simply by increasing the osmotic strength of the medium or the inoculation site (16, 105). In contrast, a *chvB* mutant grown in media containing 100 mM NaCl plus 7 mM $CaCl_2$ attached normally to pea root hairs and was virulent on leaves of *Kalanchoe daigremontiana* (105); $CaCl_2$ alone did not have this effect.

An explanation for the role of calcium in suppressing the *chvB* mutation may lie in the earlier finding (106) that attachment of wild-type *Agrobacterium* and *Rhizobium* species to pea root hairs requires production of "active" rhicadhesin, a calcium-binding protein found on the surface of the bacteria (97). *chvB* mutants appear to make an "inactive" rhicadhesin (106), although the biochemical basis for this protein's lack of activity is unknown. Pretreating pea roots or wound sites on *K. daigremontiana* leaves with active rhicadhesin partially restored attachment and virulence of a *chvB* mutant, whereas inactive rhicadhesin had no effect (106). Thus the failure of high-osmotic-strength media to restore virulence appears to be due to the lack of calcium necessary for synthesis of active rhicadhesin. In low-osmotic-strength media, however, the requisite concentration of calcium is toxic to *chvB* mutants, probably because they lack the cyclic β-(1,2)-glucans necessary for the osmoadaptation required to maintain normal structure and function of the cell envelope.

Although appealing, several aspects of this model remain uncertain. It has not been explained how the absence of calcium results in inactive rhicadhesin being made by the *chvB* mutant. Given that calcium originally was proposed to bind rhicadhesin to the bacterial cell (97), it is surprising that any rhicadhesin protein could be recovered from the *chvB* mutants. In addition, although wild-type *A. tumefaciens* grown without calcium (i.e. ≤ 0.15 mM) did not attach to pea root hairs (104, 105), bacteria grown in media that similarly lack calcium attach to other types of plant cells and induce tumors (e.g. Refs. 27, 67, 104, 106). In fact, cells grown without calcium and treated with 1 mM EDTA attached to carrot suspension culture cells (38). The role of the cyclic β-(1,2)-glucans, calcium, and rhicadhesin should be clarified when mutants specifically lacking the rhicadhesin protein are studied.

CONCLUSIONS AND FUTURE DIRECTIONS

Significant progress has been made in characterizing polysaccharides produced by phytopathogenic bacteria and in understanding their possible involvement in pathogenesis and symptom development. It is now clear that EPSs are necessary for several pathogens to cause normal disease symptoms. However, whether polysaccharides are directly responsible for inducing symptoms or act indirectly by facilitating pathogenesis (or both) is not clear. Evidence is accumulating that EPSs and cell-associated polysaccharides may be more important for phytopathogenic bacteria than previously thought, either by promoting colonization or by enhancing survival. To firmly establish the role of polysaccharides during pathogenesis will require thoroughly characterized mutants, "natural" inoculation procedures, and careful sampling to assess the fate of bacteria within plants. Besides using polysaccharide-minus mutants, engineered strains that produce polymers with different sizes or compositions should be tested. Subsequently, the molecular mechanisms by which polysaccharides affect host-pathogen interactions can be addressed. It will be especially interesting to study how polysaccharides affect the production or presentation of cell-surface and extracellular pathogen-produced molecules like harpins that may damage the plant.

It is also important to continue characterizing the polysaccharides made by phytopathogenic bacteria. The recent discovery of a cell-associated, Kdo-rich polysaccharide in *Rhizobium* and *Agrobacterium,* and its possible production by other plant pathogens (BL Ruehs, personal communication), illustrates that we still have much to learn. One should also consider whether EPSs might be important for pathogens that do not appear to produce appreciable amounts in culture. It may be that these bacteria only produce EPS in planta or that the small amount produced is present at a critical time and place to influence the interaction with its host. Finally, we should all remember that polysaccharides likely serve essential functions for phytopathogenic bacteria when they inhabit niches other than inside plants. Only by considering the multiple functions of polysaccharides in all phases of a pathogen's existence will we truly understand the importance of these complex polymers.

ACKNOWLEDGMENTS

I thank all colleagues who provided reprints, preprints, and unpublished information. Special thanks also to A Collmer, D Coplin, S Lindow, S McCarter, and M Schell for their helpful review of an earlier version of this manuscript. I gratefully acknowledge the financial support of the USDA and the Georgia Agricultural Experiment Station.

Literature Cited

1. Baer D, Gudmestad NC. 1993. Serological detection of nonmucoid strains of *Clavibacter michiganensis* subsp. sepedonicus in potato. *Phytopathology* 83:157–63
2. Barton-Willis PA, Wang MC, Staskawicz B, Keen NT. 1987. Structural studies on the O-chain polysaccharides of lipopolysaccharides from *Pseudomonas syringae* pv. *glycinea*. *Physiol. Mol. Plant. Pathol.* 30:187–97
3. Bellemann P, Bereswill S, Berger S, Geider K. 1994. Visualization of capsule formation by *Erwinia amylovora* and assays to determine amylovoran synthesis. *Int. J. Biol. Macromol.* 16:290–96
4. Bellemann P, Geider K. 1992. Localization of transposon insertions in pathogenicity mutants of *Erwinia amylovora* and their biochemical characterization. *J. Gen. Microbiol.* 138:931–40
5. Benhamou N. 1991. Cell surface interactions between tomato and *Clavibacter michiganensis* subsp. *michiganensis*: localization of some polysaccharides and hydroxyproline-rich glycoproteins in infected host leaf tissues. *Physiol. Plant Pathol.* 38:15–38
6. Bernhard F, Coplin DL, Geider K. 1993. A gene cluster for amylovoran synthesis in *Erwinia amylovora*: characterization and relationship to *cps* genes in *Erwinia stewartii*. *Mol. Gen. Genet.* 239:158–68
7. Bills D, Kung S-D, eds. 1994. *Biotechnology and Plant Protection: Bacterial Pathogenesis and Disease Resistance*. River Edge, NJ: World Sci.
8. Bishop AL. 1985. *Bacterial ring rot of potato: disease development, diagnosis, and host water relations*. PhD thesis. Univ. Wisc, Madison, WI
9. Bishop AL, Slack SA. 1992. Effect of infection with *Clavibacter michiganensis* subsp. *sepedonicus* Davis et al. on water relations in potato. *Potato Res.* 35:59–63
10. Bradley DE, Douglas CJ, Peschon J. 1984. Flagella-specific bacteriophages of *Agrobacterium tumefaciens*: demonstration of virulence of nonmotile mutants. *Can. J. Microbiol.* 30:676–81
11. Braun EJ. 1990. Colonization of resistant and susceptible maize plants by *Erwinia stewartii* strains differing in exopolysaccharide production. *Physiol. Mol. Plant Pathol.* 36:363–79
12. Breedveld MW, Miller KJ. 1994. Cyclic β-glucans of members of the family Rhizobiaceae. *Microbiol. Rev.* 58:145–61
13. Brown I, Mansfield J, Irlam I, Conrads-Strauch J, Bonas U. 1993. Ultrastructure of interactions between *Xanthomonas campestris* pv. *vesicatoria* and pepper, including immunocytochemical localization of extracellular polysaccharides and the AvrBs3 protein. *Mol. Plant-Microbe Interact.* 6:376–86
14. Buddenhagen I, Kelman A. 1964. Biological and physiological aspects of bacterial wilt caused by *Pseudomonas solanacearum*. *Annu. Rev. Phytopathol.* 2:203–30
15. Cangelosi GA, Hung L, Puvanesarajah V, Stacey G, Ozga DA, et al. 1987. Common loci for *Agrobacterium tumefaciens* and *Rhizobium meliloti* exopolysaccharide synthesis and their roles in plant interactions. *J. Bacteriol.* 169: 2086–91
16. Cangelosi GA, Martinetti G, Leigh JA, Lee CC, Theines C, Nester EW. 1989. Role for *Agrobacterium tumefaciens* ChvA protein in export of β-1,2-glucan. *J. Bacteriol.* 171:1609–15
17. Cangelosi GA, Martinetti G, Nester EW. 1990. Osmosensitivity phenotypes of *Agrobacterium tumefaciens* mutants that lack periplasmic β-1,2-glucan. *J. Bacteriol.* 172:2172–74
18. Chatterjee AK, Vidaver AK. 1986. *Advances in Plant Pathology, Vol. 4. Genetics of Pathogenicity Factors: Application to Phytopathogenic Bacteria*. New York: Academic
19. Cook D, Sequeira L. 1991. Genetic and biochemical characterization of a *Pseudomonas solanacearum* gene cluster required for extracellular polysaccharide production and for virulence. *J. Bacteriol.* 173:1654–62
20. Coplin DL, Bernhard F, Majerczak D, Geider K. 1994. Capsular polysaccharide production in Erwinias. See Ref. 52, pp. 341–56
21. Coplin DL, Frederick RD, Majerczak DR, Tuttle LD. 1992. Characterization of a gene cluster that specifies pathogenicity in *Erwinia stewartii*. *Mol. Plant-Microbe Interact.* 5:81–88
22. Coplin DL, Majerczak DR. 1990. Extracellular polysaccharide genes in *Erwinia stewartii*: directed mutagenesis and complementation analysis. *Mol. Plant-Microbe Interact.* 3:286–92
23. Daniels MJ, Barber CE, Dow JM, Han B, Liddle SA, et al. 1993. Bacterial genes required for pathogenicity: interactions between *Xanthomonas* and crucifers, In *Mechanisms of Plant Defense*

Responses, ed. B Fritig, M Legrand, pp. 53–63. Dordrecht: Kluwer Academic

24. Denny TP, Baek SR. 1991. Genetic evidence that extracellular polysaccharide is a virulence factor of *Pseudomonas solanacearum. Mol. Plant-Microbe Interact.* 4:198–206
25. Denny TP, Schell MA. 1994. Virulence and pathogenicity of *Pseudomonas solanacearum:* genetic and biochemical perspectives. See Ref. 7, pp. 127–48
26. Douglas CJ, Halperin W, Nester EW. 1982. *Agrobacterium tumefaciens* mutants affected in attachment to plant cells. *J. Bacteriol.* 152:1265–75
27. Douglas CJ, Staneloni RJ, Rubin RA, Nester EW. 1985. Identification and genetic analysis of an *Agrobacterium tumefaciens* chromosomal virulence region. *J. Bacteriol.* 161:850–60
28. Fett WF, Dunn MF. 1989. Exopolysaccharides produced by phytopathogenic *Pseudomonas syringae* pathovars in infected leaves of susceptible hosts. *Plant. Physiol.* 89:5–9
29. Fett WF, Osman SF, Dunn MF, Panopoulos NJ. 1992. Cell surface properties of *Pseudomonas syringae* pv. *phaseolicola* wild-type and *hrp* mutants. *J. Phytopathol.* 135:135–52
30. Fett WF, Osman SF, Wijey C, Singh S. 1993. Exopolysaccharides of rRNA group I pseudomonads, In *Carbohydrates and Carbohydrate Polymers,* ed. M Yalpani, pp. 52–61. Mt. Prospect, IL: ATL
31. Fiedler W, Rotering H. 1988. Properties of *Escherichia coli* mutants lacking membrane-derived oligosaccharides. *J. Biol. Chem.* 263:14684–89
32. Fuqua WC, Winans SC, Greenberg EP. 1994. Quorum sensing in bacteria: the *luxR-luxI* family of cell density-responsive transcriptional regulators. *J. Bacteriol.* 176:269–75
33. Geier G, Geider K. 1993. Characterization and influence on virulence of the levansucrase gene from the fireblight pathogen *Erwinia amylovora. Physiol. Mol. Plant Pathol.* 42:387–404
34. Goodman RN, White JA. 1981. Xylem parenchyma plasmolysis and vessel wall disorientation caused by *Erwinia amylovora. Phytopathology* 71:844–52
35. Gorin PA, Spencer JFT, Lindberg B, Lindh F. 1980. Structure of the extracellular polysaccharide from *Corynebacterium insidiosum. Carbohydr. Res.* 79: 313–15
36. Grignon C, Sentenac H. 1991. pH and ionic conditions in the apoplast. *Annu. Rev. Plant Physiol.* 42:103–28
37. Gross M, Geier G, Rudolph K, Geider K. 1992. Levan and levansucrase synthesized by the fireblight pathogen *Erwinia amylovora. Physiol. Mol. Plant Pathol.* 40:371–81
38. Gurlitz RHG, Lamb PW, Matthysse AG. 1987. Involvement of carrot cell surface proteins in attachment of *Agrobacterium tumefaciens. Plant. Physiol.* 83:564–68
39. Harding NE, Raffo S, Raimondi A, Cleary JM, Ielpi L. 1993. Identification, genetic and biochemical analysis of genes involved in synthesis of sugar nucleotide precursors of xanthan gum. *J. Gen. Microbiol.* 139:447–57
40. Henningson PJ, Gudmestad NC. 1993. Comparison of exopolysaccharides from mucoid and nonmucoid strains of *Clavibacter michiganensis* subspecies *sepedonicus. Can. J. Microbiol.* 39:291–96
41. Hibberd AM, Stall RE, Bassett MJ. 1987. Different phenotypes associated with incompatible races and resistance genes in bacterial spot disease. *Plant Dis.* 71:1075–78
42. Hignett RC, Roberts AL. 1985. A possible regulatory function for bacterial outer surface components in fireblight disease. *Physiol. Plant Pathol.* 27:235–43
43. Hisamatsu M, Abe J, Amemura A, Harada T. 1980. Structural elucidation on succinoglycan and related polysaccharides from *Agrobacterium* and *Rhizobium* by fragmentation with two special β-D-glycanases and methylation analysis. *Agric. Biol. Chem.* 44:1049–55
44. Hisamatsu M, Amemura A, Matsuo T, Matsuda H, Harada T. 1982. Cyclic (1–2)-β-D-glucan and the octasaccharide repeating-unit of succinoglycan produced by *Agrobacterium. J. Gen. Microbiol.* 128:1873–79
45. Hooykaas PJJ, Beijersbergen AGM. 1994. The virulence system of *Agrobacterium tumefaciens. Annu. Rev. Phytopathol.* 32:157–79
46. Höltje J-V, Fiedler W, Rotering H, Walderich B, van Duin J. 1988. Lysis induction of *Escherichia coli* by the cloned lysis protein of the phage MS2 depends on the presence of osmoregulatory membrane-derived oligosaccharides. *J. Biol. Chem.* 263:3539–41
47. Huang J, Carney BF, Denny TP, Weissinger AK, Schell MA. 1995. A complex network regulates expression of *eps* and other virulence genes of *Pseudomonas solanacearum. J. Bacteriol.* 177:1259–67
48. Huang J, Schell MA. 1995. Molecular characterization of the *eps* locus of *Pseudomonas solanacearum* and its

transcriptional regulation at a single promoter. *Mol. Microbiol.* In press
49. Huang P, Goodman RN. 1976. Ultrastructural modifications in apple stems induced by *Erwinia amylovora* and the fire blight toxin. *Phytopathology* 66: 269–76
50. Huang P, Huang JS, Goodman RN. 1975. Resistance mechanisms of apple shoots to an avirulent strain of *Erwinia amylovora. Physiol. Plant Pathol.* 6: 283–87
51. Iñon de Iannino N, Ugalde RA. 1989. Biochemical characterization of avirulent *Agrobacterium tumefaciens chvA* mutants: synthesis and excretion of β-(1,2)glucan. *J. Bacteriol.* 171:2842–49
52. Kado CI, Crosa JH, eds. 1994. *Molecular Mechanisms of Bacterial Virulence.* Dordrecht: Kluwer Academic
53. Kao CC, Barlow E, Sequeira L. 1992. Extracellular polysaccharide is required for wild-type virulence of *Pseudomonas solanacearum. J. Bacteriol.* 174:1068–71
54. Kao CC, Sequeira L. 1991. A gene cluster required for coordinated biosynthesis of lipopolysaccharide and extracellular polysaccharide also affects virulence of *Pseudomonas solanacearum. J. Bacteriol.* 173:7841–47
55. Kasapis S, Morris ER, Gross M, Rudolph K. 1994. Solution properties of levan polysaccharide from *Pseudomonas syringae* pv. *phaseolicola,* and its possible primary role as a blocker of recognition during pathogenesis. *Carbohydr. Polym.* 23:55–64
56. Kennedy EP. 1987. Membrane-derived oligosaccharides. In *Escherichia coli and Salmonella typhimurium: Cellular and Molecular Biology,* ed. FC Neidhardt, pp. 672–79. Washington, DC: Am. Soc. Microbiol.
57. Kingsley MT, Gabriel DW, Marlow GC, Roberts PD. 1993. The *opsX* locus of *Xanthomonas campestris* affects host range and biosynthesis of lipopolysaccharide and extracellular polysaccharide. *J. Bacteriol.* 175:5839–50
58. Leigh JA, Coplin DL. 1992. Exopolysaccharides in plant-bacterial interactions. *Annu. Rev. Microbiol.* 46: 307–46
59. Leigh JA, Walker GC. 1994. Exopolysaccharides of *Rhizobium*: synthesis, regulation and symbiotic function. *Trends. Genet.* 10:63–67
60. Loubens I, Debarbieux L, Bohin A, Lacroix JM, Bohin JP. 1993. Homology between a genetic locus (*mdoA*) involved in the osmoregulated biosynthesis of periplasmic glucans in *Escherichia coli* and a genetic locus (*hrpM*) controlling pathogenicity of *Pseudomonas syringae. Mol. Microbiol.* 10:329–40
61. Mansfield J, Brown I, Maroofi A. 1994. Bacterial pathogenicity and the plant's response: ultrastructural, biochemical and physiological perspectives. See Ref. 7, pp. 85–105
62. Matthysse AG. 1986. Initial interactions of *Agrobacterium tumefaciens* with plant host cells. *CRC Crit. Rev. Microbiol.* 13:281–307
63. Matthysse AG, Holmes KV, Gerlitz RHG. 1981. Elaboration of cellulose fibrils by *Agrobacterium tumefaciens* during attachment to carrot cells. *J. Bacteriol.* 145:583–95
64. Matthysse AG, Wagner VT. 1994. Attachment of *Agrobacterium tumefaciens* to host cells. See Ref. 52, pp. 79–92
65. May TB, Chakrabarty AM. 1994. *Pseudomonas aeruginosa*: genes and enzymes of alginate synthesis. *Trends Microbiol.* 2:151–57
66. Meletzus D, Bermpohl A, Dreier J, Eichenlaub R. 1993. Evidence for plasmid-encoded virulence factors in the phytopathogenic bacterium *Clavibacter michiganensis* subsp. *michiganensis* NCPPB382. *J. Bacteriol.* 175:2131–36
67. Metts J, West J, Doares SH, Matthysse AG. 1991. Characterization of three *Agrobacterium tumefaciens* avirulent mutants with chromosomal mutations that affect induction of *vir* genes. *J. Bacteriol.* 173:1080–87
68. Metzger M, Bellemann P, Bugert P, Geider K. 1994. Genetics of galactose metabolism of *Erwinia amylovora* and its influence on polysaccharide synthesis and virulence of the fire blight pathogen. *J. Bacteriol.* 176:450–59
69. Miller KJ, Kennedy EP, Reinhold VN. 1986. Osmotic adaptation by gram-negative bacteria: possible role for periplasmic oligosaccharides. *Science* 231:48–51
70. Morris VJ, Chilvers GR. 1984. Cold setting alginate-pectin mixed gels. *J. Sci. Food Agric.* 35:1370–76
71. Mukhopadhyay PJ, Williams I, Mills D. 1988. Molecular analysis of a pathogenicity locus in *Pseudomonas syringae* pv. *syringae. J. Bacteriol.* 170:5479–88
72. Nester EW, Verma DPS, eds. 1993. *Advances in Molecular Genetics of Plant-Microbe Interactions,* Vol. 2. Dordrecht: Kluwer Academic
73. Newman M-A, Conrads-Strauch J, Scofield G, Daniels MJ, Dow JM. 1994. Defense-related gene induction in *Brassica campestris* in response to defined

mutants of *Xanthomonas campestris* with altered pathogenicity. *Mol. Plant-Microbe Interact.* 7:553–62

74. Norelli JL, Gilbert MT, Aldwinckle HS, Zumoff CH, Beer SV. 1990. Population dynamics of nonpathogenic mutants of *Erwinia amylovora* in apple host tissue. *Acta Hortic.* 273:239–40
75. O'Connell KP, Handelsman J. 1989. *chvA* locus may be involved in export of neutral cyclic β-1,2-linked D-glucan from *Agrobacterium tumefaciens. Mol. Plant-Microbe Interact.* 2:11–16
76. Orgambide G, Montrozier H, Servin P, Roussel J, Trigalet-Demery D, Trigalet A. 1991. High heterogeneity of the exopolysaccharides of *Pseudomonas solanacearum* strain GMI 1000 and the complete structure of the major polysaccharide. *J. Biol. Chem.* 266:8312–21
77. Osman SF, Fett WF, Fishman ML. 1986. Expolysaccharides of the phytopathogen *Pseudomonas syringae* pv. *glycinea. J. Bacteriol.* 166:66–71
78. Ovod V, Ashorn P, Yakovleva L, Krohn K. 1995. Classification of *Pseudomonas syringae* with monoclonal antibodies against the core and O-side chains of the lipopolysaccharide. *Phytopathology.* 85:226–32
79. Paschke MC, Van Alfen NK. 1993. Extracellular polysaccharide impaired mutants of *Clavibacter michiganensis* subsp. *insidiosum. Physiol. Mol. Plant Pathol.* 42:309–19
80. Petrovics G, Putnoky P, Reuhs B, Kim J, Thorp TA, et al. 1993. The presence of a novel type of surface polysaccharide in *Rhizobium meliloti* requires a new fatty acid synthase-like gene cluster involved in symbiotic nodule development. *Mol. Microbiol.* 8:1083–94
81. Pierce ML, Essenberg M, Mort AJ. 1993. A comparison of the quantities of exopolysaccharide produced by *Xanthomonas campestris* pv. *malvacearum* in susceptible and resistant cotton cotyledons during early stages of infection. *Phytopathology* 83:344–49
82. Pueppke SG. 1984. Adsorption of bacteria to plant surfaces, In *Plant-Microbe Interactions,* ed. T Kosuge, EW Nester, pp. 215–61. New York: Macmillan
83. Puvanesarajah V, Schell FM, Stacey G, Douglas CJ, Nester EW. 1985. Role for 2-linked-β-D-glucan in the virulence of *Agrobacterium tumefaciens. J. Bacteriol.* 164:102–6
84. Rai PV, Strobel GA. 1969. Phytotoxic glycopeptides produced by *Corynebacterium michiganense* II: biological properties. *Phytopathology* 59:53–57
85. Ramírez ME, Fucikovsky L, Garcia-Jimenez F, Quintero R, Galindo E. 1988. Xanthan gum production by altered pathogenicity variants of *Xanthomonas campestris. Appl. Microbiol. Biotechnol.* 29:5–10
86. Reuhs BL, Carlson RW, Kim JS. 1993. *Rhizobium fredii* and *Rhizobium meliloti* produce 3-deoxy-D-*manno*-2-octulosonic acid-containing polysaccharides that are structurally analogous to group II K antigens (capsular polysaccharides) found in *Escherichia coli. J. Bacteriol.* 175:3570–80
87. Romeiro R, Karr A, Goodman R. 1981. Isolation of a factor from apple that agglutinates *Erwinia amylovora. Plant Physiol.* 68:772–77
88. Rudolph KWE, Gross M, Ebrahim-Nesbat F, Nöllenburg M, Zomorodian A, et al. 1994. The role of extracellular polysaccharides as virulence factors for phytopathogenic pseudomonads and xanthomonads. See Ref. 52, pp. 357–78
89. Rudolph KWE, Gross M, Neugebauer M, Hokawat S, Zachowski A, et al. 1989. Extracellular polysaccharides as determinants of leaf spot diseases caused by pseudomonads and xanthomonads, In *Phytotoxins and Plant Pathogenesis,* (NATO ASI ser.), ed. A Graniti, RD Durbin, A Bollio, pp. 177–218. Berlin: Springer-Verlag
90. Salmond GPC. 1994. Secretion of extracellular virulence factors by plant pathogenic bacteria. *Annu. Rev. Phytopathol.* 32:181–200
91. Schouten HJ. 1989. A possible role in pathogenesis for the swelling of extracellular slime of *Erwinia amylovora* at increasing water potential. *Neth. J. Plant Pathol.* 95:169–74
92. Seemuller EA, Beer SV. 1976. Absence of cell wall polysaccharide degradation by *Erwinia amylovora. Phytopathology* 66:433–36
93. Sequeira L. 1985. Surface components involved in bacterial pathogen-plant host recognition. *J. Cell. Sci. Suppl.* 2:301–16
94. Sijam K, Goodman RN, Karr AL. 1985. The effect of salts on the viscosity and wilt-inducing capacity of the capsular polysaccharide of *Erwinia amylovora. Physiol. Plant Pathol.* 26:231–39
95. Sijam K, Karr AL, Goodman RN. 1983. Comparison of the extracellular polysaccharides produced by *Erwinia amylovora* in apple tissue and culture medium. *Physiol. Plant Pathol.* 22:221–31
96. Sjulin TM, Beer SV. 1978. Mechanism of wilt induction by amylovorin in *Cotoneaster* shoots and its relation to wilt-

ing of shoots infected by *Erwinia amylovora. Phytopathology* 68:89–94
97. Smit G, Swart S, Lugtenberg BJJ, Kijne JW. 1992. Molecular mechanisms of attachment of *Rhizobium* bacteria to plant roots. *Mol. Microbiol.* 6:2897–903
98. Smith ARW, Munro SM, Wait R, Hignett RC. 1994. Effect on lipopolysaccharide structure of aeration during growth of a plum isolate of *Pseudomonas syringae* pv. *morsprunorum. Microbiology-UK.* 140:1585–93
99. Smith ARW, Rastall RA, Rees NH, Hignett RC, Wait R. 1990. Structure of the extracellular polysaccharide of *Erwinia amylovora*: a preliminary report. *Acta Hortic.* 273:211–19
100. Steinberger EM, Beer SV. 1988. Creation and complementation of pathogenicity mutants of *Erwinia amylovora. Mol. Plant-Microbe Interact.* 1: 135–44
101. Suhayda CG, Goodman RN. 1981. Early proliferation and migration and subsequent xylem occlusion by *Erwinia amylovora* and the fate of its extracellular polysaccharide (EPS) in apple shoots. *Phytopathology* 71:697–707
102. Sutherland IW. 1988. Bacterial surface polysaccharides: structure and function. *Int. Rev. Cytol.* 113:187–231
103. Sutton JC, Williams PH. 1970. Relation of xylem plugging to black rot lesion development in cabbage. *Can. J. Bot.* 48:391–401
104. Swart S, Logman TJJ, Lugtenberg BJJ, Smit G, Kijne JW. 1994. Several phenotypic changes in the cell envelope of *Agrobacterium tumefaciens chvB* mutants are prevented by calcium limitation. *Arch. Microbiol.* 161:310–15
105. Swart S, Lugtenberg BJJ, Smit G, Kijne JW. 1994. Rhicadhesin-mediated attachment and virulence of an *Agrobacterium tumefaciens chvB* mutant can be restored by growth in a highly osmotic medium. *J. Bacteriol.* 176:3816–19
106. Swart S, Smit G, Lugtenberg BJJ, Kijne JW. 1993. Restoration of attachment, virulence and nodulation of *Agrobacterium tumefaciens chvB* mutants by rhicadhesin. *Mol. Microbiol.* 10:597–605
107. Talaga P, Fournet B, Bohin JP. 1994. Periplasmic glucans of *Pseudomonas syringae* pv. *syringae. J. Bacteriol.* 176: 6538–44
108. Tharaud M, Menggad M, Paulin JP, Laurent J. 1994. Virulence, growth, and surface characteristics of *Erwinia amylovora* mutants with altered pathogenicity. *Microbiology-UK* 140:659–69
109. Uttaro AD, Cangelosi GA, Geremia RA, Nester EW, Ugalde RA. 1990. Biochemical characterization of avirulent *exoC* mutants of *Agrobacterium tumefaciens. J. Bacteriol.* 172:1640–46
110. Van Alfen NK. 1989. Reassessment of plant wilt toxins. *Annu. Rev. Phytopathol.* 27:533–50
111. Van Alfen NK, McMillan BD. 1982. Macromolecular plant-wilting toxins: artifacts of the bioassay method? *Phytopathology* 72:132–35
112. Van Alfen NK, McMillan BD, Dryden P. 1987. The multi-component extracellular polysaccharide of *Clavibacter michiganense* subsp. *insidiosum. Phytopathology* 77:496–501
113. Van Alfen NK, McMillan BD, Wang Y. 1987. Properties of the extracellular polysaccharides of *Clavibacter michiganense* subsp. *insidiosum* that may affect pathogenesis. *Phytopathology* 77: 501–5
114. van den Bulk RW, Löffler HJM, Dons JJM. 1989. Effect of phytotoxic compounds produced by *Clavibacter michiganensis* subsp. *michiganensis* on resistant and susceptible tomato plants. *Neth. J. Plant Pathol.* 95:107–17
115. van den Bulk RW, Zevenhuizen LPTM, Cordewener JHG, Dons JJM. 1991. Characterization of the extracellular polysaccharide produced by *Clavibacter michiganensis* subsp. *michiganensis. Phytopathology* 81:619–23
116. Vidaver AK. 1982. The plant pathogenic corynebacteria. *Annu. Rev. Microbiol.* 36:495–517
117. Wei ZM, Beer SV. 1993. HrpI of *Erwinia amylovora* functions in secretion of harpin and is a member of a new protein family. *J. Bacteriol.* 175:7958–67
118. Weibgen U, Russa R, Yokota A, Mayer H. 1993. Taxonomic significance of the lipopolysaccharide composition of the three biovars of *Agrobacterium tumefaciens. Syst. Appl. Microbiol.* 16: 177–82
119. Westra AAG, Slack SA. 1992. Isolation and characterization of extracellular polysaccharide of *Clavibacter michiganesis* subsp. *sepedonicus. Phytopathology* 82:1193–99
120. Yabuuchi E, Kosako Y, Oyaizu H, Yano I, Hotta H, et al. 1992. Proposal of *Burkholderia* gen. nov. and transfer of seven species of the genus *Pseudomonas* homology group II to the new genus, with the type species *Burkholderia cepacia* (Palleroni and Holmes 1981) comb. nov. *Microbiol. Immunol.* 36: 1251–75

121. Zhan DF, Ridout MJ, Brownsey GJ, Morris VJ. 1993. Xanthan-locust bean gum interactions and gelation. *Carbohyd. Polym.* 21:53–58
122. Zorreguieta A, Cavaignac S, Geremia RA, Ugalde RA. 1990. Osmotic regulation of β-(1-2) glucan synthesis in members of the family Rhizobiaceae. *J. Bacteriol.* 172:4701–4
123. Zorreguieta A, Ugalde RA. 1986. Formation in *Rhizobium* and *Agrobacterium* spp. of a 235-kilodalton protein intermediate in β-D(1-2) glucan synthesis. *J. Bacteriol.* 167:947–51

Annu. Rev. Phytopathol. 1995. 33:199–221

CONCEPTUAL AND PRACTICAL ASPECTS OF VARIABILITY IN ROOT-KNOT NEMATODES RELATED TO HOST PLANT RESISTANCE

P. A. Roberts
Department of Nematology, University of California, Riverside, California 92521

KEY WORDS: *Meloidogyne*, plant breeding, integrated pest management, virulence

ABSTRACT

Root-knot nematodes are the leading cause of crop loss due to plant parasitic nematodes. A few species (*Meloidogyne arenaria, M. chitwoodi, M. hapla, M. incognita, M. javanica*) have very wide host ranges that limit nonhost rotation options. Numerous host resistance genes in diverse crops are identified and could be used in cropping sequences to manage root-knot. Approaches to their combined use that maximize effectiveness and preserve durability, including pyramiding genes in a crop and in a cropping system, are considered with resistance gene specificity to nematode species and populations. (A)virulence specificity and selection in *Meloidogyne* for matching host resistance genes is examined, with particular reference to tomato and cowpea. Analogies are made with bacterial and fungal pathogen–host gene-for-gene interactions. A biotype scheme is offered for characterizing, categorizing, and communicating *Meloidogyne* (a)virulence.

INTRODUCTION

Root-knot nematodes (*Meloidogyne* spp.) are the leading cause of crop losses by plant parasitic nematodes (81). *Meloidogyne* contains more than fifty described species, of which a few (particularly *M. arenaria*, *M. chitwoodi*, *M.*

0066-4286/95/0901-0199$05.00

hapla, M. incognita, and *M. javanica*) are common and of primary agricultural importance. These *Meloidogyne* species have characteristically wide host ranges that include more than 2000 plant species and most crops worldwide (80, 81) for which new and improved management tools are urgently needed for their control (23, 31, 70, 71).

Host-plant resistance suppresses nematode development and reproduction, and although initial infection occurs, most resistant genotypes have tolerance to the injury potential of infective second-stage juveniles and produce acceptable yields on infested ground (i.e. self-protection) (18, 25, 67–69, 91). In addition, suppression of nematode multiplication results in lower residual population densities that have less damage potential to the next crop in the rotation (32, 67–69, 91).

Host-plant resistance has been used successfully to manage root-knot on a few diverse crops (17, 18, 27, 28, 67–69, 86, 88, 91), although relatively few breeding programs have targeted root-knot resistance in cooperation with nematologists to optimize the breadth of resistance required. In complex annual systems, such as in intensive vegetable production, each of several crops in the rotation is likely to be a root-knot host, vulnerable to attack, and requiring protection. Even though a rich resource of host-plant resistance genes has been identified, few if any attempts have been made to develop resistance in each of the diverse crops in an approach to protect the whole system (cropping sequence).

For the purpose of discussion, assume that resistance gene transfer into suitable crop cultivars will not in itself be a major limiting factor in the future, i.e. all resistance genes are truly "accessible." Then most important decisions for both resistance breeding and implementation in the field will be guided by the control action spectrum of resistance genes and their recipient cultivars. The considerable variability within and between *Meloidogyne* species and populations for host range and (a)virulence determinants is not well defined. No practical or conceptual scheme for characterizing this large variation has been developed and put into use. Resistance development and application will remain an ill-directed hit-or-miss approach unless the control target is clearly defined.

This review focuses on *Meloidogyne* species and population variability for host range and (a)virulence characteristics. A practical scheme for characterizing and categorizing *Meloidogyne* spp. and their (a)virulence variants is offered, with the aim of encouraging testing and adoption, and to provide a conceptual framework for clarifying the extent and nature of (a)virulence in *Meloidogyne*. The potential and some directions are offered for integrating host-plant resistance (HPR) genes in multiple crops to the extent that a "shield" of resistance can be used to effectively manage multiple species and biotypes on multiple crops. Advances in these areas should enable crop producers worldwide to capitalize on the natural resource of HPR genes, which are numerous with distinct specificities.

GENETIC VARIABILITY IN *MELOIDOGYNE* SPECIES AND POPULATIONS

Host Range and Virulence Characteristics

Because the important common *Meloidogyne* species reproduce asexually by obligate mitotic parthenogenesis (*M. hapla* is facultatively parthenogenetic) (90), traditional Mendelian-based genetic analyses of the factors determining host range and (a)virulence to resistance genes are not feasible (89). Currently, there is no transformation system for *Meloidogyne* to facilitate molecular analysis of cloned genetic determinants when this is achieved, although this may be feasible since *Caenorhabditis elegans* can be transformed (34). The genetics of host range and (a)virulence in *Meloidogyne* is complicated by a cytogenetics that includes variation in chromosome number (within and between species) due to polyploidy and aneuploidy (90). Thus our understanding of these mechanisms must be inferred from analysis of genetic determinants of resistance in the plant (99), and by analysis of the phenotype for these characters in the nematode, including comparisons of traits at the population level. Clonal selection experiments, isofemale line comparisons, and temporal (generations) and spatial shifts in host range and (a)virulence phenotypes are important approaches to studying root-knot nematode host range and (a)virulence determinants. Important inferences may also be drawn from apparently analogous host plant-parasite/pathogen systems involving amphimictic nematodes such as cyst nematodes (1, 43, 45, 92, 96), and phytopathogenic fungi (19, 21, 40, 46) and bacteria (2, 48, 49, 52, 56, 57).

HOST RANGE The common species have extensive, overlapping host ranges and sympatric distributions, often with mixed species infestations in the same field (61, 80). This places severe constraints on the choices of nonhost crops available for most crop rotation systems. Single-species infestations occur and offer a more manageable system. They probably reflect introduction patterns of infected materials into otherwise noninfested areas. For example, in the cotton and vegetable growing areas of New Mexico only *M. incognita* is found (SH Thomas, personal communication). Nonhost crops of economic value broadly effective against all species occurring in infested areas are generally unavailable. However, some can be manipulated to act as nonhosts by, for instance, planting small grain cereals during cool season periods to avoid nematode activity and root infection (77). The differential host status of some crops, i.e. nonhost for some but not all *Meloidogyne* spp., is the best that can be utilized in most situations. Opportunities exist to use combinations of differential nonhosts in rotation sequences. For example, cotton is a nonhost for *M. arenaria* and *M. javanica*, but a host and vulnerable to many common populations of *M. incognita*, whereas most alfalfa types do not host *M. incognita*. Thus rotations containing alfalfa and cotton can help to suppress infections by all three species (36). Difficulties arise because most growers plant additional crops, particularly vegetable or agronomic cash crops, and thereby reduce the frequency of nonhosts in the rotation. Development and implementation of resistant cultivars must be considered against this backdrop of multiple host rotations with few nonhost crops. Major challenges include the availability, heritability, and action spectra of resistance traits for cultivar improvement in multiple crops, and their rational deployment.

(A)VIRULENCE IN *MELOIDOGYNE* By simple definition, virulent populations are able to reproduce significantly on resistant host plants that prevent or suppress reproduction of avirulent populations of the parasite species (1, 18, 68, 84, 89). The spectrum of (a)virulence within and between *Meloidogyne* species and populations for reaction to resistance genes is complex, but not so variable or genetically unstable as to be prohibitive for effective root-knot management with resistance. The most studied (a)virulence system is for reaction to the dominant resistance gene *Mi* in tomato (4, 9, 11–13, 20, 44, 66, 76). At least two types of virulent populations have been identified within *M. incognita*, whose populations typically are avirulent to gene *Mi*. One type includes populations, or isolates of them, commonly referred to as "natural virulent" that are highly reproductive on plants carrying gene *Mi*, reproducing at levels equivalent to isolates on susceptible control plants (11, 12, 72). Natural virulent isolates have been found in field populations that apparently have not been subjected to selection by repeated exposure to *Mi*, at least in recent agricultural time, although previous exposure in evolutionary time must be considered a possibility (72, 76).

The second type of virulent population is one that can be selected from avirulent wild-type isolates under controlled greenhouse conditions by repeated inoculation on plants with resistance gene *Mi*. This laboratory-selected virulence can occur rapidly, in as few as 6–12 generations (4). Laboratory-selected virulent isolates do not reproduce as well on plants carrying *Mi* as they do on susceptible control plants. After more than 25 generations of selection, two laboratory-selected isolates produced only 60–64% of the egg masses on *Mi*-plants that they produced on susceptible plants (12). The maximum level of reproduction tends to be about 75% of that in compatible responses on susceptible plants (4, 76, 89). (A)virulence affects the infection frequency of an isolate, as measured by the numbers (proportion) of infective second-stage juveniles that can develop into gravid females. It also affects the rate of development of primary infection, and hence fecundity (eggs per egg mass), of successful females. In avirulent populations few or no infective juveniles mature and produce egg masses on plants carrying gene *Mi* under field and greenhouse conditions (13, 74, 75). Selected virulent isolates on plants with *Mi* have reduced rates of infection frequency and in some but not all cases may produce fewer eggs per egg mass compared to that on susceptible plants. Gene *Mi* apparently has a residual potency that may not be overcome by the nematode (4). This potency appears analogous to some fungal pathogen-host resistance interactions such as the *Pm4* gene for resistance in wheat to *Erysiphe graminis* f. sp. *tritici* (53). Inoculation of resistant and susceptible segregants in an F_2 population from a tomato cross provided evidence that differences in virulence were in response to gene *Mi* alone, and not due to any residual factors in the tomato genotype. Virulence reaction was the same in tomato segregants carrying gene *Mi* in the homozygous and heterozygous condition, indicating a lack of resistance gene dosage effect (4).

Recent experiments by Dalmasso and colleagues (13) have confirmed earlier work demonstrating that not all avirulent field populations of a species such as *M. incognita* can be selected for virulence, and only a small proportion of individual females produce progeny from which virulent lines can be selected (13, 59). Mother-daughter comparisons from single egg mass descent (isofemale lineages) studies provide evidence for both genetic variability and inheritance as a basis of this virulence (9, 11, 13). Selected isolates appear to be genetically stable for virulence over at least 18 generations (11, 66). Comparisons of virulence on resistant plants were made of single egg mass–derived, selected virulent and avirulent wild-type *M. incognita* isolates that were cultured on resistant or susceptible near-isogenic tomato lines. When resistant plants were inoculated with either isolate every three generations, no clear-cut differences in virulence were obtained (11). Thus, reversion to avirulence did not occur in the absence of selection pressure from gene *Mi* (11), indicating that this selected virulence is genetically stable within this time-frame. A natural virulent population of *M. incognita* was consistently more virulent on *Mi*-bearing plants when cultured continuously on plants with gene *Mi* than on susceptible plants. This suggests that there may be an adverse cost to fitness (defined as the relative contribution that an individual makes to the next generation) in the virulence trait. However, in these tests a progressive loss of virulence in the absence of selection pressure was not observed.

No clear pattern of relative fitness has emerged between laboratory-selected virulent isolates and their nonselected wild-type parents, either for primary infection capacity or for fecundity rates on susceptible plants (4, 12). Bost & Triantaphyllou (4) reported that a selected virulent isolate had an average infection frequency on susceptible genotypes of 69% of its wild-type parent culture, even though the fecundity of the successful females did not differ. In other studies, selected virulent isolates had a higher fecundity than their wild-type progenitor on susceptible plants (44). Lack of evidence for virulence selection in intensive tomato production areas raises questions about the cost to fitness of selection. In tomato monoculture or short rotations in warm climate areas (e.g. California where more than 80% of the US processing tomato crop is grown), enough heat units occur per growing season to allow several nematode generations (approximately four or five in California's Central Valley). Nevertheless, rapid field selection for virulence like that in laboratory-based conditions, in which virulence might be expected over two to three years (about 8 to 15 generations), has not occurred. Caution must be exercised when extending results obtained from experiments under controlled conditions in greenhouse pot cultures to natural events under field conditions. Fitness measurements and effects could be buffered and distorted in optimal conditions of host root culture in pathogen-free medium in pots receiving frequent irrigation and adequate fertilization.

The complexity of virulence selection and fitness processes in *Meloidogyne* species was revealed further by Castagnone and colleagues (10), who found that the same laboratory-selected *Mi*-virulent *M. incognita* isolates (referenced above) had lost the ability to reproduce on susceptible pepper (*Capsicum annuum*) plants. Those tests were designed to determine if selection for virulence on tomato gene *Mi* had altered (a)virulence to resistance genes in pepper (39). This partial loss of host range would certainly over time have broad deleterious effects on nematode fitness and survival potential. One can speculate that such a change might result from complex and multiple functions of (a)virulence loci involved in determining host range, similar to that implicated for genetic regulation of virulence and host range in phytopathogenic bacteria (21, 48). Chromosome loss in a polyploid parasite could be another possible event that would influence (a)virulence and host range.

In our studies, field populations of *M. incognita* (normally avirulent) have been identified that are virulent to the single dominant resistance gene *Rk* in cowpea (*Vigna unguiculata*) (33, 73). The virulent populations occur in fields frequently planted with resistant (gene *Rk*) cowpea, so field selection of this virulence is likely. However, examination of one of these virulent isolates cultured on susceptible tomato host plants in greenhouse pots over four years (representing 20 or more generations) revealed a significant decrease in virulence to gene *Rk* (PA Roberts, unpublished). The virulence, measured by production of egg masses on resistant (gene *Rk*) compared to susceptible control plants, decreased progressively from about 80% of those produced on suscepts initially down to about 15% after four years. No attempt was made to single egg-mass culture the isolate initially. It was a field population subsample and presumably represented some of the variability in that population. Whatever the frequency of virulent individuals initially, loss of virulence has occurred over time on susceptible tomato, i.e. in the absence of selection pressure from gene *Rk*. This suggests that the maintenance of virulence has an adverse genetic cost to fitness with virulent individuals being eliminated. This result is in direct contrast to the apparent stability of virulence to tomato resistance gene *Mi* discussed above. The avirulence condition may be linked to or serve some additional function (gene product) of advantage or necessity to the nematode. Maintenance of avirulence genes occurs in other plant pathogens in the absence of contact with the relevant resistant plant genotypes. For example, retention of avirulence function is known in disparate isolates of the rice blast fungus (*Magnaporthe grisea*) (93) and in pathovars of the bacterium *Xanthomonas campestris* (95), indicative of avirulence genes encoding important functions in the pathogen other than the incompatible reaction capacity on host genotypes with which they have no contact (21).

GENE-FOR-GENE INTERACTIONS The most comprehensive genetic analysis of nematode (a)virulence is that in the cyst nematodes, *Globodera rostochiensis* and *G. pallida*, matching single dominant resistance genes *H1* and *H2*, respectively, in potato (*Solanum tuberosum* or ex-*S. vernei*) lines (1, 43, 45). Proof for Mendelian inheritance has been obtained for (a)virulence being controlled by a single nuclear gene, with virulence recessive to avirulence, and fitting a gene-for-gene relationship for (in)compatibility between nematode parasite and plant host (1, 43, 45). Because traditional genetic analysis of (a)virulence is not possible in parthenogenetic *Meloidogyne* species, inference must be based on analysis of phenotype. If one follows the basic requirements for a nongenetic demonstration of a gene-for-gene relationship between host and pathogen (parasite) as discussed by Robinson (78) and based on the work of Flor (35), Person (63), and Habgood (37), current phenotype data for *Meloidogyne* do not provide the essential set of differential reactions.

In applying the rule that differentials must belong to a single host species and a single parasite species (78), *M. incognita* isolates and tomato resistance genes (all derived from *Lycopersicon peruvianum*) and *M. incognita* isolates and cowpea resistance genes would come closest to indicating existence of gene-for-gene interaction. However, both cases fall short of the requirement for a transagonal quadratic check within a minimum 3 × 2 (or 2 × 3) set of differential interactions (78). In order to have a transagonal fit, a minimum requirement is that at least one nematode isolate is virulent to resistance gene 1 but avirulent to resistance gene 2, and a second isolate is virulent to gene 2 but avirulent to gene 1, in addition to uniform virulence and/or avirulence on a third host genotype. Within *M. incognita*, some isolates are avirulent to both resistance genes *Mi* and *Mi3* of tomato, some isolates are virulent to both genes, and some others are virulent to *Mi* but avirulent to *Mi3* (8, 72, 101). However, isolates that are virulent to *Mi3* but avirulent to *Mi* have not yet been identified. Another study offered a similar incomplete transagonal quadratic check for *M. incognita* populations and gene *Mi* and a second resistance factor reported to reside close to *Mi* (60). Thus a true gene-for-gene relationship is indicated but not currently demonstrated.

On cowpea, we have identified *M. incognita* isolates avirulent to both resistance genes *Rk* and *Rk2* [tightly linked to or an allele of gene *Rk* (54; PA Roberts, unpublished results)], and other isolates virulent to *Rk* but avirulent to *Rk2* (54). However, isolates that are virulent to *Rk2* but avirulent to *Rk* have not yet been identified. If one steps outside the single species rule, the interaction between *M. incognita* and *M. javanica* (or *M. arenaria*) isolates with resistance genes in common bean (*Phaseolus vulgaris*) suggests a gene-for-gene relationship because a transagonal interaction is present. *M. incognita* isolates are avirulent to gene combination *Me2me3* but virulent to gene *Me1*, whereas *M. javanica* isolates are virulent to *Me2me3* but avirulent to *Me1* (62). Thus classical (*sensu* Flor, 35) gene-for-gene interactions are indicated but not yet proven to exist for *Meloidogyne* species and their resistant hosts. These common differential interactions between *Meloidogyne* biotypes and resistance genes must be considered in both the breeding and implementation of resistant cultivars.

Attempts to Discriminate Species and Biotypes

MOLECULAR, BIOCHEMICAL, AND MORPHOLOGICAL MARKERS Traditional approaches to diagnosing root-knot nematodes use morphological characters (mostly of adult females) and differential host and resistance bioassays (38). Both approaches are time consuming and expensive, and results from these approaches may not be unequivocal. Esterase and malate dehydrogenase isozyme polymorphisms distinguish *Meloidogyne* spp. and their use has become adopted by several research and advisory groups (24). Their electrophoretic separation is a highly useful, reliable diagnostic tool. Diagnosis of *Meloidogyne* spp. is further hampered by the stage-specific characters required. Field soils are frequently sampled between crops when only second-stage juveniles and eggs are present, and their morphology is not diagnostic. The development of techniques that can discriminate species and even races based on protein or DNA polymorphisms using single eggs or juveniles would be very useful.

New diagnostic approaches are focused on application of biochemical and molecular (DNA) markers that are species or race specific (7, 16, 41, 64, 100). Molecular techniques may facilitate development of fast and accurate diagnostic tools that could alleviate many of the problems associated with traditional diagnostic methods. Use of the polymerase chain reaction (79) to amplify DNA from a single nematode juvenile or egg shows great promise (14, 64,

98). To be successful, DNA marker–based diagnosis requires identification of stable, reproducible polymorphic differences between closely related and sympatrically distributed species. Such differences are present to differentiate the major common *Meloidogyne* spp. (15, 16, 26, 64, 100). Immunoassay (ELISA) based on species-specific monoclonal antibodies and a dot-blot procedure using repetitive DNA sequences can differentiate the potato cyst nematode sibling species *Globodera pallida* and *G. rostochiensis*, and can use single cysts and juveniles (1, 82).

Differentiation of host races or biotypes within a species of *Meloidogyne* has not yet been achieved (64, 100). DNA polymorphisms within a species have been detected, but they are not linked to the genetic determinants (loci) of (a)virulence. Thus they are not markers of host race or biotype differences determined by differential parasitism on specific host-plant resistance genes. Discrimination of *Meloidogyne* race and biotype differences will require linkage of the DNA or enzyme polymorphic marker(s) to the determinant(s) of race or biotype, or detection of the actual DNA coding sequence differences between virulent and avirulent nematode phenotypes. Attempts to clone avirulence genes in nematodes, such as in *G. rostochiensis* (1) and *M. incognita* (9), could provide a direct DNA probe capability for pathotypes differentiated by that specific virulence/avirulence phenotype. The gene pool similarity concept proposed by Bakker and colleagues (1) for distinguishing pathotypes of potato cyst nematode, introduced as discrete founding populations into northern Europe, may not be feasible for the complex array of *Meloidogyne* species and biotypes infecting numerous host crops and differentiated on many resistance genes.

Races or biotypes of *Meloidogyne* species generally do not show additional phenotypic differences linked with (a)virulence. Rational choice of resistant cultivars and rotation sequences can only be made when the race or biotype constitution of a field population is known. Rapid diagnosis of race or biotype without the need to bioassay would be an important advance for integrating natural HPR into rotation-based management programs.

MELOIDOGYNE SPP. RACE AND BIOTYPE SCHEMES Clearly a need exists for a characterization scheme of (a)virulence in *Meloidogyne*. In addition to the *Meloidogyne* race scheme of Sasser (38) (see below), other schemes for differentiating (a)virulence traits within and among closely related species have been primarily for some cyst nematodes (22). An international pathotype scheme for potato cyst nematodes recognizes three *G. pallida* and five *G. rostochiensis* pathotypes differentiated on potato clones with resistance genes (1, 50). Currently 16 races of soybean cyst nematode (*Heterodera glycines*) are differentiated according to reaction on four distinct soybean resistance backgrounds (65). Cereal cyst nematode (*H. avenae*) pathotypes are differentiated on small grain cereals including barley, oat, rye, and wheat cultivars (see 18). A scheme to differentiate beet cyst nematode (a)virulence characters for reaction to distinct resistance factors in sugar beet will be necessary as resistant sugar beets are released (58). Five citrus nematode (*Tylenchulus semipenetrans*) biotypes are distinguished on differential hosts including citrus and olive germplasm (42). These pathotype (biotype or race) schemes are relatively simple, based on one or very few host crop species, and can be expanded to incorporate new resistance differentials as they are identified.

Lack of a functional characterization scheme has hindered understanding, interpretation, and testing of the (a)virulence traits of *Meloidogyne* species. Currently, the only scheme with intraspecific delineations is the widely used differential host test devised by Sasser (38). This was designed primarily to aid in the species differentiation of *M. arenaria, M. hapla, M. incognita,* and *M. javanica.* The six differential hosts (designated cultivars of cotton, peanut, pepper, tobacco, tomato, and watermelon) also distinguish "races" within *M. incognita* and *M. arenaria.* Four host races of *M. incognita* are separated by parasitic ability on tobacco (cv. NC95 possessing a single dominant resistance gene to *M. incognita*) and cotton (cv. Deltapine 61 with no or unknown resistance factors); reactions of host races 1–4 to cotton/tobacco are –/–, –/+, +/–, and +/+, respectively. Two races of *M. arenaria* are separated by parasitic ability on peanut (cv. Florunner) (38). These races are not aligned with any detectable morphological, biochemical, or additional (a)virulence and host range factors or markers. The nematode management utility of these designations for *M. incognita* is restricted to cotton and tobacco production areas.

Lack of correlation of *M. incognita* races with (a)virulence factors matching a wide range of host-plant resistance genes is illustrated by tomato and cowpea resistance gene systems. Both virulent and avirulent isolates for tomato gene *Mi* occur within *M. incognita* race 1, and both virulent and avirulent isolates for cowpea gene *Rk* occur within each of *M. incognita* races 1 and 3 (73). Where additional resistance genes in the host have been identified, e.g. *Mi3* in tomato and *Rk2* in cowpea, further differentiation of these *Meloidogyne* isolates can be made. Thus an isolate can have differential (a)virulence interaction with two genes of the same host species. Similarly, a resistance gene can have a differential (in)compatibility with two isolates of not only the same *Meloidogyne* species but the same traditional host race. This is no doubt the common pattern to be expected for distribution of (a)virulence factors matching a broad range of resistance genes. Thus it is unlikely that resistance gene homologies and hence virulence gene homologies are that close at the host/parasite recognition level across the spectra of *Meloidogyne*-host interactions, even if underlying mechanisms of interaction may be similar. Interactions of a nematode isolate with resistance genes from different plant taxa can be expected to vary widely. For example, we know from routine use of certain root-knot nematode isolates in our resistance program that *M. incognita* isolates virulent on cowpea gene *Rk* are avirulent on tomato genes *Mi* and *Mi3*, and vice versa. Resistance in Lima bean is effective against isolates virulent on cowpea gene *Rk*. Gene *Me1* in common bean is effective against *M. javanica* isolates that are virulent on cowpea gene *Rk*, and so on.

A theory that emerges from these patterns is that each *Meloidogyne* species isolate (or field population) has an extensive series or bank of (a)virulence genes and alleles. These match the array of resistance genes found within an extensive host range of diverse plant taxa. Presumably each nematode isolate has a unique combination of the genes and alleles determining (a)virulence. The particular (a)virulence spectrum present in an individual will be the result of inheritance and the mutation and selection forces acting upon the initial or ancestral complement of genes and alleles, and changes in ploidy level. It is interesting to consider that a single base-pair substitution can change a fungal tomato pathogen (*Cladosporium fulvum*) gene from avirulent to virulent, resulting in loss of resistance in the tomato cultivar possessing the matching resistance gene and a new race designation (combination) for that pathogen isolate (46). Changes from avirulence to virulence resulting in loss of resistance induction also have been shown to occur by complete loss of the fungal avirulence gene, and in phytopathogenic bacteria by deletion, mutation, or insertional inactivation (see 46, 49). Similar events may occur in phytoparasitic nematodes, and in polyploid *Meloidogyne* species (a)virulence expression could be modified by gene action on homologous chromosomes. Progressive increase in virulence during selection experiments described for *M. incognita* and tomato gene *Mi* has been compared (9, 13) to pesticide resistance selection in apomictic aphid clones that results from gene (esterase) amplification (6).

A Proposed Biotype Scheme for Characterizing Meloidogyne spp. (A)virulence Phenotypes

Numerous combinations of (a)virulence/host resistance interactions are possible for each of the several common *Meloidogyne* spp. and their hundreds of host crops. A characterization scheme must include a designation of both the nematode species and the resistant host plant species (or crop recipient of wild species–donated resistance genes). Based on our work with multiple resistance genes to *M. arenaria, M. incognita,* and *M. javanica* in several diverse host crops, and incorporating some earlier suggestions put forward by Triantaphyllou (89), the following scheme for characterizing *Meloidogyne* spp. (a)virulence variants is offered. I choose the term "biotype" for these variants, as a means to circumvent the debate over pathogenicity (implied by "pathotype") and to avoid confusion with the existing "race" scheme for *Meloidogyne* species. Triantaphyllou (89) also argued a case for the term biotype. Biotype is commonly used for insect populations differentiated by host resistance, but thus far only for citrus nematode. Biotype is applied here to *Meloidogyne* spp. isolates distinguished by (a)virulence reaction to host-plant resistance factor(s). It is based on a measurement of nematode reproduction (parasitic ability) on a host plant in comparative tests with known susceptible and resistant host plant genotypes. The index of reproduction is measured by numbers of egg masses (gravid females) or preferably of eggs produced on a differential plant genotype, as a proportion of the numbers produced on a standard susceptible host genotype.

A simple code description indicating virulence or avirulence with respect to a resistance factor in a crop is given, and this would be a multiple code if more than one resistance factor (allele, gene, or combination of genes when resistance conferred by two or more genes) occurs. For example, in Table 1 a partly hypothetical series of reactions of four *M. incognita* isolates to tomato resistance genes is given. The biotype code of isolate (or strain) "DEF" is given as:

M. incognita biotype *L.esc* 1,2/3,4,5

where *L.esc* is an abbreviation of the Latin binomial of the host plant species (i. e. *L.esc* abbreviated from *Lycopersicon esculentum* for tomato); the numbers preceding the slash symbol (/) represent compatible interactions of the nematode (i.e. virulence) on resistance factors of the same number, in this example virulence on resistance factors 1 and 2; and the numbers following the slash symbol (/) represent incompatible interactions of the nematode (i.e. avirulence) on resistance factors of the same number, in this example avirulence on resistance factors 3, 4, and 5. Isolate ABC is avirulent on all five tomato resistance factors, and so a 0 numeral is placed before the slash line, to indicate that tomato is a host and the designations are for avirulence to all resistance factors. This prevents possible confusion with a nonhost scenario, where all reactions would also be placed after the slash, but no preceding 0 is applied. Isolate GHI has a combination of virulence and avirulence different from isolate DEF, whereas isolate JKL would be virulent on all resistance factors (Table 1). The isolate names used here are merely for convenience. In practice they have a laboratory- or researcher-designated number, collection site name, etc, and the name should be retained as unique to that isolate (see also 3).

In Table 2 combinations of biotypes with multiple host designations are shown for different *Meloidogyne* spp. isolates, indicating the extent and flexibility of the scheme. It accommodates expansion as new resistance factors are identified within a crop host species, and by addition of any number of host species. The potentially large numbers of biotype designations per isolate (because of multiple hosts with resistance) would require a computerized database system, for data storage and updating, from which subsets of data could be selected and published or otherwise circulated. A central collation

Table 1 Hypothetical examples of biotype designations for *M. incognita* isolates on tomato resistance genes

Isolate	Resistance factor							Biotype designation
	S	Rf1	Rf2	Rf3	Rf4	Rf5	Rfn	
ABC	+	−	−	−	−	−	?	*L.esc* 0/1,2,3,4,5
DEF	+	+	+	−	−	−	?	*L.esc* 1,2/3,4,5
GHI	+	+	−	+	−	+	?	*L.esc* 1,3,5/2,4
JKL	+	+	+	+	+	+	?	*L.esc* 1,2,3,4,5/

S = susceptible genotype; Rf1 is gene *Mi;* Rf2 is gene *Mi2;* Rf3 is gene *Mi3;* Rf4–Rf5 are genes being described (PA Roberts & JC Veremis, unpublished results); Rfn = genes yet to be identified
+ = compatible interaction (matching virulence or no R gene)
− = incompatible interaction (matching avirulence to R gene)

and organization of biotype data would provide an important framework for analysis of possible trends of homology, divergence, and relationships to host systematics, agricultural histories, geographical distributions, etc. Although the entire biotype designation for any one nematode isolate could be extensive on numerous hosts, in most situations only a small relevant portion of the total biotype designation may be necessary for communication. For example, one may need only to convey the designations relevant to the host plant(s) reported in a research paper (e.g. *L.esc* biotypes for tomato research). Only biotypes discriminated on those crops studied or involved in nematode management decisions on a cropping sequence would be desirable, e.g. *H.vul, S.tub* and *M.sat* designations for *M. chitwoodi* in a barley (*Hordeum vulgare*), potato (*Solanum tuberosum*), alfalfa (*Medicago sativa*) rotation in the Pacific Northwest (30). Breeders, seed companies, and nurseries could report biotypes of the common *Meloidogyne* species designated on a particular host genotype (e.g. seed of a tomato cultivar with gene *Mi* would be indicated to control any biotypes designated *L.esc* 0/1 but none designated *L.esc* 1/).

Adoption of the proposed scheme will require standardization of the differential tests using acceptable methods (88) and their interpretation to designate biotypes. An accepted numerical assignment of resistance factors must be offered/created and adhered to. A mechanism of making available standard differential genotypes will also be desirable. A weakness of the potato cyst nematode pathotype scheme is the inclusion of differentials with polygenically controlled, quantitatively expressed resistance that do not distinguish pathotypes unequivocally and may not represent the resistance gene spectra within derived commercial cultivars (1, 18, 50, 91). Determination of *Meloidogyne* resistance gene inheritance would strengthen the proposed biotype scheme and identify quantitatively inherited resistance that may give intermediate levels of resistance expression. These could still be incorporated into the designation

Table 2 Hypothetical examples of *Meloidogyne* spp. biotype designations based on (a)virulence to resistance genes in host crops

Species and	Tobacco		Cotton		Tomato			Cowpea			Common bean		
isolate	S	Rf1	Rf1	Rf2	S	Rf1	Rf2	S	Rf1	Rf2	S	Rf1	Rf2
M. incognita	+	−	−	−	+	−	−	+	+	−	+	+	+
ABC	*N.tab* 0/1		*G.hir* 0/1,2		*L.esc* 0/1,2			*V.ung* 1/2			*P.vul* 1,2/		
M. incognita	+	+	+	−	+	+	−	+	−	−	+	+	−
GHI	*N.tab* 1/		*G.hir* 1/2		*L.esc* 1/2			*V.ung* 0/1,2			*P.vul* 1/2		
M. javanica	+	−	−	−	+	−	+	+	+	−	+	−	+
PAR	*N.tab* 0/1		*G.hir* /1,2		*L.esc* 2/1			*V.ung* 1/2			*P.vul* 2/1		

S = susceptible genotype; Rf = resistance factor (gene or gene combination)—see text for reference to identified resistance factors in these and additional crops
+ = compatible interaction (matching virulence or no R gene)
− = incompatible interaction (matching avirulence to R gene)

scheme as a "resistance factor." The protocol and interpretation of biotype indexing with inoculation tests would also require standardization and acceptance. However, the global application of the existing *Meloidogyne* differential host test suggests that these challenges could be met by nematologists and plant breeders.

MANAGING RESISTANCE GENES AND (A)VIRULENCE

Resistance is one of several tools for use in an integrated approach for root-knot nematode management (23, 31, 69–71, 91). Briefly I consider aspects of resistance deployment influenced by the variation in (a)virulence and host range as described. Two primary attributes of host resistance for nematode management are relevant: (*a*) the value of resistance in crop self-protection, based on the level of resistant plant tolerance to injury caused by initial infection, and (*b*) the rotational value of resistance in cropping systems for protecting subsequent crops, based on the ability to suppress nematode population densities in soil by restricting nematode reproduction (18, 67–71, 91). These two attributes underpin most nematode resistance breeding and management decisions.

Wide Host Ranges in Diversified Cropping Systems

Mixed *Meloidogyne* species and population infestations combined with extensive and overlapping host ranges confound management approaches based on selective tactics like host plant resistance. This is particularly apparent in light of the diminished availability of broadly effective preplant soil fumigant nematicides (23, 70). Infested areas may contain any combination of two or more *Meloidogyne* species and have a cropping system in which all crops are susceptible hosts (36, 61). Intensive vegetable and field crop production systems of warm climates such as California and Florida are examples within the United States, as are warm temperate-to-tropical regions globally. In the interior valleys of California common vegetable and field crop production includes some combination of alfalfa hay, carrot, common bean, corn, cotton, cowpea, lettuce, Lima beans, melons, peppers, sugar beets, and tomatoes. Cruciferous crops and small grain cereals are also common, and additional crops are grown depending on locality. On any one farm or field, a subset of these crops is produced in a rotation sequence, usually dictated by economics rather than pest management concerns. Both *M. incognita* and *M. javanica* are common in these areas, and some populations of *M. arenaria* and *M. hapla* are present. Several nematode generations may be completed in a growing season and very large final populations often develop (29, 32, 36).

How Much Resistance Is Enough?

A general working assumption for *Meloidogyne* resistance in a complex host cropping sequence like the California example is the more crops with resistant cultivars, the better root-knot can be controlled. The challenge under this assumption is to identify resistance traits in the crop or wild relative species, and breed resistance into superior or acceptable cultivar genotypes. In the California example, *Meloidogyne* resistance has been identified in several major crops, including multiple resistance factors with different specificities in some crops (Table 3). This current assortment of about 20 resistance factors (Table 3) is substantial, and no doubt more resistance genes will be identified within these crops and in additional crops. Although transfer into commercial cultivars is not yet complete, being at different stages of progress and involving different degrees of breeding complexity, availability of all these traits in commercial production is likely in the next few years.

The targeted resistance factors (Table 3) collectively could enable a comprehensive management of root-knot in most annual cropping systems. The overall objective is to have a broad-based spectrum of resistance within the cropping system providing protection to all crops from all root-knot species and biotypes infesting that production region. Some crops will be self-protected by their resistance gene constitution, and some that lack resistance genes will be protected by effects of resistance genes in the preceding crop(s). However, few guidelines exist to determine the most suitable approaches to breeding for multiple resistance traits within one crop, and to assembling them in a multiple cropping system.

Pyramiding Resistance Genes in a Crop

In single pathogen–single crop systems, the aim is usually to add resistance genes into the crop to counteract the spectrum of biotypes or races that are present or prevalent. The choices include placement of resistance genes sepa-

Table 3 *Meloidogyne*-resistance traits in use or in breeding programs for California vegetable and field crops

Crop	HPR factor	*M. incognita*	*M. javanica*	*M. hapla*
Common bean	Rf1	R	S	S
	Rf2	S	R	R
Lima bean	Rf1 and 2	R	S	?
	Rf3	R	R	?
Cowpea	Rf1	R/S	R/S	R
	Rf2,3	R	R	R
Tomato	Rf1	R/S	R/S	S
	Rf2,3	R/S	R/S	?
Carrot	Rf1	R/S	R	S
	Rf2	S	S	R
Pepper	Rf1	R	S	S
	Rf2-5	R	R	R
Cotton	Rf1	R	(R)	(R)
Wheat	Rf1	R	R	(R)

Rf = resistance factor (gene or gene combination); R = resistant; S = susceptible; R/S = resistant to some but not all populations of the species; (R) = nonhost status

rately into different cultivars, i.e. cultivar X carries resistance gene *Rg1*, cultivar Y carries resistance gene *Rg2*, and so on. Alternatively, resistance genes can be pyramided in the same genotype, i.e. cultivar Z carries resistance genes *Rg1* and *Rg2*, to provide a cultivar with the maximum breadth of resistance protection.

The multiple *Meloidogyne* resistance traits in any crop identified thus far tend to be genetically independent and can be combined or separated. Tomato genes *Mi* and *Mi3* are genetically independent and could be combined (8). In fact the two loci are present together in some wild tomato genotypes (8, 101). Gene *Mi* is already bred into many commercial cultivars, but because *Mi3* apparently controls most *Meloidogyne* isolates that are controlled by *Mi*, in addition to some that are *Mi*-virulent, *Mi3* may be useful separately in cultivars, as would *Mi2*, which is heat stable (8, 72, 101). Combining *Mi2* or *Mi3* with a resistance gene that controls *M. hapla* would be most desirable. The current approach in carrot is to combine the Brazilia-derived resistance to *M. arenaria*, *M. incognita*, and *M. javanica* with the *M. hapla* resistance present in some USDA inbred lines (PA Roberts & PW Simon, unpublished). Similar attempts are under way in the University of California dry bean breeding programs to combine resistance traits with different spectra of activity into the same cultivar. These include combining common bean genes *Me1* (*M. javanica* and *M. arenaria* resistant) with *Me2me3* (*M. incognita*), and combining Lima bean resistance (genetics not determined) in line L-76 (*M. incognita*) and/or line

L-7727 (*M. incognita*) with that in L-136 (*M. javanica*) (55). Resistance to *M. incognita* (85) is being added into cotton genotypes that have a basic resistance or nonhost status to other *Meloidogyne* spp. Similarly, resistance to *M. incognita* and *M. javanica* can be added to wheat (47), a nonhost to *M. hapla*.

Nematode resistance genes with multiple alleles matching different biotype (a)virulence-specificities have not been identified, although a candidate gene is the *Meloidogyne* resistance locus *Rk* in cowpea (33, 54). Recently we identified a dominant resistance trait in cowpea that confers a high level of broad-spectrum resistance to *M. arenaria*, *M. incognita*, and *M. javanica* (54). It is tightly linked to or an allele of resistance gene *Rk*, which is used widely but controls only some isolates of these species (PA Roberts & WC Matthews, unpublished results). Multiple allele- or gene family–based resistance is common for some fungal pathogen-host resistance systems such as the rust and mildew resistance genes (19, 21, 40). This complicates efforts for pyramiding resistance genes in a cultivar, and may necessitate combining codominantly expressed alleles with the broadest spectrum of effect. Like *Mi3* to *Mi* in tomato, this new cowpea resistance gene is as effective as or more than *Rk*, and hence may be preferred separately in new cultivars of blackeye dry beans. However, our preliminary analysis of F_1 combinations between the two resistance factors suggests that combined resistance is higher against *M. javanica* than the new resistance factor alone. The reversion of *Rk*-virulent *M. incognita* to avirulence on tomato in the laboratory suggests that virulence to gene *Rk* could be managed by a crop rotation plan.

Combining independently inherited resistance genes with different biotype compatibilities could promote durability of resistance in a crop cultivar. However, this would seem unnecessary unless a unique spectrum of resistance were achieved (102). Because *Mi* and *Mi3* in tomato control many of the same common *Meloidogyne* spp. populations, there apparently is little justification for using them together in the same genotype, although resistance durability could be extended. A virulent biotype selected on *Mi* should still be avirulent to *Mi3* (72, 101), and vice versa. On a resistance gene combination that blocks nematode regeneration, virulence to both genes would need to be present simultaneously to enable an individual to reproduce. Otherwise, one gene would prohibit virulence selection on the other gene via reciprocal lethality effects.

The potential impact on resistance durability of alternating, in some chosen frequency, cultivars that each possess one resistance gene was studied using a computer simulation model (87). The model included two potato cyst nematode virulence loci and two independent potato resistance gene loci with several alternative resistance mechanisms (e.g. partial/complete; pathotype specific/nonspecific). No data exist to substantiate the simulation results, so only hypothetical inferences may be drawn. The simulation indicated that the placement of resistance genes singly in sequentially grown cultivars would shorten resistance durability compared to "pyramiding" the genes into one cultivar, which would have to be grown more frequently than either single resistance cultivar. Potato cyst nematodes complete only one generation per year, and in theory they can maintain avirulence alleles in populations via males; males do not need to feed (thus are not blocked by resistance), are favored over females by environmental sex determination under adverse conditions, and can mate with virulent females to pass on avirulence alleles to the next generation (83). Parthenogenetic *Meloidogyne* spp. with multiple generations would be more likely to result in rapid selection for virulence. Thus the importance of pyramiding resistance traits in diverse crop plant taxa within a cropping system should be considered.

Pyramiding Resistance Genes in a Cropping System

The differential virulence reactions of biotypes to resistance genes in different crop plants could enhance the durability of all resistance genes in the cropping system. Selection of virulence to one resistance gene will be unlikely to result in virulence to a different resistance gene. Furthermore, each resistance factor would need to be grown relatively infrequently, because many options exist within the rotation and this will reduce virulence selection pressure. This is a reasonable speculation arising from the observed differences in gene action on virulent biotypes between resistance in different crops. No multiple-crop-multiple-resistance systems are available yet for analysis. However, the maintenance of functional avirulence genes in some bacterial pathogens for resistance genes that are in plants not considered in the normal or recent host range (see 21, 48, 49, 93) implies that functional and genetic homologies of resistance/avirulence and susceptibility/host range could reside and be maintained across even distantly related plant taxa. Among the recently cloned resistance genes, *N* to mosaic virus in tobacco (97), *Pto* and *RPS2* to *P. syringae* in tomato (52) and *Arabidopsis* (2, 56), respectively, *Cf9* to *Cladosporium fulvum* in tomato and L^6 to *Melampsora lini* in flax (see 57), some lack gene product homology (*Pto* and *RPS2,* see 2, 56), but others (L^6, *N,* and *RPS2*) have common sequence patterns (57). *N* and *RPS2* protein product sequences are 24% identical and 49% similar (97). Thus common basic mechanisms may underlie resistance gene function to different pathogen groups (including nematodes) within the plant kingdom.

SUMMARY REMARKS

The concept of multiple natural resistance genes in several crops to manage *Meloidogyne* in multiple cropping systems is unique in pest and disease resistance use in crop production. Most plant pests and pathogens (insects, cyst and some other nematodes, fungi, bacteria, and viruses) for which significant resistance exists have generally narrow host ranges at the species or genus level. Or in the case of pathogen species with wide host range (e.g. *Pseudomonas syringae* or *Fusarium oxysporum*), the species has diverged pathogenically in the form of host-specific pathovars or formae speciales, respectively.

Resistance in multiple crops to manage *Meloidogyne* is a real possibility in the near future, as new resistance factors are made available. The conceptual focus is development of a shield of resistance. By a combination of direct action and residual impact on soil population densities, the resistance shield could be expected to effectively control the complete spectrum of common *Meloidogyne* species and their (a)virulent biotypes. Compared to important bacterial and fungal plant pathogens, *Meloidogyne* (a)virulence appears more stable, less variable, perhaps due to clonal, polyploid biology, and certainly less impacted by local (short-distance) and regional (long-distance) migration and spread as a soil-borne plant parasite. Thus rapid appearance of virulent biotypes typical of many bacteria and fungi is unlikely in *Meloidogyne*. However, biotypes with specific (a)virulence constitutions do occur, and may be selected on resistant plant genotypes. A biotype scheme is offered for characterizing, categorizing, and communicating *Meloidogyne* spp. (a)virulence. It is open-ended to facilitate addition and modification as new resistance traits and new biotypes are identified.

Literature Cited

1. Bakker J, Folkertsma RT, Rouppe Van Der Voort JNAM, De Boer JM, Gommers FJ. 1993. Changing concepts and molecular approaches in the management of virulence genes in potato cyst nematodes. *Annu. Rev. Phytopathol.* 31: 169–90
2. Bent AF, Kunkel BN, Dahlbeck D, Brown KL, Schmidt R, et al. 1994. *RPS2* of *Arabidopsis thaliana*: a leucine-rich repeat class of plant disease resistance genes. *Science* 265:1856–60
3. Bird DM, Riddle DL. 1994. A genetic nomenclature for parasitic nematodes. *J. Nematol.* 26:138–43
4. Bost SC, Triantaphyllou AC. 1982. Genetic basis of the epidemiologic effects of resistance to *Meloidogyne incognita* in the tomato cultivar Small Fry. *J. Nematol.* 14:540–44
5. Brown RH, Kerry BR, eds. 1987. *Principles and Practice of Nematode Control in Crops*. Orlando, FL: Academic. 447 pp.
6. Bunting S, Van Emden HF. 1980. Rapid response to selection for increased esterase activity on small populations of an apomictic clone of *Myzus persicae*. *Nature* 285:502–03
7. Burrows PR. 1990. The use of DNA to identify plant parasitic nematodes. *Helminth. Abstr.* 59:1–8
8. Cap GB, Roberts PA, Thomason IJ. 1993. Inheritance of heat-stable resistance to *Meloidogyne incognita* in *Lycopersicon peruvianum* and its relationship to the *Mi* gene. *Theor. Appl. Genet.* 85:777–83
9. Castagnone-Sereno P. 1994. Genetics of *Meloidogyne* virulence against resis-

tance genes from solanaceous crops. See Ref. 51, pp. 261–76
10. Castagnone-Sereno P, Bongiovanni M, Dalmasso A. 1992. Differential expression of root-knot nematode resistance genes in tomato and pepper: evidence with *Meloidogyne incognita* virulent and avirulent near-isogenic lineages. *Ann. Appl. Biol.* 120:487–92
11. Castagnone-Sereno P, Bongiovanni M, Dalmasso A. 1993. Stable virulence against the tomato resistance *Mi* gene in the parthenogenetic root-knot nematode *Meloidogyne incognita*. *Phytopathology* 83:803–5
12. Castagnone-Sereno P, Bongiovanni M, Dalmasso A. 1994. Reproduction of virulent isolates of *Meloidogyne incognita* on susceptible and *Mi*-resistant tomato. *J. Nematol.* 26:324–28
13. Castagnone-Sereno P, Wajnberg E, Bongiovanni M, Leroy F, Dalmasso A. 1994. Genetic variation in *Meloidogyne incognita* virulence against the tomato *Mi* resistance gene: evidence from isofemale line selection studies. *Theor. Appl. Genet.* 88:749–53
14. Caswell-Chen EP, Williamson VM, Wu FF. 1992. Random amplified polymorphic DNA analysis of *Heterodera cruciferae* and *H. schachtii* populations. *J. Nematol.* 24:343–51
15. Cenis JL. 1993. Identification of four major *Meloidogyne* spp. by random amplified polymorphic DNA (RAPD-PCR). *Phytopathology* 83:76–78
16. Chacón MR, Parkhouse M, Burrows PR, Gárate T. 1993. The use of a digoxigenin-labelled synthetic DNA oligonucleotide for the rapid and sensitive identification of *Meloidogyne incognita*. *Fundam. Appl. Nematol.* 16:495–99
17. Cook R. 1991. Resistance in plants to cyst and root-knot nematodes. *Agric. Zool. Rev.* 4:213–40
18. Cook R, Evans K. 1987. Resistance and tolerance. See Ref. 5, pp. 179–231
19. Crute IR. 1992. From breeding to cloning (and back again?): a case study with lettuce downy mildew. *Annu. Rev. Phytopathol.* 30:485–506
20. Dalmasso A, Castagnone-Sereno P, Bongiovanni M, De Jong A. 1991. Acquired virulence in the plant parasitic nematode *Meloidogyne incognita*. 2. Two-dimensional analysis of isogenic isolates. *Rev. Nematol.* 14:277–83
21. De Wit PJGM. 1992. Molecular characterization of gene-for-gene systems in plant-fungus interactions and the application of avirulence genes in control of plant pathogens. *Annu. Rev. Phytopathol.* 30:391–418
22. Dropkin VH. 1988. The concept of race in phytonematology. *Annu. Rev. Phytopathol.* 26:145–61
23. Duncan LW. 1991. Current options for nematode management. *Annu. Rev. Phytopathol.* 29:469–90
24. Esbenshade PR, Triantaphyllou AC. 1990. Isozyme phenotypes for the identification of *Meloidogyne* species. *J. Nematol.* 22:10–15
25. Evans K, Haydock PPJ. 1990. A review of tolerance by potato plants of cyst nematode attack, with consideration of what factors may confer tolerance and methods of assaying and improving it in crops. *Ann. Appl. Biol.* 117:703–40
26. Fargette M, Blok VC, Phillips MS, Trudgill DL. 1994. Genetic variation in tropical *Meloidogyne* species. See Ref. 51, pp. 91–96
27. Fassuliotis G. 1982. Plant resistance to root-knot nematodes. In *Nematology in the Southern Region of the United States*, ed. RD Riggs, pp. 33–49. *South. Coop. Ser. Bull.* 276
28. Fassuliotis G. 1987. Genetic basis of plant resistance to nematodes. See Ref. 94, pp. 364–71
29. Ferris H, Ball DA, Beem LW, Gudmunson LA. 1986. Using nematode count data in crop management decisions. *Calif. Agric.* 40:12–14
30. Ferris H, Carlson HL, Westerdahl BB. 1994. Nematode population changes under crop rotation sequences: consequences for potato production. *Agron. J.* 86:340–48
31. Ferris H, Castro CE, Caswell EP, Jaffee BA, Roberts PA, et al. 1992. Biological approaches to the management of plant-parasitic nematodes. In *Beyond Pesticides: Biological Approaches to Pest Management in California*, ed. JP Madden, pp. 68–101. *Univ. Calif. Div. Agric. Natl. Res. Publ.* No. 3354
32. Ferris H, Noling JW. 1987. Analysis and prediction as a basis for management decisions. See Ref. 5, pp. 49–85
33. Fery RL, Dukes PD. 1980. Inheritance of root-knot resistance in the cowpea (*Vigna unguiculata* (L.) Walp.). *J. Am. Soc. Hortic. Sci.* 105:671–74
34. Fire A, Harrison SW, Dixon D. 1990. A modular set of *lacZ* fusion vectors for studying gene expression in *Caenorhabditis elegans*. *Gene* 93:189–98
35. Flor HH. 1942. Inheritance of pathogenicity in *Melampsora lini*. *Phytopathology* 32:653–69
36. Griffin GD, Roberts PA, eds. 1994. *Quantifying Nematode Control*. *Utah State Univ. Res. Rep.* No. 149. 59 pp.

37. Habgood RM. 1970. Designation of physiologic races of plant pathogens. *Nature* 227:1268–69
38. Hartman KM, Sasser JN. 1985. Identification of *Meloidogyne* species on the basis of differential host test and perineal pattern morphology. In *An Advanced Treatise on Meloidogyne, vol. 2. Methodology*, ed. KR Barker, CC Carter, JN Sasser, pp. 69–77. Raleigh, NC: NC State Univ. Graphics
39. Hendy H, Pochard E, Dalmasso A. 1985. Transmission héréditaire de la résistance aux nématodes *Meloidogyne* Chitwood (Tylenchida) portée par 2 lignées de *Capsicum annuum* L.: étude de descendances homozygotes issues d'androgenèse. *Agronomie* 5:93–100
40. Hulbert SH, Sudupak MA, Hong KS. 1993. Genetic relationships between alleles of the *Rp1* rust resistance locus of maize. *Mol. Plant-Microb. Interact.* 6: 387–92
41. Hyman BC. 1990. Molecular diagnostics of *Meloidogyne* species. *J. Nematol.* 22:24–31
42. Inserra RN, Vovlas N, O'Bannon JH. 1980. A classification of *Tylenchulus semipenetrans* biotypes. *J. Nematol.* 12: 283–87
43. Janssen R, Bakker J, Gommers FJ. 1991. Mendelian proof for a gene-for-gene relationship between virulence of *Globodera rostochiensis* and the H_1 resistance gene in *Solanum tuberosum* ssp. *andigena* CPC 1673. *Rev. Nématol.* 14: 214–19
44. Jarquin-Barberena H, Dalmasso A, de Guiran G, Cardin M-C. 1991. Acquired virulence in the plant-parasitic nematode *Meloidogyne incognita*. 1. Biological analysis of the phenomenon. *Rev. Nématol.* 14:299–303
45. Jones FGW, Parrott DM, Perry JN. 1981. The gene-for-gene relationship and its significance for potato cyst nematodes and their solanaceous hosts. In *Plant Parasitic Nematodes*, ed. BM Zuckerman, RA Rohde, 3:23–36. New York: Academic
46. Joosten MHAJ, Cozijnsen TJ, De Wit PJGM. 1994. Host resistance to a fungal tomato pathogen lost by a single base-pair change in an avirulence gene. *Nature* 367:384–86
47. Kaloshian I, Roberts PA, Waines JG, Thomason IJ. 1990. Inheritance of resistance to root-knot nematodes in *Aegilops squarrosa* L. *J. Hered.* 81:170–72
48. Keen NT. 1990. Gene-for-gene complementarity in plant-pathogen interactions. *Annu. Rev. Genet.* 24:447–63
49. Keen NT, Bent AF, Staskawicz BJ. 1993. Plant disease resistance genes: interactions with pathogens and their improved utilization to control plant diseases. In *Biotechnology in Plant Disease Control*, ed. I Chet, pp. 65–88. New York: Wiley
50. Kort J, Ross H, Rumpenhorst, HJ, Stone AR. 1977. An international scheme for identifying and classifying pathotypes of potato cyst-nematodes *Globodera rostochiensis* and *G. pallida*. *Nematologica* 23:333–39
51. Lamberti F, De Georgi C, Bird DM, ed. 1994. *Advances in Molecular Plant Nematology*. New York/London: Plenum. 309 pp.
52. Martin GB, Brommonschenkel SH, Chunwongse J, Frary A, Ganal MW, et al. 1993. Map-based cloning of a protein kinase gene conferring disease resistance in tomato. *Science* 262:1432–36
53. Martin TJ, Ellingboe AH. 1976. Differences between compatible parasite/host genotypes involving the *Pm4* locus of wheat and the corresponding genes in *Erysiphe graminis* f. sp. *tritici*. *Phytopathology* 66:1435–38
54. Matthews WC, Roberts PA. 1994. Resistance in cowpea to *Meloidogyne incognita* isolates and *M. javanica* virulent to the *Rk* gene. *J. Nematol.* 26:108 (Abstr.)
55. McGuire DC, Allard RW, Harding JA. 1961. Inheritance of root-knot nematode resistance in Lima beans. *J. Am. Soc. Hortic. Sci.* 78:302–7
56. Mindrinos M, Katagiri F, Yu G-L, Ausubel FM. 1994. The *A. thaliana* disease resistance gene *RPS2* encodes a protein containing a nucleotide-binding site and leucine-rich repeats. *Cell* 78: 1089–99
57. Moffat AS. 1994. Mapping the sequence of disease resistance. *Science* 265:1804–5
58. Muller J. 1992. Detection of pathotypes by assessing the virulence of *Heterodera schachtii*. *Nematologica* 38:50–64
59. Netscher C. 1977. Observations and preliminary studies on the occurrence of resistance breaking biotypes of *Meloidogyne* spp. on tomato. *Cah. ORSTOM, Sér. Biol.* 11:173–78
60. Netscher C. 1978. Morphological and physiological variability of species of *Meloidogyne* in West Africa and implications for their control. *Meded. LandbHoogesch. Wageningen* 78–3:1–46
61. Netscher C, Sikora RA. 1990. Nematode parasites of vegetables. In *Plant Parasitic Nematodes In Subtropical and*

Tropical Agriculture, ed. M Luc, RA Sikora, J Bridge, pp. 273–83. Wallingford, UK: CAB Int.

62. Omwega CO, Roberts PA. 1992. Inheritance of resistance to *Meloidogyne* spp. in common bean and the genetic basis of its sensitivity to temperature. *Theor. Appl. Genet.* 83:720–26
63. Person CO. 1959. Gene-for-gene relationships in host-parasite systems. *Can. J. Bot.* 37:1101–30
64. Powers TO, Harris TS. 1993. A polymerase chain reaction method for identification of five major *Meloidogyne* species. *J. Nematol.* 25:1–6
65. Riggs RD, Schmitt DP. 1988. Complete characterization of the race scheme for *Heterodera glycines*. *J. Nematol.* 20: 392–95
66. Riggs RD, Winstead NN. 1959. Studies on resistance in tomato to root-knot nematodes and on occurrence of pathogenic biotypes. *Phytopathology* 49:716–24
67. Roberts PA. 1982. Plant resistance in nematode pest management. *J. Nematol.* 14:24–33
68. Roberts PA. 1990. Resistance to nematodes: definitions, concepts and consequences. See Ref. 88, pp. 1–15
69. Roberts PA. 1992. Current status of the availability, development and use of host plant resistance to nematodes. *J. Nematol.* 24:213–27
70. Roberts PA. 1993. Future of nematology: integration of new and improved management strategies. *J. Nematol.* 25: 383–94
71. Roberts PA. 1994. Integration and protection of novel nematode management strategies. *Fundam. Appl. Nematol.* 17: 203–10
72. Roberts PA, Dalmasso A, Cap GB, Castagnone-Sereno P. 1990. Resistance in *Lycopersicon peruvianum* to isolates of *Mi* gene-compatible *Meloidogyne* populations. *J. Nematol.* 22:585–89
73. Roberts PA, Frate CA, Matthews WC. 1995. Interactions of virulent *Meloidogyne incognita* and Fusarium wilt on resistant cowpea genotypes. *Phytopathology*. In press
74. Roberts PA, May DM. 1986. *Meloidogyne incognita* resistance characteristics in tomato genotypes developed for processing. *J. Nematol.* 18:353–59
75. Roberts PA, Thomason IJ. 1986. Variability in reproduction of isolates of *Meloidogyne incognita* and *M. javanica* on resistant tomato genotypes. *Plant Dis.* 70:547–51
76. Roberts PA, Thomason IJ. 1989. A review of variability in four *Meloidogyne* spp. measured by reproduction on several hosts including *Lycopersicon*. *Agric. Zool. Rev.* 3:225–52
77. Roberts PA, Van Gundy SD, McKinney HE. 1981. Effects of soil temperature and planting date of wheat on *Meloidogyne incognita* reproduction, soil populations and grain yield. *J. Nematol.* 13:338–45
78. Robinson RA. 1987. *Host Management in Crop Pathosystems*. New York: Macmillan. 263 pp.
79. Saiki RK, Gelfand DH, Stoffel S, Scharf SJ, Higuchi R, et al. 1988. Primer-directed enzymatic amplification of DNA with a thermostable DNA polymerase. *Science* 239:487–91
80. Sasser JN. 1980. Root-knot nematodes: a global menace to crop production. *Plant Dis.* 64:36–41
81. Sasser JN, Freckman DW. 1987. A world perspective on nematology: the role of the society. See Ref. 94, pp. 7–14
82. Schots A, Gommers FJ, Bakker J, Egberts E. 1990. Serological differentiation of plant-parasitic nematode species with polyclonal and monoclonal antibodies. *J. Nematol.* 22:16–23
83. Schouten HJ. 1994. Preservation of avirulence genes of potato cyst nematodes through environmental sex determination: a model involving complete, monogenic resistance. *Phytopathology* 84:771–73
84. Shaner G, Lacy GH, Stromberg EL, Barker KR, Pirone TP. 1992. Nomenclature and concepts of pathogenicity and virulence. *Annu. Rev. Phytopathol.* 30:47–66
85. Shepherd RL. 1986. Cotton resistance to the root-knot-Fusarium wilt complex. 2. Relation to root-knot resistance and its implications on breeding for resistance. *Crop Sci.* 26:233–37
86. Sidhu GS, Webster JM. 1981. The genetics of plant-nematode parasitic systems. *Bot. Rev.* 47:387–419
87. Spitters CJT, Ward SA. 1988. Evaluation of breeding strategies for resistance to potato cyst nematodes using a population dynamic model. *Euphytica* S:87–98
88. Starr JL, ed. 1990. *Methods for Evaluating Plant Species for Resistance to Plant-Parasitic Nematodes*. Hyattsville, MD: Soc. Nematol.
89. Triantaphyllou AC. 1987. Genetics of nematode parasitism on plants. See Ref. 94, pp. 354–63
90. Triantaphyllou AC. 1985. Cytogenetics, cytotaxonomy and phylogeny of root-

knot nematodes. In *An Advanced Treatise on Meloidogyne. Biology and Control*, ed. JN Sasser, CC Carter, 1: 113–24. Raleigh, NC: NC State Univ. Graphics

91. Trudgill DL. 1991. Resistance to and tolerance of plant parasitic nematodes in plants. *Annu. Rev. Phytopathol.* 29: 167–92
92. Turner SJ. 1990. The identification and fitness of virulent potato cyst nematode populations (*Globodera pallida*) selected on resistant *Solanum vernei* hybrids for up to eleven generations. *Ann. Appl. Biol.* 117:385–97
93. Valent B, Farrall L, Chumley FG. 1991. *Magnaporthe grisea* genes for pathogenicity and virulence identified through a series of backcrosses. *Genetics* 127:87–101
94. Veech JA, Dickson DW, ed. 1987. *Vistas on Nematology*. Hyattsville, MD: Soc. Nematol.
95. Whalen MC, Stall RE, Staskawicz BJ. 1988. Characterization of a gene from a tomato pathogen determining hypersensitive resistance in a nonhost species and genetic analysis of this resistance in bean. *Proc. Natl. Acad. Sci. USA* 85:6743–47
96. Whitehead AG. 1991. Selection for virulence in the potato cyst-nematode, *Globodera pallida*. *Ann. Appl. Biol.* 118: 395–402
97. Whitham S, Dinesh-Kumar SP, Choi D, Hehl R, Corr C, Baker B. 1994. The product of the tobacco mosaic virus resistance gene N: similarity to Toll and the interleukin-1 receptor. *Cell* 78: 1101–15
98. Williamson VM, Caswell-Chen EP, Wu FF, Hanson D. 1994. PCR for nematode identification. See Ref. 51, pp. 119–27
99. Williamson VM, Ho J-Y, Ma HM. 1992. Molecular transfer of nematode resistance genes. *J. Nematol.* 24:234–41
100. Xue B, Baillie DL, Webster JM. 1993. Amplified fragment length polymorphisms of *Meloidogyne* spp. using oligonucleotide primers. *Fundam. Appl. Nematol.* 16:481–87
101. Yaghoobi J, Kaloshian I, Wen Y, Williamson VM. 1995. Mapping a new nematode resistance locus in *Lycopersicon peruvianum*. *Theor. Appl. Genet.* In press
102. Young LD. 1992. Problems and strategies associated with long-term use of nematode resistant cultivars. *J. Nematol.* 24:228–33

Annu. Rev. Phytopathol. 1995. 33:223–49

TRANSMISSION OF VIRUSES BY PLANT NEMATODES

D. J. F. Brown, W. M. Robertson, and D. L. Trudgill
Zoology Department, Scottish Crop Research Institute, Invergowrie, Dundee, DD2 5DA, United Kingdom

KEY WORDS: Longidoridae, Nepoviruses, Trichodoridae, tobraviruses, virus-vectors

ABSTRACT

Longidorid and trichodorid nematodes transmit nepo- and tobraviruses, respectively. A specific association exists between vectors and their viruses which is a consequence of the nature, site and mechanism of virus particle retention within the vector. It is correlated with the serological properties of the virus coat protein and determined by the RNA-2 segment of the virus genome and by an inherited character of the vector. The virus coat protein is probably involved in the recognition process between vector and virus but is not the sole determinant for transmission of tobraviruses. Genetic changes made to proteins present in the RNA-2 segment of pea early-browning tobravirus have been used to reveal the probable involvement of several proteins in vector transmission. "Protruding" C-terminal amino acid sequences of tobaviruses possibly link, with the aid of a viral determined helper factor, to the site of retention.

INTRODUCTION

In the early part of the twentieth century the hop cyst nematode, *Heterodera humuli,* was implicated in the spread of nettlehead disease of hops in England (101) but was subsequently exonerated (34). However, the disease was eventually determined as being caused by a nematode-transmitted nepovirus (11, 157). Several soil-inhabiting nematodes were examined as possible vectors of other soil-borne diseases (1, 75, 127), but it was not until 1958 that *Xiphinema index* was confirmed as the natural vector of grapevine fanleaf nepovirus in

0066-4286/95/0901-0223$05.00

vineyards in California (64). In 1959, *X. diversicaudatum* was shown to be the vector of arabis mosaic nepovirus in strawberry crops in southern England (55, 74), and during the next two years *Longidorus elongatus* was identified as the vector of tomato black ring nepovirus (56) and *Paratrichodorus minor* and *P. pachydermus* as vectors of tobacco rattle tobravirus in North America and Europe, respectively (130, 168). These reports of transmission of viruses by nematodes stimulated a search for other associations.

A characteristic feature of virus transmission by nematodes is the high degree of specificity that exists between the vector species and their associated viruses. This specificity is determined by the nematodes' ability to retain virus particles at specific sites within their feeding apparatus, in a semipersistent manner (91, 109, 113, 141–143). During the past decade, research into the transmission of viruses by nematodes has increased with the development of specialized techniques and the availability of molecular biology methods to investigate the mechanisms involved in nematode-virus-plant interactions. We review the complexity and subtlety of these interactions and describe recent advances in determining and understanding the mechanisms involved in vector-nematode and plant-virus associations.

VIRUS-VECTOR NEMATODES

Four orders of nematodes contain members that feed upon plants but only a relatively few species of the orders Dorylaimida, family Longidoridae, and Triplonchida, family Trichodoridae, are vectors of plant viruses. Longidoridae and Trichodoridae nematodes are vermiform, soil-inhabiting ectoparasites that feed on roots of a wide range of plants. Most species have long life-cycles (2 to 5 years) with slow rates of multiplication (80). Detailed descriptions of the more important virus-vector species have been published (17, 18, 33, 65, 66–70, 72, 103, 128, 129, 170).

Longidorids are relatively long nematodes (2 to 12 mm) with long hollow feeding stylets (60–250 μm) that enable them to feed deep within root-tips. The stylet is formed anteriorly by an odontostyle that is used to penetrate root cells, and posteriorly by an odontophore that contains nerve tissues adjacent to the food canal, which probably enable the nematode to discriminate between feeding sites within plant roots (114, 118, 148, 155). Protractor muscles are attached to the rear of the odontophore (to the characteristic flanges present in *Xiphinema* spp) and the odontostyle may be almost entirely protracted due to the length and rigidity of the odontophore.

The esophagus consists of a long, narrow anterior tube that connects the stylet with a prominent, muscular cylindrical bulb that provides the pumping action used to withdraw plant cell contents and force them, against the nematodes' hydrostatic body-pressure, through a one-way valve into the gut. The bulb contains three large gland cells. The dorsal gland cell, and its associated gland nucleus, is connected by a duct and valve to the food canal at the anterior end of the bulb and secretions are intermittently passed forward into the plant during feeding. (117, 118, 147, 148, 155). A pair of subventral gland ducts and gland nuclei are situated halfway along the bulb, and occasionally a second pair of ducts, without gland nuclei, are present that open into the posterior of the bulb. The subventral gland ducts open into the bulb chamber, and presumably the gland cell contents are moved backwards to be ingested during feeding.

Feeding at root tips by most longidorids, but not by *X. americanum*-group nematodes, induces the formation of galls. Galls induced by *X. index* contain enlarged, multinucleate cells with dense cytoplasm (124, 166, 169, 171, 173, 174), those caused by *X. diversicaudatum* contain increased amounts of DNA, RNA, and protein (48–50), and *Longidorus* spp induce root-tip galls containing enlarged, amoeboid-shaped cells that contain increased amounts of DNA, but not multinucleate cells (8, 47, 49). *Xiphinema* spp, which induce root-tip galls, have two distinct feeding behaviors. Most frequently they feed on a column of progressively deeper cells. The contents of each cell penetrated are removed during short periods of ingestion, each lasting several seconds to several minutes, interspersed with brief pauses of 1–10 s during which dorsal gland cell secretions are injected into the cell. The second, less frequent, type of feeding resembles that usually observed with *Longidorus* spp, namely deep stylet insertion followed by 15–60 min inactivity during which salivation probably occurs and then by 1–3 h continuous ingestion (154).

There appear to be two feeding phases: an initial, inductive gall-inducing stage followed by an exploitive phase. The inductive stage, especially in *Xiphinema* spp, is probably associated with the second type of feeding behavior and the exploitive stage involves repeated bouts of salivation causing liquefaction of the cytoplasm, followed by ingestion. Feeding by *Longidorus* spp involves long periods of ingestion during which a volume approximately equivalent to 40 normal root-tip cells is removed each hour. During feeding in older root-tip galls the walls between cells develop holes, believed to be caused by the secretions from the dorsal gland, and the contents of a group of modified cells are ingested during each bout of feeding (8, 117).

Trichodorids have a solid onchiostyle that they use to penetrate root cells, and the esophagus consists of a long narrow tube that expands posteriorly to form a spathulate bulb that contains one dorsal, two anterior, and two posterior ventrosublateral glands. They typically feed on root hairs and epidermal and subepidermal cells adjacent to the zone of root elongation. These nematodes can distinguish between living and dead cells. The feeding process, which has been intensively studied with the aid of time-lapse cinephotography and computer-enhanced video imaging (171, 172), comprises five distinct phases:

exploration, perforation of the cell wall, salivation, ingestion, and withdrawal from the cell. The nematode presses its lips firmly against the cell wall, the stoma is drawn forward, and the cell wall is perforated, usually in less than 1 min, by rapid thrusts (6/s) of the onchiostyle. Upon penetration the onchiostyle is repeatedly thrust (1/s) through the puncture hole to a depth of 2 to 3 μm. The nematode then injects saliva into the cell, some of which hardens to form a hollow feeding tube. The cytoplasm of the penetrated cell rapidly aggregates at the site of penetration, the cell nucleus also moves toward the feeding site losing its granulated appearance, and, together with the cytoplasm, is ingested (171, 172). The saliva appears to be produced in the gland cells present in the esophageal bulb.

NEMATODE-TRANSMITTED VIRUSES

Nematode-transmitted viruses belong to two taxonomic groups with wide geographic distributions, mainly in the northern hemisphere. The 36 reported nepoviruses have isometrical particles whereas the three tobraviruses each have tubular particles. All three tobraviruses are transmitted naturally by trichodorid nematodes but only about one third of the nepoviruses are transmitted by longidorid nematodes. Both groups of viruses have bipartite genomes, each containing single-stranded RNA. With nepoviruses the larger molecule, RNA-1, carries genetic determinants for host range, some types of symptom expression in herbaceous hosts, and seed transmissibility. The smaller molecule, RNA-2, carries determinants for serological specificity (coat protein), nematode transmissibility, and other types of symptom expression in herbaceous hosts. Both RNAs of tobraviruses contain determinants for symptom expression and the RNA-2 contains determinants for serological specificity and vector transmission. Infectivity for plants is associated with both RNAs of nepoviruses but only with RNA-1 of tobraviruses (59, 62, 93, 95, 106).

Nematode-transmitted nepo- and tobraviruses cause mild to severe symptoms in many cultivated crops; however, they are considered to be primarily pathogens of wild plants, which they infect usually without causing symptoms. Tobacco rattle tobravirus has the most extensive host range of any plant virus. Most are seed transmitted, a feature important in their ecology for survival, persistence, and distribution. Many nematode-transmitted viruses, especially tobacco rattle tobravirus, occur as a range of serological and/or symptomatological variants. The major serological variants often have different nematode species as their natural vectors.

The taxonomy and nomenclature of plant viruses remain controversial (94), and the nepoviruses are identified primarily on serological and, to a lesser extent, on physicochemical properties. Differences in physicochemical properties have been used to separate nepoviruses into two, three, or four subgroups

(38, 82, 84, 85, 98). Furthermore, recent nucleotide sequence data of the coat protein of SLRSV (see Table 1 for explanation of virus acronyms) have indicated the virus to be only distantly related to other members of the nepovirus group (36). To overcome these taxonomic problems the name "longiviruses" has been suggested when referring to viruses transmitted by Longidorid nematodes (27).

Tobraviruses occur naturally as several serologically distinguishable strains but little attempt has been made to identify their serological affinities. Strains of TRV were thought to form three groups (63), but one is now referred to as PRV and has only been reported from Brazil, where it is transmitted by *P. minor*. The other two groups subsequently were considered as being a single heterogeneous group (61), but nucleic acid hybridization techniques using cDNA copies of unfractionated viral RNA separated 15 strains of tobraviruses into three distinct groups corresponding to PEBV, PRV, and TRV (Table 1) (120)

Pseudorecombinant nepo- and tobraviruses, with the RNA-1 or RNA-2 molecules exchanged between different strains of the same virus, have been produced in the laboratory. Such exchanges were not successful when the RNAs came from different viruses. However, several natural tobravirus recombinants have been identified, each with the pathogenicity of TRV and the serological properties of PEBV (105, 119)

VECTOR NEMATODE AND VIRUS RELATIONS

Criteria for Demonstrating Nematode Transmission of Viruses

During the 1960s and 1970s, considerable effort was directed to identifying nematode and virus associations. However, many of these reports contained insufficient evidence to justify the conclusions that particular viruses were transmitted by the nematode species involved. To overcome this problem a set of criteria was established for assessing reports of nematodes transmitting viruses, initially for longidorid and nepoviruses (152) and subsequently for trichodorids and tobraviruses (24). The criteria were broadly similar for both nematode and virus groups: (*a*) infection of the bait plant must be demonstrated; (*b*) experiments should be done with handpicked nematodes; (*c*) appropriate controls should be included to show unequivocally that the nematode is the vector; (*d*) the nematode should be fully identified; and (*e*) the virus should be fully characterized. Publications reporting nematodes as vectors of viruses were re-evaluated according to these criteria; the vector and virus associations that satisfy these criteria are listed in Table 1 (24, 152).

Table 1 Specific associations between *Longidorus, Paralongidorus, Paratrichodorus, Trichodorus,* and *Xiphinema* virus-vector nematode species and nepo- and tobraviruses

Vector species	Viruses	Acronym	Reference[a]
	NEPOVIRUSES		
Longidorus			
apulus	artichoke Italian latent (Italian strain)	AILV	108
arthensis	cherry rosette disease	CRosV	19
attenuatus	tomato black ring (German/English strain)	TBRV	53
diadecturus	peach rosette mosaic	PRMV	35
elongatus	raspberry ringspot (Scottish strain)	RRSV	132
	tomato black ring (Scottish strain)	TBRV	56
fasciatus	artichoke Italian latent (Greek strain)	AILV	122
macrosoma	raspberry ringspot (English strain)	RRSV	53
martini	mulberry ringspot	MRSV	175
Paralongidorus			
maximus	raspberry ringspot (German grapevine strain)	RRSV	76
Xiphinema			
americanum	cherry rasp leaf	CRLV	100
(senso lato[b])	peach rosette mosaic	PRMV	78
	tobacco ringspot	TRSV	42
	tomato ringspot	ToRSV	12
americanum	cherry rasp leaf	CRLV	20
(sensu stricto[c])	tobacco ringspot	TRSV	20
	tomato ringspot	ToRSV	20
bricolensis	tomato ringspot	ToRSV	20
californicum	cherry rasp leaf	CRLV	20
	tobacco ringspot	TRSV	20
	tomato ringspot	ToRSV	71
diversicaudatum	arabis mosaic	ArMV	55, 74
	strawberry latent ringspot	SLRSV	81
index	grapevine fanleaf	GFLV	64
italiae	grapevine fanleaf	GFLV	31
rivesi	cherry rasp leaf	CRLV	20
	tobacco ringspot	TRSV	20
	tomato ringspot	ToRSV	37
	TOBRAVIRUSES		
Paratrichodorus			
anemones	pea early-browning	PEBV	54
	tobacco rattle	TRV	163
minor	pepper ringspot	PRV	125
(syn. *christiei*)	tobacco rattle	TRV	168
nanus	tobacco rattle	TRV	163
pachydermus	pea early-browning	PEBV	160
	tobacco rattle	TRV	130

tansaniensis (syn. *allius*)	tobacco rattle	TRV	6
teres	pea early-browning	PEBV	160
	tobacco rattle	TRV	162
tunisiensis	tobacco rattle	TRV	121
Trichodorus			
cylindricus	tobacco rattle	TRV	163
primitivus	pea early-browning	PEBV	45
	tobacco rattle	TRV	126
similis	tobacco rattle	TRV	32
viruliferus	pea early-browning	PEBV	45
	tobacco rattle[d]	TRV	165

[a] Several original references of vector and virus associations, listed here, do not fulfil the criteria for assessing such reports (24, 152), but these associations have been confirmed in subsequent tests;
[b] Unequivocal identification of species not available or prior to the review of the *X. americanum*-group (79);
[c] Species identification determined by using individual nematodes in virus transmission studies (20, 21);
[d] Three isolates including one (No 6) that was originally identified as pea early-browning tobravirus and subsequently reclassified as an atypical isolate of tobacco rattle tobravirus (119).

Virus Transmission Test Procedures

Transmission of viruses by nematodes in laboratory and glasshouse experiments may be affected by numerous factors (90, 135, 138). Specialized techniques involving the use of individual or small numbers of handpicked nematodes have been established to examine the complex interactions present between vector nematodes and their associated viruses (24 ,136, 152). In tests to establish a nematode as a vector of a virus the nematode must: (*a*) ingest virus from a virus source plant; (*b*) retain virus that it can subsequently release into the bait plant; (*c*) infect the bait plant; and (*d*) of particular importance, it must be shown that contamination of the bait plant with the virus or transmission by an alternative vector has not occurred. With longidorid nematodes root-tip galls present on virus source plants and bait plants are counted to assess the feeding activity of the nematodes. The proportion of nematodes ingesting virus is determined by crushing whole nematodes and examining the resultant suspension by immunosorbent electron microscopy (112) or, with *Longidorus* spp, rubbing the suspension directly onto suitable indicator plants (149, 176). Tobacco rattle tobravirus can be detected in trichodorids by a reverse transcription and polymerase chain reaction method (159). These methods mainly detect virus present in the nematode gut and have not been shown to detect virus retained in the nematode head. They therefore do not provide evidence that the virus is transmissible by the nematode. Virus particles retained in the nematodes' feeding apparatus, and thus capable of being transmitted, are detected by electron microscope examination of ultra-thin sections through the specific sites of retention in vector species.

With handpicked groups of nematodes the proportion of nematodes transmitting virus may be estimated using a maximum-likelihood equation (44). When sufficient numbers of nematodes are used, these procedures provide evidence of the probable inability of a nematode population to transmit a virus.

VECTOR AND VIRUS ASSOCIATIONS

Only a few nematode species are natural vectors of a small number of plant viruses. Of approximately 375 species in the virus-vector genera of the Longidoridae and 80 in the Trichodoridae, only eight *Longidorus*, seven *Xiphinema*, seven *Paratrichodorus*, and four *Trichodorus* species have been shown to be natural vectors of viruses (Table 1) (16, 19–21, 24, 28). Recently, field studies and limited laboratory tests have provided evidence that *Paralongidorus maximus* is the natural vector of an unusual strain of RRSV (see Table 1 for explanation of virus acronyms) (76).

Except for MRSV, transmitted by *L. martini* in Japan, and PRV, transmitted by *P. minor* in Brazil, all vector and virus associations have been described from Europe and North America (16, 19, 21, 24). Several vector and virus associations have widespread geographical distributions, namely *X. index* with GFLV, *X. diversicaudatum* with ArMV, the North American nepoviruses with their vector species, and *P. pachydermus* with TRV. However, in Europe most vector and virus associations have discrete, localized distributions that may be associated with a particular crop, e.g. *L. arthensis* with CRosV in cherry orchards in northern Switzerland, *L. apulus* and *L. fasciatus* with AILV in artichoke fields in southeastern Italy and western Greece, respectively. Similarly, serologically distinguishable strains of tobraviruses are specifically associated with different trichodorid species (105), but some species are the vector of different strains of virus at different geographical localities. These different strains of virus are possibly associated with different crops, e.g. *T. similis* in tobacco fields in northern Greece transmits TRV, which is serologically different from isolates causing diseases in flower-bulbs and potatoes in The Netherlands and Belgium, respectively.

Several nepoviruses are transmitted by more than one vector. The Scottish and English strains of RRSV can each be transmitted in the laboratory by both *L. elongatus* and *L. macrosoma*, but in the field only the Scottish strain has been found associated with *L. elongatus* and the English strain only with *L. macrosoma*. *Xiphinema diversicaudatum* is the natural vector of both ArMV and SLRSV, *L. elongatus* transmits both RRSV and TBRV, and *T. primitivus* transmits both PEBV and TRV. Several species in the genera *Trichodorus* and *Paratrichodorus* transmit both TRV and PEBV, although nepoviruses, with two exceptions, are not transmitted by nematode species of more than one genus. The known exceptions are RRSV, which in vineyards in Germany occurs as two serologically distinct strains transmitted by *L. macrosoma* and *P. maximus*, respectively. In Michigan, *X. americanum* sensu lato was reported as the vector of PRMV (78), and in a peach orchard with PRMV-infected trees in Mersea Township, Essex County, Ontario, Canada, both *X. americanum* sensu lato and *L. diadecturus* are reported as vectors (4, 5, 35) However, *L. diadecturus* is widespread in North America (111), but no other population of this species has been reported to transmit virus. A population of *X. italiae* from Israel was reported to transmit GFLV (31), and subsequently a single transmission of GFLV has been reported from one experiment with *X. italiae* conducted in southern Italy. However, in an independent, extensive survey of viticultural areas in southern Italy, *X. index* transmitted GFLV in 117 of 119 samples, whereas no virus transmission occurred in 41 samples in which *X. italiae* was present (30).

SPECIFICITY OF TRANSMISSION

Field and laboratory results indicate a high degree of specificity between vector nematodes and their associated viruses. This concept is now basic to current investigations, control strategies, and statutory measures. Further support for such specificity of association is provided by the evidence that populations of a vector species differ in their ability to transmit different isolates of a virus (13, 14, 21, 25, 26, 90, 164).

With the recognition that different nematode species are vectors of distinct viruses and virus strains came the realization that the more two strains of a nepoviruses differed serologically, the greater were the differences between their nematode vectors. This led to the hypothesis that vector and virus specificity was determined by particular characters associated with the virus coat protein (29, 53, 60). For tobraviruses it was suggested that geographical association between virus isolate and vector nematode species was responsible for a specific association between virus and vector (163). Subsequently, it has been determined that, as with nepoviruses and their associated vectors, specific relationships between tobraviruses and their vectors are determined by the RNA2 segment of the virus genome and correlated with the serological properties of the virus coat protein (59, 57, 106). Therefore, specificity is a consequence of the nature, site, and mechanism of virus particle retention within the vector.

Vector specificity and efficiency of transmission is often not affected by minor serological variation; sometimes such variation occurs at individual sites (16, 28, 77, 123, 131, 137). Three antigenic variants of ToRSV were recovered from a population of *X. bricolensis* from Washington State; three variants of the Scottish strain of RRSV occur in Scotland in association with the vector *L. elongatus*; and two symptomatological variants of ArMV were each associated with *X. diversicaudatum* from Norway and Scotland (28, 77, 97). However, in contrast, several English and German isolates of TBRV, which differed in their transmissibility by a population of *L. attenuatus* from England, were not serologically distinguishable (23).

Not all vector and virus associations for which there is inadequate evidence may be incorrect, but many seem unlikely, e.g. six species in three genera and seven species in three genera are vectors of ArMV and RRSV, respectively (39–41, 53, 55, 73, 74, 76, 87, 132, 156). Such apparent nonspecific transmission could be due to unattached virus particles occasionally observed in the stoma of *Longidorus* and *Xiphinema* spp. More likely to occur in laboratory experiments are nematodes containing ingested virus or nematode feces containing virus becoming entangled in or adhering to bait plant roots. Both provide potent sources of contamination with viruses of bait plants in laboratory tests (92). However, nonspecific transmission was reported with an isolate of PRMV transmitted by one of 52 handpicked groups of ten *L. breviannulatus* (2) and one of 46 groups of two *L. elongatus* (3). In each experiment the bait plants were systemically infected with PRMV and the criteria for virus transmission listed above were fulfilled. It was concluded that the viruses had been transmitted in a nonspecific manner (3). Nonspecific transmission may occur in laboratory experiments in which nematodes are quickly transferred between virus-infected and healthy plants growing in small pots and thus only a brief period lapses between the nematodes feeding on each plant. Under natural conditions this type of feeding, and thus nonspecific transmission, is unlikely to occur.

In the foregoing section specificity of transmission has generally referred to European nepo- and tobraviruses. Populations of *X. americanum*, now considered *X. americanum* sensu lato, were reported as vectors of CRLV, PRMV, TRSV, and ToRSV (12, 42, 78, 100). Subsequently, after a taxonomic review of populations and species placed in the *X. americanum*-group, it was concluded that the group comprised 25 morphologically similar, parthenogenetic species, of which 15 were new to science (79). There are currently 43 putative species in the *X. americanum*-group; 21 species are present in North America, of which 12 are considered to be indigenous (110, 111). Prior to the taxonomic review of the species, all reports of *X. americanum* including those of virus transmission must be referred to *X. americanum* sensu lato.

Xiphinema rivesi and *X. californicum* have each subsequently been shown to be vectors of ToRSV (37, 71). In a separate study, CRLV, TRSV, and two strains of ToRSV were each transmitted by *X. californicum* and *X. rivesi*; *X. bricolensis* transmitted ToRSV strain PSP more frequently than strain PBL and did not transmit the other two viruses, and nematodes from each of three populations of *X. americanum* sensu stricto transmitted TRSV but not ToRSV strain PBL, those from Arkansas and California populations transmitted

ToRSV strain PSP, and those from a Pennsylvania population transmitted CRLV (20–22).

The transmission of three North American nepoviruses by populations of *X. americanum*, *X. californicum* and *X. rivesi* contrasts with the high degree of specificity described for European vectors and their associated viruses. However, differences in the transmissibility of strains of ToRSV by *X. californicum* (71) and *X. bricolensis* and of TRSV and strains of ToRSV between different populations of *X. americanum* sensu stricto (20–22) suggest that some specificity of transmission occurs with these nematodes. The mechanism(s) whereby three unrelated viruses can each have a specific relationship with a vector species and yet be transmitted by several other species has yet to be determined. There may be specific associations present between "local" isolates of virus and "local" *X. americanum*-group populations of the vector. The specificity of transmission of North American nepoviruses by their respective vectors may therefore be more subtle and complex than generally evident with European vectors and their associated viruses.

Species within the *X. americanum*-group are separated by relatively minor morphometrical and morphological differences, but the species used in the virus transmission experiments referred to above were readily distinguished (21). In a separate study that included the same populations, genetic discontinuities were identified between the populations, using DNA restriction fragment length polymorphism (167) to confirm the morphological discrimination between the populations.

Nematoda typically have four juvenile stages. However, several *X. americanum*-group populations from North America, including all of those used in the virus transmission experiments described above (21), have only three such stages (51, 52). *Xiphinema pachtaicum*, a species from Europe that is also within the *X. americanum*-group, had four juvenile stages. If it is confirmed that *X. americanum*-group populations with three juvenile stages are capable of transmitting virus whereas those with four are not virus-vectors, a rapid, objective method could be established to distinguish those populations/species in the *X. americanum*-group capable of transmitting virus.

VECTOR EFFICACY

A feature of nematode-transmitted viruses is their long-term persistence once established in association with their natural vector, for example the distribution of *X. diversicaudatum*, in association with ArMV, remained virtually unchanged during a period of 30 years at a site in eastern Scotland (139). Also, vector transmission enables a virus to exploit different hosts, and high rates of vector transmission are unnecessary because a virus infecting a long-term perennial need be only infrequently transmitted to ensure infection of the host

(85). Consequently, only a small proportion of the vector population need be capable of transmitting virus, e.g. only 2% to 15% of individual *L. arthensis* from the rhizosphere of CRosV-diseased cherry trees (60–100-year lifespan) transmitted virus (19). On short-term crops there is a greater need for the vector population to be more efficient, e.g. 14% to 49% of individual *L. fasciatus* from fields with AILV-infected artichokes (3–5-year lifespan) transmitted virus (DJF Brown, unpublished data). Also, in perennial crops small population densities of the vector can effectively spread virus.

The hop strain of ArMV is spread in soils containing less than one *X. diversicaudatum* per 200 g soil (102); as a result of the extensive root-system of the hop plant, this can represent approximately 20,000 individual nematodes per plant. Similarly, in vineyards GFLV can spread rapidly even when *X. index* are so scarce as to be almost undetectable (M Rudel, personal communication).

Many individuals in *Xiphinema* populations can transmit virus. In small pot experiments 80% and 30% of *X. diversicaudatum* transmitted ArMV and SLRSV, respectively. By contrast, *Longidorus* species are more variable; 27% of *L. attenuatus* transmitted TBRV, 2% and 8% of *L. elongatus* transmitted the Scottish strains of RRSV and TBRV, respectively, and only 5% of *L. macrosoma* transmitted the English strain of RRSV (153). However, frequency of transmission is not the only criterion for considering a species to be an efficient vector. *Xiphinema americanum* sensu lato is considered an efficient vector as 3% to 5% of individuals transmitted TRSV during a 24-h access to bait plants. Furthermore, individual nematodes transmitted virus to more than one plant without having to re-acquire virus and were capable of transmitting both TRSV and ToRSV (43, 88, 89). Trichodorids also are efficient vectors; one male *P. pachydermus* infected three *Nicotiana rustica* bait plants consecutively with TRV within a period of 4 days (161).

The length of time a nematode can retain virus particles will, in certain circumstances, affect its efficiency as a vector. Vector trichodorids and *Xiphinema* species can retain their associated viruses for periods of 8 to 12 months when stored at low temperatures in moist, plant-free soil, which is sufficient time for the virus to overwinter in its vector (7, 133, 134, 140). Viruses are less persistent in *Longidorus* spp, usually remaining transmissible only for about 8 weeks (96, 133, 134).

In laboratory studies environmental factors such as temperature, soil moisture, size of pot, and numbers and developmental stages of nematodes used can affect the rate of transmission (138). All nematode developmental stages can transmit but virus is not retained during the molt, nor is it passed through the egg. Transmission can also be affected by the bait plant; substitution of *Nicotiana tabacum* with *Aster sinensis* plants increased rates of transmission of TRV by *P. nanus* (163). As already noted, the virus isolate can affect transmission rates; a population of *L. attenuatus* from England transmitted seven isolates of TBRV with frequencies ranging from 3% to 78% (23). Moreover, the proportion of nematodes transmitting virus may rapidly increase in natural systems; the rate of transmission of TRV by *P. pachydermus*, naturally associated with the virus, increased from 11% to 69% after feeding for 4 weeks on initially healthy *Petunia hybrida* plants. This result suggests that the initial small transmission rate was not due to the test system but rather it reflected a small proportion of the nematodes being naturally viruliferous (104).

In general, increasing the acquisition and transmission access periods increases the frequency of transmission as nematodes access to the plant roots is increased. Although 1 h was sufficient for *X. americanum* sensu lato to acquire ToRSV, the rate of transmission during the first 48 h was not proportional to the length of time the nematodes had access to the bait plant roots, and 100% transmission did not occur until the nematodes had access for 4 days (146).

INGESTION, RETENTION, AND TRANSMISSION OF VIRUSES

For a nematode to function as a virus-vector virus particles must be ingested from an infected plant, the particles must associate and subsequently dissociate from the site of retention within the vector, and dissociated particles must be introduced into a cell of the recipient plant.

Sites of Virus Retention within Vector Nematodes

Sites of retention of virus particles were identified by electron microscopy of ultra-thin sections through the feeding apparatus of *Longidorus, Paratrichodorus, Trichodorus* and *Xiphinema* species taken from virus-infected plants (91, 109, 113, 141–143). In *L. apulus, L. arthensis, L. elongatus, L. fasciatus*, and *L. macrosoma* virus particles are associated with the inner surface of the odontostyle (Figure 1*A*). However, in early investigations with *L. elongatus* particles of RRSV and TBRV were observed between the odontostyle and the guiding sheath and this was considered to be the effective site of retention (141, 144, 145; WM Roberston, unpublished data). In *X. americanum* sensu lato, *X. diversicaudatum* and *X. index* virus particles are associated only with the cuticle lining the lumen of the odontophore and the esophagus (Figure 1*B*) (91, 109, 141–144). In *P. pachydermus* and *T. similis*, TRV particles have been observed throughout the length of the esophageal lumen adsorbed to the cuticular lining of the lumen (Figure 1*C, D*, respectively) (143, 144; WM Robertson, unpublished data). In all of these vector species, despite extensive examination, virus particles have not been observed in any of the glands or surrounding tissues. However, virus particles were observed in the anterior

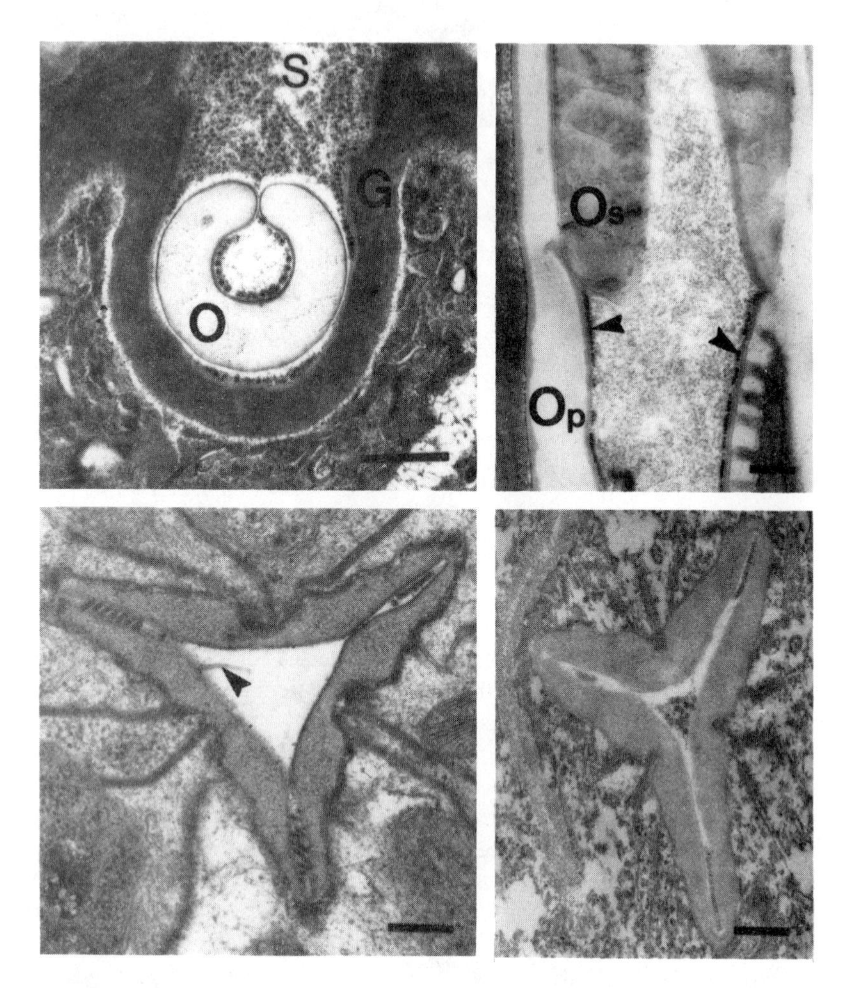

Figure 1 Nematode-transmitted viruses specifically retained at sites of retention within the feeding apparatus of four virus-vector nematode species. *A*: a transverse section through the anterior odontostyle and stoma of *Longidorus elongatus* showing RRSV virus particles specifically adsorbed to the food canal within the odontostyle (*O*) and on the surface of the guiding sheath (*G*). Also, a large group of particles lying free within the stoma (S). *B*: a longitudinal section through the odontostyle and esophageal junction of *Xiphinema index* showing groups of GFLV particles (arrowed) adsorbed to the food canal of the odontophore. Note the demarcation of particles present on the odontophore (*Op*) but absent from the odontostyle (*Os*). *C*: a transverse section through the anterior esophagus of *Paratrichodorus pachydermus* showing particles of TRV located in the apices of the triangular lumen. Note that the short particles attach predominantly by their ends and the long particles attach lengthwise; however, in this section an exceptionally long particle is attached by its end in the open part of the lumen (arrowed). *D*: a transverse section through the anterior esophagus of *Trichodorus similis* showing particles located in the apices as a monolayer and a group of particles trapped within a matrix in the open part of the lumen. Bars represent 200 nm.

mouth parts of *L. elongatus* that had been allowed to feed on ArMV-infected plants, but this virus was not transmitted by *L. elongatus* in concurrent experiments (141). Also, clumps of virus particles have been observed in the triangular section of the lumen immediately adjacent to the anterior end of the guide ring in *L. macrosoma* and *X. diversicaudatum* in sufficient quantity to provide an infectious inoculum when the nematodes next fed (144).

Genetic Determinants for Virus Retention within Vector Nematodes

During feeding most of the virus particles ingested by the nematode pass directly into the gut, and only a small proportion are retained. The capacity of *X. diversicaudatum* to retain and transmit ArMV and SLRSV is an inherited character (15), probably involving the surface of the virus coat protein and surface factors at the sites of retention (60). The RNA2 segment of the virus genomes of RRSV, TBRV, and TRV, which include the coat protein gene, contains the genetic determinants for vector transmissibility (59, 57, 106).

Although the virus coat protein is probably involved in the recognition process between vector and virus, it is not the sole determinant for transmission of tobraviruses. When the coat protein gene of a nematode-nontransmissible isolate of PEBV (SP5) was replaced with that of a highly transmissible isolate of TRV, the recombinant virus was not transmitted (SA McFarlane, DJF Brown, JF Bol, unpublished data). Full-length, infectious cDNA clones of SP5 and of an isolate of PEBV transmitted by *T. primitivus* (TPA56) were prepared. Virus containing RNA2 derived from the cDNA clone of TPA56 was transmitted by *T. primitivus* but virus containing RNA2 from the cDNA clone SP5 was not transmitted. Thus, as with TRV (106), the genetic determinants of PEBV are present on the RNA2 segment of the virus genome. However, sequencing revealed that only 11 of 3374 nucleotides differed between the nematode-transmissible and the nontransmissible clone of PEBV, but that three of these base changes would affect the amino acid sequences of the virus gene products. An analysis of these amino acid changes suggests that the 29.6-kDa protein, expressed as an open reading frame immediately following the coat protein, has a function in the transmission process and that changes in this protein prevent vector transmission of the PEBV SP5 isolate (SA McFarlane, DJF Brown, FJ Legorburu, & DJ Robinson, unpublished data). The hypothesis has been advanced, but not tested, that a 16-kDa protein produced by the RNA1 of some TRV isolates may also be involved in the transmission process as an alternative function for this protein is not known (10).

Mechanisms of Virus Retention within Vector Nematodes

Longidorus macrosoma transmits RRSV only infrequently (~5%) but nematodes given access to RRSV-infected plants contain large numbers of virus

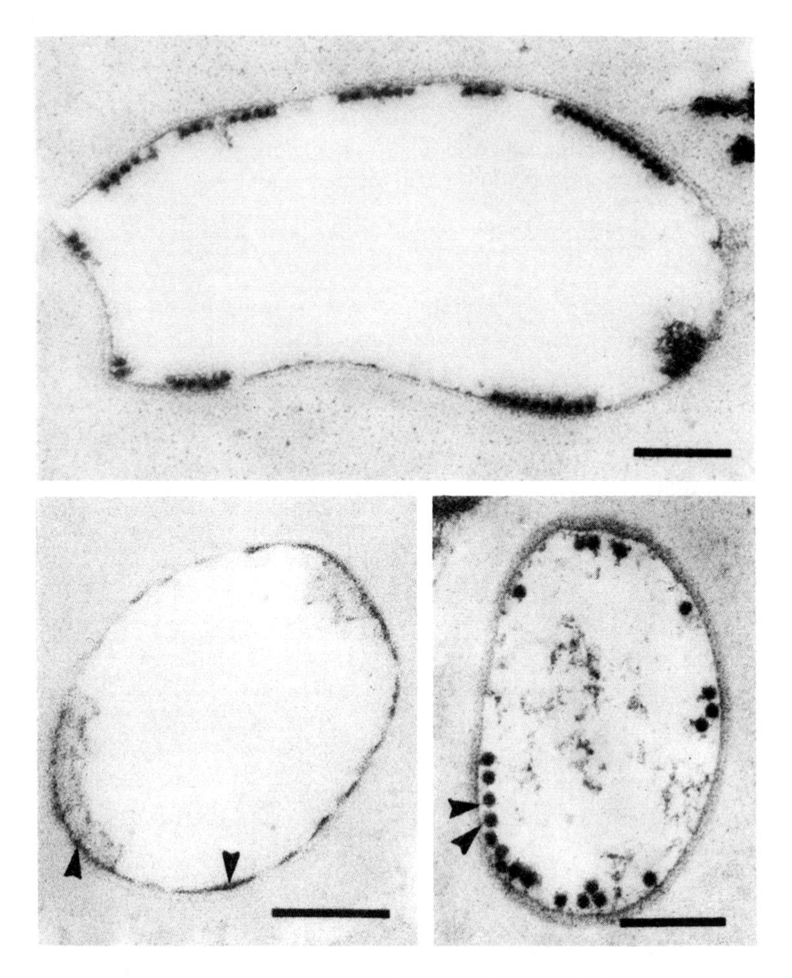

Figure 2 Transverse sections through the anterior odontophores of *Xiphinema diversicaudatum* specimens. *A*: specimen fixed in paraformaldehyde and not post-fixed in osmium but stained with uranyl acetate and lead citrate to show groups of ArMV particles lining the lumen. *B*: specimen stained for carbohydrate deposits (arrowed) on the surface of the lumen some of which have "cloud" material indicating the presence of ArMV particles (*V*). *C*: specimen SLRSV particles lining the surface of the lumen of the food canal with electron-dense structures (arrowed) linking the body of the particles to one another and to the surface of the lumen. Bars represent 200 nm.

particles within their odontostyle. Infrequent transmission is not, therefore, associated with a lack of retention but may be due to a lack of dissociation of virus particles from the specific sites of retention (150). Overall, with European virus-vector *Longidorus* spp it appears that the more efficient the nematode is as a vector, the fewer virus particles that are observed at the sites of retention (WM Robertson, unpublished data). A similar correlation is not evident with *Xiphinema* spp, which may indicate a fundamental difference in the mechanism of retention of viruses by these two groups of nematodes. A rapid retention and slow dissociation of viruses within the vector could provide an ecologically advantageous mechanism for successful virus transmission by nematodes and may occur with *Longidorus* spp (86).

The mechanisms of virus retention are unknown but it has been suggested that surface charges may be involved, because virus particles are retained during ingestion and released by pH changes induced by salivation (60, 138). The sites of virus retention on the wall of the food canal in *X. diversicaudatum* and *P. pachydermus* stain for carbohydrate (Figure 2*A*, *B*) whereas similar staining did not occur in *Longidorus* species or in the odontostyle region in *X. diversicaudatum* and *X. index*. Additionally, particles of ArMV retained within the odontophore of *X. diversicaudatum* were surrounded by a cloud of carbohydrate-staining material (115, 116). An alternative mechanism suggested by these observations is that the virus coat protein may have lectin-like properties, with carbohydrates being involved in both retention and dissociation of virus particles.

The structure of tobravirus particles has only recently been determined and affords new insight into how vector trichodorids may recognize these viruses. The tobravirus protein subunits build in a tight spiral to form a hollow rod-shaped structure. By aligning the amino acid sequences of TRV and tobacco mosaic tobamovirus [the detailed construction of the coat protein of tobacco mosaic tobamovirus has been determined by X-ray crystallography (99)], the coat protein subunits of both viruses fold in a similar manner and the N-termini and C-termini are both located on the outside of the virus particle (46). However, tobravirus coat proteins have relatively long "protruding" C-terminal amino acid sequences that, NMR measurements suggest, form mobile "arms". Consequently, the C-terminal sequence is likely involved in the retention of tobraviruses by their associated vector nematodes (86). Close examination of TRV particles present at sites of retention within *P. pachydermus* revealed a space of approximately 5 to 7 nm between the surface of the particles and the cuticle lining the esophageal lumen. This space is larger than would be occupied by a peptide of 25 amino acids associated with the C-terminal of the coat protein. However, as explained above, a second gene is probably involved in the transmissibility of tobraviruses, and its product may link the virus particle and the esophageal wall. Such a link could be formed by carbohydrate-containing material attaching to the protruding mobile C-terminal part of the virus coat protein.

Less is known about the structure of nepoviruses but TBRV may also possess an "arm-like" C-terminal peptide (86). Such a peptide, if present in nepoviruses, could have a similar role to that proposed for tobraviruses. Close examination of SLRSV particles retained in *X. diversicaudatum* has revealed electron-dense structures apparently linking particles to one another and to the surface of the lumen of the food canal (Figure 2*C*). However, it is not known whether these structures are associated with a C-terminal peptide. There are fundamental differences between the vector genera and between the viruses they transmit. Although both are transmitted by *X. diversicaudatum,* ArMV has only one whereas SLRSV has two coat protein genes. Similarly, each of three *X. americanum*-group species transmits both CRLV and TRSV, which have three and one coat proteins, respectively. However, areas of similarity may exist between unrelated viruses with the same vector. Polyproteins in the N-terminal side of the coat proteins of several nepoviruses contain substantial similarities in their amino acid sequences. Considerable similarity was evident between TBRV and RRSV (transmitted by *Longidorus* spp) and between GFLV and ToRSV (transmitted by *Xiphinema* spp) but little similarity was present between these pairs of viruses. These results may indicate a role for these virus gene products in vector transmission (9).

FUTURE TRENDS

Progress in understanding interactions between vector nematodes, their associated viruses, and plant hosts has been dependent upon the development of techniques and methods for examining and assessing virus transmission by the vectors. Also, molecular virology techniques are providing new approaches for characterizing the nematode-transmitted viruses and for identifying the genetic determinants involved in the recognition between the viruses and their vectors. However, molecular techniques have not yet been widely applied to identifying the genetic determinant(s) involved in the recognition process between the nematode and virus. Studies on aphid transmission of viruses have been considerably aided by the development of systems whereby aphids can be made to feed through artificial membranes (58). This technique has been used recently to show that several proteins from the vector aphid *Myzus persicae* bind to particles of potato leaf roll virus. Symbionin, which is synthesized by endosymbiotic bacteria in the aphid gut, was the most readily detected protein from *M. persicae*, and it was concluded that the bacteria, and its associated protein, are involved in the persistence of the virus in its vector (158). An artificial membrane feeding system is not available for studying nematode transmission of viruses but a similar approach may identify protein(s) involved in the recognition between vector nematodes and their associated viruses.

Nonspecific soil sterilants and oxime-carbamates are used to control virus-vector nematodes, but little research has been done to identify plant resistance genes effective against these nematodes. Engineered resistance involving the insertion of a TRV coat protein gene into *Nicotiana tabacum* plants provided quantitative resistance when the plants were mechanically challenged with the homologous virus. However, transgenic plants became infected when the virus was transmitted to roots and leaves by the natural vector *P. pachydermus* (107). A similar approach with coat protein constituents of ArMV and SLRSV inserted in *Nicotiana tabacum* and *N. xanthii* plants also gave protection when transgenic plants were mechanically challenged with the homologous viruses. However, unlike TRV coat protein resistance, the transgenic plants carrying constituents of ArMV and SLRSV coat proteins did not become infected when exposed to nematodes carrying the homologous viruses (S Kreiah, ML Edwards, WS Hawes, AT Jones, DJF Brown, WG McGavin, C Prehaud & JI Cooper, unpublished data). Other potentially useful engineered resistance, e.g. replicase resistance, remain to be examined for their applicability to naturally transmitted nepo- and tobraviruses.

Environmentally benign strategies for controlling nematode-transmitted viral diseases are dependent on identifying the factors involved in nematode-virus-plant interactions. In laboratory tests *X. index* readily acquired GFLV from manually infected *Chenopodium amaranticolor* plants but did not transmit the virus to healthy *C. amaranticolor* plants, although the nematodes did infect grapevine seedlings with the virus (151). In nature GFLV has only been recovered from grapevine and, despite intensive sampling, has not been recovered from weed species growing in vineyards in which GFLV was widespread (83). In other laboratory experiments *L. elongatus* and *X. diversicaudatum* readily infected *Petunia hybrida* seedlings with RRSV and ArMV, respectively, but did not infect plantlets of nine raspberry cultivars (77). Also, *X. diversicaudatum* readily transmitted ArMV to seven strawberry cultivars but the virus was present only in the roots of the plants. However, after subjecting the strawberry plants to a physiological shock such as chilling, partial defoliation, or a period of darkness, virus moved systemically into the aerial parts of the plants. A similar phenomena occurred when strawberry and raspberry plants were planted in early spring at field sites where ArMV and RRSV were present in association with their respective vectors. The viruses were recovered from the roots of growing plants during the summer but were not detected in aerial parts until the following year, after the plants had broken their winter dormancy.

Research into virus-vector nematodes and their associated viruses is still a relatively new area, but one that is progressing rapidly. Each vector and virus association has unique elements that provide the stimulus for further biological, ecological, and molecular research to identify the fundamental mechanisms that govern these dynamic processes.

ACKNOWLEDGMENT

We thank our colleagues and collaborators who have contributed in many and varied ways to our understanding of nematode transmission of plant viruses. Research at the Scottish Crop Research Institute is grant-aided by the Scottish Office Agriculture and Fisheries Department (SOAFD). Nonindigenous nematode cultures and virus isolates were held and experiments with genetically modified viruses and plants were conducted under license from the SOAFD.

Literature Cited

1. Allen MW. 1948. Relation of soil fumigation, nematodes and inoculation techniques to big vein disease of lettuce. *Phytopathology* 38:612–27
2. Allen WR. 1986. Effectiveness of Ontario populations of *Longidorus diadecturus* and *L. breviannulatus* as vectors of peach rosette mosaic and tomato black ring viruses. *Can. J. Plant Pathol.* 8:49–53
3. Allen WR, Ebsary BA. 1988. Transmission of raspberry ringspot, tomato black ring and peach rosette mosaic viruses by an Ontario population of *Longidorus elongatus. Can. J. Plant Pathol.* 10:1–5
4. Allen WR, Van Schagen JG, Ebsary BA. 1984. Comparative transmission of the peach rosette mosaic virus by Ontario populations of *Longidorus diadecturus* and *Xiphinema americanum* (Nematoda: Longidoridae). *Can. J. Plant Pathol.* 6:29–32
5. Allen WR, Van Schagen JG, Eveleigh ES. 1982. Transmission of peach rosette mosaic virus to peach, grape, and cucumber by *Longidorus diadecturus* obtained from diseased orchards in Ontario. *Can. J. Plant Pathol.* 4:16–18
6. Ayala A, Allen MW. 1966. Transmission of the Californian tobacco rattle virus by three species of the nematode genus *Trichodorus. Nematologica* 12:87 (Abstr.)
7. Bergeson GB, Athow KL, Laviolette FA, Thomasine M. 1964. Transmission, movement and vector relationships of tobacco ringspot virus in soybean. *Phytopathology* 54:723–28
8. Bleve-Zacheo T, Zacheo G, Lamberti F, Arrigoni O. 1977. Cell wall breakdown and cellular response in developing galls induced by *Longidorus apulus. Nematol. Mediterr.* 5:305–11
9. Blok VC, Wardell J, Jolly CA, Manoukian A, Robinson DJ, et al. 1992. The nucleotide sequence of RNA-2 of raspberry ringspot nepovirus. *J. Gen. Virol.* 73:2189–94
10. Boccara M, Hamilton WDO, Baulcombe DC. 1986. The organisation and interviral homologies of the genes at the 3′ end of tobacco rattle virus RNA-1. *EMBO J.* 5:223–29
11. Bock KR. 1966. Arabis mosaic and prunus necrotic ringspot viruses in hop (*Humulus lupulus L.*). *Ann. Appl. Biol.* 57:131–40
12. Breece JR, Hart WH. 1959. A possible association of nematodes with the spread of peach yellow bud mosaic virus. *Plant Dis. Rep.* 43:989–90
13. Brown DJF. 1985. The transmission of two strains of strawberry latent ringspot virus by populations of *Xiphinema diversicaudatum* (Nematoda: Dorylaimida). *Nematol. Mediterr.* 13:217–23
14. Brown DJF. 1986. The transmission of two strains of arabis mosaic virus from England by populations of *Xiphinema*

diversicaudatum from ten countries. *Rev. Nematol.* 9:82–87

15. Brown DJF. 1986. Transmission of virus by the progeny of crosses between *Xiphinema diversicaudatum* (Nematoda: Dorylaimoidae) from Italy and Scotland. *Rev. Nematol.* 9:71–74
16. Brown DJF. 1989. Viruses transmitted by nematodes. *EPPO/OEPP Bull.* 19: 453–61
17. Brown DJF, Boag B. 1975. *Longidorus macrosoma. Commonw. Inst. Helminthol. Descr. Plant Parasitic Nematodes,* 5, No. 67. 4 pp.
18. Brown DJF, Boag B. 1977. *Longidorus attenuatus. Commonw. Inst. Helminthol. Descr. Plant Parasitic Nematodes,* 7, No. 101. 4 pp.
19. Brown DJF, Grunder J, Hooper DJ, Klingler J, Kunz P. 1994. *Longidorus arthensis* sp.n. (Nematoda: Longidoridae) a vector of cherry rosette disease caused by a new nepovirus in cherry trees in Switzerland. *Nematologica* 40: 133–49
20. Brown DJF, Halbrendt JM. 1992. The virus vector potential of *Xiphinema americanum* and related species. *J. Nematol.* 24:584 (Abstr.)
21. Brown DJF, Halbrendt JM, Jones AT, Vrain TC, Robbins RT. 1994. Transmission of three North American nepoviruses by populations of four distinct *Xiphinema americanum*-group species (Nematoda: Longidoridae). *Phytopathology* 84:646–49
22. Brown DJF, Halbrendt JM, Robbins RT, Vrain TC. 1993. Transmission of nepoviruses by *Xiphinema americanum*-group nematodes. *J. Nematol.* 25: 349–54
23. Brown DJF, Murant AF, Trudgill DL. 1989. Differences between isolates of the English serotype of tomato black ring virus in their transmissibility by an English population of *Longidorus attenuatus* (Nematoda: Dorylaimoidae). *Rev. Nematol.* 12:51–56
24. Brown DJF, Ploeg AT, Robinson DJ. 1989. A review of reported associations between *Trichodorus* and *Paratrichodorus* species (Nematoda: Trichodoridae) and tobraviruses with a description of laboratory methods for examining virus transmission by trichodorids. *Rev. Nematol.* 12:235–41
25. Brown DJF, Taylor CE. 1981. Variazioni nella trasmissione di virus tra popolazioni di nematodi vettori Longidoridae. *Atti della Soc. Ital. Nematol., Giorante Nematol., Firenze, 1979,* pp. 191–204
26. Brown DJF, Trudgill DL. 1983. Differential transmissibility of arabis mosaic and strawberry latent ringspot viruses by three populations of *Xiphinema diversicaudatum* (Nematoda: Dorylaimida) from Scotland, Italy and France. *Rev. Nematol.* 6:229–38
27. Brown DJF, Trudgill DL. 1989. The occurrence and distribution of nepoviruses and their associated vector *Longidorus* and *Xiphinema* nematodes in Europe and the Mediterranean Basin. *EPPO/OEPP Bull.* 19:479–89
28. Brown DJF, Vrain TC, Jones AT, Robertson WM, Halbrendt JM, Robbins RT. 1994. *Xiphinema bricolensis*—a natural vector of three serologically distinguishable strains of tomato ringspot nepovirus. *J. Nematol.* 26:94 (Abstr.)
29. Cadman CH. 1963. Biology of soil-borne viruses. *Annu. Rev. Phytopathol.* 1:143–72
30. Catalano L, Savino V, Lamberti F. 1992. Presence of grapevine fanleaf nepovirus in populations of longidorid nematodes and their vectoring capacity. *Nematol. Mediterr.* 20:67–70
31. Cohn E, Tanne E, Nitzany FE. 1970. *Xiphinema italiae,* a new vector of grapevine fanleaf virus. *Phytopathology* 60:181–82

31a. Converse RH. 1977. Rubus virus diseases important in the United States. *HortScience* 12:471–75

32. Cremer MC, Kooistra G. 1964. Investigations on notched leaf (Kartelblad) of *Gladiolus* and its relation to tobacco rattle virus. *Nematologica* 10:69–70 (Abstr.)
33. De Waele D, Alphey TJW, Barbez D. 1985. *Paratrichodorus pachydermus. Commonw. Inst. Helminthol. Descr. Plant Parasitic Nematodes,* 8, No. 112. 4 pp.
34. Duffield CA. 1925. Nettlehead in hops. *Ann. Appl. Biol.* 12:536–43
35. Eveleigh ES, Allen WR. 1982. Description of *Longidorus diadecturus* n. sp. (Nematoda: Longidoridae), a vector of the peach rosette mosaic virus in peach orchards in southwestern Ontario, Canada. *Can. J. Zool.* 60:112–15
36. Everett KR, Milne KS, Forster RLS. 1994. Nucleotide sequence of the coat protein genes of strawberry latent ringspot virus: lack of homology to the nepoviruses and comoviruses. *J. Gen. Virol.* 75:1821–25
37. Forer LB, Hill N, Powell CA. 1981. *Xiphinema rivesi,* a new tomato ringspot vector. *Phytopathology* 71:874 (Abstr.)
38. Francki RIB, Milne RG, Hatta T. 1985. *Atlas of Plant Viruses,* 2:23–38. Boca Raton, FL: CRC. 240 pp.

39. Fritzsche R. 1964. Untersuchungen uber die Virusubertragung durch Nematoden. *Wiss. Z. Univ. Rostock Math-Naturwiss. Reihe* 13:343–37
40. Fritzsche R, Kegler H. 1968. Nematoden als Vektoren von Viruskrankheiten der Obstgewachse. *Tagungsber. Dtsch. Akad. Landwirtschftswiss. Berlin* 97:289–95
41. Fritzsche R, Thiele S. 1979. Eignung von *Xiphinema*—Herkunften aus der UdSSR und der DDR zur Ubertragung des Arabis-Mosaik-Virus. *Nachrichtenbl. Pflanzenschutzdienst DDR* 33: 103–4
42. Fulton JP. 1962. Transmission of tobacco ringspot virus by *Xiphinema americanum. Phytopathology* 52:375
43. Fulton JP. 1967. Dual transmission of tobacco ringspot and tomato ringspot virus by *Xiphinema americanum. Phytopathology* 57:535–36
44. Gibbs AJ, Gower JC. 1960. The use of a multiple-transfer method in plant virus transmission studies—some statistical points arising in the analysis of results. *Ann. Appl. Biol.* 48:75–83
45. Gibbs AJ, Harrison BD. 1964. A form of pea early-browning virus found in Great Britain. *Ann. Appl. Biol.* 54:1–11
46. Goulden MG, Davies JW, Wood KR, Lomonossoff GP. 1992. Structure of tobraviral particles: a model suggested from sequence conservation in tobraviral and tobamoviral coat proteins. *J. Mol. Biol.* 227:1–8
47. Griffiths BS, Robertson WM. 1983. Nuclear changes induced by the nematode *Longidorus elongatus* in root-tips of ryegrass *Lolium perenne. Histochem. J.* 15:927–34
48. Griffiths BS, Robertson WM. 1984. Nuclear changes induced by the nematode *Xiphinema diversicaudatum* in root-tips of strawberry. *Histochem. J.* 16:265–73
49. Griffiths BS, Robertson WM, Trudgill DL. 1982. Nuclear changes induced by the nematodes *Xiphinema diversicaudatum* and *Longidorus elongatus* in root-tips of perennial ryegrass *Lolium perenne. Histochem. J.* 14:719–30
50. Griffiths BS, Trudgill DL. 1983. A comparison of the generation times and rates of growth of *Xiphinema diversicaudatum* and *Longidorus elongatus* on a good and a poor host. *Nematologica* 29:78–87
51. Halbrendt JM, Brown DJF. 1992. Morphometric evidence for three juvenile stages in some species of *Xiphinema americanum* sensu lato. *J. Nematol.* 24: 305–9
52. Halbrendt JM, Brown DJF. 1993. Aspects of biology and development of *Xiphinema americanum* and related species. *J. Nematol.* 25:355 (Abstr.)
53. Harrison BD. 1964. Specific nematode vectors for serologically distinctive forms of raspberry ringspot and tomato black ring viruses. *Virology* 22:544–50
54. Harrison BD. 1967. Pea early-browning virus (PEBV). *1966 Rep. Rothamsted Exp. Stn.*, p. 115
55. Harrison BD, Cadman CH. 1959. Role of a dagger nematode (*Xiphinema* sp.) in outbreaks of plant disease caused by arabis mosaic virus. *Nature* 184:1624–26
56. Harrison BD, Mowat WP, Taylor CE. 1961. Transmission of a strain of tomato black ring virus by *Longidorus elongatus* (Nematoda). *Virology* 14:480–85
57. Harrison BD, Murant AF. 1978. Nematode transmissibility of pseudorecombinant isolates of tomato black ring virus. *Ann. Appl. Biol.* 86:209–12
58. Harrison BD, Murant AF. 1984. Involvement of virus-coded proteins of plant viruses by vectors. In *Vectors in Virus Biology*, ed. MA Mayo, KA Harrap, pp. 1–36. London: Academic. 188 pp.
59. Harrison BD, Murant AF, Mayo MA, Roberts IM. 1974. Distribution and determinants for symptom production, host range and nematode transmissibility between the two RNA components of raspberry ringspot virus. *J. Gen. Virol.* 22:233–47
60. Harrison BD, Robertson WM, Taylor CE. 1974. Specificity of retention and transmission of viruses by nematodes. *J. Nematol.* 6:155–64
61. Harrison BD, Robinson DJ. 1978. The tobraviruses. *Adv. Virus Res.* 23:27–77
62. Harrison BD, Robinson DJ. 1986. Tobraviruses. In *The Plant Viruses. The Rod-Shaped Plant Viruses*, ed. MHV Van Regenmortel, H Fraenkel-Conrat, pp. 339–69. New York: Plenum. 424 pp.
63. Harrison BD, Woods RD. 1966. Serotypes and particle dimensions of tobacco rattle viruses from Europe and America. *Virology* 28:610–20
64. Hewitt WB, Raski DJ, Goheen AC. 1958. Nematode vector of soil-borne fanleaf virus of grapevine. *Phytopathology* 48:586–95
65. Heyns J. 1975. *Paratrichodorus christei. Commonw. Inst. Helminthol. Descr. Plant Parasitic Nematodes*, 5, No. 69. 4 pp.
66. Heyns J. 1975. *Paralongidorus maximus. Commonw. Inst. Helminthol. Descr. Plant Parasitic Nematodes*, 5, No. 75. 4 pp.

67. Hooper DJ. 1973. *Longidorus elongatus, Commonw. Inst. Helminthol. Descr. Plant Parasitic Nematodes,* 2, No. 30. 4 pp.
68. Hooper DJ. 1976. *Trichodorus viruliferus, Commonw. Inst. Helminthol. Descr. Plant Parasitic Nematodes,* 6, No. 86. 3 pp.
69. Hooper DJ. 1977. *Paratrichodorus (Nanidorus) minor, Commonw. Inst. Helminthol. Descr. Plant Parasitic Nematodes,* 7, No. 103. 3 pp.
70. Hooper DJ, Siddiqi MR. 1972. *Trichodorus primitivus, Commonw. Inst. Helminthol. Descr. Plant Parasitic Nematodes,* 1, No. 15. 3 pp.
71. Hoy JW, Mircetich SM, Lownsbery BF. 1984. Differential transmission of prunus tomato ringspot virus strains by *Xiphinema californicum. Phytopathology* 74:332–35
72. Hunt DJ. 1993. *Aphelenchida, Longidoridae and Trichodoridae: Their Systematics and Bionomics.* Wallingford, England: CAB Int. 352 pp.
73. Iwaki M, Komuro Y. 1974. Viruses isolated from narcissus (*Narcissus* spp.) in Japan. V. Arabis mosaic virus. *Ann. Phytopathol. Soc. Jpn.* 40:344–53
74. Jha A, Posnette AF. 1959. Transmission of a virus to strawberry plants by a nematode (*Xiphinema* sp.). *Nature* 184: 962–63
75. Johnson F. 1945. The effect of chemical soil treatments on the development of wheat mosaic. *Ohio. J. Sci.* 45:125–28
76. Jones AT, Brown DJF, McGavin W, Rudel M, Altmayer B. 1994. Properties of an unusual isolate of raspberry ringspot virus from grapevine in Germany and evidence for its possible transmission by *Paralongidorus maximus. Ann. Appl. Biol.* 124:283–300
77. Jones AT, Mitchell MJ, Brown DJF. 1989. Infectibility of some new raspberry cultivars with arabis mosaic and raspberry ringspot viruses and further evidence for variation in British isolates of these two viruses. *Ann. Appl. Biol.* 115:57–69
78. Klos EJ, Fronek F, Knierim JA, Cation D. 1967. Peach rosette mosaic transmission and control studies. *Q. Bull. Mich. State Univ. Exp. Stn.* 49:287–93
79. Lamberti F, Bleve-Zacheo T. 1979. Studies on *Xiphinema americanum sensu lato* with description of fifteen new species (Nematoda, Longidoridae). *Nematol. Mediterr.* 7:51–106
80. Lamberti F, Taylor CE, Seinhorst JW, eds. 1975. *Nematode Vectors of Plant Viruses.* New York: Plenum. 460 pp.
81. Lister RM. 1964. Strawberry latent ringspot: a new nematode-borne virus. *Ann. Appl. Biol.* 54:167–76
82. Martelli GP. 1975. Some features of nematode-borne viruses and their relationships with the host plants. See Ref. 80, pp. 223–52
83. Martelli GP. 1978. Nematode-borne viruses of grapevine, their epidemiology and control. *Nematol. Mediterr.* 6:1–27
84. Martelli GP, Quacquerelli A, Gallitelli D, Savino V, Piazzolla P. 1978. A tentative grouping of nepoviruses. *Phytopathol. Mediterr.* 17:145–47
85. Martelli GP, Taylor CE. 1989. Distribution of viruses and their nematode vectors. In *Advances in Disease Vector Research,* ed. KF Harris, 6:151–89. New York: Springer-Verlag. 363 pp.
86. Mayo M, Robertson WM, Legorburu FJ, Brierley KM. 1995. Molecular approaches to an understanding of the transmission of plant viruses by nematodes. In *Advances in Molecular Plant Nematology,* ed. F Lamberti, C De Giorgi, DMcK Bird, pp. 277–93. New York: Plenum. 309 pp.
87. McElroy FD, Brown DJF, Boag B. 1976. The virus-vector and damage potential, morphometrics and distribution of *Paralongidorus maximus. J. Nematol.* 9:122–30
88. McGuire JM. 1964. Efficiency of *Xiphinema americanum* as a vector of tobacco ringspot virus. *Phytopathology* 54:799–801
89. McGuire JM. 1964. Serial transfer of *Xiphinema americanum* as a tool for studying transmission of tobacco ringspot virus. *Phytopathology* 54:900 (Abstr.)
90. McGuire JM. 1982. Nematode transmission of viruses. In *Nematology in the Southern Region of the United States,* ed. RD Riggs, pp. 190–92. *South. Coop. Ser. Bull.* 276. Fayetteville: Univ. Arkansas. 206 pp.
91. McGuire JM, Kim KS, Douthit LB. 1970. Tobacco ringspot virus in the nematode *Xiphinema americanum. Virology* 42:212–16
92. McNamara DG. 1978. *Studies on the ability of the nematode Xiphinema diversicaudatum (Micol.) to transmit raspberry ringspot virus and to survive in plant-free soil.* PhD thesis. Univ. Reading, Reading, England. 190 pp.
93. Murant AF. 1981. Nepoviruses. In *Handbook of Plant Virus Infections and Comparative Diagnosis,* ed. E Kurstak, pp. 197–238. Amsterdam: Elsevier/North Holland. 943 pp.
94. Murant AF. 1985. Taxonomy and no-

menclature of virus. *Microbiol. Sci.* 2: 218–20
95. Murant AF. 1989. Nepoviruses. In *Plant Protection and Quarantine. Selected Pests and Pathogens of Quarantine Significance*, ed. RP Khan, 2:44–57. Boca Raton, FL: CRC. 943 pp.
96. Murant AF, Taylor CE. 1965. Treatment of soil with chemicals to prevent transmission of tomato black ring and raspberry ringspot viruses by *Longidorus elongatus* (De Man). *Ann. Appl. Biol.* 55:227–37
97. Murant AF, Taylor CE, Chambers JC. 1968. Properties, relationships and transmission of a strain of raspberry ringspot virus infecting raspberry cultivars immune to the common Scottish strain. *Ann. Appl. Biol.* 61:175–86
98. Murant AF, Taylor M. 1978. Estimates of molecular weights of nepovirus RNA species by polyacrylamide gel electrophoresis under denaturing conditions. *J. Gen. Virol.* 41:53–61
99. Namba K, Casper DLD, Stubbs GJ. 1985. Computer graphics representation of levels of organization in tobacco mosaic virus structure. *Science* 227: 773–76
100. Nyland G, Lownsbery BF, Lowe BK, Mitchell JF. 1969. The transmission of cherry rasp leaf virus by *Xiphinema americanum*. *Phytopathology* 59:1111–12
101. Percival J. 1912. The eelworm disease of hops. "Nettle-headed" or "shrinkly" plants. *J. South East Agric. College, Wye* 1:5–9
102. Pitcher RS. 1975. Chemical and cultural control of nettlehead and related virus diseases of hop. See Ref. 80, 447–48
103. Pitcher RS, Siddiqi MS, Brown DJF. 1974. *Xiphinema diversicaudatum. Commonw. Inst. Helminthol. Descr. Plant Parasitic Nematodes*, 4, No. 60. 4 pp.
104. Ploeg AT, Brown DJF, Robinson DJ. 1989. Transmission of tobravirus by trichodorid nematodes. *EPPO/OEPP Bull.* 19:605–10
105. Ploeg AT, Brown DJF, Robinson DJ. 1990. The association between species of *Trichodorus* and *Paratrichodorus* vector nematodes and serotypes of tobacco rattle tobravirus. *Ann. Appl. Biol.* 121:619–30
106. Ploeg AT, Brown DJF, Robinson DJ. 1992. RNA-2 of tobacco rattle virus encodes the determinants of transmissibility by trichodorid vector nematodes. *J. Gen. Virol.* 74:1463–66
107. Ploeg AT, Mathis A, Bol JF, Brown DJF, Robinson DJ. 1993. Susceptibility of transgenic tobacco plants expressing tobacco rattle virus coat protein to nematode-transmitted and mechanically inoculated tobacco rattle virus. *J. Gen. Virol.* 74:2709–15
108. Rana GL, Roca F. 1975. Trasmissione con nematodi del virus latente italiano del carciofo (AILV). *Atti II Congr. Int. Carcifo, Bari,* 1973:139–40
109. Raski DJ, Maggenti AR, Jones NO. 1973. Location of grapevine fanleaf and yellow mosaic virus particles in *Xiphinema index*. *J. Nematol.* 5:208–11
110. Robbins RT. 1993. Distribution of *Xiphinema americanum* and related species in North America. *J. Nematol.* 25: 344–48
111. Robbins RT, Brown DJF. 1991. Comments on the taxonomy, occurrence and geographical distribution of Longidoridae (Nematoda) in North America. *Nematologica* 37:395–419
112. Roberts IM, Brown DJF. 1980. Detection of six nepoviruses in their nematode vectors by immunosorbent electron microscopy. *Ann. Appl. Biol.* 96:187–92
113. Robertson WM. 1975. Ultrastructure of nematode vectors of plant viruses with reference to their feeding apparatus. *1974 Annu. Rep. Scott. Hortic. Res. Inst., Invergowrie, Scotland,* pp. 76–77
114. Robertson WM. 1976. A possible gustatory organ associated with the odontophore in *Longidorus leptocephalus* and *Xiphinema diversicaudatum*. *Nematologica* 21:443–48
115. Robertson WM, Hendry CE. 1986. A possible role of carbohydrates in the retention of nematode-transmitted viruses. *1985 Annu. Rep. Scott. Crop Res. Inst., Invergowrie, Scotland,* p. 113
116. Robertson WM, Hendry CE. 1986. An association of carbohydrate with particles of arabis mosaic virus retained within *Xiphinema diversicaudatum*. *Ann. Appl. Biol.* 109:299–305
117. Robertson WM, Trudgill DL, Griffiths BS. 1985. Feeding of *Longidorus elongatus* and *L.leptocephalus* on root-tip galls of perennial ryegrass *Lolium perenne*. *Nemtologica* 30:222–29
118. Robertson WM, Wyss U. 1983. Feeding processes of virus-transmitting nematodes. In *Current Topics in Vector Research*, ed. KF Harris, pp. 271–95. New York: Praeger. 326 pp.
119. Robinson DJ, Hamilton WDO, Harrison BD, Baulcombe DC. 1987. Two anomalous tobravirus isolates. Evidence for RNA recombination in nature. *J. Gen. Virol.* 68:2551–61
120. Robinson DJ, Harrison BD. 1985. Evidence that broad bean yellow band virus

is a new serotype of pea early-browning virus. *J. Gen. Virol.* 66:2003–9
121. Roca F, Rana GL. 1981. *Paratrichodorus tunisiensis* (Nematoda: Trichodoridae). A new vector of tobacco rattle virus in Italy. *Nematol. Mediterr.* 9:217–19
122. Roca F, Rana GL, Kyriakopoulou PE. 1982. *Longidorus fasciatus* Roca et Lamberti vector of a serologically distinct strain of artichoke Italian latent virus in Greece. *Nematol. Mediterr.* 10: 65–69
123. Rudel M, Alebrand M, Altmayer B. 1983. Untersuchungen uber den Einsatz der ELISA-Test zum Nachweis verschiedener Rebeviren. *Die Wien-Wissenschaft* 38:177–85
124. Rumpenhorst HJ, Weischer B. 1978. Histopathological and histochemical studies on grapevine roots damaged by *Xiphinema index. Rev. Nematol.* 1:217–25
125. Salomao TA. 1975. Soil transmission of artichoke yellow band virus. *Atti II Congr. Int. Carcifo, Bari,* 1973:831–54
126. Sanger HL. 1961. Untersuchungen uber schwer ubertragbare Formen des Rattlevirus. *Proc. Conf. Potato Virus Dis., 4th.* Braunschweig, Germany, pp. 228–31
127. Schindler AF. 1958. Attempts to demonstrate the transmission of plant viruses by plant parasitic nematodes. *Plant Dis. Reptr.* 42:1348–50
128. Siddiqi MR. 1973. *Xiphinema americanum. Commonw. Inst. Helminthol. Descr. Plant Parasitic Nematodes,* 2, No. 29. 4 pp.
129. Siddiqi MR. 1974. *Xiphinema index. Commonw. Inst. Helminthol. Descr. Plant Parasitic Nematodes,* 3, No. 45. 4 pp.
130. Sol HN, Van Heuven JC, Seinhorst JW. 1960. Transmission of rattle virus and atropa belladonna mosaic virus by nematodes. *Tijdschr. Plantenziekten* 66: 228–31
131. Stellmach G, Berres R. 1985. Investigations on mixed infections of nepoviruses in *Vitis* spp. and *Chenopodium quinoa* Willd. by means of ELISA. *Phytopathol. Mediterr.* 24:125–28
132. Taylor CE. 1962. Transmission of raspberry ringspot virus by *Longidorus elongatus* (de Man) (Nematoda: Dorylaimida). *Virology* 17:493–94
133. Taylor CE. 1968. The association of ringspot viruses with their nematode vectors. *C. R. 8th Symp. Int. Nematologie, Antibes, 1965,* pp. 109–10
134. Taylor CE. 1972. Nematode transmission of plant viruses. *PANS* 18:269–82
135. Taylor CE. 1980. Nematodes. In *Vectors of Plant Pathogens,* ed. KF Harris, K Maramorosch, pp. 375–416. New York: Academic. 467 pp.
136. Taylor CE, Brown DJF. 1974. An adaptable temperature controlled cabinet. *Nematol. Mediterr.* 2:171–75
137. Taylor CE, Brown DJF. 1976. The geographical distribution of *Xiphinema* and *Longidorus* nematodes in the British Isles and Ireland. *Ann. Appl. Biol.* 84: 383–402
138. Taylor CE, Brown DJF. 1981. Nematode-virus interactions. See Ref. 177, pp. 281–301
139. Taylor CE, Brown DJF, Neilson R, Jones AT. 1994. The persistence and spread of *Xiphinema diversicaudatum* (Nematoda: Dorylaimida) in cultivated and uncultivated biotopes. *Ann. Appl. Biol.* 124:469–77
140. Taylor CE, Raski DJ. 1964. On the transmission of grape fanleaf by *Xiphinema index. Nematologica* 10: 489–95
141. Taylor CE, Robertson WM. 1969. The location of raspberry ringspot and tomato black ring viruses in the nematode vector, *Longidorus elongatus* (De Man). *Ann. Appl. Biol.* 64:233–37
142. Taylor CE, Robertson WM. 1970. Sites of virus retention in the alimentary tract of the nematode vectors, *Xiphinema diversicaudatum* (Micol.) and *X. index* (Thorne & Allen). *Ann. Appl. Biol.* 66:375–80
143. Taylor CE, Robertson WM. 1970. Location of tobacco rattle virus in the nematode vector *Trichodorus pachydermus* Seinhorst. *J. Gen. Virol.* 6:179–82
144. Taylor CE, Robertson WM. 1975. Acquisition, retention and transmission of viruses by nematodes. See Ref. 80, pp. 253–76
145. Taylor CE, Robertson WM, Roca F. 1976. Specific association of artichoke Italian latent virus with the odontostyle of its vector, *Longidorus apulus. Nematol. Mediterr.* 4:23–30
146. Teliz D, Grogan RG, Lownsbery BF. 1966. Transmission of tomato ringspot, peach yellow bud mosaic and grape yellow vein viruses by *Xiphinema americanum. Phytopathology* 56:658–63
147. Towle A, Doncaster CC. 1978. Feeding of *Longidorus caespiticola* on ryegrass, *Lolium perenne. Nematologica* 24:277–85
148. Trudgill DL. 1976. Observations on

the feeding behaviour of *Xiphinema diversicaudatum*. *Nematologica* 22: 417–23

149. Trudgill DL, Brown DJF. 1978. Frequency of transmission of some nematode-borne viruses, In *Plant Disease Epidemiology*, ed. PR Scott, A Bainbridge, pp. 283–89. London: Blackwell Sci. 329 pp.
150. Trudgill DL, Brown DJF. 1978. Ingestion, retention and transmission of two strains of raspberry ringspot virus by *Longidorus macrosoma*. *J. Nematol.* 10: 85–89
151. Trudgill DL, Brown DJF. 1980. Effect of the bait plant on transmission of virus by *Longidorus* and *Xiphinema* spp. *1979 Annu. Rep. Scott. Hortic. Res. Inst., Invergowrie, Scotland*, p. 120
152. Trudgill DL, Brown DJF, McNamara DG. 1983. Methods and criteria for assessing the transmission of plant viruses by longidorid nematodes. *Rev. Nematol.* 6:133–41
153. Trudgill DL, Brown DJF, Robertson WM. 1981. A comparison of the effectiveness of the four British virus vector species of *Longidorus* and *Xiphinema*. *Ann. Appl. Biol.* 99:63–70
154. Trudgill DL, Robertson WM. 1982. Feeding and salivation behaviour of *Xiphinema diversicaudatum* and *Longidorus elongatus*. *Nematologica* 28:177 (Abstr.)
155. Trudgill DL, Robertson WM, Wyss U. 1991. Feeding behaviour of *Xiphinema diversicaudatum*. *Rev. Nematol.* 14: 107–12
156. Valdez RB. 1972. Transmission of raspberry ringspot virus by *Longidorus caespiticola*, *L. leptocephalus* and *Xiphinema diversicaudatum* and of arabis mosaic virus by *L. caespiticola* and *X. diversicaudatum*. *Ann. Appl. Biol.* 71:229–34
157. Valdez RB, McNamara DG, Ormerod PJ, Pitcher RS, Thresh JM. 1974. Transmission of the hop strain of arabis mosaic by *Xiphinema diversicaudatum*. *Ann. Appl. Biol.* 76:113–22
158. Van Den Heuvel JFLM, Verbeek M, Van Der Wilk F. 1994. Endosymbiotic bacteria associated with circulative transmission of potato leafroll virus by *Myzus persicae*. *J. Gen. Virol.* 75:2559–65
159. Van Der Wilk F, Korsman M, Zoon F. 1994. Detection of tobacco rattle virus in nematodes by reverse transcription and polymerase chain reaction. *Eur. J. Plant Pathol.* 100:109–22
160. Van Hoof HA. 1962. *Trichodorus pachydermus* and *T. teres*, vectors of the early-browning virus of peas. *Tijdschr. Plantenziekten* 68:391–96
161. Van Hoof HA. 1964. Serial transmission of rattle virus by a single male of *Trichodorus pachydermus* Seinhorst. *Nematologica* 10:141–44
162. Van Hoof HA. 1964. *Trichodorus teres* a vector of rattle virus. *Neth. J. Plant Pathol.* 70:187
163. Van Hoof HA. 1968. Transmission of tobacco rattle virus by *Trichodorus* species. *Nematologica* 14:20–24
164. Van Hoof HA. 1971. Viruses transmitted by *Xiphinema* species in the Netherlands. *Neth. J. Plant Pathol.* 77: 30–31
165. Van Hoof HA, Maat DZ, Seinhorst JW. 1966. Viruses of the tobacco rattle virus group in Northern Italy: their vectors and serological relationships. *Neth. J. Plant Pathol.* 72:253–58
166. Vovlas N, Inserra RN, Martelli GP. 1978. Modificazioni anatomiche indotte da *Xiphinema index* e *Meloidogyne incognita* in radici di un ibrido di *Vitis vinifera* x *V. rotundifolia*. *Nematol. Mediterr.* 6:67–75
167. Vrain TC, Wakarchuk DA, Levesque CA, Hamilton RI. 1992. Intraspecific rDNA restriction fragment length polymorphism in the *Xiphinema americanum* group. *Fundam. Appl. Nematol.* 15:563–73
168. Walkinshaw CH, Griffin GD, Larson RH. 1961. *Trichodorus christei* as a vector of corky ringspot (tobacco rattle) virus. *Phytopathology* 51:806–8
169. Weischer B, Wyss U. 1976. Feeding behaviour and pathogenicity of *Xiphinema index* on grapevine roots. *Nematologica* 22:319–25
170. Wyss U. 1974. *Trichodorus similis*. *Commonw. Inst. Helminthol. Descr. Plant Parasitic Nematodes*, 4, No. 59. 4 pp.
171. Wyss U. 1981. Ectoparasitic root nematodes: feeding behaviour and plant cell responses. See Ref. 177, pp. 325–51
172. Wyss U. 1987. Video assessment of root cell responses to Dorylaimid and Tylenchid nematodes. In *Vistas on Nematology*, ed. JA Veech, DW Dickson, pp. 211–20. De Leon Springs, FL: EO Painter Printing. 509 pp.
173. Wyss U, Lehmann H, Jank-Ladwig R. 1980. Ultrastructure of modified root-tip cells in *Ficus carica*, induced by the parasitic nematode *Xiphinema index*. *J. Cell Sci.* 41:193–208
174. Wyss U, Robertson WM, Trudgill DL. 1988. Oesophageal bulb function of *Xiphinema index* and associated root

cell responses, assessed by video enhanced light microscopy. *Rev. Nematol.* 11:253–61

175. Yagita H, Komuro Y. 1972. Transmission of mulberry ringspot virus by *Longidorus martini* Merny. *Ann. Phytopathol. Soc. Jpn.* 38:275–83

176. Yassin AM. 1968. Transmission of viruses by *Longidorus elongatus*. *Nematologica* 14:419–28

177. Zuckerman BM, Rohde RA, eds. 1981. *Plant Parasitic Nematodes*, Vol. 3. New York: Academic. 508 pp.

Annu. Rev. Phytopathol. 1995. 33:251–74

BIOCHEMICAL AND BIOPHYSICAL ASPECTS OF WATER DEFICITS AND THE PREDISPOSITION TO DISEASE

John S. Boyer

College of Marine Studies and College of Agriculture, University of Delaware, Lewes, Delaware 19958

KEY WORDS: enzymes, pores, photosynthesis, embryo development, protein synthesis

---

ABSTRACT

A predisposition to disease is often observed in host plants during water deficiencies. While there are no instances in which the biochemical and biophysical causes are known with certainty, there are changes in host plants that alter their interactions with other organisms and that suggest possible mechanisms of predisposition. Gradients in water potential can be altered by water deficits and prevent growth in the host without altering growth of the pathogen. There is a decrease in photosynthetic activity and protein synthetic activity when water deficits develop that could decrease the synthesis of metabolites and enzymes important for disease resistance. A method of supplying metabolites has been used to change the flux of these molecules in intact plants and might provide tests of the causes of host predisposition to disease during water deficits.

INTRODUCTION

This review explores how water affects the growth and metabolism of plants in order to gain insight into the development of plant diseases, particularly those that become more severe when the host encounters limited water. The emphasis is on how the host might become predisposed to attack when it has been weakened by the water deficiency. The biochemical and biophysical

0066-4286/95/0901-0251$05.00

causes of this form of disease predisposition are unknown, and this review is thus restricted to drawing analogies from general interactions between organisms during water deficiencies and applying them to the predisposition response. I hope the reader will see the analogies as tentative but suggesting possibly fruitful areas for future research.

Despite the increasing understanding of water relations in disease organisms, there is relatively more known about vascular plants and it is tempting to apply this knowledge to disease. The diversity of disease and the interaction of many environmental factors make extrapolation from vascular plants a risky activity and the temptation will be avoided here. However, comparisons can be revealing between vigorous and weakened hosts, and we explore these comparisons in this review. There is little known about the biochemistry of recovery from water deficits and this area is almost entirely untouched. Also, much literature on the water relations of vascular plants while rich in insight is too extensive to include. Interested readers may wish instead to consult Close & Bray (22) or Kramer & Boyer (71) for more detailed recent treatments of plant water relations. Specific features of fungal water relations are treated in Ayres & Boddy (2).

PREDISPOSITION TO DISEASE

Water availability can alter the "proneness" of a host to attack by a pathogen, and this predisposition usually results in an increase in susceptibility to the invading organism. Examples are found mostly among the root rots, stem rots, and stem canker organisms, although a few vascular wilts and some seedling pathogens also fit this description. The pathogens usually are facultative, causing disease in some conditions but able to grow as saprophytes in others. The disease-causing organisms are present in and around the plant and attack living tissues under particular conditions that weaken the host. For example, the hyphae of *Botryosphaeria dothidea* often are present in the bark of birch (*Betula alba*) and several other tree species and grow on dead stem or branch tissues saprophytically when the tree is in a vigorous condition (106). However, when the tree is subjected to a water deficit, cankers form in living stem tissues (36). When the water deficit is relieved, the cankers stop developing and *B. dothidea* reverts to a saprophyte in the killed tissue. There is a similar development of cankers on white ash (*Fraxinus americana* L.) from attack by *Cytophoma pruinosa* and *Fusicoccum* species, which are part of the normal microflora of the tree but attack when the trees are subjected to drought (100, 119).

Fusarium roseum f. sp. *cerealis* (Cke.) Snyd. & Hans. 'culmorum' attacks wheat roots in the Pacific Northwest when high-nitrogen plants deplete soil water to a water potential of –3.5 to –4.0 MPa (1 megapascal = 10 bars), but smaller low-nitrogen plants deplete soil water only to –2.8 to –3.2 MPa and are resistant to attack by the fungus (24). The disease appears to become important because the host plants are weakened by the dry conditions and the fungus can grow while other soil organisms antagonistic to it cannot (24). A similar rapid disease progression during water deficit has been described for charcoal rot of sorghum (49) and cotton (52).

Each of these examples indicates that the host condition is modified by water deficit prior to the appearance of the disease. The host is predisposed to the disease by the decreased availability of water, and in these examples a normally benign organism becomes pathogenic. With more aggressive diseases that attack plants growing vigorously, there also is evidence that water deficit can enhance aggressiveness. *Sclerotinia* dollar spot of Kentucky bluegrass attacks plants growing in soil supplied with adequate water but the disease becomes more severe as water is depleted (34). *Cytospora* canker is more aggressive in water-deficient sugar maple but also attacks vigorous trees (79). Flooding of plant roots can cause a predisposition to infection by *Phytophthora* (44). Predisposition is thus a widespread phenomenon although it can be seen most clearly with facultative pathogens. Several reviews of these diseases are available and the reader is directed particularly to those by Cook & Papendick (25), Cook (24), and Schoeneweiss (104–6) for overviews and good examples.

Enzymes and Pore Diameters

Microbial activity depends on the secretion of extracellular enzymes that hydrolyze polymers and allow access for the invading cells and release of substrates for microbial metabolism. Diseases attacking vascular plants generally enter the host cells by dissolving the cuticle (68, 69) and/or cell wall (1, 23, 27), and one of the first questions to consider is how water affects the extracellular enzymes required for these steps.

An important theory is that the pore size in the substrate determines activity (53). Most substrates used by microorganisms are porous and the pores are often water filled. The pore size determines which pores are water filled. Most wood-rotting fungi grow at matric potentials (ψ_m) between –0.1 and –4 MPa (53). From the ψ_m, the diameter of the water-filled pores can be calculated from the equation $d = -4\sigma_w/\psi_m = -0.29 \times 10^{-6}/\psi_m$, where σ_w is the surface tension of water in the pore (7.20×10^{-8} MP a · m at 25°) and d is the diameter of the pore (m). According to this relation, fungal growth at potentials between –0.1 and –4 MPa represent pore diameters between 2.9 μm and 72 nm. It was suggested that pores as small as 72 nm might restrict the diffusion of large molecules such as extracellular enzymes (53), although some wood-decay fungi (*Geastrum* sp.) are able to grow at much lower ψ_m and thus have smaller pore diameters (53).

After this theory was proposed, pore sizes were measured in primary walls of living plant cells. There are two kinds of pores in the wall. The plasmodesmata are the most obvious because they are large, connect adjacent cells, are filled with protoplasm, and are lined with the plasmalemma. The second kind are the small pores in the composite, multifibrillar structure of the wall material. These are much more numerous than the plasmodesmata and transmit most of the molecular traffic to the plasmalemma from the cell exterior, including many pathogen enzymes. Using molecular exclusion methods, pores of the second kind are variously estimated to have diameters of 4 to 6.5 nm (4, 18, 19, 118). They freely transmit water with its diameter of only about 0.4 nm, sugars and amino acids with their diameters of 1 to 1.5 nm, and smaller proteins up to molecular weights of 60,000 with diameters up to 8.5 nm. The passage of molecules with weights larger than about 60,000 is generally blocked.

The pores permitting wood decay at the lower limit of the matric potential (–4 MPa equivalent to 72 nm) are thus nearly 10 times larger than the 8.5 nm exclusionary diameter in primary walls of living cells, and it seems unlikely that this property defines the lower limit for growth of most wood-decay fungi. On the other hand, the enzyme-exclusion theory may apply to *Geastrum* sp. that grow at matric potentials significantly below –4 MPa (53). In soils, the root rot disease of cereals caused by *Fusarium roseum* f. sp. *cerealis* (Cke.) Snyd. & Hans. 'culmorum' grew at potentials as low as –9 MPa (26), which corresponds to a pore diameter of 10 nm. Enzyme exclusion might occur at this pore diameter, and the exclusion from soil pores might determine the lower limit of growth for this organism. On the other hand, *Ophiobolus graminis* grew at soil matric potentials only as low as –4 MPa (26) and enzyme exclusion is unlikely to be the reason for this limitation.

Importantly, the molecular exclusion in living cells is controlled mostly by pectins, and removal of the pectins allows larger molecules to penetrate (4). Various pectin-degrading enzymes (polymethylgalacturonases, polygalacturonases, pectin methylesterases, pectin lyases, polygalacturonide lyases) are among the first enzymes to be induced in invading pathogens (1, 23, 27). They often have molecular weights of less than 60,000 that would allow them to enter wall pores. A few are very large (more than 200,000 Daltons) but are relatively ineffective in wall breakdown (27). As the smaller pectin-degrading enzymes enter the wall and enlarge the pores, cell death occurs rapidly, probably because turgor pressure bursts the plasmalemma at the enlarged pores, although specific products of wall degradation may also be toxic (27). The enlargement of the pores also increases the rate of diffusion of enzymes through the wall (4), and degradation products move more rapidly in and out of the cell.

Enzymes and Dehydration

Apart from the ability of enzymes to enter water-filled pores, can such low potentials directly inhibit enzyme activity? It has long been proposed that low water availability may directly modify enzyme activity because of changes in the free energy of water around the enzymes (70, 114), in spatial relationships of membrane systems, in volume, and in concentration of enzymes resulting from losses of water, or because of decreases in the water of hydration surrounding macromolecules (59, 122). These are direct effects because they involve the water molecules acting directly on the enzyme or cell structure, and some of them are undoubtedly important in some systems. For example, an isolated enzyme such as urease decreases in activity when dehydrated. The enzyme reaction produces ammonia and CO_2, both of which escape into the gas phase where they are easily detected without disturbing the hydration of the enzyme/substrate mixture. Activity is detected when the enzyme is dehydrated in equilibrium with relative humidities as low as 60% (113), which is equivalent to a matric potential of about –80 MPa (more precisely, we should speak of the chemical potential of the matrix here because at these low water contents the molal volume of water $\overline{V}_w$ used to define the matric potential (see 71) cannot be considered to be 18 $cm^3 \cdot mol^{-1}$ as required by the definition, but for comparative purposes we will overlook the problem because the conclusions will be the same).

Lysozyme similarly displays activity at very low water contents and the activity depends on the internal motion of the peptide molecule (101). The activity begins to decrease only after most of the water is removed and a monolayer of surface water remains. As the surface water is removed, the internal motion of the peptide becomes restricted and the activity disappears when only enough water remains to bind to a few exposed polar groups of the peptide.

As a consequence, dehydration directly affects enzyme activity only when very little water is present, much less than is necessary to affect most enzymes in cells. For example, nitrate reductase activity is nearly completely lost at a water potential of about –2.5 MPa when inside the cell in maize leaves (85, 112). Similarly, nitrogenase activity disappears at a water potential around –2.1 MPa in soybean nodules in soil (60). About half of the cell water remains under these conditions, which is far more than monolayer coverage. The difference seems to be that by functioning inside cells these enzymes are affected by enzyme regulators that also are altered by dehydration (see 71). Such regulation includes the synthesis of the enzyme controlled by gene action, the breakdown of the enzyme as part of the normal peptide turnover of the cell, the feedback regulation of the enzyme by substrates and products, the effects of cofactors and activators, and the solute environment around the enzyme. These regulatory factors affect the enzymes before dehydration reduces the water content to a monolayer. Extracellular enzymes would be similarly affected by the solute environment around them. Thus, for microbes in wood or soil, the lower limit of water potential that can support growth is unlikely to be determined by direct effects of water on the enzymes but rather by the regulator environment around them, whether inside or outside the microbial cells. In only a few extreme cases would enzyme dehydration be directly involved.

ENZYME REGULATION AND HOST RESISTANCE AT LOW WATER POTENTIALS

Evidence suggests that some host enzymes play a role in keeping weak pathogens from attacking healthy tissues, and the loss of these enzymes could be important in the predisposition response. Chitinase produced by the host appears to participate in the lysis of hyphae of *Verticillium albo-atrum* during tomato infection (98), and the synthesis of lysozyme, chitinase, and β–1,3-glucanase occurs frequently in plants in response to pathogen invasion (1). Chitinase and β–1,3-glucanase were detected in healthy stem and root tissues of sugar maple (*Acer saccharum*), red oak (*Quercus rubra*), black oak (*Q. velutina*), and white oak (*Q. alba*), and extracts from these tissues lysed the hyphal walls of *Armillaria mellea* (Vahl ex Fr.) Kummer, which usually attacks these trees when they are weakened by an unfavorable environmental condition (123).

For *B. dothidea*, McPartland & Schoeneweiss (80) found that hyphae in healthy European white birch showed many abnormalities that might be attributed to enzymic defenses from the host. The mycelium in healthy trees was restricted to regions close to the inoculation site and hyphae were swollen at the apices and often displayed evidence of bursting and cytoplasmic extrusion. The cell walls appeared eroded and contained large holes. In contrast, when the host was water deficient, the mycelium extended a large distance from the inoculation site and the hyphae were large with intact, unswollen tips. The cell walls did not show evidence of erosion but instead were surrounded by a sheath of mucilaginous materials. The presence of enzymes in healthy trees that can lyse the walls of an invading pathogen is consistent with the eroded walls observed in hyphae of *B. dothidea* in healthy white birch trees (80).

Accordingly, decreased protein synthesis could be a factor in the predisposition to disease in dehydrated plants. Decreased protein synthesis is a central feature of plant dehydration. Synthesis often slows during mild dehydration (58, 74, 75, 85, 110) and can cease entirely in severely desiccated tissues (6). Usually, the decreases are detected as losses in the polyribosomal content of the tissue because most other methods of measuring protein synthesis require aqueous media that rehydrate the tissue. Polyribosomes are complexes of messenger RNA and ribosomal RNA synthesizing the protein, and their disappearance is evidence for a general slowdown in protein synthesis.

Growing plant cells typically have large numbers of polyribosomes that undergo large losses during dehydration, whereas mature tissues contain fewer polyribosomes and lose fewer (7, 85). The losses in growing tissues take place soon after growth is inhibited by the water deficit (74, 75), and losses were detected within 4 h after stem growth was inhibited by transplanting soybean seedlings to vermiculite of low water content (75). The general loss in protein synthesis did not cause the growth inhibition because growth already had slowed, which also was noted in wheat leaves (110). After a time, growth resumed in the soybean stems and the polyribosomal content recovered (75). The polyribosome loss was not a whole-plant response but rather a tissue-specific response, since it was seen particularly in shoot tissues but not root tissues (35). The roots continued to grow after the transplanting and thus there was a relationship between the factors causing growth inhibition and those causing less protein synthesis.

The possibility that abscisic acid (ABA) could be a signal regulating protein synthesis began to attract attention when it was observed that treating tomato leaves with ABA or dehydration gave similar patterns of change in protein synthesis (16). However, there was no loss of polyribosomes with ABA treatment of growing tissues but a large loss with dehydration (35), which appears to exclude ABA as a cause of the general loss in polyribosomes.

The loss in polyribosomes was not due to the loss in synthesis of a particular protein, although some synthetic differences were found (7, 38). The loss was not caused by increases in ribonuclease (42), and mRNA was conserved (43). This indicated that the losses in polyribosomes were not caused by a general breakdown of mRNA.

There is evidence that the polyribosomes translate some mRNAs differently during dehydration (35, 93). A few mRNAs show increased rates of peptide synthesis and others show decreased rates. Some of the changes probably lead to enzyme changes such as observed with nitrate reductase, whose synthesis is decreased at low water potentials (85, 112). Synthesis appears to be inhibited because the nitrate flux to the leaves decreases (112). Nitrate acts as an inducer of nitrate reductase synthesis but the nitrate content of the leaf bears little relation to synthesis because most of the nitrate appears to be stored in the vacuole, and synthesis depends instead on the new cytoplasmic nitrate arriving in the transpiration stream from the roots (111). Thus the flux rather than the total cell nitrate is controlling. This is the only instance where a molecular signal has been identified that controls the synthesis of an enzyme in dehydrated plant tissue and it indicates that, as the signal from the roots decreases during dehydration, synthesis diminishes in the shoots. With diminished synthesis, the activity becomes less as the enzyme is naturally degraded by the normal turnover of the protein in the cell.

It may be important that nitrate reductase diminishes in activity in dehydrated tissue because this reductase controls the incorporation of N into amino acids for many plants. A decrease in activity might decrease the availability of amino acid substrates for overall protein synthesis by the dehydrated cells. There is evidence that cell ATP content also diminishes in growing leaf tissues of dehydrated plants (3), and ATP is required for active protein synthesis.

Therefore, during dehydration-induced loss in protein synthesis, two phenomena appear to be occurring: The synthesis of certain proteins such as nitrate reductase is more affected than for others; and the overall rate of protein synthesis tends to decrease. Although there is some understanding of the molecular regulation of individual protein synthesis such as nitrate reductase, the molecular control of the overall rate remains obscure. Because the protection of host tissues against disease may depend on proteins synthesized in response to pathogen invasion (1, 98, 123), it seems important to know whether the ability to synthesize defensive enzymes diminishes as plants become predisposed to disease. However, no such work is known to the author.

CELL ENERGETICS AND DISEASE PREDISPOSITION AT LOW WATER POTENTIALS

Photosynthesis

Figures 1*A* and 1*B* show a comparison of stem canker from *B. dothidea* in European white birch [*Betula alba*; (36)] and host photosynthetic activity in a related European white birch [*Betula pendula*; (99)] subjected to various water potentials by withholding water from the soil. The similarity in disease progression and photosynthesis inhibition is striking. This correlation is consistent with the possibility that a loss in photosynthetic activity could contribute to the onset of disease predisposition. Rewatering the plants stopped the progression of canker development after one or two days (36), and photosynthetic activity would be expected to recover similarly. Defoliation enhanced disease progression (36), adding more evidence that the lack of photosynthesis could be involved in the predisposition to the disease.

Photosynthate is required for the biosynthesis of hydrolytic enzymes and secondary metabolites that confer disease resistance, and without the required substrates, resistance may not develop or be maintained. In contrast to pathogens, invading symbionts do not elicit this type of resistance but nevertheless

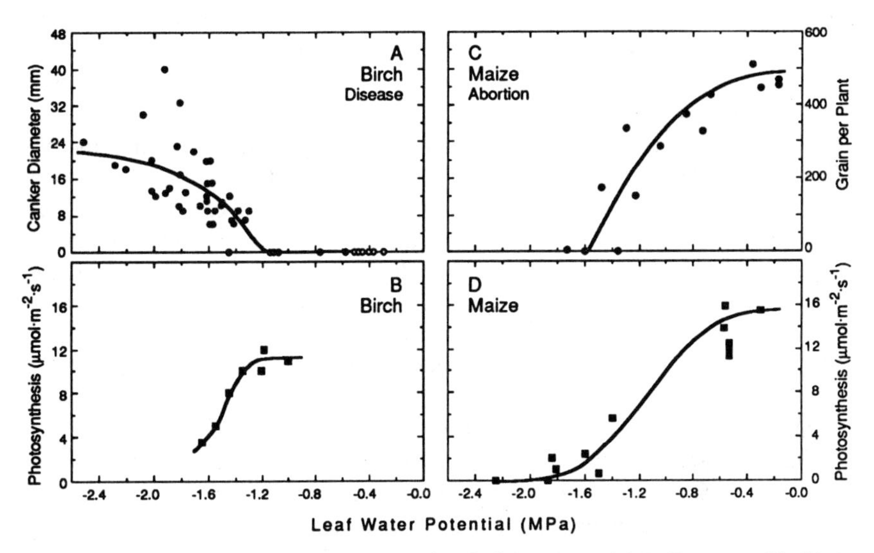

Figure 1 *A*: Increased canker development at low leaf water potentials in European white birch (*Betula alba*) inoculated with *Botryosphaeria dothidea* (data from 36); *B*: decreased photosynthesis at low leaf water potentials in European white birch [*Betula pendula*; (data from 99)]; *C*: decreased embryo development at low leaf water potentials in maize (data from 125); and *D*: decreased photosynthesis at low leaf water potentials in maize (data from 126).

are markedly susceptible to diminished supplies of photosynthate. They may therefore serve as a model from which some insights are possible. For example, *Rhizobium* invades legume roots and engages in nitrogen fixation using carbohydrate from photosynthetic activity of the host. In addition to carbohydrate, the fixation process requires N_2 gas from the atmosphere and enough O_2 so that respiration can occur in the host and bacteroid tissues but not so much that the fixation enzyme nitrogenase becomes inhibited. The delicate balance between the supply of each substrate is dynamically controlled. Removing the source of photosynthetic products by detopping the plant or girdling the stem rapidly inhibits nitrogenase (40, 56, 60, 61, 120, 121) even though substantial photosynthates (sugars and starches) are stored in the nodule tissues (56, 120, 121). Thus a continued flux of recently formed photosynthate seems necessary for N_2 fixation and there is evidence that it also may play a role in the control of the O_2 concentration inside the nodule (40, 41).

Decreasing soil water at first increases the availability of N_2 and O_2 because the soil pores drain and become filled with air (60). As dehydration becomes more severe, nitrogenase activity is markedly inhibited (60). The inhibition is accompanied by a similar inhibition in photosynthesis. Decreasing photosynthesis by decreasing the CO_2 concentration around the shoot of hydrated plants has the same effect on nitrogenase as decreasing photosynthesis by dehydration (61). Increasing photosynthesis with high CO_2 around the shoot reverses some of the inhibition of nitrogenase caused by dehydration (61). These results indicate that decreased photosynthesis is limiting or near limiting for N_2 fixation in dehydrated soil. Huang et al (61) further showed that nitrogenase activity depended on the availability of recently formed photosynthate. Although others (39, 50, 64, 97) showed that there are significant amounts of sugars and starches in nodules of dehydrating plants and argued against a limitation of N_2 fixation by limited availability of photosynthate, the dependency of N_2 fixation on recently formed carbohydrate (56, 61, 120, 121) indicates that the flux probably is more important than the amount stored (62). In effect, during dehydration, the recently formed photosynthate may be a signal from the shoot to the root that controls N_2 fixation.

Increasing the concentration of O_2 around the nodules also can reverse some of the inhibition of nitrogenase activity caused by dehydration (95, 96). The recovery is complete in nodules that are slightly dehydrated but is not complete when dehydration is more severe. Nodule respiration shows a similar response to high O_2 (54, 95). The enhancement by O_2 could be overcome by slicing the nodules (95), indicating that there was an O_2 barrier in intact nodules that was broken by slicing, and the barrier had become more effective when the nodules were dehydrated. Weisz et al (124) directly demonstrated the barrier by showing that dehydration decreased the gas permeability of nodules. Irigoyen et al (63, 65) showed that enzyme activity shifted toward hypoxic metabolism in dehydrated nodules. Thus, in addition to an inhibition by a lack of photosynthate (61), a lack of O_2 can also be limiting in some conditions, presumably because there is an inhibition of conversion of photosynthate to the substrates needed by nitrogenase (95, 96).

These results indicate that water availability has two main effects on N_2 fixation by symbiotic bacteroids. In one, gas diffusion is increased as water drains from the soil pores, and N_2 fixation tends to increase. In the other, photosynthesis and nodule gas diffusion are decreased as dehydration becomes more severe, and N_2 fixation decreases. The exact conditions where the shift occurs probably depend on soil conditions, the amount of photosynthate being produced by the plant, and the structure and condition of the nodules. Mycorrhizal symbiosis can also alter the nodule response (103).

These results with *Rhizobium* indicate that the complex metabolic relationships between the microorganism and host need to be studied in ways that distinguish between the fluxes of metabolites and the pool sizes of the same metabolites. Thus it may prove fruitful to explore whether increased predisposition to disease could result from inability to develop resistance at the enzyme level because photosynthetic flux is diminished in the host.

Embryo Growth: A Case Study

The effects of dehydration on embryo development in vascular plants in some ways resemble host predisposition to disease: An associated organism shows different growth when a water deficit occurs. Embryos can abort during a water deficit, and recent work on embryo abortion is yielding increased understanding of the abortion mechanism. A good example is maize where part of the problem is caused by a high susceptibility of floral development to inhibition of cell enlargement (21, 57, 102, 127). This susceptibility exists in part because the cells enlarge dramatically in the floral tissues during normal development, and water deficit can prevent the enlargement. However, more than cell enlargement is involved because inflorescence development can be decreased by treating maize plants with ABA during floral initiation before most enlargement of reproductive structures begins (37). Losses in reproductive activity also were reported because of megagametophyte sterility (86), asynchronous floral development (57), and nonreceptive silks (72), depending on when dehydration occurred, but when these phases were normal and plants were hand-pollinated, reproductive failure still occurred and was apparent after only a few days of dehydration (128). The loss was caused by an irreversible embryo abortion (128). This indicates that, provided there is good floral development, pollination and fertilization can be successful but a complete block in embryo growth can persist.

The block in embryo development was correlated with low photosynthetic reserves in the maternal plant around anthesis and immediately following zygote formation and it was proposed that because photosynthesis was inhibited during the treatment, the lack of reserves could have starved the embryos (126). Figure 1*C* and 1*D* show that photosynthesis was inhibited at about the same water potential as embryo development. However, the sugar content of maize embryos was not significantly different in hydrated and dehydrated plants, but the flux of sugar could have differed (129). The uptake of sugars was less in maize ovules isolated from dehydrated plants (108, 109) even though the sugar content was high, which further confirms the hypothesis that the flux of sugars could be more important than the sugar content of the developing grain.

Boyle et al (13, 14) took advantage of the finding (128) that low water potentials for a few days could prevent embryo growth, and they developed a system to vary the flux of nutrients including photosynthates (carbohydrates) to the embryos. A complete medium for embryo culture was fed to the stems of the intact plants while the water deficit was developing. The nutrients were supplied without rehydrating the plants. Figure 2 shows that the controls yielded well, but withholding water for a few days (low ψ_w) virtually eliminated grain production because of embryo abortion, which decreased grain

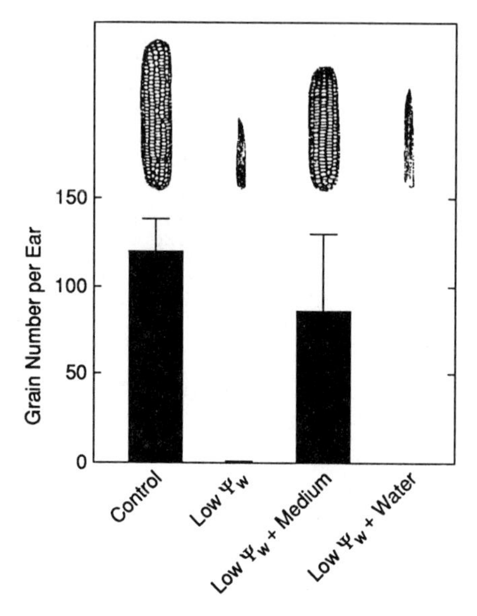

Figure 2 Drought-induced embryo abortion is prevented by feeding maize stems with a medium for embryo culture. Soil water was supplied to controls or was withheld from the plants in the low water potential treatments (low ψ_w) for 6 days prior to and during pollination. Stems of intact plants were infused with the medium (low ψ_w + medium) or the same amount of water (low ψ_w + water). Embryos were present but aborted in low ψ_w and low ψ_w + water treatments, causing small undeveloped ears. The low ψ_w + medium treatment prevented the abortion and allowed grain to grow despite dehydration of the parent plant. Data are from Reference 14.

number. Nevertheless, production was considerably restored (about 70%) when the stems were infused with the nutrient medium while low ψ_w developed (low ψ_w + medium). Infusing the same amount of water alone showed no restorative activity (low ψ_w + water). This response indicated that embryo growth failed because the parent plant was unable to supply an essential nutrient during the water deficit. It demonstrated that sufficient water was available to the embryos so that, when nutrients were supplied, reproduction could proceed even though the plants remained dehydrated.

This type of experiment offers the promise of identifying components that are required for reproductive development and may be lacking when dehydration occurs. By supplying individual components of the medium or deleting components, sucrose was identified as the active ingredient (134). Similar feeding experiments may help identify the biochemical causes of other environmentally induced losses such as those that predispose plants to disease. In effect, whole plant behavior can be studied with the power of in vitro culture.

GROWTH AND DISEASE PREDISPOSITION AT LOW WATER POTENTIALS

The pathogenesis of a microorganism depends on its ability to grow in and around living cells of the host. When the host plant becomes predisposed to invasion by a disease-causing organism, some property of the host or pathogen must change to allow increased growth of the pathogen cells and decreased growth of the host. Most of what we call plant growth is actually increased cellular water content. However, cells can change in size not only by growing but also by swelling and shrinking reversibly as they hydrate and dehydrate. The reversible swelling and shrinking is caused by the elastic nature of the cell walls as they are stretched, causing the cells to become turgid when water enters but flaccid as water departs. In both host and pathogen, gains in water content for growth need to be carefully distinguished from these reversible elastic effects by defining growth to be an irreversible increase in size.

Turgor

Because turgor pressure stretches the wall elastically, there has been much interest in whether turgor also is responsible for irreversible growth. The prevailing theory is that turgor pressure drives growth (see 29, 31, 115, 116) and, in plants exposed to drought, it might be expected that lowering the water potential might lower the turgor and thus decrease growth.

Until work by Ortega et al (94), experiments favoring turgor as the driving force for irreversible growth mostly used osmotica or soil water deficits to control turgor, or measured effects of tensions on isolated walls or killed cells. Each of these treatments modifies conditions in the wall at the same time the turgor is changed. In *Phycomyces*, a fungal saprophyte, Ortega et al (94) attempted to control turgor without altering the wall, by injecting silicon oil into the cells. The experiments allowed turgor to be increased above normal levels. The increase had little effect on growth—as though the wall compensated for the turgor change.

Zhu & Boyer (133) avoided this problem by working with the alga *Chara*, and injecting or removing cell solution to alter the turgor. When turgor was lowered moderately, there was no change in growth rate. When turgor was lowered farther, growth ceased abruptly, indicating that a threshold turgor was required. When turgor was increased, the growth resumed at the original but constant rate. Zhu & Boyer (133) concluded that turgor above the threshold was necessary for growth but did not drive growth. Importantly, they observed an immediate and large effect of inhibitors of energy metabolism, which decreased growth rates even though turgor remained constant. Thus, the limited evidence to date (94) indicates that hyphae may respond to turgor in much the same way as other plant cells (133). If so, moderate decreases in hyphal turgor may not significantly inhibit pathogen growth.

Osmotic Adjustment

Osmotic adjustment is defined as a change in the osmotic potential of the cell, usually because of a change in the cell's solute content. The adjustment maintains cell water content and turgor and protects against the concentrating effects of dehydration by increasing the force that can be exerted by cells on water in the surroundings, thus increasing water uptake. Maintaining hyphal turgor above a threshold is probably important for the progress of plant diseases, and osmotic adjustment can occur in fungi. Eamus & Jennings (46) described in detail the osmotic adjustment in mycelium of the marine saprophyte *Dendryphiella salina* exposed to high concentrations of organic solute or inorganic salts, and they listed many other fungi that show a similar response. The accumulated solutes were mostly the cytoplasmic polyols glycerol, mannitol, erythritol, arabitol, and inositol, although some other organic solute and inorganic salts also participated. Thus, as the host dehydrates and the pathogen invades the host cells, it is likely that the extracellular concentrations of solute could become high because of solute released from the host and because of the lack of water for the host, and osmotic adjustment probably would help maintain turgor in the pathogen.

However, osmotic adjustment also occurs in vascular plants (see 84, 87). Initially, it was thought to occur only in plants subjected to high salinities (5, 47, 48, 88) but it later was found in plants in drying soils (82). Much work was done to determine the effect on plant growth (84, 87). Osmotic adjustment allows growth to continue at a faster rate than it would without adjustment, albeit still at a slow rate (82, 83). The adjustment results from compatible organic solutes that can be present in the cytoplasm without inhibiting enzyme activity and are balanced in the vacuole by salts and other solutes. Compatible solutes for vascular plants are sugars, glycerol, amino acids such as proline or glycinebetaine, sugar alcohols such as mannitol, and some other low-molecular-weight metabolites (84, 87, 131, 132). If large amounts of inorganic salts are present externally, they may be accumulated but generally are stored in the vacuole (51, 55).

Osmotic adjustment solves several problems for cells undergoing dehydration (84). Because compatible solutes accumulate in the cytoplasm, enzyme function is maintained. The water content of the cell remains high and some regulatory inorganic ions do not change concentrations. Turgor is maintained and allows a moderate amount of growth where none would occur otherwise (82, 83).

However, the process is limited by how much solute can be accumulated. Maximum osmotic adjustment was 0.8 to 1 MPa in grass leaves (130), and solute accumulation was more pronounced in growing cells than in mature ones (83). Osmotic adjustment depends to a large extent on photosynthesis or stored reserves to supply substrate for compatible solute. As dehydration becomes severe, photosynthesis becomes inhibited and reserves are depleted. With a smaller solute supply, osmotic adjustment is curtailed. In the vascular plant exposed to continued water limitation, osmotic adjustment delays but cannot completely prevent dehydration. Thus, because osmotic adjustment is dependent on photosynthetically derived products in the host, the adjustment is likely to be more limited than in an attacking fungus. In effect, the fungus probably adjusts osmotically by using the released constituents from the host cells whether or not new substrates from photosynthesis are being supplied to the host.

Water Potential Gradients

Water deficits slow the growth of vascular plants, beginning first with shoot tissues and eventually extending to root tissues. The loss in growth is the first physiological response to the water deficit (10, 127), and some tissues may cease growth completely. The pathogen is not constrained in the same way and grows rapidly in the host tissues.

Compared to most pathogens, the vascular plant depends on certain specialized processes to support growth. Among them is the need for the cells in a tissue to compete with each other for water. Those cells closest to the vascular supply must conduct water fast enough to provide for all the other cells that are often at some distance from the supply, especially in growing tissues where the vascular system may be relatively undeveloped. Decreasing gradients in water potential must extend to the outlying cells in order for water to move. Figure 3*A* shows that a gradient measured in a rapidly growing soybean stem decreased smoothly from the xylem outward to the cortical cells and inward to the pith (92). As a water deficit occurs, the portion of the gradient closest to the xylem reverses (90) and becomes the wrong direction for moving water into the distant cells (Figure 3*B*). The gradient is reversed because the xylem water potential falls in response to the water deficit around the roots. As a result, the distant cells cannot grow even though their part of the gradient has not changed (Figure 3*B*). In a hypha surrounded by a solution of degradation products somewhere in this outlying part of the gradient, water availability would change very little and over long times would change slowly. Most hyphal growth can be maintained with water that crosses only a single cell wall/plasmalemma, thus eliminating the need for large gradients in water potential. More work on the importance of the xylem water potential may be useful to explain the lack of growth of the host tissues while the pathogen grows vigorously.

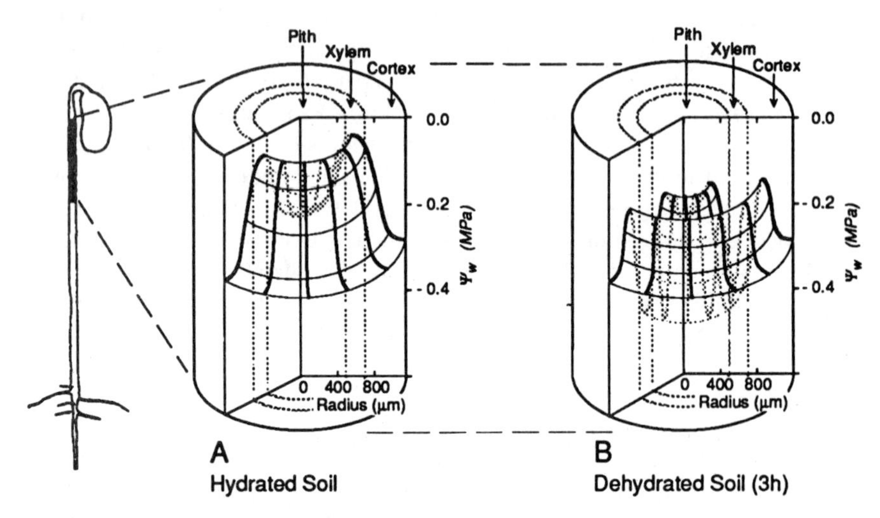

Figure 3 Section of growing soybean stem showing three-dimensional gradient in water potential (*A*) during rapid growth and (*B*) after 3 h of water deficit. Note in (*B*) that the water potential changes around the xylem but not in the outer cortical cells or the inner pith cells. With longer times, the water potentials of these cells also decrease slowly. Data for (*A*) are from (92) and for (*B*) from (90, 92).

GROWTH-INDUCED WATER POTENTIALS When transpiration is prevented so that the only water movement is for growth, water potential gradients are observed in growing tissue (Figure 3) but not mature tissue of the same organ (20, 92). They exist because the osmotic potential of the growing host cells is not balanced completely by the turgor, even though complete balance may occur in mature cells of the same organ. This unbalanced potential was suggested to arise from the enlargement of the cell walls that would prevent turgor pressure from becoming as high as it otherwise would be if the walls were rigid (9). The low pressure results in a low water potential favoring water uptake by the cells because the potential is transmitted to the cell wall (apoplast) solution as a tension (negative pressure) that moves water out of the xylem and into the cells. Tensions of an appropriate magnitude were detected in the apoplast of growing tissues but were near zero in the adjacent mature tissues (89). The tensions were able to mobilize water not only from xylem but also from the mature tissues and cells could grow on this water (77, 78)

When individual cells of plants such as *Chara* are surrounded by the external solution, growth-induced water potentials are too small to detect because water enters the cell by crossing only one cell wall/plasmalemma barrier, and similar situations often occur in hyphae in host tissue. However, long hyphae exposed to air might be expected to display growth-induced water potentials when the distance between the growing tips and the external medium is large. Eamus & Jennings (45) detected lower tip water potentials than base water potentials when the tips of several basidiomycete hyphae grew in air saturated with water vapor, and it thus appears that hyphae generate growth-induced water potentials similar to those of vascular plants when water must be transported large distances in hyphae that do not contact the external medium. One organism with which Eamus & Jennings (45) worked (*Armillaria mellea*) is a known weak pathogen of trees. The low water potentials at the hyphal tips were involved in transporting water and solute from the basal hyphae to the tips (17). Therefore, similar long-distance transport appears to occur in multicellular hyphae and vascular plants when the cell assemblages are large.

Cosgrove & Cleland (32) proposed that growth-induced water potentials might originate from high concentrations of solute in the apoplast of the growing tissues of vascular plants and detected high concentrations in extracts. Others (81) similarly proposed that large and rapid changes in solute concentrations could occur in the apoplast, but did not measure them. However, in repeating the experiments by Cosgrove & Cleland (32), Nonami & Boyer (89) found that solutes were released from phloem and cell membranes disturbed by the experimental procedures. When these problems were avoided, concentrations were too small to account for growth-induced water potentials (89). Others also detected only low solute concentrations in the apoplast of complex tissues (8, 66, 67, 107). The only exception was in the placental tissues of developing seeds where concentrations can be significant (15, 73).

Cosgrove et al (33) also suggested that growth-induced water potentials are artifacts of excision because the cell walls might relax as growth continued without water entry, and turgor would decrease. However, growth-induced water potentials were detected in plants that were completely intact where no excision artifacts were possible (9, 11, 12, 20, 89–91). After excision, Cosgrove et al (33) and Cosgrove (28, 30) found large decreases in water potential and considered them to be caused by wall relaxation. However, these experiments (28, 30, 33) were done with excised tissues that had some mature or slowly growing tissue attached. Matyssek et al (76) showed that under these conditions water was moved to the growing cells, and relaxation was delayed and became much larger than would occur when excision truly prevented water entry. In the study by Matyssek et al (76), the true relaxation was about 0.1 MPa and thus could not account for growth-induced water potentials, which are typically 0.2 to 0.4 MPa (Figure 3).

From the above, it can be seen that for multicellular tissues that have a remote water source, the growth-induced potentials appear to arise from the growth activity of the cells. The enlargement of the cell walls keeps turgor from rising to a level that would be achieved if the walls were rigid, and a gradient in growth-induced water potential develops. Water appears to be pulled into the growing cells by this potential. This suggests that the growth of the wall is the cause of the gradient, and the underlying mechanism may be tied to the metabolism of the wall. However, growth is rapidly prevented by a decrease in xylem water potential, reversing the gradient close to the xylem and inhibiting water movement to the outlying host cells. The pathogen is relatively less affected by the xylem conditions and has negligible growth-induced water potentials when the hyphae are surrounded by the solutions released from the host.

CONCLUSIONS

The complexity of the host response to water limitation opens an array of possible explanations of the increased predisposition to disease often observed in plants weakened by water deficits. It is striking that dehydration-induced losses in photosynthetic activity may be triggered at the same water potential as increased predisposition to disease. Studies of related interactions between vascular plants and microorganisms such as *Rhizobium,* and between parent plants and developing embryos show that the photosynthate flux is often more important than the total photosynthetic reserves in the tissue. Methods of feeding photosynthate from an external source can be used to vary the flux and test the contribution of the photosynthate stream. A possible explanation of why flux can be more important than total cellular pool size is offered by the response to inorganic ions such as nitrate that act as root/shoot signals of soil dehydration. Some pools of cellular nitrate appear to be sequestered in the vacuole and cause the total cell content of nitrate to be high. Nevertheless, vacuolar nitrate is relatively unavailable for signaling the induction of nitrate reductase synthesis, which depends instead on the nitrate flux from the roots. Studies are needed that distinguish between the flux of controlling metabolites such as photosynthate and the total pool size of the metabolite in order to adequately test whether the metabolite is important in the predisposition response.

Another possibly important effect is the decrease in protein synthesis in dehydrated tissues that might contribute to increased host susceptibility by preventing the synthesis of important enzymes for disease resistance in the host. Protein synthesis requires a source of carbon skeletons and amino nitrogen that may be scarce because of losses in photosynthesis and nitrate reductase activity in the host. There are losses in overall activity for protein synthesis but also changes in synthesis of particular enzymes and thus more than substrate availability is likely to be involved. Enzyme synthesis for disease resistance appears not to have been studied under dehydrating conditions.

Enzyme penetration of the pore structure of the host and direct hydration of enzymes are generally adequate for disease initiation in dehydrated hosts and become limiting only at extreme dehydrations where enzyme entrance may be blocked by the small dimensions of the remaining water-filled pores or by insufficient water for monolayer coverage of the enzyme.

Gradients in growth-induced water potentials are present in the healthy host and overall biosynthesis occurs rapidly. In the dehydrated host, decreases in xylem water potential alter part of the gradient and deprive the outlying cells of water for growth. It appears that growth is initially inhibited by this hydraulic signal, and metabolic changes follow that contribute to long-term growth inhibition. In the ensuing changes, osmotic adjustment causes an accumulation of sugars, amino acids, and related solute that might enhance the availability of substrate for the pathogen but evidence for a clear effect is lacking.

The reasons for pathogen growth while the host fails to grow during a water deficit are basically unknown but probably are related to those changes affecting the host but not the pathogen. Changes in photosynthesis, metabolite synthesis, and physical aspects of water uptake for growth in the host may be inconsequential for the pathogen while they may inhibit the maintenance of defense mechanisms in the host. Among these effects may lie some of the explanations for the increased disease susceptibility of dehydrated plants.

Literature Cited

1. Albersheim P, Anderson-Prouty AJ. 1975. Carbohydrates, proteins, cell surfaces, and the biochemistry of pathogenesis. *Annu. Rev. Plant Physiol.* 26: 31–52
2. Ayres PG, Boddy L, eds. 1986. *Water, Fungi, and Plants.* Cambridge: Cambridge Univ. Press. 413 pp.
3. Barlow EWR, Ching TM, Boersma L. 1976. Leaf growth in relation to ATP levels in water stressed corn plants. *Crop Sci.* 16:405–7
4. Baron-Epel O, Gharyal PK, Schindler M. 1988. Pectins as mediators of wall porosity in soybean cells. *Planta* 175: 389–95
5. Bernstein L. 1961. Osmotic adjustment of plants to saline media. I. Steady state. *Am. J. Bot.* 48:909–18
6. Bewley JD. 1979. Physiological aspects of desiccation tolerance. *Annu. Rev. Plant Physiol.* 30:195–238
7. Bewley JD, Larsen KM. 1982. Differences in the responses to water stress of growing and non-growing regions of maize mesocotyls: protein synthesis on total, free and membrane-bound polyribosome fractions. *J. Exp. Bot.* 33:406–15
8. Boyer JS. 1967. Leaf water potentials measured with a pressure chamber. *Plant Physiol.* 42:133–37
9. Boyer JS. 1968. Relationship of water potential to growth of leaves. *Plant Physiol.* 43:1056–62
10. Boyer JS. 1970. Leaf enlargement and metabolic rates in corn, soybean, and sunflower at various leaf water potentials. *Plant Physiol.* 46:233–35
11. Boyer JS. 1974. Water transport in plants: mechanism of apparent changes in resistance during absorption. *Planta* 117:187–207
12. Boyer JS, Cavalieri AJ, Schulze E-D.

1985. Control of cell enlargement: effects of excision, wall relaxation, and growth-induced water potentials. *Planta* 163:527–43
13. Boyle MG, Boyer JS, Morgan PW. 1991. Stem infusion of maize plants. *Crop Sci.* 31:1241–45
14. Boyle MG, Boyer JS, Morgan PW. 1991. Stem infusion of liquid culture medium prevents reproductive failure of maize at low water potentials. *Crop Sci.* 31:1246–52
15. Bradford KJ. 1994. Water stress and the water relations of seed development: a critical review. *Crop Sci.* 34:1–11
16. Bray EA. 1988. Drought- and ABA-induced changes in polypeptide and mRNA accumulation in tomato leaves. *Plant Physiol.* 88:1210–14
17. Brownlee C, Jennings DH. 1982. Long distance translocation in *Serpula lacrimans*, velocity estimates and the continuous monitoring of induced perturbations. *Trans. Br. Mycol. Soc.* 79: 143–48
17a. Callow JA, ed. 1983. *Biochemical Plant Pathology*. New York: Wiley. 484 pp.
18. Carpita NC. 1982. Limiting diameters of pores and the surface structure of plant cell walls. *Science* 218:813–14
19. Carpita NC, Sabularse D, Montezinos D, Delmer DP. 1979. Determination of the pore size of cell walls of living plant cells. *Science* 205:1144–47
20. Cavalieri AJ, Boyer JS. 1982. Water potentials induced by growth in soybean hypocotyls. *Plant Physiol.* 69:492–96
21. Claassen MM, Shaw RH. 1970. Water deficit effects on corn. II. Grain components. *Agron. J.* 62:652–55
22. Close TJ, Bray EA, eds. 1993. *Plant Responses to Cellular Dehydration During Environmental Stress*. Rockville, MD: Am. Soc. Plant Physiol. Ser. 295 pp.
23. Collmer A, Keen NT. 1986. The role of pectic enzymes in plant pathogenesis. *Annu. Rev. Phytopathol.* 24:383–409
24. Cook RJ. 1973. Influence of low plant and soil water potentials on diseases caused by soilborne fungi. *Phytopathology* 63:451–58
25. Cook RJ, Papendick RI. 1972. Influence of water potential of soils and plants on root disease. *Annu. Rev. Phytopathol.* 10:349–74
26. Cook RJ, Papendick RI, Griffin DM. 1972. Growth of two root-rot fungi as affected by osmotic and matric water potentials. *Soil Sci. Soc. Am. Proc.* 36: 78–82
27. Cooper RM. 1983. The mechanisms and significance of enzymic degradation of host cell walls by parasites. See Ref. 17a, pp. 101–35
28. Cosgrove DJ. 1985. Cell wall yield properties of growing tissue: evaluation by in vivo stress relaxation. *Plant Physiol.* 78:347–56
29. Cosgrove DJ. 1987. Wall relaxation and the driving forces for cell expansive growth. *Plant Physiol.* 84:561–64
30. Cosgrove DJ. 1987. Wall relaxation in growing stems: comparison of four species and assessment of measurement techniques. *Planta* 171:266–78
31. Cosgrove DJ. 1993. How do plant cell walls extend? *Plant Physiol.* 102:1–6
32. Cosgrove DJ, Cleland RE. 1983. Solutes in the free space of growing stem tissues. *Plant Physiol.* 72:326–31
33. Cosgrove DJ, Van Volkenburgh E, Cleland RE. 1984. Stress relaxation of cell walls and the yield threshold for growth: demonstration and measurement by micropressure probe and psychrometer techniques. *Planta* 162: 46–54
34. Couch HB, Bloom JR. 1960. Influence of environment on diseases of turfgrasses. II. Effect of nutrition, pH, and soil moisture on *Sclerotinia* dollar spot. *Phytopathology* 50:761–63
35. Creelman RA, Mason HS, Bensen RJ, Boyer JS, Mullet JE. 1990. Water deficit and abscisic acid cause differential inhibition of shoot versus root growth in soybean seedlings. *Plant Physiol.* 92: 205–14
36. Crist CR, Schoeneweiss DF. 1975. The influence of controlled stresses on susceptibility of European white birch stems to attack by *Botryosphaeria dothidea*. *Phytopathology* 65:369–73
37. Damptey HB, Coombe BG, Aspinall D. 1978. Apical dominance, water deficit, and axillary inflorescence growth in *Zea mays*: the role of abscisic acid. *Ann. Bot.* 42:1447–58
38. Dasgupta J, Bewley JD. 1984. Variations in protein synthesis in different regions of greening leaves of barley seedlings and effects of imposed water stress. *J. Exp. Bot.* 35:1450–59
39. Davey AG, Simpson RJ. 1990. Nitrogen fixation by subterranean clover at varying stages of nodule dehydration. I. Carbohydrate status and short-term recovery of nodulated root respiration. *J. Exp. Bot.* 41:1175–87
40. Denison RF, Hunt S, Layzell DB. 1992. Nitrogenase activity, nodule respiration and O_2 permeability following detopping of alfalfa and birdsfoot trefoil. *Plant Physiol.* 98:894–900
41. Denison RF, Smith DL, Legros T,

Layzell DB. 1991. Noninvasive measurement of internal oxygen concentration of field-grown soybean nodules. *Agron. J.* 83:166–69
42. Dhindsa RS, Bewley JD. 1976. Plant desiccation: polysome loss not due to ribonuclease. *Science* 191:181–82
43. Dhindsa RS, Bewley JD. 1978. Messenger RNA is conserved during drying of the drought-tolerant moss *Tortula ruralis*. *Proc. Natl. Acad. Sci. USA* 75: 842–46
44. Duniway JM, Gordon TR. 1986. Water relations and pathogen activity in soil. See Ref. 2, pp. 119–37
45. Eamus D, Jennings DH. 1984. Determination of water, solute and turgor potentials of mycelium of various basidiomycete fungi causing wood decay. *J. Exp. Bot.* 35:1782–86
46. Eamus D, Jennings DH. 1986. Water, turgor and osmotic potentials of fungi. See Ref. 2, pp. 27–48
47. Eaton FM. 1927. The water requirement and cell-sap concentration of Australian saltbush and wheat as related to the salinity of the soil. *Am. J. Bot.* 14:212–26
48. Eaton FM. 1942. Toxicity and accumulation of chloride and sulfate salts in plants. *J. Agric. Res.* 64:357–99
49. Edmunds LK. 1964. Combined relation of plant maturity, temperature, and soil moisture to charcoal stalk rot development in grain sorghum. *Phytopathology* 54:514–17
50. Fellows RJ, Patterson RP, Raper CD, Harris D. 1987. Nodule activity and allocation of photosynthate of soybean during recovery from water stress. *Plant Physiol.* 84:456–60
51. Flowers TJ, Troke PF, Yeo AR. 1977. The mechanism of salt tolerance in halophytes. *Annu. Rev. Plant Physiol.* 28: 89–121
52. Ghaffar A, Erwin DC. 1969. Effect of soil water stress on root rot of cotton caused by *Macrophomina phaseolina*. *Phytopathology* 59:795–97
53. Griffin DM. 1977. Water potential and wood-decay fungi. *Annu. Rev. Phytopathol.* 15:319–29
54. Guerin V, Trinchant J-C, Rigaud J. 1990. Nitrogen fixation (C_2H_2 reduction) by broad bean (*Vicia faba* L.) nodules and bacteroids under water-restricted conditions. *Plant Physiol.* 92: 595–601
55. Hajibagheri MA, Flowers TJ. 1989. X-ray microanalysis of ion distribution within root cortical cells of the halophyte *Suaeda maritima* (L.) Dum. *Planta* 177:131–34
56. Hartwig U, Boller BC, Baur-Hoch B, Nosberger J. 1990. The influence of carbohydrate reserves in the response of nodulated white clover to defoliation. *Ann. Bot.* 65:97–105
57. Herrero MP, Johnson RR. 1981. Drought stress and its effects on maize reproductive systems. *Crop Sci.* 21:105–10
58. Hsiao TC. 1970. Rapid changes in levels of polyribosomes in *Zea mays* in response to water stress. *Plant Physiol.* 46:281–85
59. Hsiao TC. 1973. Plant responses to water stress. *Annu. Rev. Plant Physiol.* 24:519–70
60. Huang CY, Boyer JS, Vanderhoef LN. 1975. Acetylene reduction (nitrogen fixation) and metabolic activities of soybean having various leaf and nodule water potentials. *Plant Physiol.* 56:222–27
61. Huang CY, Boyer JS, Vanderhoef LN. 1975. Limitation of acetylene reduction (nitrogen fixation) by photosynthesis in soybean having low water potentials. *Plant Physiol.* 56:228–32
62. Hunt S, Layzell DB. 1993. Gas exchange of legume nodules and the regulation of nitrogenase activity. *Annu. Rev. Plant Physiol. Plant Mol. Biol.* 44:483–511
63. Irigoyen JJ, Emerich DW, Sánchez-Díaz M. 1992. Phosphoenolpyruvate carboxylase, malate and alcohol dehydrogenase activities in alfalfa (*Medicago sativa*) nodules under water stress. *Physiol. Plant.* 84:61–66
64. Irigoyen JJ, Emerich DW, Sánchez-Díaz M. 1992. Water stress induced changes in concentrations of proline and total soluble sugars in nodulated alfalfa (*Medicago sativa*) plants. *Physiol. Plant.* 84:55–60
65. Irigoyen JJ, Sánchez-Díaz M, Emerich DW. 1992. Transient increase of anaerobically-induced enzymes during short-term drought of alfalfa root nodules. *J. Plant Physiol.* 139:397–402
66. Jachetta JJ, Appleby AP, Boersma L. 1986. Use of the pressure vessel to measure concentrations of solutes in apoplastic and membrane-filtered symplastic sap in sunflower leaves. *Plant Physiol.* 82:995–99
67. Klepper B, Kaufmann MR. 1966. Removal of salt from xylem sap by leaves and stems of guttating plants. *Plant Physiol.* 41:1743–47
68. Kolattukudy PE. 1985. Enzymatic penetration of the plant cuticle by fungal pathogens. *Annu. Rev. Phytopathol.* 23: 223–50

69. Kolattukudy PE, Köller W. 1983. Fungal penetration of the first line defensive barriers of plants. See Ref. 17a, pp. 79–100
70. Kramer PJ. 1969. *Plant and Soil Water Relationships: A Modern Synthesis.* New York: McGraw-Hill. 482 pp.
71. Kramer PJ, Boyer JS. 1995. *Water Relations of Plants and Soils.* San Diego: Academic. In press
72. Lonnquist JH, Jugenheimer RW. 1943. Factors affecting the success of pollination in corn. *J. Am. Soc. Agron.* 35:923–33
73. Maness NO, McBee GG. 1986. Role of placental sac in endosperm carbohydrate import in sorghum caryopses. *Crop Sci.* 26:1201–7
74. Mason HS, Matsuda K. 1985. Polyribosome metabolism, growth and water status in the growing tissues of osmotically stressed plant seedlings. *Physiol. Plant.* 64:95–104
75. Mason HS, Mullet JE, Boyer JS. 1988. Polysomes, messenger RNA and growth in soybean stems during development and water deficit. *Plant Physiol.* 86: 725–33
76. Matyssek R, Maruyama S, Boyer JS. 1988. Rapid wall relaxation in elongating tissues. *Plant Physiol.* 86:1163–67
77. Matyssek R, Maruyama S, Boyer JS. 1991. Growth-induced water potentials may mobilize internal water for growth. *Plant Cell Environ.* 14:917–23
78. Matyssek R, Tang A-C, Boyer JS. 1991. Plants can grow on internal water. *Plant Cell Environ.* 14:925–30
79. McPartland JM. 1983. *Stress predisposition and histopathology of canker diseases in woody hosts.* MS thesis, Univ. Ill., Urbana. 60 pp.
80. McPartland JM, Schoeneweiss DF. 1984. Hyphal morphology of *Botryosphaeria dothidea* in vessels of unstressed and drought-stressed stems of *Betula alba. Phytopathology* 74:358–62
81. Meshcheryakov A, Steudle E, Komor E. 1992. Gradients of turgor, osmotic pressure, and water potential in the cortex of the hypocotyl of growing *Ricinus* seedlings. *Plant Physiol.* 98:840–52
82. Meyer RF, Boyer JS. 1972. Sensitivity of cell division and cell elongation to low water potentials in soybean hypocotyls. *Planta* 108:77–87
83. Michelena VA, Boyer JS. 1982. Complete turgor maintenance at low water potentials in the elongating region of maize leaves. *Plant Physiol.* 69:1145–49
84. Morgan JM. 1984. Osmoregulation and water stress in higher plants. *Annu. Rev. Plant Physiol.* 35:299–319
85. Morilla CA, Boyer JS, Hageman RH. 1973. Nitrate reductase activity and polyribosomal content of corn (*Zea mays* L.) seedlings having low leaf water potentials. *Plant Physiol.* 51:817–24
86. Moss GI, Downey LA. 1971. Influence of drought stress on female gametophyte development in corn (*Zea mays* L.) and subsequent grain yield. *Crop Sci.* 11: 368–72
87. Munns R. 1988. Why measure osmotic adjustment? *Aust. J. Plant Physiol.* 15: 717–26
88. Munns R. 1993. Physiological processes limiting plant growth in saline soils: some dogmas and hypotheses. *Plant Cell Environ.* 16:15–24
89. Nonami H, Boyer JS. 1987. Origin of growth-induced water potential: solute concentration is low in apoplast of enlarging tissues. *Plant Physiol.* 83:596–601
90. Nonami H, Boyer JS. 1989. Turgor and growth at low water potentials. *Plant Physiol.* 89:798–804
91. Nonami H, Boyer JS. 1990. Primary events regulating stem growth at low water potentials. *Plant Physiol.* 93: 1601–9
92. Nonami H, Boyer JS. 1993. Direct demonstration of a growth-induced water potential gradient. *Plant Physiol.* 102: 13–19
93. Oliver MJ. 1991. Influence of protoplasmic water loss on the control of protein synthesis in the desiccation-tolerant moss *Tortula ruralis. Plant Physiol.* 97: 1501–11
94. Ortega JKE, Zehr EG, Keanini RG. 1989. In vivo creep and stress relaxation experiments to determine the wall extensibility and yield threshold for the sporangiophores of *Phycomyces. Biophys. J.* 56:465–75
95. Pankhurst CE, Sprent JI. 1975. Effects of water stress on the respiratory and nitrogen-fixing activity of soybean root nodules. *J. Exp. Bot.* 26:287–304
96. Pankhurst CE, Sprent JI. 1976. Effects of temperature and oxygen tension on the nitrogenase and respiratory activities of turgid and water-stressed soybean and French bean root nodules. *J. Exp. Bot.* 27:1–9
97. Pararajasingham S, Knievel DP. 1990. Nitrogenase activity, photosynthesis and total nonstructural carbohydrates in cowpea during and after drought stress. *Can. J. Plant Sci.* 70:1005–12
98. Pegg GF, Vessey JC. 1973. Chitinase activity in *Lycopersicon esculentum* and

its relationship to the in vivo lysis of *Verticillium albo-atrum* mycelium. *Physiol. Plant Pathol.* 3:207–22

99. Ranney TG, Bir RE, Skroch WA. 1991. Comparative drought resistance among six species of birch (*Betula*): influence of mild water stress on water relations and leaf gas exchange. *Tree Physiol.* 8:351–60
100. Ross EW. 1964. Cankers associated with ash dieback. *Phytopathology* 54: 272–75
101. Rupley JA, Gratton E, Careri G. 1983. Water and globular proteins. *Trends Biochem. Sci.* 8:18–22
102. Salter PJ, Goode JE. 1967. *Crop Responses to Water at Different Stages of Growth.* Farnham Royal, Bucks, UK: Commonw. Agric. Bur. 246 pp.
103. Sánchez-Díaz M, Pardo M, Antolín M, Peña J, Aguirreolea J. 1990. Effect of water stress on photosynthetic activity in the *Medicago-Rhizobium-Glomus* symbiosis. *Plant Sci.* 71:215–21
104. Schoeneweiss DF. 1975. Predisposition, stress, and plant disease. *Annu. Rev. Phytopathol.* 13:193–211
105. Schoeneweiss DF. 1978. Water stress as a predisposing factor in plant disease. In *Water Deficits and Plant Growth,* ed. TT Kozlowski, 5:61–99. New York: Academic
106. Schoeneweiss DF. 1986. Water stress predisposition to disease—an overview. See Ref. 2, pp. 157–74
107. Scholander PF, Hammel HT, Bradstreet ED, Hemmingsen EA. 1965. Sap pressure in vascular plants. *Science* 148: 339–46
108. Schussler JR, Westgate ME. 1991. Maize kernel set at low water potential: I. Sensitivity to reduced assimilates during early kernel growth. *Crop Sci.* 31: 1189–95
109. Schussler JR, Westgate ME. 1991. Maize kernel set at low water potential: II. Sensitivity to reduced assimilates at pollination. *Crop Sci.* 31: 1196–203
110. Scott NS, Munns R, Barlow EWR. 1979. Polyribosome content in young and aged wheat leaves subjected to drought. *J. Exp. Bot.* 30:905–11
111. Shaner DL, Boyer JS. 1976. Nitrate reductase activity in maize (*Zea mays* L.) leaves. I. Regulation by nitrate flux. *Plant Physiol.* 58:449–504
112. Shaner DL, Boyer JS. 1976. Nitrate reductase activity in maize (*Zea mays* L.) leaves. II. Regulation by nitrate flux at low leaf water potentials. *Plant Physiol.* 58:505–9
113. Skujins JJ, McLaren AD. 1967. Enzyme reaction rates at limited water activities. *Science* 158:1569–70
114. Slatyer RO. 1967. *Plant-Water Relationships.* New York: Academic. 366 pp.
115. Taiz L. 1984. Plant cell expansion: regulation of cell wall mechanical properties. *Annu. Rev. Plant Physiol.* 35:585–657
116. Taiz L, Métraux J-P, Richmond PA. 1981. Control of cell expansion in the *Nitella* internode. In *Cell Biology Monographs,* Vol. 8: *Cytomorphogenesis in Plants,* ed. O Kiermayer, pp. 231–64. Wien: Springer-Verlag
117. Deleted in proof
118. Tepfer M, Taylor IEP. 1981. The permeability of plant cell walls as measured by gel filtration chromatography. *Science* 213:761–63
119. Tobiessen P, Buchsbaum S. 1976. Ash dieback and drought. *Can. J. Bot.* 54:543–45
120. Vance CP, Heichel GH, Barnes DK, Bryan JW, Johnson LE. 1979. Nitrogen fixation, nodule development, and vegetative regrowth of alfalfa (*Medicago sativa* L.) following harvest. *Plant Physiol.* 64:1–8
121. Walsh KB, Vessey JK, Layzell DB. 1987. Carbohydrate supply and N_2 fixation in soybean. The effect of varied daylength and stem girdling. *Plant Physiol.* 85:137–44
122. Walter HD. 1931. *Die Hydratur der Pflanze und ihre physiologische-ökologische Bedeutung,* pp. 118–21. Jena: Fischer
123. Wargo PM. 1975. Lysis of the cell wall of *Armillaria mellea* by enzymes from forest trees. *Physiol. Plant Pathol.* 5:99–105
124. Weisz PR, Denison RF, Sinclair TR. 1985. Response to drought stress of nitrogen fixation (acetylene reduction) rates by field-grown soybeans. *Plant Physiol.* 78:525–30
125. Westgate ME. 1984. *Water transport, osmotic adjustment, and growth of the vegetative and reproductive structures of maize.* PhD thesis, Univ. Ill., Urbana. 157 pp.
126. Westgate ME, Boyer JS. 1985. Carbohydrate reserves and reproductive development at low water potentials in maize. *Crop Sci.* 25:762–69
127. Westgate ME, Boyer JS. 1985. Osmotic adjustment and the inhibition of leaf, root, stem and silk growth at low water potentials in maize. *Planta* 164:540–49
128. Westgate ME, Boyer JS. 1986. Reproduction at low silk and pollen water potentials in maize. *Crop Sci.* 26:951–56
129. Westgate ME, Thomson Grant DL.

1989. Water deficits and reproduction in maize: response of the reproductive tissue to water deficits at anthesis and mid-grain fill. *Plant Physiol.* 91:862–67

130. Wilson JR, Ludlow MM. 1984. Time trends of solute accumulation and the influence of potassium fertilizer on osmotic adjustment of water-stressed leaves of three tropical grasses. *Aust. J. Plant Physiol.* 10:523–37

131. Wyn Jones RG. 1980. An assessment of quaternary ammonium and related compounds as osmotic effectors in crop plants. In *Genetic Engineering of Osmoregulation*, ed. DW Rains, RC Valentine, A Hollaender, pp. 155–70. New York: Plenum

132. Yancey PH, Clark ME, Hand SC, Bowlus RD, Somero GN. 1982. Living with water stress: evolution of osmolyte systems. *Science* 217:1214–22

133. Zhu GL, Boyer JS. 1992. Enlargement in *Chara* studied with a turgor clamp: growth rate is not determined by turgor. *Plant Physiol.* 100:2071–80

134. Zinselmeier C, Lauer MJ, Boyer JS. 1995. Reversing drought-induced losses in grain yield: sucrose maintains embryo growth. *Crop Sci.* In press

Annu. Rev. Phytopathol. 1995. 33:275–97

PHYTOALEXINS, STRESS METABOLISM, AND DISEASE RESISTANCE IN PLANTS

Joseph Kuć

Department of Plant Pathology, University of Kentucky, Lexington, Kentucky 40546

KEY WORDS: defense compounds, elicitors, active defense, gene expression

> The reasonable man adapts himself to the world and the unreasonable one persists in trying to adapt the world to himself. Therefore, all progress depends upon the unreasonable man.
>
> George Bernard Shaw

ABSTRACT

Phytoalexins are low-molecular-weight antimicrobial compounds that accumulate in plants as a result of infection or stress. The rapidity and extent of their accumulation is determined by their release or the release of immediate precursors from conjugates and/or de novo synthesis, as well as detoxification as a result of plant or microbial enzymes. The rapidity of phytoalexin accumulation is associated with resistance in plants to diseases caused by fungi and bacteria, although the genetic information for phytoalexin synthesis is found in susceptible and resistant plants. Phytoalexins are only one component of the complex mechanisms for disease resistance in plants. This review considers the contribution of phytoalexins to disease resistance and evaluates molecular and nonmolecular strategies for their biosynthesis as potential technologies for disease control.

INTRODUCTION

The artist and research scientist have much in common. The scientist introduces an idea and the artist an image, both contributions may be unique to the scientist or artist and new to the world. Therefore, the research scientist and artist are different from the technologists who utilize rather than create concepts. How-

0066-4286/95/0901-0275$05.00

ever, those who create concepts and those who utilize concepts make equally valuable contributions to society. In general, the value of "good" scientific contributions increases with time as they serve as nuclei for new ideas and concepts. The value of "good" art also increases with time and serves to generate new art forms and perspectives. However, both new art and science are subject to shifts in public perception of what is in vogue. Styles dictate the direction of art as well as science.

Phytoalexins are no longer the trendsetters as defense compounds against diseases caused by fungi and bacteria. That doesn't mean that they no longer are important, that the early research was flawed, or that they may not return to center stage. It merely means that new defense-associated compounds, among them chitinases, β-1,3-glucanases, pathogenesis-related proteins in general, salicylic acid, active oxygen species, and jasmonate, have been discovered and are in style. Clearly, new information and technology are necessary to approach truth and understanding. However, to realize the maximum benefit from the new, it is necessary to incorporate that which is valid from the old, and not simply exclude all that is old because it is old and accept all that is new because it is new. It is difficult to be integrative in science, but it becomes increasingly evident that no single compound or mechanism reported to date explains disease resistance in plants and it is highly unlikely that a single compound or mechanism will suffice.

With this philosophical introduction, I hope to present in this chapter my views of what we currently know, what we think we know, and what we need to know about phytoalexins as they relate to plant disease resistance. The subject of phytoalexins has been frequently reviewed from different perspectives (2, 3,17, 25–27, 32, 45, 49, 56, 72). I do not attempt to consider all phytoalexin structures or their biosynthesis, but rather present a view of central issues in the field.

PHYTOALEXIN CONCEPT

The concept of a response phase in plant defense against pathogens has been alluded to from the early observations of plant diseases and the establishment of infectious agents as their cause. Early experiments and observations were collected and reported by Chester (13). However, not until the reports by Meyer & Muller and their associates during the period 1939–1961 was the importance of a plant response in resistance established.

Muller & Borger (62) published the work that caused a fundamental conceptual change concerning disease resistance in plants. Using techniques codeveloped with Meyer (58), Muller & Borger presented strong evidence that resistance of potato to *Phytophthora infestans* is caused by the production of fungitoxic compounds by the host. In disease (susceptibility), the production of these compounds is either suppressed or they are detoxified. The race-specific resistance they studied was not due to the absence of infection. All potatoes were as readily infected by all races of the fungus, and resistance or compatibility was expressed subsequent to infection. Muller and coworkers proposed, therefore, that toxic compounds are formed after infection in resistant interactions. This concept was expanded and formed the basis of the phytoalexin theory (58, 62, 63). It became evident that the rapidity and magnitude with which antifungal compounds were produced, rather than the magnitude alone, were important in disease resistance and that potato tubers could be protected against disease caused by a compatible race of *P. infestans* by prior inoculation with an incompatible race. At least with respect to phytoalexins and the potato, it was also evident that susceptibility was not due to a lack of genes for a defense compound in the host but rather the timing and intensity of their expression.

In later years, Muller (59, 60) designed the classical method for the elicitation of phytoalexins in plants. The epidermal tissue lining the seed cavities of legume pods was inoculated with drops containing spores of a nonpathogen (initially *Monolinia fructicola* or *P. infestans*). After less than 8 h at 20°C, the drop contained strongly fungitoxic compounds. These compounds were not detected when pods were treated with water. The compounds produced were low molecular weight and heat stable. Based on these studies and his earlier observations with potato, Muller (61) defined phytoalexins as "compounds produced after infection under the influence of two metabolic systems, that of the host and that of the parasite, and inhibitory to the parasite."

The research of Cruickshank & Perrin, however, was vital to establish the phytoalexin concept on the molecular level. They isolated the phytoalexin pisatin from inoculum droplets on opened pea pods and established its structure (19, 20, 67). These reports were followed shortly after by the characterization of phaseollin from bean pods inoculated with fungi (23, 66). At approximately the time these reports appeared, the structures of other phytoalexins were published: ipomeamarone, chlorogenic acid, umbelliferone, and scopoletin in sweet potato (1, 44, 80); 6-methoxymellein from carrot (15, 16); orchinol, hircinol, and loroglossol from orchids (39, 40). The establishment of chemical structures for phytoalexins opened the door to subsequent studies involving their biosynthesis, degradation, role in disease resistance, regulation of genes for their synthesis, and transfer of genes for their synthesis from one plant to another.

As more information became available, however, it became apparent that the definition for phytoalexins proposed by Muller was no longer sufficiently inclusive since phytoalexin accumulation is not limited to infection, although it occurs after infection. As phytoalexin research progressed, many modified definitions for phytoalexins have been proposed. The one that appeals to me most, but is not original to me, is "phytoalexins are low molecular weight antimicrobial compounds produced by plants in response to infection or stress." The term antimicrobial is perhaps too general since the existence of antiviral phytoalexins has not been established, and antimicrobial activity has most often been reported against fungi and considerably less frequently against bacteria.

It is important to emphasize that phytoalexins are not unique for infection and their accumulation after infection may merely reflect their accumulation in response to stress caused by infection. This is an important conceptual point and I, therefore, include the following discussion in this section. Phytoalexin synthesis and accumulation does not require specific structural components of fungi, bacteria and viruses, or microbial metabolites as elicitors. The three groups of pathogens, as well as nematodes, have been reported to elicit accumulation of the same phytoalexins in some plants. The list of compounds that elicit phytoalexin accumulation is awesome because of its length and the diversity of chemical structures comprising the list. Elicitors include compounds as diverse as inorganic salts (68), oligoglucans (73), ethylene (9, 10), fatty acids (7), chitosan oligomers (42), and a polypeptide (22). Over 200 compounds, microorganisms, and physiological stresses have been reported to elicit pisatin accumulation in pea, phaseollin and kievitone accumulation in green bean, and the glyceollins in soybean (48). Plant constituents released after injury or infection can also function as elicitors (25, 27, 41, 48). To cloud the issue of elicitation further, some fungicides, low temperature, and ultraviolet radiation also elicit accumulation of phytoalexins (35, 57), and some elicitors are antifungal at concentrations at which they elicit (42). Therefore, the determinant for phytoalexin synthesis and accumulation is likely to be metabolic perturbation and not the structures of elicitors except as they cause metabolic perturbation. This does not preclude the importance of microbial or plant components released during infection as elicitors of defense compounds. Based on the above, it seems appropriate, however, to shift research emphasis from the structures of elicitors to what they do and how. Somehow this awesome number of microorganisms and chemicals cause a temporary deregulation of metabolism and this is chemically translated as an alarm signal or signals. The signal(s) initiates a cascade of events leading to, among many other things, the localized synthesis and localized accumulation of phytoalexins. This can include enhanced activities and regulation of as many as 20 genes for the synthesis of a phytoalexin. This level of direction, regulation, and organization exists in the midst of apparent cellular chaos. Though many apparently unrelated changes occur in plants after infection, there is organization and purpose to cellular activity as it is directed to a return to normalcy and, in the long term, survival. A current review of the literature pertaining to the early events in the activation of plant-defense responses, including phytoalexins, is available (27).

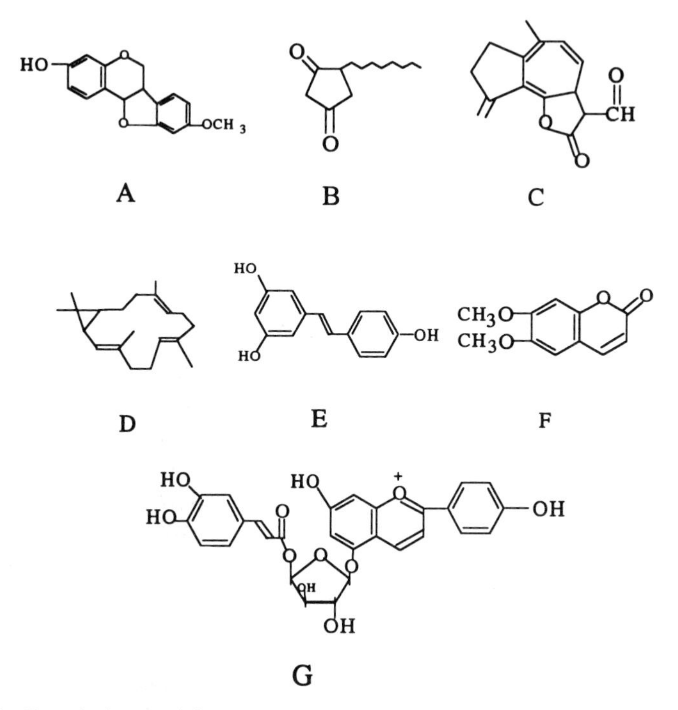

Figure 1 Phytoalexins containing only carbon, hydrogen and oxygen. *A*: Medicarpin-red clover, white clover, alfalfa, chickpea (*Phytochemistry* 1972, 62:235); *B*: Tsibulin 1d-onion (*Physiol. Mol. Plant Pathol.* 1990, 37:235–44); *C*: Lettucenin A-lettuce (*Chem. Comm.* 1985, 10:621–22); *D*: Casbene-castor bean (*Phytopathology* 1975, 14:1921–25); *E*: Resveratrol-peanut, grape (*Phytochemistry* 1976, 15:1791-93); *F*: Scoparone-citrus (*Phytochemistry* 1986, 25:1855–56); *G*: Caffeic acid ester of arabinosyl-5-0-apogeninidin-sorghum (*Physiol. Mol. Plant Pathol.* 1990, 36:381–96).

DISTRIBUTION, STRUCTURE, BIOSYNTHESIS

Distribution and Structure

Equally as striking as the number and diversity of elicitors is the number and diversity of structures for phytoalexins. More than 350 phytoalexins have been chemically characterized from approximately 30 plant families. The greatest number, 130, have been characterized from the *Leguminosae*. Most phytoalexins have been isolated from dicotyledons, but they have also been isolated from monocotyledons including rice, corn, sorghum, wheat, barley, onions, and lilies. Phytoalexins have been isolated from stems, roots, leaves, and fruits though they have not always been reported produced, or the same compounds produced, by all organs of a plant. The distribution of phytoalexins and the

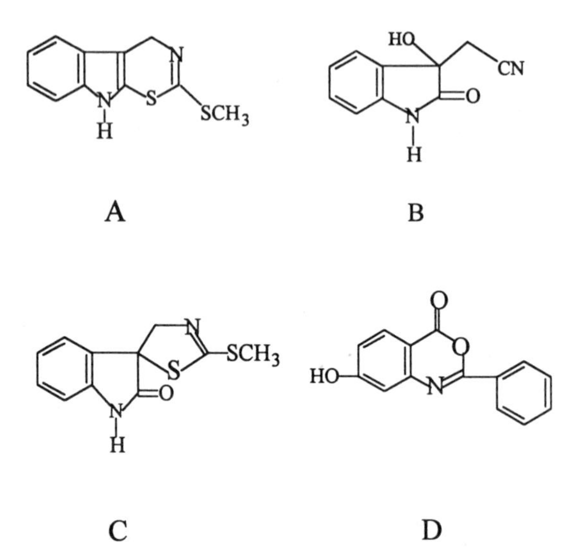

Figure 2 Phytoalexins containing atoms in addition to carbon, hydrogen, and oxygen. *A.* Cyclobrassinin-cabbage, chinese cabbage, rape, turnips. *B.* 3-Cyanomethyl-3-hydroxy-oxindole-cabbage. *C.* Spirobrassinin-turnip (*J. Plant Dis. Prot.* 1993, 4:433–42). *D.* Dianthalexin-carnation (*J. Phytopathol.* 1989, 126:281–92).

responsiveness of plant organs as related to phytoalexin formation must be determined in order to assess their contribution as defense compounds to what often are organ-specific diseases. Some pathogens of roots will not cause disease if inoculated on leaves. The isolation of a phytoalexin from an infected pod is no assurance that it can accumulate in all organs. In general, phytoalexins are lipophilic, localized at and immediately around sites of infection, and because of their great structural diversity, a relationship is not evident between their antimicrobial activity and structure. The diversity of phytoalexins is illustrated in Figures (1–3).

The phenylpropanoid phytoalexins are distributed among families as diverse as the *Leguminosae, Solanaceae, Convolvulaceae, Umbelliferae*, and *Gramineae*. With other groups of phytoalexins, however, some structural specificity is apparent in plants within a family. Isoflavonoid phytoalexins are common in the *Leguminosae*, but have not been reported in the *Solanaceae*, whereas sesquiterpenoid phytoalexins are common in the *Solanaceae* but have not been reported in the *Leguminosae*. Within a family, a level of specificity also is evident, e.g. the norsesquiterpenoid phytoalexin, rishitin, is found in potato and tobacco but not pepper, whereas the sesquiterpenoid phytoalexin, capsidiol, is found in tobacco and pepper but not potato.

Biosynthesis

The major biosynthetic pathways (shikimate, acetate-malonate, acetate-mevalonate) that provide phytoalexin precursors are common to all plants. They are responsible for the synthesis of housekeeping compounds vital for all plants. Phytoalexins arise from a diversion, in part, of housekeeping precursors. This diversion is also often associated with an enhancement of the activity of enzymes in the biosynthetic pathways for housekeeping compounds and the appearance of enzymes for steps closely associated with the biosynthesis of the phytoalexin. These latter enzymes may be at an early regulatory site and be key enzymes in a committed sequence for the biosynthesis of the phytoalexin, but they may also include enzymes late in the biosynthetic pathway. An example of the complexity of regulation can be illustrated by the biosynthesis of relatively simple sesquiterpenoid phytoalexins in potato tuber slices. Hydroxymethylglutaryl coenzyme A reductase (HMGR) is an early regulatory enzyme in the acetate-mevalonate pathway. HMGR activity increases in sliced and infected tubers, but the isozymes that arise after slicing are different from those formed after infection (14). After infection or treatment with arachidonic acid, but not after slicing alone, sesquiterpene cyclase activity increases and squalene synthetase activity is suppressed (92). Both enzymes are at a branch in the acetate-mevalonate pathway. The cyclase catalyzes the conversion of farnesyl pyrophosphate to cyclic precursors for the sesquiterpenoid phytoalexins, whereas squalene synthetase is on the path to the synthesis of sterols and sterol glycoalkaloids. Compartmentalization of biosynthesis and differential regulation of isozymes could permit increased phytoalexin synthesis without disrupting the synthesis of needed housekeeping compounds.

Phytoalexin precursors can be derived from one of the three biosynthetic pathways or a combination of two or three of the pathways (Figure 3). This adds an additional complication to regulation. For phaseollin, for example, precursors from the three pathways are required for its biosynthesis (Figure 3). Thus, not only must the three pathways be regulated to provide sufficient precursors, but enzymes for the incorporation of the precursors into suitable intermediates must also be activated/synthesized and regulated. All of this must be accomplished without disturbing for an extended period the flow of housekeeping compounds.

Experiments with inhibitors of RNA and protein synthesis and analyses for appropriate mRNA demonstrated that both the hypersensitive response and phytoalexin accumulation include the de novo synthesis of proteins (11, 25, 26, 30). The use of cell suspension cultures and purified elicitor molecules greatly facilitated research on the enzymology and molecular regulation of phytoalexin accumulation. For phenylpropanoid-derived phytoalexins, 3-de-

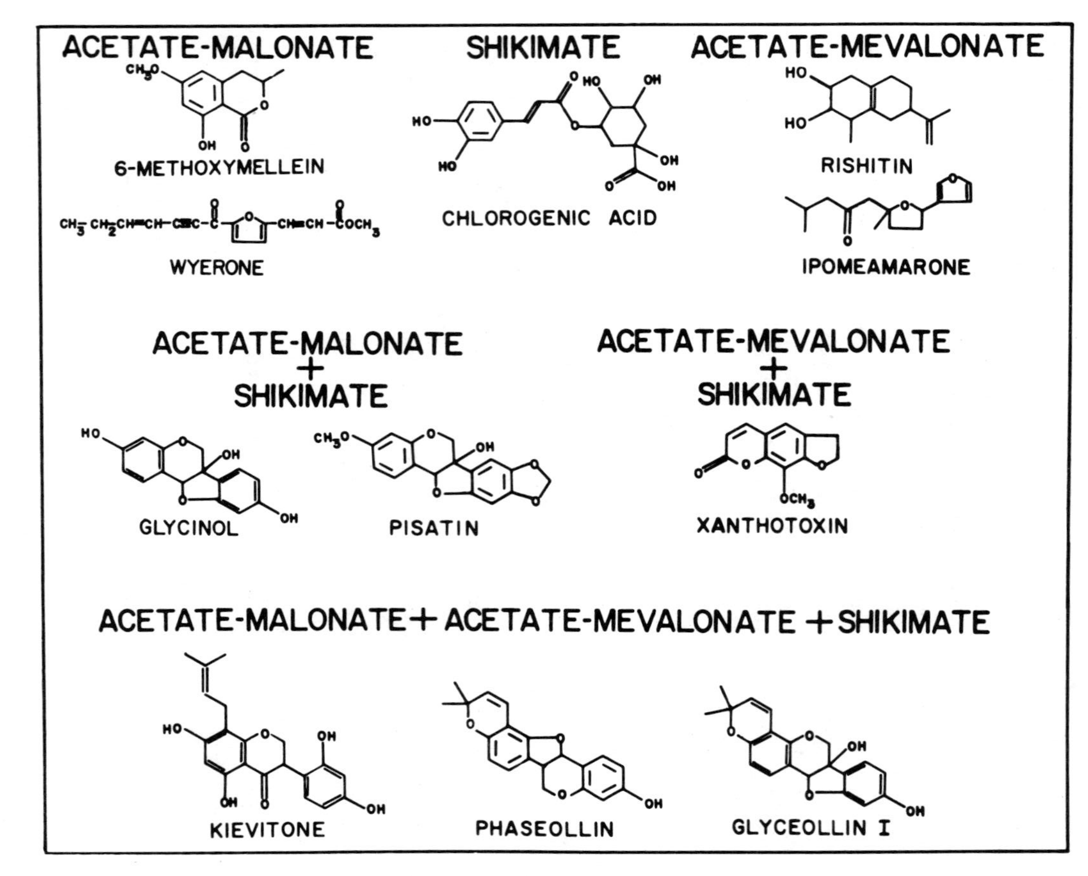

Figure 3 Pathways for the biosynthesis of some phytoalexins.

oxy-arabinoheptulosonate-7 P synthase, phenylalanine ammonia-lyase, cinnamate 4-hydroxylase, 4-coumarate coenzyme A ligase, chalcone synthase, and chalcone isomerase are regulated enzymes. Regulation is principally at the transcriptional level, although some posttranslational control is also reported. The above key enzymes and other later enzymes of phytoalexin biosynthesis appear to be coordinately regulated. In general, phytoalexin accumulation after infection or treatment with elicitors is consistent with mRNA production and enzyme activity. For mevalonate-derived sesquiterpenoid phytoalexins, HMGR (CoA) reductase, squalene synthase, and sesquiterpene cyclase appear to be key regulated enzymes (12, 14, 78, 84, 92).

The critical experiments of Graham and associates, however, provided a new perspective in assessing the need of phytoalexin synthesis de novo to account for the rapid accumulation of phytoalexins after infection in resistance (33, 34). Conjugates of the closely related isoflavones daidzein and genistein are present constitutively in large quantities in soybean. Daidzein is an immediate precursor of the glyceollins, and conjugates of daidzein are rapidly hydrolyzed to free daidzein during the incompatible reactions of soybean with *Phytophthora megasperma* f. sp. *glycinea.* High levels of glyceollins subsequently accumulate and fungal development is arrested within 48 h after infection. In compatible reactions, release of the precursors is delayed and low levels of the glyceollins accumulate only after the infection front has passed. The results suggest that glyceollin biosynthesis early after infection is not dependent on the induction of early enzymes of phenylpropanoid and flavonoid biosynthesis, and resistance is at least partly dependent on the rapid release of the aglyca from their conjugates. Similar evidence has been provided for the accumulation of medicarpin and maackiain in chickpea.

Chickpea cell suspension cultures treated with a yeast glucan elicitor accumulate the pterocarpan phytoalexins medicarpin and maackiain and their corresponding 3-0-glucoside-6′-0-malonates (53). Both pterocarpan conjugates are also constitutive metabolites of the cell cultures and are localized in the vacuoles (85). The ratio of accumulated aglyca to conjugates was reported to depend on the concentration of elicitor used. Low or moderate amounts of elicitor favor conjugate formation, whereas high doses predominantly lead to aglycone formation. Clearly, the two pterocarpan phytoalexins are not totally synthesized de novo because elicitation leads to consumption of preformed conjugates for phytoalexin aglyca formation. The release of vacuolar conjugates introduces another interesting aspect of specificity and regulation for phytoalexin accumulation. Numerous isoflavone and isoflavanone conjugates are kept in the vacuoles at a time when the glucoside malonate conjugates of medicarpin and maackiain are released (5).

One interpretation of the data is that the critical factor for resistance is the incompatibility of plant and pathogen caused by a still elusive incompatibility factor(s). Incompatibility results in rapid release of phytoalexins from conjugates, and the subsequent de novo synthesis of phytoalexins is to replace the loss of the conjugates. A key site for regulation may be at the level of the conjugates. Whether this phenomenon is general for phytoalexin accumulation remains to be resolved.

PHYTOALEXINS IN DISEASE RESISTANCE

Fungitoxicity and Association with Pathogen Development

Since phytoalexins accumulate at infection sites and since they inhibit the growth of fungi and bacteria in vitro, it is logical to consider them as possible plant-defense compounds against diseases caused by fungi and bacteria.

Depending upon the phytoalexin, fungus and bioassay, the EC_{50} for fungi is generally 10^{-3} to 10^{-5} M (3, 49). Thus, they are comparatively weak as antifungal agents. Evidence is not available that phytoalexins are translocated in plants. However, the speed of their accumulation and localization at the infection site may permit the pathogen to encounter concentrations far in excess of the EC_{50} at early stages in the infection process. Evidence for this presumption is available. In the numerous reports by Cruickshank concerning the accumulation of pisatin in pea and phaseollin in green bean, it was apparent that the phytoalexins accumulated to fungitoxic concentrations not only in inoculum droplets placed on opened pea or bean pods but also in the tissues immediately below the inoculum droplets (18, 21, 23). When conceptualizing the role of phytoalexins in disease resistance, Cruickshank and coworkers emphasized that numerous factors could influence the effectiveness of a phytoalexin as a defense compound: the antimicrobial activity as measured in vitro; rate of synthesis; localization in reference to the developing pathogen; the presence of compounds in vivo that influence antimicrobial activity; and the vulnerability of the phytoalexin to detoxification by microorganisms or the plant. Cruickshank's data in toto supported a role for phytoalexins in plant disease resistance, but there were and still are exceptions.

To provide further information as to whether phytoalexins were important to disease resistance, Yoshikawa and coworkers studied in great detail the soybean's response to *P. megasperma* f. sp. *glycinea* (88, 90). They approached the question by evaluating the time when quantitative differences in fungal growth occur in resistant and susceptible soybean cultivars and determining the glyceollin concentration in localized infection sites at various times after inoculation. In the resistant cultivar, glyceollin levels exceeding the EC_{90} and EC_{50} values were detected at and immediately adjacent to the site of infection, respectively, as early as 8 h after inoculation and increased at 9–10 h after inoculation. In contrast, little glyceollin accumulated in the susceptible cultivar during these time periods. The inhibition of fungal development in the resistant cultivar occurred at a time when levels of glyceollin were sufficient to account for the inhibition based on EC values. By 24 h after infection, however, glyceollin levels were much higher than the EC_{90} value in the susceptible cultivar, especially in tissue layers close to the immediate site of inoculation. This suggests that delayed phytoalexin accumulation is not effective in inhibiting fungal growth and that glyceollin present only in the immediate vicinity of advancing hyphae is important for restricting development of the fungus. This is supported further by the observation that glyceollin levels in newly invaded tissue of the susceptible cultivar did not exceed the EC_{50} value at any time after inoculation. The continued growth of *P. megasperma* from tissues in a susceptible soybean cultivar where levels of glyceollin exceeded the EC_{90} also raised several important questions. Were the old hyphae and young hyphae equally sensitive to glyceollin? Was glyceollin translocated from old to young hyphae? Was the inhibitory effect of glyceollin based primarily on a surface phenomenon? Were other compounds present before or produced during the interaction that reduced the toxicity of glyceollin, and were these compounds in the environment around the old but not young hyphae? Should in vitro assays of glyceollin activity utilize mycelia, sporangia, or zoospores as inoculum?

Factors Influencing Activity

In some respects, the experiments of VanEtten and coworkers (70, 76) indirectly addressed the above questions. In his interpretation of the phytoalexin theory, Cruickshank (17) proposed that in incompatible interactions, phytoalexin accumulation halts pathogen growth, thereby conferring resistance to the plant. In compatible interactions, the pathogen apparently either tolerates the accumulated phytoalexins, detoxifies them, suppresses phytoalexin accumulation, and/or avoids eliciting phytoalexin production. The results of some studies on phytoalexin involvement in host-pathogen interactions have been consistent with these proposals. However, the *Aphanomyces euteiches* Drechs.–*Pisum sativum* L. interaction is an exception. The phytoalexin, pisatin, reaches high levels in infected pea tissue during the susceptible interaction with *A. euteiches*, yet this fungus exhibits exceptional sensitivity to pisatin in in vitro bioassays (70). The fungus evidently encounters high pisatin concentrations in infected tissue, casting doubt on the argument that the anomaly could be explained by a physical separation of the fungus and the phytoalexin. The fungus does not metabolize pisatin, nor does it become tolerant to pisatin by prior exposure to low concentrations of the phytoalexin. In further studies, Sweigard & VanEtten (76) reported that incorporation of a polar lipid extract from pea into a growth medium decreased the pisatin sensitivity of *A. euteiches* in both semisolid and liquid culture bioassays. This extract also decreased sensitivity to the phytoalexins maackiain and phaseollin. Lipids extracted from squash and bean were also effective. Pea lipids also decreased the pisatin sensitivity of *Aphanomyces euteiches* f. sp. *phaseoli*, a fungus not pathogenic on pea, but did not decrease the pisatin sensitivity of *Fusarium solani* f. sp. *cucurbitae* or *Neurospora crassa*, also nonpathogens of pea. Commercially obtained phosphatidylcholine decreased the pisatin sensitivity of the four fungi tested. Experiments on the mechanism of the increased pisatin tolerance in *A. euteiches*, the pea pathogen, suggested that phosphatidylcholine decreased the uptake of pisatin from the medium. The pea lipid extract, however, did not have this effect. Neither phosphatidylcholine nor the pea extract stimulated pisatin demethylation. Thus, the effectiveness of phytoalexins as defense compounds may be determined by the presence or absence of compounds that influence their antimicrobial activity in different plants. However, whether the compounds exert this effect in vivo and how general the effect is in influencing disease resistance is not evident. What is evident is that with the information available, it is not possible to generalize about a role for such compounds or phytoalexins in disease resistance without qualifications.

Studies by Cruickshank and colleagues were principally concerned with nonhost resistance, whereas the studies of Yoshikawa and coworkers were with race-specific resistance. With respect to the phytoalexins accumulated and the rate of their accumulation after inoculation, non host and race-specific resistance cannot be differentiated. Clearly, phytoalexin levels per se are not reliable biochemical markers for disease resistance without qualifications. These qualifications would be expected to vary with the pathogen, plant, host organ or tissue inoculated, age of the plant, and environmental factors. In biology, qualifications are likely to be the rule rather than the exception.

The danger of determining whether a compound is or is not a defense compound based on association is illustrated by an analogy with the human immune system. A high white–blood cell count is indicative of a severe infection and it is routinely included in diagnostic procedures where infection is suspected. Two completely false conclusions could be reached concerning the association between the white blood cells and infection: One is that they have no role in defense because their levels are highest with infection; the second, even more absurd, is that they are the cause of the infection. The association is absolute but the conclusions based on the association are false, even granted that certain symptoms associated with the infection may be due to white blood cells or defense compounds.

Detoxification

It is well established that pathogens and plants can degrade, metabolize, and sequester phytoalexins (3, 77). Phytoalexin accumulation, whether biotically or abiotically induced, is clearly not only dependent on the rate of synthesis:

Conjugation, compartmentalization, release from conjugates, sequestering in walls or vacuoles, and degradation all influence phytoalexin level. The biochemical and subsequent genetic studies of VanEtten and his group supported the importance of pisatin detoxification by pisatin demethylase in virulence of *Nectria haematococca* (54, 77). Correlations of virulence with the production of kievitone hydratase were also reported (52, 75). It appears likely to this author that phytoalexin detoxification is related to virulence in some fungi under some conditions. Careful biochemical/genetic analyses need to be made for other fungi before the phenomenon can be accepted as a general mechanism for enhanced virulence. Phytoalexin degradation, however, does not explain race specificity in plant-pathogen interactions. A report of the transfer of pisatin demethylase to a nonpathogen of peas and the subsequent limited virulence of the nonpathogen on peas is cited as evidence of a role for phytoalexins in disease resistance (71). However, the presence of genes for detoxification of the phytoalexins produced by two hosts of *Nectria haematococca*, pea and chickpea, on dispensable chromosome segments of the fungus and the retention of virulence by the fungus in the absence of the genes, indicates that the role of detoxification of phytoalexins in virulence needs reevaluation (81). The report by VanEtten and colleagues (81) can be interpreted to mean that phytoalexins do not have a role in resistance or at least they are not the only determinants of resistance in pea and chickpea. The matter of qualification again enters into the consideration of whether phytoalexins have a role in disease resistance. The apparent exceptions to the rule may not be exceptions at all but merely an indication that we do not understand the qualifications and have not been creative enough to design the critical experiments to resolve the issue.

Multiple Phytoalexins and Stereoisomers

It is unclear whether the accumulation of different but structurally related phytoalexins in a plant is important for resistance, e.g. phaseollin, kievitone, phaseollidin in green beans; rishitin, lubimin, phytuberin in potato tuber. Is this multiplicity of phytoalexin response an insurance for survival? Different sensitivities of fungi to related phytoalexins have not been sufficiently studied and relatively little has been reported concerning the mode of action of phytoalexins.

In addition to differences in the antifungal activity of different phytoalexins for a fungus, it has been reported that stereoisomers of a phytoalexin can markedly differ in antifungal activity. The pterocarpan phytoalexins are found as (+) or (–) stereoisomers. Most legumes analyzed to date accumulate the (–) forms, with the exception of pea, which produces (+) pisatin, and *Sophora japonica*, which produces both (+) and (–) maackiain. (83). The antifungal activity of maakiain and pisatin to some fungi is stereospecific. Several pathogens isolated from alfalfa and red clover, which produce (–) pterocarpans, were more inhibited by (+) maackiain than by (–) maackiain (24). The differences in sensitivity were apparently due to the inability of the pathogens to degrade the (+) isomer. These data led to the suggestion of a novel strategy to enhance a legume's disease resistance by a one- or two-gene modification that would yield the hosts natural phytoalexin with a different steric configuration (82). This relatively simple genetic modification could clearly have profound effects on disease resistance if phytoalexin degradation by the pathogen is important in determining resistance. This has not yet been established. A further complication to this reasoning is that the conjugates of some of the pterocarpan phytoalexins (see previous section of this chapter), which may be very important for the rapid release of free phytoalexin after infection, may not be synthesized or hydrolyzed as readily in the presence of the stereoisomer unnatural to the host. Nevertheless, this is an exciting area of investigation that requires a thorough understanding of phytoalexin biosynthesis, factors influencing accumulation, and mode of action. Excellent progress has been made and continues to be made in elucidating steps in the biosynthesis of pterocarpan phytoalexins with regard to specific genes and enzymes determining stereo-configuration (4, 28, 65).

Missing Phytoalexins

Clearly, the extent to which phytoalexins may determine resistance is complicated by many factors. Though sesquiterpenoid phytoalexins accumulate in potato tubers, few or none of the sesquiterpenoid phytoalexins accumulate in potato foliage. Recent research confirms the lack of sesquiterpenoid phytoalexins in foliage inoculated with incompatible or compatible races of *P. infestans*, and evidence is presented that unsaturated fatty acids and/or their oxidation products function as phytoalexins in the foliage (31). Lipid peroxidative products, including volatile aldehydes, would be lost during the commonly followed extraction procedures for phenolic and sesquiterpenoid phytoalexins. More research is necessary to find and fit the missing pieces into the puzzle of phytoalexin distribution within a plant and in different plants. There are plants in which the presence of classical phytoalexins has not been established though their possible presence has been actively investigated, e.g. cucumber, watermelon, muskmelon (46, 74).

Suppression of Accumulation

A basic question facing plant and animal scientists is, why do the highly effective mechanisms for disease resistance occasionally fail? Why is there disease? Unfortunately, not enough research has been conducted to answer this question in the plant and, probably, in the animal field. Progress has been made in identifying factors contributing to the elicitation of resistance and the relationship of genes for avirulence to genes for resistance in plants (the gene-for-gene concept). Some argue that suppression of resistance mechanisms and the specificity of suppressors does not fit the gene-for-gene concept. Nevertheless, suppression of defense responses, including phytoalexin accumulation, is well documented, and a considerable body of evidence is available that it participates in determining susceptibility/resistance and race-specificity (29, 43, 64, 69, 86, 91). It has been reported for plants as different as potato and chickpea and fungi as different as *P. infestans* and *Ascochyta rabiei*. In all systems studied to date, the suppressors are either low-molecular-weight polysaccharides or glycopeptides. In chickpea, the suppressor reduced the accumulation not only of medicarpin and maackiain but also of the isoflavones and isoflavone glucosylconjugates (43). In potato, low-molecular-weight branched glucans may determine race-specificity in the interaction of tubers with *P. infestans* in part by suppressing the accumulation of sesquiterpenoid phytoalexins (29, 69). Other branched glucans, however, enhance arachidonic acid and other C_{20} unsaturated fatty acid–elicited accumulation of sesquiterpenoid phytoalexins (69). Careful genetic analyses relating specific genes for pathogenicity to specific suppressor molecules, to modes of action in a host, and ultimately to susceptibility are needed. One message seems to be apparent in considering resistance/susceptibility. Though there are generalizations that are valid with qualifications and exceptions, each plant-pathogen interaction, each host, pathogen, and disease is different and needs a fine tuning of research effort.

DIRECTIONS FOR FUTURE RESEARCH

Phytoalexins as Pesticides

A logical consequence of the information presented concerning phytoalexins would be their utilization to protect plants against disease. One scenario would be to spray plants and treat seeds or soak soil with naturally occurring compounds such as phytoalexins. The assumptions underlying such actions are that the compounds are safe because they are naturally occurring, and that they would be effective because they are the compounds in nature for plant protection, and existing plants have survived the selection pressure of evolution. There are flaws in this reasoning. Naturally occurring compounds are not necessarily safe. Many of the world's most potent poisons are derived from plants. Some plant tissues are extremely toxic to animals because they contain protective compounds, e.g. potato, tomato, and tobacco foliage, potato peel, and potato sprouts. The phytoalexins are not notably antimicrobial, though they do accumulate at sites of infection to levels sufficient to inhibit the development of some fungi and bacteria. Since phytoalexins are not translocated, plants would have to be frequently sprayed not only to account for losses due to rainfall, but also to protect emerging foliage or developing fruit. The degradation of phytoalexins by microorganisms and plants would further necessitate frequent application. Phytoalexins on seed surfaces would likely be unstable once seeds were planted in soil, and soil drenches would also prove unsatisfactory for the same reason. The use of naturally occurring defense compounds to protect plants would also likely be uneconomical. The structures of most of these compounds are complex, synthesis would be difficult, and the cost of isolation from natural sources would probably be high.

Despite these possible flaws in the scenario, I concede that a natural defense compound may be found that is highly active, can be easily and economically synthesized, and is safe to use. Even if such a compound could not be used directly, it might serve as a model for the synthesis of new pesticides. If we use these compounds as pesticides, however, we reintroduce all the problems associated with the current use of pesticides. I suggest it is well to study plant-defense compounds to understand the natural mechanisms for disease resistance rather than for their use directly in disease control. There may be more promising approaches to the problem. Why not regulate the expression of plant genes and enable plants to protect themselves?

Gene Expression for Phytoalexin Accumulation and Disease Control

An alternative to the direct use of defense compounds to protect plants is the elicitation within the plant of such compounds by regulating gene expression and, hence, gene products leading to their synthesis and accumulation. Since some cell wall components from fungi, bacteria, and plants elicit phytoalexin accumulation, foliar sprays of elicitor molecules were tested as possible disease-control agents. The use of such elicitors, however, proved ineffective for many of the reasons that sprays containing phytoalexins are ineffective. In addition, the repeated application of elicitors of phytoalexin accumulation resulted in resistant, but stunted, plants, perhaps because the elicitors caused a marked diversion of energy and carbon precursors from vital processes. Plants that constitutively accumulate phytoalexins also were stunted and nonproductive. In nature, phytoalexins are produced at or around sites of infection when and where needed.

To circumvent the disadvantages of using elicitors of phytoalexins directly on the plant, a strategy was put forth to engineer enhanced and rapid elicitor release only after infection and at the site of infection. Soybean β-1,3-endoglucanase cDNA was expressed in tobacco plants and the transformed plants contained β-1,3-glucanase activity at levels fourfold higher than those found in untransformed plants (89). The transformed plants with high glucanase activities showed a degree of disease resistance to a diverse group of fungal pathogens and there was a positive correlation between glucanase activity and disease resistance. The results support a role for endoglucanase in the expression of disease resistance in plants. Purified soybean endoglucanase, however, did not show any fungitoxicity to the fungi used in the study when assayed in vitro under conditions analogous to those used by others. The data suggest the possibility that disease-resistant transgenic plants can be developed by expressing genes encoding nonfungitoxic proteins. These proteins may modify subtle resistance-expression mechanisms and enhance the release of active elicitor molecules. Other fungal cell wall degradation products, chitin and chitosan oligomers, also have antifungal activity, and, like glucan oligomers released by β-1,3-glucanase, some serve as elicitors of phytoalexin accumulation (42). A plausible scenario, suggested by some investigators, would be that chitinases and deacetylases (produced by the plant as a result of fungal infection) hydrolyze fungal cell walls, release chitin and chitosan oligomers; the oligomers participate in slowing development of the fungi because of their antifungal activity as well as their ability to elicit phytoalexin accumulation, which further restricts fungal development. This one-two punch, along with glucan release from some fungi, may be part of a much bigger picture involving the synchronization and activation of multiple mechanisms for disease resistance including hydroxyproline-rich glycoproteins, lignin, callose, pathogenesis-related hydrolases and antifungal proteins, lipoxygenases, cell wall cross-linking phenolics, and active oxygen species. I believe it is time for those studying phytoalexins to take greater account of other putative defense compounds, and for those currently studying pathogenesis-related proteins to resurrect phytoalexins and other defense mechanisms. Are we at the "blind men describing the elephant" stage of this research where, depending upon one's position, one's conclusions can be markedly different, in themselves correct, but in toto wrong.

Suggestions have been made to enhance the resistance in plants by incorporating genes for phytoalexin synthesis from unrelated plants. The reasoning is that phytoalexins from a carrot, which is not susceptible to many diseases of beans, would protect beans against several of its pathogens. The suggestions are unlikely to prove useful for practical disease control since phytoalexins have a broad spectrum of biological activity. Those from carrot often inhibit carrot pathogens as well as pathogens of bean, and vice versa. The pathways for phytoalexin synthesis are also quite complex and differ in numerous enzymes. Thus, it is unlikely that the transfer of a single gene would often be sufficient for the synthesis of most foreign phytoalexins. However, the transfer of stilbene synthase from grapevine to tobacco with the resultant synthesis and accumulation of resveratrol in tobacco may be an exception (36). The transgenic tobacco synthesizing resveratrol had enhanced resistance to disease caused by *Botrytis cinerea.* This result is difficult to interpret, however, since the increased resistance may have been due less to the ability to synthesize resveratrol and more to the ability to accumulate the phytoalexin rapidly after infection. If the phytoalexins found in tobacco, e.g. capsidiol and scopoletin, were also produced quickly enough after infection, resistance to disease might also be increased. The report by Hahn et al (36) does not provide data for phytoalexin accumulation other than resveratrol.

The use of phytoalexin stereoisomers is another approach for control. Such stereoisomers may be resistant to microbial and plant enzymes that degrade and detoxify the normal phytoalexins. They could be applied directly as sprays; however, this methodology has all the inherent problems of current pesticides. A second and potentially more effective method would be to genetically engineer plants by introducing a key enzyme in a biosynthetic pathway to produce the desired stereoisomer or its precursor (see previous section; 82). Thus the natural defense mechanism would be utilized to produce pathogen-resistant phytoalexins when and where needed.

Since disease resistance in plants is likely not determined by the presence or absence of genes for gene products that restrict pathogen development and since resistance is in large part determined by the rapidity and magnitude of gene expression and the activity of such gene products, it is probable that all plants have the genetic potential for resistance. This potential can be expressed systemically by restricted inoculation with pathogens, attenuated pathogens, and selected nonpathogens, and by treatment with chemicals that are signals produced by inoculated or treated plants, signals unrelated to natural signals or chemicals that release signals. The SIR is effective against diseases caused by some fungi, bacteria, and viruses; and it has been successfully tested in the laboratory and field (46, 47, 50, 51, 55). Immunization systemically enhances the levels of some putative defense compounds, e.g. chitinases, β-1,3-glucanases and peroxidases (6, 8, 55, 79, 87). Another major effect, however, is to sensitize the plant to respond rapidly after infection by accumulating phytoalexins, hydroxyproline-rich glycoproteins, and lignin at the site of infection, and by further enhancing systemic levels of chitinases, β-1,3-glucanases, other PR-proteins, and peroxidases (37, 38, 47, 50, 51, 55, 79, 87). Thus, SIR activates multiple mechanisms for resistance rather than any one compound or mechanism, as is the case with suggested genetic engineering strategies. In addition, it does not require the laborious procedures to produce high-yielding, uniform genetically engineered plants and eliminates the objections raised by those who are concerned about the introduction of "foreign genes" into plants.

FINAL REMARKS

I do believe phytoalexins are one of the mechanisms for disease resistance in plants. However, even with fungi, where the case is strongest for their contribution, there are qualifications. An understanding of the contribution of phytoalexins to resistance would facilitate their exploitation for disease control. Such exploitation, however, is likely to depend upon an understanding and integration of the multiple mechanisms for disease resistance and factors such as signals, signal transduction, elicitors, suppressors, and detoxification. More effort and money should be directed to integrate information and synthesize models to help explain the complexity of interaction. A major complex problem facing researchers in this field is that, although often a single gene in a host determines resistance, many mechanisms for resistance and many genes coding for such mechanisms have been reported and, although a single gene may determine pathogenicity, many mechanisms to explain pathogenicity have been described.

For every complex question, there is a simple answer—and it's wrong.

Attributed to Mark Twain

Literature Cited

1. Akazawa T, Wada K. 1961. Analytical study of ipomeamarone and chlorogenic acid alterations in sweet potato roots infected by *Ceratocystic fimbriata*. *Plant Physiol.* 36:139–44

1a. Asada Y, Bushnell WR, Ouchi S, Vance CP, eds. 1982. *Plant Infection—The Physiological and Biochemical Basis.* Berlin/New York: Jpn. Soc. Press

2. Bailey J, Mansfield J, eds. 1982. *Phytoalexins.* Glasgow: Blackie. 334 pp.

3. Barz W, Bless W, Borger-Papendorf G, Gunia W, Mackenbrock U, et al. 1990. Phytoalexins as part of induced defense reactions in plants: their elicitation, function and metabolism. In *Bioactive Compounds from Plants*, pp. 140–56. Ciba Found. Symp. 154. New York: Wiley

4. Barz W, Mackenbrock U. 1995. Constitutive and elicitation induced metabolism of isoflavones and pterocarpans in chickpea (*Cicer arietinum*) cell suspension cultures. *Plant Cell Tissue Organ Cult.* In press

5. Barz W, Welle R. 1992. Biosynthesis and metabolism of isoflavones and pterocarpan phytoalexins in chickpea, soybean and phytopathogenic fungi. *Recent Adv. Phytochem.* 26:139–64

6. Boller T. 1987. Hydrolytic enzymes in plant disease resistance. In *Plant Microbe Interactions, Molecular and Genetic Perspectives*, ed. T Kosuge, E Nester, 2:385–413. New York: Macmillan

7. Bostock R, Kuć J, Laine R. 1981. Eicosapentaenoic and arachidonic acids from *Phytophtora infestans* elicit fungitoxic sesquiterpenes in potato. *Science* 212:67–69

8. Carr J, Klessig D. 1989. The pathogenesis-related proteins of plants. In *Genetic Engineering—Principles and Methods*, ed. J Setlow, 11:65–109. New York: Plenum

9. Chalutz E, DeVay J, Maxie E. 1969. Ethylene induced isocoumarin formation in carrot root tissues. *Plant Physiol.* 44:235–41

10. Chalutz E, Stahmann M. 1969. Induction of pisatin by ethylene. *Phytopathology* 59:1972–73

11. Chappell J, Hahlbrock K. 1984. Transcription of plant defense genes in response to UV light or fungal elicitor. *Nature* 311:76–78

12. Chappell J, Von Lankin C, Vogeli U, Bhatt P. 1989. Sterol and sesquiterpenoid biosynthesis during a growth cycle of tobacco cell suspension cultures. *Plant Cell Rep.* 8:48–52

13. Chester K. 1933. The problem of acquired physiological immunity in plants. *Q. Rev. Biol.* 8:129–54, 275–324
14. Choi D, Bostock R, Avdiushko S, Hildebrand D. 1994. Lipid-derived signals that discriminate wound- and pathogen-responsive isoprenoid pathways in plants: methyl jasmonate and the fungal elicitor arachidonic acid induce different 3-hydroxy-3-methylglutaryl-coenzyme A reductase genes and antimicrobial isoprenoids in *Solanum tuberosum* L. *Proc. Natl. Acad. Sci. USA* 91:2329–33
15. Condon P, Kuć J. 1960. Isolation of a fungitoxic compound from carrot root tissue inoculated with *Ceratocystis fimbriata. Phytopathology* 50:267–70
16. Condon P, Kuć J. 1962. Confirmation of the identity of a fungitoxic compound produced by carrot root tissue. *Phytopathology* 52:182–83
17. Cruickshank IAM. 1963. Phytoalexins. *Annu. Rev. Phytopathol.* 1:351–74
18. Cruickshank IAM. 1966. Defense mechanisms in plants. *World Rev. Pest Control* 5:161–75
19. Cruickshank IAM, Perrin DR. 1960. Isolation of a phytoalexin from *Pisum sativum* L. *Nature* 187:799–800
20. Cruickskank IAM, Perrin DR. 1961. Studies on phytoalexins III. The isolation assay and general properties of a phytoalexin from *Pisum sativum* L. *Aust. J. Biol. Sci.* 14:336–48
21. Cruickshank IAM, Perrin DR. 1965. Studies on phytoalexins. VIII. The effect of some further factors on the formation, stability and localization of pisatin in vitro. *Aust. J. Biol. Sci.* 18: 817–28
22. Cruickshank IAM, Perrin DR. 1968. The isolation and partial characterization of monilicolin A, a polypeptide with phaseollin-inducing activity from *Monilinia fructicola. Life Sci.* 7:449–58
23. Cruickshank IAM, Perrin DR. 1971. Studies on phytoalexins. XI. The induction, antimicrobial spectrum and chemical assay of phaseollin. *Phytopathol. Z.* 70:209–22
24. Delserone LM, Matthews DE, Van Etten HD. 1992. Differential toxicity of enantiomers of maackiain and pisatin to phytopathogenic fungi. *Phytochemistry* 31: 3813–19
25. Dixon R. 1986. The phytoalexin response: elicitation, signalling, and control of host gene expression. *Biol. Rev.* 61:239–91
26. Dixon RA, Harrison MJ. 1990. Activation, structure and organization of genes involved in microbial defense in plants. *Adv. Genet.* 28:165–233
27. Dixon RA, Harrison MJ. 1994. Early events in the activation of plant defense responses. *Annu. Rev. Phytopathol.* 32: 479–501
28. Dixon RA, Harrison MJ, Paiva NL. 1995. The isoflavonoid phytoalexin pathway: from enzymes to genes to transcription factors. *Physiol. Plant.* In press
29. Doke N, Chai H, Kawaguichi A. 1987. Biochemical basis of triggering and suppression of hypersensitive cell response. See Ref. 63a, pp. 235–51
30. Ebel J. 1986. Phytoalexin synthesis: the biochemical analysis of the induction process. *Annu. Rev. Phytopathol.* 24: 235–64
31. Ertz SD, Friend J. 1993. Fungitoxic fatty acids in potato leaves inoculated with *Phytophthora infestans. Abstr. Int. Congr. Plant Pathol., 6th,* p. 231
32. Friend J. 1991. The biochemistry and cell biology of interaction. *Adv. Plant Pathol.* 7:85–129
33. Graham TL, Graham MY. 1991. Gyceollin elicitors induce major but distinctly different shifts in isoflavonoid metabolism in proximal and distal soybean cell populations. *Mol. Plant-Microbe Interact.* 4:60–68
34. Graham TL, Kim JE, Graham MY. 1990. Role of constitutive isoflavone conjugates in the accumulation of glyceollin in soybean infected with *Phytophthora megasperma. Mol. Plant-Microbe Interact.* 3:157–66
35. Hadwiger L, Schwochaw M. 1971. Ultraviolet light-induced formation of pisatin and phenylalanine ammonia lyase. *Plant Physiol.* 47:588–90
36. Hahn R, Reif H, Krause E, Langebarteis R, Kindl H, et al. 1993. Disease resistance results from foreign phytoalexin expression in a novel plant. *Nature* 361: 153–56
37. Hammerschmidt R, Kuć J. 1982. Lignification as a mechanism for induced systemic resistance in cucumber. *Physiol. Plant Pathol.* 20:61–71
38. Hammerschmidt R, Nuckles E, Kuć J. 1982. Association of peroxidase activity with induced systemic resistance in cucumber. *Physiol. Plant Pathol.* 20:73–82
39. Hardegger E, Biland H, Carrodi H. 1963. Synthese von 2,4-dimethoxy-6-hydroxyphenanthren und Konstitution des Orchinols. *Helv. Chim. Acta* 46:1354–60
40. Hardegger E, Schellenbaum M, Corrodi H. 1963. Welstoffe und Antibiotica. Uber induqierte Abwehrstoffe bei Orchideen II. *Helv. Chim. Acta* 46:1171–80
41. Hargreaves J, Bailey J. 1978. Phytoalexin production by hypocotyls of *Phaseolus vulgaris* in response to con-

stitutive metabolites released by damaged bean cells. *Physiol. Plant Pathol.* 13:89–100

42. Kendra DF, Hadwiger LA. 1984. Characterization of the smallest chitosan oligomer that is maximally antifungal to *Fusarium solani* and elicits pisatin formation in *Pisum sativum*. *Exp. Mycol.* 8:276–82
43. Kessmann H, Barz W. 1986. Elicitation and suppression of phytoalexin and isoflavone accumulation in cotyledons of *Cicer arietinum* L. as caused by wounding and polymeric components from the fungus *Ascochyta rabiei*. *J. Phytopathol.* 117:321–35
44. Kubota T, Matsuura T. 1953. Chemical studies on the black rot disease of sweet potatoes. *J. Chem. Soc. Jpn.* 74:101–9, 197–99, 248–51, 668–70
45. Kuć J. 1972. Phytoalexins. *Annu. Rev. Phytopathol.* 10:207–32
46. Kuć J. 1982. Induced immunity to plant disease. *Bioscience* 32:854–60
47. Kuć J. 1987. Plant immunization and its applicability for disease control. In *Innovative Approaches to Plant Disease Control*, ed. I Chet, pp. 255–74. New York: Wiley
48. Kuć J. 1991. Phytoalexins: perspectives and prospects. See Ref. 72, pp. 595–603
49. Kuć J. 1992. Antifungal compounds in plants. In *Phytochemical Resources for Medicine and Agriculture*, ed. H Nigg, D Seigler, pp. 159–84. New York: Plenum
50. Kuć J. 1993. Nonpesticide control of plant disease by immunization. In *Proc. Int. Symp. Syst. Fungic. Antifung. Compd., 10th*, ed. H Lyr, C Polter, pp. 225–37. Stuttgart: Ulmer
51. Kuć J, Strobel N, 1992. Induced resistance using pathogens and nonpathogens. In *Biological Control of Plant Diseases*, ed. E Tjamos, G Papavisas, RJ Cook, pp. 295–303. New York: Plenum
52. Kuhn P, Smith D. 1979. Isolation from *Fusarium solani* f. sp. *phaseoli* of an enzyme system responsible for kievitone and phaseollidin detoxification. *Physiol. Plant Pathol.* 14:179–80
53. Mackenbrock U, Gunia W, Barz W. 1993. Accumulation and metabolism of medicarpin and maackiain malonylglucosides in elicited chickpea (*Cicer prietinum* L.) cell suspension cultures. *J. Plant Physiol.* 142:385–91
54. Mackintosh S, Matthews D, VanEtten H. 1989. Two additional genes for pisatin demethylation and their relationship to the pathogenicity of *Nectria haematococca* on pea. *Mol. Plant-Microbe Interact.* 2:354–62
55. Madamanchi NR, Kuć J. 1991. Induced systemic resistance in plants. In *The Fungal Spore and Disease Initiation in Plants and Animals*, ed. G Cole, H Hoch, pp. 347–62. New York: Plenum
56. Mahadevan A. 1991. *Post Infectional Defense Mechanisms*. New Delhi: Today and Tomorrow Printers. 871 pp.
57. Mercier J, Arul J, Chantal J. 1993. Effect of UV-C on phytoalexin accumulation and resistance to *Botrytis cinerea* in stored carrots. *J. Phytopathol.* 139:17–25
58. Meyer G. 1939. Zellphysiologische und anatomische Untersuchungen uber die Reaktion der Kartoflelknolle auf den Angriff der *Phytophthora infestans* bei Sorten verschiedener Resistenz. *Arb. Biol. Vers. Land Forstwirtsch.* 23:97–132
59. Muller KO. 1956. Einige einfache Versuche zum Nachweis von Phytoalexinen. *Phytopathol. Z.* 27:237–54
60. Muller KO. 1958. Studies on phytoalexins. The formation and immunological significance of phytoalexins produced by *Phaseolus vulgaris* in response to infections with *Sclerotinia fructicola* and *Phytophthora infestans*. *Aust. J. Biol. Sci.* 11:275–300
61. Muller KO. 1961. The phytoalexin concept and its methodological significance. *Recent Adv. Bot.* 1:396–400
62. Muller K, Borger H. 1940. Experimentelle Untersuchungen uber die *Phythophthora*-Resistenz der Kartoffel. *Arb. Biol. Reichsanst. Land Forstwirtsch.* 23:189–231
63. Muller KO, Meyer G, Klinkowski M. 1939. Physiologische genetische Untersuchungen uber die Resistenz der Kartoffel gegenuber *Phytophthora infestans*. *Naturwissenschaften* 27:765–68

63a. Nishimura S, Vance C, Doke N, eds. 1987. *Molecular Determinants of Plant Disease*. Berlin/Tokyo: Springer-Verlag

64. Ouchi S, Oku H. 1982. Physiological basis of susceptibility induced by pathogens. See Ref. 1a, pp. 117–36
65. Paiva NL, Sun Y, Dixon RA, VanEtten HD, Hrazdina G. 1994. Molecular cloning of isoflavone reductase from pea (*Pisum sativum* L.): evidence for a 3R-isoflavanone intermediate in (+) pisatin biosynthesis. *Arch. Biochem. Biophys.* 312:501–10
66. Perrin DR. 1964. The structure of phaseollin. *Tetrahedron Lett.* 1:29–35
67. Perrin DR, Bottomly W. 1962. Studies

on phytoalexins. V. The structure of pisatin from *Pisum sativum* L. *J. Am. Chem. Soc.* 84:1919–22

68. Perrin DR, Cruickshank IAM. 1965. Studies on phytoalexins. VII. The chemical stimulation of pisatin formation in *Pisum sativum* L. *Aust. J. Biol. Sci.* 18:803–16
69. Preisig C, Kuć J. 1987. Phytoalexins, elicitors, enhancers, suppressors and other considerations in the regulation of R-gene resistance to *Phytophthora infestans* in potato. See Ref. 63a, pp. 203–27
70. Pueppke S, VanEtten H, 1976. The relation between pisatin and development of *Aphanomyces euteiches* in diseased *Pisum sativum*. *Phytopathology* 66: 1174–85
71. Schafer W, Straney D, Cuiffetti L, VanEtten H, Yoder O. 1989. One enzyme makes a fungal pathogen, but not a saprophyte, virulent on a new host plant. *Science* 246:247–49
72. Sharma R, Salunkhe D, eds. 1991. *Mycotoxins and Phytoalexins*. Boca Raton: CRC. 775 pp.
73. Sharp J, Albersheim P, Ossowski O, Pilotti A, Garegg P, Lindberg B. 1984. Comparison of the structures and elicitor activities of a synthetic and mycelial-wall-derived hexa-(B-D-glucopyranosyl)-D-glucitol. *J. Biol. Chem.* 259: 11341–345
74. Siegrist J, Jeblick W, Kauss H. 1994. Defense responses in infected and elicited cucumber (*Cucumis sativus* L.) hypocotyl segments exhibiting acquired resistance. *Plant Physiol.* 105: 1365–74
75. Smith D, Wheeler H, Banks S, Cleveland T. 1984. Association between lowered kievitone hydratase activity and reduced virulence to bean in variants of *Fusarium solani* f. sp. *phaseoli*. *Physiol. Plant Pathol.* 25:135–47
76. Sweigard J, VanEtten H. 1987. Reduction in pisatin sensitivity of *Aphanomyces euteiches* by polar lipid extracts. *Phytopathology* 77:771–75
77. Tegtmeier, K, VanEtten H. 1982. The role of pisatin tolerance and degradation in virulence of *Nectria haematococca* on peas: a genetic analysis. *Phytopathology* 72:608–12
78. Tjamos E, Kuć J. 1982. Inhibition of steroid glycoalkaloid accumulation by arachidonic and eicosapentaenoic acids in potato. *Science* 217:542–44
79. Tuzun S, Rao N, Vogeli U, Schardl C, Kuć J. 1989. Induced systemic resistance to blue mold: early induction of β-1,3-glucanases, chitinases and other pathogenesis-related (b) proteins in immunized tobacco. *Phytopathology* 79: 979–83
80. Uritani I. 1963. The biochemical basis for disease resistance induced by infection. *Conn. Agric. Exp. Stn. Bull.* 663:4–19
81. VanEtten HD, Funnel-Baerg D, Wasman C, Covert S. 1993. Pathogenicity genes on nonessential elements of the genome in *Nectria haematococca*. *Int. Congr. Plant Pathol., 6th.* Abstr. S11.2: 13
82. VanEtten HD, Matthews DE, Matthews PS. 1989. Phytoalexin detoxification: importance for pathogenicity and practical implications. *Annu. Rev. Phytopathol.* 27:143–64
83. VanEtten HD, Matthews PS, Mercer EH. 1983. (+)Maackiain and (+) medicarpin as phytoalexins in *Sophora japonica* and identification of the (–) isomers by biotransformation. *Phytochemistry* 22:2291–95
84. Vogeli U, Chappell J. 1988. Induction of sesquiterpene cyclase and suppression of squalene synthetase activities in plant cell cultures treated with fungal elicitor. *Plant Physiol.* 88:1291–96
85. Weidemann C, Tenhaken R, Hohe U, Barz W. 1991. Medicarpin and maackiain-3-0-glucoside-6′-0 malonate conjugates are constitutive compounds of chickpea (*Cicer arietinum* L.) cell cultures. *Plant Cell Rep.* 10:371–74
86. Yamada T, Hashimota H, Shiraishi T, Oker H. 1989. Suppression of pisatin, phenylalanine ammonia-lyase mRNA, and chalcone synthase mRNA accumulation by a putative pathogenicity factor from the fungus *Mycosphaerella pinodes*. *Mol. Plant-Microbe Interact.* 2: 256–61
87. Ye X, Pan S, Kuć J. 1989. Pathogenesis-related proteins and systemic resistance to blue mold and tobacco mosaic virus induced by tobacco mosaic virus, *Peronospora tabacina* and aspirin. *Physiol. Mol. Plant Pathol.* 35: 161–75
88. Yoshikawa M, Masago H. 1982. Biochemical mechanism of glyceollin accumulation in soybean. See Ref. 1a, pp. 265–80
89. Yoshikawa M, Tsuda M, Takeuchi Y. 1993. Resistance to fungal diseases in transgenic tobacco plants expressing the phytoalexin elicitor factor, β-1,3-endoglucanase from soybean. *Naturwissenschaften* 80:417–20
90. Yoshikawa M, Yamauchi K, Masago H. 1978. Glyceollin: its role in restricting fungal growth in resistant soybean hy-

pocotyls infected with *Phytophthora megasperma* var. *sojae*. *Physiol. Plant Pathol.* 12:73–82

91. Ziegler E, Pontzen R. 1982. Specific inhibition of glucan-elicited glyceollin accumulation in soybeans by an extracellular mannan-protein of *Phytophthora megasperma* f.sp. *glycinea*. *Physiol. Plant Pathol.* 20:321–31
92. Zook M, Kuć J. 1991. Induction of sesquiterpene cyclase and suppression of squalene synthetase activity in elicitor-treated or fungal-infected potato tuber tissue. *Physiol. Mol. Plant Pathol.* 39:377–90

Annu. Rev. Phytopathol. 1995. 33:299–321

ACTIVE OXYGEN IN PLANT PATHOGENESIS[1]

C. Jacyn Baker[*] and Elizabeth W. Orlandi[**]

[*]US Department of Agriculture, Agricultural Research Service, Molecular Plant Pathology Laboratory, Beltsville, Maryland 20705; [**]University of Maryland, Department of Botany, College Park, Maryland 20742-5815

KEY WORDS: catalase, hypersensitive response, peroxidase, NAD(P)H oxidase, lipoxygenase

ABSTRACT

Plant cells produce active oxygen during interactions with potential pathogens. Active oxygen species, including superoxide, hydrogen peroxide, and the hydroxyl radical, could potentially affect many cellular processes involved in plant/pathogen interactions. Active oxygen can be difficult to monitor in plant cells because many of the species are short-lived and are subject to cellular antioxidant mechanisms such as superoxide dismutases, peroxidases, the ascorbate/glutathione cycle, and catalase. Modifications of the luminol-dependent chemiluminescence assay have facilitated studies on both the production and scavenging of active oxygen that occurs during incompatible plant/bacteria interactions. Many potential sources for active oxygen production have been identified such as NADPH oxidases and peroxidases, but it is still unclear which mechanisms predominate during plant/pathogen interactions. The active oxygen produced in response to pathogens and elicitors has been hypothesized to have direct antimicrobial effects and to play a role in other defense mechanisms including lignin production, lipid peroxidation, phytoalexin production, and the hypersensitive response.

INTRODUCTION

In recent years an exciting area of disease physiology has emerged reporting the involvement of active oxygen (AO) species in plant pathogenesis. As in any new field there has been a flurry of initial reports linking AO species with key events in plant-pathogen interactions including signal transduction, antimicrobial effects, membrane lipoxidation, cell wall modification, induced resistance, and hypersensitve cell death. Much of the work in the field has been descriptive, identifying when and where AO production occurs. The slow transition of this phase is in part due to the complexity of AO metabolism, the transient nature of the reactants and products, and the lack of techniques appropriate for the measurement of these processes in biological systems.

Because of the extraordinary potential impact of AO on several aspects of plant-pathogen interactions, it is important to be aware of its existence as well as to better understand the cellular mechanisms that govern its metabolism. In this chapter we provide a critical review and update on some of the current hypotheses regarding the production and role of AO during plant-pathogen interactions and comment on the future direction and needs in AO research. To accomplish this, a brief overview of AO metabolism and detection is discussed along with an updated description of its production during plant-pathogen interactions.

ACTIVE OXYGEN IN PLANTS

For the purpose of this review, the term "active oxygen" refers primarily to species that result from the reduction of molecular oxygen (O_2) (Figure 1). This general term is useful because several species of AO may be present in a biological system at one time and it is often difficult to distinguish between them. The characteristics, cellular locations, and common methods of detection of AO species important in plant-pathogen interactions have been previously reviewed (14, 95) and are only briefly discussed here.

Active Oxygen Species

Molecular oxygen, itself relatively unreactive and nontoxic, becomes reactive and, in some cases, dangerously reactive to biological systems when its electron structure is altered. The first one-electron reduction resulting in the superoxide radical (O_2^-) requires a slight input of energy that is often provided by NAD(P)H in biological systems (Figure 1*A*). The following three reductions of O_2^- yielding hydrogen peroxide (H_2O_2), the hydroxyl radical (OH•), and, finally, water occur spontaneously or in the presence of an appropriate reaction partner. No extra energy is required for these reductions to occur.

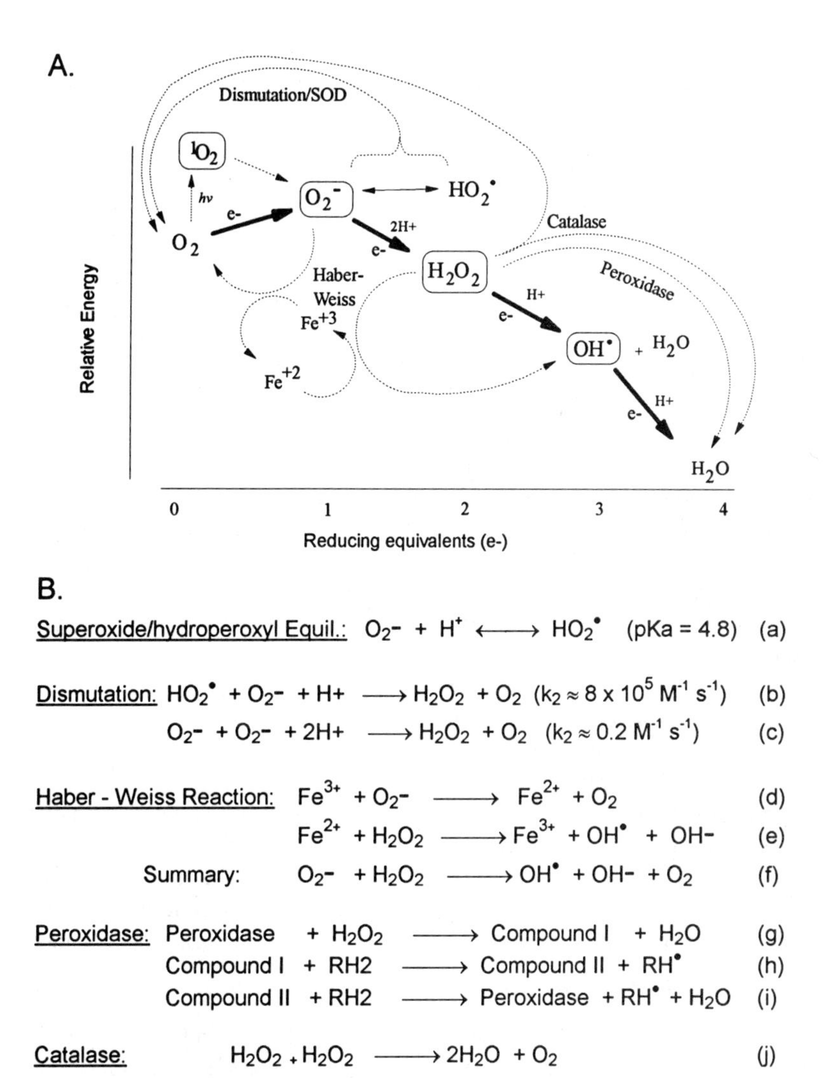

Figure 1 *A*. Active oxygen species derived from molecular oxygen and interconversion pathways that are likely to exist in plants. *B*. Chemical equations depicting key reactions involving active oxygen species in plants.

In the plant, superoxide exists in equilibrium with its conjugate acid, the hydroperoxyl radical ($HO_2\bullet$) (Figure 1*B*, reaction *a*). The hydroperoxyl radical is more lipophilic than superoxide and, therefore, more capable of lipid peroxidation. The superoxide in either form dismutates to hydrogen peroxide and oxygen either spontaneously at neutral or slightly acidic pH or with superoxide dismutase as a catalyst (Figure 1*B*, equations *b, c*). Therefore, whenever superoxide is formed significant amounts of hydrogen peroxide are also formed.

Hydrogen peroxide is a relatively stable oxidant and is slower in reacting with biological molecules than O_2^-. Unlike the other AO species discussed here, H_2O_2 readily crosses the lipid bilayer of cell membranes. Hydrogen peroxide can directly oxidize transition metals such as Fe^{2+} (Figure 1*B*, reaction *e*) or oxidize organic molecules, generally via peroxidases (Figure 1*B*, reactions *g–i*).

The hydroxyl radical is believed to be formed via the Haber-Weiss reaction (Figure 1*A*, equations *d–f*). Because of the role of transition metals as a catalyst, the site of OH• formation appears to be determined by the location of metal ion complexes in the cell. The hydroxyl radical has a half life in the range of microseconds and therefore it reacts with materials close to its generation site. Because of its short half life, it has thus far been difficult to conduct extensive precise studies of this radical in biological materials. Recent attempts to monitor OH• during plant/bacteria interactions were unsuccessful (85). However, the involvement of OH• in plant-pathogen interactions cannot be ruled out for the reasons discussed above.

Another AO species worth mentioning here is singlet oxygen ($^1O_2$), which generally results from a photochemical excitation of oxygen (Figure 1*A*). Although singlet oxygen has been demonstrated in plants (65), it does not appear to be a major product during the interconversions depicted in Figure 1 and has not been found to be involved in AO production associated with plant-pathogen interactions (89).

In Figure 1 we have included the primary interactions between these different species that are of interest in plant-pathogen interactions. Other mechanisms of interconversion have been demonstrated in the test tube or under unique conditions in specific organelles but, based on our present understanding, are not likely to impact upon AO production during pathogenesis.

Antioxidant Mechanisms

Active oxygen species occur during normal metabolism in healthy plant cells and would be toxic if allowed to accumulate. As a result the cell has several mechanisms in place for detoxification of these species. These include small antioxidant molecules, single enzymes, and more complex systems that scavenge AO and efficiently protect the cellular compartments where they are localized. Many of these systems have been extensively reviewed elsewhere (36, 40, 49). The following is a brief overview of the antioxidant mechanisms that may play a role in plant-pathogen interactions.

Superoxide dismutases (SOD), which catalyze the dismutation of O_2^- and HO_2• to H_2O_2 as described above (Figure 1*B*, reactions *b, c*), are metal-containing enzymes that are found in the cytosol, chloroplasts, and mitochondria of plants. Some forms of the enzyme are also found in prokaryotes (42, 51). The H_2O_2 formed through dismutation of O_2^- can then be scavenged and detoxified by the mechanisms described below. SOD is generally viewed as an important antioxidant. However, elevated concentrations of the enzyme in animal and bacterial cells have been reported to induce cell disfunction and death (92, 106). A recent study reported the capacity of SOD to catalyze the generation of OH• from H_2O_2 and hypothesized that this mechanism would allow OH• to cause oxidative damage away from membrane-bound transition metal sites (106).

Ascorbate is an excellent antioxidant found throughout the plant cell. It directly scavenges O_2^- and OH• (49) and detoxifies H_2O_2 via ascorbate peroxidase (40). Ascorbate is regenerated by dehydroascorbate reductase at the expense of glutathione, which is in turn regenerated by glutathione reductase at the expense of NADPH. Glutathione also reacts directly with OH•. Another scavenger molecule important in oxygen detoxification is α-tocopherol (vitamin E), which is located in the lipid bilayer region and protects against lipid peroxidation by scavenging AO-generated lipid peroxides. The resulting oxidized radical of α-tocopherol can then be regenerated by the ascorbate/glutathione cycle described above. In addition to the above antioxidant systems, numerous flavonoid compounds in plant cells have been demonstrated to scavenge O_2^- and may help to prevent lipid peroxidation (54).

Catalase, which converts H_2O_2 to H_2O and O_2 (Figure 1*B*, reaction *j*), is found in most aerobic organisms and plays a significant role in reducing high levels of H_2O_2 in the peroxisomes. However, its importance as a H_2O_2-scavenger during plant-pathogen interactions may be overestimated. Due to its high K_m, catalase is very inefficient at scavenging low levels of H_2O_2 produced in cells. For example, 1 mg of catalase placed in 10 mM H_2O_2 will initially neutralize 4000 μmoles H_2O_2/min; the same amount of catalase placed in 10 μM H_2O_2 only neutralizes 2 μmoles H_2O_2/min, more than 1000-fold less (CJ Baker, unpublished results). For this reason, the scavenging systems described above are believed to play a more important role than catalase in reducing cellular levels of H_2O_2 during early plant-pathogen interactions. An exception might be if the ascorbate/glutathione system becomes overloaded in a stress situation and NADPH levels are depleted. Unlike this system, catalase does not require NADPH and so might be a more important antioxidant under these conditions.

Detection of AO Species

Several assay systems have been used to detect AO species in biological systems. Some of the most common assays used in plant tissues have been reviewed previously (14). Spectrophotometric assays for O_2^-, such as the reduction of ferricytochrome c (41), and nitroblue tetrazolium (NBT) (10) or the oxidation of epinephrine (75), have been used extensively. NBT has also been used as a microscopic probe for O_2^- production in vivo (1).

Several assays have been developed that measure the peroxidase-catalyzed oxidation of various fluorescent dyes by H_2O_2. Some of the fluorescent compounds used include pyranine (5), scopoletin (87), and homovanillin (47). These assays are somewhat limited in their usefulness as they monitor the destruction of a substrate in a very narrow concentration range and under nonsaturating conditions. They are also subject to fluctuations in peroxidase activity.

Chemiluminescent methods for measuring AO production have become popular in recent years (56, 59, 72, 101). The peroxidase-catalyzed oxidation of luminol by H_2O_2 results in the emission of photons that can be immediately measured in a luminometer (48):

$$\text{Luminol} + H_2O_2 \xrightarrow{\text{peroxidase}} \text{Aminophthalate} + N_2 + h\nu.$$

Because the luminol-mediated chemiluminescent (LDC) assay depends upon the measurement of emitted photons instead of a product that accumulates, measurement must be made instantaneously with the addition of luminol so as not to miss the reaction completely. Also, exogenously added peroxidase makes the assay less subject to changes in endogenous peroxidase activities (45). The LDC assay allows the detection of very low levels of active oxygen in particulate samples and is extremely rapid and reproducible.

When interpreting the results of these assays, one must keep in mind that the complexity of oxygen metabolism in cells (*a*) reduces the specificity of these assays for any one AO species; (*b*) reduces the specificity of added enzyme inhibitors; and (*c*) makes it difficult to accurately quantify levels of AO without first evaluating changes in the levels of competing reactants and enzymes. For example, epinephrine, which is commonly used as a "specific" indicator for O_2^-, can also be oxidized by H_2O_2 via peroxidase (GL Harmon, unpublished observation). Therefore, assays for O_2^- must incorporate the use of SOD to determine how much of the reduction or oxidation of the indicator molecule is due to other materials in the sample. Similarities between the many enzymes involved in AO metabolism also complicate the interpretation of assay results. For example, the inhibitor aminotriazole, which is commonly used to destroy catalase activity, also inhibits the peroxidase that serves as catalyst for several of the most common assays for H_2O_2 (CJ Baker, unpublished results).

As described in the preceding section, antioxidant and conversion mechanisms in both plants and pathogens continuously lower the concentrations of AO species. Therefore, rapid assays are imperative in these experimental systems. Changes in AO scavenging must be considered when production of AO is quantified (16). Recently, we modified the LDC assay so that changes in peroxidase and H_2O_2-scavenging activities can be monitored concomitantly with the measurement of AO (16). By monitoring these changes we were able to more accurately interpret the changes in AO production measured during plant-pathogen interactions. This study is discussed in greater detail in the following sections.

ACTIVE OXYGEN PRODUCTION DURING PLANT-PATHOGEN INTERACTIONS

This review focuses on AO metabolism during the early stages of pathogenesis, although AO production also presumably occurs during later stages associated with lipid peroxidation, lignin formation, and senescence. Much of our knowledge of AO production in plant cells comes from studies of plant suspension cells treated with live pathogens or various AO-eliciting compounds. Suspension cells have proven invaluable for examining plant-pathogen interactions at the biochemical level because they allow for easier and more accurate monitoring of AO production, ion fluxes, pH changes, and changes in other stress-related processes. In particular, the treatment of suspension cells with bacteria has allowed the direct measurement of changes in AO production and the various cellular processes involved in AO metabolism during the induction of the hypersensitive response (HR).

AO Production During the Hypersensitive Response

In 1983, Doke published his first report of AO production in potato tubers undergoing a HR (30). He demonstrated that O_2^- production occurred in potato tissues upon inoculation with an incompatible (HR-causing) race of *Phytophthora infestans* or its hyphal wall components, but not after treatment with a compatible (disease-causing) race.

Other laboratories subsequently reported AO production occurring early in incompatible plant-bacteria interactions (1, 58, 59). The AO production was associated with changes in membrane permeability and lipid peroxidation associated with incompatible reactions to *Pseudomonas syringae* (*P.s.*) pathovars. Further studies in our laboratory using plant suspension cells more precisely characterized the AO response during incompatible interactions as consisting of two distinct phases (19, 59) (Figure 2). Phase I is a relatively short-lived, nonspecific response that occurs immediately after the addition of either compatible or incompatible pathovars. Phase II is a relatively long-lived response occurring 1.5 to 3 h after inoculation and appears to be specific to incompatible interactions. Concomitant with the phase II AO response is a K^+/H^+ exchange response (XR) first described by Atkinson et al (9). The XR and phase II AO responses were termed "recognition responses" as they signified the earliest detectable reaction of plant cells specific to incompatible pathogens.

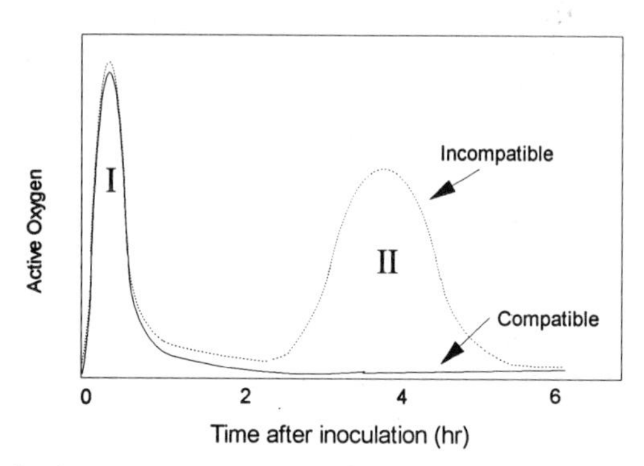

Figure 2 Phases of active oxygen production in plant-bacterial interactions.

The phase II AO response is tightly correlated with nonhost-pathogen interactions at the species level (15, 19) and incompatible race-cultivar interactions at the subspecies level (18, 81). For example, *P.s.* pv. *tabaci*, a tobacco pathogen, does not elicit a phase II AO response on tobacco but does elicit the response in nonhost soybean cells (Figure 3*A*, *B*). Likewise, *P.s.* pv. *glycinea* races 4 and 6 both elicited a phase II AO response in tobacco, a nonhost. However, in soybean (*Glycine max* L. cv. Mandarin) cells, race-specificity was demonstrated with these pathogens: Phase II AO production occurred only after treatment with the incompatible race 6 and not the compatible race 4 (Figure 3*B*). The hypersensitive death of suspension cells was monitored using a spectrophotometric technique (17, 18) with the histological stain Evan's blue (100). The results of the cell death assays indicated that the interactions that resulted in the production of phase II AO subsequently resulted in hypersensitive cell death several hours later (Figure 3*C*, *D*).

During the studies discussed above we found, as expected, that increases in the level of bacterial inoculum from 10^7 to 10^8 cfu/ml corresponded to increases in phase I AO production in suspension cells. However, we were surprised to find that the phase II AO response was actually lower when higher amounts of bacteria were used. This discrepancy was found to be due to increases in AO-scavenging activity induced early during treatment with the higher inoculum levels (16). Therefore, higher inoculum levels of bacteria ($> 10^7$ cfu/ml) actually led to less measurable phase II AO production than lower levels of inoculum. This may help to explain the report by Devlin et al (29) that phase II active oxygen did not occur in suspension cells of white clover inoculated with high innoculum (0.4×10^8 cfu/mL) of the incompatible *Pseudomonas corrugata*.

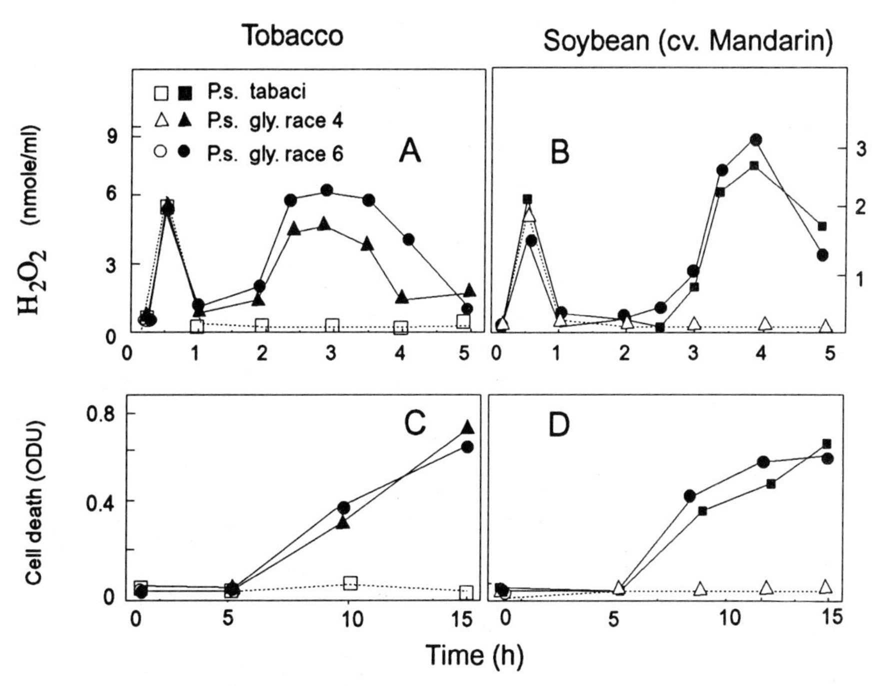

Figure 3 Active oxygen production (*A, B*) and cell death (*C, D*) in suspension cells inoculated with bacterial pathogens. Filled symbols = incompatible; unfilled = compatible.

Genetic Basis for the Bacterial Elicitation of the Phase II AO Response

The two-phased AO response has been a consistent characteristic of incompatible plant-bacteria interactions in several systems tested, including soybean (18), tobacco (19), potato (CJ Baker & KL Deahl, unpublished results), and alfalfa (CJ Baker & NR O'Neill, unpublished results) cells treated with various *Pseudomonas sp.* and in tobacco treated with *Erwinia amylovora* (20). Studies were undertaken to determine if the AO response was controlled by the same genetic determinants responsible for the HR.

The phase II AO production monitored during race-cultivar specific interactions in the soybean system (Figure 3*B*) is apparently due to interactions between the avirulence gene *avr*A (94) in *P.s.* pv. *glycinea* race 6 and the corresponding resistance gene in soybean cultivars. In one study, incorporation of the *avr*A gene transformed the compatible race 4 into an incompatible pathogen that induced the phase II AO response in Mandarin suspension cells (81). Race 4 strains containing only the vector plasmid without the *avr*A gene did not induce this response.

Phase II AO production during incompatible interactions on the nonhost-pathogen species level between *P.s.* pv. *syringae* and tobacco suspension cells has been shown to be under the control of the genes contained in the bacterial *hrp/hrm* region. Mutations in these *hrp* genes (53) blocked the HR and the phase II AO response (44). Insertion of a cosmid (*pHIR11*) containing the *hrp/hrm* region of *P.s.* pv. *syringae* into the saprophyte *P. fluorescens* (53) transformed it into an incompatible bacterium that induces the phase II AO response in suspension cells (44). Recently, proteinaceous products of the *hrp* regions of *E. amylovora* and *P.s.* pv. *syringae*, termed harpins, have been shown to elicit the HR (52, 105) and appear to be the bacterial elicitors of the phase II AO response (20; EW Orlandi, unpublished results).

Recently, studies have been performed to identify the elicitor of the phase I AO response seen with compatible, incompatible, and saprophytic bacteria (81a). The putative phase I elicitor was isolated from loosely bound extracellular products of *Pseudomonas* pathovars. Elicitor activity was destroyed by protease treatment, indicating that the phase I elicitor is comprised, at least partially, of a protein. Inhibitors of cellular Ca^{++} influx and protein phosphorylation that inhibit the phase II AO response also inhibited phase I, indicating that both phases may be triggered by traditional signal transduction mechanisms. This conflicts with earlier hypotheses (14) suggesting that extracellular peroxidase was involved in the nonspecific phase I response as these enzymes would not be expected to require Ca^{++} influx or protein phosphorylation.

As mentioned previously, most of the work described above was done with suspension cell cultures. The more precise assays used in the detection of AO production and metabolism do not adapt easily to whole-tissue studies. Although cell death assays assure us that the suspension cells are undergoing the same type of pathogen recognition and subsequent hypersensitive cell death, it would be preferable to monitor these parameters in whole tissues. This remains an important goal for future studies.

Other Elicitors of the AO response

In addition to the whole-pathogen studies described above, the AO response has been monitored in plant cells treated with eliciting compounds that are not associated with the bacterial HR response. Polygalacturonases purified from fungal pathogens were demonstrated to induce H_2O_2 production in cotton leaves (78). In more recent years, other elicitors extracted from fungal cell walls have proved to be effective AO elicitors (3, 5, 6, 72, 91), as have oligogalacturonides isolated from plant cell walls (5, 57, 68). Most of these elicitors induce AO production in plant cells within a few minutes of treatment.

Cell-free elicitors have been useful in several studies concerning the mechanisms by which pathogens trigger AO production; the signal transduction pathways activated in the plant cell; the likely sources for AO production; and the possible effects of AO production on other defense responses. Studies using cell-free elicitors have the obvious advantage over those with viable pathogens in that they allow the researcher to focus on the plant side of the interaction. For example, inhibitors can be added and the effects on plant cells studied without worrying about undesired effects on pathogen cells. However, by simplifying the system and eliminating the pathogen these studies cannot portray an accurate picture of AO metabolism during plant-pathogen interactions. The impact of the pathogen on AO metabolism including pathogen-produced elicitors and suppressors of elicitation (31) and antioxidant systems must be considered.

PROBABLE SOURCES OF AO PRODUCTION DURING PLANT-PATHOGEN INTERACTIONS

Although many potential sources of AO species in plant cells have been identified, it is still unclear which mechanisms predominate during plant-pathogen interactions. Several enzyme systems have been proposed as the source of AO production, including NADPH oxidases, peroxidases, and lipoxygenases (LOX). The following section briefly describes AO production by these systems and discusses the evidence for their involvement during the early stages of plant-pathogen interactions.

NAD(P)H Oxidases

The best characterized example of active oxygen involvement in disease resistance is the "respiratory burst" of animal phagocytes that is responsible for the killing of foreign bacteria (see 12 for a review). The respiratory burst is attributed to the activation of an NADPH oxidase that transfers electrons from NADPH on the inside of the membrane to molecular oxygen on the outside of the membrane, leading to the generation of O_2^- (93) (Figure 4).

Doke hypothesized that a membrane-bound NADPH oxidase was responsible for the generation of superoxide in potato tissues undergoing a HR to incompatible races of *Phytophthora infestans* or their hyphal wall components (32). Although some of the methods and conclusions of his earlier studies were criticized and later defended by Doke (43), there is still valid evidence that AO production in plants may closely resemble the respiratory burst in animal phagocytes. The signaling pathway elucidated for AO production in plants has, so far, closely paralleled that established for the animal respiratory burst enzyme. This includes the hypothesized involvement of GTP-binding proteins (67, 76, 103), phospholipase C activation (69), and a need for continuous Ca^{++}-influx and protein kinase activity (20, 34, 91). In addition, some of the compounds found to elicit the AO burst in animal cells also elicit AO production in plant cells (67, 91; EW Orlandi, unpublished results). More direct

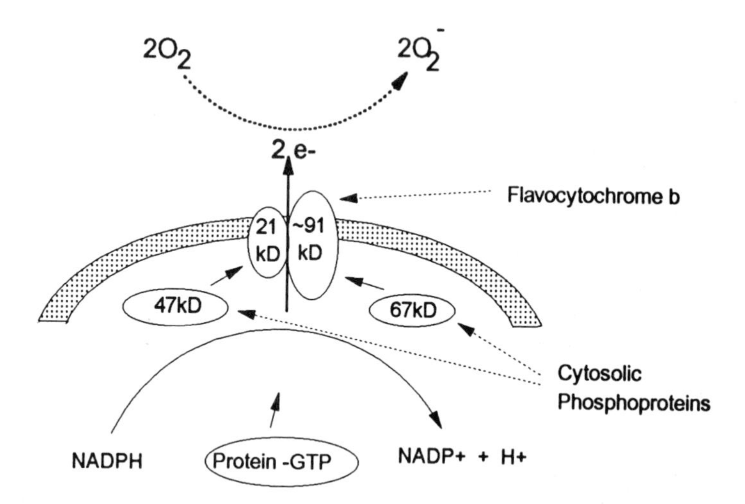

Figure 4 Activation of the respiratory burst oxidase in animal phagocytes. Activation involves a GTP-binding cytoplasmic protein in conjunction with two cytosolic proteins that associate and activate the cytochrome b heterodimer. Two electrons from NADPH are thereby transported across the membrane to molecular oxygen.

evidence for the presence of an NADPH respiratory burst oxidase in plants may be forthcoming. Antibodies raised against components of the human neutrophil respiratory burst oxidase have recently been demonstrated to cross-react with proteins of the same molecular weight from several plant cell lines (PS Low, personal communication).

Peroxidases

Peroxidases have been demonstrated to act as NAD(P)H oxidases leading to the formation of AO in addition to their role in AO scavenging (Figure 1*B*, equations *g–i*). In the cell wall, peroxidase was reported to oxidize NAD(P)H, leading to the formation of O_2^- and H_2O_2, the latter of which is used in the formation of lignin (37). AO production by peroxidases was also found in purified plasma membranes in the presence of various phenolic compounds and added NADH (7). The proposed source of extracellular NAD(P)H for peroxidases is cell wall–bound malate dehydrogenase using malate transported across the cell membrane (46). However, the physiological significance of this mechanism was thoughtfully criticized by Rubinstein & Luster (88) who pointed out that the relatively high pH and mM concentrations of NAD(P)H required for the reaction were not likely in the cellular apoplast.

Peroxidases have been proposed as the source of AO production during plant-pathogen interactions (2, 84). Although it appears possible that membrane-bound or cell wall peroxidases could play a role in pathogen-induced AO production, no definitive evidence for this scenario has been presented. Recently, peroxidase activity has been demonstrated to increase during incompatible reactions (2, 101). However, the increases are often after the time period of AO production and may well be part of the AO-scavenging activity induced during the interaction. A recent report that known peroxidase inhibitors greatly inhibited AO production measured by the LDC assay after treatment with fungal elicitors was interpreted as an indication that peroxidases are involved in the AO production (101). However, it is not clear if these authors took into account the effect of the inhibitors on the assay itself, as peroxidase is the catalyst for the LDC assay.

Lipoxygenases

Purified lipoxygenase (LOX) has been reported to produce AO in vitro when supplied with free linoleic acid and peroxides of linoleic acid. Chamulitrat et al (26) detected superoxide production by LOX incubated with fatty acid, fatty acid hydroperoxides, and molecular oxygen. In another study, low amounts of singlet oxygen were detected upon addition of linoleic acid and 13-hydroperoxylinoleic acid to purified soybean LOX (55). However, a recent study using a highly sensitive polychromatic spectrometer found no $^1O_2$ production during LOX activity (79).

Although numerous studies have suggested that LOX plays an important role in AO production during the HR (8, 27, 83) they have been primarily corollary. The monitored increases in extractable LOX enzyme activity during the HR do not necessarily translate to increased LOX reactions in vivo. As pointed out by Thompson et al (97), LOX reactions are likely to be determined more by availability of free fatty acid substrate than by changes in the total activity of the enzyme. These fatty acids could be released by phospholipases (70) or via the action of AO species on plasma membranes.

AO scavengers inhibit membrane deterioration during the bacteria-induced HR, indicating that AO production precedes lipid peroxidation (1, 60). Increases in LOX activity were, therefore, hypothesized to be in response to fatty acids released by AO-induced lipid peroxidation of membranes (60). This view is supported by recent studies on senescence hypothesizing that LOX may actually play a role in preserving membrane integrity rather than in propagating its destruction (4, 104).

Additional evidence that AO production precedes LOX activity has been published recently. In soybean cell suspensions treated with fungal elicitors AO production was detected within the first few minutes whereas lipid hydroperoxides, the products of LOX, were not detected until 6 h after treatment.

In tobacco suspension cells treated with fungal elicitors, LOX activity did not increase until 3 h after treatment (39).

In our laboratory we recently studied the effects of eicosatetranoic acid (ETYA), an analog of arachadonic acid that reportedly inhibits lipoxygenase, on AO production in tobacco suspension cells. We found that ETYA alone was actually a weak elicitor of AO and that it synergistically enhanced the elicitation of AO production by *P.s.* pv. *syringae* or oligogalacturonides (EW Orlandi, unpublished results). These results indicate that the AO production that occurs early in the plant-bacteria interaction is not likely due to LOX reactions. On the contrary, the AO-eliciting activity of ETYA is actually consistent with the presence of an AO-producing NAD(P)H oxidase since fatty acid stimulation of these oxidases has been reported in both animals (13) and plants (23, 77).

POSSIBLE ROLES FOR AO DURING PLANT PATHOGENESIS

Active oxygen species appear to be able to affect nearly all aspects of biological systems, including direct effects on proteins, lipids, polysaccharides, and nucleic acids (98, 99). For this reason, several roles have been hypothesized for AO species during plant-pathogen interactions. Some of these roles are well established such as the involvement of H_2O_2 in lignin formation discussed below. Others have been proposed more recently based on corollary evidence, and more definitive studies are needed.

Antimicrobial Activities of AO

The antimicrobial role of active oxygen species in animals has long been recognized. The engulfment of bacteria during phagocytosis followed by production of AO in the enclosed area around the bacteria allow concentrations of AO and its derivatives to reach antimicrobial levels. However, since plant cells lack the mobility to surround or engulf invading pathogens, it is not as clear whether AO species reach sufficient levels for antimicrobial effects to play a significant role in plant pathogenesis. The antimicrobial effects of exogenously added O_2^- and H_2O_2 have been demonstrated against fungi (11, 82, 84) and bacteria (33, 61). However, it is still not clear whether the antimicrobial effects of AO play a role in vivo (58, 61, 74).

In related studies, recent reports have demonstrated that stationary-phase bacteria cells produce several-fold more catalase than log-phase cells and suggest that they are more resistant to high levels of H_2O_2.(62–64). However, the studies did not monitor active oxygen concentrations and did not take into account the differences in bacterial populations at these two stages of growth. Recent results in our laboratory and others (73) demonstrate that at the high

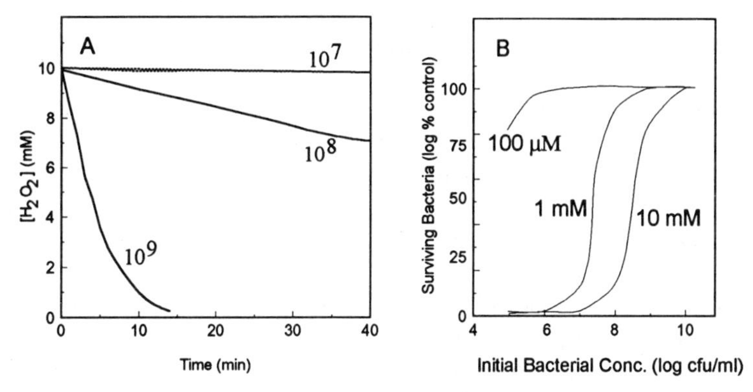

Figure 5 *A*. Scavenging of 10 mM hydrogen peroxide by *P. syringae* pv. *syringae* at different bacterial concentrations (log cfu/ml). *B*. Ability of different initial bacterial concentrations to survive 30 min exposure to different concentrations of hydrogen peroxide.

concentrations of H_2O_2 (1.0–50 mM) used in the previous studies, the primary factor in determining survival is the bacterial concentration rather than the scavenging activity of an individual cell. For example, in Figure 5*A*, the higher the bacterial concentration, the more rapid the reduction of the 10 mM H$_2$O2 concentration to tolerable levels in the μM range. Figure 5*B* demonstrates that bacteria can survive in increasingly higher levels of H_2O_2 if the initial inoculum also increases. We found that bacteria can tolerate H_2O_2 concentrations if they can reduce it to tolerable levels within the first 15 to 30 min. The tolerable level of H_2O_2 is defined by the individual bacterial cell and its scavenging ability, which includes catalase and the ascorbate-glutathione cycle. Unlike antibiotics, which a single bacterial cell can resist because of their semipermeable membrane, H_2O_2 is neutral and is able to easily pass through biological membranes. Therefore, a bacterial population must reduce the environmental H_2O_2 concentration rapidly to a tolerable level in order for individual cells to survive.

AO as a Direct Cause of Hypersensitive Cell Death

Over 20 years ago, Mussell (78) demonstrated that H_2O_2 damaged plant tissues either when supplied exogenously or when generated by added oxidases and their substrates. Exogenously supplied catalase blocked the cell death. Although $[H_2O_2]$ in leaf tissue treated with endopolygalacturonase elicitor was estimated to be 150 μM, no concentration was given for the added H_2O_2, making it difficult to determine if physiological concentrations of H_2O_2 were used. More recently, Adam et al (1) demonstrated that exogenously supplied O_2^- prepared from riboflavin-EDTA infiltrated into leaf tissue would also cause cell death symptoms similar to those of HR. These symptoms are presumably due to AO-induced lipid peroxidation and subsequent membrane damage leading to loss of integrity and cell death.

Recently, Levine et al (71) attempted to demonstrate that H_2O_2 produced during early plant-pathogen interactions is the direct cause of cellular death during the HR. In soybean cell suspensions incubated with *P.s.* pv. *glycinea*, they found the two-phased AO response described earlier (18, 45, 81). In their studies designed to monitor the effects of H_2O_2 on the development of the HR, the authors added exogenous H_2O_2 in the 2 to 10 mM range, much above that found in vivo during the two-phased AO response (Figure 3*B*). To monitor this level of H_2O_2 they relied on the iodine-starch assay (80), which is orders of magnitude less sensitive than the scopoletin assay (87) used for their bacterial studies. The high levels of exogenously added H_2O_2 made cell death due to oxidative stress inevitable.

Studies conducted over the past several years do not support a role of the early AO responses in directly causing plant cell death. First, as described above, all bacteria added to cells elicit an immediate and often very large phase I burst of AO, yet not all bacteria elicit hypersensitive cell death. Heat-killed bacteria elicit AO production several times greater than the phase II response elicited by viable bacteria, and yet they do not elicit the HR (81a). One could argue that there is something unique about the timing of the phase II AO response that triggers cell death. However, *P.s.* pv. *syringae* with mutations in the *hrm* region of the *hrp/hrm* gene cluster described earlier are able to fully elicit both phase I and II in cell suspensions yet do not cause cell death in either cell suspensions or leaf tissue (JA Glazener, unpublished results). Therefore, although AO supplied in nonphysiological concentrations elicits defense responses and cell death, the AO production found during the first few hours after bacterial treatment may be one of a series of events leading to HR but is not itself the direct cause of hypersensitive cell death. AO production during the later stages of the HR response may be sufficient to directly cause cell death, but AO production has not been well-documented beyond the first few hours.

AO-Induced Cell Wall Strengthening

For many years evidence has been published that lignification of the cell wall occurs during pathogenesis (50). An increase in H_2O_2 and peroxidase was demonstrated and, depending on the timing, this could strengthen the cell wall and slow pathogen ingress. Similarly, Bradley et al (21) and Brisson et al (24) observed what they hypothesized to be oxidative cross-linking of SDS-extractable proteins from the cell walls of bean (*Phaseolus vulgaris* L.) and soybean (*Glycine max* L.) suspension cells after treatment with crude fungal elicitors. Presumably, the cross-linking could play a role in discouraging or slowing pathogen ingress and apparently also decrease protoplastibility of cells. The authors hypothesized that proteins were made insoluble by peroxidase-catalyzed cross-linking mediated by the burst of H_2O_2 induced by the elicitor treatments. Although this explanation of the responses seems reasonable, a concern in this work is that they also found cross-linking with glutathione, which is considered an antioxidant and has been associated with stimulation of a wider range of defense responses (35). This suggests that mechanisms other than active oxygen may play a part in the observed cross-linking induced by fungal elicitors. Further direct evidence is needed to confirm this.

The Role of AO in Phytoalexin Elicitation

A connection between the production of AO and phytoalexins was first suggested by the work of Doke (30, 31) when he found that known fungal phytoalexin elicitors also stimulated O_2^- formation. Direct application of superoxide-generating systems or H_2O_2 to potato tuber slices resulted in induction of phytoalexin production (25). Since these early studies, the connection between AO production and phytoalexin synthesis has been explored in bean (86), soybean (5, 66), white clover (29), tomato (102), and other plant species. Several studies have found a tight correlation between AO and phytoalexin production and have demonstrated that treatment of plant cells with AO scavengers prevents elicitation of phytoalexins. Other studies have found phytoalexin production in the absence of measurable AO production (29, 102). However, the failure to detect AO production after elicitation of cells may be due to increases in AO-scavenging activity as described above.

Apostol et al (5) reported a correlation between AO production and glyceollin synthesis in soybean suspension cells treated with a crude fungal elicitor. They found that relatively high amounts of exogenously added H_2O_2 (0.5–1.0 mM) were sufficient to induce glyceollin production, and hypothesized that H_2O_2 acted as a second messenger in phytoalexin elicitation. However, a recent study from their laboratory demonstrated that the crude elicitor used in the first study was composed of a protein moiety that elicits phytoalexin induction and a carbohydrate portion that is apparently responsible for the elicitation of AO production (28). These conflicting studies suggest that, although they are closely correlated, it may be possible to separate the two phenomena, indicating that AO production may not be directly responsible for the elicitation of phytoalexin production in vivo.

SUMMARY

Although we have been aware of active oxygen species for some time, the study of their involvement in many physiological processes is still an emerging field. The chemical principles, techniques, and strategies traditionally used for carbon-based analyses often cannot be applied directly to the study of AO. For this reason, advances in understanding of the importance of AO metabolism have been and still are dependent on the development of techniques both directly and indirectly related to the measurement of AO species.

Transgenic plants lacking or overproducing antioxidant enzymes may prove to be valuable tools in understanding the biochemistry of AO metabolism (40, 90, 96). Interestingly, these studies have demonstrated that overproduction of antioxidant enzymes alone does not necessarily protect plants from oxidative stress. In the case of SOD-overproduction, for example, one might actually expect an increase in oxidative stress due to increased production of H_2O_2 or OH• by the mechanisms described previously (106). Another concern is that increased production of antioxidant enzymes, which are predominantly metaloenzymes, would increase the amount of soluble transition metal in the plant and could catalyze increased production of OH• by the Haber-Weiss reaction (Figure 1*B*, reactions *d–f*). Also, it must be kept in mind that increased enzyme concentration does not necessarily mean increased activity of the pathway; activity is also dependent on the availability of substrates and the reaction conditions.

In summary, AO is clearly important in plant-pathogen interactions, but it likely plays its role in concert with other mechanisms. As the role of AO and associated phenomena in plant-pathogen interactions are examined from different perspectives and with improved techniques a more accurate and realistic understanding is emerging.

Literature Cited

1. Adam A, Farkas T, Somlya G, Hevesi M, Kiraly Z. 1989. Consequence of O_2^- generation during a bacterially induced hypersensitive reaction in tobacco: deterioration of membrane lipids. *Physiol. Mol. Plant Pathol.* 34:13–26
2. Adam AL, Bestwick CS, Galal AA, Manninger K, Barna B, et al. 1993. What is the putative source of free radical generation during hypersensitive response in plants? See Ref. 77a, pp. 35–43
3. Anderson AJ, Rogers K, Tepper CS, Blee K, Cardon J. 1991. Timing of molecular events following elicitor treatment of plant cells. *Physiol. Mol. Plant Pathol.* 38:1–13
4. Anderson JM. 1989. Membrane-derived fatty acids as precursors to second messengers. In *Second Messengers in Plant Growth and Development*, ed. WF Boss, DJ Morre, 6:181–212. New York: Liss. 348 pp.
5. Apostol I, Heinstein PF, Low PS. 1989. Rapid stimulation of an oxidative burst during elicitation of cultured plant cells. *Plant Physiol.* 90:109–16
6. Arnott T, Murphy TM. 1991. A comparison of the effects of fungal elicitor and ultraviolet radiation on ion transport and hydrogen peroxide synthesis by rose cells. *Environ. Exp. Bot.* 31:209–16
7. Askerlund P, Larsson C, Widell S, Moller IM. 1987. NAD(P)H oxidase and

peroxidase activities in purified plasma membranes from cauliflower inflorescences. *Physiol. Plant.* 71:9–19

8. Atkinson MM. 1993. Molecular mechanisms of pathogen recognition by plants. *Adv. Plant Pathol.* 10:35–64
9. Atkinson MM, Huang JS, Knopp JA. 1985. The hypersensitive reaction of tobacco to *Pseudomonas syringae* pv. *pisi*: activation of plasmalemma K^+/H^+ exchange mechanism. *Plant Physiol.* 79: 843–47
10. Auclair C, Boisin E. 1985. Measurement of superoxide anion: nitroblue tetrazolium reduction. In *Handbook of Methods for Oxygen Radical Research,* ed. RA Greenwald, pp. 123–32. Boca Raton, FL: CRC. 447 pp.
11. Aver'yanov AA, Lapikova VP, Djawakhia VG. 1993. Active oxygen mediates heat-induced resistance of rice plant to blast disease. *Plant Sci.* 92:27–34
12. Babior BM. 1987. The respiratory burst oxidase. *Trends Biochem. Sci.* 12:241–43
13. Badwey JA, Karnovsky ML. 1986. Production of superoxide by phagocytic leukocytes: a paradigm for stimulus-response phenomena. *Curr. Top. Cell. Regul.* 28:183–208
14. Baker CJ. 1994. Active oxygen metabolism during plant/bacterial recognition. In *Biotechnology and Plant Protection,* ed. DD Bills, S Kung, pp. 275–94. River Edge, NJ/London: World Sci. 337 pp.
15. Baker CJ, Atkinson MM, Collmer A. 1987. Concurrent loss in T*n*5 mutants of *Pseudomonas syringae* pv *syringae* of the ability to induce the hypersensitive response and host plasma membrane K^+/H^+ exchange in tobacco. *Phytopathology* 77:1268–72
16. Baker CJ, Harmon GL, Glazener JA, Orlandi EW. 1995. A non-invasive technique for monitoring peroxidative and H_2O_2-scavenging activities during interactions between bacterial plant pathogens and suspension cells. *Plant Physiol.* In press
17. Baker CJ, Mock NM. 1994. An improved method for monitoring cell death in cell suspension and leaf disc assays using Evans Blue. *Plant Cell Tissue Organ Cult.* 39;7–12
18. Baker CJ, Mock NM, Glazener JA, Orlandi EW. 1993. Recognition responses in pathogen/nonhost and race/cultivar interactions involving soybean (*Glycine max*) and *Pseudomonas syringae* pathovars. *Physiol. Mol. Plant Pathol.* 43:81–94
19. Baker CJ, O'Neill NR, Keppler LD, Orlandi EW. 1991. Early responses during plant-bacteria interactions in tobacco cell suspensions. *Phytopathology* 81: 1504–7
20. Baker CJ, Orlandi EW, Mock NM. 1993. Harpin, an elicitor of the hypersensitive response in tobacco caused by *Erwinia amylovora,* elicits active oxygen production in suspension cells. *Plant Physiol.* 102:1341–44
21. Bradley DJ, Kjellbom P, Lamb CJ. 1992. Elicitor- and wound-induced oxidative cross-linking of a proline-rich plant cell wall protein: a novel, rapid defense response. *Cell* 70:21–30
22. Deleted in proof
23. Brightman AO, Zhu Z, Morre DJ. 1991. Activation of plasma membrane NADH oxidase activity by products of phospholipase A. *Plant Physiol.* 96:1314–20
24. Brisson LF, Tenhaken R, Lamb CJ. 1994. The function of oxidative cross-linking of cell wall structural proteins in plant disease resistance. *Plant Cell.* 6:1703–12
25. Chai HB, Doke N. 1987. Activation of the potential of potato leaf tissue to react hypersensitively to *Phytophthora infestans* by cytospore germination fluid and the enhancement of this potential by calcium ions. *Physiol. Mol. Plant Pathol.* 30:27–37
26. Chamulitrat W, Hughes MF, Eling TE, Mason RP. 1991. Superoxide and peroxyl radical generation from the reduction of polyunsaturated fatty-acid hydroperoxides by soybean lipoxygenase. *Arch. Biochem.* 290:153–59
27. Croft KPC, Voisey CR, Slusarenko AJ. 1990. Mechanisim of hypersensitive cell collapse: correlation of increased lipoxygenase activity with membrane damage in leaves of *Phaseolus vulgaris* (L.) inoculated with an avirulent race of *Pseudomonas syringae* pv. *phasicola. Physiol. Mol. Plant Pathol.* 36:49–62
28. Davis D, Merida J, Legendre L, Low PS, Heinstein P. 1993. Independent elicitation of the oxidative burst and phytoalexin formation in cultured plant cells. *Phytochemistry* 32:607–11
29. Devlin WS, Gustine DL. 1992. Involvement of the oxidative burst in phytoalexin accumulation and the hypersensitive reaction. *Plant Physiol.* 100: 1189–95
30. Doke N. 1983. Involvement of superoxide anion generation in the hypersensitive response of potato tuber tissues to infection with an incompatible race of *Phytophthora infestans* and to the hyphal wall components. *Physiol. Plant Pathol.* 23:345–57

31. Doke N. 1983. Generation of superoxide anion by potato tuber protoplasts during the hypersensitive response to hyphal wall components of *Phytophthora infestans* and specific inhibition of the reaction by suppressors of hypersensitivity. *Physiol. Plant Pathol.* 23:359–67
32. Doke N. 1985. NADPH-dependent O^- generation in membrane fractions isolated from wounded potato tubers inoculated with *Phytophthora infestans. Physiol. Plant Pathol.* 27:311–22
33. Doke N. 1987. Release of Ca^{++} bound to plasmamembrane in potato tuber tissue cells in the initiation of hypersensitive reaction. *Ann. Phytopathol. Soc. Jpn.* 53:391 (Abstr.)
34. Doke N, Miura Y. 1995. In vitro activation of NADPH-dependent O-generating system in a plasmamembrane-rich fraction of potato tuber tissues by treatment with an elicitor from *Phytophthora infestans* or with digitonin. *Physiol. Mol. Plant Pathol.* 46:17–28
35. Edwards R, Blount JW, Dixon RA. 1991. Glutathione and elicitation of the phytoalexin response in legume cell cultures. *Planta* 184:403–9
36. Elstner EF. 1987. Metabolism of activated oxygen species. In *The Biochemistry of Plants,* ed. DD Davies, 11:253–315. San Diego: Academic
37. Elstner EF, Heupel A. 1976. Formation of hydrogen peroxide by isolated cell walls from horseradish (*Armoracia lapathifolia* Gilib.). *Planta* 130:175–80
38. Elstner EF, Wagner GA, Schutz W. 1988. Activated oxygen in green plants in relation to stress situations. *Curr. Top. Plant Biochem. Physiol.* 7:159–87
39. Fournier J, Pouenat M-L, Rickauer M, Rabinovitch-Chable H, Rigaud M, et al. 1993. Purification and characterization of elicitor-induced lipoxygenase in tobacco cells. *Plant J.* 3:63–70
40. Foyer CH, Descourvieres P, Kunert KJ. 1994. Protection against oxygen radicals: an important defence mechanism studied in transgenic plants. *Plant Cell Environ.* 17:507–23
41. Fridovich I. 1985. Measurement of superoxide anion: cytochrome c. In *Handbook of Methods for Oxygen Radical Research,* ed. RA Greenwald, pp. 121–22. Boca Raton, FL: CRC. 447 pp.
42. Fridovich I. 1986. Superoxide dismutases. *Adv. Enzymol.* 58:61–97
43. Friend J. 1991. The biochemistry and cell biology of interaction. *Adv. Plant Pathol.* 7:85–129
44. Glazener JA, Huang HC, Baker CJ. 1991. Active oxygen induction in tobacco cell suspensions treated with *Pseudomonas fluorescens* containing the cosmid *pHIR11* and with strains containing *TnPho*A mutations in the *hrp* cluster. *Phytopathology* 81:1196 (Abstr.)
45. Glazener JA, Orlandi EW, Harmon GL, Baker CJ. 1991. An improved method for monitoring active oxygen in bacteria-treated suspension cells using luminol-dependent chemiluminescence. *Physiol. Mol. Plant Pathol.* 39:123–33
46. Gross GG, Janse C, Elstner EF. 1977. Involvement of malate, monophenols, and the superoxide radical in hydrogen peroxide formation by isolated cell walls from horseradish (*Armoracia lapathifolia* Gilib.). *Planta* 136:271–76
47. Guilbault GG. 1973. *Practical Fluorescence: Theory, Methods, and Techniques,* pp. 1256–62. New York: Dekker
48. Guilbault GG, Brignac PJ Jr, Juneau M. 1968. New substrates for the fluorometric determination of oxidative enzymes. *Anal. Chem.* 40:1256–62
49. Halliwell B. 1974. Superoxide dismutase, catalase and glutathione peroxidase: solutions to the problems of living with oxygen. *New Phytol.* 73:1075–86
50. Hammerschmidt R, Kuc J. 1982. Lignification as a mechanism for induced systemic resistance in cucumber. *Physiol. Plant Pathol.* 20:61–71
51. Hassan HM. 1989. Microbial superoxide dismutases. *Adv. Genet.* 26:65–97
52. He SY, Huang HC, Collmer A. 1993. *Pseudomonas syringae* pv. *syringae* harpinPss: a protein that is secreted via the hrp pathway and elicits the hypersensitive response in plants. *Cell* 73:1–20
53. Huang HC, Schuurink R, Denny TP, Atkinson MM, Baker CJ, et al. 1988. Molecular cloning of a *Pseudomonas syringae* pv. *syringae* gene cluster that enables *Pseudomonas fluorescens* to elicit the hypersensitive response in tobacco. *J. Bacteriol.* 170:4748–56
54. Huguet AI, Manez S, Alcaraz MJ. 1990. Superoxide scavenging properties of flavonoids in a non-enzymic system. *Z. Naturforsch. Teil C* 45:19–24
55. Kanofsky JR, Axelrod B. 1986. Singlet oxygen production by soybean lipoxygenase isozymes. *J. Biol. Chem.* 261:1099–104
56. Kauss H, Jeblick W, Ziegler J, Krabler W. 1994. Pretreatment of parsley (*Petroselinum crispum* L.) suspension cultures with methyl jasmonate enhances elicitation of activated oxygen species. *Plant Physiol.* 105:89–94
57. Keppler LD, Atkinson MM, Baker CJ. 1988. Association of "Active Oxygen" with induction of increased extracellular

pH in tobacco suspension cultures by pectic oligosaccharides. *Curr. Top. Plant Biochem. Physiol.* 7:230

58. Keppler LD, Baker CJ. 1989. O^- initiated lipid peroxidation in a bacteria-induced hypersensitive reaction in tobacco cell suspensions. *Phytopathology* 79: 555–62
59. Keppler LD, Baker CJ, Atkinson MM. 1989. Active oxygen production during a bacteria-induced hypersensitive reaction in tobacco suspension cells. *Phytopathology* 79:974–78
60. Keppler LD, Novacky A. 1987. The initiation of membrane lipid peroxidation during bacteria-induced hypersensitive reaction. *Physiol. Mol. Plant Pathol.* 30:233–45
61. Kiraly Z, El-Zahaby H, Galal A, Abdou S, Adam A, et al. 1993. Effect of oxy free radicals on plant pathogenic bacteria and fungi and on some plant diseases. See Ref. 77a, pp. 9–19
62. Klotz MG. 1993. The importance of bacterial growth phase for in planta virulence and pathogenicity testing: coordinated stress response regulation in fluorescent pseudomonads. *Can. J. Microbiol.* 39:948–57
63. Klotz MG, Anderson AJ. 1994. The role of catalase isozymes in the culturability of the root colonizer *Pseudomonas putida* after exposure to hydrogen H_2O_2 and antibiotics. *Can. J. Microbiol.* 40: 382–87
64. Klotz MG, Hutcheson SW. 1992. Multiple periplasmic catalases in phytopathogenic strains of *Pseudomonas syringae. Appl. Environ. Microbiol.* 58: 2468–73
65. Knox JP, Dodge AD. 1985. Singlet oxygen and plants. *Phytochemistry* 24:889–96
66. Kondo Y, Hanawa F, Miyazawa T, Mizutani J. 1993. Detection of rapid and transient generation of activated oxygen and phospholipid hydroperoxide in soybean after treatment with fungal elicitor by chemiluminescence assay. In *Mechanisms of Plant Defense Responses*, ed. B Fritig, M Legrand, pp. 148–51. Dordrecht, Netherlands: Kluwer
67. Legendre L, Heinstein PF, Low PS. 1992. Evidence for participation of GTP-binding proteins in elicitation of the rapid oxidative burst in cultured soybean cells. *J. Biol. Chem.* 267: 20140–47
68. Legendre L, Rueter S, Heinstein PF, Low PS. 1993. Characterization of the oligogalacturonide-induced oxidative burst in cultured soybean (*Glycine max*) Cells. *Plant Physiol.* 102:233–40
69. Legendre L, Yueh YG, Crain R, Haddock N, Heinstein PF, et al. 1993. Phospholipase C activation during elicitation of the oxidative burst in cultured plant cells. *J. Biol. Chem.* 268:24559–63
70. Leshem YY. 1987. Membrane phospholipid catabolism and Ca^{++} activity in control of senescence. *Physiol. Plant.* 69:551–59
71. Levine A, Tenhaken R, Dixon R, Lamb C. 1994. H_2O_2 from the oxidative burst orchestrates the plant hypersensitive disease resistance response. *Cell* 79:583–93
72. Lindner WA, Hoffmann C, Grisebach H. 1988. Rapid elicitor-induced chemiluminescence in soybean cell suspension cultures. *Phytochemistry* 27:2501–3
73. Ma M, Eaton JW. 1992. Multicellular oxidant defense in unicellular organisms. *Proc. Natl. Acad. Sci. USA* 89: 7924–28
74. Minardi P, Mazzucchi U. 1988. No evidence of direct superoxide anion effect in hypersensitive death of *Pseudomonas syringae* van Hall in tobacco leaf tissue. *J. Phytopathol.* 122:351–58
75. Misra HP, Fridovich I. 1972. The role of superoxide anion in the autoxidation of epinepherine and a simple assay for superoxide dismutase. *J. Biol. Chem.* 247:3170
76. Miura Y, Kawakita K, Yoshioka H, Doke N. 1992. In vitro activation of NADPH-O^- generating reaction by elicitor using microsomal fraction from potato tuber tissues. *Ann. Phytopathol. Soc. Jpn.* 58:564 (Abstr.)
77. Morre DJ, Brightman AO. 1991. NADH oxidase of plasma membranes. *J. Bioenerg. Biomembr.* 23:469–89

77a. Mozsik G, Emerit I, Feher J, Matkovics B, Vincze A, eds. 1993. *Oxygen Free Radicals and Scavengers in the Natural Sciences.* Budapest: Akademiai Kiado. 356 pp.

78. Mussell HW. 1973. Endopolygalacturonase: evidence for involvement in verticillium wilt of cotton. *Phytopathology* 63:62–70
79. Nagoshi T, Watanabe N, Suzuki S, Usa M, Watanabe H, et al. 1992. Spectral analyses of low level chemiluminescence of a short lifetime using a highly sensitive polychromatic spectrometer incorporating a two dimensional photoncounting type detector. *Phytochem. Photobiol.* 56:89–94
80. Olson PD, Varner JE. 1993. Hydrogen peroxide and lignification. *Plant J.* 4:887–92
81. Orlandi EW, Hutcheson SW, Baker CJ. 1992. Early physiological responses associated with race-specific recognition

in soybean leaf tissue and cell suspensions treated with *Pseudomonas syringae* pv. *glycinea*. *Physiol. Mol. Plant Pathol.* 40:173–80

81a. Orlandi EW, Mock NM, Baker CJ. 1995. The elicitation and signal transduction pathways involved in the two-phased active oxygen response during plant/bacteria interactions. *J. Cell Biochem.* 21A:489(Suppl.) (Abstr.)

82. Ouf MF, Gazar AA, Shehata ZE, Abdou E, Kiraly Z, et al. 1993. The effect of superoxide anion on germination and infectivity of wheat stem rust (*Puccinia graminis pers.* F. Sp. *tritici* Eriks and Henn.) Uredospores. *Cereal Res. Commun.* 21:31–37

83. Peever TL, Higgins VJ. 1989. Electrolyte leakage, lipoxygenase, and lipid peroxidation induced in tomato leaf tissue by specific and nonspecific elicitors from *Cladosporium fulvum*. *Plant Physiol.* 90:867–75

84. Peng M, Kuc J. 1992. Peroxidase-generated hydrogen peroxide as a source of antifungal activity in vitro and on tobacco leaf disks. *Phytopathology* 82: 696–99

85. Popham PL, Novacky A. 1991. Use of dimethyl sulfoxide to detect hydroxyl radical during bacteria-induced hypersensitive reaction. *Plant Physiol.* 96: 1157–60

86. Rogers KR, Albert F, Anderson J. 1988. Lipid peroxidation is a consequence of elicitor activity. *Plant Physiol.* 86:547–53

87. Root RK, Metcalf J, Oshino N, Chance B. 1975. H_2O_2 release from human granulocytes during phagocytosis: documentation, quantitation and some regulating factors. *J. Clin. Invest.* 55:945–55

88. Rubinstein B, Luster DG. 1993. Plasma membrane redox activity: components and role in plant processes. *Annu. Rev. Plant Physiol. Plant Mol. Biol.* 44:131–55

89. Salzwedel J, Daub M, Huang J. 1989. Effects of singlet oxygen quenchers and pH on the bacterially induced hypersensitive reaction in tobacco suspension cell cultures. *Plant Physiol.* 90:25–28

90. Scandalios JG. 1993. Oxygen stress and superoxide dismutases. *Plant Physiol.* 101:7–12

91. Schwacke R, Hager A. 1992. Fungal elicitors induce a transient release of active oxygen species from cultured spruce cells that is dependent on Ca^{++} and protein-kinase activity. *Planta* 187: 136–41

92. Scott MD, Meshnick SR, Eaton JW. 1987. Superoxide dismutase-rich bacteria: paradoxical increase in oxidant toxicity. *J. Biol. Chem.* 262:3640–45

93. Segal AW, Abo A. 1993.The biochemical basis of the NADPH oxidase of phagocytes. *Trends Biochem. Sci.* 18: 48–52

94. Staskawicz BJ, Dahlbeck D, Keen NT. 1984. Cloned avirulence gene of *Pseudomonas syringae* pv. *glycinea* determines race-specific incompatibility on *Glycine max* (L.) Merr. *Proc. Natl. Acad. Sci. USA* 81:6024–28

95. Sutherland MW. 1991.The generation of oxygen radicals during host plant responses to infection. *Physiol. Mol. Plant Pathol.* 39:79–83

96. Tepperman JM, Dunsmuir P. 1990. Transformed plants with elevated levels of chloroplastic SOD are not more resistant to superoxide toxicity. *Plant Mol. Biol.* 14:501–11

97. Thompson JE, Brown JH, Gopinadhan PT, Yao K. 1991. Membrane phospholipid catabolism primes the production of activated oxygen, in senescing tissues. In *Active Oxygen/Oxidative Stress and Plant Metabolism*, ed. EI Pell, KL Steffen, pp. 57–66. Rockville, MD: Am. Soc. Plant Physiol.

98. Thompson JE, Legge RL, Barber RF. 1987. The role of free radicals in senescence and wounding. *New Phytol.* 105: 317–44

99. Trippi VS, Gidrol X, Pradet A. 1989. Effects of oxidative stress caused by oxygen and hydrogen peroxide on energy metabolism and senescence in oat leaves. *Plant Cell Physiol.* 30: 157–62

100. Turner JG, Novacky A. 1974. The quantitative relationship between plant and bacterial cells involved in the hypersensitive reaction. *Phytopathology* 64:885–90

101. Vera-Estrella R, Blumwald E, Higgins VJ. 1992. Effect of specific elicitors of *Cladosporium fulvum* on tomato suspension cells: evidence for the involvement of active oxygen species. *Plant Physiol.* 99:1208–15

102. Vera-Estrella R, Blumwald E, Higgins VJ. 1993. Non-specific glycopeptide elicitors of *Cladosporium fulvum:* evidence for involvement of active oxygen species in elicitor-induced effects on tomato cell suspensions. *Physiol. Mol. Plant Pathol.* 42:9–22

103. Vera-Estrella R, Higgins VJ, Blumwald E. 1994. Plant defense response to fungal pathogens. *Plant Physiol.* 106:97–102

104. Wang J, Fujimoto K, Miyazawa T, Endo Y, Kitamura K. 1990. Sensitivities of

lipoxygenase-lacking soybean seeds to accelerated aging and their chemiluminescence levels. *Phytochemistry* 29: 3739–42

105. Wei ZM, Laby RJ, Zumoff CH, Bauer DW, He SY, et al. 1992. Harpin, elicitor of the HR produced by the plant pathogen *Erwinia amylovora*. *Science* 257: 85–88

106. Yim MB, Chock PB, Stadtman ER. 1990. Copper, zinc superoxide dismutase catalyzes hydroxyl radical production from hydrogen peroxide. *Proc. Natl. Acad. Sci. USA* 87:5006–10

Annu. Rev. Phytopathol. 1995. 33:323–43

PATHOGEN-DERIVED RESISTANCE TO PLANT VIRUSES

George P. Lomonossoff
Department of Virus Research, John Innes Centre, Colney Lane, Norwich NR4 7UH, United Kingdom

KEY WORDS: virus, resistance, transgenic plant, regeneration, cosuppression

---

ABSTRACT

Transformation of plants with portions of viral genomes frequently gives rise to lines of plants that are resistant to the virus from which the sequence was derived. This phenomenon has been termed "pathogen-derived resistance." The nature of the resistance obtained is variable and can be either protein- or RNA-mediated. RNA-mediated resistance often protects against very high levels of inoculum but is highly specific; protein-mediated resistance generally offers lower level but broader spectrum resistance. The mechanism of protein-mediated resistance is probably specific to the protein concerned, while RNA-mediated resistance appears to operate by a mechanism similar to that of cosuppression. The type of resistance obtained is governed, at least in part, by the way the transgene is inserted into the plant chromosomes.

INTRODUCTION

Virus diseases cause serious losses worldwide in horticultural and agricultural crops. For example, virus diseases of rice in Southeast Asia can be devastating, causing losses of more than $\$1 \times 10^9$ per year (31). Conventional breeding programs to develop resistance are effective, but also protracted and expensive, and the resistance can be circumvented by virus variation. Alternative and additional approaches to protecting plants from infection with viruses have made use of the techniques of plant transformation and regeneration. These techniques have enabled the introduction of DNA sequences from "foreign"

0066-4286/95/0901-0323$05.00

organisms into the plant genome. The number of plant species amenable to transformation and regeneration has increased greatly in recent years (14), and it will likely be possible to produce fertile transformants of most, if not all, significant crops in the not-too-distant future.

There are a number of ways in which resistance could potentially be achieved by using plant transformation and regeneration; several of these methods have been explored in considerable detail. For example, plant transformation might allow the natural virus resistance genes present in one plant species to be introduced into another species. However, the problems in identifying and characterizing such resistance genes have prevented this approach from being used so far. Other approaches have included introducing sequences either corresponding to ameliorative satellite RNAs, encoding antiviral proteins, or antisense RNAs corresponding to segments of viral genomes. Details of such strategies can be found in recent reviews (5, 83, 84) and are not discussed further here.

An alternative approach to engineering resistance was proposed by Sanford & Johnson (75); they suggested the possibility of engineering resistance by transforming a susceptible plant with genes derived from the pathogen itself. This form of resistance, which they termed "parasite-derived resistance" (subsequently termed "pathogen-derived resistance": PDR[1]) was envisaged to operate through the expression of the viral gene product at either an inappropriate time, in inappropriate amounts, or in an inappropriate form during the infection cycle, thereby perturbing the ability of the pathogen to sustain an infection. The first illustration that PDR is indeed a viable way of producing virus-resistant plants was provided by the experiments of Powell-Abel et al (69), who demonstrated that tobacco plants transgenic for, and expressing, the TMV coat protein were resistant to infection with the virus. This discovery opened a new field of plant science research, and many papers have been published in the past decade illustrating the utility of PDR. This review aims to use the available information to draw some conclusions as to how the phenomenon might operate.

APPLICATION OF PDR

Viruses To Which PDR Has Been Successfully Applied

The majority of viruses against which PDR has been successfully developed have genomes that consist of positive-strand RNA. These viruses include

[1]Abbreviations used: PDR, pathogen-derived resistance; AlMV, alfalfa mosaic virus; BMV, brome mosaic virus; CMV, cucumber mosaic virus; CyRSV, cymbidium ringspot virus; PEBV, pea early browning virus; PLRV, potato leafroll virus; PVS, potato virus S; PVX, potato virus X; PVY, potato virus Y; SHMV, sunn-hemp mosaic virus; TEV, tobacco etch virus; TMV, tobacco mosaic virus; TVMV, tobacco vein mottling virus; TSWV, tomato spotted wilt virus.

members of the tobamo-, cucumo-, potex-, poty-, luteo-, carla-, ilar-, tobra-, nepo-, and alfalfa mosaic virus groups. In addition, the concept has also been applied to the ambisense RNA tospovirus group (18, 67). There is currently a single report of the successful application of PDR to a member of the geminivirus group of single-stranded DNA–containing plant viruses (41). Although plants transgenic for portions of caulimovirus genomes have been produced (4, 76), there have been no reports to date of the effectiveness of PDR against double-stranded DNA–containing plant viruses. For a recent list of viruses against which PDR has been developed, the reader should consult Table 3.2 in Reference 29.

Viral Sequences That Can Confer Resistance

Initially, all attempts to use PDR to produce virus-resistant plants followed the lead of Powell-Abel et al (69) by making plants transgenic for viral coat protein genes. In a large number of cases transgenic lines could be regenerated that showed greater or lesser degrees of resistance when challenged with the virus from which the gene was derived. (For detailed accounts of the production and properties of such plants, see References 7, 24, 29.)

The second portion of a viral genome shown to be capable of conferring resistance was the gene encoding the viral replicase. This was, as in coat protein–mediated protection, first demonstrated with TMV (27). It was subsequently found that the equivalent sequences from tobra-, cucumo-, potex-, tombus-, and potyviruses could also confer resistance, and the phenomenon was termed replicase-mediated resistance (for recent reviews of the phenomenon, see References 13, 49). Other portions of plant viruses also can give rise to a resistant phenotype. Such portions include genes encoding the NIa proteases of potyviruses (53, 82), the helper component (HC) of potyviruses (unpublished data quoted in Reference 82), the movement proteins of tobamoviruses, bromoviruses, and potexviruses (8, 42, 54), and the 3′ noncoding region of a tymovirus (87). Probably the simplest conclusion is that any part of a plant viral genome can potentially give rise to PDR.

CHARACTERISTICS OF PDR

Level of Resistance Obtained

Variable levels of resistance have been observed in plants transgenic for viral coat proteins, ranging from resistance to only low levels of inoculum (approximately 1μg/ml virus), as seen with plants transgenic for the TMV coat protein (64), to resistance to quite high levels (25–50μg/ml virus) of inoculum, as seen in the case of CMV (17). Even different constructs derived from the same coat protein gene can give different levels of protection. For example, lines of plants transgenic for a full-length copy of the TEV coat protein showed little or no delay of symptoms when challenged with TEV, though there appeared to be a slight attenuation of symptoms (44). By contrast, plants harboring some truncated forms of the coat protein showed significant delays in the appearance of symptoms. The ability of coat protein–mediated resistance to protect plants against inoculation with viral RNA is also variable. For example, plants transgenic for the TMV coat protein are not resistant when the inoculum is applied as RNA (64). This situation is also found with plants transgenic for the coat protein of AlMV (47). By contrast, plants transgenic for the coat proteins of a number of other viruses, such as PVX (30) and PVS (52), are equally resistant to infection with virions or RNA.

Transformation of a plant with sequences derived from the replicase gene can have widely differing consequences. In many cases plants that are highly resistant to the virus from which the sequence was derived can be regenerated (2, 3, 9, 19, 27, 50, 51, 74). However, plants transgenic for sequences derived from the replicases of AlMV and BMV (58, 79) were not only fully susceptible to subsequent challenge but could also support the replication of incomplete mixtures of the viral RNAs, i.e. the transgene-derived products could complement functions missing from the viral genome. This dramatic difference in phenotype was initially ascribed to the fact the resistant plant usually contained aberrant versions of the replicase gene. However, the finding that an intact version of the replicase gene from PVX could confer high levels of resistance (9) makes this explanation unlikely. The possible reasons for the remarkable variation in phenotype in plants transgenic for replicase sequences are discussed later.

Specificity of Resistance

Generally, PDR tends to be fairly specific for the virus from which the transgene was isolated. For example, tobacco plants transgenic for the coat protein of the U_1 strain of TMV are most resistant to that strain of TMV and its close relatives, show less resistance to more distantly related tobamoviruses (63), and display little or no resistance to unrelated viruses (1). Likewise, plants transgenic for the AlMV coat protein were not protected against the unrelated virus, TMV (47). However, there are instances where the resistance conferred by a coat protein gene is quite wide; for example, Namba et al (60) found that the coat protein from a subgroup II strain of CMV (strain WL) provides protection against several subgroup I strains, and Maiti et al (53) found that plants transgenic for the TVMV coat protein protected plants against infection by other potyviruses. The resistance obtained in plants transgenic for viral replicases or portions thereof is, if anything, more specific than that found with coat protein transgenics. For example, Golemboski et al (27) found that plants transgenic for the 54-K gene of the U_1 strain of TMV, although highly resistant to that strain and mutants derived from it, were not resistant to other tobamoviruses or unrelated viruses. Likewise, plants transgenic for portions of the replicases of PEBV, PVX, and CMV are only resistant to the strain of the virus, and its close relatives, from which the transgene was isolated (50, 51, 88). An exception to this high degree of specificity, however, was found in plants transgenic for the TMV 183-kDa replicase gene which is interrupted by a bacterial insertion element (19). A number of lines of plants transgenic for the modified replicase were highly resistant not only to the strain of TMV from which the gene was isolated but also to distantly related tobamoviruses. The reason for this broadening of specificity is unclear.

The broadest resistance reported to date is seen in plants transgenic for viral movement proteins. Plants transgenic for the movement protein of BMV are resistant to infection with TMV (54), and plants transgenic for the movement protein of PVX have broad-spectrum resistance against systemic infection by plant viruses with a triple gene block (8). Even more strikingly, plants transgenic for a deleted version of the TMV movement protein are resistant to a wide range of plant viruses (16, 42).

Site in the Plant at Which PDR Operates

Protoplasts isolated from plants displaying PDR generally show a resistance phenotype, although the way the resistance is manifested can differ. For example, TMV inoculation of tobacco protoplasts isolated from plants transgenic for the viral coat protein resulted in infection of a lower proportion of cells than is seen with protoplasts isolated from nontransgenic tobacco (71). However, the protoplasts that did become infected accumulated virus to the same level as the nontransgenic control. Similarly, protoplasts isolated from alfalfa plants transgenic for the AlMV coat protein were markedly less susceptible to infection by AlMV (33), and protoplasts from plants transgenic for an untranslatable version of the TEV coat protein gene did not support virus replication (45). Protoplasts isolated from plants transgenic for replicase genes of TMV, CMV, PVX, and CyRSV appear to be highly resistant to infection by the respective virus (10, 12, 50, 74). In these cases it appears that viral replication is suppressed in every inoculated protoplast. The only documented instance of resistance in whole plants but not in protoplasts is in the case of the resistance to TMV found in tobacco transgenic for, and expressing, the BMV movement protein (54). However, since the protein concerned is involved in cell-to-cell movement, which does not occur in protoplasts, this result is perhaps not surprising.

From the above, it appears safe to conclude that PDR against plant viruses generally works at the single cell level. However, this is an oversimplification. For example, the presence of the TMV coat protein gene in transgenic plants appears to limit the spread of the virus after initial infection with TMV RNA

(85). Evidence also suggests that the presence of the transgene in plants transgenic for a portion of the CMV replicase, as well as suppressing replication of the virus at the single cell level, also inhibits systemic movement of the virus (10).

THE MECHANISM OF PDR

Problems in the Elucidation of the Mechanism

A major problem encountered in determining the mechanism of PDR is the essentially random nature of the transformation process. In most cases, the transformation is mediated by *Agrobacterium*, although particle bombardment is increasingly being used. Both methods result in the incorporation of a variable number of copies of the gene of interest into one or several different locations in the host genome in various different conformations. A large number of independent lines of plants can thus be regenerated from a single transformation experiment; each line may have the transgene incorporated in a different copy number and in different locations. Lines of plants with very different phenotypes may result, even though each contains at least a single copy of the transgene. These phenotypic differences are demonstrated in the case of plants transformed with a portion of the replicase gene from PEBV (51). Twenty progeny plants obtained by the self-fertilization of an original transgenic line that showed a high level of resistance to PEBV were assessed for resistance and their chromosomal DNA was analyzed by southern blotting to assess the copy number and arrangement of the transgene. The results (Figure 1) showed that resistance was correlated with the presence of a 5.5-Kb *HindIII* restriction enzyme fragment, and the two lines [9 and 13] lacking this fragment were fully virus susceptible.

The random nature of transgene insertion into the host chromosome also makes it very difficult to carry out revealing studies with mutants. For example, if one wishes to study the effect of a deletion in a transgene on its ability to evoke a resistance phenotype, it is necessary to repeat the whole transformation/regeneration procedure to produce entirely new lines of plants harboring the deleted gene. It is highly improbable that any of the new lines will contain the new form of the transgene inserted in precisely the same manner as in the original line. Thus any phenotypic difference between plants transgenic for the original and mutant transgene are as likely to be due to differences in the way the genes are integrated as they are to be due to the presence of the mutation.

Protein or RNA-Mediated Protection?

One of the original rationales for constructing plants transgenic for the TMV coat protein was that the plants may be protected from subsequent challenge

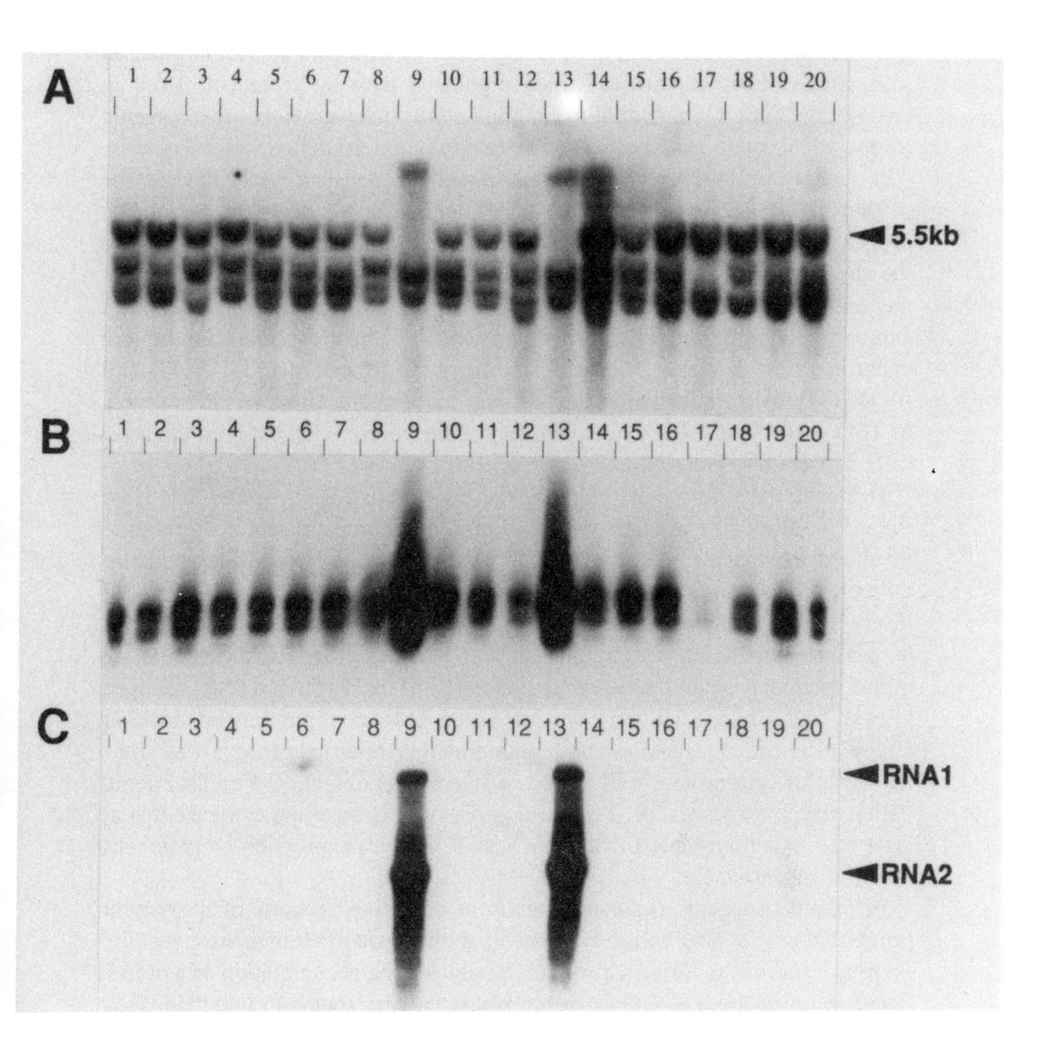

Figure 1 Analysis of 20 self-fertilized progeny plants derived from a line of *Nicotiana benthamiana* plants transgenic for the PEBV 54-K gene. *A*: Southern blot analysis of genomic DNA isolated from the plants. The DNA was digested with *HindIII* and the blot probed with a PEBV 54-K gene-specific probe. The position of the 5.5-kb fragment associated with resistance is indicated. Panel *B*: Southern blot analysis of RT-PCR amplified 54-K-specific mRNA. *C*: Northern blot analysis of plants inoculated with PEBV. The blot was probed with sequences specific to the two viral RNAs, the positions of which are marked. Only those plants susceptible to infection give a detectable signal. (Reproduced, with modification, from Reference 51.)

with the virus by a mechanism similar to that which occurs in cross-protection (69). Since previous work (77) had implicated the coat protein of TMV in cross-protection, it was natural to examine plants that were not only transgenic for the coat protein gene but were also expressing detectable levels of protein. Although Powell-Abel et al (69) examined the resistance to TMV of the progeny of a self-fertilized transgenic line and concluded that only those progeny lines expressing the coat protein ("expressors") showed resistance, the nonexpressing lines probably did not harbor any copies of transgene. Thus the analysis did not provide direct evidence for the requirement for protein expression as opposed to the mere presence of the transgene. Subsequent experiments with plants transgenic for the coat proteins of AlMV (47) and PVX (30) only examined plants expressing coat protein. However, in the cases of TMV, AlMV, and PVX, there appears to be a correlation between the level of coat protein expression and resistance (30, 47, 70).

Evidence that the accumulation of the coat protein itself, rather than, say, the mRNA encoding it, is responsible for at least some cases of coat protein–mediated resistance has been provided by examining the effects of temperature on protein levels and resistance. In both TMV (62) and CMV (65) the decreased levels of coat protein found at elevated temperatures are correlated with a decrease in resistance. To examine further whether protein expression is required for coat protein–mediated resistance to TMV, Powell et al (70) produced lines of plants transgenic for a version of the TMV coat protein gene lacking an initiation codon. Transformants containing this modified gene showed similar levels of coat protein mRNA to transformants containing the intact gene, but displayed no resistance. This finding provides compelling evidence that at least in plants transgenic for the TMV coat protein, expression of protein is required for resistance.

In plants transgenic for heterologous or defective versions of movement proteins there is also some evidence that the protein itself is responsible. Lapidot et al (42) reported a correlation between the accumulation of a defective version of the TMV movement protein and resistance to infection. Furthermore, it has been shown that only a dysfunctional form of the protein can give rise to resistance, whereas the presence of wild-type protein actually potentiates virus infection. Thus the defect in the movement protein has the properties of a dominant negative mutation (32).

The first indications that expression of protein from a transgene is not a prerequisite for resistance came from the finding of lack of correlation between protein levels and resistance found in lines of plants transgenic for, and expressing, the coat proteins of PLRV and PVY (38, 43). Indeed, some lines of plants transgenic for the PLRV coat protein and showing high levels of resistance to the virus did not contain detectable levels of coat protein (39). Subsequently, it was demonstrated that resistance to TSWV could be induced by transforming plants with a translationally defective version of the N gene of the virus (18). Perhaps the most striking evidence that protein synthesis is not a prerequisite for resistance is provided by the case of TEV. As mentioned previously, plants transgenic for, and expressing, an intact or N-terminally deleted form of the coat protein of TEV show little or no resistance to virus infection (44). However, the plants showed a "recovery" phenotype in which newly emerging leaves were devoid of symptoms (the analysis of such tissue is discussed later). By contrast, plants transgenic for an untranslatable version of the coat protein showed a high level of resistance to viral infection (44, 45). This result is completely the opposite of that obtained with TMV (70) and it appears difficult, if not impossible, to reconcile the two observations. This inconsistency has led to the conclusion that there may be two types of resistance that can be induced in transgenic plants—protein- or RNA-mediated protection.

Additional evidence that there are indeed two forms of resistance was provided by studies on plants transgenic for versions of the nucleocapsid (N) gene of TSWV. In lines of plants transgenic for a translatable form of the N gene, resistance to the virus from which the gene had been isolated (the homologous virus) was inversely correlated with the level of protein expression, whereas resistance to a distantly related virus was positively correlated (67). Pang et al (68) subsequently demonstrated that several lines of plants transgenic for an untranslatable form of the TSWV N gene showed a high degree of resistance against the homologous virus but no resistance to the more distantly related ones. Analysis of the N gene–specific mRNA levels in various lines transgenic for the untranslatable version of the gene showed that the greatest resistance was found in the plants accumulating the lowest levels of transcript. From these results it was concluded that two types of resistance were induced by the presence of the same transgene: one that requires expression of the protein and gives low level but fairly broad-spectrum resistance and the second that is RNA-mediated and provides high-level and highly specific resistance. In the case of replicase-mediated resistance, it has generally proved impossible to detect expression of any protein from the transgene (see Table 1 in Reference 6). The only exception was found in plants transgenic for a mutated form of the replicase of PVX, where the amount of protein found correlated inversely with the level of resistance (50). The only evidence that protein expression is required for replicase-mediated resistance was provided by transient expression studies (11). In these studies, protoplasts from non-transgenic tobacco were simultaneously transfected with TMV RNA and intact or mutated forms of the TMV replicase sequence (the 54-K gene). Although the presence of the wild-type sequence markedly inhibited replication, a frameshift mutant that could direct the synthesis of a much truncated protein had no effect. Since the wild-type and frameshift mutants differed only by a single nucleotide, the most obvious interpretation is that expression of a protein is required for resistance. However, it is possible or even probable (see below) that transient expression studies are not good models for what is occurring in transgenic plants, and it has been suggested (6) that all cases of replicase-mediated resistance may, in fact, be examples of RNA-mediated protection.

Is Transcription of the Transgene Required?

Though PDR that does not require the expression of protein is now generally referred to as "RNA-mediated" resistance, there is little direct evidence that transcription is actually required. It is possible, at least theoretically, that the presence of transgene DNA per se in a plant could be responsible for resistance, a possibility originally raised in connection with TMV replicase-mediated resistance (48). The only direct evidence that transcription is required to obtain resistance was provided by studies in which plants were transformed with a promoterless construct containing the TSWV N gene (68). All 12 lines transgenic for this construct failed to produce detectable amounts of transgene-specific mRNA, and none of the lines tested was resistant. Though this result is indicative that transcription is necessary, it is very common to produce lines that are not resistant even when plants are transformed with promoter-containing constructs. Thus, to provide compelling evidence that a promoterless construct cannot give resistance requires analysis of a very large number of transformed lines.

An additional difficulty in using promoterless constructs is that it is always possible that a promoterless sequence might integrate into the host genome in such a way that transcripts are produced off a nearby endogenous promoter. Conceivably, an apparently promoterless transgene could give rise to a resistant phenotype even if transcription is required. To eliminate this possibility, it will be necessary to carry out a detailed analysis of the transcripts derived from the "promoterless" transgene.

Potential Mechanism(s) for Protein-Mediated Resistance

By far the most extensively studied example of protein-mediated resistance is that provided by the coat protein of TMV. The early observation that the resistance is effective against virus particles but not against viral RNA suggested that the resistance operates by inhibiting a step in the viral replication cycle prior to the release of viral RNA from the virus particle. Evidence that it is an early step in viral disassembly which is inhibited by the presence of coat protein was provided by experiments in which coat protein–expressing plants or protoplasts prepared from them were inoculated with TMV virions pretreated at pH8.0 (71). This treatment results in the removal of a few molecules of coat protein from the 5′ terminus of the viral RNA, allowing ribosomes to bind to the 5′ noncoding region of the RNA (59). The pretreated virions were able to initiate an infection of both whole plants and protoplasts as efficiently as viral RNA; this suggests that it is the initial removal of viral subunits that is inhibited by the presence of coat protein. Support for this hypothesis is provided by the observation that inoculation of coat protein–expressing protoplasts with TMV resulted in the appearance of far fewer disassembly complexes ("striposomes") than were found when protoplasts not expressing coat protein were inoculated (86). Furthermore, it has been shown, using promoters with differing tissue specificities, that the coat protein must accumulate in the initially infected tissue of a plant for TMV coat protein–mediated resistance to be effective (unpublished data quoted in Reference 73).

Two models have been proposed to explain how the presence of transgene-derived coat protein might inhibit the early stages of virus disassembly (72). First, the presence of transgene-derived coat protein in the cytoplasm of a cell may tip the disassembly-assembly equilibrium of incoming virus particles in favor of assembly and thus effectively inhibit uncoating. Second, there may be some kind of receptor or uncoating site within cells that is responsible for initiating the disassembly of virus particles. Transgene-derived coat protein is then envisaged as blocking this site prior to infection, preventing virions from attaching to it. At present there is no clear evidence pointing to which, if either, of these mechanisms is correct.

There is evidence that TMV coat protein–mediated resistance also operates at a stage in the virus infection cycle subsequent to uncoating. Osbourn et al (66) reported that protoplasts from TMV coat protein–expressing tobacco are susceptible to infection by the related tobamovirus SHMV even if the viral RNA is encapsidated by the U_1 coat protein. This susceptibility was interpreted as implying that as well as inhibiting viral disassembly, the presence of coat protein inhibits a later step in replication that requires the presence of the homologous (U_1) RNA. As discussed earlier, the presence of transgene-derived coat protein also appears to have an inhibitory effect on the long-distance spread of the virus (85). The nature of this inhibition has been investigated by grafting experiments in which segments of stem from coat protein–expressing plants were inserted between the rootstock and apical region of plants not expressing coat protein. Following inoculation of the rootstock, the development of an infection in the apical region was monitored. Curiously, the presence of the section of coat protein–expressing stem only inhibited the development of such an infection if the section had a leaf attached. The reason for this is unclear.

Assuming that the resistance found in plants transgenic for a defective form of the TMV movement protein is protein mediated, several explanations can be advanced (42). The mutant, dysfunctional protein might retain its ability to bind to sites (plasmodesmata?) in the cells and compete with the functional protein encoded by the incoming virus. In support of this notion, it was reported that the defective movement protein was able to increase the size exclusion limit of plasmodesmata to a lesser extent that wild-type protein. Alternatively, the mutation might affect the nucleic acid binding properties of the protein (15). Thus the dysfunctional protein might bind nonproductively to the viral RNA and thus compete with the functional protein.

In cases where the protection afforded by the N gene of tospoviruses is believed to be due to the protein itself, it has been proposed that this might operate by causing aberrant replication (26). During the replication of negative-strand viruses the amount of free N protein determines whether the viral polymerase acts to transcribe mRNA from the genome (at low N protein levels) or acts to replicate the genome (high N protein levels). The presence of transgene-derived N protein early in infection could lead to a premature switch of the polymerase from the transcription to the replicative mode, leading to an abortive infection.

Potential Mechanism(s) of RNA-Mediated Resistance

Lindbo & Dougherty (46) carried out a detailed analysis of the "recovered" tissue from TEV-infected tobacco plants transgenic for a full-length or N-terminally truncated form of the TEV coat protein. The recovered tissue did not contain any virus and could not be infected by subsequent inoculation with the virus; protoplasts isolated from such tissue were also unable to support TEV replication. Prior infection of the transgenic plants with TEV was found to be necessary to induce the virus-resistant state, even though equivalent tissue from transgenic plants not previously inoculated with TEV was fully susceptible. Examination of the expression of the transgene in recovered tissue revealed the absence of detectable coat protein and much (12–22-fold) reduced levels of transcript as compared with that found in uninoculated transgenic tissue. A similarly reduced level of transcript has been reported in other systems where extreme resistance was manifested. For example, in plants transgenic for the N gene of TWSV the highest level of resistance was found in plants accumulating low levels of transcript (68). The effect is illustrated in Figure 1 in the case of plants transgenic for the 54-K gene of PEBV. Lines that are highly resistant to the virus contain lower amounts of 54-K-specific transcripts than plants that are not resistant (lines 9 and 13).

The lower level of transcripts found in resistant tissue could be due either to a reduced rate of transcription of the transgene or to a higher rate of RNA degradation. To discriminate between these two possibilities, Lindbo et al (46) performed runoff transcription assays. These assays showed that the rate of transcription of the transgenes in recovered tissue was not reduced compared to the rate found in uninoculated transgenic tissue. These observations led Lindbo et al (46) to postulate that the antiviral state found in recovered tissue is due to the induction of a specific cellular RNA-degradation mechanism.

This mechanism would be activated by the presence of "unacceptably" high levels of a particular transcript and act to reduce the level of such a transcript. Since the transgene is comprised of virus-specific sequences, the induced mechanism would also operate against the genomic RNA from the virus from which the transgene was derived. The mechanism proposed above explains why, in many cases, it has been found that virus resistance is inversely correlated with the steady-state levels of transcripts in transgenic plants. It also explains the specificity of the resistance, since the mechanism requires a high degree of homology between the transgene and the incoming viral nucleic acid.

The need for prior infection to induce the antiviral state in recovered tissue is presumably a consequence of the inability of the original transgene on its own to produce high enough levels of transcript to provoke the degradation mechanism. By transforming plants with an untranslatable version of the TEV coat protein linked to the 5′ untranslated region of TEV, Dougherty et al (22) isolated lines of plants that were intrinsically highly resistant to TEV infection in addition to lines displaying the recovery phenotype. Untranslatable transcripts may therefore be more effective at provoking the degradation mechanism than translatable transcripts.

Smith et al (78) produced double-haploid (DH) lines of tobacco expressing translatable and untranslatable versions of the PVY coat protein. Analysis of the lines provided confirmation that the transcript level was inversely correlated with resistance and that untranslatable constructs are more effective at conferring high levels of resistance. By using DH plants, Smith et al (78) produced isogenic lines of tobacco that were identical in terms of the copy number and organization of the transgene. Unexpectedly, the different lines showed differing levels of resistance. The level of resistance was inversely correlated with transcript level and appears to be related to the methylation state of the transgene DNA. This result implies that one should not necessarily expect a consistent resistance phenotype even in homozygous transgenic plants. Indeed, in the DH isogenic lines analyzed by Smith et al (78), the lines displaying high levels of resistance always gave rise to a proportion of susceptible progeny.

The nature of the RNA-degrading mechanism is unclear. Lindbo et al (46) suggested the possible involvement of a 130-kDa RNA-dependent RNA-polymerase present in plant cells. This enzyme has no known specificity, at least in vitro, and copies added template RNA into small complementary fragments. It is stimulated during virus infection and its presence has long bedeviled the characterization of virus-encoded polymerases (20, 21, 81). It was hypothesized that the enzyme might use overexpressed RNAs as a template and that the resultant small fragments of negative-sense RNA might hybridize with further copies of the RNA making them potential targets for degradation by a specific RNAse.

Relationship of PDR to Cosuppression

Cosuppression is the phenomenon in which the presence of a transgene leads to the suppression of expression ("silencing") of both the transgene and its homologous endogenous counterpart. It first came to prominence when petunia plants were transformed with an additional copy of the gene encoding chalcone synthase, a key enzyme in the synthesis of floral pigments. Instead of resulting in the anticipated increase in the level of pigment, many (up to 50%) of the transformed plants showed reduced levels or the absence of pigment in the flowers (56, 61, 80). This lack of pigment was correlated with low steady-state levels of transcripts derived from both the inserted and endogenous gene. The rates of transcription were, however, found to be unaffected (40, 57). This phenomenon has subsequently been described for many other transgene/plant combinations (for reviews, see References 23, 25). Numerous mechanisms have been postulated to explain cosuppression, including alterations to the structure of the chromatin, competition between genes for nondiffusible factors in the nucleus, methylation of specific DNA sequences, and perturbations to RNA metabolism. The cellular processes that could conceivably be affected so as to cause cosuppression are illustrated in Figure 2.

Among the hypotheses advanced to explain cosuppression is the "biochemical switch" mechanism proposed Matzke & Matzke (55). This mechanism proposes the existence of genetic regulatory circuits that can exist in alternative stable states. The conversion between states would be very sensitive to a threshold level of a gene product such as the accumulated levels of transcribed RNA. When the threshold level is exceeded, a highly specific RNA-degradation mechanism would be activated, dramatically reducing the levels of a specific mRNA in the cell. In the case of cosuppression, the presence of extra copies of a gene would allow the critical threshold to be exceeded, leading to degradation of transcripts derived from all copies of the gene. This mechanism is clearly similar to that proposed by Lindbo et al (46) to explain the high level of resistance found in plants transgenic for nontranslatable copies of the TEV coat protein gene. Evidence for the relationship between cosuppression and PDR is provided by analysis of the organization of transgenes in plants. Generally, cosuppression appears to be correlated with the insertion of multiple copies of a transgene. In work with tobacco plants transgenic for the GUS gene, Hobbs et al (34, 35) showed that cosuppression was correlated with the presence of inverted repeats of T-DNA, whereas plants with single copies of T-DNA at a locus expressed high levels of GUS. Consistent with the existence of a relationship between cosuppression and PDR, Smith et al (78) provided evidence that high levels of resistance to PVY were correlated with the insertion of multiple copies of T-DNA; all lines of plants containing only a single copy of the T-DNA were susceptible. Strikingly, the 5.5-kb fragment that

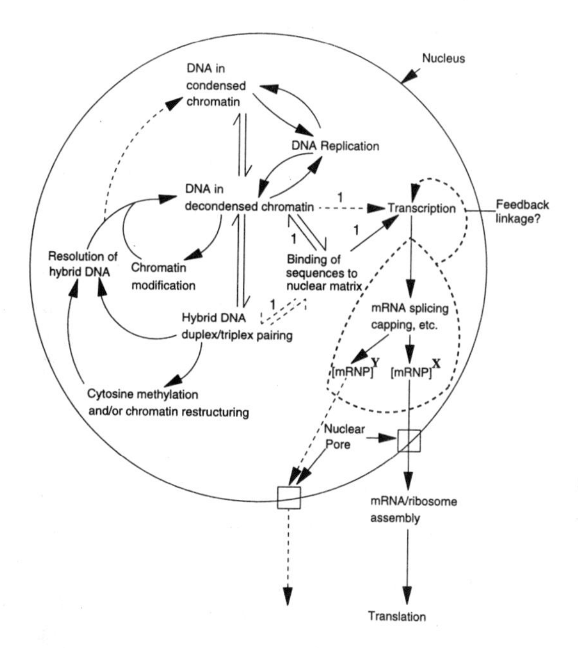

Figure 2 Illustration of the processes that could be affected so as to cause cosuppression. Steps involving an alteration to the structure of the chromatin are labeled 1. Posttranscriptional effects are envisaged as giving rise to aberrant ($mRNP^Y$) rather than normal ($mRNP^X$) ribonucleoprotein complexes. These aberrant complexes may be targets for RNA degradation. (Reproduced, with modification, from Reference 25.)

cosegregates with resistance and low levels of transcript accumulation in plants transgenic for the PEBV 54-K gene (Figure 1) has the correct size to represent an inverted repeat of that gene.

Hobbs et al (35) demonstrated that a cross between plants expressing high levels of GUS and those expressing low levels gave rise to progeny that were of the low-expressing phenotype. Thus the low-expression, cosuppressed phenotype is dominant. A similar phenomenom has recently been observed in plants transgenic for the replicase of PVX (D Baulcombe, personal communication). Plants exhibiting an extreme resistance to PVX and accumulating low levels of transgene transcript were crossed with plants containing the same transgene but exhibiting a less resistant phenotype and which accumulated higher levels of transcript. All progeny plants that contained copies of both genes showed the extreme resistance/low RNA phenotype, a result similar to that of Hobbs et al (35).

Control of Type of Resistance Obtained

Upon transformation with a virus-derived sequence, what determines whether one will obtain a protein- or RNA-mediated phenomenon? The results obtained with the TSWV (68) and TEV (44, 45) suggest that both types of resistance can be obtained in a single transformation experiment. Since the inserted sequence is constant, the different effects must be due to the manner in which the DNA is inserted.

The variability in GUS expression noted by Hobbs et al (34, 35) explains why different types of resistance might be obtained by transformation with the same viral sequence. If the sequence integrates as a single copy at a given locus, large amounts of mRNA will accumulate, leading, provided the mRNA is translatable, to the synthesis of a protein. The presence of the protein can then affect the ability of an incoming virus to initiate an infection either positively, as in plants expressing the replicase genes of AlMV and BMV (58, 79), or negatively, as in the case of viral coat proteins. In lines containing multiple insertions of the gene at the same locus, cosuppression can be induced leading to only low levels of transcripts and protein accumulating. These plants will probably manifest an RNA-mediated resistance.

Can the type of resistance be controlled? At a simple level, the answer is yes. If a plant is transformed with an untranslatable sequence from a virus, presumably a protein-mediated phenomenon is unlikely to be found. It also appears that untranslatable sequences are more efficient at evoking RNA-mediated protection than translatable sequences (46). Dougherty et al (22) suggested that untranslatable RNAs may in some way be recognized as aberrant and may be more efficient in stimulating a degradative pathway than translatable RNAs. One possible mechanism is that translatable mRNAs will bind ribosomes and be sequestered in polysomes and thus are removed from the pool of RNA sensed by the degradation mechanism. Assuming that the type of resistance obtained is closely correlated with the organization and copy number of the transgene, controlling the way a transgene integrates is potentially a means of determining whether one obtains protein- or RNA-mediated protection.

FUTURE PROSPECTS

The ability to produce virus-resistant plants is clearly of agronomic significance. A number of field trials are currently taking place to assess the usefulness of the resistance obtained with PDR (28, 36, 37). To date, these trials have been mainly carried out with plants transgenic for, and expressing, viral coat proteins, and the results have been encouraging. At first sight, the higher levels of resistance obtained with RNA-mediated resistance might appear to be of more practical use. However, this type of resistance is usually more specific than that mediated by protein and hence may be overcome more easily in a field situation owing to variations in the virus population. Furthermore, the lower levels of resistance obtained with, for example, expressed coat protein may not be particularly disadvantageous, given the low levels of inoculum likely to be encountered in the field situation. Clearly, comparative field tests are needed to resolve this question.

ACKNOWLEDGMENTS

I would like to thank David Baulcombe and Dick Flavell for helpful discussions, David Baulcombe for allowing me to quote unpublished data, and John Shaw for critically reading the manuscript.

Literature Cited

1. Anderson EJ, Stark DM, Nelson RS, Turner NE, Beachy RN. 1989. Transgenic plants that express the coat protein gene of TMV or AlMV interfere with disease development of non-related viruses. *Phytopathology* 12:1284–90
2. Anderson JM, Palukaitis P, Zaitlin M. 1992. A defective replicase gene induces resistance to cucumber mosaic virus in transgenic tobacco plants. *Proc. Natl. Acad. Sci. USA* 89:8759–63
3. Audy P, Palukaitis P, Slack SA, Zaitlin M. 1994. Replicase-mediated resistance to potato virus Y in transgenic tobacco plants. *Mol. Plant-Microbe Interact.* 7: 15–22
4. Baughman, GA, Jacobs, JD, Howell, SH. 1988. Cauliflower mosaic virus gene VI produces a symptomatic phenotype in transgenic tobacco plants. *Proc. Natl. Acad. Sci. USA* 85:733–37
5. Baulcombe D. 1994. Novel strategies for engineering virus resistance in plants. *Curr. Opin. Biotechnol.* 5:117–24
6. Baulcombe D. 1994. Replicase-mediated resistance: a novel type of virus resistance in transgenic plants? *Trends Microbiol.* 2:60–63
7. Beachy RN, Loesch-Fries S, Turner NE. 1990. Coat protein-mediated resistance against virus infection. *Annu. Rev. Phytopathol.* 28:451–74
8. Beck DL, van Dolleweerd CJ, Lough TJ, Balmori E, Voot DM, et al. 1994. Disruption of virus movement confers broad-spectrum resistance against systemic infection by plant viruses with a triple gene block. *Proc. Natl. Acad. Sci. USA* 91:10310–14
9. Braun CJ, Hemenway CL. 1992. Expression of amino-terminal portions or full-length viral replicase genes in transgenic plants confers resistance to potato virus X infection. *Plant Cell* 4:735–44
10. Carr JP, Gal-On A, Palukaitis P, Zaitlin M. 1994. Replicase-mediated resistance to cucumber mosaic virus in transgenic plants involves suppression of both virus replication in the inoculated leaves and long-distance movement. *Virology* 199: 439–47
11. Carr JP, Marsh LE, Lomonossoff GP, Sekiya ME, Zaitlin M. 1992. Resistance to tobacco mosaic virus induced by the 54-kDa gene sequence requires expression of the 54-kDa protein. *Mol. Plant-Microbe Interact.* 5:397–404
12. Carr JP, Zaitlin M. 1991. Resistance in transgenic plants expressing a nonstructural gene sequence of tobacco mosaic virus is a consequence of markedly reduced virus replication. *Mol. Plant-Microbe Interact.* 4:579–85
13. Carr JP, Zaitlin M. 1993. Replicase-mediated resistance. *Semin. Virol.* 4:339–47
14. Christou P. 1994. Genetic engineering

of crop legumes and cereals: current status and recent advances. *Agro Food Ind. Hi-Tech* 5(2):17–27
15. Citovsky V, Knorr D, Schuster G, Zambryski P. 1990. The P30 movement protein of tobacco mosaic virus is a single-strand nucleic acid binding protein. *Cell* 60:637–47
16. Cooper B, Lapidot M, Heick JA, Dodds JA, Beachy RN. 1995. A defective movement protein of TMV in transgenic plants confers resistance to multiple viruses whereas the functional analog increases susceptibility. *Virology* 206: 307–13
17. Cuozzo M, O'Connell K, Kaniewski W, Fang R-X, Chua N-H, Tumer NE. 1988. Viral protection in transgenic plants expressing the cucumber mosaic virus coat protein or its antisense RNA. *Bio/Technology* 6:549–57
18. de Haan P, Gielen JJL, Prins M, Wijkamp IG, van Schepen A, et al. 1992. Characterisation of RNA-mediated resistance to tomato spotted wilt virus in transgenic tobacco plants. *Bio/Technology* 10:1133–37
19. Donson J, Kearney CM, Turpen TH, Khan IA, Kurath G, et al. 1993. Broad resistance to tobamoviruses is mediated by a modified tobacco mosaic virus replicase transgene. *Mol. Plant-Pathogen Interact.* 6:635–42
20. Dorssers L, van der Meer J, van Kammen A, Zabel P. 1983. The cowpea mosaic virus RNA replication complex and the host-encoded RNA-dependent RNA polymerase-template complex are functionally different. *Virology* 125: 155-74
21. Dorssers L, Zabel P, van der Meer J, van Kammen A. 1982. Purification of a host-encoded RNA-dependent RNA polymerase from cowpea mosaic virus-infected cowpea leaves. *Virology* 116: 236–49
22. Dougherty WG, Lindbo JA, Smith HA, Parks TD, Swaney S, Proebsting WM. 1994. RNA-mediated virus-resistance in transgenic plants: exploitation of a cellular pathway possibly involved in RNA degradation. *Mol. Plant-Pathogen Interact.* 7:544–52
23. Finegan J, McElroy D. 1994. Transgene inactivation: plants fight back! *Bio/Technology* 12:883–88
24. Fitchen JH, Beachy RN. 1993. Genetically engineered protection against viruses in transgenic plants. *Annu. Rev. Microbiol.* 47:739–63
25. Flavell RB. 1994. Inactivation of gene expression in plants as a consequence of specific sequence duplication. *Proc. Natl. Acad. Sci. USA* 91:3490–96
26. Goldbach R, de Haan P. 1993. Prospects of engineered forms of resistance against tomato spotted wilt virus. *Semin. Virol.* 4:381–87
27. Golemboski DB, Lomonossoff GP, Zaitlin M. 1990. Plants transformed with a tobacco mosaic virus nonstructural gene sequence are resistant to the virus. *Proc. Natl. Acad. Sci. USA* 87:6311–15
28. Gonsalves D, Slightom JL. 1993. Coat protein-mediated protection: analysis of transgenic plants for resistance in a variety of crops. *Semin. Virol.* 4:397–405
29. Grumet R. 1994. Development of virus resistant plants via genetic engineering. *Plant Breed. Rev.* 12:47–79
30. Hemenway C, Fang R-X, Kaniewski W, Chua N-H, Tumer NE. 1988. Analysis of the mechanism of protection in transgenic plants expressing the potato virus X coat protein or its antisense RNA. *EMBO J* 7:1273–80
31. Herdt RW. 1991. Research priorities for rice biotechnology. In *Rice Biotechnology*, ed. GS Khush, GH Toenissen, pp. 19–54. Wallingford, UK: CAB Int.
32. Herskovitz I. 1987. Functional inactivation of genes by dominant negative mutations. *Nature* 229:219–22
33. Hill KK, Jarvis-Eagan N, Halk EL, Krahn KJ, Liao LE, et al. 1991. The development of virus-resistant alfalfa, *Medicago sativa* L. *Bio/Technology* 9: 373–77
34. Hobbs SLA, Kpodar P, DeLong CMO. 1990. The effect of T-DNA copy number, position and methylation on reporter gene expression in tobacco transformants. *Plant Mol. Biol.* 15:851–64
35. Hobbs SLA, Warkentin TD, DeLong CMO. 1993. Transgene copy number can be positively or negatively associated with transgene expression. *Plant Mol. Biol.* 21:17–26
36. Jongedijk E, Huisman MJ, Cornelissen BJC. 1993. Agronomic performance and field resistance of genetically modified, virus-resistant potato plants. *Semin. Virol.* 4:407–16
37. Kaniewski WK, Thomas PE. 1993. Field testing of virus resistant transgenic plants. *Semin. Virol.* 4:389–96
38. Kawchuk LM, Martin RR, McPherson J. 1990. Resistance in transgenic potato expressing the potato leafroll virus coat protein gene. *Mol. Plant-Microbe Interact.* 3:301–7
39. Kawchuk LM, Martin RR, McPherson J. 1991. Sense and antisense RNA-mediated resistance to potato leafroll virus

in Russet Burbank potato plants. *Mol. Plant-Microbe Interact.* 4:247–53
40. Kooter JM, Mol JNM. 1993. Trans-inactivation of gene expression in plants. *Curr. Opin. Biotechnol.* 4:166–71
41. Kunik T, Salomon R, Zamir D, Navot N, Zeidan M, et al. 1994. Transgenic tomato plants expressing the tomato yellow leaf curl capsid protein are resistant to the virus. *Bio/Technology* 12:500–4
42. Lapidot M, Gafney R, Ding B, Wolf S, Lucas WJ, Beachy RN. 1993. A dysfunctional movement protein of tobacco mosaic virus that partially modifies the plasmodesmata and limits virus spread in transgenic plants. *Plant J.* 4:959–70
43. Lawson C, Kaniewski W, Haley L, Rozman R, Newell C, et al. 1990. Engineering resistance to mixed virus infection in a commercial potato cultivar. *Bio/Technology* 8:127–34
44. Lindbo JA, Dougherty WG. 1992. Pathogen-derived resistance to a potyvirus: immune and resistant phenotypes in transgenic tobacco expressing altered forms of a potyvirus coat protein nucleotide sequence. *Mol. Plant-Pathogen Interact.* 5:144–53
45. Lindbo JA, Dougherty WG. 1992. Untranslatable transcripts of the tobacco etch virus coat protein gene sequence can interfere with tobacco etch virus replication in transgenic plants and protoplasts. *Virology* 189:725–33
46. Lindbo JA, Silva-Rosales L, Proebsting WM, Dougherty WG. 1993. Induction of a highly specific antiviral state in transgenic plants: implications for regulation of gene expression and virus resistance. *Plant Cell* 5:1749–59
47. Loesch-Fries LS, Merlo D, Zinnen T, Burhop L, Hill K, et al. 1987. Expression of alfalfa mosaic RNA 4 in transgenic plants confers virus resistance. *EMBO J* 6:1845–51
48. Lomonossoff GP. 1993. Virus resistance mediated by a non-structural gene sequence. In *Transgenic Plants: Fundamentals and Applications,* ed. A Hiatt, pp. 79–91. New York: Marcel Dekker
49. Lomonossoff GP, Davies JW. 1992. Replicase-mediated resistance to plant viruses. *Sci. Prog.* 76:537–51
50. Longstaff M, Brigneti G, Boccard F, Chapman S, Baulcombe D. 1993. Extreme resistance to potato virus X infection in plants expressing a modified component of the putative viral replicase. *EMBO J.* 12:379–86
51. MacFarlane SA, Davies JW. 1992. Plants transformed with a region of the 201-kiloDalton replicase gene of pea early browning virus are resistant to virus infection. *Proc. Natl. Acad. Sci. USA* 89:5829–33
52. MacKenzie DJ, Tremaine JH. 1990. Transgenic *Nicotiana debneyii* expressing viral coat protein are resistant to potato virus S infection. *J. Gen. Virol.* 71:2167-70
53. Maiti IB, Murphy JF, Shaw JG, Hunt AG. 1993. Plants that express a potyvirus proteinase gene are resistant to virus infection. *Proc. Natl. Acad. Sci. USA* 90:6110–14
54. Malyshenko SI, Kondakova OA, Navarova JV, Kaplan IB, Taliansky ME, Atabekov JG. 1993. Reduction of tobacco mosaic virus accumulation in transgenic plants producing non-functional viral transport proteins. *J. Gen. Virol.* 74:1149–56
55. Matzke MA, Matzke AJM. 1993. Genomic imprinting in plants—parental effects and transinactivation phenomena. *Annu. Rev. Plant Physiol. Plant Mol. Biol.* 44:53–76
56. Mol JNM, Stuitje AR, van der Krol A. 1989. Genetic manipulation of floral pigmentation genes. *Plant Mol. Biol.* 13:287–94
57. Mol J, van Blokland R, Kooter J. 1991. More about co-suppression. *Trends Biotechnol.* 9:182–83
58. Mori M, Mise K, Okuno T, Furusawa I. 1992. Expression of brome mosaic virus-encoded replicase genes in transgenic tobacco plants. *J. Gen. Virol.* 73:169–72
59. Mundry KW, Watkins PAC, Ashfield T, Plaskitt KA, Eiseler-Walter S, Wilson TMA. 1991. Complete uncoating of the 5′ leader sequence of tobacco mosaic virus RNA occurs rapidly and is required to initiate co-translational virus disassembly in vitro. *J. Gen. Virol.* 72:769–77
60. Namba S, Ling K, Gonsalves C, Gonsalves D, Slightom JL. 1991. Expression of a gene encoding the coat protein of cucumber mosaic virus (CMV) strain WL appears to provide protection to tobacco plants against infection by different CMV strains. *Gene* 107:181–88
61. Napoli C, Lemieux C, Jorgensen R. 1990. Introduction of a chimeric chalcone synthase gene into petunia results in reversible co-suppression of homologous genes in trans. *Plant Cell* 2:279–89
62. Nejidat A, Beachy RN. 1989. Decreased levels of TMV coat protein in transgenic tobacco plants at elevated temperatures reduces resistance to TMV infection. *Virology* 173:531–38

63. Nejidat A, Beachy RN. 1990. Transgenic tobacco plants expressing a tobacco mosaic virus coat protein gene are resistant to some tobamoviruses. *Mol. Plant Microbe Interact.* 3:247–51
64. Nelson RS, Powell-Abel P, Beachy RN. 1987. Lesions and virus accumulation in inoculated transgenic tobacco plants expressing the coat protein gene of tobacco mosaic virus. *Virology* 158:126–32
65. Okuno T, Nakayama M, Furusawa I. 1993. Cucumber mosaic virus coat protein-mediated resistance. *Semin. Virol.* 4:357–61
66. Osbourn JK, Watts JW, Beachy RN, Wilson TMA. 1989. Evidence that nucleocapsid disassembly and a later step in virus replication are inhibited in transgenic tobacco protoplasts expressing TMV coat protein. *Virology* 172:370–73
67. Pang SZ, Nagpala P, Wang M, Slightom JL, Gonsalves D. 1992. Resistance to heterologous isolates of tomato spotted wilt virus in transgenic tobacco expressing its nucleocapsid gene. *Phytopathology* 82:1223–29
68. Pang SZ, Slightom JL, Gonsalves D. 1993. Different mechanisms protect transgenic tobacco against tomato spotted wilt and impatiens necrotic spot tospoviruses. *Bio/Technology* 11:819–24
69. Powell-Abel P, Nelson RS, De B, Hoffmann N, Rogers SG, et al. 1986. Delay of disease development in transgenic plants that express the tobacco mosaic virus coat protein. *Science* 232: 738–43
70. Powell PA, Sanders PR, Tumer N, Fraley RT, Beachy RN. 1990. Protection against tobacco mosaic virus infection in transgenic plants requires accumulation of coat protein rather than coat protein RNA sequences. *Virology* 175: 124–30
71. Register JC, Beachy RN. 1988. Resistance to TMV in transgenic plants results from interference with an early event in infection. *Virology* 166:524–32
72. Register JC, Powell PA, Nelson RS, Beachy RN. 1989. Genetically engineered cross-protection against TMV interferes with initial infection and long distance spread of the virus. In *Molecular Biology of Plant-Pathogen Interactions*, ed. B Staskawicz, P Ahlquist, O Yoder, pp. 269–81. New York: Liss
73. Reimann-Philipp U, Beachy RN. 1993. The mechanism(s) of coat protein-mediated resistance against tobacco mosaic virus. *Semin. Virol.* 4:349-56
74. Rubino L, Lupo R, Russo M. 1993. Resistance to cymbidium ringspot tombusvirus infection in transgenic *Nicotiana benthamiana* plants expressing a full-length viral replicase gene. *Mol. Plant-Pathogen Interact.* 6:729–34
75. Sanford JC, Johnson SA. 1985. The concept of parasite-derived resistance: deriving resistance genes from the parasites own genome. *J. Theor. Biol.* 115: 395–405
76. Schoelz JE, Wintermantel WM. 1993. Expansion of viral host range through complementation and recombination in transgenic plants. *Plant Cell* 5:1669–79
77. Sherwood JL, Fulton RW. 1982. The specific involvement of coat protein in tobacco mosaic virus cross protection. *Virology* 119:150–58
78. Smith HA, Swaney SL, Parks TD, Wernsman EA, Dougherty WG. 1994. Transgenic plant virus resistance mediated by untranslatable sense RNAs: expression, regulation and fate of non- essential RNAs. *Plant Cell* 6:1441–53
79. Taschner PEM, van der Kuyl A, Neeleman L, Bol JF. 1991. Replication of an incomplete alfalfa mosaic virus genome in plants transformed with viral replicase genes. *Virology* 181:445–50
80. van der Krol AR, Mur LA, Beld M, Mol JNM, Stuitje AR. 1990. Flavonoid genes in petunia: addition of a limited number of gene copies may lead to a suppression of gene expression. *Plant Cell* 2:291–99
81. van der Meer J, Dorssers L, van Kammen A, Zabel P. 1984. The RNA-dependent RNA polymerase of cowpea is not involved in cowpea mosaic virus RNA replication: immunological evidence. *Virology* 132:413–25
82. Vardi E, Sela I, Edelbaum O, Livneh O, Kuznetsova L, Stram Y. 1993. Plants transformed with a cistron of a potato virus Y protease (NIa) are resistant to virus infection. *Proc. Natl. Acad. Sci. USA* 90:7513–17
83. Wilson TMA. 1993. Strategies to protect crop plants against viruses: pathogen derived resistance blossoms. *Proc. Natl. Acad. Sci. USA* 90:3134–41
84. Wilson TMA, Davies JW. 1994. New roads to crop protection against viruses. *Outlook Agric.* 23:33–39
85. Wisniewski LA, Powell PA, Nelson RS, Beachy RN. 1990. Local and systemic spread of tobacco mosaic virus in transgenic tobacco. *Plant Cell* 2:559–67
86. Wu X, Beachy RN, Wilson TMA, Shaw JG. 1990. Inhibition of uncoating of tobacco mosaic virus particles in protoplasts from transgenic tobacco plants

that express the viral coat protein gene. *Virology* 179:893–95

87. Zaccomer B, Cellier F, Boyer J-C, Haenni A-L, Tepfer M. 1993. Transgenic plants that express genes including the 3′ untranslated region of the turnip mosaic virus (TYMV) genome are partially protected against TYMV infection. *Gene* 136:87–94
88. Zaitlin M, Anderson JM, Perry KL, Zhang L, Palukaitis P. 1994. Specificity of replicase mediated resistance to cucumber mosaic virus. *Virology* 201: 200–5

Annu. Rev. Phytopathol. 1995. 33:345–68

THE MOLECULAR BASIS OF INFECTION AND NODULATION BY RHIZOBIA: The Ins and Outs of Sympathogenesis

Herman P. Spaink
Clusius Laboratory, Institute of Molecular Plant Sciences, Leiden University, Leiden, The Netherlands

KEYWORDS: host specificity, plant-microbe interactions, signal molecules, symbiosis, organogenesis, plant defense

---

ABSTRACT

Bacteria belonging to the genera *Rhizobium, Bradyrhizobium,* and *Azorhizobium,* collectively known as rhizobia, penetrate the roots (or adventitious roots) of their leguminous host plants via tubular structures, the infection threads. During infection of the host plant they trigger the formation of a new organ, the root nodule, in which a differentiated form of rhizobia, the bacteroid, fixes nitrogen into ammonia, which can then be used by the plant. This review presents an update of the recent literature on the molecular biology of the infection and nodulation of plants by rhizobia, with special emphasis on results pertinent to other plant-microbe interactions. Particular attention is given to determinants of host specificity such as flavonoid and lipo-chitin oligosaccharide signal molecules.

INTRODUCTION

Bacteria belonging to the genera *Rhizobium, Bradyrhizobium,* and *Azorhizobium,* collectively known as rhizobia, penetrate the roots (or adventitious roots)

0066-4286/95/09015-0345$05.00

of their leguminous host plants via tubular structures, the infection threads. During infection of the host plant they trigger the formation of a new organ, the root nodule, in which a differentiated form of rhizobia, the bacteroid, fixes nitrogen into ammonia, which can then be used by the plant. This symbiosis can be very host specific; cross-inoculation groups of bacterial species are defined by host range. Examples of cross-inoculation groups include *R. leguminosarum* biovar *viciae* with peas, vetches, lentils, and sweet peas as hosts; *R. leguminosarum* biovar *trifolii* with clovers as hosts; *R. meliloti* with alfalfa and sweet clovers as hosts; or *B. japonicum* with soybean and a few other (sub)tropical leguminous plant species as hosts. By contrast, other rhizobial strains have a very broad host range of infection and nodulation. For example, *Rhizobium* strain NGR234 can nodulate at least 70 genera of legumes as well as the nonlegume *Parasponia andersonii* (111).

This review presents an update of the recent literature on infection and nodulation of plants by rhizobia, with special emphasis on results pertinent to other plant-microbe interactions. In previous reviews the properties of rhizobia were compared to those of other plant-invasive microorganisms (35, 43, 90, 154). Arguments were advanced that in the early stages of the rhizobium-plant interaction, before the host plant derives any benefit from the symbiosis, rhizobia can act as parasites or potential plant pathogens. Only in later stages of the infection process is it clear that the interaction is symbiotic, not strictly parasitic, which prompted Djordjevic et al (35) to state that rhizobia are "the refined parasites of legumes." The conclusion that this is a highly coevolved relationship is supported by the existence of certain rhizobitoxine-producing strains of *Bradyrhizobium* that are pathogenic on their host and may represent ancestral forms not yet "refined" to a symbiotic interaction. Nontoxin-producing rhizobia also represent a potential pathogenic threat to a host plant until the nodule is formed and beneficial nitrogen fixation is insured. Hence, to avoid nonsymbiotic infections, the plant likely evolved intricately regulated mechanisms to assess the infection process at various stages and to activate resistance strategies if an infection is not moving toward symbiosis. An interesting parallel evolutionary issue regards the benefit of symbiosis to the bacteria, at least in terms of reproduction. Once the bacteroids are formed in the nodule, the rhizobia are effectively locked into this state since there is no evidence that the bacteroids dedifferentiate to free-living bacteria and emerge from a senescent nodule. This apparent dilemma that follows infection of root hairs by rhizobia has been described by Kijne et al (77) as the "rhizobium trap." Given the potential risks of engaging in symbiosis, the selection pressure for symbiosis to evolve was strong. The reason why this form of "sympathogenesis" has evolved exclusively in the family of the leguminous plants remains mysterious.

This review describes the mechanisms by which rhizobia and host plants regulate each other's gene expression. I also discuss how the linkage between the mechanisms for plant defense and those for normal plant (organ) development complicates study of the specific mechanisms involved in regulating symbiosis. Constraints of space prevent my dealing with the stages of the symbiosis after the release of the bacteria from the infection threads (for reviews, see References 50, 74, 158, 159).

REVIEW OF REVIEWS

The molecular basis of the infection and nodulation process has received widespread attention as a model system to study host-specific infection and organogenesis. Recent progress in molecular and genetic technology and understanding of signal transduction have fostered rapid advances in the fundamental study of plant responses to bacterial signals. Numerous research papers have appeared over the past 10 years on the rhizobium-plant interaction and the reported data have been regularly and widely reviewed. Recent reviews have focused on the following:

1. The genetics of the rhizobial genes involved in infection and nodulation that are inducible by plant signal molecules. Subjects covered include the transcriptional regulation of these genes (called *nod*, *nol*, or *noe* genes (Figure 1) (58, 79, 126, 131) and the biochemical function of their translation products (19, 32, 51, 58, 128, 131).
2. The exchange of molecular signals between rhizobia and plants leading to root nodule formation, particularly the plant flavonoids, which induce the transcription of the *nod*, *nol*, and *noe* genes, and the rhizobial lipo-chitin oligosaccharides (LCOs), which can induce nodule organogenesis. Evidence that these signal molecules are the basis of the host specificity of nodulation and, at least in part, infection (17, 31, 32, 37, 42, 43, 51, 54, 58, 86, 86, 90, 128, 131, 135, 141, 154a) has attracted great attention.
3. Factors other than LCOs that also are thought to be involved in the infection process include extracellular polysaccharides (EPS) (85), lipopolysaccharides (LPS) (74), cyclic β-glucans (16), and capsular polysaccharides (CPS) (15).
4. Factors involved in rhizobial competition for infection and nodulation (148), including those involved in the attachment of rhizobia to root hairs (130).
5. The response of plants to rhizobial infection and nodulation signals, especially nodulins, i.e. plant genes differentially expressed during infection and nodulation (18, 52, 64, 158, 160).

FUNCTION OF RHIZOBIAL GENES REGULATED BY PLANT SIGNALS

Regulation of the nod, nol, *and* noe *Genes*

Flavonoids or isoflavonoids excreted by the plant induce transcription of the rhizobial *nod, nol,* or *noe* genes (Figure 1). The promoter of the (iso)flavonoid-inducible operons usually, but not always (14), contains a consensus sequence called the *nod* box. In all cases, the induction process requires the gene *nodD*, which is a member of the *LysR* family of transcriptional regulators. Multiple allelic forms of the *nodD* gene are present in different rhizobia. In *B. japonicum* gene regulation by isoflavonoids also involves a set of genes of the two-component regulatory family, *nodV* (sensor) and *nodW* (regulator) (122). The possibility of cross talk between other members of this family, *nwsA* and *nwsB*, has been demonstrated (59). In addition, (iso)flavonoids also have been shown to exert negative regulation on *nodD*-mediated expression. Repressors were found in *R. meliloti*, called *nolR* (79), and in *B. japonicum*, called *nolA* (40). In the later stages of symbiosis, transcription of the *nod* and *nol* genes is turned off by an as yet unknown mechanism that does not involve the reported capacity of some (iso)flavonoids to inhibit *nod* or *nol* gene expression. For a more detailed overview of *nod* and *nol* gene transcriptional regulation the reader is referred to several recent, detailed reviews (51, 58, 79, 126).

Biosynthesis and Secretion of LCO

STRUCTURES OF LCO In response to induction by flavonoids, rhizobia produce LCO signals that induce various symbiosis-related responses (see below). As indicated in Figure 2, LCOs produced by all rhizobia consist of an oligosaccharide backbone of β-1,4-linked *N*-acetyl-D-glucosamine, varying in length between 3 and 5 sugar units. A fatty acid group, of variable structure, is attached to the nitrogen of the nonreducing sugar moiety. A special α,β-unsaturated fatty acid moiety can be present in LCOs produced by *R. meliloti* and *R. leguminosarum* biovars *viciae* and *trifolii* (Figure 3) (87, 128a, 134, 136). This polyunsaturated fatty acyl group is not present in the LCOs of other rhizobial species where the fatty acyl moieties consist of classes commonly found in cell membrane phospholipids. Other substitutions on the chitin backbone are rhizobial strain specific. *O*-linked substitutions include sulphate (*R*5), acetyl (*R*4 or *R*5), carbamoyl (*R*2–4), glycerol (*R*6), and sugar moieties (*R*5) such as arabinose, fucose, or various derivatives of fucose (11, 20, 49, 86, 91, 97, 106, 107, 121). An *N*-linked methyl group (*R*2) also occurs in the LCOs of several species (20, 91, 97, 106, 107).

BACKBONE SYNTHESIS Proteins encoded by the rhizobial *nod* and *nol* genes play a crucial role in the biosynthesis of the LCOs (Figure 1). The NodA,

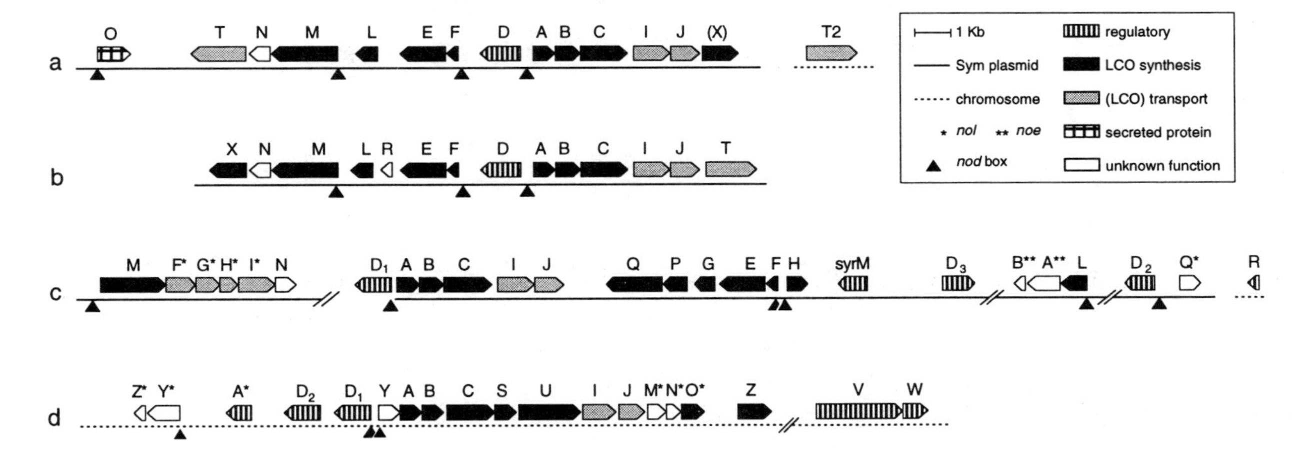

Figure 1 Genetic map and proposed role of *nod*, *nol*, and *noe* genes of rhizobia. Arrows depict the open reading frames of the genes of (*a*) *R. leguminosarum* biovar *viciae* (strain 248, except *nodX*, which is present in strain TOM), (*b*) *R. leguminosarum* biovar *trifolii* (strain ANU843), (*c*) *R. meliloti* (strains 2011 and AK41), and (*d*) *B. japonicum* (strain USDA110). [Data are from References 128, and 131, except those for *noeAB* (3; J Dénarié, personal communication), *nodT2* (116), and the genes of *B. japonicum* (40, 41, 92).] Other rhizobial genes include *nolB, nolE, nolJ, nolT, nolU, nolV, nolW,* and *nolX* of *R. fredii*, which have been indicated to determine cultivar specificity (14, 96), *nodK* of *B. elkanii* (homologous to *nodY*) (39), *nolC* of *R. leguminosarum* biovar *phaseoli* and *nolP*, and *nolK* of *A. caulinodans* (57).

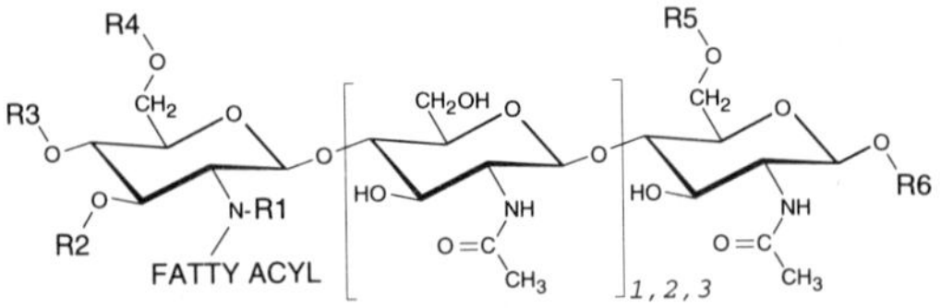

Figure 2 General structure of lipo-chitin oligosaccharides (LCOs). The nature of possible substituents (indicated by *R*) is described in the text. In the absence of host-specific modifications, *R*1–6 stand for hydrogen groups.

NodB, and NodC proteins are sufficient to produce a basic LCO structure. Recent results indicate that the NodC and NodB proteins function as a chitin synthase and a chitin deacetylase, respectively, providing a key intermediate in the synthesis of LCO, called the NodBC metabolite (Figure 4) (56, 71, 73, 139). The NodA protein has been implicated in the addition of the fatty acyl moiety to the NodBC metabolites (4, 118, 139).

LCO-MODIFYING ENZYMES The acetyl transferase NodL (12) and the methyltransferase NodS (53a) are responsible for the presence of an *O*-acetyl and *N*-methyl moiety on the nonreducing terminal saccharide, respectively. Both appear to use the NodBC metabolite as a preferred substrate (Figure 4) (53; GV Bloemberg, personal communication). The NodH protein, which acts as a sulphotransferase responsible for the sulphate moiety on the reducing terminal sugar can use chitin oligomers as substrates (4). However, NodH has been indicated to prefer the complete LCO molecules as a substrate (128b) over chitin oligosaccharides and NodBC metabolites. In this respect, the NodH protein is different from the NodS and NodL proteins.

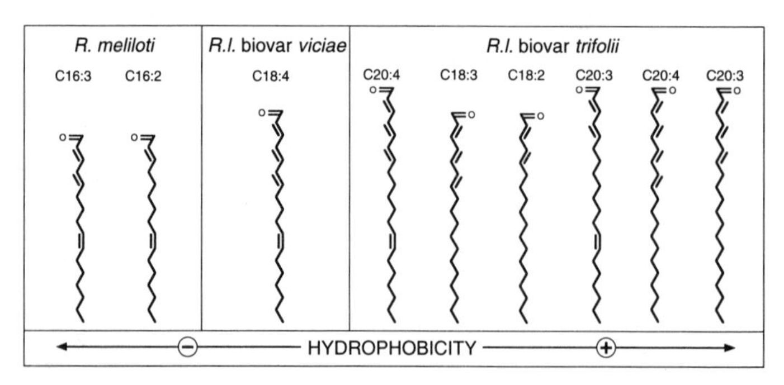

Figure 3 Chemical structures of polyunsaturated fatty acyl moieties of LCOs of *R. meliloti* and *R. leguminosarum* biovars *viciae* and *trifolii*.

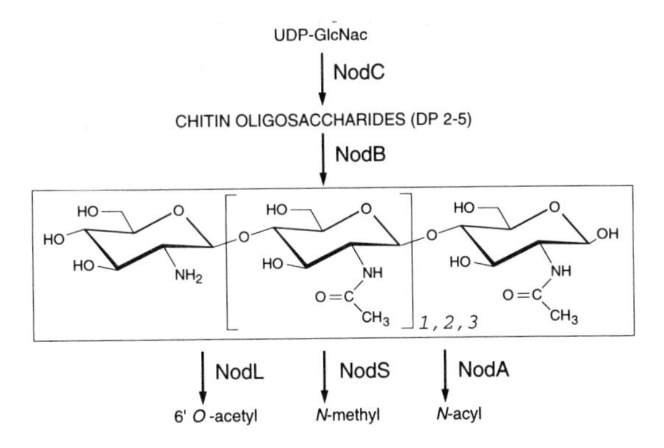

Figure 4 **Model for the biosynthesis of lipo-chitin oligosaccharides.**

ENZYMES INVOLVED IN THE SYNTHESIS OF BUILDING BLOCKS The NodF and NodE proteins probably function as an acyl carrier protein and a β-ketoacyl-synthase, respectively, and are required for the biosynthesis of the special polyunsaturated fatty acid moiety shown in Figure 3 (30, 55, 115). These fatty acyl moieties are incorporated into LCO by means of the putative acyltransferase NodA and can also be found as substituents of the phospholipids (54, 55). The NodP and NodQ proteins, encoded by members of a family of three sets of genes in *R. meliloti*, function together as a sulphurylase and a kinase, leading to the production of the sulphate donor 3′-phosphoadenosine 5′-phosphosulphate (PAPS), which is the substrate for NodH (4, 129). The NodM protein is a glucosamine synthase homologous to the chromosomally encoded glucosamine synthase (GlmS) counterpart (5, 94).

OTHER ENZYMES Synthesis of the LCOs and various host-specific substituents requires other enzymes as well. Some are encoded by genes that also are required for other nonsymbiotic functions, such as fatty acid biosynthesis. Other functions could be encoded by *nod* or *nol* genes that are involved in the biosynthesis of LCO, but whose biochemical function is still unclear (19, 58).

SECRETION OF LCO Little is known about the mechanisms of secretion of LCOs. Sequence homologies with genes of ATP-dependent transport systems suggest that the NodI protein (which contains an ATP-binding cassette), NodJ (which is probably a dimer and integrated in the cytoplasmic membrane), and NodT protein (which is integrated in the outer membrane and contains a fatty acyl substituent) may be involved in transport of LCOs (43, 58, 156). Some experimental results point to the involvement of these gene products in secretion of LCOs in *R. leguminosarum* biovar *trifolii* (95). Recent results (HP Spaink, AHM Wijfjes & BJJ Lugtenberg, submitted) show that NodI and NodJ are involved in the efficiency of LCO secretion but are not required for secretion of LCOs in *R. leguminosarum* biovars *viciae* and *trifolii*. It also has been suggested that the genes *nolF* and *nolGHI* may play a role in LCO secretion, based on their sequence homology with putative membrane fusion proteins and efflux pump proteins, respectively (119).

Other Functions

The gene *nodO* from *R. leguminosarum* biovar *viciae*, which also is positively regulated by *nodD* and flavonoids, encodes an excreted calcium-binding protein (27, 43–45). The NodO protein forms ion channels in membranes that would allow the movement of monovalent cations across the membrane (147). It was suggested that NodO plays a role in nodulation signaling by forming specific channels in the plasma membrane of the host plants. Other rhizobia also excrete proteins after flavonoid induction (82); however, the genes encoding these proteins and their possible functions are not known. Other flavonoid-inducible phenotypes whose the regulatory mechanisms have not yet been documented include: modulation of cell-associated polysaccharides (114); the accumulation of diglycosyl diacylglycerol membrane glycolipids (99); indole-3-acetic acid production (108); and a leucine-responsive regulatory protein (102). Rhizobial genes, encoding putative outer membrane proteins, are down-regulated in the plant, but the signals responsible have not been identified. Interestingly, one gene, *ropA*, is very similar to a gene of the animal cell-invading pathogen *Brucella abortus* (28, 48).

DETERMINANTS OF GUEST-HOST RECOGNITION

Flavonoids

Flavonoids may not seem to be likely candidates to function as host-specific signals to *Rhizobium*. First, flavonoids are not unique to the roots of leguminous plants, but are present in several organs of a wide range of plant species (93). Second, several leguminous plants exude flavonoid compounds that also activate rhizobia that are unable to nodulate these plants. Nevertheless, there is strong evidence that recognition of specific flavonoids by rhizobia is an important basis of host specificity (67, 79, 131, 138). The host specificity of this induction process is determined by the regulatory bacterial NodD protein, which presumably interacts directly with the flavonoids. Recognition of flavonoids is complex on several grounds. (*a*) All rhizobia recognize broad spectra of (iso)flavonoids as inducers or antiinducers. This ability may reflect an adaptation to the range of compounds produced by different species of host plant. Some bacterial species have single *nodD* genes that recognize some structural aspects of flavonoids or have multiple copies of the *nodD* gene with different, more-defined, flavonoid-specificities (79). (*b*) In the latter case, there may be a differential response towards certain plant signals, allowing qualitative response of bacterial signal production to plant signals during the symbiotic process (29). (*c*) After induction by bacterial signals the plant modulates the expression of the genes involved in (iso)flavonoid biosynthesis (see below).

The Lipo-Chitin Oligosaccharides

LCOs are major mediators of host specificity in nodule induction. Different substituents of the LCOs account for their host-specific characteristics. The NodH-determined sulphate moiety of the LCOs from *R. meliloti* is required to induce responses in host plants and to prevent activity on individual nonhost plants (3, 87, 117). The NodE-determined polyunsaturated fatty acyl moieties of the *R. leguminosarum* biovars *viciae* and *trifolii* and *R. meliloti* (Figure 3) are also important effectors or determinants of host specificity (137, 150). For *R. leguminosarum* biovars *viciae* and *trifolii*, the difference in the hydrophobicity of polyunsaturated fatty acyl moieties (Figure 3) determines the difference in their respective host ranges, *Vicia* or *Trifolium* (134). With respect to their occurrence in the LCOs of various rhizobia (Figure 2), the polyunsaturated fatty acyl moieties seem to play a role only in the initiation of indeterminate root nodule primordia (Figure 5). A role of the specialized fatty acids in targeting of LCO to the inner cortex could be proposed on the basis of these observations. The LCOs from bacteria that associate mainly with plants that form determinate root nodules often contain an NodS-determined *N*-linked methyl group at the nonreducing saccharide or an additional 6-linked saccharide moiety. The presence of the 2-*0*-methylfucose moiety of the LCO from *B. japonicum* has been linked to the *nodZ* gene, which is not flavonoid inducible (140). However, the LCOs of broad host range bacteria such as *R. tropicii*, which nodulate a broad range of determinate nodule-forming plants species, seem not to contain an additional saccharide moiety (106). Furthermore, the presence of the 2-*0*-methylfucose moiety is apparently not essential to induce nodules on the hosts of *B. japonicum* (NK Peters, personal communication). Thus the role of additional saccharide substituents is unclear.

In addition to a role in nodule formation, LCOs also determine the specificity of the infection process. Since externally added LCOs from homologous rhizobia allow the infection of *Vigna* and *Glycine* plants by heterologous rhizobia, Relić et al (111) characterized LCO as "a key to the legume door."

Bacterial Exopolysaccharides

The exopolysaccharides (EPSs) have also been suggested to play a role in host specificity of the infection process (35). The observation that nodule invasion

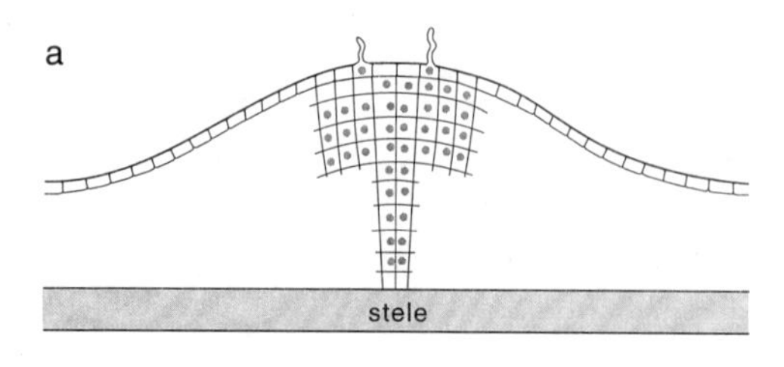

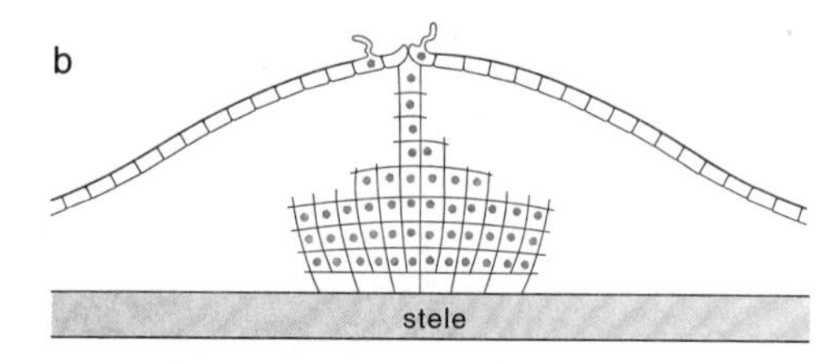

Figure 5 Schematic description of the induction of early stages of determinate (*A*) or indeterminate (*B*) nodule formation by LCOs. This figure is based upon microscopic studies of LCO-induced roots of *Vicia sativa* (indeterminate nodules), *Glycine soja* (145), *Lotus preslii* (91), and *Phaseolus vulgaris* (determinate nodules). [For further description of the development of indeterminate and determinate nodules the reader is referred to references (17, 18, 52).] In some plant species, such as *Acacia*, the induction of determinate primordia (*A*) lead to the formation of indeterminate type nodules (IM López-Lara, personal communication).

by EPS mutants could be restored by exogenously added EPS from the guest bacteria but not by EPS from incompatible rhizobia (7, 38) supports this proposed role. LPSs or CPSs, such as the recently described polysaccharide that resembles group II K antigens of *E. coli*, may also be determinants of host specificity during the infection process (15, 74, 113).

Plant Determinants

Plant lectins are the only known example of plant factors that determine host specificity. Diaz et al (33) constructed transgenic white clover plants containing the pea lectin gene *psl*, which could be nodulated by the heterologous pea-nodulating bacteria. Although their results initially pointed to a function of pea lectin as a receptor of LCO, more recent findings raise questions about this function. A more likely explanation is that the pea lectin facilitates infection by the rhizobia and thereby makes the transgenic plants susceptible to the heterologous nodulation signals (78). Various lectins might also play a role during the late stages of symbiosis (8, 75, 76). Plant chitinases, which can specifically degrade LCOs to their disaccharidic inactive forms (144), also have been suggested to play a role in specificity. However, a general role in autoregulation of the nodule induction process is more likely (63, 144). One possible explanation for the role of LCO hydrophobicity in host specificity could lie in the difference in fatty acid composition in the host plants. However, Bloemberg & Thomas-Oates (personal communication) demonstrated that such differences are not detectable by gas chromatographic analysis of total fatty acyl extracts of the roots of host plants.

REACTION OF PLANTS TO RHIZOBIAL INFECTION

Root Hair Curling

After rhizobia attach to the tips of emerging root hairs in the host plant, the infection process is initiated by curling of the root hairs. Rhizobia are thus trapped inside so-called shepherd's crook structures, and infection thread formation is initiated within the crook (76, 77). External application of LCOs, at concentrations between 10^{-8} and 10^{-12} M, elicits responses such as depolarization of membrane potential (46), modulation of proton and calcium ion fluxes (2), or curling, branching, and swelling (20, 63, 87, 91, 105, 112, 136) of root hairs of the respective host plants. These responses are probably all related to the process of root hair curling. The earliest detectable response to LCO is induction of membrane depolarization, which occurs within 10 min (46). On *Vicia sativa*, 5–10 min interaction between LCO and the root is sufficient to induce root hair deformation that is visible within 1 h (63). After 3 h, most root hairs in a small susceptible zone of the roots can be deformed. LCOs apparently do not need to contain host-specific substituents for the induction of these responses. For instance, root hair deformation on *V. sativa* can be induced by LCOs derived from the compatible *R. leguminosarum* biovar *viciae* but also by the LCOs from *B. japonicum* (20) or *R. loti* (91). By contrast, the sulphate group of the LCOs from *R. meliloti*, which is important for curling of root hairs in *Medicago sativa* (87), substantially decreases the capacity of the LCO to induce root hair deformation on *V. sativa* (32, 63, 144). LCOs can induce the formation of shepherd crook's curling on *Macroptilium atropurpureum* (112). Induction of this response appears not to be dependent on host-specific substituents of the LCO, although it takes 10–100 times higher concentrations of LCOs from heterologous rhizobia to achieve the same degree of curling as that given by the LCOs from *Rhizobium* strain NGR234. On *V. sativa* plants, shepherd crook's structures cannot be induced by purified LCOs. For this plant, the physical presence of rhizobia attached to root hairs is required in addition to the LCOs.

Little is known about the genetic basis of root hair deformation. Krause et al (81) recently identified a gene encoding a lipid transfer-like protein, the expression of which increased in root hairs treated with LCOs or rhizobia. It was speculated that this gene may have a role in the transport of the LCOs.

The Infection Thread

Rhizobial infection proceeds in many host plants via the formation of infection threads (17, 76). The mechanisms underlying the growth of these tubular structures through the root hair cell and subsequently through the root cortical tissue are poorly understood. Cellular changes include nuclear movements mediated by microtubuli, the breakdown of cell wall material, the formation of new cell walls, and the production of an infection thread matrix. The matrix contains the bacteria as well as plant-derived glycoproteins, as was shown for *Pisum sativum* (74). Evidence indicates that infection thread formation is a result of activation of the cell cycle in cells of the outer cortex. This activation was suggested by the host-specific induction of preinfection thread structures in the outer cortex of *Vicia* roots by LCOs (151). These preinfection thread structures are characterized by the formation of so-called cytoplasmic bridges (6); they are radially aligned, giving the impression of cytoplasmic threads crossing the outer cortex. The formation of these structures, which are indistinguishable from those observed after infection with bacteria from *R. leguminosarum* biovar *viciae*, always precedes the formation of infection threads (6, 151, 155). The formation of cytoplasmic bridges in vacuolated cells is preceded by cell polarization; the nucleus moves to the center of the cell just as in cells that are about to divide. Yang et al (164) showed that cells of the preinfection threads have indeed entered the cell cycle but do not divide as a result of arrest in the G2 phase. The process of preinfection thread formation is accompanied by local cell wall modifications that are probably related to the induction of tip growth. However, complete cell wall degradation was not observed in the absence of rhizobia. Based on these observations, Van Spronsen et al (155) proposed a two-step process of cell wall degradation for infection thread formation: local cell wall modification by plant enzymes induced by LCO, followed by complete cell wall degradation in the presence of rhizobia. Given the developmental similarities between infection thread growth and polar tip growth, these findings suggest that other related processes such as pollen tube growth may be directed by parallel signal transduction pathways. Interestingly, a genetic locus has been described in *Arabidopsis* that is involved in both root hair and pollen tube expansion (125).

At the molecular genetic level, LCO signals produce some of the same responses as those observed during the process of rhizobial infection, including the induction of nodulin gene expression, e.g. *enod12* from which the expression in time and place is strongly correlated with the early steps in the symbiosis. *Pisum sativum* and *Medicago sativa* appear to contain two copies of this nodulin gene, *enod12a* and *enod12b* (9, 52, 68, 124). In *M. sativa*, the two copies of *enod12* appear to be differentially regulated: *enod12a* expression in roots is associated with the infection process, whereas *enod12B* expression is related to root nodule organogenesis (9). Transgenic *M. varia* plants have been constructed that contain gene fusions of the single *enod12* gene of *M. truncatula* with the reporter gene *uidA*. These fusions provide a valuable molecular marker for studying LCO signal transduction in the plant (104). The results obtained with this system show that rhizobial infections as well as purified LCOs elicit *enod12* expression in the epidermal cells in the zone of emerging and maturing root hairs within 3 h of inoculation (72). Infection studies of these transgenic reporter plants with *R. meliloti exoA* mutants indicate that the transcription of the *enod12* gene is activated at a very low level in noninfected regions of *Rhizobium*-elicited nodules (84, 103). This conclusion is consistent with the results of Hirsch et al (65), who did not detect *enod12* expression in *M. sativa* nodules induced by bacterial infection mutants. By contrast, expression studies of *M. truncatula enod12* in another genotype of *M. varia*, which can develop nodules in the absence of rhizobia (Nar^+) (149), show that *enod12* is expressed in spontaneous nodules and that its expression can therefore be uncoupled from the infection process (103). Although the expression of the *enod12* gene is closely correlated with the infection process in *Rhizobium*-induced nodules, the *enod12* gene apparently is not required for infection and nodulation of *Medicago* plants (24).

The Root Nodule

At micromolar concentrations, externally applied purified LCOs elicit the formation of nodule primordia in the cortex that are indistinguishable from the nodule primordia in the first stage of normal nodule organogenesis. Furthermore, as in plants infected by rhizobia, the primordia are only induced at certain positions in the plant root, i.e. where young root hairs emerge, opposite (or almost opposite) the protoxylem poles of the central cylinder. This holds true for indeterminate nodule-forming plants such as *Vicia sativa* (136) and *Medicago sativa* 150), as well as for determinate nodule-forming plants such as *Glycine soja* (145), *Lotus preslii* (91), and *Phaseolus vulgaris* (IM Lopez-Lara, personal communication) (Figure 5). In *Medicago* and *Glycine soja*, the nodule primordia can develop into full-grown nodules with the anatomical and histological features of genuine rhizobia-induced nodules. This development has not been observed in other plants such as *Vicia*, rather nodules stop developing at a stage at which small outgrowths are externally visible on the roots. Induction of root nodule primordia on *Vicia* by LCO is strongly inhibited by ethylene (153) in agreement with its inhibitory effect on nodulation by *Rhizobium* (83). Furthermore, root nodulation by *Rhizobium* is reduced in *Vicia* roots treated with LCOs, a phenotype called JAN (jamming of nodulation) (AAN van Brussel, personal communication). These observations cannot yet be linked with findings that rhizobial infection strongly autoregulates the number of root nodules (18). Caetano-Anollés & Gresshof (18) showed that organized nodular structures trigger the feedback regulatory mechanisms even in the absence of bacterial infection. By contrast, nodules formed spontaneously on *Medicago* plants with a Nar^{+} phenotype apparently have a different autoregulatory mechanism or lack a key controlling element in the process (103).

Little is known about the chemical signals determining the position of root nodulation. The gradient hypothesis offers one explanation for the local reaction of individual inner cortical cells to rhizobial signals; this postulates that variations in concentration of a signal determines that only certain cortical cells respond (88). A hormone from the central stele, which stimulated cell division in pea root explants at nanomolar concentrations, has been purified and shown to be uridine. The strong correlation between the activity of this root signal and autoregulation of nodulation suggests that uridine plays an essential role in initiating root nodule formation (G Smit, C de Koster, J Schripsema, et al, submitted). Microtargeting experiments of signal molecules confirmed the important role of uridine; these results showed that induction of root cortical cell divisions by chitin oligosaccharides was dependent on the cointroduction of uridine (133). In *Medicago*, the nodule primordia-inducing effect of LCO can be mimicked by the addition of other compounds such as the auxin transport inhibitor 2,3,5-triiodobenzoic acid (TIBA) (52, 64). Furthermore, mutant strains of *R. meliloti* that are unable to produce LCO can be restored to their nodulation phenotype by introducing the *tzs* gene from *Agrobacterium*, which results in the production of *trans*-zeatin (22). These results suggest that nodule initiation is regulated by common plant hormonal mechanisms.

At the genetic level, the induction of root nodule primordia coincides with the activation of various genes involved in cell cycle regulation (164). Savouré et al (123) showed that the cognate LCOs also have host-specific responses on gene expression in suspension cultures of *Medicago microcallus*. At nanomolar concentrations, increased expression of histone H3-1, the *cdc2* homologue from *Schizosaccharomyces pombe*, and a cyclin-encoding gene demonstrated the host-specific activation of the cell cycle. Stimulation of the cell cycle also was indicated by enhanced thymidine incorporation, elevated numbers of S-phase cells, and increased kinase activity of *cdc2*-related complexes.

Several (nodulin) genes that are differentially expressed during the early stages of nodule formation have been reported (34, 52, 80, 161, 165). The nodulin gene *enod40*, a good marker for nodule primordium formation (80, 165), is activated early after treatment of roots with wild-type compatible rhizobia, purified LCOs, TIBA, or *Rhizobium* LCO-strains containing a zeatin gene from *Agrobacterium* (K Pavlowski, H Franssen, T Bisseling, personal communication). However, expression of *enod40* was detectable in the cells of the protoxylem poles bordering the nodule primordia before it could be detected in the cells of the nodule primordium itself. The function of the *enod40* transcript is unknown. Because no significant open reading frames were found in *enod40* cDNAs of several plant species, it was suggested that *enod40* encodes a nontranslatable RNA (23). The introduction into *M. varia* of constructs that overexpress *enod40* resulted in overactive cell proliferation of the transformed explants. Introduction of the antisense *enod40* construct resulted in an arrest of growth of the transformed explants (23). These results indicate that *enod40* plays an important role in plant development.

Responses Related To Plant Defense

There are several reports on the induction of phenotypes by wild-type or mutant rhizobia that could be related to a defense response. The induced responses include: phytoalexin accumulation (143), chitinase and peroxidase enzyme activity (120, 143, 157), de novo (iso)flavonoid production and excretion (25, 109, 127, 152, 163), expression of genes of the (iso)flavonoid pathway (36, 47, 60, 89, 110), accumulation of intercellular matrix material (101), secondary cell wall modifications (101), and the hypersensitive response, as defined by the observation of local cell death (157). The suggestion that the observed reactions, except for the latter phenotype, serve as a defense response should be treated with caution. Alternatively, these phenotypes may merely reflect the fact that, in plants, regulation of several (presumed) defense-related mechanisms is integrated with regulation of normal developmental processes, and vice versa (1, 13, 66, 135). However, since induction of plant defense responses is not necessarily correlated with the induction of the hypersensitive response (69), all the observed phenotypes might well have a function in defense against (sym)pathogens.

Differences in the type of rhizobial strain or mutants, plant species, and the experimental conditions used in most test systems make it difficult to draw general conclusions from published results. In some cases, the rhizobial mutants that were used have not been characterized at a molecular level and thus the genetic, molecular, or biochemical bases of their actions cannot be determined. Induction of the reported phenotypes, however, can be linked to a condition or a specific bacterial trait in the following results:

1. Activation of the hypersensitive response as a result of abortive infections, even in *Medicago* plants infected by the compatible wild-type *R. meliloti* bacteria (157).
2. Triggering of some of the defense-related responses described above by EPS-defective mutants of *R. meliloti* and *B. japonicum* and LPS-defective mutants of *R. leguminosarum* that can nodulate the host plant but not invade the nodules (98, 100, 101). Exogenously supplied EPS forms appear to be

able to restore the defect in nodule invasion of EPS-defective mutants (7, 38). A low molecular weight form of the EPS of *R. meliloti* has been shown to be responsible for the complementing effect (7, 85). These EPSs might be the equivalent of known "silencers" of a defense response (70).

3. Induction of flavonoid synthesis genes in the host plant. These genes include those encoding phenylalanine ammonia-lyase (PAL) and chalcone synthase (CHS) (110). The observed induction of PAL and CHS gene expression in *Vicia* by wild-type *R. leguminosarum* biovar *viciae* is correlated with the production of new flavonoids that induce transcription of the *nod* genes (109). These responses are condition dependent since they are only detectable in roots that are not shielded from light, and are probably dependent on the presence of ethylene (152, 153). It was also shown that other rhizobia induce the (local) synthesis of (iso)flavonoids in the host plants, but the role of ethylene was not tested (25, 36, 47, 89, 127). CHS gene expression was higher for rhizobial mutants that were unable to infect nodules than it was in normal wild-type nodulation (60, 163). These results suggest that induction of nodule primordia, combined with the absence of nodule invasion, triggers a regulatory mechanism involving flavonoids, although its function is not clear.

The resemblance of LCOs to other reported elicitors of defense responses makes them obvious candidates as elicitors responsible for the above-mentioned responses (10, 21, 142). A positive correlation with LCO production has indeed been demonstrated in the induction of the (iso)flavonoid biosynthesis pathway. Interestingly, the structural requirements of LCOs to induce flavonoid biosynthesis in *Vicia* plants are identical to the requirements to induce nodule primordia. Possible functions of the inducible flavonoids include:

1. Modulation of *nod* gene expression to enforce "better" LCO production (higher or lower levels, or structurally different molecules) by the rhizobia. Such a continuous feed-back mechanism could enable the plant to differentially regulate rhizobial signal production at different stages of the infection and nodulation process. These different stages could have different requirements with respect to quantity or quality of LCOs.
2. Induction of a direct hormonal effect by acting as auxin transport inhibitors. This could lead to the stimulation or inhibition of further nodule initiation or nodule outgrowth. An inhibitory effect would seem to contrast with the nodule-inducing capacity of (externally supplied) auxin transport inhibitors such as TIBA.
3. Provision of a defense mechanism against co-invading microorganisms, which are not potentially beneficial. Toxicity in inducible flavonoids may be higher than in the constitutively expressed flavonoids. The effect of

ethylene would be consistent with its synergistic function in triggering plant defense mechanisms (21, 162).

4. Finally, a similar mechanism also may play a role in the infection of plants by other microorganisms. One example is the interaction between roots of *Medicago* and mycorrhizal fungi, which is also accompanied by differential expression of (iso)flavonoid pathway genes (61, 62).

ACKNOWLEDGMENTS

I am grateful to Dr. Helmi Schlaman for critically reading the manuscript and for many stimulating discussions. I also thank my colleagues from Leiden University and Drs. D Geelen, M Schultze, NK Peters, K Pavlowski, H Franssen, T Bisseling, and JT Thomas-Oates for communicating results prior to publication. The author is supported by the Royal Netherlands Academy of Arts and Sciences and the Netherlands Organization for Scientific Research.

Literature Cited

1. Albrecht C, Laurent P, Lapeyrie F. 1994. Eucalyptus root and shoot chitinases, induced following root colonization by pathogenic versus ectomycorrhizal fungi, compared on one- and two-dimensional activity gels. *Plant Sci.* 100:157–64
2. Allen NS, Bennett MN, Cox DN, Shipley A, Ehrhardt DW, et al. 1994. Effects of Nod factors on alfalfa root hair Ca^{++}, and H^+ currents and on cytoskeletal behavior. See Ref. 26, pp. 107–14
3. Ardourel M, Demont N, Debellé F, Maillet F, de Billy F, et al. 1994. *Rhizobium meliloti* lipo-oligosaccharide nodulation factors: different structural requirements for bacterial entry into target root hair cells and induction of plant symbiotic developmental responses. *Plant Cell* 6:1357–74
4. Atkinson EM, Palcic MM, Hindsgaul O, Long R. 1994. Biosynthesis of *Rhizobium meliloti* lipooligosaccharide Nod factors: NodA is required for an N-acetyltransferase activity. *Proc. Natl. Acad. Sci. USA* 91:8418–22
5. Baev N, Endre G, Petrovics G, Banfalvi Z, Kondorosi A. 1991. Six nodulation genes of *nod* box locus 4 in *Rhizobium meliloti* are involved in nodulation signal production: *nodM* codes for D-glucosamine synthetase. *Mol. Gen. Genet.* 228:113–24
6. Bakhuizen R. 1988. *The plant cytoskeleton in the Rhizobium-legume symbiosis.* PhD thesis. Leiden Univ., Leiden, The Netherlands. 148 pp.
7. Battisti L, Lara JC, Leigh JA. 1992. Specific oligosaccharide form of the *Rhizobium meliloti* exopolysaccharide promotes nodule invasion in alfalfa. *Proc. Natl. Acad. Sci. USA* 89:5625–29
8. Bauchrowitz MA, Barker DG, Lescure B, Truchet G. 1994. Promoter activities of medicago lectin genes during the symbiotic interactions between *R. meliloti* and transgenic alfalfa. See Ref. 26, pp. 135–38
9. Bauer P, Crespi MD, Szécsi J, Allison LA, Schultze M, et al. 1994. Alfalfa *Enod12* genes are differentially regulated during nodule development by Nod factors and *Rhizobium* invasion. *Plant Physiol.* 105:585–92
10. Baureithel K, Felix G, Boller T. 1994. Specific, high affinity binding of chitin fragments to tomato cells and membranes. Competitive inhibition of binding by derivatives of chitooligosaccharides and a Nod factor of *Rhizobium*. *J. Biol. Chem.* 269:17931–38

11. Bec-Ferté MP, Krishnan HB, Promé D, Savagnac A, Pueppke SG, et al. 1994. Structures of nodulation factors from the nitrogen-fixing soybean symbiont *Rhizobium fredii* USDA257. *Biochemistry* 33:11782–88
12. Bloemberg GV, Thomas-Oates JE, Lugtenberg BJJ, Spaink HP. 1994. Nodulation protein NodL of *Rhizobium leguminosarum* *O*-acetylates lipo-oligosaccharides, chitin fragments and *N*-acetylglucosamine in vitro. *Mol. Microbiol.* 11:793–804
13. Boot K. 1993. *Regulation of auxin-induced genes in cell-suspension cultures from Nicotiana tabacum*. PhD thesis. Leiden Univ., Leiden, The Netherlands. 161 pp.
14. Boundy-Mills KL, Kosslak RM, Tully RE, Pueppke SG, Lohrke S, Sadowsky MJ. 1994. Induction of the *Rhizobium fredii nod* box-independent nodulation gene *nolJ* requires a functional *nodD1* gene. *Mol. Plant-Microbe Interact.* 7: 305–08
15. Breedveld MW. 1992. *Oligo- and polysaccharide synthesis by Rhizobium leguminosarum and Rhizobium meliloti*. PhD thesis. Wageningen Univ., Wageningen, The Netherlands. 127 pp.
16. Breedveld MW, Miller KJ. 1994. Cyclic β-glucans of members of the family *Rhizobiaceae*. *Microbiol. Rev.* 58:145–61
17. Brewin NJ. 1991. Development of the legume root nodule. *Annu. Rev. Cell Biol.* 7:191–226
18. Caetano-Anollés, Gresshoff PM. 1991. Plant genetic control of nodulation. *Annu. Rev. Microbiol.* 45:345–82
19. Carlson RW, Price NPJ, Stacey G. 1994. The biosynthesis of rhizobial lipo-oligosaccharide nodulation signal molecules. *Mol. Plant-Microbe Interact.* 7: 684–95
20. Carlson RW, Sanjuan J, Bhat R, Glushka J, Spaink HP, et al. 1993. The structures and biological activities of the lipo-oligosaccharide nodulation signals produced by type I and type II strains of *Bradyrhizobium japonicum*. *J. Biol. Chem.* 268:18372–81
21. Collinge DB, Kragh KM, Mikkelsen JD, Nielsen KK, Rasmussen U, et al. 1993. Plant chitinases. *Plant J.* 3:31–40
22. Cooper JB, Long SR. 1994. Morphogenetic rescue of *Rhizobium meliloti* nodulation mutants by *trans*-zeatin secretion. *Plant Cell* 6:215–25
23. Crespi MD, Jurkevitch E, Poiret M, d'Aubeton-Carafa Y, Petrovics G, et al. 1994. *enod40*, a gene expressed during nodule organogenesis, codes for a non-translatable RNA involved in plant growth. *EMBO J.* 13:5099–112
24. Csanádi G, Szécsi J, Kaló P, Kiss P, Endre G, et al. 1994. *ENOD12*, an early nodulin gene is not required for nodule formation and efficient nitrogen fixation in alfalfa. *Plant Cell* 6:201–13
25. Dakora FD, Joseph CM, Phillips DA. 1993. Alfalfa (*Medicago sativa* L.) root exudates contain isoflavonoids in the presence of *Rhizobium meliloti*. *Plant Physiol.* 101:819–24
26. Daniels MJ, Downie JA, Osbourne AE, eds. 1994. *Advances in Molecular Genetics of Plant-Microbe Interactions*, Vol. 3. Dordrecht: Kluwer. 414 pp.
27. de Maagd RA, Wijfjes AHM, Spaink HP, Ruiz-Sainz JE, Wijffelman CA, et al. 1989. *nodO*, a new *nod* gene of the *Rhizobium leguminosarum* biovar *viciae* Sym plasmid pRL1JI, encodes a secreted protein. *J. Bacteriol.* 171:6764–70
28. de Maagd RA, Yang W-C, Goosen-de Roo L, Mulders IHM, Roest HP, et al. 1994. Down regulation of expression of the *Rhizobium leguminosarum* outer membrane protein gene *ropA* occurs abruptly in interzone II-III of pea nodules and can be uncoupled from *nif* gene activation. *Mol. Plant-Microbe Interact.* 7:276–81
29. Demont N, Ardourel M, Maillet F, Promé D, Ferro M, et al. 1994. The *Rhizobium meliloti* regulatory *nodD3* and *syrM* genes control the synthesis of a particular class of nodulation factors *N*-acylated by (w-1)-hydroxylated fatty acids. *EMBO J.* 13:2139–49
30. Demont N, Debellé F, Aurelle H, Dénarié J, Promé JC. 1993. Role of the *Rhizobium meliloti* *N*-acylated and *N*-acylated genes in the biosynthesis of lipo-oligosaccharidic nodulation factors. *J. Biol. Chem.* 268:20134–42
31. Dénarié J, Cullimore J. 1993. Lipo-oligosaccharide nodulation factors: a new class of signaling molecules mediating recognition and morphogenesis. *Cell* 74: 951–54
32. Dénarié J, Debellé F, Rosenberg C. 1992. Signaling and host range variation in nodulation. *Annu. Rev. Microbiol.* 46:497–531
33. Diaz CL, Melchers LS, Hooykaas PJJ, Lugtenberg BJJ, Kijne J. 1989. Root lectin as a determinant of host-plant specificity in the *Rhizobium*-legume symbiosis. *Nature* 338:579–81
34. Dickstein R, Prusty R, Peng T, Ngo W, Smith ME. 1993. ENOD8, a novel early nodule-specific gene is expressed in empty alfalfa nodules. *Mol. Plant-Microbe Interact.* 6:715–21

35. Djordjevic MA, Gabriel DW, Rolfe BG. 1987. *Rhizobium*—The refined parasite of legumes. *Annu. Rev. Phytopathol.* 25: 145–68
36. Djordjevic MA, Lawson CGR, Mathesius U, Weinman JJ, Arioli T, et al. 1994. Developmental and environmental regulation of chalcone synthase expression in subterranean clover. See Ref. 26, pp. 131–34
37. Djordjevic MA, Weinman JJ. 1991. Factors determining host recognition in the clover-*Rhizobium* symbiosis. *Aust. J. Plant Physiol.* 18:543–57
38. Djordjevic SP, Chen H, Batley M, Redmond JW, Rolfe BG. 1987. Nitrogen fixation ability of exopolysaccharide synthesis mutants of *Rhizobium* sp. strain NGR234 and *Rhizobium trifolii* is restored by the addition of homologous exopolysaccharides. *J. Bacteriol.* 169: 53–60
39. Dobert RC, Brei BT, Triplett EW. 1994. DNA sequence of the common nodulation genes of *Bradyrhizobium elkanii* and their phylogenetic relationship to those of other nodulating bacteria. *Mol. Plant-Microbe Interact.* 7:564–72
40. Dockendorff TC, Sanjuan J, Grob P, Stacey G. 1994. NolA represses nod gene expression in *Bradyrhizobiun japonicum. Mol. Plant-Microbe Interact.* 7:596–602
41. Dockendorff TC, Sharma AJ, Stacey G. 1994. Identification and characterization of the *nolYZ* genes of*Bradyrhizobiun japonicum. Mol. Plant-Microbe Interact.*7:173–80
42. Downie JA. 1991. A *nod* of recognition. *Curr. Opin. Biol.* 1:382–84
43. Downie JA. 1994. Signalling strategies for nodulation of legumes by rhizobia. *Trends Microbiol.* 2:318–24
44. Downie JA, Surin BP. 1990. Either of two *nod* loci can complement the nodulation defect of a *nod* deletion mutant of *Rhizobium leguminosarum* bv *viciae. Mol. Gen. Genet.* 222:81–86
45. Economou A, Davies AE, Johnston AWB, Downie JA. 1994. The *Rhizobium leguminosarum* biovar *viciae nodO* gene can enable a *nodE*-mutant of *Rhizobium leguminosarum* biovar *trifolii* to nodulate vetch. *Microbiology* 140:2341–47
46. Ehrhardt DW, Atkinson EM, Long SR. 1992. Depolarization of alfalfa root hair membrane potential by *Rhizobiun meliloti* Nod factors. *Science* 256:998–1000
47. Estabrook EM, Sengupta-Gopalan C. 1990. Differential expression of phenylalanine ammonia-lyase and chalcone synthase during soybean nodule development. *Plant Cell* 3:299–308
48. Ficht TA, Bearden SW, Sowa BA, Adams G. 1989. DNA sequence and expression of the 36-kilodalton outer membrane protein gene of *Brucella abortus. Infect. Immun.* 57:3281–91
49. Firmin JL, Wilson KE, Carlson RW, Davies AE, Downie JA. 1993. Resistance to nodulation of cv. Afghanistan peas is overcome by *nodX*, which mediates an *O*-acetylation of the *Rhizobium leguminosarum* lipo-oligosaccharide nodulation factor. *Mol. Microbiol.* 10: 351–60
50. Fischer H-M. 1994. Genetic regulation of nitrogen fixation in rhizobia. *Microbiol. Rev.* 58:352–86
51. Fisher RF, Long SR. 1992. *Rhizobium*-plant signal exchange. *Nature* 357:655–60
52. Franssen HJ, Vijn I, Yang WC, Bisseling T. 1992. Developmental aspects of the *Rhizobium*-legume symbiosis. *Plant Mol. Biol.* 19:89–107
53. Geelen D, Leyman B, Mergaert P, Klarskov K, van Montagu M, et al. 1995. NodS is an *S*-adenosyl-L- methione-dependent methyltransferase that methylates chitin fragments deactylated at the non-reducing end. *Mol. Microbiol.* In press
53a. Geelen D, Mergaert P, Geremia RA, Goormachtig S, van Montagu M, et al. 1993. Identification of *nodSUIJ* genes in locus *1* of *Azorhizobium caulinodans*: evidence that *nodS* encodes a methyltransferase involved in Nod factor modification. *Mol. Microbiol.* 9:145–54
54. Geiger O, Ritsema T, van Brussel AAN, Tak T, Wijfjes AHM, Bloemberg GV, et al. 1994. Role of rhizobial lipo-oligosaccharides in root nodule formation on leguminous plants. *Plant Soil* 161: 81–89
55. Geiger O, Thomas-Oates JE, Glushka J, Spaink HP, Lugtenberg BJJ. 1994. Phospholipids of *Rhizobium* contain *nodE*-determined highly unsaturated fatty acid moieties. *J. Biol. Chem.* 269:11090–97
56. Geremia RA, Mergaert P, Geelen D, van Montagu M, Holsters M. 1994. The NodC protein of *Azorhizobium caulinodens* is an *N*-acetylglucosaminyltransferase. *Proc. Natl. Acad. Sci. USA* 91: 2669–73
57. Goethals K, Mergaert P, Gao M, Geelen D, van Montagu M, Holsters M. 1992. Identification of a new inducible nodulation gene in *Azorhizobium caulinodans. Mol. Plant-Microbe Interact.* 5: 405–11
58. Göttfert M. 1993. Regulation and func-

tion of rhizobial nodulation genes. *FEMS Microbiol. Rev.* 104:39–64
59. Grob P, Hennecke H, Göttfert M. 1994. Cross-talk between the two-component regulatory systems NodVW and NwsAB of *Bradyrhizobium japonicum. FEMS Microbiol. Lett.* 120:349–53
60. Grosskopf E, Ha DTC, Wingender R, Röhrig H, Szecsi J, et al. 1993. Enhanced levels of chalcone synthase in alfalfa nodules induced by a Fix$^-$ mutant of *Rhizobium meliloti. Mol. Plant-Microbe Interact.* 6:173–81
61. Harrison MJ, Dixon RA. 1993. Isoflavonoid accumulation and expression of defense gene transcripts during the establishment of vesicular-arbuscular mycorrhizal associations in roots of *Medicago truncatula. Mol. Plant-Microbe Interact.* 6:643–54
62. Harrison MJ, Dixon RA. 1994. Spatial patterns of expression of flavonoid/isoflavonoid pathway genes during interactions between roots of *Medicago truncatula* and the mycorrhizal fungus *Glomus versiforme. Plant J.* 6:9–20
63. Heidstra R, Geurts R, Franssen H, Spaink HP, van Kammen A, et al. 1994. Root hair deformation activity of nodulation factors and their fate on *Vicia sativa. Plant Physiol.* 105:787–97
64. Hirsch AM. 1992. Developmental biology of legume nodulation. *New Phytol.* 122:211–37
65. Hirsch AN, McKhann HI, Löbler M. 1992. Bacterial-induced changes in plant form and function. *Int. J. Plant Sci.* 153:S171–78
66. Hoekstra SS. 1993. *Accumulation of indole alkaloids in plant-organ cultures.* PhD thesis. Leiden Univ., Leiden. The Netherlands. 127 pp.
67. Horvath B, Bachem CW, Schell J, Kondorosi A. 1987. Host-specific regulation of nodulation genes in *Rhizobium* is mediated by a plant-signal, interacting with the *nodD* gene product. *EMBO J.* 6:841–48
68. Horvath B, Heidstra R, Lados M, Moerman M, Spaink HP, et al. 1993. Induction of pea early nodulin expression by Nod factors of *Rhizobium. Plant J.* 4:727–33
69. Jakobek JL, Lindgren PB. 1993. Generalized induction of defense responses in bean is not correlated with the induction of the hypersensitive reaction. *Plant Cell* 5:49–56
70. Jakobek JL, Smith JA, Lindgren PB. 1993. Suppression of bean defense response by *Pseudomonas syringae. Plant Cell* 5:57–63
71. John M, Röhrig H, Schmidt J, Wieneke U, Schell J. 1993. *Rhizobium* NodB protein involved in nodulation signal synthesis is a chitooligosaccharide deacetylase. *Proc. Natl. Acad. Sci. USA* 90:625–29
72. Journet EP, Pichon M, Dedieu A, Debilly F, Truchet G, Barker DG. 1994. *Rhizobium meliloti* Nod factors elicit cell-specific transcription of the ENOD12 gene in transgenic alfalfa. *Plant J.* 6:241–49
73. Kafetzopoulos D, Thireos G, Vournakis JN, Bouriotis V. 1993. The primary structure of a fungal chitin deacetylase reveals the function for two bacterial gene products. *Proc. Natl. Acad. Sci. USA* 90:8005–8
74. Kannenberg EL, Brewin NJ. 1994. Host-plant invasion by *Rhizobium*: the role of cell surface components. *Trends Microbiol.* 2:277–83
75. Kardailsky IV, Brewin NJ. 1994. A new lectin-type glycoprotein identified in the peribacteroid fluid of pea nodules. See Ref. 26, pp. 139–42
76. Kijne JW. 1992. The *Rhizobium* infection process. In *Biological Nitrogen Fixation*, ed. G Stacey, RH Burris, HJ Evans, pp. 349–98. New York: Chapman & Hall. 943 pp.
77. Kijne JW, Bakhuizen R, van Brussel AAN, Canter Cremers HCJ, Diaz CL, et al. 1992. The *Rhizobium* trap: root hair curling in the root nodule symbiosis. In *Perspectives in Plant Cell Recognition*, ed. JA Callow, JR Green, 1: 267–84. Cambridge: Cambridge Univ. Press
78. Kijne JW, Diaz C, van Eijsden R, Booij P, Demel R, et al. 1994. Lectin and Nod factors in *Rhizobium*-legume symbiosis. In *Proc. Eur. Nitrogen Fixation Congr., 1st*, ed. G Kiss, G Endre, pp. 106–10. Szeged, Hungary: Officina Press
79. Kondorosi A. 1992. Regulation of nodulation genes in rhizobia. In *Molecular Signals in Plant-Microbe Communications*, ed. DPS Verma, pp. 325–40. Boca Raton: CRC Press
80. Kouchi H, Hata S. 1993. Isolation and characterization of novel nodulin cDNAs representing genes expressed at early stages of soybean nodule development. *Mol. Gen. Genet.* 238:106–19
81. Krause A, Sigrist CJA, Dehning I, Sommer H, Broughton WJ. 1994. Accumulation of transcripts encoding a lipid transfer-like protein during deformation of nodulation-competent *Vigna unguiculata* root hairs. *Mol. Plant-Microbe Interact.* 7:411–18
82. Krishnan HB, Pueppke SG. 1993. Flavonoid inducers of nodulation genes

stimulate *Rhizobium fredii* USDA257 to export proteins into the environment. *Mol. Plant-Microbe Interact.* 6:107–13

83. Lee KH, Larue TA. 1992. Exogenous ethylene inhibits nodulation of *Pisum sativum* L. cv Sparkle. *Plant Physiol.* 100:1759–63
84. Leigh JA, Barker DG, Journet EP, Truchet G. 1994. Role of surface factors in plant-microbe interactions: involvement of *Rhizobium meliloti* exopolysaccharide during early infection events in alfalfa. See Ref. 26, pp. 143–50
85. Leigh JA, Walker GC. 1994. Exopolysaccharides of *Rhizobium*: synthesis, regulation and symbiotic function. *Trends Genet.* 10:63–67
86. Lerouge P. 1994. Symbiotic host specificity between leguminous plants and rhizobia is determined by substituted and acylated glucosamine oligosaccharide signals. *Glycobiology* 4:127–34
87. Lerouge P, Roche P, Faucher C, Maillet F, Truchet G, et al. 1990. Symbiotic host-specificity of *Rhizobium meliloti* is determined by a sulphated and acylated glucosamine oligosaccharide signal. *Nature* 344:781–84
88. Libbenga KR, van Iren F, Bogers RJ, Schraag-Lamers MF. 1973. The role of hormones and gradients in the initiation of cortex proliferation and nodule formation in *Pisum sativum* L. *Planta* 114: 19–39
89. Lawson CGR, Djordjevic MA, Weinman JJ, Rolfe BG. 1994. *Rhizobium* inoculation and physical wounding result in the rapid induction of the same chalcone synthase copy in *Trifolium subterraneum*. *Mol. Plant-Microbe Interact.* 7:498–507
90. Long SR, Staskawicz BJ. 1993. Prokaryotic plant parasites. *Cell* 73:921–35
91. López-Lara IM, van den Berg JDJ, Thomas Oates JE, Glushka J, Lugtenberg BJJ, et al. 1995. Structural identification of the lipo-chitin oligosaccharide nodulation signals of *Rhizobium loti*. *Mol. Microbiol.* 15:627–38
92. Luka S, Sanjuan J, Carlson RW, Stacey G. 1993. *nolMNO* genes of *Bradyrhizobium japonicum* are co-transcribed with *nodYABCSUIJ*, and *nolO* is involved in the synthesis of the lipo-oligosaccharide nodulation signal. *J.Biol. Chem.* 268: 27053–59
93. Maxwell EA, Harrison MJ, Dixon RA. 1993. Molecular characterization and expression of alfalfa isoliquiretigenin 2′-*O*-methyltransferase, an enzyme specifically involved in the biosynthesis of an inducer of *Rhizobium meliloti* nodulation genes. *Plant J.* 4:971–81
94. Marie C, Barny M-A, Downie JA. 1992. *Rhizobium leguminosarum* has two glucosamine synthases, GlmS and NodM, required for nodulation and development of nitrogen fixing nodules. *Mol. Microbiol.* 6:843–51
95. Mckay IA, Djordjevic MA. 1993. Production and excretion of Nod metabolites by *Rhizobium leguminosarum* bv. *trifolii* are disrupted by the same environmental factors that reduce nodulation in the field. *Appl. Env. Microbiol.* 59: 3385–92
96. Meinhardt LW, Krishnan HB, Balatti PA, Pueppke SG. 1993. Molecular cloning and characterization of a sym plasmid locus that regulates cultivar-specific nodulation of soybean by *Rhizobium fredii* USDA257. *Mol. Microbiol.* 9:17–29
97. Mergaert P, van Montagu M, Promé J-C, Holsters M. 1993. Three unusual modifications, a D-arabinosyl, a *N*-methyl, and a carbamoyl group, are present on the Nod factors of *Azorhizobium caulinodans* strain ORS571. *Proc. Natl. Acad. Sci. USA* 90:1551–55
98. Niehaus K, Kapp D, Pühler A. 1993. Plant defense and delayed infection of alfalfa pseudonodules induced by an exopolysaccharide (EPS I)-deficient *Rhizobium meliloti* mutant. *Planta* 190:415–25
99. Orgambide GG, Philip Hollingsworth S, Hollingsworth RI, Dazzo FB. 1994. Flavone-enhanced accumulation and symbiosis-related biological activity of a diglycosyl diacylglycerol membrane glycolipid from *Rhizobium leguminosarum* biovar *trifolii*. *J. Bacteriol.* 176:4338–47
100. Parniske M, Schmidt PE, Kosch K, Müller P. 1994. Plant defense responses of host plants with determinate nodules induced by EPS-defective *exoB* mutants of *Bradyrhizobium japonicum*. *Mol. Plant-Microbe Interact.* 7:631–38
101. Perotto S, Brewin NJ, Kannenberg EL. 1994. Cytological evidence for a host defense response that reduces cell and tissue invasion in pea nodules by lipopolysaccharide-defective mutants of *Rhizobium leguminosarum* strain 3841. *Mol. Plant-Microbe Interact.* 7:99–112
102. Perret X, Fellay R, Bjourson AJ, Cooper JE, Brenner S, et al. 1994. Substraction hybridisation and shot-gun sequencing: a new approach to identify symbiotic loci. *Nucleic Acid Res.* 22:1335–41
103. Pichon M, Journet EP, de Billy F, Dedieu A, Huguet T, et al. 1994. *ENOD12* gene expression as a molecular marker for comparing *Rhizobium*-de-

pendent and -independent nodulation in alfalfa. *Mol. Plant-Microbe Interact.* 7: 740–47

104. Pichon M, Journet EP, Dedieu A, de Billy F, Truchet G, et al. 1992. *Rhizobium meliloti* elicits transient expression of the early nodulin gene *ENOD12* in the differentiating root epidermis of transgenic alfalfa. *Plant Cell* 40:1199–211
105. Plazinski J, Ridge RW, McKay IA, Djordjevic MA. 1994. The *nodABC* genes of *Rhizobium leguminosarum* biovar *trifolii* confer root-hair curling ability to a diverse range of soil bacteria and the ability to induce novel root hair swellings on beans. *Aust. J. Plant Physiol.* 21:311–25
106. Poupot R, Martinez-Romero E, Promé J-C. 1993. Nodulation factors from *Rhizobium tropici* are sulphated or non-sulphated chitopentasaccharides containing an *N*-methyl-*N*-acylglucosaminyl terminus. *Biochemistry* 32:10430–35
107. Price NPJ, Relić B, Talmont F, Lewin A, Promé D, et al. 1992. Broad-host-range *Rhizobium* species strain NGR234 secretes a family of carbamoylated, and fucosylated, nodulation signals that are *O*-acetylated or sulphated. *Mol. Microbiol.* 6:3575–84
108. Prinsen E, Chauvaux N, Schmidt J, John M, Wieneke U, et al. 1991. Stimulation of indole-3-acetic acid production in *Rhizobium* by flavonoids. *Fed. Eur. Biochem. Soc.* 282:53–55
109. Recourt K, Schripsema J, Kijne JW, van Brussel AAN, Lugtenberg BJJ. 1991. Inoculation of *Vicia sativa* subsp. *nigra* roots with *R. leguminosarum* biovar *viciae* results in release of *nod* gene activating flavanones and chalcones. *Plant Mol. Biol.* 16:841–52
110. Recourt K, Van Tunen AJ, Mur LA, Van Brussel AAN, Lugtenberg BJJ, et al. 1992. Activation of flavonoid biosynthesis in roots of *Vicia sativa* subsp. *nigra* plants by inoculation with *Rhizobium leguminosarum* biovar *viciae*. *Plant Mol. Biol.* 19:411–20
111. Relić B, Perret X, Estradagarcia MT, Kopcinska J, Golinowski W, et al. 1994. Nod factors of *Rhizobium* are a key to the legume door. *Mol. Microbiol.* 13: 171–78
112. Relić B, Talmont F, Kopcinska J, Golinowski W, Promé J-C, et al. 1993. Biological activity of *Rhizobium* sp. NGR234 Nod-factors on *Macroptilium atropurpureum*. *Mol. Plant-Microbe Interact.* 6:764–74
113. Reuhs BL, Carlson RW, Kim JS. 1993. *Rhizobium fredii* and *Rhizobium meliloti* produce 3-deoxy-*D*-manno-2-octulosonic acid-containing polysaccharides that are structurally analogous to group II K antigens (capsular polysaccharides) found in *Escherichia coli*. *J.Bacteriol.* 175:3570–80
114. Reuhs BL, Kim JS, Badgett A, Carlson RW. 1994. Production of cell associated polysaccharides of *Rhizobium fredii* USDA205 is modulated by apigenin and host root extract. *Mol. Plant-Microbe Interact.* 7:240–47
115. Ritsema T, Geiger O, van Dillewijn P, Lugtenberg BJJ, Spaink HP. 1994. Serine residue 45 of nodulation protein NodF from *Rhizobium leguminosarum* bv. *viciae* is essential for its biological function. *J. Bacteriol.* 176:7740–43
116. Rivilla R, Downie JA. 1994. Identification of a *Rhizobium leguminosarum* gene homologous to *nodT* but located outside the symbiotic plasmid. *Gene* 144:87–91
117. Roche P, Debellé F, Maillet F, Lerouge P, Faucher C, et al. 1991. Molecular basis of symbiotic host specificity in *Rhizobium meliloti*: *nodH* and *nodPQ* genes encode the sulphation of lipo-oligosaccharide signals. *Cell* 67:1131–43
118. Röhrig H, Schmidt J, Wieneke U, Kondorosi E, Barlier I, et al. 1994. Biosynthesis of lipooligosaccharide nodulation factors-*Rhizobium* NodA protein is involved in *N*-acylation of the chitooligosaccharide backbone. *Proc. Natl. Acad. Sci.* 91:3122–26
119. Saier MH Jr, Tam R, Reizer A, Reizer J. 1994. Two novel families of bacterial membrane proteins concerned with nodulation, cell division and transport. *Mol. Microbiol.* 11:841–47
120. Salzwedel JL, Dazzo FB. 1993. pSym *nod* gene influence on elicitation of peroxidase activity from white clover and pea roots by rhizobia and their cell-free supernatants. *Mol. Plant-Microbe Interact.* 6:127–34
121. Sanjuan J, Carlson RW, Spaink HP, Bhat UR, Barbour WM, et al. 1992. A 2-*O*-methylfucose moiety is present in the lipo-oligosaccharide nodulation signal of *Bradyrhizobium japonicum*. *Proc. Natl. Acad. Sci. USA* 89:8789–93
122. Sanjuan J, Grob P, Göttfert M, Hennecke H, Stacey G. 1994. NodW is essential for full expression of the common nodulation genes in *Bradyrhizobium japonicum*. *Mol. Plant-Microbe Interact.* 7:364–69
123. Savouré A, Magyar Z, Pierre M, Brown S, Schultze M, et al. 1994. Activation of the cell cycle machinery and the isoflavonoid biosynthesis pathway by active *Rhizobium meliloti* Nod signal

molecules in *Medicago microcallus* suspensions. *EMBO J.* 13:1093–102
124. Scheres B, van de Wiel C, Zalensky A, Horvath B, Spaink HP, et al. 1990. The *ENOD12* gene product is involved in the infection process during the pea-*Rhizobium* interaction. *Cell* 60:281–94
125. Schiefelbein J, Galway M, Masucci J, Ford S. 1993. Pollen tube and root-hair tip growth is disrupted in a mutant of *Arabidopsis thaliana. Plant Physiol.* 103:979–85
126. Schlaman HRM, Okker RJH, Lugtenberg BJJ. 1992. Regulation of nodulation gene expression by NodD in rhizobia. *J. Bacteriol.* 174:5177–82
127. Schmidt PE, Broughton WJ, Werner D. 1994. Nod factors of *Bradyrhizobium japonicum* and *Rhizobium* sp. NGR234 induce flavonoid accumulation in soybean root exudate. *Mol. Plant-Microbe Interact.* 7:384–90
128. Schultze M, Kondorosi E, Ratet P, Buiré M, Kondorosi A. 1994. Cell and molecular biology of *Rhizobium*-plant interactions. *Int. Rev. Cytol.* 156:1–75
128a. Schultze M, Quiclet-Sire B, Kondorosi E, Virelizier H, Glushka JN, et al. 1992. *Rhizobium meliloti* produces a family of sulphated lipo-oligosaccharides exhibiting different degrees of plant host specificity. *Proc. Natl. Acad. Sci. USA* 89:192–96
128b. Schultze M, Staehelin C, Röhrig H, John M, Schmidt J, et al. 1995. In vitro sulfotransferase activity of *Rhizobium meliloti* NodH protein: lipochitooligosaccharide nodulation signals are sulfated after synthesis of the core structure. *Proc. Natl. Acad. Sci. USA*. In press
129. Schwedock J, Long SR. 1992. *Rhizobium meliloti* genes involved in sulphate activation: the two copies of *nodPQ* and a new locus *saa. Genetics* 132:899–909
130. Smit G, Swart S, Lugtenberg BJJ, Kijne JW. 1992. Molecular mechanisms of attachment of *Rhizobium* bacteria to plant roots. *Mol. Microbiol.* 6:2897–903
131. Spaink HP. 1994. The molecular basis of the host specificity of *Rhizobium* bacteria. *Antonie van Leeuwenhoek* 65:81–98
132. Spaink HP, Aarts A, Stacey G, Bloemberg GV, Lugtenberg BJJ, et al. 1992. Detection and separation of *Rhizobium* and *Bradyrhizobium* Nod metabolites using thin layer chromatography. *Mol. Plant-Microbe Interact.* 5:72–80
133. Spaink HP, Bloemberg GV, Wijfjes AHM, Ritsema T, Geiger O, et al. 1994. The molecular basis of host specificity in the *Rhizobium leguminosarum*-plant interaction. See Ref. 26, pp. 91–98
134. Spaink HP, Bloemberg GV, van Brussel AAN, Lugtenberg BJJ, van der Drift KMGM, et al. 1995 Host-specificity of *Rhizobium leguminosarum* is determined by the hydrophobicity of highly unsaturated fatty acyl moieties of the nodulation factors. *Mol. Plant-Microbe Interact.* 8:155–64
135. Spaink HP, Lugtenberg BJJ. 1994. Role of rhizobial lipo-chitin oligosaccharide signal molecules in root nodule organogenesis. *Plant Mol. Biol.* 26:1413–22
136. Spaink HP, Sheeley DM, van Brussel AAN, Glushka J, York WS, et al. 1991. A novel highly unsaturated fatty acid moiety of lipo-oligosaccharide signals determines host specificity of *Rhizobium. Nature* 354:125–30
137. Spaink HP, Weinman J, Djordjevic MA, Wijffelman CA, Okker RJH, et al. 1989. Genetic analysis and cellular localization of the *Rhizobium* host specificity-determining NodE protein. *EMBO J.* 8:2811–18
138. Spaink HP, Wijffelman CA, Pees E, Okker RJH, Lugtenberg, BJJ. 1987. *Rhizobium* nodulation gene *nodD* as a determinant of host specificity. *Nature* 328:337–40
139. Spaink HP, Wijfjes AHM, van der Drift KMGM, Haverkamp J, Thomas-Oates JE, et al. 1994. Structural identification of metabolites produced by the NodB and NodC proteins of *Rhizobium leguminosarum. Mol. Microbiol.* 13:821–31
140. Stacey G, Luka S, Sanjuan J, Banfalvi Z, Nieuwkoop AJ, et al. 1994. *nodZ*, a unique host-specific nodulation gene, is involved in the fucosylation of the lipooligosaccharide nodulation signal of *Bradyrhizobium japonicum. J. Bacteriol.* 176:620–33
141. Stacey G, Sanjuan J, Spaink H, van Brussel T, Lugtenberg BJJ, et al. 1993. *Rhizobium* lipo-oligosaccharides; novel plant growth regulators. In *Plant Responses to the Environment*, ed. PM Gresshoff, pp. 45–58. Boca Raton: CRC Press. 184 pp.
142. Staehelin C, Granado J, Müller J, Wiemken A, Mellor RB, et al. 1994. Perception of *Rhizobium* nodulation factors by tomato cells and inactivation by root chitinases. *Proc. Natl. Acad. Sci. USA* 91:2196–200
143. Staehelin J, Mellor RB, Wiemken A, Boller T. 1992. Chitinase and peroxidase in effective (fix^+) and ineffective (fix^-) soybean nodules. *Planta* 187:295–300
144. Staehelin C, Schultze M, Kondorosi E, Mellor RB, Boller T, et al. 1994. Structural modifications in *Rhizobium meliloti* Nod factors influence their sta-

bility against hydrolysis by root chitinases. *Plant J.* 5:319–30
145. Stokkermans TJW, Peters NK. 1994. *Bradyrhizobium elkanii* lipo-oligosaccharides signals induce complete nodules structures on *Glycine soja* Siebold et Zucc. *Planta* 193:413–20
146. Stokkermans TJW, Sanjuan J, Ruan X, Stacey G, Peters KN. 1992. *Bradyrhizobium japonicum* rhizobitoxine mutants with altered host-range on *Rj4* soybeans. *Mol. Plant-Microbe Interact.* 5:504–12
147. Sutton JM, Lea EJA, Downie JA. 1994. The nodulation-signaling protein NodO from *Rhizobium leguminosarum* biovar *viciae* forms ion channels in membranes. *Proc. Natl. Acad. Sci. USA* 91:9990–94
148. Triplett EW, Sadowsky MJ. 1992. Genetics of competition for nodulation of legumes. *Annu. Rev. Microbiol.* 46:399–428
149. Truchet G, Barker DG, Camut S, de Billy F, Vasse J, et al. 1989. Alfalfa nodulation in the absence of *Rhizobium. Mol. Gen. Genet.* 219:65–68
150. Truchet G, Roche P, Lerouge P, Vasse J, Camut S, et al. 1991. Sulphated lipo-oligosaccharide signals of *Rhizobium meliloti* elicit root nodule organogenesis in alfalfa. *Nature* 351:670–73
151. van Brussel AAN, Bakhuizen R, van Spronsen P, Spaink HP, Tak T, et al. 1992. Induction of pre-infection thread structures in the host plant by lipo-oligosaccharides of *Rhizobium. Science* 257:70–72
152. van Brussel AAN, Recourt K, Pees E, Spaink HP, Tak T, et al. 1990. A biovar-specific signal of *Rhizobium leguminosarum* bv. *viciae* induces increased nodulation gene-inducing activity in root exudate of *Vicia sativa* subsp. *nigra. J. Bacteriol.* 172:5394–401
153. van Brussel AAN, Tak T, Spaink HP, Kijne JW. 1992. Light and ethylene influence the expression of nodulation phenotypes induced by *Rhizobium* Nod factors on *Vicia sativa* ssp. *nigra. Int. Symp. Mol. Plant-Microbe Int., 6th, Seattle*:137 (Abstr.)
154. Vance CP. 1983. *Rhizobium* infection and nodulation: a beneficial plant disease? *Ann. Rev. Microbiol.* 37:399–424
154a. van Rhijn P, Vanderleyden J. 1995. The *Rhizobium-* plant symbiosis. *Microbiol. Rev.* 59:124–42
155. van Spronsen PC, Bakhuizen R, van Brussel AAN, Kijne JW. 1994. Cell wall degradation during infection thread formation by the root nodule bacterium *Rhizobium leguminosarum* is a two-step process. *Eur. J. Cell Biol.* 64:88–94
156. Vásquez M, Santana O, Quinto C. 1993. The NodI and NodJ proteins from *Rhizobium* and *Bradyrhizobium* strains are similar to capsular polysaccharide secretion proteins from gram-negative bacteria. *Mol. Microbiol.* 8:369–77
157. Vasse J, de Billy F, Truchet G. 1993. Abortion of infection during the *Rhizobium meliloti*-alfalfa symbiotic interaction is accompanied by a hypersensitive reaction. *Plant J.* 4:555–66
158. Verma DPS. 1992. Signals in root nodule organogenesis and endocytosis of *Rhizobium. Plant Cell* 4:373–82
159. Verma DPS, Hong Z, Gu X. 1994. Signal transduction and endocytosis of rhizobia in the host cells. See Ref. 26, pp. 123–30
160. Vijn I, das Neves L, van Kammen A, Franssen H, Bisseling T. 1993. Nod factors and nodulation in plants. *Science* 260:1764–65
161. Wilson RC, Long FX, Maruoka EM, Cooper JB. 1994. A new proline-rich early nodulin from *Medicago truncatula* is highly expressed in nodule meristematic cells. *Plant Cell* 6:1265–75
162. Xu Y, Chang P-FL, Liu D, Narasimhan ML, Raghothama KG, et al. 1994. Plant defense genes are synergistically induced by ethylene and methyl jasmonate. *Plant Cell* 6:1077–85
163. Yang WC, Canter Cremers HCJ, Hogendijk P, Katinakis P, Wijffelman CA, et al. 1992. In situ localization of chalcone synthase mRNA in pea root nodule development. *Plant J.* 2:143–51
164. Yang WC, de Blank C, Meskiene I, Hirt H, Bakker J, et al. 1994. Nod factors reactivate the cell cycle during infection and nodule primordium formation, but the cycle is only completed in primordium formation. *Plant Cell* 6: 1415–26
165. Yang WC, Katinakis P, Hendriks P, Smolders A, de Vries F, et al. 1993. Characterization of *GmENOD40*, a gene showing novel patterns of cell-specific expression during soybean nodule development. *Plant J.* 3:573–85

Annu. Rev. Phytopathol. 1995. 33:369–91

CLONALITY IN SOILBORNE, PLANT-PATHOGENIC FUNGI

James B. Anderson and Linda M. Kohn
Department of Botany, Erindale College, University of Toronto, Mississauga, Ontario, Canada L5L 1C6

KEY WORDS: *Armillaria, Sclerotinia,* genet, asexual reproduction

ABSTRACT

Understanding patterns of mating and recombination, gene flow, and drift in fungal populations requires that both sexual and clonal components of reproduction be identified. After considering the genetic consequences of clonality, we focus on two species of soilborne, plant-pathogenic fungi with strikingly different modes of clonal propagation for which there is a combination of genetic and spatial data. At one extreme, clones of the root-infecting basidiomycete *Armillaria gallica* are territorial; each clonal individual arises in a unique mating event and from this point of origin then ramifies vegetatively as rhizomorphs to colonize territories often including the root systems of many adjacent trees in forests. At the other extreme, clones of the plant pathogenic ascomycete *Sclerotinia sclerotiorum* are dispersive; they are disseminated as sclerotia and ascospores with the result that clones are disconnected from their points of origin, are spatially mixed within agricultural fields, and are capable of movement between widely separated locations.

INTRODUCTION

With the development of molecular-genetic markers that can be interpreted as Mendelian determinants, interest in the population genetics of plant-pathogenic fungi has grown substantially in the past several years. Recent articles have reviewed general aspects of population genetic theory and its application to fungi (10, 23, 49, 54, 56). Almost every study of fungal populations to date asks questions about genetic structure: How is genetic variability distributed

0066-4286/95/0901-0369$05.00

spatially, temporally, or with respect to host species or cultivar? From this kind of investigation, the ultimate goal is to draw conclusions about past or present processes, including patterns of mating and recombination, gene flow, and drift. Understanding these processes requires that both sexual and clonal components of the population genetic structure be clearly identified and taken into consideration. Because asexual reproduction plays such a prominent role in fungal life cycles, the clonal component is likely to be significant (40).

Clonality is at least as important in soilborne fungal pathosystems as it is in pathosystems where aerial plant parts are infected by a "spore shower" of airborne inoculum, which includes propagules from a much wider area than does inoculum in soils. By definition, soilborne pathogens can penetrate the host plant directly from the soil, characteristically causing seed rot, damping off, root rot, or wilt diseases, although some can also infect aerial plant parts by means of airborne or waterborne spores (for a comprehensive review of soilborne pathogens, see ref. 9). However, all fungi termed "soilborne" produce a resistant propagative structure that persists in soil. Structures that can persist in soil are usually protected by thick walls against the effects of cycles of saturation and desiccation, freezing and thawing, as well as microbial predation and other destructive elements. Although some pathogens, such as *Pythium*, may produce thick-walled sexual oospores, the resistant structures are most often asexual reproductive propagules or hyphal aggregates, such as chlamydospores and other conidia, rhizomorphs, or sclerotia. A persistent structure with the hyphal characteristics of radial growth and physiological continuity, such as a rhizomorph, will extend through the soil, increasing the territory of one genotype for potentially many years. Persistent structures that are small, discrete propagules, such as chlamydospores and sclerotia, occupy small, discrete territories and can be spatially rearranged when soil is cultivated or irrigated, but may remain stationary in uncultivated native soils. Asexual reproductive propagules, including these soilborne resistant structures, will propagate clonal lineages, either briefly or for many seasons, either locally or over a wide geographical area. Although extensive systematic, genetic, and epidemiological data exist for important soilborne pathogens such as *Phytophthora* spp. (6, 7), *Verticillium albo-atrum* (17), *Gibberella fujikuroi* (47), *Rhizoctonia solani* (78), and *Sclerotium rolfsii* (62), the extent of clonality, the spatial distribution of genotypes, and the origins of genotypic variability in populations of these systems have not yet been elucidated.

In this review we first consider the genetic consequences of sexual and asexual reproduction on populations, then discuss a fundamental unit of the population, the genetic individual, and how genetic individuals are identified, how they are distributed spatially, and how they arise. We focus on two species of soilborne plant-pathogenic fungi with strikingly different modes of clonal propagation for which there is a combination of genetic and spatial data on samples from populations. At one extreme, clones of the root-infecting basidiomycete *Armillaria gallica* (synonym *A. bulbosa*) are territorial; they grow vegetatively as rhizomorphs to colonize spatial territories that often include the root systems of many adjacent trees over substantial areas in managed and unmanaged forests. At the other extreme, clones of the plant pathogenic ascomycete *Sclerotinia sclerotiorum* are dispersive; they are disseminated as sclerotia and ascospores, with the result that clones are spatially mixed locally and a clone may be found in widely separated locations. Such differences in patterns of clonal reproduction have been discussed elsewhere with reference to plants (32).

Genetic Consequences of Sexual and Asexual Reproduction

Although many intricate genetic mechanisms regulating sexual and somatic incompatibility in fungi have been described (2), the life histories and reproductive strategies of most fungi remain incompletely known. In these cases, making inferences about patterns of sexual and asexual reproduction in nature remains problematic. Regardless of current knowledge about the life cycle of a fungus in the laboratory, a useful starting point in examining natural populations is to ask whether observed patterns of genetic variability are consistent with strict clonality, random mating (panmixia), or a mixture of clonality and mating. This approach is well reviewed (45, 53, 76). Detecting clonality or genetic exchange and recombination generally hinges on determining whether associations among alleles at several loci, chosen as a sample of the genome, are random or nonrandom. Clonality produces a clear pattern in populations. For a series of loci known to be polymorphic, repeated recovery of the same multilocus genotype, especially over long distances or periods of time, can be taken as a strong indication of clonal reproduction. In clonal reproduction, the entire genome is effectively linked. The higher the proportion of clonality to genetic exchange and recombination in a population, the more strongly the alleles at different loci will be associated. In contrast, when all reproduction is sexual and mating is random, multilocus genotypes are not repeatedly recovered, and there is no association between alleles at one locus and alleles at another locus. This approach indicated that one population of *Candida albicans* pathogenic on humans is clonal (60).

Another approach to distinguishing clonality and recombination asks whether phylogenetic trees inferred for different segments of DNA in the genome are congruent or incongruent (20, 28a, 59). If a population is purely clonal, then all individual trees for different sequences should be congruent. If genetic exchange and recombination have occurred, then incongruence between gene trees will result as sequences with different patterns of descent are reshuffled. This phylogenetic approach clearly indicated a recombined population structure for *Coccidioides immitis*, another human pathogen (11), even though there have been no direct observations of sexuality in this fungus. In reality, most fungal populations probably occupy an intermediate position between the extremes of complete clonality and panmixia, with some clonal reproduction and some genetic exchange and recombination. When the clonal component of reproduction is subtracted from a mixed system, genetic exchange may be random or may be partitioned, for example, by host or geographical separation.

The Genetic Individual

In studying fungal populations, it is necessary to define the individual in genetic terms, in physical or physiological terms, or both. All fungi exist as physical units, or ramets, that can be defined as any asexual, anamorphic structure, such as a discrete patch of mycelium, a sclerotium, a conidium, or an individual hypha. In studies concerned with patterns of mating, the fundamental unit is the genetic individual, or genet, a term with ample precedent in plants (31) and, more recently, in fungi (8, 65). In plants, the genet includes all vegetative derivatives of a seed, whether physically connected or detached. We define the fungal genet to include all derivatives of a mitotic cell lineage that originates in either of two different ways. In fungi with dikaryotic or diploid mycelia, i.e. basidiomycetes and oomycetes, a new genet is created by the mating of two genetically unlike gametic nuclei. The definition of the genet in these fungi is therefore completely analogous to that in plants. The genet concept requires reinterpretation for fungi with haploid mycelia, i.e. ascomycetes and zygomycetes, because they have no diploid vegetative phase equivalent to that in seed-bearing plants. In fungi with haploid mycelia, we propose that a genet includes all vegetative derivatives of the fusion of two genetically unlike gametic nuclei and the ensuing meiosis.

In referring to either plants or fungi, the genet represents one particular kind of clone: one whose origin is associated with genetic exchange, for example in mating. The term "clone" is used in a broader context than the "genet" to include all asexual derivatives from a designated point of origin. Clones for which the earliest origin in natural populations is not known cannot automatically be considered to be genets. In practical terms, this means that all isolates with the same multilocus genotype must have arisen through asexual propagation from some common origin; these isolates belong to the same clone. Furthermore, it may be clear in some cases that highly similar, but not identical, clonal genotypes are unlikely to have arisen independently in sexual reproduction. In these cases, the highly similar clonal genotypes can be inferred to represent the same "clonal lineage" (26, 50) . In our view, the clonal lineage is really a clone whose origin is defined to reflect genetic changes that have accumulated during successive cycles of asexual reproduction.

An important aspect of the genet is that derivatives of the mitotic lineage need not be identical throughout the genome; a mutation or mitotic-recombination event does not create a new genet. In characterizing many fungal populations, a fundamental question is whether genotypic differences found in a sample of isolates are attributable to (*a*) genetic change accumulated during asexual propagation of a genet, especially if indications are that the genet is old or widely distributed geographically, or to (*b*) the presence of more than one genet. Base substitutions or deletions may occur during the mitotic cycle, albeit at a low frequency. Changes may occur more frequently with the potentially interrelated processes of chromosome rearrangement and movement of transposable elements (37). A good example of change within mitotic lineages is the extensive variation in electrophoretic karyotypes in clonal lineages of *Magnaporthe grisea* (75); some of these changes are even detected after serial transfer in the laboratory. In addition to karyotypic variability, tandemly repeated elements may increase or decrease in number during mitotic growth (61). Over time, many such genetic changes might accrue among the surviving members of a mitotic lineage. Changes accumulated during asexual propagation may even affect the phenotype (37). Deeply divergent members of the lineage may be difficult to recognize as such and could be mistakenly assumed to represent more than one genet.

Identification of the Genetic Individual

A common misconception is that a population with high levels of allelic diversity at individual loci must be predominantly sexual. However, even entirely asexual populations can harbor high levels of allelic diversity at a given locus or loci. The key expectation associated with clonal reproduction is a lower level of multilocus, genotypic diversity than would be the case in an entirely sexual population. The extent to which the genets in a sample of isolates can be distinguished will depend upon the number of genetic loci assayed, the frequencies of the alleles at these loci, and the extent to which mating in the population approximates a random pattern. If levels of gene diversity are high, and genets arise by more-or-less random mating, then it will be relatively easy to find a sufficient number of polymorphic markers to identify most or all genets in a sample of isolates with a high degree of confidence.

Under some conditions, genets may be difficult or impossible to identify. If there is significant inbreeding in small or isolated populations, and levels of genetic diversity relative to some larger population are low, then finding sufficient markers to identify genets may be difficult. Distinguishing actual genets under these conditions, however, may not be as critical as in an outbreeding population since most members of a highly inbred group are expected to share a preponderance of genes and combinations of genes that are identical by descent. In functional terms, the most highly inbred population might approximate a clone.

Patterns of Genetic Exchange and the Origins of Genetic Individuals in Populations

In attempting to distinguish genets, it is necessary to consider the antithesis of clonality: genetic exchange and recombination. Most importantly, patterns of genetic exchange and recombination giving rise to genets in a population may be strongly influenced by the presence or absence of a sexual incompatibility system. In homothallic species, a single haploid nucleus contained in a spore, hypha, or yeast cell carries the ability to complete the life cycle, without another nucleus of independent origin (33, 43). Meiosis resulting from selfed, homothallic fruiting is homozygous at each locus, no segregation occurs, and the intact genome is transmitted from one generation to the next. Because there is no fusion of genetically unlike gametes, no new genets are created. The prevalence of selfing in a homothallic fungus would therefore maintain a strongly clonal population structure. Although selfing is common in homothallic fungi, there is no genetic mechanism preventing outcrossing in such species. For example, although *Sordaria fimicola* and *Emericella nidulans* are homothallic, strains of each species are readily crossed in the laboratory (22). In the latter species, there is now strong evidence for genetic exchange and recombination in nature (24). Also, one fungal species, *Cryphonectria parasitica,* undergoes a mix of selfing and crossing in nature (58).

In heterothallic fungi, a single haploid nucleus must combine with another nucleus carrying a compatible mating type in order to complete the life cycle. The expected proportion of compatible matings among siblings and nonsiblings depends on the number of mating-type loci and the number of mating-type specificities, commonly termed alleles, at each locus in the population. One adaptive interpretation of sexual incompatibility systems is that they constrain levels of potential inbreeding among siblings, while allowing equal or higher levels of potential outbreeding (42). Despite the attention given to the genetics of sexual incompatibility systems, however, the actual proportion of sibling vs nonsibling matings among gametes of a heterothallic species in nature is not easily estimated because there is no physical connection between generations, as there is in an array of seeds resulting from many independent fertilizations in a single plant. One way to address this difficulty in basidiomycetes is to resolve the two component nuclear genotypes of dikaryons and then use the principle of genetic exclusion to identify which dikaryons in the local population could not possibly be the parents of those nuclei. Among the possible parents, the relative likelihood of parentage can then be calculated (77; see ref. 70a for analysis of male contribution in plants). In addition to the uncertainty regarding the fate of gametes produced by individual fruit-bodies, the actual extent to which inbreeding and outbreeding affect fitness of fungal genets has not been studied. Although an interesting variety of recessive mutations that deleteriously affect the sexual diplophase have been described in wild-collected strains of *Neurospora crassa* (48), we know of no quantitative studies of inbreeding depression in fungi comparable to the many such studies in plants and animals for which the evolutionary consequences of such systems have been considered in detail (14).

In addition to sexual mating, somatic exchange and parasexuality have the potential to create what should be considered new genets. Although parasexual genetic exchange may be limited in nature due to the prevalence of somatic incompatibility mechanisms in fungi (25, 46), its widespread occurrence cannot be ruled out. Unfortunately, it will be difficult using indirect genetic means to distinguish parasexual recombination from sexual recombination in those natural populations with any potential at all for sexual fruiting. Both parasexual and sexual exchange result in frequent reassortment of chromosomes. With only a few unlinked markers, the two processes might be indistinguishable. One expected difference is that parasexuality should result in lower rates of crossing over than would sexual recombination. An estimate of the relative contributions of parasexual vs sexual genetic exchange to recombination would therefore require detailed linkage maps, numerous markers, and a realistic model of the population structure. In the laboratory, parasexuality has been followed in many ascomycetes, but usually when strong selection for heterokaryosis and diploidy is applied to strains that are isogenic except for their marker mutations. In basidiomycetes, somatic recombination is often observed without any artificial selection for heterokaryosis or diploidy. For example, a heterokaryon can contribute a fertilizing nucleus to a haploid homokaryon in a process known as the Buller phenomenon. This process has been described as a Pandora's box, a prolific source of genetic complications that include somatic recombination of nuclear genotypes and, in one case, specific transfer of mating-type genes from one nuclear background to another (63). Also, there is surprising new evidence from the basidiomycetes, *Heterobasidion annosum* and *Agaricus bisporus,* that somatic recombination occurs at strikingly high frequencies when unrelated heterokaryons are confronted (30, 80). In *A. bisporus,* this somatic recombination even includes both reassociation of chromosomes and crossing over (80). Matings between diploid and haploid strains of *Armillaria gallica* can also result in recombinant nuclei (13), which are effectively new genets.

ARMILLARIA GALLICA: A ROOT-INFECTING BASIDIOMYCETE WITH TERRITORIAL CLONES

Clonal genotypes of *A. gallica* and other root-infecting basidiomycetes are usually found in discrete patches in forest soils and are not repeatedly recovered from different localities. Because of this pattern of spatial distribution, clones of these fungi can safely be inferred to represent genets, each of which originated in a unique mating event and then grew vegetatively from a single point of origin over a period of time to reach its present distribution.

Identification of Genets

Genets of *A. gallica* and other species of root-infecting basidiomycetes have been identified according to three criteria: somatic incompatibility reactions, mating types, and neutral molecular markers such as allozymes and DNA polymorphisms. Somatic incompatibility is the visible reaction line that usually results when isolates of different genets are paired in culture. In all cases, isolates of the same genet show no somatic incompatibility when confronted in culture. In nature, somatic compatibility may be the determining mechanism limiting spatial mixing, anastomosis, and physiological exchange among mycelia of different genets (64, 65). Unfortunately, nothing is known about the cellular basis for the recognition system underlying somatic incompatibility in basidiomycetes.

As a genetic criterion for identifying genets of root-infecting basidiomycetes, somatic compatibility has been extremely useful, but with limitations. Somatic incompatibility is a simple phenotype with a complex genetic basis. While recent studies of *H. annosum* (29) and *Pleurotus ostreatus* (51) show up to several segregating loci from a given fruit-body, the total number of genetic loci in the population and the degree of polymorphism at these loci are unknown. As would be expected from these and other studies (36, 44), the frequency and strength of somatic incompatibility reactions is lower among related than among unrelated individuals. Despite these limitations, most genets would be expected to be somatically incompatible in outbreeding populations, although occasionally two distinct genets might be somatically compatible by chance inheritance of determinants. Given these limitations, how effective are somatic incompatibility tests in identifying the genets in a sample of isolates? There are reports of (*a*) perfect agreement among the groups identified by somatic incompatibility testing and the groups identified by their composite phenotypes with respect to molecular or mating-type markers (68, 69), (*b*) an overall pattern of agreement with a few exceptions (28, 36, 44), and (*c*) an overall pattern of disagreement (35). Generalizations about the usefulness of somatic compatibility in identifying genets based on a comparison of somatic compatibility tests and other markers may be presumptuous. Some disagreement among the groups identified with somatic compatibility tests and the groups identified with other markers is expected by chance alone in randomly mating populations, but disagreement is expected to occur more frequently in populations with significant inbreeding and where only a few determinants are segregating. To fully assess how effectively somatic incompatibility will identify genets, knowledge of the underlying genetics of the phenomenon and of the structure of the population is required; these characteristics may well vary among populations and species.

Another biologically important criterion used to identify genets of basidiomycetes is mating-type polymorphism. Although mating types of basidiomycetes can be genetically interpreted very clearly as one or two unlinked "factors," each with a series of multiple specificities or "alleles," they have one important limitation in identifying genets. Mating-type loci are constrained to heterozygosity in dikaryons or diploids because only haploids with different mating-type alleles at the one or two factors can mate successfully. Therefore, mating-type alleles should function well in identifying genets of unrelated origin, but cannot distinguish genets arising from the mating of sibling spores, since all such genets would carry exactly the same set of mating-type alleles. Studies relying only on mating-type alleles cannot distinguish whether a set of isolates carrying the same mating-type alleles and arising from a distinct spatial patch represent (*a*) one genet that reached its present size by vegetative growth or (*b*) several closely related genets. Because basidiospores show heavy, local deposition, even producing a visible sporeprint near the fruit-body, the second possibility cannot be automatically excluded.

Given the limitations of both somatic incompatibility and mating-type tests, additional markers that are selectively neutral become necessary to distinguish vegetative growth from local dissemination of basidiospores followed by sibling mating. The two patterns of colonization lead to different genetic expectations. Essentially, vegetative growth should preserve heterozygosity, whereas mating of spore germlings that are related by descent should result in loss of heterozygosity at some loci in some of the resulting genets. The probability that a diploid arising from mating of sibling spores would retain heterozygosity at a given locus is 0.5. When enough loci have been assayed such that the joint probability falls below a suitable threshold, the null hypothesis of sibling mating can be rejected. This approach was used to exclude the possibility that several isolates of a genet of *A. gallica* from an area of more than 15 hectares of forest in Northern Michigan resulted from sibling mating among basidiospores (68).

Spatial Distribution of Genets

Numerous studies have spatially mapped genets of root-infecting basidiomycetes in the field by taking samples of isolates from defined locations and then characterizing the isolates with respect to one or more genetic criteria. These fungi include both root-rotting (3, 15, 19, 28, 36, 44, 67–69, 72) and ectomycorrhizal species (18). From these studies, a clear pattern emerges: Genets occupy spatial territories ranging from very small, represented by only a single isolate, to very large, represented by many isolates from many adjacent root-systems over several hectares of forest. Each genet appears to occupy a territory to the exclusion of other genets of the same species. In *A. gallica,* inoculum of a genet, the network of rhizomorphs, is resident in both roots and fallen logs. After a substrate is colonized, rhizomorphs radiate out into the surrounding soil, extending the network into the humus layer, enabling the fungus to colonize other suitable substrates that come into contact with the humus layer. Although other root-infecting basidiomycetes spread by a variety of means, including root-to-root contacts, the territoriality of clones is always evident. In a recent study of *A. gallica,* isolates were collected from eight sampling sites separated by 20 km to 2000 km, with 50 m to 200 m between individual isolates within a site. Among 250 isolates, 123 genets were identified on the basis of unique multilocus genotypes (JB Anderson, unpublished). For each of 35 genets, more than one isolate was recovered; isolates of these genets were invariably from adjacent or nearby collection points. Such a pattern of distribution of genotypes is consistent with vegetative spread of genets within sites and is inconsistent with the spread of genets from place to place by dispersal of mitotic propagules.

Although the distributions of the genets of many species of basidiomycetes have been mapped, none of these studies could pinpoint the exact spatial limits of the genet, since samples were taken only from selected points within a locality. Resolving the spatial borders of genets depends upon the scale of sampling. For example, Guillaumin et al (28) sampled forests at two different spatial scales. When collections were made at intervals of 20 m, no spatial mixing of genets was detected. But when an area 2.5 m square at the border of two genets was selected for further, intensive sampling at a much finer scale, rhizomorphs of the two adjacent genets were shown to be spatially mixed. Although the territories of the two genets were distinct, there was some overlap of the genets at their borders.

Although there is usually little spatial overlap of genets of the same species, there can be considerable overlap of the genets of different species, which in effect are partitioned in different ecological niches. For example, one study examined an area of mixed hardwood forest in Michigan that was clear cut and replanted with red pine in 1984 (69). The area is occupied by two genets of *A. gallica,* which extend throughout the entire clear cut and many hectares of surrounding forest. *A. gallica* was present as an extensive network of rhizomorphs throughout the humus layer, fruited abundantly on nearly every hardwood stump (68, 70), but was not associated with any of the red pine seedlings that were dying from *Armillaria* root rot each year. By 1989, in addition to the two genets of *A. gallica,* 20 genets of *A. ostoyae* were identified among isolates from dead or dying red pine seedlings (Figure 1). Although some of the same genets of *A. ostoyae* fruited on hardwood stumps in some years, most were recovered only from infected red pines and never as rhizomorphs in the soil on this site. Several of these genets had killed adjacent red

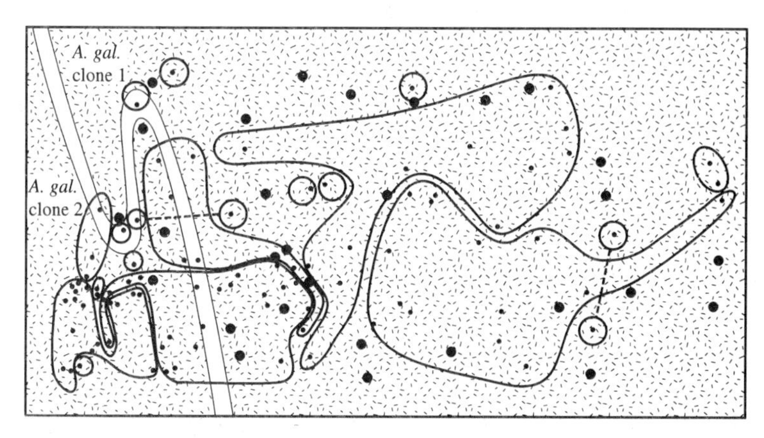

Figure 1 Territoriality of genets of *Armillaria ostoyae* and *A. gallica* in 75 × 140 m area of mixed hardwood forest that was clear cut and replanted with red pine. Large dots represent collections of *A. gallica*. The two large genets of *A. gallica*, shown by the stippled areas and marked as clone 1 or clone 2 (see ref. 70), extend for several hundred meters out into the surrounding uncut hardwood forest (see ref. 68). Small dots represent collections of *A. ostoyae* from dead or dying pine seedlings, which are encircled to show the approximate distribution of genets (see ref. 69).

pine seedlings in centers that seemed to be expanding from year to year. Similar results of spatial overlap among the genets of different species have also been reported by others (28, 66, 79). Such examples of spatial overlap between species may be an artifact of the coarse scale of sampling. At a fine scale, the species must occupy different spaces.

While genets and their spatial distributions have been relatively easy to describe, the physical or physiological units of a genet are not obvious because the mycelium cannot be seen in its entirety within its opaque substrate and can only be sampled destructively. The entire mycelium within a large territory occupied by a genet is unlikely to function as one physical or physiological unit. Within the territory, there is continual turnover of fungal biomass; cells are not immortal. There are likely to be physical gaps in the mycelium constituting the genet. In fact, man-made obstructions (road cuts) are known to bisect genets covering large territories (36), including the largest genet of *A. gallica* identified on the Michigan site (JB Anderson, unpublished). Furthermore, what might constitute a distinct physiological unit at one time could become subdivided into a number of units through famine-induced thinning or mechanical fragmentation of the vegetative network of rhizomorphs and mycelium. Some evidence for such thinning is found in the few cases where a genet occurs in noncontiguous patches within a site (19, 28, 36). Conversely, adjacent mycelia of the same genet could coalesce freely during periods of growth.

How did genets of *A. gallica* and other root-infecting basidiomycetes come to occupy such large territories in forests? All estimates of genet age in basidiomycetes are based on measurements of growth rate under field conditions (36, 67, 68). These estimates all assume that (*a*) growth rate is uniform from year to year and (*b*) growth occurs in a simple radial manner. The colonization of root systems by a genet may not actually proceed in this way in nature. Given the discrete territories commonly observed, it is likely that linear expansion is impeded where two genets meet one another. Furthermore, living mycelium within the territory occupied by a genet is continually replaced as food bases decay and new ones become available. There are two main hypotheses that could explain the present large size of genets: (*a*) an area was initially sparsely colonized by a few genets each growing unimpeded to its present size, together filling the entire habitat, or (*b*) an area was initially colonized by many genets, some of which gradually expanded their territory at the expense of others, perhaps by colonizing new food bases more efficiently than their neighbors. Recent observations in *H. annosum* (74) are most easily reconciled with the latter hypothesis. Freshly cut stump surfaces of pine have been shown to contain many genets of *H. annosum*. The stump surface presumably acts as a favorable infection court for basidiospores, some of which mate to form heterokaryons, whereas others remain homokaryotic (73). While many genets are found on the stump surfaces, the root-systems of trees that have been heavily infected for a longer time generally harbor only a single genet. This observation is consistent with the possibility that the many initial genets of *H. annosum* are "weeded out" in a process driven by either selection for the fittest genoytpes or chance. Similarly, the size distributions of genets of *Armillaria* species may be the result of such a process (79). Unlike *Heterobasidion*, the infection pathway taken by basidiospores of *A. gallica* or other *Armillaria* species is unknown.

Origin of Genets

If each genet arose through a mating event and then reached its present distribution through vegetative growth alone, what was the nature of the original mating event? We have been investigating the possibility that mtDNA can answer this question because its mode of transmission is different from that of nuclear markers and it is highly polymorphic in the population. In laboratory matings between haploid strains, hyphae anastomose and nuclei migrate extensively through the two mycelia. Each strain usually acts simultaneously as donor and recipient of nuclei. After reciprocal nuclear migration, there is pairing between the two types of nuclei, resulting in a uniformly dikaryotic or diploid mycelium. While the nuclei migrate extensively, the mitochondria, as marked by mtDNA haplotypes, do not migrate. Any mixing of mtDNA types must be restricted to the immediate area where the hyphae of mates anastomose. This pattern of nuclear and mitochondrial transmission leads to a colony with a uniform nuclear genotype that is a mosaic of mtDNA types (1, 34, 52, 70, 71).

The many reports on mitochondrial transmission in laboratory experiments led us to ask whether a diploid genet of *Armillaria* is a mosaic of mtDNA types in nature. At present, no mtDNA mosaics have been found among the 35 genets of *A. gallica* (68, 70; JB Anderson, unpublished) and the 7 genets of *A. ostoyae* (69) for which 2 or more isolates were recovered from field collections. Why are mitochondrial mosaics prevalent in laboratory matings, but not in genets found in nature? We see three possibilities. First, there may be unequal input of cytoplasm during mating in the field. For example, when a spore fertilizes a haploid mycelium, the mtDNA haplotype of the spore is unlikely ever to be recovered given the vastly more numerous mtDNA molecules of the haploid mycelium. In the laboratory matings, generally two equal-sized haploid colonies are confronted, making it much more likely that each mtDNA type will be transmitted. Second, one of the two mtDNA types may have a selective or replicative advantage in a given diploid background. So far there is no experimental evidence for or against this possibility. Third, one mtDNA type may be lost randomly as portions of the mycelium of a genet die and other small portions serve as inoculum for new colonization. These possibilities are not mutually exclusive. That there is mixing of mtDNA types at some point during the origin of genets in nature is supported by preliminary data suggesting that the high level of mtDNA diversity in *A. gallica* and *A. ostoyae* is due at least in part to recombination. We are currently testing the recombination hypothesis in *A. gallica.*

SCLEROTINIA SCLEROTIORUM: A STEM-INFECTING ASCOMYCETE WITH DISPERSIVE CLONES

In both *S. sclerotiorum* and *A. gallica,* clonal propagation can yield a large amount of genetically uniform biomass for each clone. When, in the course of a population survey, this biomass is sampled randomly, in transects, or in grids, isolates with the same multilocus genotype will be recovered. Although clonal propagation is important in both species, the spatial distribution of clones is entirely different. In *A. gallica* clones are the equivalent of genets occupying discrete territories. In contrast, in agricultural populations of *S. sclerotiorum,* clones are scattered in small, noncontiguous units or patches that are spatially disconnected from their points of origin and are highly mixed within fields. For this reason, it is not clear whether clones of *S. sclerotiorum* represent genets.

Identification of clones

In *S. sclerotiorum*, as in *A. gallica*, somatic compatibility and molecular markers are linked in clonal genotypes. Unlike *A. gallica*, however, *S. sclerotiorum* has no sexual incompatibility system and therefore mating types are unavailable for identification of clones. Clones of *S. sclerotiorum* are distinguished by the repeated recovery of genotypes sharing two features. First, in canola, sunflower, and weed species associated with agriculture, all members of a clone are somatically compatible with all other members of the same clone and are incompatible with members of other clones (38, 41). Somatic compatibility in *S. sclerotiorum* has been termed mycelial compatibility. Mycelial incompatibility is evident in a reaction line between mycelia that are paired in culture. While mycelial incompatibility is a clearly expressed binary trait, additional genetic criteria are needed to determine whether mycelial compatibility groups (MCGs) are clones. This is because the underlying genetics of the phenomenon are not known. Because of the high diversity in MCGs, however, we can conclude that the trait must be determined by several loci and/or alleles. The actual mechanisms of somatic incompatibility in *S. sclerotiorum*, as in basidiomycetes and other ascomycetes, are not known. The second distinguishing feature is that each clone has a unique set of molecular markers including a unique DNA fingerprint (38, 41), one of two RFLP phenotypes associated with the large mitochondrial rRNA gene (41), and the presence among all members of a clone or the absence among all members of a clone of a Group IC intron within the small, mitochondrial rRNA gene (12).

Of these molecular criteria, DNA fingerprinting has been used the most extensively in *S. sclerotiorum* because the chance of a match in fingerprints in individuals of independent, sexual origin is several orders of magnitude lower than for the other criteria, except for mycelial compatibility, where the chance of a match cannot be estimated (38). The repeated element used for DNA fingerprinting is known to be present on most of the chromosomes of *S. sclerotiorum* (41). Part of the dispersed, repeated element has been sequenced and this is similar to the 3′ end of a non-LTR-retrotransposon (4).

Spatial Distribution of Clones

Because *S. sclerotiorum* is homothallic, airborne ascospores produced as a result of selfing can propagate clonal genotypes. The asexually produced sclerotia also propagate clonal genotypes and can survive for documented periods of 4–5 years in the soil, dispersing through cultivation or movement of infested seed, irrigation water, or agricultural machinery. The distribution of clones of *S. sclerotiorum* has been studied most extensively in canola fields (38, 41). In one study of four fields in western Canada (39), 594 clonal genotypes were found among 2747 isolates. Each of the fields was infested

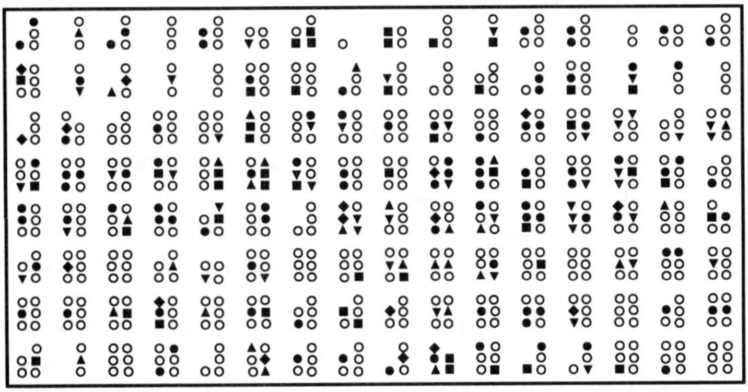

Figure 2 Spatial mixing of clones of *Sclerotinia sclerotiorum* in a canola field. Data are from field 4 in the study of Kohli et al (39). The five most frequent clones are designated with solid symbols. Other, less frequent clones are designated with an unfilled circle. Up to six isolates in each of 64 sampling stations in the 400 × 800 grid divided at 50 m intervals are grouped. The isolates on the left of each group were isolated from the petals of flowers; those on the right were from lesions on stems. All six isolates within a group came from approximately the same location; the order of isolates within each column of three is arbitrary.

and infected by many clones. Clone frequencies in each field ranged from high to low, with 5% of the clones collected more than seven times and 75% of the clones collected only once. Nine of the 30 most frequently collected clones from this study were previously recovered in a macrogeographical sample from western Canada in 1990 (38) and one frequently sampled clone had also been recovered in Ontario in 1989 (41). Within each of the four fields, clonal genotypes were distributed randomly among the collection points (Figure 2). Despite the random distribution of clones within fields, however, clone frequencies differed between some of the fields. Although clone frequencies were not significantly different between the two fields sampled in Meadow Lake, Saskatchewan, they were marginally significantly different between the two fields sampled in Olds, Alberta. Between the Alberta and Saskatchewan fields, however, clone frequencies were highly significantly different, indicating a clonal population that is subdivided geographically.

In addition to studying the spatial distribution of clones, clone frequencies were compared during the course of the disease cycle of sclerotinia stem rot. Soilborne sclerotia produce apothecia during the flowering period of canola. When petals infested by airborne ascospores dehisce, fall, and adhere to canola stems, the ascospores germinate and initiate disease lesions in the stems. In the two fields in Saskatchewan, there was no difference in clone frequencies among isolates taken from infested petals when compared with those from disease lesions. This indicates that strong selection for clonal genotypes is not always present. In sharp contrast, clone frequencies were highly significantly different between isolates from petals and isolates from lesions in the two fields in Alberta. Some, but not all, of the difference was due to the appearance of two clones that had not previously been recovered in sampling on canola but were recovered at high frequency in these fields. Both clones increased in frequency during the disease cycle. It is tempting to speculate that selection was responsible for observed changes in clone frequency. For example, differences in fecundity, measured as the number of apothecia per unit sclerotium over time, a trait possibly under selection, could result in an increase in the frequency of certain clones. Alternatively, chance cannot be ruled out. For example, prolonged immigration of a clone from other fields could result in an increase in frequency between inoculum and lesion phases of the disease cycle, although this could also be related to fecundity. The lack of change in clone frequencies in the 1991 season and the change in the 1992 season indicate that processes effecting change in clone frequencies are not uniform from place to place and from year to year.

The overall pattern of dispersive clonal reproduction in *S. sclerotiorum* contrasts with the patterns in some other fungal plant pathogens that are not soilborne, but have dispersive clonal reproduction in at least three important ways. First, clone diversity in *S. sclerotiorum* is high relative to that in *Phytopthora infestans* (23) and *Magnaporthe grisea* (50) in most agricultural fields. Although the different levels of clone diversity may be in part an artifact of how the clones are identified in these species, the distribution of genotypes is very different. Clonal genotypes of *P. infestans* and *M. grisea* cluster into a small number of groups within which genotypes are highly similar; in *S. sclerotiorum* clonal genotypes do not cluster into a small number of distinct groups. Second, while clones of *S. sclerotiorum* have apparently migrated between agricultural fields in different regions, their geographic origins are unknown because archival population samples do not exist. In contrast, there is strong genetic and epidemiological evidence that a single clone of *P. infestans* caused the Irish potato famine after an early migration from Central Mexico to Europe (26). This clonal lineage now has a worldwide distribution. Further data indicate that there have been subsequent, migrations of other clonal lineages of *P. infestans* from central Mexico to potato growing regions around the world. Third, while clones of *S. sclerotiorum, P. infestans*, and *M. grisea* are highly dispersive, clones of some phytopathogenic fungi are less dispersive. In *Septoria tritici,* clones arise within a predominantly sexual population and appear to exist for only a single season during which they are limited in their dispersal, resulting in highly clustered distribution within a field (16, 55).

Origin of Clones

A fundamental question arising from studies of clonality in *S. sclerotiorum* is whether clones represent genets. If so, this would mean that most clonal genotypes originated as ascospores resulting from crossing. The alternative possibility is that all of the observed clones arose from a single source without genetic exchange and recombination through an accumulation of mutations. Although all 50 apothecia of *S. sclerotiorum* sampled to date from canola fields were the products of selfing (as evidenced by the absence of recombinant DNA fingerprint or mycelial compatibility phenotypes among sibling monospore isolates), the existence of outcrossing at some low, but evolutionarily significant level cannot be ruled out. This question can be addressed indirectly by examining whether the alleles at different loci are associated with one another. The problem is finding suitable marker loci. Unfortunately, restriction polymorphism in single-copy DNA is not very common in this species. Of 20 such sequences examined for 10 enzymes no variants were found (Y Kohli & LM Kohn, unpublished). The only DNA known to be associated with a high level of variation is the sequence used for DNA fingerprinting (41). We therefore tested whether associations among fragments in isolates from the study of four fields in western Canada (40) are random. In this sample of isolates, the fragments were assigned to 53 distinct size classes (39). Of these, the 24 fragment classes present in frequencies between 10 and 90% were analyzed for association. Fragments with frequencies outside this range were not tested for association in this preliminary analysis because the numbers in many of the expected genotypic classes were fewer than five, ruling out the use of the χ-square test. When all 2747 isolates were included in the analysis, only 16% of the pairs of fragment size classes were randomly associated, while the remainder were not. When the data were censored to include only one representative of each of the 594 clones, however, the frequency of pairs found to be randomly associated rose to 62% (Y Kohli, unpublished). That the null hypothesis of random association could not be rejected in so many comparisons cannot be due merely to the loss of statistical power associated with the reduction in sample size. We interpret this preliminary result as consistent with at least some genetic exchange and recombination that would result in the origin of new clones. We cannot rule out the possibility that the observed genotypes arose solely by repeated, independent loss or gain of certain fingerprint fragments. This possibility would be heightened if the insertion of the transposable element into genomic DNA is not random. Independent loss or gain of fragments in the complete absence of recombination, however, is difficult to reconcile with the random association of so many of the fingerprint fragments. Furthermore, comparison of all pairs of fingerprint fragments, including those that are not randomly associated, revealed the presence of all four possible combinations of presence/absence. In the absence of recombination, repeated gain or loss of at least one fingerprint fragment would therefore be required for each pair of the 53 fragment size classes to explain the observed fingerprints.

Population Structure of S. sclerotiorum in the Wild

In addition to its occurrence on most agricultural crops, with the notable exception of cereals, *S. sclerotiorum* infects many wild plants in natural plant populations not associated with agriculture (5). The population structure of *S. sclerotiorum* in the wild appears, however, to be very different from that of *S. sclerotiorum* in agricultural systems and is included here as an example of small, isolated populations in which the impact of clonal reproduction is difficult to assess.

In two sites in Norway, abundant apothecia of *S. sclerotiorum* have been observed under dense canopies of *Ranunculus ficaria,* over several seasons. *R. ficaria* is a common, perennial, woodland understory plant that spreads clonally by rhizomes, presenting a recurring, stable host population. Apothecia are usually produced for about two weeks when *R. ficaria* is in bloom. Isolates of *S. sclerotiorum* from apothecia under *R. ficaria* are more variable in phenotypic traits such as growth rate, pigmentation, and the amount of aerial mycelium than the comparatively uniform agricultural isolates from both Canada and Norway. On the basis of a study of *S. sclerotiorum* in the two *R. ficaria* sites, and a comparative sample of *S. sclerotiorum* from agricultural crops of canola and potato in Norway (40), agricultural and wild populations of *S. sclerotiorum* can be compared.

The structure of the two wild populations of *S. sclerotiorum* differed from that of agricultural populations studied to date in four major ways. First, in the wild populations, unique DNA fingerprints and mycelial compatibility groupings were not congruent; in agricultural populations in North America, Norway, and presumably elsewhere, each clone had a unique fingerprint and all individual members of a clone were mycelially compatible with all members of the same clone, but incompatible with members of other clones. Second, in the wild populations there were relatively few fingerprints; in agricultural fields, fingerprint diversity is much higher, e.g. five apothecia from a crop would likely have four to five different fingerprints. Third, within what are apparently isolated wild populations, strong spatial substructuring was evident as a localized distribution of fingerprints within a site; in agricultural populations, clones have been observed to be randomly dispersed spatially. Last, there was evidence that some outcrossing had occurred because sibling ascospores from each of several apothecia showed different mycelial compatibility reactions, indicative of crossing and segregation; in studies of agricultural populations, no segregation has been observed in studies of sibling monosporous isolates from field-collected or in vitro–grown apothecia in attempted crosses between clones. The data suggest that the two wild populations of *S. sclerotium* on *Ranunculus* are each isolated and inbred. Within each of these populations only a small number of alleles determining mycelial compatiblity and DNA fingerprint are segregating; most are fixed. It is possible that identical fingerprint or mycelial compatiblity types can arise repeatedly, in independent crosses within inbreeding, spatial subgroups within each site. As a result, clonality would not be easily differentiated from groups arising via inbreeding with the markers presently available. A similar situation has been described in *C. parasitica* (57)

The highly local, site-specific, near fixation of compatibility or incompatibility and of DNA fingerprint in the wild populations *of S. sclerotium* on *Ranunculus* is unlike the spatial mixing observed in agricultural populations of *S. sclerotiorum* and is actually more reminiscent of the territoriality observed among clones of *A. gallica.* Gordon et al (27) had earlier reported a highly aggregated distribution of mitochondrial haplotypes of *Fusarium oxysporum* in native soil, compared to a random distribution in cultivated soil, suggesting that agricultural practices may mix and homogenize soilborne pathogen populations. Our data on *S. sclerotiorum* have been consistent with this hypothesis.

CONCLUSIONS

The focus of this review is variation in patterns of clonality among soilborne, plant pathogenic fungi. In *A. gallica* and other root-infecting basidiomycetes in forests, clones exist as discrete patches that can be interpreted as genets, each of which is spatially connected to a mating event at its point of origin. In these fungi, each clone colonizes only through the relatively slow process of vegetative growth. Since clones cannot disperse from place to place, no clone has the opportunity to become frequent in the larger population, which is completely sexual.

The clonal biology of *S. sclerotiorum* and other soilborne agricultural pathogens is vastly different from that of *A. gallica.* In many species, including *S. sclerotiorum,* the potential for genetic exchange and recombination has not been documented. In these cases, it is not known whether clones (*a*) originated in sexual reproduction and therefore represent genets, or (*b*) evolved without genetic exchange. Furthermore, clones of *S. sclerotiorum* exist in spatially noncontiguous patches that are disconnected from their place of origin. The consequence of this is that clones of *S. sclerotiorum,* unlike clones of *A. gallica,* have the opportunity to increase in frequency in the population well beyond their place of origin.

Studies of the spatial patterns of clonality raise several questions for future research. Can the high frequency and wide distribution of some clones of *S.*

sclerotiorum in agriculture be explained by their relative fitness? Or is the prevalence of some clones more a result of the random fluctuation in clone frequency that is characteristic of a small founding population? The comparable question can be asked for *A. gallica.* Do some genets occupy large territories because they are fitter than their competitors? Or does chance determine which genets ultimately colonize large territories?

An experimental approach to questions of selection in soilborne pathogens with clonal reproduction is eminently feasible (21). In clonal propagation, all portions of the genome, including both neutral genetic markers and determinants of phenotypic traits, are linked. In both *S. sclerotiorum* and *A. gallica,* the relative fitness of specific genetic individuals could be readily measured in a given environment by following the frequency of any existing genetic marker. At the level of large populations, however, genetic markers may or may not correlate with phenotypic traits under selection, depending on the nature of clonal propagation. Because clones of *S. sclerotiorum* are highly dispersive within a largely asexual population in agricultural fields, genetic markers would be expected to correlate with phenotypic traits under selection. In contrast, because clones of *A. gallica* are nondispersive within a sexual population, traits under selection would not be expected to correlate with known genetic markers due to the prevalence of recombination.

Finally, although the spatial origin of clones of *A. gallica* is known, that of clones of *S. sclerotiorum* remains an open question. Did the clonal genotypes of *S. sclerotiorum,* which are now prevalent in agriculture, originally come from one or more genetically diverse and possibly sexual populations on wild plant hosts? If so, where are these wild populations located? Migration of clones from specific regions is not only possible, but has been amply documented in *P. infestans* (26). An unknown quantity of basic biological interest and practical importance is the long-term potential of soilborne fungal pathogens now in the wild to enter agricultural systems (40). To assess this potential, we need to know much more about populations of fungal pathogens in the wild.

ACKNOWLEDGMENTS

We are grateful to Michael G. Milgroom, Barry J. Saville, and Johann N. Bruhn for many helpful comments on earlier drafts of this article. Our work with *S. sclerotiorum* and *A. gallica* is supported by the Natural Sciences and Engineering Research Council of Canada.

Literature Cited

1. Ainsworth M, Rayner ADM, Broxholme SJ, Beeching JR. 1990. Occurrence of unilateral genetic transfer and genomic replacement between strains of *Stereum hirsutum* from non-outcrossing and outcrossing populations. *New Phytol.* 115: 119–28
2. Anderson JB, Kohn LM, Leslie JF. 1992. Genetic mechanisms in fungal adaptation. In *The Fungal Community: Its Organization and Role in the Ecosystem,* ed. GC Carroll, DT Wicklow, pp. 73–98. New York: Dekker. 2nd ed.
3. Bae H, Hansen EM, Strauss SH. 1994. Restriction fragment length polymorphisms demonstrate single origin of infection centers in *Phellinus weirii. Can. J. Bot.* 72:440–47
4. Baller LM. 1992. *Analysis of a dispersed repetitive DNA element in Sclerotinia sclerotiorum.* MSc. thesis. Univ. Toronto. 63 pp.
5. Boland G, Hall R. 1994. Index of plant hosts of *Sclerotinia sclerotiorum. Can. J. Plant Pathol.* 16:93–108
6. Brasier CM. 1992. Evolutionary biology of Phytophthora: Part I: Genetic system, sexuality and the generation of variation. *Annu. Rev. Phytopathol.* 30: 153–71
7. Brasier CM, Hansen EM. 1992. Evolutionary biology of Phytophthora: Part II: Phylogeny, speciation, and population structure. *Annu. Rev. Phytopathol.* 30: 173–200
8. Brasier CM, Rayner ADM. 1987. Whither terminology below the species level in fungi? In *Evolutionary Biology of the Fungi,* ed. ADM Rayner, CM Brasier, D Moore, pp. 379–88. Cambridge, UK: Cambridge Univ. Press
9. Bruehl GW. 1987. *Soilborne Plant Pathogens.* New York: Macmillan. 368 pp.
10. Burdon JJ. 1993. The structure of pathogen populations in natural plant communities. *Annu. Rev. Phytopathol.* 31: 305–23
11. Burt A, Carter DA, Taylor JW, White TJ. 1994. *Genetic epidemiology of the human pathogen Coccidioides immitis.* Presented at Int. Mycol. Congr., 5th, Vancouver, Canada
12. Carbone I, Anderson JB, Kohn LM. 1995. A group I intron in the mitochondrial small subunit ribosomal RNA gene of *Sclerotinia sclerotiorum. Curr. Genet.* 27:166–76
13. Carvalho DB, Smith ML, Anderson JB. 1995. Genetic exchange between diploid and haploid mycelia of *Armillaria gallica. Mycol. Res.* In press
14. Charlesworth D, Charlesworth B. 1987. Inbreeding depression and its evolutionary consequences. *Annu. Rev. Ecol. Syst.* 18:237–68
15. Chase TE, Ullrich RC. 1983. Sexuality, distribution, and dispersal of *Heterobasidion annosum* in pine plantations of Vermont. *Mycologia* 75:825–31
16. Chen R-S, Boeger JM, McDonald BA. 1994. Genetic stability in a population of a plant pathogenic fungus over time. *Mol. Ecol.* 3:209–18
17. Correll JC, Gordon TR, McCain AM. 1988. Vegetative compatibility and pathogenicity of *Verticillium alboatrum. Phytopathology* 78:1017–21
18. Dahlberg A, Stenlid J. 1990. Population structure and dynamics in *Suillus bovinus* reflected by spatial distribution of fungal clones. *New Phytol.* 115:487–95
19. Dickman A, Cook S. 1989. Fire and fungus in a mountain hemlock forest. *Can. J. Bot.* 67:2005–16
20. Dykhuizen DE, Green L. 1991. Recombination in *Escherichia coli* and the definition of biological species. *J. Bacteriol.* 173:7257–68
21. Ennos RA, McConnell KC. 1995. Using genetic markers to investigate natural selection in fungal populations. *Can. J. Bot.* In press
22. Fincham JRS, Day PR, Radford A. 1979. *Fungal Genetics.* Berkeley: Univ. Calif. Press. 636 pp. 4th ed.
23. Fry WE, Goodwin SB, Matuszak JM, Spielman LJ, Milgroom MG. 1992. Population genetics and intercontinental migrations of *Phytophthora infestans. Annu. Rev. Phytopathol.* 30:107–29
24. Geiser DM, Arnold ML, Timberlake WE. 1994. Sexual origins of British *Aspergillus nidulans* isolates. *Proc. Natl. Acad. Sci. USA* 91:2349–52
25. Glass NL, Kuldau GA. 1992. Mating type and vegetative incompatibility in filamentous ascomycetes. *Annu. Rev. Phytopathol.* 30:201–24
26. Goodwin SB, Cohen BA, Fry WE. 1994. Panglobal distribution of a single clonal lineage of the Irish potato famine fungus. *Proc. Natl. Acad. Sci. USA* 91:11591–95
27. Gordon TR, Okamoto D, Milgroom M. 1992. The structure and interrelationship of fungal populations in native and cultivated soils. *Mol. Ecol.* 1:241–49
28. Guillaumin JJ, Anderson JB, Legrand P, Ghahari S. 1994. Use of different

methods for mapping clones of *Armillaria* spp in four forests of central France. In *Proc. IUFRO Int. Conf. Root Butt Rots, 8th*, ed. M Johansson, J Stenlid, pp. 437–58. Uppsala: Swedish Univ. Agric. Sci.

28a. Guttman DS, Dykhouizen DE. 1994. Clonal divergence in *Escherichia coli* as a result of recombination, not mutation. *Science* 266:1380–83

29. Hansen EM, Stenlid J, Johansson M. 1993. Genetic control of somatic incompatibility in the root-rotting basidiomycete *Heterobasidion annosum. Mycol. Res.* 97:1229–33

30. Hansen EM, Stenlid J, Johansson M. 1993. Somatic incompatibility and nuclear reassortment in *Heterobasidion annosum. Mycol. Res.* 97:1223–28

31. Harper JL. 1977. *Population Biology of Plants.* New York: Academic

32. Harper JL. 1985. Modules, branches and the capture of resources. In *Population Biology and Evolution of Clonal Organisms*, ed. JBC Jackson, LW Buss, RE Cook, pp. 1–33. New Haven, CT: Yale Univ. Press

33. Herskowitz I. 1988. Life cycle of the budding yeast *Saccharomyces cerevisiae. Microbol. Rev.* 52:536–53

34. Hintz WEA, Anderson JB, Horgen PA. 1988. Nuclear migration and mitochondrial inheritance in the mushroom *Agaricus bitorquis. Genetics* 119:35–42

35. Jacobson KM, Miller OK, Turner BJ. 1993. Randomly amplified polymorphic DNA markers are superior to somatic incompatibility tests for discriminating genotypes in natural populations of the ectomycorrhizal fungus *Suillus granulatus. Proc. Natl. Acad. Sci. USA* 90: 9159–63

36. Kile GA. 1983. Identification of genotypes and the clonal development of *Armillaria luteobubalina* Watling & Kile in eucalypt forests. *Aust. J. Bot.* 31:657–71

37. Kistler HC, Miao VPW. 1992. New modes of genetic change in filamentous fungi. *Annu. Rev. Phytopathol.* 30:131–52

38. Kohli Y, Morrall RAA, Anderson JB, Kohn LM. 1992. Local and trans-Canadian clonal distribution of *Sclerotinia sclerotiorum* on canola. *Phytopathology* 82:875–80

39. Kohli YL, Brunner LJ, Yoell H, Milgroom MG, Anderson JB, Morrall RA, Kohn LM. 1995. Clonal dispersal and spatial mixing in populations of the plant pathogenic fungus, *Sclerotinia sclerotiorum. Mol. Ecol.* 4:69–77

40. Kohn LM. 1995. The clonal dynamic in wild and agricultural plant pathogen populations. *Can. J. Bot.* In press

41. Kohn LM, Carbone I, Stasovski E, Royer J, Anderson JB. 1991. Mycelial incompatibility and molecular markers identify genetic variability in field populations of *Sclerotinia sclerotiorum. Phytopathology* 81:480–85

42. Koltin Y, Stamberg J, Lemke PA. 1972. Genetic structure and evolution of the incompatibility factors in higher fungi. *Bacteriol. Rev.* 36:156–71

43. Korf RP. 1952. The terms homothallism and heterothallism. *Nature* 170:534–36

44. Korhonen K. 1978. Interfertility and clonal size in the *Armillaria mellea* complex. *Karstenia* 18:31–42

45. Lenski RE. 1993. Assessing the genetic structure of microbial populations. *Proc. Natl. Acad. Sci. USA* 90:4334–36

46. Leslie JF. 1993. Fungal vegetative compatibility. *Annu. Rev. Phytopathol.* 31: 127–51

47. Leslie JF. 1995. *Gibberella fujikuroi:* available populations and variable traits. *Can. J. Bot.* In press

48. Leslie JF, Raju NB. 1985. Recessive mutations from natural populations of *Neurospora crassa* that are expressed in the sexual diplophase. *Genetics* 111: 759–77

49. Leung H, Nelson RJ, Leach JE. 1993. Population structure of plant pathogenic fungi and bacteria. *Adv. Plant Pathol.* 10:157–205

50. Levy M, Romao J, Marchetti MA, Hamer JE. 1991. DNA fingerprinting with a dispersed repeated sequence resolves pathotype diversity in the rice blast fungus. *Plant Cell* 3:95–102

51. Malik M, Vilgalys R. 1994. *Towards the genetic basis of somatic incompatibility in Pleurotus Ostreatus.* Presented at Int. Mycol. Congr., 5th. Vancouver, Canada

52. May G, Taylor JW. 1988. Patterns of mating and mitochondrial DNA inheritance in the agaric basidiomycete *Coprinus cinereus. Genetics* 118:213–20

53. Maynard Smith J, Smith NH, O'Rourke M, Spratt BG. 1993. How clonal are bacteria? *Proc. Natl. Acad. Sci. USA* 90:4384–88

54. McDermott JM, McDonald BA. 1993. Gene flow in plant pathosystems. *Annu. Rev. Phytopathol.* 31:353–73

55. McDonald BA, Martinez JP. 1990. DNA restriction fragment length polymorphisms among *Mycosphaerella graminicola* (anamorph *Septoria tritici*) isolates collected from a single wheat field. *Phytopathology* 80:1368–73

56. Milgroom MG. 1995. Analysis of popu-

lation structure in fungal plant pathogens. In *Disease Analysis Through Genetics and Biotechnology: Interdisciplinary Bridges to Improved Sorghum and Millet Crops*, ed. JF Leslie, RA Frederiksen, pp. 203–19. Ames, IA: Iowa State Univ. Press

57. Milgroom MG. 1995. Population biology of the chestnut blight fungus *Cryphonectria parasitica*. *Can. J. Bot.* In press
58. Milgroom MG, Lipari SE, Ennos RA, Liu YC. 1993. Estimation of the outcrossing rate in the chestnut blight fungus *Cryphonectria parasitica*. *Heredity* 70:385–92
59. Milkman R, Bridges MM. 1993. Molecular evolution of the *Escherichia coli* chromosome. IV. sequence comparisons. *Genetics* 133:455–68
60. Pujol C, Reynes J, Renaud F, Raymond M, Tibayrenc M, et al. 1993. The yeast *Candida albicans* has a clonal mode of reproduction in a population of infected human immunodeficiency virus-positive patients. *Proc. Natl. Acad. Sci. USA* 90:9456–59
61. Pukkila PJ, Skrzynia C. 1993. Frequent changes in the number of reiterated ribsomal RNA genes throughout the life cycle of the basidiomycete *Coprinus cinereus*. *Genetics* 133:203–11
62. Punja ZK. 1988. *Sclerotium (Aethalia) rolfsii*, a pathogen of many plant species. *Adv. Plant Pathol.* 6:523–34
63. Raper JR. 1966. *Genetics of Sexuality in Higher Fungi*. New York: Ronald
64. Rayner ADM. 1991. The challenge of the individualistic mycelium. *Mycologia* 83:48–71
65. Rayner ADM. 1991. The phytopathological significance of mycelial individualism. *Annu. Rev. Phytopathol.* 29:305–23
66. Rizzo DM, Harrington TC. 1993. Delineation and biology of clones of *Armillaria ostoyae*, *A. gemina*, and *A. calvescens*. *Mycologia* 85:164–74
67. Shaw CGI, Roth LF. 1976. Persistence and distribution of a clone of *Armillaria mellea* in a ponderosa pine forest. *Phytopathology* 66:1210–13
68. Smith ML, Bruhn JN, Anderson JB. 1992. The fungus *Armillaria bulbosa* is among the largest and oldest living organisms. *Nature* 356:428–31
69. Smith ML, Bruhn JN, Anderson JB. 1994. Relatedness and distribution of *Armillaria* genets infecting red pine seedlings. *Phytopathology* 84:822–29
70. Smith ML, Duchesne L, Bruhn J, Anderson JB. 1990. Mitochondrial genetics in a natural population of the plant pathogen *Armillaria*. *Genetics* 126:575–82

70a. Smouse PE, Meagher TR. 1994. Genetic analysis of male reproductive contributions in *Chamaeliirium luteum* (L) Gray (Liliaceae). *Genetics* 136:313–22

71. Specht CA, Novotny CP, Ullrich RC. 1992. Mitochondrial DNA of *Schizophyllum commune:* restriction map, genetic map, and mode of inheritance. *Curr. Genet.* 22:129–34
72. Stenlid J. 1985. Population structure in *Heterobasidion annosum* as determined by somatic incompatibility, sexual incompatibility, and isoenzyme patterns. *Can. J. Bot.* 63:2268–73
73. Stenlid J. 1994. Homokaryotic *Heterobasidion annosum* mycelia in stumps of Norway spruce. In *Proc. IUFRO Int. Conf. Root Butt Rots, 8th*, ed. M Johansson, J Stenlid, pp. 243–48. Uppsala: Swedish Univ. Agric. Sci.
74. Swedjemark G, Stenlid J. 1993. Population dynamics of the root rot fungus *Heterobasidion annosum* following thinning of *Picea abies*. *Oikos* 66:247–54
75. Talbot NJ, Salch YP, Ma M, Hamer JE. 1993. Karyotypic variation within lineages of the rice blast fungus *Magnaporthe grisea*. *Appl. Environ. Microbiol.* 59:585–93
76. Tibayrenc M, Kjellberg F, Arnaud J, Oury B, Breniere SF, et al. 1991. Are eukaryotic microorganisms clonal or sexual? A population genetics vantage. *Proc. Natl. Acad. Sci. USA* 88:5129–33
77. Vilgalys R. 1994. *Patterns of basidiospore dispersal and gene flow in the oyster mushroom Pleurotus ostreatus*. Presented at Int. Mycol. Congr., 5th, Vancouver, Canada
78. Vilgalys R, Cubeta MA. 1994. Molecular systematics and population biology of Rhizoctonia. *Annu. Rev. Phytopathol.*32:135–55
79. Worrall JJ. 1984. Population structure of *Armillaria* species in several forest types. *Mycologia* 86:401–7
80. Xu J, Horgen PA, Anderson JB. 1994. *Somatic recombination in Agaricus bisporus*. Presented at Int. Mycol. Congr., 5th, Vancouver, Canada

Annu. Rev. Phytopathol. 1995. 15:393–427

MOLECULAR APPROACHES TO MANIPULATION OF DISEASE RESISTANCE GENES

R. Michelmore
Department of Vegetable Crops, University of California, Davis, California 95616

KEY WORDS: germplasm, marker-aided selection, molecular marker, quantitative trait locus, transgene

ABSTRACT

Various molecular approaches are greatly increasing our ability to characterize and manipulate disease resistance genes in plants. Molecular markers are allowing the dissection of monogenic and quantitative resistance. The utility of marker-aided selection in breeding for disease resistance is being evaluated. Markers will soon enhance the effective deployment of resistance genes to provide more stable resistance. Several resistance genes have been characterized at the molecular level. This has led to the identification of several mechanistic classes and provides the opportunity for transgenic disease control strategies. As the mechanisms of determining specificity and generating variation at resistance loci are understood in greater detail, they will be manipulated to generate new specificities, probably using microbial selection systems. Disease resistance strategies utilizing classical resistance genes will complement the use of novel transgenes.

INTRODUCTION

Molecular techniques are providing the tools to characterize all stages of the interaction between plants and potential pathogens, from the earliest recognition events to the changes that occur during resistant and susceptible responses. Although many genes are involved in determining the outcome of the interaction between plants and potential pathogens, only one or a few loci have

0199-9885/95/0715-0393$05.00

been shown to determine the genetic variation observed in a specific plant-pathogen interaction.

This review focuses on how various molecular approaches are increasing our ability to characterize and manipulate plant genes that exhibit the natural or induced variation in resistance detectable by the plant breeder and geneticist. Studies on such loci are increasing our understanding of how resistance genes function, how they evolve to have new specificities, and how they might be manipulated to provide useful levels of disease control. Resistance generated by using novel transgenes has been reviewed earlier (e.g. 2, 60, 113, 140) and is considered here only to the extent that it complements or substitutes for natural resistance genes.

Molecular approaches are having a similar impact on our understanding of plant pathogens, and complementary studies on pathogens are of comparable importance to studies on resistance genes. Parallel investigations of the molecular determinants of pathogenicity and virulence are critical to our understanding of resistance gene function. Studies on pathogen variation and epidemiology are integral to characterizing and manipulating the evolution of different types of resistance genes. However, such studies are beyond the scope of this review and have been described elsewhere (e.g. 17, 19, 52, 66).

Some of the impact of molecular genetics will be almost immediate. Molecular markers will greatly assist in the preservation and exploitation of germplasm, allow marker-aided selection, and facilitate in generating particular combinations of resistance genes and in resistance gene deployment. Increasing success in cloning resistance genes from numerous species will allow us to characterize the diversity of mechanistic classes that exist and the genetic changes that generate variation in specificity. In the intermediate term, molecular markers will allow the characterization and manipulation of genes determining quantitative resistance. Also, cloned resistance genes will be used for novel transgenic approaches. In the long term, we can anticipate generating new resistance specificities in vitro or in microorganisms to increase the efficiency of selecting rare recombinant sequences.

MAPPING MONOGENIC RESISTANCE

Disease resistance genes are major components of many breeding programs and genetic mapping studies. The advent of several technologies to identify numerous molecular markers (4, 130) allows all regions of the genome to be assayed for linkage to resistance. This has recently resulted in the identification of an increasing number of molecular markers linked to resistance genes using a variety of approaches (Table 1) (64). Near-isogenic lines (NILs) that have resulted from backcrossing programs to introgress resistance have been used extensively (143). Bulked segregant analysis (BSA) (82) is being used increasingly as it allows the rapid mapping of monogenic resistance genes using segregating populations. Marker analyses have located many disease resistance genes to clusters in the genome, supporting data from classical segregation analyses of resistance genes (96). As many resistance genes are clustered, it is prudent to begin mapping new resistance genes by BSA with markers linked to known resistance genes. Less than 20 individual PCRs were required to locate R18 within the major cluster of resistance genes in lettuce (67). In the absence of linkage to such characterized markers, BSA subsequently can be used to screen large numbers of arbitrary loci and to identify markers linked to the targeted resistance gene. These new markers then can be mapped using previously characterized mapping populations to determine the genetic position of the resistance gene. The map position provides a further source of linked markers (14, 67, 122). When the disease screen is difficult or unreliable, it is possible to use flanking molecular markers to identify those F_2 individuals or F_3 families that are recombinant in the region of the resistance gene; this allows screening efforts to be focused on the informative individuals or families and the gene order and map distances to be determined more accurately (54, 135).

Although markers linked to many resistance genes are now known, often the closest markers are not tightly linked (Table 1); therefore, identification of additional markers is required before marker-aided selection or initiation of map-based cloning. In species with high-density maps, such as tomato and species syntenic to it, many additional markers are now potentially available. A similar situation exists within the syntenic Gramineae species. However, many of these markers may not be informative and therefore available within the breeding gene pool because most of the maps were developed using wide crosses. Consequently, identification of additional markers may still be necessary. It is now relatively facile to saturate particular regions of the genome with molecular markers using RAPD (Random Amplified Polymorphic DNA) (130) or AFLP (Amplified Fragment Length Polymorphism) (145) markers. AFLP analysis provides many loci per PCR reaction and allows the genome to be screened efficiently. BSA of a cross within the targeted gene pool is appropriate for identification of markers to be used in marker-aided selection. To identify markers that are extremely close to resistance genes as starting points for map-based cloning, screening of deletion mutants is more efficient. This is because all markers in the region will be detected, not just those that are polymorphic between the parents. We were able to screen for markers spaced on average every 42 Kb through the genome of lettuce using deletion mutants of *Dm* genes that had been induced by fast neutron bombardment (D Lavelle, R Michelmore, unpublished observations).

Table 1 Recent examples of mapping and identification of markers linked to disease resistance genes[a]

Host	Pathogen/Pest	Gene	Method of identification[b]	Type of markers[c]	Closest markers[d]	Reference
Tomato	*Meloidogyne incognita*	*Mi*	NILs	RAPD	0	132
		Mi3	BSA	RAPD, RFLP	0	139
	Leveillula taurica	*Lv*	F_2 seg.	RFLP, RAPD	5.5 interval	14
	Oidium lycopersicon	*Ol1*	BSA	RAPD, RFLP	*	122
	Verticillium dahliae	*Ve*	NILs	RAPD	0	51
	Yellow leaf curl virus	*Ty1*	BC_1S_1 seg.	RFLP	4 interval	147
Potato	*Globodera rostochiensis*	*H1*	F_1seg.	RFLP	0	29
	Phytophthora infestans	*R1* & *R3*	P_1seg.	RFLP	4 & 2	24
Lettuce	*Bremia lactucae*	*Dm17* & *18*	BSA	RAPD, SCAR	4, 6 & 0.6, 3	67
		Dm8 & *10*	BSA	RAPD, SCAR	1, 2 & 1, 2	135
	Plasmopara lactucae-radicis	*plr*	BSA	RAPD, RFLP	2, 3	53
	Turnip mosaic virus	*Tu*	F_2 seg.	RAPD, RFLP	0.4, 0.7	102
Soybean	Soybean mosaic virus	*Rsv*	F_2 seg.	μsat., RFLP	0.5, 35.9	144
Common bean	*Uromyces appendiculatus*	*Up2*	BC_6F_2 BSA	RAPD	0	83
	Common bean mosaic virus	*I*	NILs	RAPD	1 to 5	35

Pea	Pea seed-borne mosaic virus	*sbm-1*	F_2 seg.	RFLP, RAPD	8	120
	Pea common mosaic virus	*mo*	F_2 seg.	RFLP,	15	22a
	Erysiphe polygoni	*er*		RAPD,	11	
	Fusarium oxysporum	*Fw*		& μsat	6	
Maize	*Bipolaris maydis*	*rhm*	F_2 seg.	RFLP	0.5, 1	146
Barley	*Rhynchosporium secalis*	*Rh*	NILs, BSA	RAPD	7	1
	Erysiphe graminis f.sp. hordei	*Ml(La)*	DH seg.	RFLP	1,8	31
	Barley yellow mosaic virus	*ym4*	DH seg.	RFLP	1,1	33
Oats	*Puccinia graminis*	*Pg3*	NILs	RAPD	0	93
	Puccinia coronata	*Pc68*	BSA	RAPD	5	94
Wheat	*Heterodera avenae*	*Cre*	NILs	RFLP	6, 8	131
	Puccinia recondita	*Lr9*	NILs	RAPD	0	107

[a] Earlier examples are tabulated in Lefebrve & Chevre (64).
[b] NIL = Marker first identified using near-isogenic lines; F_2 seg. = Marker identified by conventional segregation analysis of F_2 population; DH seg. = Segregation analysis of doubled haploid population; BSA = Bulked Segregant Analysis.
[c] RFLP = Restriction Fragment Length Polymorphism; RAPD = Random Amplified Polymorphic DNA; SCAR = Sequence Characterized Amplified Region (91); μsat = microsatellite.
[d] Distances in cM to closest markers each side reported in paper. For several of these genes, there are now potentially closer markers; see text. *Could not be localized precisely.

CHARACTERIZATION OF QUANTITATIVE RESISTANCE

The advent of numerous molecular markers throughout the genome provides the opportunity to analyze the Mendelian factors determining quantitative traits (Quantitative Trait Loci, QTL). Over the past three years, several extensive studies to dissect quantitative resistance have been reported (Table 2).

For most aspects, identification of disease resistance QTL is no different than genetic dissection of other quantitative traits (reviewed in 116). Genomic regions contributing to resistance have usually been identified using regression analysis, interval mapping, or both (Table 2). Although there is no single optimal method for data analysis, consensus is emerging as to the required progeny size, marker density, degree of replication, and resolution of analyses. When the effect of a QTL is large, it will be detected by most of the methods. It is less clear what is the optimal method for avoiding Type I (false positives) or Type II (failure to detect real a QTL) errors when the effect is small (42). Initially, the threshold for a significant association between a marker and a phenotype was based on theoretical considerations; valid significance thresholds can now be determined empirically for each dataset using simulation (13). Simulations are also increasingly being used to indicate the resolution of experimental designs. Also, the initial analysis only leads to the hypothesis that a QTL exists at a particular position; the QTL has to be confirmed in subsequent experiments. This is only beginning for disease resistance (9, 16, 126). To investigate the effects of individual QTLs in multiple environments and against multiple pathogens, it is now necessary to generate near-isogenic lines for disease resistance QTLs using backcross programs based on linked markers.

Some aspects of plant-pathogen interactions complicate QTL dissection of disease resistance. Differences in pathogen virulence between screens introduce additional variation. The lack of detection of a QTL at multiple locations could be due to race-specificity of the QTL or true genotype × environment interactions. In addition, disease incidence or severity may be rated on an ordinal scale rather than a truly quantitative evaluation that has a normal distribution required by several statistical approaches (64). Furthermore, the timing and method of inoculation may alter the QTLs detected (16).

One of the primary impacts of QTL analysis is to indicate the number and effects of the major genetic factors controlling quantitative resistance. The number of QTL detected ranges from 2 to 13 (Table 2). In several cases, few loci control the majority of the genetic variation, and this quantitative resistance should be considered oligogenic rather than polygenic. One locus accounted for 77% of the genetic variance for resistance to *P. solanacearum* in tomato (16) and 58% of the genetic variance for resistance to *Plasmodiophora brassicae* in *Brassica oleracea* (62). In both of these diseases, resistance is clearly

Table 2 Dissection of quantitative trait loci determining quantitative disease resistance

Host	Pathogen/Pest Disease	No. of markers[a] Genome coverage[d]	Population size[b] Method of analysis[e]	No. of QTL[c] Effects[f]	Reference
Tomato	Insects	36 to 99 RFLP Incomplete	74/900, 96 F_2 Regression	3 $\Sigma = 38$	85
	Pseudomonas solanacearum Bacterial wilt	67 RFLP, 12 RAPD 75%	71 F_2 MMQTL	3 24 to 77, $\Sigma = 81$	16
Potato	*Globodera rostochiensis* Potato cyst-nematode	29 RFLP Partial	96 F_1 ANOVA	2 $\Sigma = 22$	57
	Phytophthora infestans Late blight	77 + 68 each parent	189 F_1 LSIM	13	65
Brassica oleracea	*Plasmodiophora brassicae* Clubroot	198 RFLP 94%	90 F_2/F_3 MMQTL	2 15, 58, $\Sigma = 61$	62
Bean	*Xanthomonas campestris* Bacterial blight	152 RFLP Complete	70 F_2/F_3 Reg., MMQTL	4^+ 17 to 32, $\Sigma = 75$	87
Mung bean	*Erysiphe polygoni* Powdery mildew	141 RFLP Majority	58 F_2/F_3 Reg., MMQTL	3 $\Sigma = 58$	141
Soybean	*Heterodera glycines* Soybean cyst-nematode	36 RFLP, 7 RAPD	56 F_2/F_3 ANOVA	3 21 to 40, $\Sigma = 52$	15
Pea	*Ascochyta pisi* Blight	56 RFLP, 6 others Ca 1/3	174 F_2 AV, MMQTL	4 $\Sigma = 74$	22
Barley	*Erysiphe graminis* Powdery mildew	155 RFLP, 3 others Complete	113 DH MMQTL	2 $\Sigma = 20$	37
		61 RFLP	28 F_1 Diallele ANOVA	5 or 6	104
	Puccinia striiformis f.sp hordei Stripe rust	78 RFLP 70%	110 DH MMQTL	2 10, 57	12

Table 2 *(Continued)*

Host	Pathogen/Pest Disease	No. of markers[a] Genome coverage[d]	Population size[b] Method of analysis[e]	No. of QTL[c] Effects[f]	Reference
Maize	*Cercospora zeae-maydis*	87, 87, 78 RFLP	139, 193, 144	F_2/F_3 9*	9
	Grey leaf spot	77, 74, 79%	ANOVA	4 to 26, Σ = 24 to 58 (3 populations)	
	Exserohilum turcicum	103 RFLP	150 F_2/F_3	7	27
	Northern corn leaf blight	Complete	MMQTL	7 to 18, Σ = 29 to 24 (three traits)	
	Ostrinia nubilalis	87 RFLP	300 F_2/F_3	7	108
	European corn borer	Complete	MMQTL	Σ = 38	
	Gibberella zeae	95 RFLP, 19 RAPD	112 F_2/F_3	4 to 5	92
	Stalk and ear rot	Complete	Reg., MMQTL	Σ = 20	
	Colletotrichum graminicola	113 RFLP	158, 151 F_2/F_3	1	49
	Anthracnose stalk rot	75%	MMQTL	16 to 75 (2 traits)	
Pearl Millet	*Sclerospora graminicola*	22 RFLP	93 to 119 F_2/F_4	5	48
	Downy mildew		MMQTL	8 to 48, Σ = 65	
Rice	*Pyricularia oryzae*	127 RFLP	131/281 RIL	10	126
	Blast	Complete		Reg., MMQTL	19 to 60

[a] The number and type of informative markers analyzed.

[b] The number of individuals and the population analyzed. F_2/F_3 = Genetic analysis made with F_2 but the disease screen made with F_3 families. DH = Doubled haploids. RIL = Recombinant inbred lines.

[c] The number of significant QTL reported above the LOD threshold used. They may not have all been detected in the same population or environment. * A relaxed LOD threshold identified at least nine QTL in three populations; only one was significant in all three environments tested.

[d] The proportion of the genome covered by informative markers.

[e] Method of statistical analysis: Reg. = Regression analysis; ANOVA, AV = Analysis of variance; MMQTL = Interval mapping using Mapmaker QTL. LSIM = Least squares interval mapping. In several cases multiple methods were used; the primary methods are shown in the table.

[f] The range in percentage of the variance explained by individual QTL and the maximum total variance (Σ=) explained in a single experiment. Missing data indicate that either they were not apparent from the paper or that they were too complicated to be included in tabular form.

quantitative and, prior to molecular marker analysis, it had been impossible to dissect the genetics of resistance. The numbers of loci detected represent minimum numbers. QTL analyses do not preclude the presence of genes of minor effect that were below the threshold of significance for detection in the experiment. In several cases, the sizes of the populations studied were small and/or the whole genome was not analyzed with informative markers (Table 2). As with all QTL analyses, demonstration of a significant association between resistance and a chromosomal region does not distinguish between a single QTL of large effect or multiple QTL of smaller effects. No resistance QTL has yet been dissected by recombination to distinguish between these possibilities.

A variety of gene actions have been detected for disease resistance QTL. Some QTL appear to be additive, for example, resistance to gray leaf spot of maize (9). Others, such as resistance to downy mildew of pearl millet, are mainly dominant or possibly overdominant (47, 104). Recessive QTLs for resistance were only occasionally indicated, as in resistance to bacterial blight in bean (87). Alleles increasing resistance usually originated from the resistant parent, although occasionally alleles increasing resistance came from the susceptible parent; this would explain the transgressive segregation that can occur (9, 48).

Individual QTLs were not always detected in repeated analyses. There were several possible reasons for this. In some cases, it was probably due to genotype × environment interactions (9). However, when LOD scores were high, QTLs seemed to be stable across environments and greenhouse tests were good predictors of field performance (47, 126). In other cases, variation was apparently due to pathogen heterogeneity and race-specificity of the QTLs. All the QTLs for resistance to downy mildew in pearl millet were race specific; no QTL contributed to resistance against all four pathogen populations tested (48). Several of the QTLs for resistance to *Phytophthora infestans* were race specific (65). Both isolate-specific and isolate-nonspecific QTLs were detected in a diallele analysis of eight barley cultivars inoculated with five isolates of powdery mildew (104). The life cycle of most pathogens requires several distinct phases of interaction with its host, and different genetic interactions may occur between plant and pathogen during each of these stages. Analysis of the pathogen proliferation at each of these stages, scoring of disease symptoms by several criteria, and the use of different inoculation procedures may reveal the genes responsible for such differences. Different QTLs conditioned disease severity and lesion size and number of Northern corn leaf blight (27). Inoculation of tomato shoots and roots revealed different QTLs for resistance to *P. solanacearum* (16).

The relationship between the determinants of quantitative resistance and the single genes that confer complete resistance ("major genes," *sensu* the plant pathology literature) is unclear. Defeated major genes may have residual effects (86). Also, some QTLs are race specific (48, 65), but this does not necessarily imply a function similar to race-specific major genes. However, QTL for several traits may be allelic with extreme alleles (3, 103), and disease resistance may be no exception. This possibility is supported by the mapping of QTLs for resistance to the vicinity of major resistance genes in several species. QTLs for resistance to *Exserohilum turcicum* in maize mapped to three monogenic resistances to the same pathogen (27). Three QTLs conditioning partial resistance to rice blast mapped to loci conditioning complete resistance in other genotypes (126). QTLs for resistance to powdery mildew in barley mapped to positions previously known for major resistance loci (104), as did some QTLs for resistance to *P. infestans* in potato (65). A QTL for resistance to potato cyst-nematode maps to the syntenous region in potato as *I2* for resistance to *Fusarium oxysporum* in tomato (57). However, mapping alone cannot definitively demonstrate similarity between major genes and QTL for resistance. The resolution of QTL analysis is too low to locate genes precisely, and linkage between functionally dissimilar genes can occur by chance. Cloning of multiple alleles of major resistance genes and the generation of transgenic, truly isogenic lines may provide definitive data, if it can be demonstrated that some alleles determine qualitative resistance while others contribute to quantitative resistance. Analysis of these isogenic transgenics with a range of isolates also will demonstrate whether defeated resistance genes still confer some level of resistance. The answer to this experiment depends on whether the differences in the pathogen result in a partial or complete loss of interaction with the resistance gene product. Partial or weakened interactions may lead to partial resistance. This will probably occur in some cases, but not in others, and is consistent with the range of dominance relationships observed for resistance genes and their mutants.

Mapping of QTLs for resistance may provide clues as to their function. Those that map within clusters of major genes can be tentatively hypothesized to have similar functions and therefore to be involved in recognition processes. Other QTLs may cosegregate with genes involved in other aspects of resistance, such as genes encoding proteins involved in resistance response (65). As more transcribed sequences are mapped, it may be increasingly possible to postulate function from cosegregation. However, again the poor genetic resolution of QTL analysis and the possibility of coincidental linkage require great caution in interpreting such inferences. Only transgenic studies will provide definitive data.

Cloning of QTLs for disease resistance is still many years away unless they are allelic with previously cloned genes such as major resistance or response genes. Techniques for cloning genes with unknown products and patterns of expression, such as map-based cloning or transposon tagging, currently are inapplicable to most disease resistance QTL. The genetic resolution is too poor and the phenotype of most individual QTL too imprecise for tagging or complementation. Map-based cloning may be applicable to a QTL that determines the majority of the genetic variance. However, this will be difficult and requires additional technical advances, particularly for complementation with large genomic fragments, before it is routinely feasible (79).

GERMPLASM CHARACTERIZATION AND EXPLOITATION

Wild germplasm is an important source of disease resistance genes. One of the major challenges in germplasm collection, conservation, and screening is measuring relevant diversity. The limited molecular marker analyses of germplasm made to date have utilized random loci (30). The development of markers tightly linked to clusters of resistance genes and the cloning of resistance genes provide the opportunity to assay variation at resistance gene loci using relevant, hyperpolymorphic loci. When collecting germplasm, it is difficult to determine the optimal sampling frequency without extensive characterization. Studies on disease resistance have indicated that large numbers of monogenic resistances may be effective in some natural populations (7, 10); molecular markers indicate that other populations may be close to genetically homogeneous because of founder effects (101). Development of PCR-based markers for resistance gene regions will provide a quick assay for allelic diversity at pertinent and variable segments of the genome, so that large numbers of individuals can be sampled within and between populations. This will allow the population genetics of resistance to be studied and collections made appropriately.

There is great diversity of disease resistance in the gene pools of many crop species. However, when accessions are identified that are resistant to all known isolates of a pathogen, it is difficult to estimate how many distinct sources of resistance are represented. Hybridization markers for the major clusters of resistance genes in a species will identify which accessions have similar haplotypes and may therefore be redundant in a collection or breeding program. Molecular markers at resistance loci will also provide a useful indication of the genetic relationship between the resistant accessions and allow the breeder to select accessions representing the greatest diversity of resistant haplotypes. This will significantly increase exploitation of the natural diversity in resistance.

Markers linked to resistance genes will be useful in fingerprinting and demonstrating the unique identity of cultivars. Traditionally, resistance genes often have been used as distinguishing characters for cultivar registration. When backcross programs are carried out in parallel, it may be difficult to distinguish the products, particularly in inbred crops that have little intraspecific variability. The existence of numerous markers flanking resistance genes allows cross-over points between the donor and recurrent chromosomes to be determined accurately. The chances are small that recombination events either side of the introgressed resistance gene would be identical during independent backcross programs. Therefore, it should be relatively easy to demonstrate the novelty of backcross material. However, the level of novelty now required for a new cultivar and limits of material "essentially derived" from other genotypes has yet to be defined at the legal level. Will novel cross-over positions be sufficient to constitute a unique cultivar?

MARKER-AIDED SELECTION

Marker-aided selection (MAS) is potentially useful for breeding for disease resistance in at least four ways: as a substitute for a disease screen, to accelerate the return to the genotype of the recurrent parent during backcrossing, to reduce linkage drag of linked deleterious genes, and to select for disease resistance QTL.

There are several situations in which molecular markers can substitute for a disease screen (77). Some disease screens are difficult to conduct because of incomplete penetrance of the resistance, variability in aggressiveness or availability of the pathogen, or sensitivity of the disease reaction to environmental conditions. As a consequence, the disease screen may be unreliable. Some disease screens are time consuming or only can be conducted at particular locations, times of year, or stages of plant development. Molecular markers can allow selection of resistant individuals at the seedling stage as soon as the plants are large enough to yield sufficient DNA or protein for the assay, independent of the environmental conditions. The isozyme marker, *Aps1*, has been used commercially for many years as a substitute for screening with nematodes to select for *Mi1* in tomato (100). A more tightly linked PCR-based marker is now available that allows *Mi* to be followed in lines that have lost the diagnostic *Aps1* allele due to recombination (132). Markers also can help in combining genes for resistance to pathogens for which the disease screens are mutually incompatible. Alternatively, infection may interfere with selection for other traits, even in resistant individuals, rendering simultaneous selection for resistance and agronomic or horticultural traits impractical. In addition, molecular markers can substitute for the use of a foreign pathogen. This allows breeding against diseases or variants that have yet to evade quarantine barriers or breeding in a country in which the pathogen does not occur naturally (48). Several genetic situations favor the use of markers. Markers allow multiple genes that are effective against all known isolates of a single pathogen to be combined (see below). If resistance is recessive, either testing of selfed progeny in each backcross generation or alteration of backcross and selfed generations is required to identify those individuals carrying the resistant allele using a disease screen. Codominant loci or dominant markers linked in cis to resistance can identify individuals carrying the recessive allele to be used for the next backcross. Codominant markers can also be used to identify homozygous resistant individuals after selfing at the end of a backcross program. In all these situations, however, molecular markers should not totally substitute for a disease screen. At least occasional testing with the pathogen is necessary to check that the resistance gene is still present and effective.

Selection on molecular markers in regions unlinked to resistance has the potential to accelerate the return to the recurrent parent genotype during backcross programs (118). However, backcrossing is an efficient method for substituting chromosomes and simulations suggest that the use of markers that are unlinked to the introgessed gene may not be as beneficial as sometimes advocated (40). This is likely to be particularly true when the use of markers is compared to programs that include selection for the recurrent parent phenotype; selection on only a few markers per chromosome is required for a rapid return to the genotype of the recurrent parent (40), and plant phenotype is probably determined by multiple loci scattered through the genome. Therefore, MAS for unlinked regions may have limited benefit in species with short generation times. It is likely to be more advantageous in species with long generation times. It will be interesting to dissect existing breeding pedigrees with molecular markers to determine a posteriori the maximum potential gains that could have been made using MAS (H Witsenboer & O Ochoa, personal communications).

Molecular markers clearly do have great potential for reducing linkage drag in chromosomal regions linked to the resistance gene being introgressed. In contrast to the rate of substitution of unlinked regions, backcrossing is very inefficient at removing linked regions. After ten generations, the theoretical average size of the introgressed segment is still 18 cM, and even after many backcross generations, the amounts of donor chromosome remaining vary greatly (112, 118). The region remaining around the *Tm-2* locus (*resistance to tobacco mosaic virus* in tomato) after introgression from *Lycopersicon peruvianum* into *L. esculentum* varied from 4 to over 50 cM (142). Similar variation has been observed around the *Mi* locus (*resistance to Meloidogyne incognita*) that was introgressed into tomato cultivars from *L. peruvianum* nearly 50 years ago (78). Introgression of resistance genes from wild species is often accompanied by the transfer of deleterious linked genes. Minimizing the donor chromosome segment is further hindered by reduced recombination within introgressed segments (39, 78). In addition, because resistance is often dominant, it is impossible to identify individuals carrying a recombinant chromosome without progeny testing; this is tedious for many diseases. Codominant PCR-based markers can be used to screen large numbers of individuals and select rare recombinants close to the resistance gene. Selection of recombinants on either side of the resistance gene in sequential generations is the most efficient strategy (40, 118).

As resistance genes become cloned they will be useful as the ultimate genetic marker for resistance. Southern hybridization with the resistance genes cloned thus far identify small, polymorphic and predominantly clustered, multigene families (25, 46, 63, 71, 84, 129). As more resistance genes are sequenced, it should be possible to design PCR-based markers specific for each allele using oligonucleotide primers that anneal to regions of sequence divergence. No recombination will be possible between such markers and the resistance specificity. Anonymous flanking markers still will be needed to screen for recombinants beside the resistance gene.

MAS has considerable potential for transferring QTLs for disease resistance. Molecular markers will allow the reliable selection of individual QTL because the heritability of the markers is one, in contrast to the QTL whose selection is obstructed by genetic and environmental noise. Markers allow the selection of individuals carrying favorable alleles from either parent and avoids the inclusion of individuals that are homozygous for unfavorable alleles. Markers for important resistance QTL are now being developed in several species (Table 2); this could allow selection for complex genotypes containing loci determining both major gene and quantitative resistance that have eluded the breeders using solely classical approaches (126). However, as with all quantitative traits, the potential for MAS of QTL in breeding programs must be considered separately from the potential of molecular markers for QTL dissection. The potential of the latter is already being realized (see above), but such experiments involve large, time-consuming analyses that may not be applicable to breeding programs. MAS should involve inexpensive, reliable, routine assays that are more cost effective than direct selection. The latest generation of computer programs allow the weighting of different QTL-markers and the simulated comparison of MAS vs phenotypic selection, or both (23, 61; B Liu, personal communication). Several factors will determine the usefulness of QTL-marker associations for MAS: prevalence in the germplasm, effectiveness against the diversity of pathogen genotypes, epistatic interactions with other loci, variations in linkage phase, and genotype × environment interactions (as well as the development of easily assayable markers). None of these factors has been studied extensively for disease resistance. In corn, the same QTLs for stalk rot resistance were identified in two populations (49), but only some of the QTLs for gray leaf spot resistance were detected in multiple populations (9). Two different experimental approaches detected overlapping sets of QTLs for resistance to powdery mildew in barley (37, 104). QTLs for pearl millet downy mildew and potato late blight resistance were race specific (48, 65), whereas those for blast resistance were not (126). Of the QTLs for gray leaf spot in maize, only one was effective in all three of the environments tested (9); in contrast, resistance to corn borer was detected in both environments (108). The usefulness of MAS for disease resistance QTL will have to be determined on a case-by-case basis.

MAS has been much advocated but has yet to be adopted for widespread use. There are several reasons for this. For many diseases, it remains easier, cheaper, and more reliable to screen with the pathogen than with molecular markers. Molecular marker technology is currently too expensive and technically demanding for routine analysis of most large breeding populations. In addition, identification of loosely linked markers in one mapping population does not necessarily provide markers that are informative in many breeding populations. Genetic maps for several, particularly inbred, crops are based on interspecific or intertype crosses (e.g. 11, 53, 99, 117). Many of the mapped markers are not polymorphic within the breeding gene pool. However, as resistance genes have often been introgressed from wild species, there is often a good probability that once closely linked markers have been identified, they will remain useful. Microsatellite markers are likely to be informative in a greater range of genetic backgrounds than other types of markers; therefore, development of microsatellite markers linked to disease resistance genes should be a high priority. As markers become identified for more genes, the effort and cost per trait diminishes and it becomes increasingly advantageous to utilize molecular markers. Also, inexpensive and rapid assays are being developed. Therefore, MAS for disease resistance will soon become a routine component of many breeding programs.

RESISTANCE GENE COMBINATIONS AND DEPLOYMENT

The identification of molecular markers linked to resistance genes provides several opportunities for resistance strategies, such as resistance gene pyramiding and gene deployment, that were proposed many decades ago but have not been implemented extensively because of practical difficulties.

Molecular markers will greatly increase our ability to select for particular combinations of resistance genes. As resistance genes to diverse pathogens are frequently clustered in the genome, there is the danger of introducing susceptibility to one disease while breeding for resistance to another (148). This can be averted if the map position of the different resistances is known. Bulked segregant analysis using markers linked to the clusters of resistance genes can very quickly determine the linkage relationships of a new resistance to known resistance genes (see above). When two resistance genes are located in the same cluster but in different parents (linked in repulsion), it is often impossible using disease screens alone to identify recombinants with the genes in coupling, without testing numerous F_3 families. Analysis of codominant markers flanking the resistance genes allows the identification of individuals carrying the desired recombinant chromosome with the resistant alleles in coupling. Resistance to both diseases or to isolates of the same disease can then be selected in subsequent generations as a single Mendelian unit using either molecular markers or a single disease screen.

Molecular markers will allow the pyramiding of genes that are effective against all known variants of a pathogen. Germplasm screens often identify accessions that are resistant to all isolates of a particular pathogen. However, it is difficult to determine how many distinct resistances are present and it is impossible to combine multiple resistances into a single genotype (unless making hybrids and then the limit is two) using disease screens because of the lack of an isolate diagnostic for each gene. Molecular markers at resistance loci will identify which accessions are likely to carry different genes (see above). Bulked segregant analysis in an early generation segregating for the individual resistances should indicate the map position of the resistance and identify linked markers to the different genes. These markers can then be used to select genotypes carrying combinations of multiple genes. Molecular markers similarly allow resistance genes of different mechanistic classes to be combined; this may increase their durability compared to their use individually (137).

Deployment of different resistance genes has been promoted for many years to diversify the resistance being used and limit the exposure of individual genes (136, 137). However, it has been applied to a limited extent owing to the restricted amount of effective variation available or the effort required to introgress multiple resistance genes. It has proved effective when it has been utilized, for example, to reduce losses to barley powdery mildew in Eastern Europe (137, 138). Molecular markers will increase the diversity of resistance genes available for deployment (see above) and increase the speed of introgression between genotypes. Gene deployment strategies should be integrated with information on pathogen variation (138). Most screens that rely on determination of a pathogen's virulence phenotype are too slow and labor intensive to allow sufficient numbers of samples to be processed to gain comprehensive data on pathogen variability. PCR-based markers that are diagnostic for the major variants, particularly if they are linked to the determinants of virulence (or avirulence genes themselves as they become available), will allow large numbers of isolates to be characterized from within and between populations. This in turn will indicate which resistance genes are likely to be effective and determine the prevalence and location of resistance-breaking variants.

MOLECULAR ISOLATION OF RESISTANCE GENES

The potential and details of various strategies for cloning resistance genes have been reviewed extensively (e.g. 5, 81). Several resistance genes have now been isolated and are being characterized at the molecular level after years of effort by several groups. This work has been thoroughly reviewed (e.g. 18, 70) and little new can be added until more resistance genes are cloned and compared to the few currently isolated. Therefore, only the salient points are summarized here.

Successful Cloning of Resistance Genes

All resistance genes cloned to date have relied on either transposon tagging or map-based cloning. Neither of these methods assumes anything about the nature of the resistance gene product. All the cloned resistance genes had unambiguous phenotypes and had been well defined genetically as single loci.

Four genes have been isolated by transposon tagging. *Hm1*, encoding resistance to *Cochliobolus carbonum* race 1, was cloned from maize using the transposable element, *Mutator* (45). *Cf9*, encoding resistance to *Cladosporium fulvum* in tomato (46), the *N* gene in tobacco for resistance to tobacco mosaic virus (129), and the *L6* gene in flax for rust resistance (63) were tagged using the heterologous corn elements, *Ac/Ds*. The biology of these diseases allowed large numbers of plants to be scored reliably for mutations. Powerful positive selection strategies were used to identify mutations in *Cf9* and *N* but such a strategy was not available for *L6*. Tagging of *Cf9* and *L6* started with elements linked to the targeted gene to increase the probability of insertion. Some mutations occurred in all of these genes that were not caused by the characterized element; however, the targeted loci were not so unstable that the background mutation rate swamped insertional inactivation events. Insertional inactivation was confirmed by mapping, the isolation of multiple mutants, and germinal or somatic instability. Only complementation with *N* has been reported to date.

Three genes, *Pto* and a candidate for *Prf* in tomato determining resistance to *Pseudomonas syringae* pv. *tomato* (71; B Staskawicz & J Salmeron, personal communications), and *RPS2* encoding resistance to *P. syringae* pv. *maculicola* in *Arabidopsis* (6, 84) have been isolated by map-based cloning. Isolation proceeded through the well-defined steps of precise genetic localization, saturation of the region with markers, isolation of large genomic YAC clones, identification of cDNAs from the area, and finally complementation. Mutants were important in the identification of *Prf* and *RSP2* but were not used in the initial cloning of *Pto*, which relied predominantly on recombinants to localize the gene. The cloning of *RSP2* occurred much faster than the cloning of *Pto*; this was due to the close proximity of *RPS2* to an ABA response gene that was the subject of a well-advanced map-based cloning effort, as well as the multiple tools that were available for the small genome of *Arabidopsis*. Complementation with candidate clones containing *RPS2* was efficiently performed either using a transient assay or whole-plant transformation. For both *Pto* and *RPS2*, very few resistant plants were obtained by stable expression of cDNA clones from the 35S promoter. Transformation with genomic clones of *RPS2* more frequently resulted in resistance.

Cloning of Numerous Additional Resistance Genes

The recent cloning of these six resistance genes provides new opportunities and perspectives for cloning numerous related genes from crop species (79). Most of the cloned genes hybridized to multigene families within the original species (25, 46, 63, 72, 84, 129), which suggests that other resistance genes can be isolated by cross-hybridization. The recurring motifs imply that cloning may be also possible using PCR with oligonucleotide primers for conserved sequences. However, there are probably many related sequences in the genome; therefore, in both cases, it will be challenging to identify the sequences encoding the targeted specificities. The primary limitation will be the genetic definition of the targeted resistance gene.

Resistance genes will continue to be cloned by transposon tagging and map-based cloning. Currently, more resistance genes have been isolated by transposon tagging than by map-based cloning. This is likely to be reversed in the near future, partly because significantly more resistance genes are currently being targeted using map-based cloning than transposon tagging (70) and partly because of technical advances for map-based cloning (79). Map-based cloning involves several distinct, labor-intensive steps. However, this approach is becoming increasingly efficient due to technical improvements for saturation of genomic regions containing resistance genes with markers, for cloning and characterization of large genomic fragments, and for efficient complementation. In particular, the resolution of genetic analysis is now approaching the size of fragments that can be used for complementation. The identification of the correct open reading frame in large genomic fragments ("the end game") remains a major hurdle.

Resistance genes have been cloned simultaneously from several crop species as well as the model species, *Arabidopsis*. This has resulted from sufficient technical advances to allow cloning of long-known resistance genes from the crop species occurring concurrently with the genetic definition of resistance genes in *Arabidopsis*. Continued technical developments for cloning from large genomes will increase the ease of gene isolation from crop species. More resistance genes are being identified in *Arabidopsis*, and its genome is being characterized in ever greater detail. Therefore, the cloning of large numbers of resistance genes can be expected from both crop and model species in the near future.

The genes cloned from crop and model species may be used rather differently. Genes cloned from crop species may have greater immediate applied value as they encode resistance to agriculturally important diseases. Also, because of their diversity and well-documented breeding history, they will be powerful tools for characterization of the mechanisms generating variation in specificity (see below). Dissection of the mechanisms of resistance, particularly the molecular events occurring during perception of a pathogen and transduction of the signal that results in the resistance response, will rely heavily on studies of *Arabidopsis* and other model species because of the efficient generation, molecular identification, and manipulation of mutants.

MECHANISTIC CLASSES OF RESISTANCE GENES

Genetic variation in disease resistance has been known since the turn of the century. Much of this variation was identified by plant breeders as the result of screening germplasm for resistant lines or selection of resistant individuals during field epidemics. Plant breeders have been more interested in the level and durability of resistance than in the underlying mechanisms. Not surprisingly, therefore, cloning of resistance genes is demonstrating that the genetic variation identified by plant breeders is due to a diversity of resistance mechanisms.

Based on genetic and physiological data as well as teleological arguments, there are likely to be at least four broad mechanistic classes of resistance genes. These classes are now being confirmed by cloning and molecular characterization. Some resistance genes encode components of receptor systems that detect, either directly or indirectly, the presence of potential pathogens. Activation of such receptors probably initiates a signal transduction pathway that results in the induction of the generic response genes (32, 59). Genetic and now molecular data suggest that genes involved in gene-for-gene interactions belong to this class. *Pto*, *Cf9*, and the *Prf* candidate from tomato (46, 71; B Staskawicz & J Salmeron, personal communications), the *N* gene from tobacco (129), *Rsp2* from *Arabidopsis* (6, 84), and the *L6* gene from flax (63) all seem to belong to this category on the basis of their sequence similarity to known signal transduction components. However, binding of a ligand derived directly or indirectly from the pathogen to any of these gene products has yet to be demonstrated.

Several classes of resistance gene are not components of signal transduction pathways. A second class of genes encodes products that detoxify and deactivate compounds that the pathogen requires to cause disease. *Hm1* from corn, which encodes a reductase that deactivates HC-toxin, is clearly in this category

(45, 76). Plants contain many catabolic capabilities. Therefore, it is likely that other biochemical processes have been adapted to inactivate molecules required by pathogens. A third class of resistance genes may encode altered targets for pathogen-derived molecules required for pathogenesis. A resistant allele would encode a product that no longer interacted with the pathogen factor and would tend to be recessive. The dominant allele providing a compatible product could be considered a susceptibility factor. The absolute dependence of viruses on the host's biochemical machinery provides particular opportunities for such resistance to evolve. Therefore, some of the recessive genes for resistance to viruses may be of this type (26). However, no such viral resistance gene has been cloned to date and characterized at the molecular level. Yet another category of resistance genes includes those encoding structural or constitutive biochemical barriers to pathogens.

Within each mechanistic class, there will be subclasses of functionally distinct resistance genes. *Pto* that encodes a protein kinase is clearly different from the other characterized receptor-related genes that contain leucine-rich regions (LRRs) but lack a protein kinase domain. Within the few LRR-containing proteins characterized so far, there seem to be two subclasses: one containing a GTP-binding motif (*RPS2, N, L6, Prf*), the other lacking it (*Cf9*). Characterization of multiple genes conferring resistance to diverse pathogens from several plant species will determine the diversity and characteristics within each category. It is interesting, however, that the majority of genes now isolated, conferring resistance to a virus (*N*), bacterium (*Pto* and *Prf*), or fungus (*Cf9, L6*), belong to the receptor-related class. This suggests that many of the resistance genes identified by plant breeders will fall into this class and that similar recognition mechanisms and signal transduction pathways operate in diverse disease interactions.

It is unlikely that any of these mechanistic classes of resistance genes have evolved *de novo*. Domains involved in a variety of cellular processes have probably been recruited to provide resistance, and related homologs undoubtedly exist that are involved in other cellular processes. The LRR motif is present in functionally diverse proteins and is thought to be involved in protein-protein interactions (56). *Cf9* has sequence similarity to antifungal polygalacturonase-inhibiting proteins as well as the nonkinase domains of plant receptor-like protein kinases (46). *Pto* has homology to several serine/threonine kinases. No resistance gene has yet been reported that has both a LRR and a protein kinase domain. Whether all receptor-type resistances comprise a two-component system with separate receptor and effector molecules (possibly allowing greater evolutionary flexibility?) remains to be seen. An open reading frame from the *Xa21* region of rice has both an LRR region and a protein kinase domain separated by a transmembrane motif (P Ronald, personal communication). It is unclear whether the LRR domain is involved in interactions between components of the signal transduction pathway and/or in binding of a ligand. Furthermore, the LRR motif may not be the only type of receptor domain utilized (assuming that specificity resides in this domain). As more resistance genes are characterized, one can anticipate the identification of resistance genes with carbohydrate or lipid binding domains. However, it is interesting that four of the first six resistance genes cloned contain an LRR motif, suggesting that this is a common feature of this class of genes.

Mutation studies have shown that several genes are necessary for resistance gene function. Mutations to susceptibility to *Pseudomonas syringae* pv. *tomato* in tomato identified equal numbers of mutations at *Pto* and a new complementary and tightly linked locus, *Prf* (105). Screening for increased susceptibility to *Cladosporium fulvum* in tomato identified six mutations at *Cf9* and one at each of two independent loci, *Rcr1* and *Rcr2* (*Required for Cladosporium resistance*) (36). Similarly, mutants in two new complementary loci, *Nar1* and *Nar2* (*Necessary for* Mla_{12} resistance), have been identified that are required for resistance determined by Mla_{12} in barley (28). Screening of lettuce for susceptibility to *Bremia lactucae* identified mutants at several *Dm* loci but no complementary loci (88). With the exception of *Prf*, the greatest frequency of mutations were identified in the loci already identified as natural variants by plant breeders. The numbers of families screened was low in all of these studies. Only one or two mutants were identified at new, complementary loci, which indicates that the mutant screens had not saturated the resistance phenotype and that not all mutable loci had been detected. It is unclear what this distribution of mutants reflects; it may be the consequence of the different stabilities of the genes. Resistance genes are known to be inherently unstable (97), and the mutagenic treatment may have increased their instability. Therefore, obtaining multiple mutants in resistance genes does not imply that saturation mutagenesis has been achieved. Not all steps in the pathway may be mutable; there may be redundancy in some parts of the pathway, some steps may be encoded by duplicated genes, or mutations in some genes may be lethal. However, if the sequence similarity of *N* and *RPS2* products to Toll in *Drosophila* and the mammalian Interleukin-1 receptor (129) implies involvement of a similar signal transduction pathway, then there may be few downstream components. This is consistent with the low number of complementation groups so far identified. As more resistance genes are cloned, it is likely that the variation identified by the breeder will be shown to represent a range of steps in this signal transduction pathway, but that one component (the receptor?) is more variable than others.

Mutation studies have also identified disease lesion mimics. These mutants have varying degrees of apparently spontaneous necrosis. They are numerous in contrast to resistance loci (124). Some of the mutations map to resistance gene loci; others apparently do not (96). It is difficult to envisage how all of such loci could be primarily involved in resistance. Alternatively, the disease lesion mimic phenotype may have a wide range of underlying causes. Some may be related to the resistance; others may involve some form of default necrosis triggered in a stressed cell by additional nonspecific stress caused by a pathogen or an abiotic factor. Now that several disease lesion mimics have been identified in *Arabidopsis* (21, 34), such genes should soon be cloned and their relationship to disease resistance determined.

TRANSGENIC STRATEGIES FOR UTILIZING CLONED RESISTANCE GENES

Cloning of resistance genes provides several opportunities for transgenic disease resistance. It will be possible to pyramid resistance genes using cassettes of genes, with each gene effective against different strains of a single pathogen or against different pathogens. Transgenic approaches overcome two objections to pyramiding resistance genes by classical methods. Unlike the sequential introduction of multiple genes by traditional backcrossing, the time required for introduction of resistance by transformation is independent of the number of transgenes in the construct. Also, molecular approaches will allow the identification of an increased number of resistance genes; therefore, multiple resistances can be combined and not all the available genetic resources will be potentially squandered in a single genotype. Once introduced, the cassette of transgenes can be manipulated as a single Mendelian unit in classical breeding programs. This would be aided by the inclusion of a whole-plant selectable marker, such as herbicide resistance. Introduction of transgenes will allow different types of resistance to be combined. Also, it will be possible to introduce transgenes conferring high levels of resistance into genotypes with polygenic resistance or empirically proven durable resistance. Alternatively, it may be possible to combine different mechanistic classes of genes. All of these strategies should increase the durability of resistance.

Transfer of resistance genes by transformation has been advocated as an efficient substitute for backcrossing (52). This will definitely be true in some cases. Transformation with a single transgene will not have deleterious effects due to linked genes (linkage drag), as is sometimes the case with classical backcrossing (e.g. *Tm-2* for virus resistance in tomato) (142). Transformation may also allow transfer across species barriers that are difficult or impossible to cross sexually. Transformation is also highly applicable to heterozygous clonally propagated crops such as potato, cassava, and several plantation crops, where any hybridization results in the loss of agronomic type. However, transformation for all crop species currently involves passage through tissue culture, which is a potentially mutagenic process. Backcrossing and/or extensive field testing are required to ensure that agronomic type has been retained and to satisfy regulatory requirements. Backcrossing is a powerful and efficient method for regaining the genotype of the recurrent parent. Therefore, it may be more practical to introduce a transgene or cassette of transgenes into one cultivar, identify lines with optimal levels of expression and insertion events, and then transfer the trait to other genotypes by classical backcrossing aided by a whole-plant selectable marker. This is more predictable and avoids detailed characterization of each new genotype that is transformed.

Interspecific transfer of transgenes has great potential, particularly if there is no effective resistance within the target species. Genes that provide resistance against pathogens or pests with wide host ranges are obvious candidates for transfer. *Mi* has provided resistance in tomato to the nematode *Meloidogyne incognita* for many years and is currently the focus of cloning efforts by several groups (39, 78). Transfer of *Mi* to other species may provide effective resistance to this important pest. Several lines of evidence suggest that interspecific transfer may provide resistance to pathogens of the recipient species. Transfer of cloned avirulence genes between bacterial pathovars of *Pseudomonas syringae* or *Xanthomonas campestris* demonstrated that the same avirulence genes were recognized in multiple plant species (55, 127, 128). For example, resistance to *avrRpt2* exists in *Arabidopsis*, bean, and soybean (58). This suggests that functionally homologous resistance genes exist in different species and that in some cases, nonhost resistance is due to the detection of multiple avirulence gene products. Therefore, transfer of resistance gene(s) from a nonhost could provide resistance. Related pathogens may produce similar ligands. Different *Phytophthora* species secrete different elicitins; all induce hypersensitivity in tobacco (50). Mutagenesis of the tightly linked genes *Lr20* and *Sr15* for resistance to leaf and stem rust in wheat resulted in simultaneous changes in both specificities, suggesting that one gene recognizes both pathogens (66a). Therefore, transfer of a resistance gene against one pathogen, for example, lettuce downy mildew (*Bremia lactucae*), may result in resistance to other downy mildews in other plant species. This strategy will only be effective occasionally, and its potency will have to be determined empirically.

The success of interspecific transfers will depend partly on how many unique components are required and how readily they interact with downstream elements in the heterologous signal transduction pathway to the resistance response. Mutagenesis of resistance to *Pseudomonas syringae* in tomato identified at least two genes, *Pto* and *Prf*, that are required (105). The success of complementation experiments with *Pto* implies that *Prf* homologs are functional in both resistant and susceptible genotypes (71). Therefore, transfers of single resistance genes between closely related species is more likely to be successful than between more distantly related species. Transfer of resistance between distantly related species may require the introduction of multiple genes.

Widespread use of the same resistance gene in numerous crops may decrease the durability of resistance. Even within a single species, there are several cases where a resistance gene has remained effective when used on a limited scale but has broken down when used extensively (137). There has been some consideration of this with regard to the use of multiple transgenic crops expressing insecticidal genes from *Bacillus thuringiensis* (20, 75, 115); however, no consensus has emerged on how to maximize the longevity of protection provided by such transgenes. Widespread use in multiple crops may hasten the failure of genes like *Mi* from tomato; the longevity of this resistance gene may have partly been the consequence of the diverse host range of the nematode, only one of which expressed *Mi*. Resistance-breaking strains have already developed in areas of high selection pressure (43).

The cloning of resistance genes and their matching avirulence genes provide the opportunity for several transgenic strategies to induce the resistance response nonspecifically. Such strategies will be effective when susceptibility is due to the lack of induction of the resistance response rather than the detoxification or inactivation of the defense response. Expression of an avirulence gene in the plant from a promoter induced either by nonspecific elicitors or as part of the normally susceptible response could result in the induction of the hypersensitive resistance response (HR) (19). Such strategies, however, require very tightly regulated promoters; otherwise, leaky expression will result in catastrophic induction of HR. Several such promoters are being sought (74, 95); a subcomponent of a promoter for putative membrane channel protein in tobacco seems to be specifically induced by nematode infection (89). The advantage of such strategies is that they rely on the natural defense mechanisms in the plant that have proved effective against many pathogens.

EVOLUTION OF NEW SPECIFICITIES

Resistance genes in many species are clustered in the genome. This led to the hypothesis that some resistance genes are members of large multigene families that diverged to have different specificities (80, 96). The sequence similarity between the majority of the genes cloned thus far and their hybridization to small multigene families support this idea. The clustering is most evident for dominant genes involved in gene-for-gene interactions. Recessive genes tend not to be clustered, although this is not always true. Recessive resistance in lettuce to *Plasmopara lactucae-radicis* mapped in the same cluster as two dominant genes for resistance to *Bremia lactucae* and a dominant gene for resistance to turnip mosaic virus (54, 135).

Distinguishing between different resistances being truly allelic or encoded by tightly linked loci is difficult genetically (110); this distinction is greatly helped by the availability of cloned genes. Sequence divergence or hemizygosity may repress recombination, and recombination between tightly linked genes may never be detected. Recombinants were never detected between the *Pto* and *Fen* genes (72); the separate specificities could only be demonstrated by mutation and complementation with separate cloned members of the multigene family (71, 73, 105). Hybridization with the cloned *L6* gene supports the genetic dissection of the *L* and *M* loci (25). Specificities at the *L* locus are considered to be allelic and only a single RFLP cosegregated with this locus; in contrast, recombinants have been frequently observed within the *M* locus and multiple RFLPs cosegregated with this locus. Detection of rare, non-Mendelian, susceptible progeny or spontaneous meiotic instability at resistance loci may reflect a number of possible genetic events, including insertional inactivation by a transposon (134), unequal crossing-over (114), intragenic recombination, or spontaneous mutation. Cloned genes and tightly linked markers are providing the tools to differentiate between these processes.

The generation of new resistance specificities may involve random or programmed genetic rearrangements within clusters of resistance genes. Gene duplication and subsequent divergence has been demonstrated for several types of genes in diverse species (8). Duplication is thought to increase the rate of evolution of new functions in the duplicated sequences (98, 125). The initial duplication event may have resulted from transposon activity or mispairing between dispersed repeats (121). Subsequent amplification of duplicated sequences could occur by unequal crossing-over between lower copy repeats (114). A thorough genetic analysis of instability at the *Rp1* locus in corn demonstrated that the loss of resistance was associated with recombination; both arrangements of flanking markers were recovered, which suggests the involvement of unequal crossing-over (114). However, markers that were sufficiently closely linked to *Rp1* were not available to demonstrate gain or loss of genetic material. Genetic markers were lost in a spontaneous *Dm3* mutant in the major *Dm* gene cluster of lettuce (P Anderson, R Arroyo-Garcia, P Okubara, R Michelmore, unpublished observations) and in mutants generated as part of the transposon-tagging experiments to clone *L6*, *N*, and *Cf9* (46, 64, 129); in these cases, polymorphic flanking markers were either not available or not analyzed to determine the role of recombination.

Duplication of sequences within and between clusters provides further evidence for functional and evolutionary relatedness between clusters (22a, 25, 90). However, the existence of duplicated sequences does not imply that they are involved in unequal crossing-over and in generating new specificities. There are many duplicated sequences in most regions of the genome of higher plants. Neither clusters of resistance genes nor other regions have been characterized in sufficient detail to determine to what extent the genomic structure of resistance gene clusters is unique. The 280-Kb region around the *Adh1* locus in corn contained a minimum of 37 repeated genomic sequences (111).

It remains an open question as to whether resistance genes predominantly evolve by programmed rearrangements, random mutation, or a mixture of both. The diversity of resistance specificities suggests that a mechanism must exist for generating new specificities faster than the spontaneous mutation rate (97). The majority of the resistance genes cloned so far hybridize to small multigene families in multiple genotypes. Sequencing multiple alleles from these families and the characterization of spontaneous mutants and natural variants will provide evidence for the mechanisms generating changes in specificity. If evolution of resistance genes is a random process, it will be difficult to detect new specificities as it will only be possible to test new alleles with a limited number of pathogen genotypes. If, however, programmed genetic changes occur in both forward and backward directions, it may be possible to detect ancestral specificities using even a small number of pathogen genotypes.

The same recognition specificity to a prevalent pathogen ligand may have evolved multiple times. The detection of the same avirulence genes in taxonomically distant species (55, 127, 128) are examples either of convergent evolution to detect the same ligand or of identity by descent and preservation of an ancient specificity. This issue will be resolved when these genes are cloned from each of these species and compared at the sequence level. Multiple origins for genes encoding the same specificity within a species is more difficult to document. There is only one report of the same specificity mapping to different genetic positions (119). Analysis of molecular marker haplotypes of the second largest cluster of resistance genes in lettuce demonstrated distinct origins for two resistance genes that detected the same avirulence gene; however, both genes mapped to the same position, and it is unclear whether they represent an ancient specificity or multiple evolution of the same specificity (135). Sequence analysis of the genes when cloned will resolve this question.

SELECTION OF NEW SPECIFICITIES

Sequence comparisons and domain swaps between cloned resistance genes will identify regions that are critical to receptor and effector functions. Studies of the structure of resistance gene clusters may reveal programmed genetic rearrangements that have generated alleles with different specificities. However, it is likely to be difficult to select new specificities that are effective against particular pathogen ligands through the use of plant genetics. It will probably require the selection of new specificities in a microbial system from very large numbers of variants generated in sequences known to be critical to specificity. Such an approach is now proving highly successful with mammalian antibodies using phage presentation to generate new specificities (68, 133). Antibody fragments have been expressed as part of bacteriophage coat proteins and phage expressing particular specificities selected on the basis of binding affinity to a specific antigen. This provides powerful enrichment for sequences encoding individual specificities. A variety of strategies are available to generate further variation and to select for increased binding affinities. Using these molecular techniques to mimic evolution occurring in the mammalian immune system, new specificities have been selected against diverse epitopes (44, 69). Adaptation of such a system to plant resistance genes would offer the opportunity to develop genes encoding specificities to any antigen. In particular, it would be possible to target resistance specificities to components of the pathogen that are vital to its survival. Such sequences would be "strong" resistance genes *sensu* van der Plank (123). It may also be possible to select resistance specificities against pathogens or pests for which there are currently no resistance genes. Whether phage presentation approaches can be adapted to generate new resistance genes, or whether some other microbial system will prove optimal remains to be seen. The critical factors will be the ability to generate sufficient variation in appropriate regions of resistance genes, the capacity for efficient positive selection of functional sequences from very large pools of variants, and the ease of reconstituting a functional resistance gene.

RESISTANCE GENES OR TRANSGENES?

Molecular biology is providing an increasing diversity of options for disease control. In addition to the strategies described above, there are growing numbers of novel transgenes, unrelated to classical resistance genes, that may provide high levels of control. The production of hydrolytic enzymes, antinutrients, enzymes for phytoalexin synthesis, and viral components have been shown to protect against pathogens that have been typically difficult to control (60, 113). It is unclear the extent to which these transgenic strategies will complement or substitute for strategies involving classical resistance genes. Will classical monogenic resistance genes remain the mainstay of resistance breeding programs or be supplanted by novel transgenes?

Classical resistance genes will continue to provide resistance using the plant's existing defense mechanisms. There are unlikely to be any increased physiological costs or potentially detrimental changes in nutritional value (perceived or real). In some of the short-term marker-aided selection strategies, the end product is not transgenic and therefore there are no public acceptance concerns. However, classical resistance is not available for some diseases, particularly for some of the more omnivorous and less-specialized pathogens and pests. Also, some classical resistance genes have proved effective for only limited time periods. The challenge is to use molecular techniques to provide more predictable and durable resistance. It should be possible to provide an increased diversity of resistance that is less likely to be overcome and to increase the speed with which new resistances are introduced. This would be an unstable equilibrium in which the plant breeder and molecular biologist play a major role in controlling the pathosystem, rather than responding to the dictates of pathogen variability.

Novel transgenes are more unknown and difficult to generalize. The full range of potential strategies is only just beginning to be realized. They have the capacity to provide resistance to pathogens such as the necrotrophs and insects that have been difficult to control through classical breeding. The key to their use will be whether there are any secondary consequences to their expression, the ease with which it can be demonstrated that there are no detrimental changes in nutritional quality or adverse ecological consequences, and their durability. In several cases, overcoming the novel resistance seems to require a major change in the pathogen (for example, the acquisition of the ability to metabolize a new phytoalexin) and therefore its durability might be predicted to be great. However, given the evolutionary flexibility shown by pathogens, it will be dangerous and naive to rely solely on one control strategy.

CONCLUSIONS AND EPILOGUE

The cloning and characterization of numerous resistance genes will be a major activity over the next few years. We are in a Renaissance period. We are no longer limited by available techniques; continued technical advances will increase the ease of gene isolation. The primary rate-limiting step will be the genetic definition of the targeted resistance gene at the classical level (as well as funding for the research).

Characterization of numerous genes will resolve several outstanding questions on the function and evolution of resistance genes. It will also address the relationship of resistance genes of various mechanistic classes to nonhost resistance. This in turn will provide new options for disease resistance strategies.

Great progress is being made in the dissection of quantitative resistance. Detailed mapping and ultimately cloning of major genes and QTL for resistance will determine the relationships between these genes.

Marker-aided selection and transgenes will become increasingly employed for disease control. The relative balance between the different strategies will be different for each disease and will depend on the relative costs, benefits, and sources of variation. It is debatable whether we should aim for unstable equilibrium or permanent fix. Several strategies allow efficient generation and manipulation of resistances that ultimately may be rendered ineffective by changes in the pathogen. This will allow the predictable management of an unstable equilibrium. Other novel transgene strategies may be more difficult for the pathogen to overcome and therefore could provide a permanent fix, but currently represent greater unknowns regarding their secondary effects and durability.

The application of molecular techniques to improve disease resistance has the potential for contrasting consequences on genetic diversity and genetic vulnerability of crop species. On the one hand, there is the potential to generate highly resistant genotypes using expensive high-technology approaches. This will tend to reduce the number of high-performing genotypes and restrict genetic diversity, particularly if transgenic approaches are involved, because few genotypes will be transformed initially. On the other hand, molecular markers provide the ability to detect and manipulate increased diversity at resistance loci. Also, marker-aided backcrossing will increase speed of transfer of transgenes between genotypes. It is important that markers are used to maximize diversity, both of the resistances used and of the background genotypes. This is particularly important in less developed countries where production may be currently buffered by greater diversity of local genotypes that should be preserved, and where the social consequences of failures in resistance are greater than in other regions.

ACKNOWLEDGMENTS

I thank numerous colleagues, particularly past and present members of my lab as well as members of the NSF Center for Engineering Plants for Resistance Against Pathogens (CEPRAP) at the University of California, Davis, for useful discussions over the years. I also acknowledge the financial support from the USDA NRICGP #s 92-37300-7547 and 93-37300-8772.

Literature Cited

1. Barua UM, Chalmers KJ, Hackett CA, Thomas WTB, Powell W, Waugh R. 1993. Identification of RAPD markers linked to a *Rynchosporium secalis* resistance locus in barley using near-isogenic lines and bulked segregant analysis. *Heredity* 71:177–84
2. Beachy RN. 1993. Virus resistance through expression of coat protein genes. See Ref. 12a, pp. 89–104
3. Beavis WD, Grant D, Albertson M, Fincher R. 1991. Quantitative trait loci for plant height in four maize populations and their association with qualitative genetic loci. *Theor. Appl. Genet.* 83:141–45
4. Beckmann JS, Soller M. 1983. Restriction fragment length polymorphisms in genetic improvement: methodologies, mapping and costs. *Theor. Appl. Genet.* 67:35–43
5. Bennetzen JL, Jones JDG. 1992. Approaches and progress in the molecular cloning of plant disease resistance genes. In *Genetic Engineering*, ed. JK Setlow, 14:99–124. New York: Plenum
6. Bent AF, Kunkel BN, Dahlbeck D, Brown KL, Schmidt R, et al. 1994. *RPS2* of *Arabidopsis thaliana*: a leucine-rich repeat class of plant disease resistance genes. *Science* 265:1856–60
7. Bevan JR, Clarke DD, Crute IR. 1993.

Resistance to *Erysiphe fisheri* in two populations of *Senecio vulgaris*. *Plant Pathol.* 42:636–46
8. Borst P, Greaves DR. 1987. Programmed gene rearrangements altering expression. *Science* 235:658–67
9. Bubeck DM, Goodman MM, Beavis WD, Grant D. 1993. Quantitative trait loci controlling resistance to gray leaf spot in maize. *Crop Sci.* 33:838–47
10. Burdon JJ. 1987. Phenotypic and genetic patterns of resistance to the pathogen *Phakopsora pachyrhizi* in populations of *Glycine canescens*. *Oceologia* 73: 257–67
11. Cause M, Fulton T, Gu Cho Y, Sag Ahn N, Wu K, et al. 1995. Saturated molecular map of the rice genome based on an interspecific backcross population. *Genetics*. In press
12. Chen FQ, Prehn D, Hayes PM, Mulrooney D, Corey A, Vivar H. 1994. Mapping genes for resistance to barley stripe rust (*Puccinia striiformis* f.sp. *hordei*). *Theor. Appl. Genet.* 88:215–19
12a. Chet I, ed. 1993. *Biotechnology in Plant Disease Control*. New York: Wiley-Liss
13. Churchill GA, Doerge RW. 1994. Empirical threshold values for quantitative trait mapping. *Genetics* 138:963–71
14. Chuwongse J, Bunn TB, Crossman C, Jiang J, Tanksley SD. 1994. Chromosomal localization and molecular marker tagging of the powdery mildew resistance gene (*Lv*) in tomato. *Theor. Appl. Genet.* 89:76–79
15. Concibido V, Denny RL, Boutin SR, Hautea R, Orf JH, Young ND. 1994. DNA marker analysis of loci underlying resistance to soybean cyst nematode (*Heterodera glycines* Ichinohe). *Crop Sci.* 34:240–46
16. Danesh D, Aarons S, McGill GE, Young ND. 1994. Genetic dissection of oligogenic resistance to bacterial wilt in tomato. *Mol. Plant-Microbe Interact.* 7: 464–71
17. Dangl JL. 1994. Bacterial pathogenesis of plants and animals. *Curr. Topics Microbiol.* 192:99–118
18. Dangl JL. 1995. Pièce de résistance: novel classes of plant disease resistance genes. *Cell*. In press
19. de Wit PJGM. 1992. Molecular characterization of gene-for-gene systems in plant fungus interactions and the application of avirulence genes in control of plant pathogens. *Annu. Rev. Phytopathol.* 30:391–418
20. Denholm I, Rowland MW. 1992. Tactics for managing pesticide resistance in arthropods: theory and practice. *Annu. Rev. Entomol.* 37:91–112
21. Dietrich RA, Delaney TP, Uknes SJ, Ward ER, Ryals JA, Dangl JL. 1994. *Arabidopsis* mutants simulating disease resistance response. *Cell* 77:565–77
22. Dirlewanger E, Isaac PG, Ranade S, Belajouza M, Cousin R, de Vienne D. 1994. Restriction fragment length polymorphism analysis of loci associated with disease resistance genes and developmental traits in *Pisum sativum* L. *Theor. Appl. Genet.* 88:17–27
22a. Dixon MS, Jones DA, Hatzixanthis K, Ganal MW, Tanksley SD, Jones JDG. 1995. High resolution mapping of the physical location of the tomato *Cf-2* gene. *Mol. Plant-Microbe Interact.* In press
23. Edwards MD, Page NJ. 1994. Evaluation of marker-assisted selection through computer simulation. *Theor. Appl. Genet.* 88:376–82
24. El-Kharbotly A, Leonards-Schippers C, Huigen DJ, Jacobsen E, Pereira A, et al. 1994. Segregation analysis and RFLP mapping of the *R1* and *R3* alleles conferring race-specific resistance to *Phytophthora infestans* in progeny of dihaploid potato plants. *Mol. Gen. Genet.* 242:749–54
25. Ellis JG, Lawrence GJ, Finnegan EJ, Anderson PA. 1995. Contrasting complexity of two rust resistance loci in flax. *Proc. Natl. Acad. Sci. USA* In press
26. Frasier RSS. 1992. The genetics of plant-virus interactions: implications for plant breeding. See Ref. 45a, pp. 175–85
27. Freymark PJ, Lee M, Woodman WL, Martinson CA. 1993. Quantitative and qualitative loci affecting host response to *Exserohilum turcicum* in maize (*Zea mays* L.). *Theor. Appl. Genet.* 87:537–44
28. Friealdenhoven A, Scherag B, Hollricher K, Collinge DB, Thordal-Christensen H, Schulze-Lefert P. 1994. *Nar-1* and *Nar-2*, two loci required for Mla_{12}-specified race-specific resistance to powdery mildew in barley. *Plant Cell* 6:983–94
29. Gebhardt C, Mugniery D, Ritter E, Salamini F, Bonnel E. 1993. Identification of RFLP markers closely linked to the *H1* gene conferring resistance to *Globodera rostochiensis* in potato. *Theor. Appl. Genet.* 85:541–44
30. Gepts P. 1993. The use of molecular and biochemical markers in crop evolution studies. *Evol. Biol.* 27:51–94
31. Giese H, Holm-Jensen AG, Jensen HP, Jensen J. 1993. Localization of the Laevigatum powdery mildew resistance gene to barley chromosome 2 by the use of RFLP markers. *Theor. Appl. Genet.* 85:897–900

32. Godiard L, Grant MR, Dietrich RA, Kiedrowski S, Dangl JL. 1994. Perception and response in plant disease resistance. *Curr. Opin. Gen. Dev.* 4:662–71
33. Graner A, Bauer E. 1993. RFLP mapping of the *ym4* virus resistance gene in barley. *Theor. Appl. Genet.* 86:689–93
34. Greenberg JT, Ausubel FM. 1993. *Arabidopsis* mutants compromised for the control of cellular damage during pathogenesis and aging. *Plant J.* 4:327–41
35. Haley SD, Afanador L, Kelly JD. 1994. Identification and application of a random amplified polymorphic DNA marker for the *I* gene (potyvirus resistance) in common bean. *Phytopathology* 84:157–60
36. Hammond-Kosack KE, Jones DA, Jones JDG. 1994. Identification of two genes required in tomato for full *Cf-9*-dependent resistance to *Cladosporium fulvum*. *Plant Cell* 6:361–74
37. Heun M. 1992. Mapping quantitative powdery mildew resistance of barley using restriction fragment length polymorphism map. *Genome* 35:1019–25
38. Deleted in proof
39. Ho JY, Weide R, Ma HM, van Wordragen MF, Lambert KN, et al. 1992. The root-knot nematode resistance gene (*Mi*) in tomato: construction of a molecular linkage map and identification of dominant cDNA markers in resistant genotypes. *Plant J.* 2:971–82
40. Hospital F, Chevalet C, Mulsant P. 1992. Using markers in gene introgression breeding programs. *Genetics* 132:1199–210
41. Islam MR, Shepherd KW, Mayo GME. 1989. Recombination among genes at the *L* group in flax conferring resistance to rust. *Theor. Appl. Genet.* 77:540–46
41a. Jacobs T, Parlevliet JE, eds. 1993. *Durability of Disease Resistance*. Dordrecht: Kluwer
42. Jansen RC. 1994. Controlling the type I and type II errors in mapping quantitative trait loci. *Genetics* 138:871–81
43. Jarquin-Barberena H, Dalmasso A, deGuiran G, Cardin MC. 1991. Acquired virulence in the plant parasitic nematode *Meloidogyne incognita*. 1. Biological analysis of the phenomenon. *Rev. Nematol.* 14:261–75
44. Jespers LS, Roberts A, Mahler SM, Winter G. 1994. Guiding selection of human antibodies from phage display repertoires to a single epitope of an antigen. *Bio/Technology* 12:899–903
45. Johal GS, Briggs SP. 1992. Reductase encoded by the *Hm1* disease resistance gene in maize. *Science* 258:985–87
45a. Johnson R, Jellis GJ, eds. 1992. *Breeding for Disease Resistance. Euphytica.* Dordrecht: Kluwer. Vol. 63
46. Jones DA, Thomas CM, Hammond-Kosack KE, Balint-Kurti PJ, Jones JDG. 1994. Isolation of the tomato *Cf-9* gene for resistance to *Cladosporium fulvum* by transposon tagging. *Science* 266:789–93
47. Jones ES. 1995. *Mapping quantitative trait loci for resistance to downy mildew in pearl millet*. PhD thesis. Univ. Wales, Bangor, UK
48. Jones ES, Liu CJ, Gale MD, Hash CT, Witcombe JR. 1995. Mapping quantitative trait loci for downy mildew resistance in pearl millet. *Theor. Appl. Genet.* In press
49. Jung M, Weldekidan T, Schaff D, Paterson A, Tingey S, Hawk J. 1994. Generation means analysis and quantitative trait locus mapping of anthracnose stalk rot genes in maize. *Theor. Appl. Genet.* 89:413–18
50. Kamoun S, Young M, Forster H, Coffey MD, Tyler BM. 1994. Potential role of elicitins in the interaction between *Phytophthora* species and tobacco. *Appl. Environ. Microbiol.* 60:1593–98
51. Kawchuk LM, Lynch DR, Hachey J, Bains PS. 1994. Identification of a codominant amplified polymorphic DNA marker linked to the verticillium resistance gene in tomato. *Theor. Appl. Genet.* 89:661–64
52. Keen NT, Dawson WO. 1992. Pathogen avirulence and elicitors of plant defense. In *Plant Gene Research*, ed. T Boller, F Meins, 8:76–103. New York: Springer
53. Kesseli RV, Paran I, Michelmore RW. 1994. Analysis of a detailed genetic linkage map of *Lactuca sativa* (lettuce) constructed from RFLP and RAPD markers. *Genetics* 136:1435–46
54. Kesseli RV, Witsenboer H, Vandemark GJ, Stanghellini ME, Michelmore RW. 1993. Recessive resistance to *Plasmopara lactucae-radicis* maps by bulked segregant analysis to a cluster of dominant resistance genes in lettuce. *Mol. Plant-Microbe Interact.* 6:722–28
55. Kobayashi DY, Tamaki SJ, Keen NT. 1989. Cloned avirulence genes from the tomato pathogen *Pseudomonas syringae* pv *tomato* confer cultivar specificity on soybean. *Proc. Natl. Acad. Sci. USA* 86:157–61
56. Kobe B, Deisenhofer J. 1994. The leucine-rich repeat: a versatile binding motif. *Trends Biol. Sci.* 19:415–21
57. Kreike CM, de Koning JRA, Vinke JH, van Ooijen JW, Geghardt C. 1993. Mapping of loci involved in quantitatively

inherited resistance to the potato cyst-nematode *Globodera rostochiensis* pathotype *Ro1*. *Theor. Appl. Genet.* 87: 464–70
58. Kunkel BN, Bent AF, Dalhbeck D, Innes RW, Staskawicz BJ. 1993. *RPS2*, an *Arabidopsis* disease resistance locus specifying recognition of *Pseudomonas syringae* strains expressing the avirulence gene *avrRpt2*. *Plant Cell* 5:865–75
59. Lamb CJ. 1994. Plant disease resistance genes in signal perception and transduction. *Cell* 76:419–22
60. Lamb CJ, Ryals JA, Ward ER, Dixon RA. 1992. Emerging strategies for enhancing crop resistance to microbial pathogens. *Bio/Technology* 19:1436–45
61. Lande R, Thompson R. 1990. Efficiency of marker assisted selection in the improvement of quantitative traits. *Genetics* 124:743–56
62. Landry BS, Hubert N, Crete R, Chang MS, Lincoln SE, Etoh T. 1992. A genetic map for *Brassica oleracea* based on RFLP markers detected with expressed DNA sequences and mapping of resistance genes to race 2 of *Plasmodiophora brassicae* (Woronin). *Genome* 35:409–20
63. Lawrence GJ, Ellis JG, Finnegan EJ. 1994. Cloning a rust-resistance gene in flax. In *Advances in Molecular Genetics of Plant-Microbe Interactions*, ed. MJ Daniels, 3:303–6. New York: Kluwer
64. Lefebvre V, Chevre A-M. 1995. Tools for marking plant disease and pest resistance genes: a review. *Agronomie*. In press
65. Leonards-Schippers C, Giefffers W, Schafer-Pregl R, Ritter E, Knapp SJ, et al. 1994. Quantitative resistance to *Phytophthora infestans* in potato: a case study for mapping in an allogamous plant species. *Genetics* 137:67–77
66. Leung H, Nelson RJ, Leach JE. 1993. Population structure of plant pathogenic fungi and bacteria. In *Advances in Plant Pathology*, ed. JH Andrews, IC Tommerup, 10:157–205. New York: Academic
66a. MacIntosh RA. 1977. *Nature of Induced Mutations Affecting Disease Reaction in Wheat. Induced Mutations for Disease Resistance in Crop Plants*, pp. 551–65. Vienna: IAEA
67. Maisonneuve B, Bellec Y, Anderson P, Michelmore RW. 1994. Rapid mapping of two genes for resistance to downy mildew from *Lactuca serriola* to existing clusters of resistance genes. *Theor. Appl.* 89:96–104
68. Marks JD, Hoogenboom HR, Griffiths AD, Winter G. 1992. Molecular evolution of proteins on filamentous phage: mimicking the strategy of the immune system. *J. Biol. Chem.* 267:16007–10
69. Marks JD, Ouwehand WH, Bye JM, Finnern R, Gorick BD, et al. 1993. Human antibody fragments specific for human blood group antigens from a phage display library. *Bio/Technology* 11:1145–49
70. Martin GB. 1995. Recent successes in cloning plant disease resistance genes. In *Plant Microbe Interactions*, ed. G Stacey, N Keen. Vol. 1. In press
71. Martin GB, Brommonschenkel SH, Chunwongse J, Frary A, Ganal MW, et al. 1993. Map-based cloning of a protein kinase gene conferring disease resistance in tomato. *Science* 262:1432–36
72. Martin GB, de Vicente MC, Tanksley SD. 1993. High-resolution linkage analysis and physical characterization of the *Pto* bacterial resistance locus in tomato. *Mol. Plant-Microbe Interact.* 6: 26–34
73. Martin GB, Frary A, Wu T, Brommonschenkel S, Chunwongse J, et al. 1994. A member of the *Pto* family confers sensitivity to Fenthion resulting in rapid cell death. *Plant Cell* 6:1543–52
74. Martini N, Egen M, Runtz I, Strittmatter G. 1993. Promoter sequences of a potato pathogenesis-related gene mediate transcriptional activation selectively upon fungal infection. *Mol. Gen. Genet.* 236: 179–86
75. McGaughey WH, Whalon ME. 1992. Managing insect resistance to *Bacillus thuringiensis* toxins. *Science* 258:1451–55
76. Meeley RB, Johal GS, Briggs SP, Walton JD. 1992. A biochemical phenotype for a disease resistance gene of maize. *Plant Cell* 4:71–77
77. Melchinger AE. 1990. Use of molecular markers in breeding for oligogenic disease resistance. *Plant Breed.* 104:1–19
78. Messeguer R, Ganal M, de Vicente MC, Young ND, Bolkan H, Tanksley SD. 1991. High resolution RFLP map around the root knot nematode resistance gene *(Mi)* in tomato. *Theor. Appl. Genet.* 82: 529–36
79. Michelmore RW. 1995. Isolation of disease resistance genes from crop plants. *Curr. Opin. Biotechnol.* 6: In press
80. Michelmore RW, Hulbert SH, Landry BS, Leung H. 1987. Towards a molecular understanding of lettuce downy mildew. In *Genetics and Plant Pathogenesis*, ed. PR Day, GJ Jellis, pp. 221–31. Cambridge, MA: Blackwell Sci.
81. Michelmore RW, Kesseli RV, Francis DM, Paran I, Fortin MG, Yang C-H.

1992. Strategies for cloning plant disease resistance genes. In *Molecular Plant Pathology—A Practical Approach*, ed. S Gurr, MJ McPherson, DJ Bowles, 2:233–88. Oxford: IRL Press

82. Michelmore RW, Paran I, Kesseli RV. 1991. Identification of markers linked to disease resistance genes by bulked segregant analysis: a rapid method to detect markers in specific genomic regions using segregating populations. *Proc. Natl. Acad. Sci. USA* 88:9828–32
83. Miklas PN, Stavely JR, Kelly JD. 1993. Identification and potential use of molecular markers for rust resistance in common bean. *Theor. Appl. Genet.* 85: 745–49
84. Mindrinos M, Katagiri F, Yu G-L, Ausubel FM. 1994. The *A. thaliana* disease resistance gene *RPS2* encodes a protein containing a nucleotide-binding site and leucine-rich repeats. *Cell* 78: 1089–99
85. Neinhuis J, Helentjaris T, Slocum M, Ruggero B, Schaeffer A. 1987. Restriction fragment length polymorphism analysis of loci associated with insect resistance in tomato. *Crop Sci.* 27:797–803
86. Nelson RR. 1978. Genetics of horizontal resistance to plant disease. *Annu. Rev. Phytopathol.* 16:359–78
87. Nodari RO, Tsai SM, Guzman P, Gilbertson RL, Gepts P. 1993. Toward an integrated map of common bean. 3. Mapping genetic factors controlling host-bacterial interactions. *Genetics* 134:341–50
88. Okubara P, Anderson P, Michelmore RW. 1994. Mutants of downy mildew resistance in *Lactuca sativa* (lettuce). *Genetics* 137:867–74
89. Opperman CH, Taylor CG, Conkling MA. 1994. Root-knot nematode-directed expression of a plant root-specific gene. *Science* 263:221–23
90. Paran I, Kesseli RV, Westphal L, Michelmore RW. 1992. Recent amplification of triose phosphate isomerase related sequences in lettuce. *Genome* 35:627–35
91. Paran I, Michelmore RW. 1994. Development of reliable PCR-based markers linked to downy mildew resistance genes in lettuce. *Theor. Appl. Genet.* 85:985–93
92. Pè ME, Gianfranceschi L, Taramino G, Tarchini R, Angelini P, et al. 1993. Mapping quantitative trait loci (QTLs) for resistance to *Gibberella zeae* infection in maize. *Mol. Gen. Genet.* 241:11–16
93. Penner GA, Chong J, Levesque-Lemay M, Molnar SJ, Fedak G. 1993. Identification of a RAPD marker linked to the oat stem rust gene *Pg3*. *Theor. Appl. Genet.* 85:702–5
94. Penner GA, Chong J, Wight CP, Molnar SJ, Fedak G. 1993b. Identification of a RAPD marker for the crown rust resistance gene *Pc68* in oats. *Genome* 36: 818–20
95. Pontier D, Godiard L, Marco Y, Roby D. 1994. *hsr203J*, a tobacco gene whose activation is rapid highly localized, and specific for incompatible plant pathogen interactions. *Plant J.* 5:507–21
96. Pryor T. 1987. Origin and structure of disease resistance genes in plants. *Trends Genet.* 3:157–61
97. Pryor T, Ellis J. 1993. The genetic complexity of fungal resistance genes in plants. *Adv. Plant Pathol.* 10:281–305
98. Purugganan MD, Wessler SR. 1994. Molecular evolution of the plant *R* regulatory gene family. *Genetics* 138:849–54
99. Reinisch AJ, Dong J, Brubaker CL, Stelly DM, Wendel JF, Paterson AH. 1994. A detailed map of cotton, *Gossypium hirsutum* x *Gossypium barbadense*: chromosome organization and evolution in a disomic polyploid genome. *Genetics* 138:829–47
100. Rick CM, Fobes JF. 1974. Association of an allozyme with nematode resistance. *Rep. Tomato Genet. Coop.* 24:25
101. Rick CM, Fobes JF. 1975. Allozyme variation in cultivated tomato and closely related species. *Bull. Torrey Bot. Club* 102:376–84
102. Robbins MA, Witsenboer H, Michelmore RW, Laliberte J-F, Fortin MG. 1994. Genetic mapping of turnip mosiac virus resistance in *Lactuca sativa*. *Theor. Appl. Genet.* 89:583–89
103. Robertson DS. 1989. Understanding the relationship between qualitative and quantitative genetics. In *Development and Application of Molecular Markers to Problems in Plant Genetics*, ed. T Helentjaris, B Burr, pp. 81–87. Cold Spring Harbor: Cold Spring Harbor Lab.
104. Saghai Maroof MA, Zhang Q, Biyashev RM. 1994. Molecular marker analysis of powdery mildew resistance in barley. *Theor. Appl. Genet.* 88:733–40
105. Salmeron JM, Barker SJ, Carland FM, Mehta AY, Staskawicz BJ. 1994. Tomato mutants altered in bacterial disease resistance provide evidence for a new locus controlling pathogen recognition. *Plant Cell* 6:511–20
106. Deleted in proof
107. Schachermayr G, Siedler H, Gale MD, Winzeler H, Winzeler M, Keller B. 1994. Identification and localization of

molecular markers linked to the *Lr9* leaf rust resistance gene of wheat. *Theor. Appl. Genet.* 88:110–15
108. Schon CC, Lee M, Melchinger AE, Guthrie WD, Woodman WL. 1993. Mapping and characterization of quantitative trait loci affecting resistance against second-generation European corn borer in maize with the aid of RFLPs. *Heredity* 70:648–59
109. Deleted in proof
110. Shepherd KW, Mayo GME. 1972. Genes conferring specific plant disease resistance. *Science* 175:375–80
111. Springer PS, Edwards KJ, Bennetzen JL. 1994. DNA class organization on maize *Adh1* yeast artificial chromosomes. *Proc. Natl. Acad. Sci. USA* 91:863–67
112. Stam P, Zeven AC. 1981. The theoretical proportion of the donor genome in near-isogenic lines of self-fertilizers bred by backcrossing. *Euphytica* 30: 227–38
113. Stiekema WJ, Visser B, Florak DEA. 1993. Is durable resistance against viruses and bacteria attainable via biotechnology? See Ref. 41a, pp. 71–81
114. Sudupak MA, Bennetzen JL, Hulbert SH. 1993. Unequal exchange and meiotic instability of disease-resistance genes in the *Rp1* region of maize. *Genetics* 133:119–25
115. Tabashnik BE. 1994. Evolution of resistance to *Bacillus thuringiensis*. *Annu. Rev. Entomol.* 39:47–79
116. Tanksley SD. 1994. Mapping polygenes. *Trends Genet.* 27:205–33
117. Tanksley SD, Ganal MW, Prince JP, de Vincente MC, Bonierbale MW, et al. 1992. High density molecular linkage maps of the tomato and potato genomes. *Genetics* 132:1141–60
118. Tanksley SD, Young ND, Paterson AH, Bonierbale MW. 1989. RFLP mapping in plant breeding: new tools for an old science. *Bio/Technology* 7:257–64
119. Teverson DM, Taylor JD, Crute IR, Kornegay J, Jenner CE, et al. 1995. Pathogenic variation in *Pseudomonas syringae* pv. *phaseolicola* III. Analysis of the gene-for-gene relationship between pathogen races and *Phaseolus vulgaris* cultivars. *Plant Pathol.* In press
120. Timmerman GM, Frew TJ, Miller AL, Weeden NF, Jermyn WA. 1993. Linkage mapping of *sbm-1*, a gene conferring resistance to pea seed-borne mosaic virus, using molecular markers in *Pisum sativum*. *Theor. Appl. Genet.* 85:609–15
121. Tsubota SI, Rosenberg H, Szostak H, Rubin D, Scheldl P. 1989. The cloning of the *Bar* region and the B breakpoint in *Drosophila melanogaster*: evidence for a transposon induced rearrangement. *Genetics* 122:881–90
122. van der Beck JG, Pet G, Lindhout P. 1994. Resistance to powdery mildew (*Oidium lycopersicon*) in *Lycopersicon hirsutum* is controlled by an incompletely-dominant gene *Ol-1* on chromosome 6. *Theor. Appl. Genet.* 89: 467–73
123. van der Plank JE. 1968. *Disease Resistance in Plants*. New York: Academic
124. Walbot V, Hoisington DA, Neuffer MG. 1983. Disease lesion mimics in maize. In *Genetic Engineering of Plants*, ed. T Kosuge, C Meredith, pp. 431–42. New York: Plenum
125. Walsh JB. 1995. How often do duplicated genes evolve new functions? *Genetics* 139:421–28
126. Wang G-L, Mackill DJ, Bonman JM, McCouch SR, Champoux MC, Nelson RJ. 1994. RFLP mapping of genes conferring complete and partial resistance to blast in a durably resistant rice cultivar. *Genetics* 136:1421–34
127. Whalen MC, Innes RW, Bent AF, Staskawicz BJ. 1991. Identification of *Pseudomonas syringae* pathogens of *Arabidopsis* and a bacterial locus determining avirulence on both *Arabidopsis* and soybean. *Plant Cell* 3:49–59
128. Whalen MC, Stall RE, Staskawicz BJ. 1988. Characterization of a gene from a tomato pathogen determining hypersensitive resistance in non-host species and genetic analysis of this resistance in bean. *Proc. Natl. Acad. Sci. USA* 85:6743–47
129. Whitham S, Dinesh-Kumar SP, Choi D, Hehl R, Corr C, Baker B. 1994. The product of tobacco mosaic virus resistance gene *N*: similarity to Toll and the interleukin-1 receptor. *Cell* 78:1101–15
130. Williams GGK, Kubelik AR, Livak KJ, Rafalski JA, Tingey SV. 1990. DNA polymorphisms amplified by arbitrary primers are useful as genetic markers. *Nucleic Acids Res.* 18:6531–35
131. Williams KJ, Fisher JM, Langridge P. 1994. Identification of RFLP markers linked to the cereal cyst nematode resistance gene (*Cre*) in wheat. *Theor. Appl. Genet.* 89:927–30
132. Williamson VM, Ho JY, Wu FF, Miller N, Kaloshian I. 1994. A PCR-based marker tightly linked to the nematode resistance gene, *Mi*, in tomato. *Theor. Appl. Genet.* 87:757–63
133. Winter G, Griffiths AD, Hawkins RE, Hoogenboom HR. 1994. Making antibodies by phage display technology. *Annu. Rev. Immunol.* 12:433–55
134. Wise RP, Ellingboe AH. 1985. Fine

structure and instability of the *Ml-a* locus in barley. *Genetics* 111:113–30
135. Witsenboer H, Kesseli RV, Fortin MG, Stanghellini M, Michelmore RW. 1995. Sources and genetic structure of a cluster of genes for resistance to three pathogens in lettuce. *Theor. Appl. Genet.* In press
136. Wolfe MS. 1985. The current status and prospects for multiline cultivars and variety mixtures for disease resistance. *Annu. Rev. Phytopathol.* 23:251–73
137. Wolfe MS. 1993. Can the strategic use of disease resistant hosts protect their inherent durability. See Ref. 41a, pp. 83–96
138. Wolfe MS, Brandle U, Koller B, Limpert E, McDermott JM, et al. 1992. Barley mildew in Europe: population biology and host resistance. See Ref. 45a, pp. 125–39
139. Yaghoobi J, Kaloshian I, Wen Y, Williamson VM. 1995. Mapping a new nematode resistance locus in *Lycopersicon peruvianum*. *Theor. Appl. Genet.* In press
140. Yoneyama K, Anzai H. 1993. Transgenic plants resistant to diseases by the detoxification of toxins. See Ref. 12a, pp. 115–37
141. Young ND, Danesh D, Menancio-Hautea D, Kumar L. 1993. Mapping oligogenic resistance to powdery mildew in mungbean using RFLPs. *Theor. Appl. Genet.* 87:243–49
142. Young ND, Tanksley SD. 1989. RFLP analysis of the size of the chromosomal segments retained around the *Tm-2* locus of tomato during backcross breeding. *Theor. Appl. Genet.* 77:95–101
143. Young ND, Zamir D, Ganal MW, Tanksley SD. 1988. Use of isogenic lines and simultaneous probing to identify DNA markers tightly linked to the *Tm-2a* gene in tomato. *Genetics* 120:579–85
144. Yu YG, Saghai Maroof MA, Buss GR, Maughan PJ, Tolin SA. 1994. RFLP and microsatellite mapping of a gene for soybean mosaic virus resistance. *Phytopathology* 84:60–64
145. Zabeau M, Vos P. 1993. Selective restriction fragment amplification: A general method for DNA fingerprinting. *Eur. Patent Appl. 92402629.7.* Publ. No. 0 534 858 A1
146. Zaitlin D, DeMars S, Ma Y. 1993. Linkage of *rhm*, a recessive gene for resistance to southern corn leaf blight, to RFLP marker loci in maize (*Zea mays*) seedlings. *Genome* 36:555–64
147. Zamir D, Ekstein-Michelson I, Zakay Y, Navot N, Zeiden M, et al. 1994. Mapping and introgression of a tomato yellow leaf curl virus tolerance gene, *TY-1*. *Theor. Appl. Genet.* 88:141–46
148. Zeven AC, Knott DR, Johnson R. 1983. Investigation of linkage drag in near-isogenic lines of wheat by testing for reaction to races of stem rust, leaf rust and yellow rust. *Euphytica* 32:319–27

Annu. Rev. Phytopathol. 1995. 33:429–43

USE OF ALIEN GENES FOR THE DEVELOPMENT OF DISEASE RESISTANCE IN WHEAT[1]

Stephen S. Jones[1], *Timothy D. Murray*[2], *and Robert E. Allan*[1]
[1]USDA-ARS and Department of Crop and Soil Sciences, Washington State University, Pullman, Washington 99164-6420; [2]Department of Plant Pathology, Washington State University, Pullman, Washington 99164-6430

KEY WORDS: *Triticum aestivum*, wild wheat, disease resistance, plant breeding

---

ABSTRACT

The genus *Triticum* contains three ploidy levels and about 30 species. Most of these species have been investigated as sources of disease-resistance genes and several have been used in successful transfers of resistance to domestic wheat (*T. aestivum*, genomes AABBDD). In addition, at least six genera from the tribe Triticeae have been used successfully as donors of disease-resistance genes for domestic wheat. The amount of alien chromatin involved in these transfers varies from a single gene to chromosome arms or entire chromosomes. No attempt was made in this review to describe all alien resistance gene transfers in wheat or to outline the various techniques involved. Alien disease resistance genes covered in detail are those that confer resistance to barley yellow dwarf virus, wheat streak mosaic virus, Cephalosporium stripe (caused by *Cephalosporium gramineum*) and eyespot (caused by *Pseudocerosporella herpotrichoides*).

INTRODUCTION

No other cultivated plant is equal to wheat (*Triticum* spp.) in the breadth of our knowledge of its genomic structures and relationships, availability of wild

[1]The US Government has the right to retain a nonexclusive, royalty-free license in and to any copyright covering this paper.

germplasm, and global importance. The genus *Triticum* contains three ploidy levels and about 30 species (24). Most of these species have been investigated as sources of disease-resistance genes, and several have been used in successful transfers of resistance to domestic wheat (24, 35, 40, 58). In addition, at least six genera from the tribe Triticeae have been used successfully as donors of disease-resistance genes for domestic wheat (24, 35, 58). The amount of alien chromatin involved in these transfers varies from a single gene (3) to chromosome arms or entire chromosomes (27). Ignoring selection, the main factor determining the amount of alien chromatin carried through generations of recombination is chromosome pairing. Recombination frequencies involving homologous chromosomes such as from *T. tauschii* (genome DD) and D genome chromosomes of wheat (*T. aestivum*, genomes AABBDD) can be very close to those observed for wheat-chromosome to wheat-chromosome exchanges. Jones et al (36, 37) recombined two different 1D chromosomes from *T. tauschii* with the 1D chromosome of wheat cultivar Chinese Spring to map the alien leaf rust (caused by *Puccinia recondita*) resistance gene *Lr21*, and stem rust (caused by *P. graminis* f. sp. *tritici*) resistance gene *Sr33* relative to other markers. They found that the recombination frequencies among previously mapped markers in these crosses were consistent with those from other wheat-by-wheat crosses. Conversely, transfers between chromsomes that do not normally pair are much more complex and usually result in translocations and a large amount of alien chromatin introgressed into the wheat genomes. The use of induced homoeologous pairing, spontaneous translocations, and irradiation to achieve gene transfer has been described by several authors (listed in Reference 35).

Recent reviews have addressed the transfer of disease-resistance genes from wild wheats to domestic wheat (24, 35, 40, 41, 58). Similarly, most of these reviews outline the techniques and complexities involved in alien transfers. Therefore, no attempt was made in this review to describe all alien resistance-gene transfers in wheat or to outline the various techniques involved. Rather, we sought to describe economically important examples of disease resistance that cover a range of diseases for which there is little or no genetic variation in domestic wheat.

THE NEED FOR AND VALUE OF ALIEN SOURCES OF DISEASE RESISTANCE

The power of alien sources of resistance is not only to expand existing genetic variation in the crop but, more importantly, to introduce truly novel variation. Although adult plant resistance (race-nonspecific) genes have been described in alien sources (25, 40), the majority have been race-specific genes for foliar fungal pathogens, especially the obligately parasitic rust fungi (58). The greater need for alien disease-resistance genes, though, is against pathogens for which there is limited or no genetic variation in domestic wheat. Examples of the latter include viruses and the facultative saprophytic and facultative parasitic fungal pathogens of the foliage and roots. Indeed, this review addresses two virus-caused diseases, barley yellow dwarf and wheat streak mosaic, and two fungal diseases, eyespot and *Cephalosporium* stripe, each of which is caused by a facultative saprophyte. Each of these diseases is economically important and widespread in various wheat-growing areas of the world. Other diseases that are not discussed, but for which resistance is inadequate or does not exist in wheat include: scab (caused by *Fusarium* sp.), Take-all (caused by *Gaeumannomyces graminis* var. *tritici*), *Septoria/Stagonospora* blotch (caused by *Septoria tritici* and *Stagonospora nodorum*), and sharp eyespot (caused by *Rhizoctonia cerealis*).

Barley Yellow Dwarf

Barley yellow dwarf disease (BYDV) was first described by Oswald & Huston (68) in 1951. The disease is caused by any one of several closely related luteoviruses (48) that are transmitted by several aphid species (11, 29). It is currently the most economically important virus disease of small grains worldwide (14). Bruehl (11) lists 97 grasses that can serve as a host for BYDV. Only low levels of resistance or tolerance have been reported for barley yellow dwarf in wheat (15, 70). Bruehl & Toko (13) were the first to show that *Agropyron* spp. (syn. *Lophopyrum* and *Thinopyrum*) were highly resistant to BYDV and, noting a lack of resistance in wheat, listed several wheat relatives in addition to *Agropyron* spp. that had immune reactions to the virus. Although single genes for resistance to BYDV have not been described in wheat genetically, chromosome and genome manipulations have provided valuable insight into the number and location of the genes controlling resistance. Transfers of genes involving *Agropyron* chromosomes are complex because they do not pair well, or at all, with their wheat homoeologues.

Surveys of *Agropyron* spp. and amphiploids derived from common wheat × *Agropyron* spp. have shown high levels of resistance to BYDV (30, 57, 75, 76, 91). Using field disease reactions McGuire et al (57) found *Lophopyrum elongatum* (Host) A. Love (2n=2X=14) and *L. ponticum* (Podp.) A. Love (2n=10X=70) were both highly resistant. In the same study McGuire et al (57) used a Chinese Spring wheat/*L. elongatum* amphiploid (genomes AABBDDEE) and disomic substitution lines (*L. elongatum* chromosomes substituted for wheat homoeologues) to determine the chromosomal location of the resistance genes. They found that although all of the alien chromosomes had an effect on resistance, major genes for BYDV resistance were located on chromosomes 2E and 5E.

Zhong et al (95) characterized a *L. ponticum* × common wheat backcross line, SW35, highly resistant to BYDV under field conditions. Based on in situ hybridization, meiotic chromosome pairing, and isozyme analysis, this 46-chromosome line had chromosomes 3E and 6E in addition, and 5E was substituted for wheat chromosome 5D. Line SW35 is very wheat-like and will be important in the eventual gene transfer because of its low number of chromosomes. Manipulations with this line to reduce the chromosome number even further are under way (G. Zhong, personal communication).

Xin et al (91) characterized Zhong 4, a wheat × *Thinopyrum intermedium* (syn. *A. intermedium*) amphiploid (2n=8X=56), which was shown by Zhang et al (94) to be highly resistant to BYDV under field conditions. Based on enzyme-linked immunosorbent assay (ELISA) readings, F_1 plants with 49 chromosomes derived from Zhong 4 × several common wheats were shown to have disease reactions intermediate to the resistant and susceptible parents: none of the F_1 plants were as resistant as Zhong 4. Although the authors were not able to reconstitute the resistance of the amphiploid in lines with fewer chromosomes, they were confident that the lines will have value in future transfers of resistance. Banks et al (8) are using this material in further attempts to reduce the alien chromatin while retaining the resistance.

Both cDNA dot blotting and ELISA were used by Goulart et al (30) to study the BYDV resistance in BC_2 lines derived from common wheat × *L. ponticum*. The backcross lines were segregating for resistance and chromosome number, but the number of chromosomes for individual lines was not reported. Chromosome numbers ranged from 44 to 55, and they reported that the lines with higher chromosome number were, in general, more resistant, including several lines that were as resistant as the *L. ponticum* parent. Although field testing was not performed on these lines, the results agree with those of McGuire et al (57) and suggest that inheritance of resistance to BYDV is complex and controlled by many genes.

Wheat Streak Mosaic

Wheat streak mosaic (WSMV) is a disease primarily of winter wheat but which can also cause damage to spring wheat (7). The vector for WSMV is the wheat curl mite (*Eriophyes tulipae* Keifer, syn. *Aceria tulipae*) (6). Although slight levels of tolerance have been observed to WSMV, there is no resistance in common wheat (26, 53). As with BYDV resistance, *Agropyron* spp. exhibit very high levels of resistance to WSMV (47, 59, 73, 75, 79, 87, 88). Strategies for developing resistant cultivars include transfer of genes for resistance to the virus, the vector, or both.

Jiang et al (34) characterized wheat lines CI 15321 and CI 15322 (74), which are highly resistant to both WSMV and the wheat curl mite (53). These lines were derived from a progeny with high chromosome number from common wheat × *L. ponticum* that was resistant to WSMV (73). Using wheat lines that were either chromosome substitutions or translocations, Jiang et al (34) determined that chromosome 1E confers high resistance to WSMV in these lines.

Chromosome 4 from *Th. intermedium* has also been found to confer resistance to WSMV in a wheat background (85). Friebe et al (27) designated this chromosome 4Ai-2 and, using translocated chromosomes, determined that the short arm conferred resistance. They identified line CI 17884 as a 4DL/4AiS substitution for chromosome 4D and line CI 17766 as the same translocated chromosome substituted for chromosome 4A.

Successful resistance to the mite vector is also an effective method of controlling disease. Resistance to the wheat curl mite was derived from lines immune to WSMV that were triple substitutions of *L. ponticum* chromosomes for wheat homoeologs (43). Larson & Atkinson (44) identified chromosome 6Ag as controlling resistance to the mite in a wheat background. Working with different material, Whelan & Hart (89) reported a wheat/*L. ponticum* translocation involving chromosome 6Ag that also confers resistance to establishment of the mite.

Although there appear to be several diverse sources of resistance available to WSMV, there are no commercial cultivars carrying any of these genes. Friebe et al (27) point out that all of the genetic stocks carrying resistance have undesirable agronomic characteristics associated with resistance.

Cephalosporium Stripe

Cephalosporium stripe, caused by *Cephalosporium gramineum* Nis. & Ika. (syn. *Hymenula cerealis* Ell. & Ev.), is a vascular wilt disease of wheat and other small grains and grasses. The disease was first described from Japan in 1933 (65, 66) and later in North America (10) and the United Kingdom (31). Early workers recognized that the pathogen had a wide host range among the small grains and grasses (10, 66). Nisikado et al (66) found *Cephalosporium* stripe on wheat, barley (*Hordeum vulgare* L.), *Avena fatua* L. and later (65) on *Alopecurus agrestis* L. Bruehl (10) found the disease occurring naturally on wheat, rye (*Secale cereale* L.), *Agropyron repens* (L.) Beauv., *Bromus marginatus* Nees., *Dactylis glomerata* L., and *Elymus glaucus* Buckl. Bruehl (10) also demonstrated that 11 other genera of grasses were hosts when greenhouse-grown plants were inoculated with conidia of the pathogen.

Nisikado et al (66) were the first to suggest control of *Cephalosporium* stripe by selection and breeding of resistant cultivars. Several different groups have screened wheat for resistance to *Cephalosporium* stripe (10, 12, 51, 54, 56, 66, 93), and most have arrived at the conclusions that immunity to *C. gramineum* does not exist in wheat, and wheat cultivars vary in the amount of disease and yield loss they sustain, with most considered moderately to highly susceptible. In 1985, however, Mathre et al (55) reported that *A. elongatum*, *A. intermedium*, and *Agrotriticum* #3525 (2n=56, AABBDDEE) were all highly resistant to *C. gramineum* in growth chamber and field screening studies. Other wheat relatives including *T. monococcum*, *T. dicoccum*, *T. timopheevii*, and *T. durum*, were susceptible. Some accessions of *T. tauschii* were moderately resistant, however.

Allan (4) demonstrated that *A. elongatum* chromatin has considerable value for resistance to *Cephalosporium* stripe and tolerance to eyespot (caused by *Pseudocercosporella herpotrichoides*). The hard red winter wheat CI 13113 (Chinese Spring 2*/*A. elongatum*/Pawnee) was initially used as a parent because it expressed resistance to leaf rust (*Puccinia recondita*), stripe rust (*P. striiformis*) and stem rust (*P. graminis*). Subsequent progeny derived from crosses involving CI 13113 were found to be highly resistant to Cephalosporium stripe and have tolerance to eyespot. One line, PI 561033 (WA7437), proved to have combined resistances to the three rusts and both soilborne pathogens (4). This line was derived from a complex cross involving CI 13113 and three club wheat parents (Paha/CI 13645/ Chinese Spring 2*/*A. elongatum*/Pawnee/3/2*Omar). Cytological characterization of PI 13113 and PI 561033 confirmed that both lines contain *A. elongatum* chromatin. Using in situ hybridization with total genomic *A. elongatum* DNA as a probe, it was determined that both lines are euploid and have chromosome 6Ag substituted for chromosome 6A (X Cai & SS Jones, submitted). Further improvement is needed in agronomic and quality characteristics of club wheat germplasm containing *A. elongatum* source resistance. PI 561033 and other sib lines have adequate baking and milling properties but cookie diameter (an important trait for predicting end-use quality) is marginal, and the maturity of plants in the field is often delayed compared to other related lines lacking *A. elongatum* chromatin (4).

Jones et al (38) reported that an amphiploid of Chinese Spring wheat and *L. elongatum* (AABBDDEE) produced by BC Jenkins (20) was resistant to *Cephalosporium* stripe. They also used a complete set of *L. elongatum* substitution lines derived from this amphiploid (21, 23, 82) to identify chromosomes that carry the major genes for resistance. Using plants inoculated in the growth chamber (38), they showed that chromosomes 2E and 3E have a significant effect on resistance in a Chinese Spring wheat background. For both critical chromosomes, substitutions for the A and B genome homoeologues had a greater effect than substitutions for the D genome homoeologues. Working with the same substitution lines, McGuire et al (57) observed a similar phenomenon: when chromosome 2E was substituted for chromosome 2A it had higher BYDV disease scores than the 2E for 2B or 2D substitutions. They suggested that the absence of chromosome 2A increases susceptibility to BYDV. An alternate, although not contrary, explanation is that chromosome 2A carries a resistance gene(s) that interacts with the chromosome 2E gene(s) to increase resistance. Differential effects of homoeologue substitution have been demonstrated in these same substitution lines for resistance to soil salinity (67). Clearly, the choice of chromosome targeted for transfer can have an effect on the ultimate level of expression.

Development of commercially acceptable cultivars with resistance to *Cephalosporium* stripe has been slow, and cultivars with high levels of resistance adapted to the Pacific Northwest are not currently available. Disease development in field plots varies among years and locations (12, 51, 54, 56), making selection of resistant genotypes difficult. Such variation may be due to differences in density of pathogen inoculum and techniques (51, 52, 54, 56). In addition, the disease is strongly influenced by environment, especially soil freezing during the winter, which may also interact with genotype (12, 50, 56). Anderegg & Murray (5) developed a growth chamber–greenhouse method for studying *Cephalosporium* stripe that has been used to screen wheat germplasm and relatives for disease resistance (38). Environmental variation is reduced and results are available in 6 months compared with 11 months for field tests. Identification and selection of resistant genotypes may be more precise if pathogen colonization of plants is measured rather than estimating disease severity. Qi & Murray (69) transformed *C. gramineum* with the β-glucuronidase (GUS) reporter gene from *Escherichia coli* and were able to measure GUS activity in plants as soon as 10 days after inoculation. Preliminary studies have indicated that colonization of resistant genotypes, reflected by GUS activity, is less than in susceptible genotypes (TD Murray & CA Blank, unpublished).

Eyespot

Eyespot, caused by the soilborne pathogen *Pseudocercosporella herpotrichoides* (Fron.) Deighton (teleomorph=*Tapesia yallundae* Wallwork and Spooner), is a stem base disease of wheat grown in cool, temperate climates. The discovery of strains of the pathogen that are resistant to benzimidazole-type fungicides (39, 64) has led to increased efforts to produce resistant cultivars. Only three genes have been described for eyespot resistance: cultivar Cappelle-Desprez (84) carries a gene of unknown source on chromosome 7A (45); breeding line VPM-1 (49) has a gene (*Pch1*) derived from *T. ventricosum* on chromosome 7D (33); and recently, Murray et al (63) mapped a new resistance gene(s) to chromosome 4V in *Dasypyrum villosum*.

Resistance to eyespot was reported in *T. ventricosum* and other wild wheats nearly 60 years ago by Sprague (78). He predicted that resistant wheat cultivars might eventually be developed from wild relatives of wheat. Simonet (77) used a bridging species, *T. persicum*, in the cross *T. persicum/T. ventricosum//T. aestivum* Marne to transfer eyespot resistance to hexaploid wheat. Maia (49) then used these lines to transfer resistance into the cultivated wheat line

VPM-1. The first commercial wheat to use the eyespot resistance of VPM-1 was the French cultivar Roazon (33). However, it was never widely grown.

The eyespot resistance of VPM-1 is believed to be controlled by a single locus (3, 17, 80). Using monosomic analysis, Jahier et al (33) showed that resistance is located on chromosome 7D. Doussinault et al (19) showed that an independently derived line, H-93-70, also having eyespot resistance from *T. ventricosum*, has a single dominant resistance gene designated *Pch1*. Worland et al (90) believed that the *Pch1* gene of H-93-70 was allelic with the gene of VPM-1, and this was subsequently confirmed by Mena et al (61).

The use of *Pch1* in cultivar development in the US Pacific Northwest is a success story on the use of wild gene resources for improvement of disease resistance. The USDA-ARS winter wheat breeding program at Washington State University initiated crosses with VPM-1 in 1975. Although some progeny of these crosses had eyespot resistance, they lacked other important traits such as commercially acceptable end-use quality, yield potential, and cold-hardiness. Yield potential of VPM-1 was 30% lower than the long-term check Nugaines based on 16 site-years of tests in Washington State. In 1974 though, eyespot resistant selections 951 and 421 of VPM/Moisson from the National Institute of Agronomic Research of LaRheu, France, were obtained. These two lines were agronomically superior to VPM-1, had yields 14 to 24% less than Nugaines in 16 site-years of tests, and had better milling quality than VPM-1. Both lines were used extensively as parents in an attempt to develop a resistant cultivar that was acceptable to the growers.

From 1978 to 1983, several thousand progeny from over 100 populations involving VPM-1/Moisson 421 and VPM-1/Moisson 951 were evaluated in head-row nurseries. Lines with promising agronomic type and resistance to endemic foliar diseases were placed in preliminary yield tests. Approximately 80 club-headed and common-headed wheat lines were selected each year from preliminary tests and placed in advanced yield tests.

Disease trials consisted of replicated inoculated and uninoculated (control) paired plots. The inoculated plots were sprayed with conidia of *P. herpotrichoides* in November each year, and the control plots were sprayed with a benzimidazole fungicide for protection. Comparisons were made between the paired plots for grain yield and symptoms of eyespot including lodged tillers and white heads (prematurely dead spikes). Lines with eyespot resistance sustained minimal grain yield loss and had low incidences of lodging and white heads.

In 1984, two promising lines, WA7163 and WA7166, with eyespot resistance were identified. In addition to eyespot resistance, both expressed resistance to stripe rust, leaf rust, and stem rust (9). These two lines had high grain yield potential, adequate cold-hardiness, excellent milling quality, and fair-to-good soft wheat flour quality. In 1988, WA7163, a soft white winter wheat with a common spike, was named Madsen and released to growers (1). A year later WA7166, a club type soft white winter wheat, was released and named Hyak (2). Both cultivars have been readily accepted by growers in the Pacific Northwest primarily because of their resistance to eyespot. In 1994, the two cultivars were grown on over 500,000 ha in the region, and use of fungicides to control eyespot has been reduced significantly. Estimates indicate over 250,000 ha planted to Madsen and Hyak did not require fungicide treatment; this reduced growers production costs by ca $40/ha (TD Murray, unpublished). In 1994, Madsen was the mostly widely grown cultivar in the Pacific Northwest (BC Miller, personal communication).

The *Pch1* resistance gene has had only limited use in Europe. According to Law et al (46), substitution of the 7D chromosome of VPM1 into several adapted United Kingdom wheats depresses yield by about 6%. These authors indicated that it should be possible to break the deleterious linkages between genes for low yield and the *Pch1* gene, as has been done in the US wheats.

Other genes may interact with *Pch1* to increase or decrease eyespot resistance. Hollins et al (32) concluded the potent eyespot resistance of Rendezvous was due to the combined resistance of Cappelle-Desprez and VPM-1. Similarly, Allan & Roberts (3) identified progeny with mean lesion indices that transgressed both resistant parents for higher resistance in a cross between VPM/Moisson 951 (resistant) and Cerco (moderately resistant).

The *Pch1* gene does not provide complete resistance to eyespot or the subsequent losses in grain yield associated with the disease. Madsen sustained significant yield losses (average 15%) in 5 of 13 tests when inoculated and control plots were compared (RE Allan, unpublished). Murray & Bruehl (62) observed similar results with VPM-1, which sustained significant loss in grain yield one year in four under favorable disease conditions. Additional sources of eyespot resistance are needed because the resistance conferred by *Pch1* is not complete and may not be durable. Increases in yield loss of Madsen relative to susceptible cultivars have been observed over a 12-year period in eyespot field evaluation tests (RE Allan, unpublished). In a search for new sources of resistance genes for eyespot, hundreds of accessions of *T. tauschii* (DD genome) have been screened and were shown to carry a high frequency of resistance (92). Work is in progress to transfer these new source genes, as well as the recently described gene from *D. villosum* (63), into Madsen, Hyak, and other adapted backgrounds (TD Murray & SS Jones, unpublished).

Recently, de la Peña & Murray (16) described an improved method for evaluating resistance to eyespot of wheat genotypes. They used a GUS-transformed eyespot strain to measure differences in disease development of 4- to 8-week-old wheat seedlings. This strain has the β-glucuronidase (GUS) reporter gene, and production of the enzyme is highly correlated with growth of the pathogen. de la Peña & Murray (16) showed that differences in GUS activity in seedling tissue of wheat genotypes was closely correlated with differences in resistance to eyespot. The method differentiated among highly resistant, resistant, and susceptible genotypes. Evaluation of resistance with this technique is reduced from about 11 months to 2 months.

Selecting for resistance to the eyespot pathogen using field or greenhouse screening procedures is labor intensive, sometimes inaccurate, and in the case of field tests, slow, taking up to 11 months. Breeding for eyespot resistance in Madsen and Hyak was greatly enhanced with the discovery of a marker for the *Pch1* gene of VPM-1 and its derivatives. Gale et al (28) suggested that the eyespot resistant gene was on 7DL and probably at the distal end because it segregated independently of an isozyme marker for alpha-amylase located near the centromere. The 7DL endopeptidase locus was shown to be about 42 map units from the centromere (60). Several progeny from a cross between Sel. 421 (eyespot resistant) to Sel. 101 (eyespot susceptible) were classified for their endopeptidase allele frequencies and eyespot resistance (60), based on lesion index scores of mature straw. A close association between the VPM-1 *Pch1* gene and the endopeptidase allele *EP-V1* (*Ep-D1b*) was detected. Others subsequently verified this close linkage using different parental lines (42, 83, 90). Summers et al (81) compared the association between the *Pch1* gene and *EP-D1b* endopeptidase allele among large numbers of progeny and confirmed the two genes were tightly linked in coupling. Worland et al (90) suggested that the *Ep-D1b* gene may confer eyespot resistance. Mena et al (61) proved that *Ep-D1b* and *Pch1* genes are only closely linked and can be separated from each other; they also showed *Pch1* was transferred from chromosome $7D^V$ of *T. ventricosum*. Traditional evaluation of eyespot disease is difficult, time consuming, and not always successful.

DISCUSSION

Except for a few cases reported in other reviews (35, 58) and the eyespot resistance transfers discussed here, most alien gene transfers for disease resistance are still not used in commercial wheats. The resistance levels achieved in adapted lines are in some examples equal to that of the wild species, although linked with these genes come undesirable traits that affect yield, end-use quality, and other agronomic traits. Continued rounds of radiation or homoeologous pairing to promote recombination are still required to remove unwanted alien chromatin before breeders can incorporate these valuable genes into commercially acceptable cultivars.

New approaches are also needed for the identification of resistant genotypes, especially when dealing with race-nonspecific resistance and pathogens that are facultative parasites. In these cases, potentially valuable genotypes can be lost owing to an inability to select resistant individuals because resistance is not complete and environmental effects may be large.

Marker-based selection, such as the *Ep-D1b* marker for the *Pch1* eyespot resistance gene, enables selection of resistant individuals with certainty and eliminates undesirable genes. Developing markers still requires the identification of resistant genotypes based on a phenotypic response following inoculation with a pathogen. Use of a GUS-transformed strain of *P. herpotrichoides* (16) along with chromosome addition lines enabled Murray et al (63) to map a new gene for resistance in *D. villosum*, which was not possible with visual disease evaluations. The relative ease of transformation of most fungi should allow the use of reporter genes for detection of resistance in other host-pathogen systems.

Chromosome manipulation or, more specifically, chromosome engineering, which has been practiced for at least 50 years (71), is the main tool used by wheat researchers for introgression of alien genes. Not only has this work led to hundreds of well-characterized alien transfers but it has also set a standard for gene transformation systems to try to equal. Wheat is unique among plants in that it allows geneticists to target gene transfers to specific chromosomes through the use of aneuploid stocks and the induction of homoeologous pairing and recombination (72). Although transformation systems have recently been successful in wheat (86), chromosome engineering still has substantial advantages and potential that should not be abandoned. When transferring alien resistance genes, wheat cytogeneticists deal with the classical definition of the gene, a unit of inheritance. Thus, not only is the coding sequence of the gene transferred but so is everything else needed to confer resistance. What the "everything else" is at this point is not clear. It certainly involves regulatory sequences but may also include positional and other undescribed effects. Genetic transformation will certainly have an impact on the introduction of novel disease resistances into wheat, but this impact will not be realized until resistance genes are cloned, characterized, and their expression limited to the correct timing and tissue. In the meantime, chromosome engineering, aided by the latest techniques of in situ hybridization, chromosome banding, and DNA manipulation, will continue to contribute significantly to our understanding and to solve the very real problems involved in wheat-pathogen interactions.

Literature Cited

1. Allan RE, Peterson CJ Jr., Rubenthaler GL, Line RF, Roberts DE. 1989. Registration of 'Madsen' wheat. *Crop Sci.* 29:1575–76
2. Allan RE, Peterson CJ Jr., Rubenthaler GL, Line RF, Roberts DE. 1990. Registration of 'Hyak' wheat. *Crop Sci.* 30:234
3. Allan RE, Roberts DE. 1991. Inheritance of reaction to strawbreaker foot rot in two wheat populations. *Crop Sci.* 31: 943–47
4. Allan RE, Rubenthaler GL, Moris CF, Line RF. 1993. Registration of three soft white winter wheat germplasm lines resistant or tolerant to strawbreaker foot rot. *Crop Sci.* 33:1111–12
5. Anderegg JC, Murray TD. 1988. Influence of soil matric potential and soil pH on Cephalosporium stripe of winter wheat in the greenhouse. *Plant Dis.* 72: 1011–16
6. Ashwoth IJ Jr, Futrell MC. 1961. Sources, transmission, symptomatology and distribution of wheat streak mosaic virus in Texas. *Plant Dis. Rep.* 45:220–24
7. Atkison TG, Slykhuis JT. 1963. Relation of spring drought, summer rains, and high fall temperatures to the wheat streak mosaic epiphytotic in southern Alberta. *Can. Plant Dis. Surv.* 43:154–59
8. Banks P, Xu SJ, Wang RC, Larkin PJ. 1993. Varying chromosome composition of 56-chromosome wheat × *Thinopyrum intermedium* partial amphiploids. *Genome* 36:207–15
9. Bariana HS, McIntosh RA. 1993. Cytogenetics studies in wheat XV. Location of rust resistance genes in VPM1 and their genetic linkage with other disease resistance genes in chromosome 2A. *Genome* 36:476–82
10. Bruehl GW. 1957. Cephalosporium stripe disease of wheat. *Phytopathology* 47:641–49
11. Bruehl GW. 1961. Barley yellow dwarf. Monogr. No. 1. *Am. Phytopathol. Soc.* 52 pp.
12. Bruehl GW, Murray TD, Allan RE. 1986. Resistance of winter wheats to Cephalosporium stripe in the field. *Plant Dis.* 70:314–16
13. Bruehl GW, Toko HV. 1957. Host range of two strains of the cereal yellow dwarf virus. *Plant Dis. Rep.* 41:730–34
14. Burnett PA, Comeau A, Qualset CO. 1994. Host plant tolerance or resistance for control of barley yellow dwarf. In *Barley Yellow Dwarf*, ed. CJ D'Arcy, PA Burnett, 1:321–43. DF Mexico: CIMMYT. 374 pp.
15. Cisar G, Brown CM, Jedlinski H. 1982b. Diallel analyses for tolerance in winter wheat to the barley yellow dwarf virus. *Crop Sci.* 22:328–33
16. de la Peña RC, Murray TD. 1994. Identifying wheat genotypes resistant to eyespot disease with a β-glucuronidase-transformed strain of *Pseudocercosporella herpotrichoides. Phytopathology* 84:972–77
17. Delibes A, Dosba F, Doussinault G, Garcigra-Olmedo F, Sanchez-Monge R. 1977. Resistance to eyespot (*Cercosporella herpotrichoides*) and distribution of biochemical markers in hexaploid lines derived from a double cross (*Triticum turgidum* × *Aegilops ventricosa*) × *T. aestivum. Proc. EUCARPIA Congr., 8th* 8:917
18. Delibes A, Sanchez-Monge R, Garcigra-Olmedo F. 1977. Biochemical and cytological studies of genetic transfer from the Mv genome of *Aegilops ventricosa* into hexaploid wheat: a progress report. *Proc. EUCARPIA Congr., 8th* 8:819
19. Doussinault G, Dosba F, Jahier J. 1983. New results on the improvement of the level of resistance to eyespot in wheat. *Proc. Int. Wheat Genet. Symp., 6th.* 6: 1938
20. Dvoràk J. 1976. The cytogenetic structure of a 56-chromosome derivative from a cross between *Triticum aestivum* and *Agropyron elongatum* (2n=70). *Can. J. Genet. Cytol.* 18:271–79
21. Dvoràk J. 1980. Homology between *Agropyron elongatum* chromosomes and *Triticum aestivum* chromosomes. *Can. J. Genet. Cytol.* 22:237–59
22. Dvoràk J. 1981. Genome relationships among *Elytrigia (=Agropyron) elongata, E. stipifolia,"E. elongata* 4x," *E. caespitosa, E. intermedia,* and "*E. elongata* 10x." *Can. J. Genet. Cytol.* 23:481–92
23. Dvoràk J, Chen KC. 1984. Phylogenetic relationships between chromosomes of wheat and chromosome 2E of *Elytrigia elongata. Can. J. Genet. Cytol.* 26:128–32
24. Dvoràk J, McGuire PE. 1991. Triticeae, the gene pool for wheat breeding. In *Genome Mapping of Wheat and Related Species: Proceedings of a Public Workshop.* Rep. 7. 7:3–8. Univ. Calif. Genet. Resour. Program, Davis, CA
25. Dyck PL, Lukow OM. 1988. The genetic analysis of two interspecific sources of

leaf rust resistance and their effect on the quality of common wheat. *Can. J. Plant Sci.* 68:633–39
26. Edwards MC, McMullen MP. 1988. Variation in tolerance to wheat streak mosaic virus among cultivars of hard red spring wheat. *Plant Dis.* 705–7
27. Friebe B, Mukai Y, Dhaliwal HS, Martin TJ, Gill BS. 1991. Identification of alien chromatin specifying resistance to wheat streak mosaic and greenbug in wheat germ plasm by C-banding and in situ hybridization. *Theor. Appl. Genet.* 81:381–89
28. Gale MD, Scott PR, Law CN, Ainsworth, Hollins TW, Worland AJ. 1984. An a-amylase gene from *Aegilops ventricosa* transferred to bread wheat together with a factor for eyespot resistance. *Heredity* 52:431–35
29. Gildow RE. 1990. Barley yellow dwarf virus-aphid vector interactions associated with virus transmission and vector specificity. In *World Perspectives on Barley Yellow Dwarf* ed. PA Burnett, pp. 111–22. DF Mexico: CIMMYT
30. Goulart LR, Mackeizie SA, Ohm HW, Lister RM. 1993. Barley yellow dwarf virus resistance in a wheat × wheatgrass population. *Crop Sci.* 33:595–99
31. Gray EG, Noble M. 1960. Cephalosporium stripe in cereals in Scotland. *Plant Prot. Bull.* 8:46
32. Hollins TW, Lockley KD, Blackman JA, Scott PR, Bingham J. 1988. Field performance of Rendezvous, a wheat cultivar with resistance to eyespot (*Pseudocercosporella herpotrichoides*) derived from *Aegilops ventricosa*. *Plant Pathol.* 37:251–60
33. Jahier J, Doussinault G, Dosba F, Bourgeois F. 1978. Monosomic analysis of resistance to eyespot in the variety 'Roazon'. *Proc. Int. Wheat Genet. Symp., 5th*, pp. 437–40
34. Jiang J, Friebe B, Dhaliwal HS, Martin TJ, Gill BS. 1993. Molecular cytogenetic analysis of *Agropyron elongatum* chromatin in wheat germplasm specifying resistance to wheat streak mosaic virus. *Theor. Appl. Genet.* 86:41–48
35. Jiang J, Friebe B, Gill BS. 1994. Recent advances in alien gene transfer in wheat. *Euphytica* 73:199–212
36. Jones SS, Dvoràk J. 1991. Use of double-ditelosomic and normal chromosome 1D recombinant substitution lines to map *Sr33* on chromosome 1DS in wheat. *Genome* 34:505–8
37. Jones SS, Dvoràk J, Qualset CO. 1990. Linkage relations of *Gli-D1*, *Rg2*, and *Lr21* on the short arm of chromosome 1D in wheat. *Genome* 33:937–40
38. Jones SS, Murray TD, Allan RE. 1993. Source of resistance to *Cephalosporium gramineum*. *Proc. Int. Wheat Genet. Symp., 8th*. Beijing, China. In press
39. King JE, Griffin MJ. 1985. Survey of benomyl resistance in *Pseudocercosporella herpotrichoides* on winter wheat and barley in England and Wales in 1983. *Plant Pathol.* 34:272–83
40. Knott DR. 1987. Transferring alien genes to wheat. In *Wheat and Wheat Improvement*, ed. EG Heyne, 2:462–71. Madison, WI: Am. Soc. Agron. 765 pp.
41. Knott DR, Dvoràk J. 1976. Alien germ plasm as a source of resistance to disease. *Annu. Rev. Phytopathol.* 14:211–35
42. Koebner RMD, Miller TW, Snape JW, Law CN. 1988. Wheat endopeptidase: genetic control, polymorphism, intrachromosomal gene location, and alien variation. *Genome* 30:186–92
43. Larson RI, Atkinson TG. 1970. Identity of the wheat chromosome replaced by *Agropyron* chromosomes in a triple alien chromosome substitution line immune to wheat streak mosaic. *Can J. Genet. Cytol.* 12:145–50
44. Larson RI, Atkison TG. 1973. Wheat-*Agropyron* chromosome substitution lines as source of resistance to wheat streak mosaic virus and its vector, *Aceria tulipae*. In *Proc. Int. Wheat Genet. Symp., 4th.*, Columbia, Mo: Univ. Missouri, pp. 173–84
45. Law CN, Scott PR, Worland AJ, Hollins TW. 1976. The inheritance of resistance to eyespot (*Cercosporella herpotrichoides*) in wheat. *Genet. Res. Camb.* 25:739
46. Law CN, Worland AJ, Hollins TW, Koebner RMD, Scott PR. 1988. The genetics of two sources of resistance to eyespot (*Pseudocercosporella herpotrichoides*) in wheat. *Proc. Int. Wheat Genet. Symp., 7th*, ed. TE Miller, RMD Koebner, pp. 835–40. Cambridge, UK: Inst. Plant Sci. Res.
47. Lay CL, Wells DG, Gardner WAS. 1971. Immunity from wheat streak mosaic virus in irradiated *Agrotriticum* progenies. *Crop Sci.* 11:431–32
48. Lister RM, Clement D, Skaria M, Foster JE. 1984. Biological differences between barley yellow dwarf viruses in relation to their epidemiology and host reactions. In *Barley Yellow Dwarf, A Proceedings of the Workshop*. DF Mexico: CYMMYT. pp. 16–25
49. Maia N. 1967. Obtention de blés tendres résistants au piétinverse par croisements interspécifiques blés × *Aegilops*. *C. R. Acad. Agric. Fr.* 53:149–54

50. Martin JM, Johnston RH, Mathre DE. 1989. Factors affecting the severity of Cephalosporium stripe of winter wheat. *Can. J. Plant Pathol.* 11:361–67
51. Martin JM, Mathre DE, Johnston RH. 1983. Genetic variation for reaction to *Cephalosporium gramineum* in four crosses of winter wheat. *Can. J. Plant Sci.* 63:623–30
52. Martin JM, Mathre DE, Johnston RH. 1986. Winter wheat genotype responses to *Cephalosporium gramineum* inoculum levels. *Plant Dis.* 70:421–23
53. Martin TJ, Harvey TL, Livers RW. 1976. Resistance to wheat streak mosaic virus and its vector, *Aceria tulipae. Phytopathology* 66:346–49
54. Mathre DE, Johnston RH. 1975. Cephalosporium stripe on winter wheat: procedures for determining host response. *Crop Sci.* 15:591–94
55. Mathre DE, Johnston RH, Martin JM. 1985. Sources of resistance to *Cephalosporium gramineum* in *Triticum* and *Agropyron* species. *Euphytica* 34:419–24
56. Mathre DE, Johnston RH, McGuire CF. 1977. Cephalosporium stripe of winter wheat: pathogen virulence, sources of resistance, and effect on grain quality. *Phytopathology* 67:1142–48
57. McGuire PE, Zhong Gan-Yuan, Qualset CO. 1995. Resistance to barley yellow dwarf virus disease in derivatives of crosses between hexaploid wheat and species of *Lophopyrum* (Triticeae; Poaceae).*Plant Breed.* In press
58. McIntosh RA. 1991. Alien sources of disease resistance in bread wheats. *Proc. Dr. H. Kihara Memorial Int. Symp. Cytoplasmic Eng. Wheat. Nuclear and Organellar Genomes of Wheat Species,* eds. T Sasakuma, T Kinoshita, pp. 320–32. Yokahoma, Jpn.
59. Mckinney HH, Sando WJ. 1951. Susceptibility and resistance to the wheat streak mosaic virus in the genera *Triticum, Agropyron, Secale,* and ceratin hybrids. *Plant Dis. Rep.* 35:476–78
60. McMillin DE, Allan RE, Roberts DE. 1986. Association of an isozyme locus and strawbreaker foot rot resistance derived from *Aegilops ventricosa* in wheat. *Theor. Appl. Genet.* 72:743–47
61. Mena M, Doussinault G, Lopez-Braña I, Aguaded S, Garcigra-Olmedo F, Delibes A. 1992. Eyespot resistance gene *Pch-1* in H-93 wheat lines. Evidence of linkage to markers of chromosome group 7 and resolution from the endopeptidase locus *Ep-D1b. Theor. Appl. Genet.* 83:1044–47
62. Murray TD, Bruehl GW. 1986. Effect of host resistance to *Pseudocercosporella herpotrichoides* on yield and yield components in winter wheat. *Plant Dis.* 70:851–56
63. Murray TD, de la Peña RC, Yildirim A, Jones SS. 1994. A new source of resistance to *Pseudocercosporella herpotrichoides,* cause of eyespot disease of wheat, located on chromosome 4V of *Dasypyrum villosum. Plant Breed.* 113:281–86
64. Murray TD, Smiley RW, Uddin W. 1990. Resistance to benzimidazole fungicides in *Pseudocercosporella herpotrichoides* in Washington and Oregon. *Phytopathology* 80:1041 (Abstr.)
65. Nisikado Y, Higuti T. 1938. Comparative studies on *Cephalosporium gramineum* Nisikado et Ikata, which caused the stripe disease of wheat, and *C. acremonium* Corda. *Ber. Ohara Inst. Landwirtsch. Forsch.* 8:283–304
66. Nisikado Y, Matsumoto H, Yamauti K. 1934. Studies on a new *Cephalosporium,* which causes the stripe disease of wheat. *Ber. Ohara Inst. Landwirtsch. Forsch.* 6:275–306
67. Omeilan JA, Epstein E, Dvoràk J. 1991. Salt tolerance and ionic relations of wheat as affected by individual chromosomes of salt tolerant *Lophopyrum elogatum. Genome* 34:961–74
68. Oswald JW, Houston BR. 1951. A new virus disease of cereals transmissible by aphids. *Plant Dis. Rep.* 35:471–75
69. Qi M, Murray TD. 1994. Transformation of *Cephalosporium gramineum* with the β-glucuronidase gene. *Phytopathology* 84:1169 (Abstr.)
70. Qualset CO. 1984. Evaluation and breeding methods for barley yellow dwarf resistance. In *Barley Yellow Dwarf. A Proc. Workshop* 1:72–80. Mexico, DF: CIMMYT
71. Sears ER. 1939. Cytogenetic studies with polyploid species of wheat. I. Chromosomal aberrations in the progeny of a haploid of *Triticum vulgare. Genetics* 24:509–23
72. Sears ER. 1977. An induced mutant with homoeologous pairing in wheat. *Can. J. Genet. Cytol.* 19:585–93
73. Sebesta EE, Bellingham RC. 1963. Wheat viruses and their genetic control. In *Proc. Int. Wheat Genet. Symp. Lund. Sweden, Hereditas, 2nd,* (Suppl.) pp. 184–201
74. Sebesta EE, Young HC, Wood EA. 1972. Wheat streak mosaic virus resistance. *Ann. Wheat Newsl.* 18:136
75. Sharma HC, Gill BS, Uyemoto JK 1984. High level of resistance in *Agropyron* species to barley yellow dwarf and

wheat streak mosaic virus. *Phytopathol. Z.* 110:143–47
76. Sharma HC, Ohm HW, Lister RM, Foster JE, Shukle RH. 1989. Responses of wheatgrasses and wheat × wheatgrass hybrid to barley yellow dwarf virus. *Theor. Appl. Genet.* 77:369–74
77. Simonet M. 1957. Hybrids interspécifiques et intergénériques. *Ann. Amél. Plantes* 4:395–411
78. Sprague R. 1936. Relative susceptibility of certain species of gramineae to *Cercosporella herpotrichoides. J. Agric. Res.* 53:659–70
79. Stoddard SL, Gill BS, Lommel SA. 1987. Genetic expression of wheat streak mosaic virus resistance in two wheat wheatgrass hybrids. *Crop Sci.* 27: 514–19
80. Strausbaugh CA, Murray TD. 1989. Inheritance of resistance to *Pseudocercosporella herpotrichoides* in three cultivars of winter wheat. *Phytopathology* 79:1048–53
81. Summers RW, Koebner RMD, Hollins TW, Forster J, Macartney DP. 1988. The use of an isozyme marker in breeding wheat (*Triticum aestivum*) resistant to the eyespot pathogen (*Pseudocercosporella herpotrichoides*). *Proc. Int. Wheat Genet. Symp., 7th,* pp. 1195–97. Cambridge, UK: Inst. Plant Sci. Res.
82. Tuleen N, Hart GE. 1988. Isolation and characterization of wheat *Elytrigia elongata* chromosome 3E and 5E and substitution lines. *Genome* 30:519–24
83. Vahl U, Muller G. 1991. Endopeptidase EP1 as a marker for the eyespot resistance gene *Pch1* from *Aegilops Ventricosa* in wheat line H9370. *Plant Breed.* 107:779
84. Vincent A, Ponchet J, Koller J. 1952. Recherche de variétés de blés tendres peu sensibles au piétinverse: résultats préliminaires. *Ann. Amél. Plantes* 2: 459–72
85. Wang RC, Liang GH. 1977. Cytogenetic location of genes for resistance to wheat streak mosaic in an *Agropyron* substitute line. *J. Hered.* 68:375–78
86. Weeks JT, Anderson OD, Blechl AE. 1993. Rapid production of multiple independent lines of fertile transgenic wheat (*Triticum aestivum*). *Plant Physiol.* 102:1077–84
87. Well DG, Kota RS, Sandhu HS, Gardner WS , Finney KF. 1982. Registration of one disomic substitution line and five translocation lines of winter wheat germ plasm resistant to wheat streak mosaic virus. *Crop Sci.* 22:1277–78
88. Well DG, Wong R, Sze-Chung, Lay CL, Gardner WS, Buchenau GW. 1973. Registration of C.I. 15092 and C.I. 15093 wheat germ plasm. *Crop Sci.* 13:776
89. Whelan EDP, Hart GE. 1988. A spontaneous translocation that transfers wheat curl mite resistance from decaploid *Agropyron elongatum* to common wheat. *Genome* 30:289–92
90. Worland AJ, Law CN, Hollins TW, Koebner RMD, Giura A. 1988. Location of a gene for resistance to eyespot (*Pseudocercosporella herpotrichoides*) on chromosome 7D of bread wheat. *Plant Breed.* 101:43–51
91. Xin ZY, Brettel RIS, Cheng ZM, Waterhouse PM, Appels P, et al. 1988. Characterization of potential source of barley yellow dwarf virus resistance for wheat. *Genome* 30:250–57
92. Yildirim A, Jones SS, Murray TD. 1994. Population and cytogenetic studies of resistance to eyespot in wheat and its wild relatives. *Agron. Abstr.* 86:133 (Abstr.)
93. Yunoki T, Sakurai Y. 1965. Control of the Cephalosporium stripe disease of wheat. *Bull. Chugoku Agric. Exp. Stn.* A-11:113–44
94. Zhang QF, Guan WN, Ren ZY, Zhu XS, Tsai JH. 1983. Transmission of barley yellow dwarf virus strains from northwestern China by four aphid species. *Plant Dis.* 67:895–99
95. Zhong GY, McGuire PE, Qualset CO, Dvoràk J. 1994. Cytological and molecular characterization of a *Triticum aestivum* × *Lophopyrum ponticum* backcross derivative resistant to barley yellow dwarf. *Genome* 37:876–81

Annu. Rev. Phytopathol. 1995. 33:445–66

EPIDEMIOLOGICAL APPROACH TO DISEASE MANAGEMENT THROUGH SEED TECHNOLOGY

Denis C. McGee
Seed Science Center and Department of Plant Pathology, Iowa State University, Ames, Iowa 50011

KEY WORDS: seed pathology, seedborne disease

ABSTRACT

This review describes how knowledge of the epidemiology of seed diseases provides opportunities for disease management through modern seed technology and how current research priorities are meeting these needs. Management strategies used to minimize seed infection in the seed production field include cultural, chemical, and disease resistance. Harvesting, drying, processing, and storage operations can be used to minimize the spread and to eradicate seedborne pathogens. A range of methods are used to detect seedborne pathogens. Effective application of test results, however, is dependent on knowledge of inoculum thresholds for transmission of pathogens in the planting field. Control of seed transmission can be achieved by physical, chemical, and biological seed treatments. The impact of regulation of seedborne diseases in the worldwide movement of seeds has significant impact on the interests of the seed industry, germ plasm banks, and international research organizations.

INTRODUCTION

The quality of planted seeds has a critical influence on the ability of crops to become established and to realize their full potential of yield and value. A complex technology is required to ensure high standards of seed quality that involves producing, harvesting, processing, storing, and planting the seed.

00066-4286/95/0901-0445$05.00

Throughout this process, careful handling to avoid mechanical injury and protection from adverse environmental conditions, pests, and diseases are imperative. No one factor is necessarily more important than another with respect to maintenance of seed quality, but almost all seed crops require some measure of disease control. Some significant changes have occurred in the global seed industry since previous reviews of seed pathology (2, 77); these include the greater importance of seed-transmitted pathogens in quality assurance programs; increased demands for international phytosanitary certificates for seed exports; concerns about dissemination of pathogens by germ plasm; improved seed treatment technology; and the application of biotechnology to seed disease control practices. This review addresses how knowledge of the epidemiology of seed diseases can promote disease management through modern seed technology and how current research priorities are meeting these needs.

DISEASE IMPACT ON SEED MANAGEMENT SYSTEMS

Seed pathology emerged as a subdiscipline of plant pathology from analysis of seed quality in the early part of this century (77). Since then, a worldwide process of cataloguing microflorae of seeds has associated approximately 2400 microorganisms with the seeds of 383 genera of plants (87). Concurrently, epidemiological studies were carried out on the seedborne phase of economically important diseases such as bacterial blights of beans, smuts of cereals, and Stewart's wilt of corn (77). Baker (4) first identified epidemiological studies with seed pathology. He described three environments in which seeds exist: the seed production field; harvesting, processing, and storage; and the planted field. He also defined categories of pathogen-seed associations within the environments and indicated how these related to control strategies. McGee (63) integrated the life cycle of a plant pathogen into the three environments and suggested that the role of a seed pathologist should be to study the seed aspects of the life cycle of the pathogen and their interactions with environmental, cultural, and genetic factors that influence the cycle. The model in Figure 1 combines the concepts proposed by Baker (4) and McGee (63) and provides a framework for discussion of an epidemiological approach to management of seed diseases.

THE SEED PRODUCTION FIELD

Diseases can have indirect effects on seeds in the production field in that the seed is not associated in any way with the pathogen but other plant parts are diseased; this renders the plant physiologically ill equipped to complete the development and maturation of the seeds. Shrunken or malformed seeds may

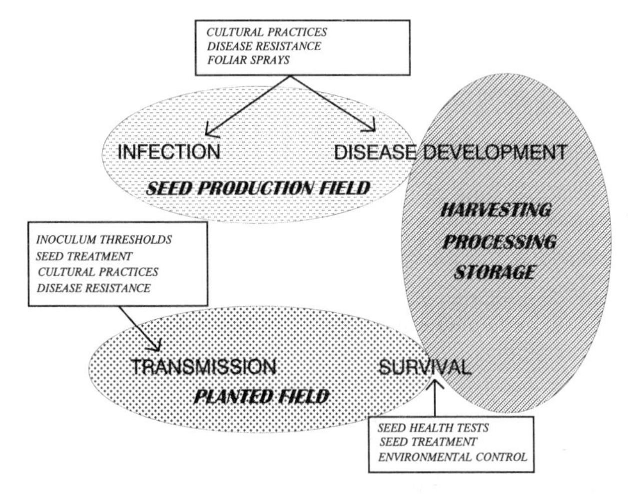

Figure 1 Schematic diagram integrating the environments in which seeds exist, the life cycle of a seedborne pathogen, and disease management strategies.

result, and germination potential may be lost or reduced. Management practices for these pathogens are essentially the same as those used in general crop production and are not considered in this review.

Direct effects means that the seed itself is diseased, thus the viability and appearance of the seed is affected and/or the pathogen is transmitted to the plant grown from the seed. Seed infection can occur during the three distinct physiological phases in the seed production field; anthesis, which covers the period from initiation of floral primordia to fertilization of the embryo; seed development, which represents the period during which the fruiting structures grow and develop to full physiological maturity; and seed maturation, which is the dry down period that continues until the seed is harvested. Each phase has unique characteristics with respect to the epidemiology and management of seed disease.

Elimination of Inoculum Sources

The first opportunity for management of seed disease is to eradicate or reduce pathogen inoculum in the seed production field. Garzonio & McGee (23) demonstrated that soybean crop residues were the major source of inoculum for Phomopsis seed decay of soybeans and that seed infection was substantially lower in soybean fields rotated annually with corn compared to those in a continuous soybean rotation. Schaad & Dianese (93) identified cruciferous weeds in California as potentially important sources of contamination of Brassica seed fields by *Xanthomonas campestris* pv. *campestris*. The primary inoculum sources for ergot (*Claviceps purpurea*) and blind seed (*Gloeotinia temulenta*) in perennial grass seed production fields in Oregon are infested seeds from previous crops. These are destroyed by burning stubble or are removed by vacuuming fields (1).

Disease Management During Anthesis

Several important groups of pathogens including smuts, ergots, viruses, and nematodes infect seeds during anthesis (77). A unique feature of infection at this growth stage is the facility for infection of embryos and other internal seed tissues. Opportunities for infection, however, are usually limited in time. Embryo invasion by viruses from the mother plant is dependent on short-lived cytoplasmic connections to the male or female gametophytes (50, 51). The optimal time for *Fusarium moniliforme* infection of maize kernels by silk inoculation occurs when silks begin to senesce (39).

The infection process also may be influenced by environment. Rain and warm temperatures following anthesis resulted in increased grain mold contamination of caryopses of sorghum (118). Knowledge of the mechanism of and environmental influences on infection at this growth stage has been used to advantage in disease management. Resistance to infection by *Claviceps fusiformis* in pearl millet has been attributed to shrivelling of the stigma induced by pollination (108). By establishing that the potato spindle viroid was primarily pollen transmitted, this pathogen was controlled in potato breeding programs by avoiding the use of pollen from infected plants (98). The potential for biological control during anthesis was demonstrated by inoculation of wheat florets with a strain of *C. purpurea* that did not biosynthesize ergot alkaloids, but had sufficient parasitic vigor to displace alkaloid-producing strains (105).

Disease Management During Seed Development

Seed infection during seed development can occur by invasion through natural openings including the funiculus and micropyle, by direct penetration of the seed or caryopsis, or from pods or fleshy fruits (4, 77). Infection also can be strongly influenced by environment. Stepwise regression analysis of *Fusarium* spp. infection of cereals seeds in relation to agroclimactic variables indicated that infection increased significantly with precipitation on days preceding ripeness when minimum temperatures on days following heading were relatively high (11). Osorio & McGee (80) showed that exposure of soybean pods to frost at –4.5 or –2.5°C immediately before physiological maturity predisposed seeds to infection by *Fusarium graminearum* and *Alternaria alternata* but reduced seed infection by *Phomopsis longicolla*.

There are numerous reports of fungicide applications in seed production fields to control seedborne pathogens such as *Fusarium* ssp. on lentils (31).

These studies rarely consider the disease epidemiology. A more strategic approach to chemical control is evident in the elucidation of the pattern of control of *Xanthomonas campestris* pv. *vignicola* in cowpea by copper sprays by determining the distribution of two lysotypes of the pathogen in the plots by bacteriophage typing (26). Extensive studies of the epidemiology of Phomopsis seed decay of soybeans have led to effective control of this disease by predictive methods that determine the need to apply benzimidazole fungicides. One method uses a points scale that accounts for the presence of conditions known to be associated with development of the disease (103). Another determines the incidence of pod infection by sampling pods in the field and subjecting them to a simple laboratory assay (64). The pod test method is based on the knowledge that pod infection can occur at any time from flowering onwards, but the fungus will not infect seeds until seed maturation begins (64). For both methods, it is necessary that the predictive measurement is made and fungicide applied before the seeds are infected. As indicated by Balducchi & McGee (5), the pathogen will move from pods to seeds only under precise conditions of temperature and humidity. The probability of the occurrence of these conditions under weather patterns in mid-west states of the U.S. indicates that both methods establish a sufficient degree of risk of severe seed disease to be of practical value (66).

Cultural practices provide options to manage seed disease in the production field: adjustment of planting time, crop rotation, elimination of weed hosts, irrigation practices, etc (4, 77). A recent study in Tanzania showed that seedborne populations of *Pseudomonas syringae* pv. *phaseolicola* were higher on beans intercropped with maize than on beans grown alone. Maize leaves did not support high populations of the pathogen, but bean leaves in the intercrop system took longer to dry after rainfall and were wetter than those in the monoculture (58). Biological control of seed infection during seed development was demonstrated by the reduction of aflatoxin B_1 in the cottonseed after simultaneous inoculation of wounded cotton bolls with toxigenic and atoxigenic strains of *Aspergillus flavus* (15).

Disease Management During Seed Maturation

Pathogens like *Fusarium moniliforme* of corn (39, 68) that infect seeds during seed development continue the infection process during seed maturation. Also, fungi such as *Botrytis, Fusarium, Alternaria,* and *Cladosporium* spp that commonly infest soil and crop residues will invade seeds under prolonged periods of wet weather at this growth stage (65) and cause seed discoloration and loss of viability. The condition can be particularly severe when harvest is delayed due to wet weather. So-called "weathered" seeds experience physiological deterioration as well as pathological damage (100). Using tetrazolium salt staining, Moore (76) demonstrated that physiological damage to soybean seeds was induced by alternate wetting and drying during simulated delayed harvest conditions. Effective control of disease during seed maturation is achieved by harvesting the seed as soon as it is sufficiently dry. Planting dates may be manipulated to avoid conditions favorable for seed infection as in the case of Phomopsis seed decay of soybeans in which the chances of temperature and humidity conditions favorable for seed infection occurring are much lower for late compared to early planted crops (5).

Resistance to seed infection by pathogens in the seed production field exists for many important diseases including various ear rots of maize and black rot of crucifers, but this resistance is usually the result of breeding to control the disease on the whole plant (68, 124). There are a few examples of breeding specifically for resistance to infection of the seeds. A genotype resistant to Phomopsis seed decay and sources of resistance to *Cercospora kikuchii*, the cause of purple seed stain of soybeans, have been identified (10, 89). Halloin (34) described several potential defense mechanisms that could be used in breeding for resistance to seed infection. Grain hardness, ergosterol content, and tannins have been implicated in resistance to molding of sorghum grains (8, 85). Evidence was also obtained for tannin extracts from peanuts to inhibit aflatoxin production by *Aspergillus parasiticus* (3).

HARVESTING, CONDITIONING AND STORING ENVIRONMENT

Disease Management During Harvesting and Processing

The harvesting process provides opportunities for pathogen structures such as sclerotia, nematode soil peds, and teliospores to contaminate seed lots. This type of contamination can be minimized by setting the harvesting equipment to avoid contact with the soil and to eliminate physically altered seeds or pathogen structures. Bacterial colonies on plant parts can become attached to seed surfaces during harvest (30). Gossen (28) showed that *Ascochyta lentis* spread in the swath of lentils and that control should be directed at reducing the drying time of the swath before seed is harvested. Seeds of eastern white pine and white spruce in closed cones were usually free of fungi while on the tree, but fungal infection spread during processing and seed extraction, particularly if cones were kept on the forest floor for prolonged periods of time after their removal from branches (75). Fermentation used routinely to extract tomato seeds from the fruit pulp greatly reduced populations of *Clavibacter michiganense* subsp. *michiganense* in tomato seeds (16).

The potential for modern seed processing to reduce seed infection was demonstrated by the separation of sclerotia of *Sclerotinia sclerotiorum* and of *Phomopsis*-infected seeds and plant debris from soybean seed lots passed over air-screen cleaners and gravity tables (82). Paulsen (81) used a computer vision system to detect purple-stained soybean seed infected with *Cercospora kikuchii* with a 91% degree of accuracy. Walcott et al (113) used an ultrasound signal to detect soybean seeds with reduced germination potential resulting from asymptomatic infection by the storage fungi (*Aspergillus* and *Penicillium* spp.). By determining that spiny prickles of fully developed fruits were the most common sites of infection of carrot by *Alternaria dauci*, Strandberg (101) reduced seed infection from 26.4 to 4.8% by removing these particles during milling.

Disease Management During Storage

Storage fungi (*Aspergillus* and *Penicillium* spp.) invade grains and seeds stored at moisture contents in equilibrium with ambient relative humidities ranging from 65–90% and can cause major losses in seed viability. Christensen & Meronuck (12) have long contended that no significant invasion of seeds by storage fungi occurs before storage, but this contention has been challenged by others (74), who suggested that the storage fungi probably invade seeds in the field and that controls should be focused at the preharvest as well as the postharvest phase. There is little dispute that effective management of storage fungi invasion is obtained by drying of seeds to below the minimum moisture content for storage fungi invasion and maintaining this moisture content by aeration (99). The effectiveness of this practice often breaks down, however, when seed is held in storage facilities with poor environmental controls.

Alternative methods of control of storage fungi have been investigated in recent years. A polyvinylidene chloride copolymer emulsion applied to maize and soybean seeds effectively controlled storage fungi invasion during storage for three months at 85% relative humidity and 25°C by reducing the rate of uptake of moisture by the seeds (71). Soybean oil, applied at a rate used to suppress grain dust, reduced growth of storage fungi in maize and soybeans during 12 months in field storage bins (72, 123). Insect and storage fungi were controlled by mineral oil and soybean oil treatment of dry beans (33). The fungicides thiabendazole and iprodione have been shown to suppress growth of storage fungi in stored corn (123). The potential for natural products (flavonoids, isoflavonoids, and their derivatives) to control storage fungi was indicated on seeds of soybean and bean (121).

Various studies on survival of seedborne pathogens including *Phytophthora nicotianae* var. *nicotianae* in bell pepper (7) and *Acremonium coenophialum* in tall fescue (122) have shown that the length of survival varied greatly, depending on pathogen type and storage conditions. This variation confirms the conclusion of Baker (4) that aging of seeds is an unreliable method to eliminate seedborne pathogens.

Interaction of Pathogens with Laboratory Seed Quality Tests

Seedborne pathogens can interfere with the ability of laboratory germination tests to establish the germination potential that most accurately predicts field performance. Zorrilla (129) showed that seed germination values for seed lots with substantial levels of infection by *Phomopsis longicolla* were not well correlated with subsequent field emergence in test protocols in which the seeds were not covered with a substrate. Good correlations were obtained, however, in tests in which seeds were covered with sand. It was concluded that seedborne inoculum was reduced in the sand test by removal of infected seed coats from cotyledons by physical restraints during seedling emergence, as would happen in the soil environment.

Germination tests also can be contaminated by airborne microorganisms in the testing laboratory. McGee (62) demonstrated that airborne conidia of *Penicillium* spp contaminated the substrate used for grass seed germination tests at the time seeds were planted. The fungus then grew on the seeds and substrate during incubation and drastically reduced seed germination. The problem was corrected by planting seeds in a sterile environment.

Seed Health Testing

Seed health testing is used primarily to manage diseases by inoculum thresholds; to determine the potential effect of seedborne inoculum on stand establishment in the planted field; and to meet the requirements for phytosanitary certification of seed lots to be exported. The subject has been extensively reviewed (2, 77), but the principles of methods currently in use and their main advantages and limitations are briefly described.

Field inspection requires that the seed production field be examined for symptoms of a disease on growing plants. The method is based on the assumption that incidence of infection on plants and seeds are related. Although there are few diseases where this relationship has been validated, the procedure remains the backbone of phytosanitary certification in many countries.

Seeds may be assayed by direct examination, during which time the pathogen is not altered and is observed directly in, on, or accompanying the seed. Seeds may be examined visually for clear signs or symptoms expressed on the seed surface. Another approach is to soften seed tissues and then examine the internal tissues of the seed microscopically for mycelium of the pathogen. Structures such as smut teliospores that accompany seeds are identified microscopically after release into a wash solution. These tests are inexpensive and rapid, but some protocols have low sensitivity or are labor intensive.

Incubation tests require that seeds be subject to conditions that select for and optimize growth of the target pathogen. Assays usually require pretreatment with a chemical to surface disinfest the seeds, followed by incubation on blotters or culture medium under precisely defined environmental conditions. After incubation, the target pathogen is identified by morphological or biochemical tests. The method is sensitive for many fungal and bacterial pathogens. Major advances in the past 15 years in the development of selective media for different groups of fungal and bacterial pathogens have greatly increased the utility of culture plate testing, particularly for bacterial assays.

For grow-out tests, seeds are planted in the field or greenhouse in the absence of other inoculum sources. Seedlings are examined for symptoms produced by the seedborne pathogen. The procedure requires much time, space, and labor. It also tends to lack sensitivity, but it can predict well the extent of seed transmission of pathogens in the planted field.

Serological assays for seedborne pathogens were first reported in 1965 with an agglutination test for *Pseudomonas phaseolicola* in beans (32) and a double diffusion assay for barley stripe mosaic virus (36). The introduction of enzyme-linked immunosorbent assay (ELISA) to plant pathology in 1976 (112) stimulated rapid advances in the use of serological assays for seedborne pathogens. With diagnostic kits now available from the private sector for ELISA and its variants, serology has become cost-effective and practical to detect seedborne pathogens throughout the world. The technique, however, requires rigorous testing to ensure specificity. Lamka et al (55), for example, tested antibodies to *Erwinia stewartii* against 167 bacterial strains isolated from maize seeds. Once obtained, a supply of high-quality antiserum must be maintained. This requirement can be a significant problem with polyclonal antibodies because batches can vary in specificity. Monoclonal antibodies theoretically should resolve this problem by re-culturing the original clonal cell line. Other considerations in optimizing and standardizing assays include insuring release of the antigen from host particles (17); competition from seed tissues for binding sites in ELISA wells; and reagents that might sterically interfere with binding of the antigen or cause conformational changes in the antibody (109). ELISA test data must be very carefully interpreted. The setting of test thresholds is a critical feature in the design, verification and routine use of the assay (104). A well-known weakness in serological tests has been the propensity to detect false positives caused by binding of the antibodies to epitopes, which may no longer be a component of a viable propagule of the pathogen. For economic reasons, it is particularly important that high-value seed be assayed by a method that detects viable propagules of the pathogen. This problem can be overcome by combining serological assays with a viability test. In the assay for *Clavibacter michiganensis* ssp. *insidiosus* in alfalfa seeds, for example, immunofluorescence (IF) staining is followed by a culture step if it is not unequivocally negative (78). Modifications to IF such as immunosorbent immunofluorescence and immuno-isolation procedures include pathogen viability checks (20).

DNA hybridization assays use a DNA probe that is complementary to DNA in the genome of the plant pathogen. The probe is applied to a DNA extract from seed and hybridized material detected by dot-blot hybridization assays. The technique has been used successfully to detect the corn downy mildews *Peronosclerospora sorghi* and *P. sacchari* (126, 127) and *Pseudomonas syringae* pv. *phaseolicola* in bean seeds (92). As with serological techniques, it is necessary to ensure specificity of the probe to the pathogen. Yao et al (127) verified that the probe developed to detect *Peronosclerospora sacchari* did not hybridize with a range of fungi isolated from the same corn seeds. The sensitivity of the test for *Pseudomonas syringae* pv. *phaseolicola* in bean seeds was increased by amplification of a segment of the gene cluster detected by the probe by polymerase chain reaction (PCR). The method was used to detect the pathogen in a commercial seed lot that failed to yield the pathogen in conventional selective medium methods (83).

THE PLANTED FIELD

Seedling Emergence and Establishment

For some seedborne pathogens, relationships to seed quality have been established on sound epidemiological bases. Examples are correlations between laboratory and field emergence for three seedborne pathogens of sorghum (125) and the finding that *Tilletia indica* infection of wheat seeds had very little effect on seed viability but did adversely affect seed vigor (119). Most research reports on this topic, however, only indicate potential associations between seedborne pathogens and germination or emergence, either by artificial inoculation of seeds by the pathogen (73) or by application of filtrates from cultures of the pathogen (96). The ability of soilborne pathogens to parasitize seeds is influenced by the structure and condition of the seeds as well as by cultural and climatic factors. Hering et al (40) showed that infection of wheat embryos by *Pythium* species during seed germination was affected by seed age and soil matrix potential. Glumes of spelt seeds seemed to protect seeds against soil-borne fungi and to increase the fitness of seeds under low temperature and water logging conditions (88). Gleason & Ferriss (27) showed that effects of soil moisture and temperature on the relative growth of soybean seeds and the pathogen were a determining factor in the development of seedling decay by seedborne *Phomopsis longicolla*.

In developed countries, almost all maize, cotton, sorghum, and vegetable seeds are treated with the broad-spectrum fungicides captan or thiram for protection against both seed- and soilborne pathogens during field emergence (45). For other seeds, the need for seed treatment has not been so clear cut. An epidemiological study in Iowa in the early 1980s showed that seed treatment to improve stand establishment of soybeans was justified under a relatively narrow spectrum of conditions that took into account seedborne pathogens, seed quality, and soil temperatures (117). The potential for more efficient use of seed-treatment fungicides has been demonstrated for maize and peas by application of fungicides in combination with polymer film coatings (53, 70). A detailed study of the respective infection patterns of sugar beet seed parts by the control agent *Pseudomonas* spp. and the pathogen *Pythium ultimum* provided valuable insight into the efficacy of control of the pathogen and required dosages of the biopesticide (79).

Transmission of Seedborne Pathogens

EPIDEMIOLOGICAL FACTORS AFFECTING SEED TRANSMISSION There are few seedborne pathogens for which the mechanism of seed transmission is as well understood as for *Peronospora ducometi* in buckwheat. Primary infection occurs by the germination of seedborne oospores, which results in systemic invasion of the seedling followed by development of secondary conidial inoculum on the cotyledons and subsequent repeated leaf infections as the disease progresses up the plant (128). Transmission can be dependent on the plant part infected. Blackgram mottle virus is transmitted through the embryo of the seeds of *Vigna mungo,* but the presence of virus in the testa alone did not result in its transmission through seed (111). Physiological factors may affect the capacity of the seeds to transmit pathogens. Downy mildew pathogens in maize can be transmitted when seeds are freshly harvested, but not once the seeds are dried (67). Arabis mosaic nepovirus is transmitted inefficiently by *Nicotiana* seeds, because the virus reduces seed germination (116). A pseudorecombinant containing RNA-1 influences seed transmissibility of cucumber mosaic virus in *Phaseolus vulgaris* (47).

Environmental factors play a major role in the efficiency of seed transmission of plant pathogens. Cabbage seedling disease caused by *Alternaria brassicicola,* for example, does not occur below 15°C in heavily infected seed lots (6). The respective roles of seedborne inoculum and other sources of the pathogen can vary depending on environment. For oil-seed rape, seedborne *Phoma lingam* is not an important inoculum source compared to crop residues (61), but it is of major importance for cruciferous vegetable crops in which a seed-bed phase efficiently spreads seedborne inoculum to the transplants (114, 124).

INOCULUM THRESHOLDS Inoculum thresholds have been established on a sound epidemiological basis for only a few pathogens, including *Phoma lingam* of crucifers (22), *Pseudomonas syringae* pv. *phaseolicola* (115), and lettuce mosaic virus (29). For many seedborne pathogens, inoculum thresholds are determined either arbitrarily or by field observation data (54, 91). To be of value, however, the thresholds should be established in well-designed experiments. The first step is to have a suitable seed health assay. The literature is replete with descriptions of test methods, but very few are thoroughly researched to determine if they are specific, accurate, reproducible, and practical. The assays also should have a degree of sensitivity that relates to the application of the test results. The next step is to plant seeds with different infection levels in the field and establish a correlation with plant infection. For diseases that have no repeating cycles of infection, such as seedling infecting smuts, strong correlations between seed infection and field disease can usually be expected. Rennie & Seaton (86) showed that the loose smut embryo test of barley seeds was highly correlated with field disease across a range of environments and in different cultivars of barley. It is much more difficult to establish inoculum thresholds for diseases for which secondary infections occur from other inoculum sources. The influence of secondary infection by aphids is the primary reason why lettuce mosaic virus is controlled with a tolerance of 0 infected seeds in 30,000 in California, whereas in the Netherlands, the tolerance is 0 in 2000. The Netherlands has a cooler climate and thus a lower aphid population than California. Furthermore, the growers break the disease cycle with other crops, whereas lettuce is grown under continuous rotation in California (54).

The final step in establishing an inoculum threshold is to apply appropriate statistical analysis to results. This often has been accomplished by interpolating the value directly from the relationships between seed infection and field infection (32, 94). Various statistical formulae have been proposed to estimate inoculum thresholds; these include the most probable number method (107), a Poisson and binomial probability system (13), and two models for determining necessary sample sizes that would allow detection of low levels of disease contamination (25). Russell (90) pointed out that establishment of acceptable threshold levels of seedborne pathogens requires sampling techniques that are representative of seed lots tested and that adhere to acceptable statistical procedures. Given that there are only a few reliable threshold levels for seedborne pathogens, little attention has been given to sample sizes. Seed-health certification programs normally use standard sampling procedures in the International Rules for Seed Testing 1993 (44), which were developed primarily for analyses of germination and mechanical purity.

REGIONAL CERTIFICATION PROGRAMS Regulatory programs exist to protect against spread of diseases by seeds within geographic regions. In Europe, seed lots must meet certain minimum standards of quality, which includes specific diseases, before seeds can be marketed (110). In the United States, there are no federal seed health standards, but states and regions may set their own. The state of Idaho, which produces a large proportion of the snap bean seed in the United States, developed strict regulations to protect the crop from bacterial blights (120). This program uses knowledge of the epidemiology of the disease that includes laboratory assays of the seeds and field controls. Periodic epidemics caused by seedborne disease that have cost the seed industry multimillion dollars in losses worldwide and lawsuits have stimulated seed companies to enact their own standards for seed health, which sometimes can be more rigorous than those mandated by official regulations (120). Watermelon fruit blotch (21) became so serious in 1994 that several US seed companies halted the sales of watermelon seeds for a period of time because of liability.

PHYTOSANITARY CERTIFICATION Seeds that move in international commerce must comply with seed health requirements as defined by a standard phytosanitary certificate recognized by most countries (97). The system has some serious problems, however. Phytosanitary regulations are determined by individual countries and often are made on the basis of a poor understanding of the economic losses that introduction of particular pathogens could potentially incur; minimal knowledge of relationships between tolerances in seed assays to risks of transmission of the pathogen to the planted crop; and lack of standardized testing protocols. In addition, there is widespread use of phytosanitary certification of seeds as trade barriers.

The primary source of information for quarantine regulations is the Annotated List of Seedborne Diseases (87), which lists all seedborne microorganisms that have been recorded as being associated with seeds of all crops. This is an excellent starting point for data on seedborne diseases, but it does not provide information on the economic importance of the listed microorganisms. The extensive world literature on seedborne diseases remains largely untapped because information is not systematically organized and is difficult to access. Recent advances in the electronic media promise efficient, low-cost access to data on seedborne disease for use in establishing more rationale and scientifically justifiable phytosanitary regulations.

GERM PLASM Concern about germ plasm collections as sources of seedborne pathogens was first expressed by the discovery of important pathogens in the seeds of 2% of the 4489 PI accessions at the USDA Plant Introduction Station in Ames, Iowa (57). Subsequently, the introduction of pea seedborne mosaic virus (PsbMV) into pea breeding lines from a germ plasm collection had severe economic consequences for commercial pea production in North America (37). Programs enacted to eliminate PsbMV and BCMV from US germ plasm collections (37, 52) have required sensitive and accurate techniques to assay for the pathogen in mother plants and seeds and a sound knowledge of the epidemiology of the disease throughout propagative cycles of seeds (37). An important consideration in devising pathogen elimination strategies is to ensure that genetic diversity of the germ plasm collection is maintained (52).

A limitation in assaying germ plasm collections for seedborne pathogens has been a lack of assays that do not destroy the seeds. This is a particular concern for accessions containing small numbers of high-value seeds. Higley et al (41) developed methods to assay seeds nondestructively for several fungal, viral, and bacterial seedborne pathogens. Excising of cores of seed tissues proved to be a particularly successful approach over the three major steps in the process: extraction of the pathogen; assay of the extract according to preexisting methods; and restoration of seeds to a safe physiological state. To minimize loss in seed viability, consideration must also be given to the seed species, the seed structure from which the tissue is removed, and the conditions for hydrating and drying seeds (42).

International Agricultural Research Centers throughout the world have taken steps during the past 10 years to minimize the introduction and spread of exotic seedborne pathogens by seed exchange (49). The detection of karnal bunt, *Tilletia indica,* of wheat from Mexico and the resulting quarantine of Mexican wheat by the United States had a major impact on the wheat breeding program at CIMMYT (9). Several international centers have implemented programs to manage seedborne pathogens that include monitoring pathogens present in seed lots, modification of seed production practices to minimize seed infection or transmission of pathogens by seeds, and use of seed treatments (60).

SEED TREATMENT Chemical, physical, and biological seed treatments have changed dramatically in the past 20 years as a result of new fungicide chemistry, advances in biological control, and environmental regulations that have either banned or restricted the use of fungicides. Fungicide seed treatment remains the most widely used practice, and established materials such as captan and thiram still are the mainstay of seed treatment chemistry. Several systemic seed treatment fungicides have been introduced in recent years, including metalaxyl, iprodione, and triadimenol. These materials control deep-seated infections in seeds and can protect seedlings against infection for a few weeks into the growing season.

Economic and environmental considerations require that chemical products developed in the future be applied at low and efficient dosages. To meet these standards, the traditional research method of testing materials at various application rates across a range of locations for several years will have to include studies that elucidate the mechanisms and epidemiology of disease control (63). The benefits of epidemiological research on seed treatment are evident in the finding that inconsistent control of common bunt of wheat resulted from failure to attain the recommended rates and random variability of the seed microenvironment (24). Also, improved control of internal infection of cabbage seed by *Alternaria brassisicola* (59) and of karnal bunt in wheat (119) was achieved by taking into account the location of pathogens within the seeds.

Chemical control of seedborne bacteria has had limited success, either because of lack of control of internal inoculum (43) or phytotoxicicity to the seeds (19). Antibiotics, applied in polyethylene glycol (PEG), reduced infection by *Xanthomonas campestris* pv. *phaseoli* in bean seeds, but were phytotoxic (56).

Heat treatment is a traditional method that is still commonly used. Hot water or oil treatments can reduce the incidence of seedborne pathogens, but can result in loss of seed viability (84, 102). Hot water also has limitations in eliminating low infections of bacteria of epidemiological significance (124). Microwave heating successfully reduced transmission of soybean mosaic virus, with little reduction in germination in seeds treated at 8.5% moisture content, but germination was considerably reduced in seeds treated at 16% (48).

Advances in biopesticide agents are attributable, in part, to the attention paid to the mechanisms of control, particularly in the soil environment (14, 38). There are numerous reports of potentially valuable biological control microorganisms, some of which are supplied as seed treatments, but the developmental process to bring these into commercial practice is long and arduous (14). Naturally occurring microorganisms also can be exploited for biological control. Seed-transmitted *Acremonium coenophialum* and *A. lolii* are toxic to animals in pastures, but endophyte-infested grasses often show superior growth due to insect and pathogen control (95). Endophytes are now introduced into turf grass varieties to improve agronomic performance.

Traditional dust or slurry applications of seed treatment fungicides are now regarded as inefficient and environmentally hazardous. Application of chemical or biopesticides in film coatings or pellets reduces the loss of materials and allows for the delivery of multiple products (35, 106). Metalaxyl, etridiazole, and captan applied by fluid drilling controlled *Pythium* damping-off of tomatoes (106). *Enterobacter* spp. proliferated in priming solutions and provided protection of beet seeds planted in *Pythium*-infested soil (106). Bioprotectants and chemical pesticides provided effective control when added together in solid matrix priming (106).

RESISTANCE Cultivar-specific resistance to seed transmission has been reported for BSMV in barley, PsbMV in peas, AMV in alfalfa, and SMV in soybean (47), but the mechanisms relate to resistance to infection of the seed tissues on the mother plant in the seed production field (47). However, some viruses are suggested to be nontransmissible because of inhibited replication during seed germination (47). No examples could be found of resistance specifically to seed transmission of fungal or bacterial pathogens in the planted field. As with viruses, varietal differences in seed transmission in the planted field reflect resistance to the disease on the mother plant (39, 124).

RESEARCH PRIORITIES

In recent years, seedborne diseases have adversely affected worldwide trading of seeds as well as research programs of international research and germ plasm centers. Brennan et al (9) pointed out that economic evaluation of seedborne diseases should incorporate the effect of trading restrictions such as quarantines or embargoes imposed by importing countries. From 1988-94, laboratory assays for *Pseudomonas syringae* pv. *glycinea* in soybean seeds exported to Europe from the U.S. have cost seed companies approximately $1.5 million, despite the fact that the pathogen exists in Europe and is considered to be of no economic importance in other regions of the world (69).

A review of the literature on seed pathology over the period 1982–94 indicated that almost a quarter of approximately 2000 citations simply catalogued the presence of microorganisms on seeds. These purely descriptive commentaries do not address the potential for crop damage by planting diseased seeds or the management of seedborne disease. Indiscriminate cataloguing of seedborne microorganisms on seeds obscures seedborne pathogens that might be of genuine economic importance. A case in point is maize chlorotic mottle, first reported in 1973 (68). No serious effort was made to determine that the pathogen was seedborne until it caused an epidemic in winter nurseries in Hawaii. Because of the concerns of importing infected seeds into the United States, seed transmission of the pathogen was investigated and shown to occur (46). Cataloguing of seedborne pathogens is necessary, particularly for viruses and bacteria that traditionally have been neglected for lack of adequate assays. Priority should be given to pathogens that meet the criteria of limited distribution and of potential economic importance, as in the case of maize chlorotic mottle (68).

Previous reviews of seed pathology have stressed the importance of establishing inoculum thresholds for seedborne pathogens (54, 77). Only a handful of the citations examined for this review addressed this topic. Research on inoculum thresholds is both complex and expensive, but it is so fundamental to realistic and effective management of seed transmission of plant pathogens that little improvement in the seed health system worldwide will be possible unless priorities in seed pathology research are changed. A positive sign of progress on inoculum thresholds is the formation of national and international committees in 1994 comprising European and US scientists from both industry and the public sectors with a mandate to collaborate with regulatory authorities in the management of economically important seedborne diseases. The technical approach has been to assemble scientific information on seedborne diseases, prioritize them based on economic importance, and develop management strategies incorporating standardized assays that relate to inoculum thresholds currently in existence or to be developed with further research. Also included will be recommendations for management throughout the complete production, processing, and planting cycle. These initiatives and projects such as the "Guidelines for Safe Movement of Germplasm" (18), sponsored by the International Board for Plant Genetic Resources, can lead to management systems for seed diseases that protect against the spread of economically important plant pathogens without posing unnecessary barriers to the movement of seeds.

Literature Cited

1. Alderman SC. 1991. Assessment of ergot and blind seed diseases of grasses in the Willamette valley of Oregon. *Plant Dis.* 75:1038–41
2. Agarwal VK, Sinclair JB. 1987. *Principles of Seed Pathology.* Boca Raton, FL: CRC Press. Vol. 1, 176 pp., 2, 168 pp.
3. Azaizeh HA, Pettit RE, Sarr BA, Phillips TD. 1990. Effect of peanut tannin extracts on growth of *Aspergillus parasiticus* and aflatoxin production. *Mycopathologia* 110:125–32
4. Baker KF. 1972. Seed pathology. In *Seed Biology*, ed. TT Kozlowski, 2:317–416. New York: Academic. 447 pp.
5. Balducchi AJ, McGee DC. 1987. Environmental factors influencing infection of soybean seeds by *Phomopsis* and *Diaporthe* species during seed maturation. *Plant Dis.* 71:209–12
6. Bassey EO, Gabrielson RL. 1983. The effects of humidity, seed infection level, temperature and nutrient stress on cabbage seedling disease caused by *Alternaria brassicicola. Seed Sci. Technol.* 11:403–10
7. Bhardwaj SS, Sharma SL, Tyagi SNS. 1987. Perpetuation of *Phytophthora nicotianae* var. *nicotianae* in bell pepper seeds. *Indian J. Plant Pathol.* 5:94
8. Bosman JL, Rabie CJ, Pretorius AJ. 1991. Fungi associated with high and low tannin sorghum grain in the Republic of South Africa. *Phytophylactica* 23: 47–52
9. Brennan JP, Warham EJ, Byerlee D, Hernandez-Estrada J. 1992. Evaluating the economic impact of quality-reducing, seed-borne diseases: lessons from Karnal bunt of wheat. *Agric. Econ.* 6: 345–52
10. Brown EA, Minor HC, Calvert OH. 1987. A soybean genotype resistant to Phomopsis seed decay. *Crop Sci.*. 27: 895–98
11. Castonguay Y, Couture L. 1983. Epidemiology of cereal grain contamination by *Fusarium* spp. *Can. J. Plant Pathol.* 5:222–28
12. Christensen CM, Meronuck RA. 1986. In *Quality Maintenance in Stored Grains & Seeds,* Minneapolis, MN: Univ. Minnesota Press. 138 pp.
13. Coleno A, Trigalet A, Digat B. 1976. Detection of seed lots contaminated by a phytopathogenic bacterium. *Ann. Phytopathol.* 8:355–64
14. Cook RJ. 1993. Making greater use of introduced microorganisms for biological control of plant pathogens. *Annu. Rev. Phytopathol.* 31:53–80
15. Cotty PJ. 1990. Effect of atoxigenic strains of *Aspergillus flavus* on aflatoxin contamination of developing cottonseed. *Plant Dis.* 74:233–35
16. Dhanvantari BN. 1989. Effect of seed extraction methods and seed treatments on control of tomato bacterial canker. *Can. J. Plant Pathol.* 11:400–8
17. Ding XS, Cockbain AJ, Govier D. 1992. Improvements in the detection of pea seed-borne mosaic virus by ELISA. *Ann. Appl. Biol.* 121:75–83
18. FAO/IBPGR. *Technical Guidelines for the Safe Movement of Germplasm*

19. Fatmi M, Schaad NW, Bolkan HA. 1991. Seed treatments for eradicating *Clavibacter michiganensis* subsp. *michiganensis* from naturally infected tomato seeds. *Plant Dis.*. 75:383–85
20. Franken AAJM, van Vuurde JWL. 1990. Problems and new approaches in the use of serology for seed-borne bacteria.*Seed Sci. Technol.* 18:415–26
21. Frankle WG, Hopkins DL, Stall RE. 1993. Ingress of the watermelon fruit blotch bacterium into fruit. *Plant Dis.* 77:1090–92
22. Gabrielson RL, Mulanax MW, Matsuoka K, Williams PH, Whiteaker GP, Maguire JD. 1977. Fungicide eradication of seedborne *Phoma lingam* of crucifers. *Plant Dis. Rep.* 61:118–21
23. Garzonio DM, McGee DC. 1983. Comparison of seeds and crop residues as sources of inoculum for pod and stem blight of soybeans.*Plant Dis.* 67:1374–76
24. Gaudet DA, Puchalski BJ, Entz T. 1992 Application methods influencing the effectiveness of carboxin for control of common bunt caused by *Tilletia tritici* and *T. laevis* in spring wheat. *Plant Dis.* 76:64–66
25. Geng S, Campbell RN, Carter M, Hills FJ. 1983. Quality-control programs for seedborne pathogens. *Plant Dis.* 67: 236–42
26. Gitaitis RD, Bell DK, Smittle DA. 1986. Epidemiology and control of bacterial blight and canker of cowpea. *Plant Dis.* 70:187–90
27. Gleason ML, Ferriss RS. 1987. Effects of soil moisture and temperature on Phomopsis seed decay of soybean in relation to host and pathogen growth rates. *Phytopathology* 77: 1152–57
28. Gossen BD, Morrall, RAA. 1984. Seed quality loss at harvest due to Ascochyta blight of lentil. *Can. J. Plant Pathol.* 6:233–37
29. Grogan RG. 1980. Control of lettuce mosaic with virus free seed. *Plant Dis.* 64:446–49
30. Groth DE, Braun EJ. 1989. Survival, seed transmission and epiphytic development of *Xanthomonas campestris* pv. *glycines* in the north-central United States. *Plant Dis.* 73:326–30
31. Gupta JC, Gupta OM, Gupta O. 1991. Influence of pre-harvest fungicide spray on *Fusarium* species associated with lentil seeds. *Indian J. Mycol. Plant Pathol.* 21:186–89
32. Guthrie JW, Huber DM, Fenwick HS. 1965. Serological detection of halo blight. *Plant Dis. Rep.* 49:297–99
33. Hall JS, Harman GE. 1991. Efficacy of oil treatments of legume seeds for control of *Aspergillus* and *Zabrotes*. *Crop Prot.* 10:315–19
34. Halloin JM. 1983. Deterioration resistance mechanisms in seeds. *Phytopathology* 73:335–39
35. Halmer P. 1988. Technical and commercial aspects of seed pelleting and film-coating. See Ref. 58a, pp. 191–204
36. Hamilton RI. 1965. An embryo test for detecting seed-borne barley stripe mosaic virus in barley. *Phytopathology* 55: 798–99
37. Hampton RO, Kraft JM, Muehlbauer FJ. 1993. Minimizing the threat of seedborne pathogens in crop germ plasm: elimination of pea seedborne mosaic virus from the USDA-ARS germ plasm collection of *Pisum sativum*. *Plant Dis.* 77:220–24
38. Harman GE, Nelson EB. 1994. Mechanisms of protection of seed and seedlings by biological seed treatments: implications for practical disease control. See Ref. 58b, pp. 283–92
39. Headrick JM, Pataky JK, Juvik JA. 1990 Relationships among carbohydrate content of kernels, condition of silks after pollination, and the response of sweet corn inbred lines to infection of kernels by *Fusarium moniliforme*. *Phytopathology* 80:487–94
40. Hering TF, Cook RJ, Tang WH. 1987. Infection of wheat embryos by *Pythium* species during seed germination and the influence of seed age and soil matrix potential. *Phytopathology* 77:1104–8
41. Higley PM, McGee DC, Burris JS. 1993. Development of methodology for non-destructive assay of bacteria, fungi and viruses in seeds of large-seeded field crops. *Seed Sci. Technol.* 21:399–409
42. Higley PM, McGee DC, Burris JS. 1994. Effects of non-destructive tissue extraction on the viability of corn, soybean, and bean seeds. *Seed Sci. Technol.* 22:245–52
43. Honervogt B, Lehmann-Danzinger H. 1992. Comparison of thermal and chemical treatment of cotton seed to control bacterial blight (*Xanthomonas campestris* pv. *malvacearum*). *J. Phytopathol.* 134:103–9
44. International Seed Testing Association. 1993. International Rules for Seed Testing 1993. *Seed Sci. Technol.* 21 (Suppl.)
45. Jeffs KA. ed. 1986. *Seed Treatment*. Thornton Heath, UK: BCPC. 332 pp. 2nd ed.
46. Jensen SG, Wysong DS, Ball EM, Higley PM. 1991. Seed transmission of

maize chlorotic mottle virus. *Plant Dis.* 75:497–98

47. Johansen E, Edwards MC, Hampton RO. 1994. Seed transmission of viruses: current perspectives. *Annu. Rev. Phytopathol.* 32:363–86
48. Jolicoeur G, Hackam R, Tu JC. 1982. The selective inactivation of seed-borne soybean mosaic virus by exposure to microwaves. *J. Microwave Power* 17: 341–44
49. Kahn RP. 1988.The importance of seed health in international seed exchange. In *Rice Seed Health*, pp. 7–20. Manila, Philippines: IRRI. 362 pp.
50. Kapil RN, Bhatnagar AK. 1981. Ultrastructure and biology of female gametophyte in flowering plants. *Int. Rev. Cytol.* 70:291–341
51. Keijzer CJ, Willemse MTM. 1988. Tissue interactions in the developing locule of *Gastera verrucosa* during microgametogenesis. *Acta Bot. Neerl.* 37: 475–92
52. Klein RE, Wyatt SD, Kaiser WJ. 1990. Effect of diseased plant elimination on genetic diversity and bean common mosaic virus incidence in *Phaseolus vulgaris* germ plasm collections. *Plant Dis.* 74:911–13
53. Kosters P. 1994. Field emergence of peas as affected by seed quality and fungicide seed treatments. See Ref. 58b, pp. 207–10
54. Kuan TL. 1988. Inoculum thresholds of seedborne pathogens: overview. *Phytopathology* 78:867–68
55. Lamka GL, Hill JH, McGee DC, Braun EJ. 1991. Development of an immunosorbent assay for seedborne *Erwinia stewartii. Phytopathology* 81:839–46
56. Liang LZ, Halloin JM, Saettler AW. 1992. Use of polyethylene glycol and glycerol as carriers of antibiotics for reduction of *Xanthomonas campestris* pv. *phaseoli* in navy bean seeds. *Plant Dis..* 76:875–79
57. Leppik EE. 1970. Gene centers of plants as sources of disease resistance. *Annu. Rev. Phytopathol.* 8:323–44
58. Mabagala RB, Saettler AW. 1992. *Pseudomonas syringae* pv. *phaseolicola* populations and halo blight severity in beans grown alone or intercropped with maize in northern Tanzania. *Plant Dis.* 76:687–92

58a. Martin T, ed. 1988. *Application to Seeds and Soil.* Mono. 39. Thornton Heath, UK: BCPC. 404 pp.

58b. Martin T, ed. 1994. *Seed Treatment: Progress and Prospects.* Mono. 57. Thornton Heath, UK: BCPC. 482 pp.

59. Maude RB, Humpherson-Jones FM. 1980. The effect of iprodione on the seed-borne phase of *Alternaria brassicicola. Ann. Appl. Biol.* 95:321–27
60. Mew TW, Gergon EB, Merca SD. 1990. Impact of seedborne pathogens in rice germplasm exchange. *Seed Sci. Technol.* 18:44–50
61. McGee DC. 1977. Blackleg (*Leptosphaeria maculans* (Desm.) Ces et de Not) of rapeseed in Victoria: sources of infection and relationships between inoculum, environmental factors and disease severity. *Aust. J. Agric. Res.* 28: 53–62
62. McGee DC. 1979. *Penicillium* contamination of grass seed germination tests. *J. Seed Technol.* 4:18–23
63. McGee DC. 1981. Seed pathology: its place in modern seed production. *Plant Dis.* 65:638–42
64. McGee DC. 1986. Prediction of Phomopsis seed decay by measuring soybean pod infection. *Plant Dis.* 70: 329–33
65. McGee DC. 1986. Environmental factors associated with preharvest deterioration of seeds. In *Physiological-Pathological Interactions Affecting Seed Deterioration,* Special Publ. 12:53–63. Madison, WI: Crop Sci. Soc. Am. 95 pp.
66. McGee DC. 1988. Evaluation of current predictive methods for control of Phomopsis seed decay of soybeans. In *Soybean Diseases of the North Central Region,* ed. TD Wyllie, DH Scott. pp. 22–25. St. Paul, MN. APS Press. 149 pp.
67. McGee DC. 1988. Seedborne and seed transmitted diseases of maize in rice-based cropping systems In *Rice Seed Health,* pp. 203–13. Manila, Philippines: IRRI. 362 pp.
68. McGee DC. 1988. *Maize Diseases: A Reference Source for Seed Technologists.* St. Paul, MN: APS Press. 150 pp.
69. McGee DC. 1992. *Soybean Diseases: a Reference Source for Seed Technologists.* St. Paul, MN. APS Press. 151 pp.
70. McGee DC, Arias-Rivas B, Burris JS. 1994. Impact of seed-coating polymers on maize seed decay by soilborne *Pythium* species. See Ref. 58b, pp. 117–21
71. McGee DC, Henning, A, Burris JS. 1988. Seed encapsulation methods for control of storage fungi. See Ref. 58a, pp. 257–64
72. McGee DC, Iles A, Misra MK. 1989. Suppression of storage fungi in grain with soybean oil. *Phytopathology* 79: 1140 (Abstr.)
73. McLaughlin RJ, Martyn RD. 1982.

Identification and pathogenicity of *Fusarium* species isolated from surface-disinfested watermelon seed. *J. Seed Technol.* 7:97–107

74. McLean M, Berjak P. 1987. Maize grains and their associated mycoflora—a micro-ecological consideration. *Seed Sci. Technol.* 15:831–50
75. Mittal RK, Wang BSP. 1987. Fungi associated with seeds of eastern white pine and white spruce during cone processing and seed extraction. *Can. J. For. Res.* 17:1026–34
76. Moore RP. 1971. Mechanisms of water damage to mature soybean seed. *Proc. Off. Seed Anal.* 61:112–18
77. Neergaard P. 1977. *Seed Pathology.* New York: Wiley. Vol. 1, 2. 1187 pp.
78. Nemeth J, Laszlo E, Emody L. 1991. *Clavibacter michiganensis* ssp. *insidiosus* in lucerne seeds. *Bull. OEPP* 21: 713–18
79. Osburn RM, Schroth MN, Hancock JG, Hendson M. 1989. Dynamics of sugar beet seed colonization by *Pythium ultimum* and *Pseudomonas* species: effects on seed rot and damping-off. *Phytopathology* 79:709–16
80. Osorio JA, McGee DC. 1992. Effects of freeze damage on soybean seed mycoflora and germination. *Plant Dis.* 75: 879–82
81. Paulsen MR. 1990. Using machine vision to inspect oilseeds. *INFORM* 1:50–55
82. Pittis JE, da Cruz Machado J. 1987. Analysis of mycoflora of inert matter removed during the processing of soybean seeds. *Fitopatol. Brasil.* 12:408–10
83. Prossen D, Hatziloukas E, Schaad NW, Panopoulos NJ. 1993. Specific detection of *Pseudomonas syringae* pv *phaseolicola* DNA in bean seed by polymerase chain reaction-based amplification of a phaseolotoxin gene region. *Phytopathology* 83:965–70
84. Pyndji MM, Sinclair JB, Singh T. 1987. Soybean seed thermotherapy with heated vegetable oils. *Plant Dis.* 71: 213–16
85. Ramamurthi J, Kherdekar MS, Stenhouse JW, Jambunathan R. 1992. Sorghum grain hardness and its relationship to mold susceptibility and mold resistance. *J. Agric. Food Chem.* 40:1403–8
86. Rennie WJ, Seaton RD. 1975. Loose smut of barley: the embryo test as a means of assessing loose smut infection in seed stocks. *Seed Sci. Technol.* 3:697–709
87. Richardson MJ. 1990. *An Annotated List of Seed-borne Diseases.* Zurich, Switzerland: Int. Seed Test. Assoc. 4th ed.
88. Riesen T, Winzeler H, Ruegger A, Fried PM. 1986. The effect of glumes on fungal infection of germinating seed of spelt (*Triticum spelta* L.) in comparison to wheat (*Triticum aestivum* L.). *J. Phytopathol.* 115:318–24
89. Roy KW. 1982. *Cercospora kikuchii* and other pigmented *Cercospora* species: cultural and reproductive characteristics and pathogenicity to soybean. *Can. J. Plant Pathol.* 4:226–32
90. Russell TS. 1988. Inoculum thresholds of seedborne pathogens: some aspects of sampling and statistics in seed health testing and the establishment of threshold levels. *Phytopathology* 78:880–81
91. Schaad NW. 1988. Inoculum thresholds of seedborne pathogens: bacteria. *Phytopathology* 78:872–75
92. Schaad NW, Azad H, Peet RC, Panopoulos NJ. 1989. Identification of *Pseudomonas syringae* pv *phaseolicola* by a DNA hybridization probe. *Phytopathology* 79:903–7
93. Schaad NW, Dianese JC. 1981. Cruciferous weeds as sources of inoculum of *Xanthomonas campestris* to black rot of crucifers. *Phytopathology* 71: 1215–20
94. Schaad NW, Sitterley WR, Humaydan H. 1980. Relationship of incidence of seedborne *Xanthomonas campestris* to black rot of crucifers. *Plant Dis.* 64:91–92
95. Siegel MR, Latch GCM, Johnson MC. 1987. Fungal endophytes of grasses. *Annu. Rev. Phytopathol.* 25:293–315
96. Shree MP. 1985. Effect of culture filtrates of species of *Drechslera, Exserohilum* and *Helminthosporium* isolated from seeds on germination and seedling growth of sorghum varieties and hybrids. *Seed Res.* 13:13–18
97. Smith IM. 1990. New techniques for detection and identification of seedborne pathogens in relation to international seed exchange and the zero tolerance concept. *Seed Sci. Technol.* 18:461–65
98. Singh RP, Boucher A, Somerville TH. 1992. Detection of potato spindle tuber viroid in the pollen and various parts of potato plant pollinated with viroid-infected pollen. *Plant Dis.* 76:951–53
99. Sinha RN, Coutts A, Tuma D, Muir WE. 1989. Seasonal changes in seedgerm protein, fat acidity and microflora in non-ventilated and ventilated binstored moist wheat. *Sci. Aliments* 9:769–84
100. Sohi HS, Aulakh KS, Randhawa HS. 1988. Effect of prolonged field exposure of seed cotton to climatic conditions on

seed quality. *Indian Phytopathol.* 41: 220–24
101. Strandberg JO. 1983. Infection and colonization of inflorescences and mericarps of carrot by *Alternaria dauci. Plant Dis.* 67:1351–53
102. Strandberg JO, White JM. 1989. Response of carrot seeds to heat treatments. *J. Am. Soc. Hortic. Sci.* 114: 766–69
103. Stuckey RE, Jacques RM, Tekrony DM, Egli DM. 1981. *Foliar Fungicides Can Improve Seed Quality.* Lexington, KY: KY Crop Improv. Assoc.
104. Sutula CL, Gillett JM, Morrissey SM, Ramsdell DC. 1986. Interpreting ELISA data and establishing the positive-negative threshold. *Plant Dis.* 70: 722–26
105. Swan DJ, Mantle PG. 1991. Parasitic interactions between *Claviceps purpurea* strains in wheat and an acute necrotic host response. *Mycol. Res.* 95: 807–10
106. Taylor AG, Harman GE. 1990. Concepts and technologies of selected seed treatments. *Annu. Rev. Phytopathol.* 28: 321–39
107. Taylor JD. 1970. The quantitative estimation of the infection of bean seed with *Pseudomonas phaseolicola* (Burkh.) Dowson. *Ann. Appl. Biol.* 66: 29–36
108. Thakur RP, Williams RJ. 1980. Pollination effects on pearl millet ergot. *Phytopathology* 70:80–84
109. Tijessen P. 1985. *Practice and Theory of Enzyme Immunoassay.* New York: Elsevier Sci. 540 pp.
110. Tonkin JHB. 1994. Seed standards in legislation: assessing seed quality and effect on seed treatment. See Ref. 58b, pp. 427–40
111. Varma A, Krishnareddy M, Malathi VG. 1992. Influence of the amount of blackgram mottle virus in different tissues on transmission through the seeds of *Vigna mungo. Plant Pathol.* 41:274–81
112. Voller A, Bartlett A, Bidwell DE, Clark MF, Adams AN. 1976. The detection of viruses by enzyme-linked immunosorbent assay (ELISA). *J. Gen. Virol.* 33:165–67
113. Walcott R, Misra MK, McGee DC. 1994. The detection of asymptomatic soybean seeds infected with *Aspergillus* and *Penicillium* spp. using ultrasound analysis. *Phytopathology* 84:1153 (Abstr.)
114. Walker JC. 1922. Seed treatment and rainfall in relation to control of cabbage black-leg. *USDA Bull. 1029*
115. Walker JC, Patel PN. 1964. Splash dispersal and wind factors in epidemiology of halo blight of bean. *Phytopathology* 54:140–41
116. Walkey DGA, Brocklehurst PA, Parker JE. 1985. Some physiological effects of two seed-transmitted viruses on flowering, seed production and seed vigour in *Nicotiana* and *Chenopodium* plants. *New Phytol.* 99:117–28
117. Wall MT, McGee DC, Burris JS. 1983. Emergence and yield of fungicide treated soybeans differing in quality. *Agron. J.* 75:969–73
118. Waniska RD, Forbes GA, Bandyopadhyay R, Frederiksen RA, Rooney LA. 1992. Cereal chemistry and grain mold resistance. In *Sorghum and Millet Diseases: a Second World Review,* ed. WAJ de Milliano, RA Frederiksen, GD Bengston, pp. 265–72. Patancheru, Andra Pradesh, India: ICRISAT. 370 pp.
119. Warham EJ. 1990. Effect of *Tilletia indica* infection on viability, germination and vigor of wheat seed. *Plant Dis.* 74:130–32
120. Webster DM, Atkin JD, Cross JE. 1983. Bacterial blights of snap beans and their control. *Plant Dis.* 67:935–40
121. Weidenborner M, Hindorf H, Jha HC. 1990. Control of storage fungi of the genus *Aspergillus* on legumes with flavonoids and isoflavonoids. *Angew. Bot.* 64:175–90
122. Welty RE, Azevedo MD, Cooper TM. 1987. Influence of moisture content, temperature, and length of storage on seed germination and survival of endophytic fungi in seeds of tall fescue and perennial ryegrass. *Phytopathology* 77: 893–900
123. White DG, Toman J. 1994. Effects of postharvest oil and fungicide application on storage fungi in corn following high-temperature grain drying. *Plant Dis.* 78: 38–43
124. Williams PH. 1980. Black rot: a continuing threat to world crucifers. *Plant Dis.* 64:736–42
125. Wu WS, Cheng KC. 1990. Relationships between seed health, seed vigour and the performance of sorghum in the field. *Seed Sci. Technol.* 18:713–19
126. Yao CL, Frederiksen RA, Magill CW. 1990. Seed transmission of sorghum downy mildew: detection by DNA hybridization. *Seed Sci. Technol.* 18:201–7
127. Yao CL, Magill CW, Frederiksen RA, Bonde MR, Wang Y, Pin-shan W. 1991. Detection and identification of *Peronosclerospora sacchari* in maize by DNA hybridization. *Phytopathology* 81:901–5
128. Zimmer RC, McKeen WE, Campbell

CG. 1992. Location of oospores in buckwheat seed and probable roles of oospores and conidia of *Peronospora ducometi* in the disease cycle on buckwheat. *J. Phytopathol.* 135:217–23

129. Zorrilla G, Knapp AD, McGee DC. 1994. Severity of Phomopsis seed decay, seed quality evaluation and field performance of soybean. *Crop Sci.* 34: 172–77

Annu. Rev. Phytopathol. 1995. 33:467–88

MODELS FROM PLANT PATHOLOGY ON THE MOVEMENT AND FATE OF NEW GENOTYPES OF MICROORGANISMS IN THE ENVIRONMENT

C. C. Mundt

Department of Botany and Plant Pathology, Oregon State University, Cordley Hall 2082, Corvallis, Oregon 97331-2902

KEY WORDS: dispersal, ecological niche, fitness, genetically engineered microorganisms, host range

ABSTRACT

Because of their long history of studying the environmental fate of nonengineered plant pathogens, plant pathologists can contribute to an evaluation of the environmental fate of genetically engineered microorganisms (GEMs). This review discusses three relevant subtopics: dispersal, change of ecological niche as exemplified by host range, and fitness. Dispersal of microorganisms in the daytime, turbulent atmosphere provides significant potential of long-distance dispersal, and matters of scale are critically important in determining this potential. Studies of plant pathogen host range confirm the commonly held view that microorganisms are plastic with regard to adaptation to environment, but suggest that change in ecological niche will be inversely proportional to the magnitude of the change. Fitness of plant pathogens is not constant with regard to either environment or time, and the same is likely to be true of genetically engineered microorganisms.

INTRODUCTION

The movement and fate of genetically engineered organisms has aroused much concern in recent years. Genetically engineered microorganisms (GEMs) have

0066-4286/95/0901-0467$05.00

received particular emphasis owing to their fast multiplication rate, and because these organisms are not easily observed in the environment. Further, the commonly held association of microorganisms with disease has generated fear over the potential harm of GEMs. Given the limited knowledge of microbiology and ecology among the general public, it is important that accurate, impartial information be made available on the potential risks associated with GEMs.

Scientists from many disciplines have a role to play in evaluating the movement and fate of GEMs, from the molecular biologist to the global ecologist; no scientific discipline can provide all of the required information. Nevertheless, plant pathologists can make a unique contribution because of our long history of studying the environmental fate of nonengineered plant pathogens. A complete review of all relevant plant pathology literature would be impossible in a single chapter, and thus I restrict my analysis to three relevant subtopics: dispersal, change of ecological niche as exemplified by pathogen host range, and fitness. I discuss only selected examples to demonstrate the relevance to GEMs and emphasize fungal pathogens in the aboveground environment. The reader is also referred to two recent, related reviews by plant pathologists (46, 108).

DISPERSAL

Dispersal of infective units, a key component of plant disease epidemics, can be divided into the processes of release, transport, and deposition (2, 89). Here I discuss only transport, as it is this process that is of most concern in evaluating the potential movement of GEMs in the environment. There are three major modes of propagule transport in the air: nocturnal dispersal, splash or rain dispersal, and diurnal dispersal in the turbulent atmosphere.

Nocturnal Dispersal

The nocturnal movement of propagules is a fascinating topic, though rarely studied. The structure of the atmosphere varies drastically between daytime and nighttime. During daytime, the atmosphere is characterized by considerable turbulence due to convective heating and interaction of moving air with surface obstructions (see below). At night, the atmosphere becomes highly structured owing to the lack of convective heating. However, air movement, though slow, can show complex patterns caused by differential cooling and topographical variation. This was demonstrated elegantly for white pine (*Pinus strobus*) blister rust (caused by *Cronartium ribicola*) by Van Arsdel (103), who released smoke during the night to determine patterns of wind movement as related to blister rust infections. One such example is given in Figure 1, which shows a circulating pattern of smoke from the lower swamplands where

Figure 1 Pattern of a nighttime smoke release from swamp ribes in relation to position of white pines infected by blister rust. Reprinted with permission from Van Arsdel (103).

Ribes (the alternate host of white pine blister rust) grows, to white pines on the upper hillsides. This pattern was caused by rapid cooling of air on the open slopes, with the dense, cooled air flowing downhill. The taller trees in the lower elevations, however, prevented further radiative cooling of this air, which was then forced upwards by the cooler air behind it. However, its upward movement was eventually blocked by an inversion layer at the top of the tree canopy, allowing for horizontal movement to the upper hillsides. Van Arsdel (103) noted that blister rust infections tended to be much more prevalent on white pine on the upper hillsides (dark trees in the left panel of Figure 1) than on white pines that were adjacent to *Ribes* in swamps (White trees of Figure 1), and suggested that this infection pattern was due to the nocturnal movement of sporidia of the pathogen. This and other patterns shown by Van Arsdel (103) are the only examples of which I am aware for nighttime dispersal of a plant pathogen. However, this work is highly important, as it demonstrates the potential for dispersal of microorganisms in unexpected patterns that might be missed by standard sampling designs close to the source of the microbe if one considers only daytime dispersal patterns.

Splash Dispersal

The transport of propagules via splash dispersal by rain or irrigation probably occurs for nearly every plant pathogen that produces propagules at or above the soil surface. However, this transport mechanism is relatively more important for pathogens that produce large propagules and/or produce propagules in some type of mucilage, as splash may provide the only mechanism for significant movement of such pathogens. Splash dispersal of plant pathogens is influenced by factors such as rain drop size and velocity; rainfall duration; and physical characteristics of the propagule, propagule substrate, and crop canopy in which infection occurs (23, 59). However, splash dispersal is a small-scale phenomenon. For example, Madden (59) noted that few splash droplets travel more than 20 cm, and Fitt et al (23) concluded that few inoculum-carrying droplets are splashed beyond 1 m from the source. Thus, this short-distance dispersal is often reflected by very steep gradients away from an inoculum source (21). Mathematically, such gradients are often described well (21) by the exponential model $y = ae^{-bx}$, in which y is the number of propagules deposited or number of infections formed per unit area at x units of distance from the source of inoculum, a is the number of propagules or infections per unit area at the source, e is the base of natural logarithms, and b is a parameter that describes the steepness of the gradient (44). This model describes an exponential decline with distance in which the proportionate decrease in propagule or infection number per unit distance is constant (21). Propagule or disease gradients would be expected to follow an exponential model when decline with distance is due entirely to deposition (62). This would seem an appropriate model for splash-dispersed pathogens, because the relatively large splash droplets would have a high impaction efficiency on foliage (34). Thus the exponential model would describe a situation where inoculum is being "filtered" from the air at a constant proportion as inoculum moves through a plant canopy.

The short-distance movement caused by splash dispersal is apparently reflected in the genetic structure of plant pathogens. McDonald et al (65) reported distinct spatial clustering of haplotypes of *Rhynchosporium secalis* within 10 m square plots of barley. Even more striking were the results of McDonald & Martinez (64) who found that, with one exception, the same haplotype of *Mycosphaerella graminicola* was never found in more than one of six sampling grids that were placed 10 m apart within a wheat (*Triticum aestivum*) field.

Turbulent Dispersal

Of more interest is microbial dispersal in the turbulent, daytime atmosphere, as such dispersal impacts a large number of microbial species and provides much potential for long-distance transport. Several models originally developed to describe the dispersion of air pollution and radioactive fallout have been applied to plant pathogens (2, 63). Such models are based on the assumption that turbulence will dilute a point source of particles through the influence of expanding eddies and calculate predicted variances of dispersion in the downwind, crosswind, and vertical directions based on physical attributes of

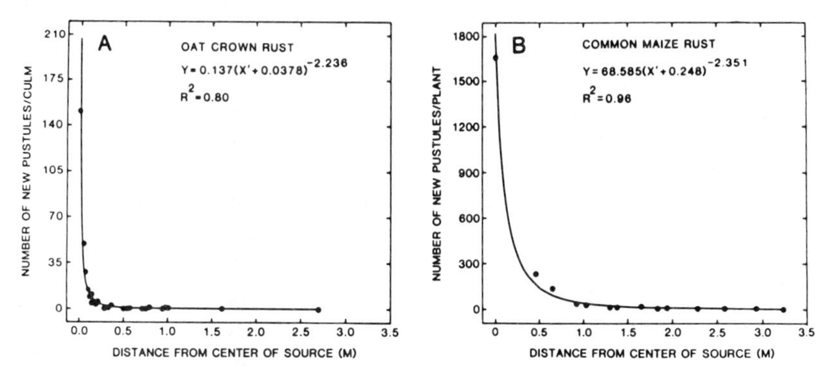

Figure 2 Disease gradients of two rust diseases after one cycle of infection from single, infected source plants. Reprinted with permission from Mundt & Leonard (72).

the particles and of the atmosphere. However, the result of these complex processes can often be visualized in simple gradient plots of number of propagules or amount of disease vs distance from the inoculum source (Figure 2). Such gradients have been reported for a large number of wind-dispersed pathogens (21), and are often described by an inverse power function of the form $y = ax^{-b}$, in which y, x, and b are as in the exponential model above and a is the number of propagules or infections per unit area at one unit of distance from the source (33, 34). The power law describes gradients for which the decline of numbers with distance is due to diffusion (62). In contrast to the exponential model, the power law describes a gradient that decreases with distance from the source (21), providing an extended tail that may contain a significant proportion of the dispersed inoculum.

Several aspects of gradients of the type shown in Figure 2 deserve mention. First, such gradient plots are often strikingly steep. However, steep dispersal gradients should be of no surprise for a point source of inoculum, because dispensing a given amount of inoculum at the source into a larger area would have to result in a gradient with an inverse power of ~2, simply based on geometric considerations, as area increases with the square of distance. Second, it is very common to incorrectly interpret such gradients to mean that the majority of inoculum is deposited near the source. In reality, the steepness of a gradient per se tells little about the proportion of inoculum dispersed close to vs far from the source. It is economically unfeasible to measure dispersal at far distances from a source because the number of propagules deposited per sampling unit is very small. However, total deposition integrated over all potential sampling units at that distance can be very large, and a significant number of deposition events may thus be missed. In fact, factors that cause steep dispersal gradients often cause a *larger* proportion of inoculum to be dispersed distant from the source (21, 34, 63). For example, increased wind speed increases turbulence and, hence, the rate of diffusion with distance from the source. This results in both a steeper gradient, but also greater potential for long-distance dispersal (34) or, in other words, a greater distance of mean dispersal and a larger proportion of inoculum in the "tail" of the gradient. Thus, disease gradients of wheat leaf rust (caused by *Puccinia recondita*) were found to be much steeper in a high-wind site than at a low-wind site (69).

The main reason for lack of information on the average dispersal distance of plant pathogens is failure to quantify the source strength, which is very difficult to measure in the field. Instead, nearly all studies have measured only the number of propagules or infections at different distances, without relating those numbers to the total release. Surprisingly few exceptions exist. Fitt & McCartney (22) considered data of two older studies where the source strength of spores was measured for three species. They concluded that the fraction of spores that remained airborne was greater for the two species with smaller spores, despite the fact that gradients for all three species were similar. The best example of accounting for source strength of which I am aware was by Lindow et al (57) in a pretest for the release of genetically engineered bacterial strains. They applied known numbers of *Pseudomonas syringae* in the center of each of three 7 × 7 m plots of oats (*Avena sativa*), with the microbe being sampled from both plants and soil around the release site. They found very steep dispersal gradients typical of that expected for wind dispersal. By integrating counts over the entire sampled area, however, they calculated that only 0.005% of the originally applied bacterial cells were actually recovered. Even allowing for a large decline in viability, there was still the potential for considerable dispersal of the bacterium out of the plot areas. Interestingly, a mathematical model developed to predict wind dispersal of bacteria (47) showed a very high correspondence with the field data of Lindow et al (57) for *relative* deposition with distance from the source, but required scaling to provide correspondence with the *actual* numbers observed.

Matters of Scale

Evidence for long-distance dispersal of plant pathogens has been reviewed (73, 85). Many studies involve long-distance dispersal of fungal spores, because they are easily identifiable, and well adapted for survival in the atmosphere. For example, Gregory (34) analyzed data from an epidemic of wheat stem rust caused by *Puccinia graminis* (92) and concluded that the pathogen was dispersed in a single jump from infected wheat in Texas, depositing over 2000 spores per sq m during a 48-hour period ~1000 km north in Wisconsin. Studies of long-distance dispersal of other pathogens, such as bacteria, have been hampered by difficulty of identification and sensitivity to environment. However, Franc et al (26) discussed evidence supporting the hypothesis that viable cells of *Erwinia caratovora* (causal agent of potato [*Solanum tuberosum*] soft rot) are dispersed over hundreds of kilometers in aerosols.

The ability of pathogens with steep dispersal gradients to be dispersed long distance has two explanations. First, massive numbers of propagules can be produced during a plant disease epidemic. For example, Schafer & Roelfs (86) calculated that 2.3×10^{14} uredospores of *P. graminis* are produced ha^{-1} day^{-1} in a wheat field with 10% of leaf area diseased. Thus even a small proportion of dispersal in the tail of a gradient could cause significant transport of inoculum at long distance. Second, an area source of inoculum produces a different dispersal gradient than a point source (33, 34). A point source of inoculum is very small relative to the area over which inoculum deposition is being measured. An area source, on the other hand, is large relative to the area where deposition is measured. Thus, a single infected wheat field in Kansas would function as a point source for states northward, but would be an area source for other fields in the local vicinity. An area source provides a shallower gradient than does a point source (33, 34), primarily because an area source actually consists of multiple point sources with overlapping gradients. Mathematically, an area source of inoculum reduces b of the power law by more than one unit (33, 34). Given that values of b for plant pathogens generally range between 1.5 to 3.0, the influence of an area source on dispersal gradient flattening can be substantial.

Dispersal from a single point source of inoculum provides one pattern, whereas dispersal from multiple sources within a region may provide another. For example, modeling studies suggested that relatively little inoculum of cereal rusts (*Puccinia* spp.) would be dispersed between any pair of agricultural fields (71). However, in a larger region with many infected fields, the "tails" from multiple inoculum sources would overlap and result in considerable amounts of inoculum exchange among fields. This concept is supported by small-scale field tests with a cereal rust (CC Mundt, LS Brophy, & SC Kohler, in preparation) and with the brown spot disease [caused by *Pseudomonas syringae* pv. *syringae* of bean (*Phaseolus vulgaris*) (54)]. Lindemann et al (54) compared brown spot severity in six plots planted in a commercial bean-growing district of Wisconsin, and five plots planted either 16 km west or 27 km east of that region, where no commercial beans are grown. Brown spot epidemics developed in four of the six plots within the bean-growing area and none of the plots planted outside of this region, despite the fact that the planted beans were inoculated with the pathogen and that gross environmental variables did not differ among the 11 sites. These results suggest that inoculum production on a regional basis may have been required to maintain brown spot epidemics.

Some of the above comments on long-range dispersal are based on a dispersal model that is very simple and highly descriptive, but probably adequate for describing basic concepts. However, more mechanistic models of long-distance dispersal based on trajectories of individual dispersal events have been described for plant pathogens (3, 16).

Survival During Dispersal

The ability of a microorganism to survive long-distance dispersal is important, but not easily assessed. Aylor (3) noted the difficulty of measuring propagule viability in large populations of fungal spores, in which even a very small percentage of viability would have practical consequences. Methods for determining viability can also influence results. For example, Minogue & Fry (68) found that sporangia of *Phytophthora infestans* exposed to humidities between 40 and 88% had germination rates of less than 1% when rapidly rehydrated on agar plates, the typical test used for assessing germinability of fungal spores. In contrast, germination was 30% when rehydrated over a 2 min period, a condition more similar to that which fungal spores are exposed to in the field. Similarly, Wilson & Lindow (107a) found that culturing methods accurately estimated sizes of actively growing populations of *P. syringae*, but underestimated by as much as 75% the number of viable cells for populations that were older than 80 h.

Implications for GEMs

Studies of the dispersal of plant pathogens provide several implications for the release of GEMs. Not surprisingly, wind dispersal under daytime, turbulent conditions has the most potential for significant dispersal of a GEM. Such dispersal will almost inevitably lead to a very steep gradient of dispersal away from a point source of release. Thus, one can probably safely assume that many fewer GEMs will be deposited per unit area nearby the site of release than within it. However, this tells little about the number or proportion of the GEMs that would be dispersed out of the release site nor how far the microorganism could be dispersed on average. Studies of plant pathogens suggest that wind dispersal can cause large proportions of a GEM to be dispersed outside of the area of release and to be dispersed very long distances. Thus, if one requires that a GEM be contained mostly within the release site, and if that microorganism is capable of being wind dispersed, it should not be released.

Most studies of the dispersal of GEMs have focused on dispersal from initial test sites, perhaps more for political and legal reasons rather than environmental and safety reasons. The more important issue is the potential impact of a GEM when it is used more extensively, e.g. in commercial agriculture. Scientists studying the environmental safety of transgenic plants have come to a similar conclusion (93a). Of utmost concern for microorganisms is the potential for overlapping sources and for area source effects to cause significantly greater dispersal than might be predicted based on initial release data.

ECOLOGICAL NICHE AS DEMONSTRATED BY HOST RANGE

The extensive studies of pathogen host range by plant pathologists may provide clues regarding the ability of GEMs to change ecological niche. Plant pathogens commonly show characteristic host specificities to genotypes within a plant species, and also to different host species and genera. These are discussed below by level of specificity.

Intraspecific Host Specialization

The ability of pathogens to adapt to different genotypes of a plant species has been studied in detail. Such studies have mostly involved adaptation to formerly resistant, cultivated varieties (cultivars) of crop species, which has been a common occurrence for many pathogens (e.g. 9). Flor's pioneering work in the 1940s and 1950s provided a genetic basis for such adaptation, and culminated in the gene-for-gene hypothesis, which states that "for each gene that conditions reaction in the host, there is a corresponding gene in the parasite that conditions pathogenicity" (25). The accumulated evidence suggests strongly that such gene-for-gene specificity is conditioned by specific matching between gene products produced by a dominant allele for resistance in the host and the corresponding, dominant allele for avirulence in the pathogen (15, 19, 24, 82). Such interactions likely result in a signaling process that "turns on" a general defense mechanism in the host. Such gene-for-gene systems have been proven or strongly suggested to operate in a variety of host-pathogen systems (27, 30, 32).

A consequence of gene-for-gene interactions is that any mutation that arrests the function of an avirulence gene will restore the ability of the pathogen to induce disease on a host genotype with the corresponding gene for resistance. Thus, for example, the average useful life of race-specific genes for resistance to rust diseases of wheat was estimated to be only five years (43). This process of pathogen adaptation may be facilitated by the extreme plasticity of fungal genomes, which are greatly influenced by deletions, duplications, and translocations (14, 67, 94, 95) and may provide mechanisms for rapidly averting the influence of host resistance genes (70). Plasmid transfer (81) and the infidelity associated with RNA replication (18) may provide similar opportunities for variability in bacterial and viral plant pathogens, respectively.

Although classical gene-for-gene interactions are common in plant pathogen systems, other mechanisms of adaptation to a host likely exist. A smaller proportion of plant pathogens produce host-specific toxins (107), which result in an "opposite" gene-for-gene interaction wherein virulence is the active function (19, 20). Studies of plant pathogens at the population level, rather than the individual isolate level, have shown that pathogens can become quantitatively adapted to specific host genotypes (11, 13, 36, 39, 50, 90). It is unclear, and difficult to determine (84), if such quantitative adaptation involves gene-for-gene interactions.

A nonpathogenic mutant of the fungal pathogen *Collectotrichum magna* derived through UV radiation (28) was compared with the pathogenic, wild-type isolate (29). Except for the ability to produce necrotic lesions, the mutant maintained all of the characteristics of the wild-type, including appressorial formation, systemic mycelial growth within the host, and host range. A cross between the mutant and wild-type strain indicated that nonpathogenicity was due to mutation at a single genetic locus. Thus, a single-gene change altered the ecological niche of *C. magna* from that of a pathogen to that of an endophytic mutualist.

Interspecific Host Specialization

Perhaps of greater relevance to issues surrounding the release of GEMs is the ability of plant pathogens to attack different species of hosts. Reports of changes in the known host range of plant pathogens are very common. However, many of these host-range changes could result from chance discoveries over time, rather than a genetic change in the pathogen. Of greater interest are examples where common, well-studied pathogens have apparently changed their host range. Three examples are given below for important pathogens, or groups of pathogens, on major cereal crops.

The cereal rusts (*Puccinia* spp.) are a group of biotrophic pathogens demonstrating a high degree of host specificity on cereal crops. In addition to races specialized on different cultivars of a cereal crop, these pathogens are also divided into different formae speciales (morphologically indistinguishable forms of a pathogen species that are specialized to different host species). Though such formae speciales are commonly recognized in agriculture, the host ranges of these rust species are much broader in natural ecosystems (8). Thus, agricultural formae speciales may be capable of genetic exchange. In fact, Burdon et al (10) showed through isoenzyme studies that scabrum rust (pathogenic on *Agropyron scabrum* and *Hordeum vulgare*) in Australia arose from somatic recombination between *P. graminis* f.sp. *tritici* (wheat stem rust) and *P. graminis* f.sp. *secalis* (rye stem rust). *P. coronata* has been known to cause significant levels of crown rust disease on cultivated oats (*Avena sativa*) for over 200 years, and is also known to be able to infect a large number of other wild and cultivated grasses (91). In 1991, however, cultivated barley (*Hordeum vulgare*) in Nebraska was found to be heavily infected by what appeared to be a new variety of *P. coronata* (38).

Gaeumannomyces graminis, causal agent of the take-all disease of wheat and other small grains, is one of the most serious soilborne pathogens of wheat throughout the world, especially where wheat is grown continuously. Initially, the pathogen was not known to cause disease on oats. Oats were often grown in rotation with wheat to control the take-all disease (97). From the late 1920s, however, take-all on oats began to be reported worldwide (12, 97). It was later determined that oats produce the compound *avenacin*, which is toxic to the wheat-attacking strains of the pathogen. However, the oat-attacking strains of *G. graminis* produce a glucosidase that inactivates *avenacin*, thus allowing the fungus to also become a pathogen of oats (98). Though rye (*Secale cereale*) is considered highly resistant to take-all, Hollins & Scott (35) recently reported continuous variation in ability to infect rye among wheat-attacking isolates of *G. graminis*, including isolates that substantially infected rye. Thus, *G. graminis* may currently be extending its host range to another cereal species. Though the rye- and wheat-attacking strains coexist within cereal fields (35, 78), molecular analyses suggests that there is little gene flow between the two groups in nature (78), and the origin of the rye-attacking strains is unknown.

Magnaporthe grisea (anamorph *Pyricularia oryzae*) is the causal agent of blast disease, the most important disease of rice (*Oryza sativa*), and is also known to infect a number of grassy weeds. Recent development of genetic systems for studying the blast fungus (101), and recent significant changes in the host range of the pathogen, allow for interesting observations regarding the adaptability of microorganisms. Though wild grasses have been postulated to be a source of inoculum for epidemics on rice (58), studies of repetitive DNA sequences in isolates collected from rice vs other grasses suggest that rice isolates represent distinct populations (4, 102). Despite this genetic differentiation, rice and non-rice isolates can be mated, but repeated backcrossing to a rice isolate is required to restore full compatibility of the pathogen with rice (53, 102). These results suggest that the pathogen is capable of genetically altering its host range, but that a larger number of genes are required to attain the high fitness required for field survival on a new host.

Known for centuries as a devastating pathogen of rice, the ability of *M. grisea* to become a pathogen of wheat was a shocking surprise to agriculturalists. Wheat blast caused by *M. grisea* was initially identified in seven counties of Brazil in 1985. By 1991, the disease had spread to all wheat-growing regions of Brazil, which encompass about 3 million ha, and caused significant economic losses in some areas (100). Though originally identified in areas producing both rice and wheat, host range (100) and molecular (101) studies suggest that the origin of the wheat blast pathogen was from strains infecting wild grasses. Lesser known, but also dramatic, was the ryegrass (*Lolium* spp.) blast epidemic caused by *M. grisea* in the southeastern U.S. in 1971 (96). Previously unknown on ryegrass, the disease caused severe economic losses. Though the 1971 epidemic was the most severe, and likely exacerbated by Hurricane Edith, the disease has remained a significant economic problem. The origin of strain(s) causing ryegrass blast is unknown (96). The wheat blast and ryegrass blast examples demonstrate that it is possible for a pathogen species to change hosts and have sufficient fitness to cause severe epidemics over extensive areas.

Other examples of fluidity in the host ranges of pathogens exist. For example, Nelson (74) reported 13 different successful interspecific crosses among *Helminthosporium* spp., many of which parasitize different grass hosts. Some of these successful crosses were surprising. For example, fertile crosses were obtained between *H. carbonum*, a foliar pathogen of maize (*Zea mays*) and *H. victoreae*, which infects the roots of oats (74, 75). Viruses can sometimes be mechanically transmitted to plant species that are not normally considered hosts in the field, suggesting that the virus could easily extend its host range if barriers to vector transmission were overcome. For example, barley mild mosaic virus was mechanically transmitted to rye, and typical symptoms were produced, despite the fact that rye had previously been considered a nonhost of the virus (80).

Intergeneric Host Specialization

The above examples represent cases of increased host range on different members of the same plant family. It is also interesting to know if plant pathogens possess the capability of adapting to more distantly related hosts. Madhosingh (60) demonstrated interspecific hybrid production through mycelial anastomosis between *Fusarium oxysporum lycopersici*, a tomato pathogen that does not attack wheat, and *Fusarium graminearum*, primarily a pathogen of cereal and grasses but which can also infect tomatoes to some degree in greenhouse inoculations. Perhaps the most interesting studies of pathogen host range have been with the pathogen *Nectria haematococca* (anamorph *Fusarium solani*). *N. haematococca* can infect roots and stems of pea (*Pisum sativum*), and infection is determined in part by the pathogen's ability to inactivate the phytoalexin pistatin through a demethylation process. Isolates lacking the ability to demethylate pistatin are unable to infect peas (106). One gene for pistatin is carried on a dispensable or "B" chromosome of *N. haematococca* (66). Although it is unknown if this dispensable chromosome can be horizontally transferred in nature (66), genetic transformation studies have been conducted to address the potential implications of such a transfer (79, 87, 88). *Cochliobolus heterostrophus* (a foliar pathogen of maize) and *Aspergillus nidulans* (a saprophyte) were transformed with pistatin demethylation genes from *N. haematococca* and tested for virulence on peas (87). *A. nidulans* was unable to induce disease on roots, stems, or leaves of pea, though the demethylation genes were expressed at levels equal to or greater than in *N. haematococca*. Not surprisingly, more than a single gene change was required to convert a saprophyte to a plant pathogen. A surprising result was the ability of the transformed *C. heterostrophus* to induce disease on stems and leaves of pea; transformed *C. heterostrophus* was unable to infect pea roots, however (88). These results suggest that *C. heterostrophus*, normally a foliar pathogen of maize, possesses all of the genes necessary to be a pathogen of foliage in general, and required only a single additional gene to make a major shift in host range. The ability to make this change and also be fit to survive and compete in nature is a more complex question, and is discussed later in this chapter.

Implications for GEMs

The above discussion of pathogen host range as an example of ecological niche provides a few points of relevance to the fate of GEMs. First, the examples confirm the commonly held view that microbial populations possess considerable plasticity with regard to adaptation to environment. Second, the frequency with which a microorganism is able to make a change in niche is likely to be inversely related to the magnitude of that change. Selection of a pathogen containing a mutation to virulence against a single, race-specific gene for disease resistance is easily attainable because only a single null mutation is required, and the pathogen already possesses genes for basic compatibility with that host species. Examples of plant pathogens "overcoming" the resistance of crop cultivars are common. Alteration of host range to attack different host species may be a less complex process than once thought, though it likely requires changes both in pathogenicity genes as well as fitness characteristics necessary to survive and compete on the new host. Host-range changes within the same plant family have occurred, sometimes with devastating consequences, but are much less frequent than adaptation for different host genotypes within the same species. Similarly, changes of host range across plant families should be even less frequent, and this would seem to be the case. However, the successful transformation of *C. heterostrophus* to become a pathogen of pea (79,87, 88) is a disturbing result. The example of host range evolution as a change in ecological niches may represent a worst case scenario, as host genotypes exert very direct and strong selection on pathogens. One might argue that saprophytic microbes would change niche less rapidly. Thus, further evolutionary studies of microbial populations are warranted to estimate the frequency and magnitude of changes in ecological niche.

A final relevant point involves the effect of null mutations on microorganisms. Plant pathologists would normally expect a nonpathogenic mutant to have lost some aspect of basic compatibility with its host. Thus, it was significant that a nonpathogenic isolate of *C. magma* became an endophytic mutualist. In this specific case, the mutant might actually provide a positive role as a biological control agent against plant pathogens (29). Nevertheless, that study should serve as an important reminder that genetic alterations can occasionally cause unexpected changes. This is an especially important point, as null mutations are often considered benign in terms of nontarget impacts.

FITNESS

Plant pathologists have emphasized evaluating the fitness of microorganisms, as such analyses are crucial to the management of plant pathogens for disease control. Particular emphasis has been given to evaluating the fitness of pathogen isolates that possess the virulence traits necessary to attack plant genotypes with race-specific genes for resistance. Vanderplank (104) hypothesized that such virulent pathogens would possess reduced fitness on susceptible hosts, as virulence would force the pathogen to use alternative metabolic pathways that are less metabolically efficient. This proposed mechanism is in conflict with strong evidence suggesting that the active function in gene-for-gene systems is between gene products of avirulence genes and resistance genes (15, 19). However, it might reasonably be postulated that dominant alleles for avirulence code for gene products that function to control traits other than avirulence, and that plants have merely evolved to recognize these products to induce the resistance response. Thus, recessive alleles for virulence would still cause reduced fitness to a pathogen in proportion to the fitness value of the gene product produced by the avirulence allele.

Measuring Fitness

As with other organisms, fitness of plant pathogens has been analyzed with studies of both realized fitness in competition studies, as well as measurement of individual components of fitness (1). One problem in evaluating the fitness of plant pathogens is that many reproduce asexually for part or all of their life cycle. It becomes very difficult to differentiate the fitness effects of a specific gene from the fitness effects of other genes in an asexual organism (51, 105). This was demonstrated very clearly by Bronson & Ellingboe (7), who studied fitness effects of virulence alleles of *Erysiphe graminis* in competition studies on wheat. They found a race possessing four known virulence genes to be approximately 24% less fit than a race possessing no known genes for virulence. However, analysis of the fitness of progeny resulting from a sexual cross between the two isolates showed that virulence and fitness segregated independently. Parlevliet (83) provided a thorough review and analysis of fitness costs associated with virulence. He concluded that existing data do not support the hypothesis that virulence is associated with reduced fitness as a general phenomenon. Instead, Parlevliet (83) proposed that the initial observations of Vanderplank (104) and others were caused by deleterious effects of mutations in recently evolved races. He suggested that there will be subsequent selection for modifying genes to restore fitness, and discussed evidence for lack of a cost of virulence in "older" races. One situation in which a cost of virulence may occur, however, is for pathogens that produce host-specific toxins. Production of such toxins may entail a significant physiological cost (45), and near-isogenic lines of *Cochliobolus heterostrophus* possessing the T1 toxin have been found to be less fit than the wild-type parent lacking toxin production (45, 52).

Molecular tools are beginning to shed additional light on fitness of plant pathogens. For example, site-directed mutagenesis proved highly useful in demonstrating that lack of ice nucleation activity in *P. syringae* and *P. fluorescens* did not reduce the fitness of these species in competition with ice-nucleating isolates. Further, inoculating with either the Ice^+ or the Ice^- phenotype in advance of or in higher concentration than the other phenotype allowed the first strain to preemptively exclude the second (55). Tn*5* mutagenesis enabled Lindow et al (56) to identify traits that condition reduced epiphytic fitness of *P. syringae* on bean leaves. More than half of the mutants with reduced fitness exhibited decreased ability to survive on dry leaf surfaces relative to the wild-type. In contrast, transformation of the fungal pathogen *C. heterostrophus* demonstrated clearly the necessity for careful interpretation of fitness studies. Protoplasts of *C. heterostrophus* were transformed with a plasmid containing a gene for hygromycin B resistance, and the resulting isolates were compared with the wild-type in competition studies (42). They found 92% of the transformed isolates to be significantly less fit than the wild-type. However, this fitness reduction was due in part to the transformation protocol itself, as 40% of isolates mock-transformed without receiving the plasmid DNA also showed a significant fitness reduction. Further, one transformant regained fitness after successive generations of asexual reproduction on the host. Genetic transformation of the saprophytic, soilborne bacterium *Bacillus subtilis* showed that fitness reductions associated with resistance to the antibiotic rifampcin can be ameliorated through replacement by alternative alleles with less fitness cost, and also by fitness modifiers present within the genome of the bacterium (13a). These results support the view expressed earlier by Parlevliet (83) that fitness costs associated with a trait may be transient.

The relative fitnesses of microbial genotypes are not necessarily constant and can vary greatly depending on experimental conditions. Competition studies of rust races have shown that pathogen density (number of infections per unit leaf area) (40, 41), temperature (41, 61), and host genetic background (61) can all change the ranking of fitness differences between pathogen genotypes under controlled conditions. Further, field studies of *P. striiformis* on wheat suggested that selection may be influenced by interactions among different pathogen genotypes (17). Martens (61) showed that relative fitnesses measured in growth chambers can be very different from those observed in the field. In competition studies on susceptible wheat genotypes, races with the fewest genes for virulence were generally more fit than races with a larger number of virulence genes in growth chambers, whereas the opposite was true in experimental field plots. It has been noted that host genotype × pathogen genotype × environment interactions can strongly influence interpretation of quantitative variation in host/parasite systems (48). Such interactions have been observed in both greenhouse (37) and field (49) studies.

Changes in Fitness

Several examples demonstrate the potential for sudden pathogen fitness changes to result in significant epidemics. For example, race 15B of *P. graminis* f. sp. was identified many years before it suddenly caused the devastating wheat stem rust epidemics of the 1950s in North America (93). Thus one might expect that an increase of fitness, perhaps mediated through differential effects of temperature (61), was required for that race to rapidly dominate the pathogen population in 1950. A clearer example is that of race T of *C. heterostrophus*, which caused the Southern corn leaf blight epidemic in 1970 (99). Key to that epidemic was the ability of race T to produce a host-specific toxin that caused large, chlorotic lesions on corn cultivars possessing the T-type, male-sterile cytoplasm. Some isolates of the pathogen collected as early as 1955 apparently produced the T-toxin (77). However, the earlier isolates were not nearly as fit as those that caused the 1970 epidemic (76). Brasier (5) described the displacement of the nonaggressive strain of *Ophiostoma ulmi*, causal agent of Dutch elm (*Ulmus* spp.) disease, in Europe. The first epidemic, which began between 1910–20, was caused by the nonaggressive strain, but this epidemic subsequently declined in severity. More recently, European elms have been decimated by two races of the aggressive strain of *O. ulmi*, one introduced from North America and spreading from west to east and the second originating from eastern Europe or Asia and spreading from east to west. It is unclear if these newer, aggressive genotypes will be maintained or ultimately replaced by attenuated types. A more recent and equally dramatic example is the worldwide spread of new, more highly fit isolates of *Phytophthora infestans*. Molecular analyses suggest that, until recently, all isolates outside of Mexico may have originated from the same clonal lineage responsible for the potato late blight epidemics of the 1840s in North America and Europe. In the 1970s, however, this lineage began to be replaced by another that is of the opposite mating type, more highly fit, and which has a high frequency of insensitivity to the fungicide metalaxyl. This new lineage may have been responsible for late blight epidemics in Europe and the Middle East during the 1980s (31), as well as the spectacular epidemics of North America in the summer of 1994. Presumably, this new lineage originated through migration from Mexico, a center of origin for *P. infestans* (31).

Implications for GEMs

The above discussion of plant pathogen fitness suggests that a GEM cannot be assumed to be of reduced fitness simply because it has been genetically modified. Fitness in the field is a complex phenomenon, and I agree with Brill (6) that simple fitness measurements under controlled environmental conditions may have limited value for field prediction. There is likely to be generation of and selection for increased fitness of GEMs once released, providing for increases of fitness over time. However, very large fitness changes are likely to follow the punctuated equilibrium model of evolution, with long periods of apparent stability, punctuated by large infrequent changes.

SUMMARY AND CONCLUSIONS

Four generalizations can be derived from the above discussion of plant pathogen populations:

1. Microbial populations are often capable of considerable aerial dispersal.
2. Matters of scale are critical to the movement and fate of microorganisms. Thus large-scale release of a microorganism will provide for considerably greater dispersal potential, and will likely provide increased opportunities for exclusion of native microbial populations.
3. Microorganisms possess the capacity to change ecological niche, but such changes will likely occur at relatively low frequencies that are inversely proportional to the magnitude of the change.
4. Fitness of a microorganism in nature is difficult to measure or predict with much accuracy, and is unlikely to be constant with respect to either environmental conditions or time.

More precise formulation of the unsatisfying generalizations described above will likely be of little help in addressing the more difficult questions associated with the use of GEMs. These questions include: What is an acceptable level of risk? Are natural microbial populations qualitatively or quantitatively superior to genetically engineered ones? Will use of GEMs reduce genetic diversity of microbial populations? Is nature a more appropriate model for the management of bioresources than is the model of industrial processes? Such questions cannot be answered with single equations or controlled experiments. The answers may become available in the future, but for the moment, I am thankful to be able to avoid addressing such issues in this chapter.

ACKNOWLEDGMENT

I greatly appreciate the excellent assistance of J Poppleton in preparing the manuscript and the comments of JE Loper on an earlier draft of this chapter.

Literature Cited

1. Antonovics J. Alexander AM. 1989. The concept of fitness in plant-fungal pathogen systems. See Ref. 52a, 2:185–214
2. Aylor DE. 1978. Dispersal in time and space: aerial pathogens. In *Plant Disease: An Advanced Treatise*, ed. JG Horsfall, EB Cowling, 5:159–80. New York: Academic
3. Aylor DE. 1986. A framework for examining inter-regional aerial transport of fungal spores. *Agric. For. Meteorol.* 38:263–88
4. Borromeo ES, Nelson RJ, Bonman JM, Leung H. 1993. Genetic differentiation among isolates of *Pyricularia* infecting rice and weed hosts. *Phytopathology* 83:393–99
5. Brasier CM. 1987. Recent genetic changes in the *Ophiostoma ulmi* population: the threat to the future of elm. In *Populations of Plant Pathogens: Their Dynamics and Genetics*, ed. MS Wolfe, CE Caten, pp. 213–26. Oxford: Blackwell Sci.
6. Brill WJ. 1985. Safety concerns and genetic engineering in agriculture. *Science* 227:381–84
7. Bronson CR, Ellingboe AH. 1986. The influence of four unnecessary genes for virulence on the fitness of *Erysiphe graminis* f.sp. *tritici*. *Phytopathology* 76:154–58
8. Browning JA. 1980. Genetic protective mechanisms of plant-pathogen populations: their coevolution and use in breeding for resistance. In *Biology and Breeding for Resistance to Arthropods and Pathogens in Agricultural Plants*, ed. MK Harris, pp. 53–75. Texas Agric. Exper. Stn. Misc. Publ. 1451
9. Browning JA, Simons MD, Frey KJ, Murphy HC. 1969. Regional deployment for conservation of oat crown rust resistance genes. In *Disease Consequences of Intensive and Extensive Culture of Field Crops*, ed. JA Browning, pp. 49–56. Iowa Agric. Home Econ. Spec. Rep. 64
10. Burdon JJ, Marshall DR, Luig NH. 1981. Isozyme analysis indicates that a virulent cereal rust pathogen is a somatic hybrid. *Nature* 293:565–66
11. Caten CE. 1974. Intra-racial variation in *Phytophthora infestans* and adaptation to field resistance for potato late blight. *Ann. Appl. Biol.* 77:259–70
12. Chambers SC, Flentje NT. 1967. Studies on oat-attacking and wheat-attacking isolates of *Ophiobolus graminis* in Australia. *Aust. J. Biol. Sci.* 20:927–40
13. Clifford BC, Clothier RB. 1974. Physiologic specialization of *Puccinia hordei* on barley hosts with nonhypersensitive resistance. *Trans. Br. Mycol. Soc.* 63:421–30

13a. Cohan FM, King EC, Zawadzki P. 1994. Amelioration of the deleterious pleiotrophic effectc of an adaptive mutation in *Bacillus subtilis*. *Evolution* 48:81–95

14. Cooley RN, Caten CE. 1991. Variation in electrophoretic karyotype between strains of *Septoria nodorum*. *Mol. Gen. Genet.* 228:17–23
15. Damman KE Jr. 1987. Where is the specificity in gene-for-gene systems? *Phytopathology* 77:55–56
16. Davis JM. 1987. Modeling the long-range transport of plant pathogens in the atmosphere. *Annu. Rev. Phytopathol.* 25:169–88
17. DiLeone JA, Mundt CC. 1994. Effect of wheat cultivar mixtures on populations of *Puccinia striiformis* races. *Plant Pathol.* 43:917–30
18. Domingo E, Martinez-Salas E, de la Torre F, Portela A, Ortin J., et al. 1985. The quasispecies (extremely heterogeneous) nature of viral RNA genome populations: biological relevance—a review. *Gene* 40:1–8
19. Ellingboe AH. 1976. Genetics of host-parasite interactions. In *Physiological Plant Pathology*, ed. R Heitefuss, PH

Williams, pp. 761–78. Berlin: Springer-Verlag
20. Ellingboe AH. 1981. Changing concepts in host-pathogen interactions. *Annu. Rev. Phytopathol.* 19:125–43
21. Fitt BDL, Gregory PH, Todd AD, McCartney HA, MacDonald OC. 1987. Spore dispersal and plant disease gradients: a comparison between two empirical models. *J. Phytopathol.* 118: 227–42
22. Fitt BDL, McCartney HA. 1986. Spore dispersal in relation to epidemic models. See Ref. 52a, 1:311–45
23. Fitt BDL, McCartney HA, Walklate PJ. 1989. The role of rain in dispersal of pathogen inoculum. *Annu. Rev. Phytopathol.* 27:241–70
24. Flor HH. 1960. The inheritance of X-ray-induced mutations to virulence in a uredeospore culture of race 1 of *Melampsora lini*. *Phytopathology* 50: 603–5
25. Flor, HH. 1971. The current status of the gene-for-gene concept. *Annu. Rev. Phytopathol.* 9:275–96
26. Franc GD, Harrison MD, Powelson ML. 1985. The dispersal of phytopathogenic bacteria. In *The Movement and Dispersal of Agriculturally Important Biotic Agents*, ed. DR MacKenzie, CS Barfield, GC Kennedy, RD Berger, pp. 37–48. Baton Rouge, LA: Claitor's
27. Fraser RSS. 1990. The genetics of resistance to plant viruses. *Annu. Rev. Phytopathol.* 28:179–200
28. Freeman S, Rodriguez RJ. 1992. A rapid, reliable bioassay for pathogenicity of *Colletotrichum magna* on cucurbits and its use in screening for nonpathogenic mutants. *Plant Dis.* 76: 901–5
29. Freeman S, Rodriguez RJ. 1993. Genetic conversion of a fungal plant pathogen to a nonpathogenic, endophytic mutualist. *Science* 260:75–78
30. Fry WE. 1982. *Principles of Plant Disease Management*. New York: Academic
31. Fry WE, Drenth A, Tooley PW, Sujkowski LS, Koh YJ, et al. 1993. Historical and recent migrations of *Phytophthora infestans*: chronology, pathways, and implications. *Plant Dis.* 77: 653–61
32. Gabriel DW, Burgess A, Lazo CR. 1986. Gene-for-gene interactions of five cloned avirulence genes from *Xanthomonas campestris* pv. *malvacearum* with specific resistance genes in cotton. *Proc. Natl. Acad. Sci. USA* 83:6415–19
33. Gregory PH. 1968. Interpreting plant disease dispersal gradients. *Annu. Rev. Phytopathol.* 6:189–212
34. Gregory PH. 1973. *The Microbiology of the Atmosphere*. London: Leonard Hill. 2nd ed.
35. Hollins TW, Scott PR. 1990. Pathogenicity of *Gaeumannomyces graminis* isolates to wheat and rye seedlings. *Plant Pathol.* 39:269–73
36. Jefferey SIB, Jinks JL, Grindle M. 1962. Intraracial variation in *Phytophthora infestans* and field resistance to potato late blight. *Genetica* 32:323–28
37. Jenns AE, Leonard KJ, Moll RH. 1982. Variation in the expression of specificity in two maize diseases. *Euphytica* 31: 269–79
38. Jin Y, Steffenson BJ. 1992. *Puccinia coronata* on barley. *Plant Dis.* 76:1283
39. Jinks JL, Grindle M. 1963. Changes induced by training in *Phytophthora infestans*. *Heredity* 18:245–64
40. Kardin MK, Groth JV. 1989. Density-dependent fitness interactions in the bean rust fungus. *Phytopathology* 79: 409–12
41. Katsuya K, Green GJ. 1967. Reproductive potentials of races 15B and 56 of wheat stem rust. *Can. J. Bot.* 45:1077–91
42. Keller, NP, Bergstrom GC, Yoder OC. 1990. Effects of genetic transformation on fitness of *Cochliobolus heterostrophus*. *Phytopathology* 80:1166–73
43. Kilpatrik RA. 1975. *New Wheat Cultivars and Longevity of Rust Resistance, 1971–1975*. USDA, ARS-NE64. 20 pp.
44. Kiyosawa S, Shiyomi M. 1972. A theoretical evaluation of the effect of mixing resistant variety with susceptible variety for controlling plant diseases. *Ann. Phytopathol. Soc. Jpn.* 38:41–51
45. Klittich CJR, Bronson CR. 1986. Reduced fitness associated with *TOX1* of *Cochliobolus heterostrophus*. *Phytopathology* 76:1294–98
46. Kluepfel DA. 1993. The behavior and tracking of bacteria in the rhizosphere. *Annu. Rev. Phytopathol.* 31:441–72
47. Knudsen GR. 1989. Model to predict aerial dispersal of bacteria during environmental release. *Appl. Environ. Microbiol.* 55:2641–47
48. Kulkarni RN, Chopra VL. 1982. Environment as the cause of differential interaction between host cultivars and pathogenic races. *Phytopathology* 72: 1384–86
49. Latin RX, MacKenzie DR, Cole H Jr. 1981. The influence of host and pathogen genotypes on the apparent infection rates of potato late blight epidemics. *Phytopathology* 71:82–85

50. Leonard KJ. 1969. Selection in heterogeneous populations of *Puccinia graminis* f.sp. *avenae*. *Phytopathology* 59:1851–57
51. Leonard KJ. 1977. Selection pressures and plant pathogens. *Ann. NY Acad. Sci.* 287:207–22
52. Leonard KJ. 1977. Virulence, temperature optima, and competitive abilities of isolines of races T and O of *Bipolaris maydis*. *Phytopathology* 67:1273–79
52a. Leonard KJ, Fry WE, eds. 1989. *Plant Disease Epidemiology*. New York: McGraw-Hill, Vols. 1, 2
53. Leung H, Borromeo ES, Bernardo MA, Notteghem JL. 1988. Genetic analysis of virulence in the rice blast fungus. *Phytopathology* 78:1227–33
54. Lindemann J, Arny DC, Upper CD. 1984. Epiphytic populations of *Pseudomonas syringae* pv. *syringae* on snapbeans and nonhost plants and incidence of bacterial brown spot in relation to cropping pattern. *Phytopathology* 74: 1329–33
55. Lindemann J, Suslow TV. 1987. Competition between ice nucleation-active wild type and ice nucleation-deficient deletion mutant strains of *Pseudomonas syringae* and *P. fluorescens* biovar I and biological control of frost injury on strawberry blossoms. *Phytopathology* 77:882–86
56. Lindow SE, Andersen G, Beattie GA. 1993. Characteristics of insertional mutants of *Pseudomonas syringae* with reduced epiphytic fitness. *Appl. Environ. Microbiol.* 59:1593–601
57. Lindow SE, Knudsen GR, Seidler RJ, Walter MV, Lambou VW, et al. 1988. Aerial dispersal and epiphytic survival of *Pseudomonas syringae* during a pretest for the release of genetically engineered strains into the environment. *Appl. Environ. Microbiol.* 54:1557–63
58. Mackill AO, Bonman JM. 1986. New hosts of *Pyricularia oryzae*. *Plant Dis.* 70:125–27
59. Madden LV. 1992. Rainfall and the dispersal of fungal spores. *Adv. Plant Pathol.* 8:39–79
60. Madhosingh C. 1992. Interspecific hybrids between *Fusarium lycopersici* and *Fusarium graminearum* by mycelial anastomoses. *J. Phytopathol.* 136:113–23
61. Martens JW. 1973. Competitive ability of oat stem rust races in mixtures. *Can. J. Bot.* 51:2233–36
62. McCartney HA, Bainbridge A. 1984. Deposition gradients near to a point source in a barley crop. *Phytopathol. Z.* 109:219–36
63. McCartney HA, Fitt BDL. 1985. Construction of dispersal models. *Adv. Plant Pathol.* 3:107–43
64. McDonald BA, Martinez JP. 1990. DNA restriction fragment length polymorphisms among *Mycosphaerella graminicola* (anamorph *Septoria tritici*) isolates collected from a single wheat field. *Phytopathology* 80:1368–73
65. McDonald BA, McDermott JM, Allard RW, Webster RK. 1989. Coevolution of host and pathogen populations in the *Hordeum vulgare-Rynchosporium secalis* pathosystem. *Proc. Natl. Acad. Sci. USA* 86:3924–27
66. Miao VP, Covert SF, Van Etten HD. 1991. A fungal gene for antibiotic resistance on a dispensable ("B") chromosome. *Science* 254:1773–76
67. Mills D, McCluskey K. 1990. Electrophoretic karyotypes of fungi: the new cytology. *Mol. Plant-Microbe Interact.* 3:351–57
68. Minogue KP, Fry WE. 1981. Effect of temperature, relative humidity, and rehydration rate on germination of dried sporangia of *Phytophthora infestans*. *Phytopathology* 71:1181–84
69. Mundt CC. 1989. Use of the modified Gregory model to describe primary disease gradients of wheat leaf rust away from area sources of inoculum. *Phytopathology* 79:241–46
70. Mundt CC. 1991. Probability of mutation to multiple virulence and durability of resistance gene pyramids: further comments. *Phytopathology* 21:240–42
71. Mundt CC, Brophy LS. 1988. Influence of number of host genotype units on the effectiveness of host mixtures for disease control: a modeling approach. *Phytopathology* 78:1087–94
72. Mundt CC, Leonard KJ. 1985. A modification of Gregory's model for describing plant disease gradients. *Phytopathology* 75:930–35
73. Nagarajan S, Singh DV. 1990. Long distance dispersion of rust pathogens. *Annu. Rev. Phytopathol.* 28:73–92
74. Nelson RR. 1960. Evolution of sexuality and pathogenicity. I. Interspecific crosses in the genus *Helminthosporium*. *Phytopathology* 50:375–77
75. Nelson RR. 1961. Evidence of gene pools for pathogenicity in species of *Helminthosporium*. *Phytopathology* 51: 736–37
76. Nelson RR. 1973. Pathogenic variation and host resistance. In *Breeding Plants for Disease Resistance*, ed. RR Nelson, pp. 40–48. University Park, PA: Pennsylvania State Univ. Press
77. Nelson RR, Ayers JE, Cole H, Peterson

DH. 1970. Studies and observations on the past occurrence and geographical distribution of isolates of race T of *Helminthosporium maydis*. *Plant Dis. Rptr.* 54:1123–26
78. O'Dell M, Flavell RB. 1992. The classification of isolates of *Gaeumannomyces graminis* from wheat, rye and oats using restriction fragment length polymorphisms in families of repeated DNA sequences. *Plant Pathol.* 41:554–62
79. Oeser B, Yoder OC. 1994. Pathogenesis by *Cochliobolus heterostrophus* transformants expressing a cutinase-encoding gene from *Nectria haematococca*. *Mol. Plant-Microbe Interact.* 7:282–88
80. Ordon F, Huth W, Friedt W. 1992. Mechanical transmission of barley mild mosaic virus (Ba MMV) to rye (*Secale cereale* L.). *J. Phytopathol.* 135:84–87
81. Panopoulos NJ, Peet RC. 1985. The molecular genetics of plant pathogenic bacteria and their plasmids. *Annu. Rev. Phytopathol.* 23:381–419
82. Parlevliet JE. 1981. Disease resistance in plants and its consequences for plant breeding. In *Plant Breeding*. II. ed. KJ Frey, pp. 309–63. Ames, IA: Iowa State Univ. Press
83. Parlevliet JE. 1981. Stabilizing selection in crop pathosystems: an empty concept or a reality? *Euphytica* 30:259–69
84. Parlevliet JE, Zadoks JC. 1977. The integrated concept of disease resistance; a new view including horizontal and vertical resistance in plants. *Euphytica* 26:5–21
85. Pedgley DE. 1986. Long distance transport of spores. See Ref. 52a, 1: 346–65
86. Schafer JF, Roelfs AP. 1985. Estimated relation between numbers of urediniospores of *Puccinia graminis* f.sp. *tritici* and rates of occurrence of virulence. *Phytopathology* 75:749–50
87. Schäfer W, Stanley D, Ciuffetti L, Van Etten HD, Yoder OC. 1989. One enzyme makes a fungal pathogen, but not a saprophyte, virulent on a new host. *Science* 246:247–49
88. Schäfer W, Yoder OC. 1994. Organ specificity of fungal pathogens on host and non-host plants. *Physiol. Mol. Plant Pathol.* 45:211–18
89. Schrodter H. 1960. Dispersal by air and water—the flight and landing. In *Plant Pathology: An Advanced Treatise*, ed. JG Horsfall, EB Cowling, 3:169–227. New York: Academic
90. Sierotzki H, Eggenschwiler M, Boillat O, McDermott JM, Gessler C. 1994. Detection of variation in virulence toward susceptible apple cultivars in natural populations of *Venturia inaequalis*. *Phytopathology* 84:1005–9
91. Simons MD. 1970. *Crown Rust of Oats and Grasses*. Monogr. No. 5, APS. Worcester, MA: Heffernan
92. Stakman EC, Hamilton, LM. 1939. Stem rust in 1938. *Plant Dis. Reptr. Suppl.* 117:69–83
93. Stakman EC, Harrar JG. 1957. *Principles of Plant Pathology*. New York: Ronald
93a. Stone R. 1994. Large plots are next test for transgenic crop safety. *Science* 266: 1472–73
94. Talbot NJ, Oliver RP, Coddington A. 1991. Pulsed field gel electrophoresis reveals chromosome length differences between strains of *Cladosporium fulvum* (syn. *Fulvia fulva*). *Mol. Gen. Genet.* 229:267–72
95. Tzeng TH, Lynjolm LK, Ford CF, Bronson CR. 1992. A restriction fragment length polymorphism map and electrophoretic karyotype of the fungal maize pathogen *Cochliobolus heterostrophus*. *Genetics* 130:81–96
96. Trevathan LE, Moss MA, Blasingame D. 1994. Ryegrass blast. *Plant Dis.* 78: 113–17
97. Turner EM. 1940. *Ophiobolus graminis* Sacc. var. *avenae* var. N., as the cause of take all or whiteheads of oats in Wales. *Trans. Br. Mycol. Soc.* 24:269–81
98. Turner EM. 1961. An enzymic basis for pathogen specificity in *Ophiobolus graminis*. *J. Exp. Bot.* 12:169–75
99. Ullstrup AJ. 1972. The impacts of the Southern corn leaf blight epidemics of 1970–71. *Annu. Rev. Phytopathol.* 10: 37–50
100. Urashima AS, Igarashi S, Kato H. 1993. Host range, mating type, and fertility of *Pyricularia grisea* from wheat in Brazil. *Plant Dis.* 77:1211–16
101. Valent B. 1990. Rice blast as a model system for plant pathology. *Phytopathology* 80:33–36
102. Valent B, Farrall L, Chumley FG. 1991. *Magnaporthe grisea* genes for pathogenicity and virulence identified through a series of backcrosses. *Genetics* 127:87–101
103. Van Arsdel EP. 1967. The nocturnal diffusion and transport of spores. *Phytopathology* 57:1221–29
104. Vanderplank JE. 1968. *Disease Resistance in Plants*. New York: Academic
105. Vanderplank JE. 1975. *Principles of Plant Infection*. New York: Academic
106. Van Etten HD, Matthews DE, Matthews PS. 1989. Phytoalexin detoxification: importance for pathogenicity and prac-

tical implications. *Phytopathology* 27: 143–64
107. Walton JD, Panaccione DG. 1993. Host selective toxins and disease specificity: perspectives and progress. *Annu. Rev. Phytopathol.* 31:275–303
107a. Wilson M, Lindow SE. 1992. Relationship of total viable and culturable cells in epiphytic populations of *Pseudomonas syringae*. *Appl. Environ. Microbiol.* 58:3908–13
108. Wilson M, Lindow SE. 1993. Release of recombinant microorganisms. *Annu. Rev. Microbiol.* 47:913–44

Annu. Rev. Phytopathol. 1995. 15:489–527

REMOTE SENSING AND IMAGE ANALYSIS IN PLANT PATHOLOGY

Hans-Eric Nilsson
Laboratory for Phytopathometry, Department of Plant Pathology, Swedish University of Agricultural Sciences, P.O. Box 7044, S-75007 Uppsala, Sweden

KEY WORDS: phytopathometry, visual assessment, infrared photography, radiometry, videography, IR-thermography, NMR, radar, laser fluorescence

---

ABSTRACT

This paper reviews various applications of remote sensing and image analysis in plant pathology. It describes technical methods and their possibilities, but also emphasizes the biological prerequisites and restrictions of practical applications. The subject area comprises many nondestructive and noninvasive methods to detect and assess plant diseases and stress objectively and cost-effectively, even though to date we cannot identify the specific cause of the damage. As a supplement to conventional methods, these new methods have great potential to facilitate and increase accuracy and precision in plant pathological research.

INTRODUCTION

Remote sensing and digital image analysis are methods of acquisition and interpretation of measurements of an object without physical contact between the measuring device and the object. The object can be analyzed many times noninvasively and without damage. The specific properties of the vegetation, healthy or diseased, influence the amount and quality of radiation reflected or emitted from the leaves and canopies; application of this technology to phytopathological research such as phytopathometry is of great interest.

A plant becomes stressed when any biotic or abiotic factor adversely affects growth and development. Stress can be acute or chronic, and can accelerate

0066-4286/95/0901-0489$05.00

many changes that resemble the senescence syndrome (264). Stress or disease is expressed in many ways. Problems in water supply or control of tissue water balance close stomata and impede photosynthesis, reduce evapotranspiration, and increase leaf surface temperature. Other symptoms include morphological changes such as leaf curling, change in leaf angle, wilting or stunting, and chlorosis, necrosis, or premature abscission of plant parts. Detection and rapid accurate quantification of early symptoms are often difficult. Remote sensing is a means of detecting and assessing changes in plants and canopies. However, there are many kinds of stress, numerous infectious diseases and physiological problems, as well as injuries caused by insects and/or nematodes. Furthermore, damage often results in thinner crop plants but more bare soil or weed infestations.

Notwithstanding the title of this paper, I do not restrict references only to applications of remote sensing methods in plant pathology and phytopathometry. Successful use of remote sensing is dependent on the accurate study of healthy and diseased plants and canopies with conventional methods. To evaluate influencing factors, we must compare remote sensing data from healthy plants with data from plants under stress. General aspects of remote sensing and conventional assessment of vegetation are both discussed.

VISUAL DISEASE ASSESSMENTS

Our visual system, the human eye and brain, is a good example of remote sensing; we acquire, analyze, enhance, and store images in enormous quantities and at high speed. A trained person can select from the mass of information that is available visually and pick out specific disease symptoms. By contrast, most remote sensing instruments integrate all the recorded data, which then must be computerized, reduced, and enhanced before it can be analyzed and evaluated.

There are various methods of making visual disease assessments. Some are based on the Weber-Fechner law: Visual acuity is proportional to the logarithm of the intensity of the stimulus (145). This law has large steps in the scale around 50% infected area and decreasing steps at both ends of the scale. Other methods are based on linear relationships between stimuli and perception, and have a constant stepwise increase from 0 to 100% visual damage (54, 140, 146). Many researchers have used provisional scales adapted to each actual project (146). The provisional scale may be recalculated to a percentage severity value called *disease index* (247), which is distinct from *frequency*.

Even though human vision is unique, there are individual differences in light or color perception and thus in the accuracy of estimating color, shape, size, and pattern of factors like disease symptoms. Accommodation and spatial contrast sensitivity are fundamental to a well-functioning vision (313). Unsuitable illumination in color and intensity, as well as the background color, may all influence visual assessments. Tiredness and lack of concentration also reduce precision. Visual interference is also important; for example, after having assessed heavily infected plots it is easy to underestimate slight infections, or after having assessed slight infections it is easy to overestimate areas that are heavily infected. Sherwood et al (320), in a test with two groups of five experienced scorers, found that overestimation was greatest when the infected area was smallest, often being two to three times the actual area. When two leaves had equal total spotted areas, the leaf with smaller spots was usually scored higher. Regression analysis showed that overestimation was inversely proportional to the natural logarithm of the disease area for all scorers and also directly proportional to the number of spots for five scorers. Substantial variation between raters at visual disease assessment has also been demonstrated by Nutter & Schultz (278), Shokes et al (322), Weber & Jörg (361, and personal information) and several of my own experiments (unpublished). A review of accuracy, reliability and illusions in disease assessment is given by Campbell & Madden (54).

One study comprised 223 color photographs of various-sized fungal infection-spots from three golf greens in 1994 (HE Nilsson, unpublished data). The experiment was originally devised by B Gerhardson to evaluate biological control of fungal infections through the use of bacterial isolates. Specific fungal spots were documented by color photography monthly from May to October, once using a digital camera (Kodak DCS420) as well as a hand-held Cropscan radiometer. The pictures were first graded by 11 plant pathologists in relation to the amount of infected area and then evaluated by multispectral image analysis. There were significant differences in assessment accuracy between the raters, in the intercept, slope, and variation. One rater with some degree of red/green color blindness overestimated slight disease infections and substantially underestimated heavy infections in comparison with the average of his colleagues. A twelfth person, who suffered from more severe red/green color blindness had, as expected, great difficulties in assessing the disease severity.

In many diseases caused both by infections (e.g. some viral infections) and physiological stresses (e.g. damage from air pollution), the symptoms are not advanced enough to be visual at the time of assessment. Fredericksen & Skelly (107) report that net photosynthetic rate decreased up to 14% before visible symptoms of ozone injury became evident on *Prunus serotina* Ehrh. and *Lirodendron tulipifera* L. and yellow poplar.

Various organs are of different relative importance for plant growth. In studies of leaf rust of wheat, Seck et al (318) estimated that the flag leaf and the two leaves immediately below contribute 26, 12, and 3% to the final grain weight, respectively. The linear regression of yield on average disease severity per tiller was significant, with a slope of 0.47 and an R^2 = 3D 0.90. When the estimated leaf rust severity of each leaf was weighted by the relative contribution of that leaf to obtain an effective severity (ES) per tiller, the regression between the ES and yield loss was highly significant, with $R^2 = 0.96$ and slope = 1.0.

Visual assessment of plant diseases may be facilitated by various disease assessment methods, published scales (1, 101, 140, 145, 170, 247, 278, 280, 328, 361), and computer-aided training programs such as DISEASE PRO and ALFALFA.PRO (269, 276, 279), DISTRAIN (345), and ESTIMATE (Gerhard Weber Software, Eutin, Germany; personal information).

Training by such computer programs can be very cost effective. Weber & Jörg (361; G Weber, personal information) documented substantial variation between trained and untrained raters in accuracy and repeatability in assessing cereal powdery mildew. After training with ESTIMATE, the accuracy of 10 raters increased, on average, from $R^2 = 0.80$ to $R^2 = 0.92$. However, even with training, variation amon g raters remains an important source of error in visual plant disease assessment (146, 320, 322). Remote sensing technology offers a more objective and consistent means of assessing crop loss and may be a valuable supplement to conventional methods (see 7, 272, 273, 321; HE Nilsson, unpublished). [For discussions on methods of sampling test plants and areas to ensure accurate representation of the intended plant population, the reader is referred to (9, 54, 69, 100, 117–119, 146, 189, 210–213, 336, 337, 360).]

Side effects like those caused by cultivar-specific properties, energy-consuming resistance reactions, simple or multiple infections, synergistic and competitive pathogens, and chemical control agents are all important considerations, even in samplings to compare visual assessment and remote sensing data.

INSTRUMENTAL REMOTE SENSING

The term remote sensing is usually restricted to instruments that measure electromagnetic radiation reflected or emitted from an object. The instruments record radiation in various parts of the electromagnetic spectrum, ultraviolet (UV, 10–390 nm), visible (ca 390–770 nm), near infrared (NIR, 770–1300 nm), mid-infrared (MIR, 1300–2500 nm), thermal infrared (2.5–15 μm), microwaves, etc (173). There are some differences in the literature on these wave ranges, and IR is often used instead of NIR. The human eye records three visual spectral ranges, red (R), green (G), and blue (B), but sensitivity to red over 650–700 nm is slight. A radiometer can record important details also in NIR and longer wavelengths.

Techniques for recording information in noncontact sensing include: cameras with films and filters in differing combinations; specialized electronic instruments like radiometers, video systems, sonar instruments, radar instruments, etc; use of various platforms, heights, and distances; satellites, space shuttles, high-altitude and other aircraft, unmanned powered platforms, balloons, helicopters, radiopiloted model aircraft; ground-based truck-mounted hydraulic arms or other platforms and hand-held systems; and close range photogrammetry and macroscopy, fiber optic macro- and microscopes.

The amount of reflected light (radiance) as a percentage of incoming light (irradiance) is usually called the reflectance factor. If the radiance from a healthy leaf is measured by a suitable radiometer, it is possible to detect a slight reflectance in blue (450–480 nm) and red (600–700 nm), a little more in green (500–550 nm), and substantially more in NIR at 750–1100 nm. The slight reflectance in the visual range is a result of the intensity of light absorption by various plant pigments such as chlorophyll and xanthophyll. Any physiological stress, disease, or reduced amount of photosynthetic pigments causes an increase in red and blue reflectance, and often also affects the yellow region. Moreover, NIR reflectance often decreases substantially. Changes in spectral signatures due to deficiencies of nutrients and damage by pathogens, pests and environmental factors have been widely reported (e.g. 113, 184, 185, 281).

It is sometimes sufficient to analyze only the relative decrease in NIR reflectance (270). However, better information can be obtained by combining data from various spectral ranges, such as IR/R or (IR-R)/(IR+R). The IR/R ratio is often closely correlated to the leaf area index (LAI), whereas the latter (the normalized difference vegetation index or NDvi) is often closely correlated to green biomass. Changes in these ratios may be a relative estimate of stress when data from different plots or parts of a field are compared—even if the specific reason of the stress cannot be identified.

Many other formulae or vegetation indices have been developed to reduce multispectral data to a single number to assess characteristics such as leaf area, biomass, and stress. Some researchers calculate NDvi specifically for certain IR-bands, whereas others use different spectral band combinations and names to distinguish between quantities or conditions. Specific examples include transformed vegetation index (Tvi), perpendicular vegetation index (Pvi), soil-adjusted vegetation index (SAvi), transformed soil-adjusted vegetation index (TSAvi), MSAvi (modified SAvi), weighted NDvi (WDvi), brightness (BN), yellowness (YN), greenness (GN), green vegetation index (Gvi), normalized pigment chlorophyll index (NPCi; using 680 and 430 nm), and physiological reflectance index (PRi). PRi is defined by Penuelas et al (294) and Gamon et al (108) as NDvi using 550 and 530 nm. Adjustments for both atmospheric effects and soil reflectance have resulted in SARvi (242). Further adjustments have been suggested by Gu & Guyot (129).

Gamon et al (108) made diurnal narrow-band spectral reflectance measurements of sunflower leaves. The PRi correlated well with the epoxidation state (EPS) of the xanthophyll cycle pigments and with the efficiency of photosynthesis in control and nitrogen stress canopies, but not in water stress canopies undergoing midday wilting. Both the control and N-stress canopies had midday declines in visible reflectance, and the reflectance change in the green (near 550 nm) was especially noticeable in the N-stress canopy. All three canopies had a substantial decline in NIR from 8:00 A.M. to noon. In the water-stress canopy there was little diurnal change in green, but a slight increase in blue and a large increase in red reflectance. PRi calculated on 550 and 531 nm gave the best correlation to EPS (calculated from molar concentrations of the three xanthophyll cycle pigments violaxanthin, antheraxanthin, and zeaxanthin) for N-stress, whereas the combination of 570 and 531nm was best for water stress. PRi is assumed to improve conventional remote sensing using the (NIR-R)/(NIR+R) ratio.

Additional information on the use of spectral ratios (mainly for measurements of biomass production) can be found in the following references: wheat (5, 12, 165, 169, 175); wheat cultivars (304); soybean (17, 148); maize (17, 175, 370); sunflower (108, 294); sorghum (311, 312); cotton (370); agave, sweet gum (148); *Amaranthus* sp. (red pigment) (77); grass biomass (222); grassland (347); trees and forests (67, 70, 148); NPCi (294); MSAvi (66); factor effects (124); other crops and diseases or general aspects (11, 17, 67, 72–74, 160, 252–254, 256, 257, 259–263).

Todd et al (343) found that NDvi, SAvi, and Gvi performed much better on grazed rangelands where the senescent vegetation component is minimized as compared with ungrazed sites, and that red reflectance values (TM3 of Landsat 5) are more useful in discriminating between senescent vegetation and high-reflecting soil backgrounds. However, for low-reflecting soils the other indices should be superior. Bausch (24) and Qi et al (308) suggested a modified SAvi (MSAvi) where a soil-factor (L-factor) supplemented the ordinary NDvi for better sensitivity to soil cover in studies of corn and cotton canopies.

Artan & Neale (10), Everitt et al (94), and Wiegand et al (366, 369) demonstrated good agreement when comparing IR/R and various vegetation indices and biophysical parameters of alfalfa, grasses, and wheat and corn, respectively, recorded via a multispectral video and radiometer systems.

In recent years, research has provided ever more evidence of the usefulness of vegetation indices involving other wavebands such as mid-infrared together with NIR and R for studies of canopy water status (18, 57, 79, 86, 109, 180, 348). Malthus et al (214) indicate that the most promising NIR reflectances were beyond the range 760–790 nm in studies of sugar beet canopies infected by beet yellows virus.

The availabilty of more advanced narrowband radiometers has spurred interest in canopy spectral properties that are also in the mid-infrared region. This region is greatly influenced by biochemical properties of the leaves, such as the contents of C, H, N, O, starch, cellulose, lignin, and water (2, 56, 75, 78, 294, 296–298, 362–364, 376).

Perry & Lautenschlager (295) reviewed the history and formulae of some four dozen vegetation indices. Many of these indices are calculated using data from older Landsats and radiometers; the formulae may need to be modified to accord with the properties of newer sensors, as well with the spectral properties, growth architecture, and growing conditions specific to newer cultivars. Such modifications should be based on well-defined plant material (e.g. data from healthy canopies supplemented by data from canopies infected to various disease levels by specific pathogens), and should be professionally assessed. Plant breeders should supply isogenic lines of various crop plant species for appropriate field experiments to be conducted on carefully selected sites. The test canopy could be designed via fertilization or irrigation, for example, for optimal variation as a test material.

An important aspect on radiometry of crop canopies is the variation in solar angle during the day and its influence on canopy reflectance that may be related to plant row orientation (166, 305). Bégué (26) in particular stressed the effect of sun angle on spectral vegetation indices for various homogenicities of the canopy. Lord et al (208) reported that sun angle affected the red more than the NIR reflectance of four crops at various row directions and distances (wheat, NNW-SSE rows 18–35 cm apart; barley, NNW-SSE rows 18 cm apart; corn, NNW-SSE rows 46 cm apart; and sunflower, N-S rows 92 cm apart). Such an effect is also related to growth stage, plant size, canopy architecture, and soil coverage. Note that a 12-cm row distance is common in cereals in Sweden, whereas 15 cm or more is common elsewhere. A mere 3-cm difference in row space can significantly affect the relationship between recorded soil and canopy reflectance and thus the reflectance of R and NIR, and must consequently be considered at hand-held radiometry.

Changes in diurnal spectral properties are also caused by variations in moisture stress (161, 305, 323; HE Nilsson, unpublished), and can be substantial. Tolerance to water stress of individual cultivars of many crop species varies, as does the influence on stress of insufficient soil water, the effects of fertilizers and diseases, or other growing conditions.

Dewfall on leaves influences the spectral reflectance (303). Pinter (302) demonstrated a significant correlation between dew density and the ratios of NIR (760–900 nm) and yellowish green (520–600 nm) as well as between MIR (2080–2350 nm) and red (630–690 nm). Hunt (148) demonstrated a significant correlation between leaf-water thickness and MIR/NIR, and Downing et al (84) found correlations within bands in the range 1400 to 2500 nm.

Dewfall, dew density, and duration are also important from the plant pathologist's point-of-view (see section below on retro-reflection photography).

The spectral signature is also influenced by the amount of pigments, leaf angle, leaf surface texture, diseases and stress, plant growth stage, and growing and measurement conditions (162). Alteration of leaf angles and circadian leaf movements in seedling sunflowers infected by *Pseudomonas tagetes* were reported by Kennedy (177), and cultivar differences in the way barley spikes turned towards the sun were observed in my field experiments in Sweden. Usually, soil nitrogen and water strongly influence canopy spectral signatures (164, 262, 263). These factors as well as varietal differences in plant growth and development may mask the effects of minor disease infections. Erectophile or planophile cereal varieties may reveal differences in spectral signatures (165, 304, 311; HE Nilsson, unpublished data). Pubescence and awns may influence reflectance (98, 110, 111, 114, 237, 262, 263). A recent review of stereophotogrammetry of plants with varying leaf angles was made by Herbert (139).

One of the field experiments carried out in Sweden in 1992 by me and a student, Anneli Carlsson, demonstrated an environmental effect of leaf spectral properties. It comprised two potato cultivars differently resistant to late blight disease (*Phytophthora infestans* (Mont.) de Bary) and had been sprayed at various times for control of late blight. However, due to the local climate conditions, there were no visual symptoms of late blight or visual physiological stress on any of the cultivars at the Cropscan radiometry on 12 August, but the spectral signatures of the plots were substantially influenced by effects from the fungicide sprayed onto the leaves two days earlier but only slightly on those sprayed two weeks before (e.g. 30% decrease in IR/R and 35% in IR/Blue of the first group).

The magnitude of the irradiance reflected as polarized light depends on the geometric arrangement of the canopy and the angle of the incident light on the leaves, as well as on the characteristics of the leaf surface. Important factors are the optical index of refraction and the surface roughness characteristics of the cuticle (354). Leaf cuticle properties vary with species and cultivars, developmental state, environmental conditions, and stress (22, 23, 125, 126, 351–355). Polarization phenomena of leaves can be interesting in, for example, studies of how virus infection influences leaf surface properties and they can often be seen via a polarizing filter.

Remote sensing of vegetation by satellites typically uses broad spectral bands of the order of 100 nm width. This spectral resolution is adequate for parts of the spectrum where reflectance only shows gradual change with wavelength and no fine structure is evident. However, vegetation shows sharp reflectance changes in some spectral ranges, notably the so-called "red edge" or the sharp change in leaf reflectance between 680 and 750 nm due to canopy properties. This change has attracted increased attention to narrow-band multispectral radiometry, notably in examining the first or second derivative of a narrow band reflectance spectral curve (50, 52, 83, 144, 192, 194, 220, 294, 309). The red edge is a unique feature of green vegetation because it results from two special optical properties of plant tissue, high internal leaf scattering causing large near infrared reflectance and chlorophyll absorption giving low red reflectance. Horler et al (144) pointed out that red edge measurements are valuable for assessments of vegetative chlorophyll status and leaf area index independently of ground cover variations, and thus are particularly suitable for early stress detection. However, the red edge is a complex contour composed of a number of components and varying according to plant growth stage (309).

PHOTOGRAPHY, AERIAL PHOTOGRAPHY, AND PHOTOGRAMMETRY

Aerial photography was first used in the 1920s (245, 246, 335) to survey infection by *Phymatotrichum omnivorum* (Shear) Duggar in Texas, and in the early 1930s infrared plate-films were used in studies of virus diseases of potatoes and tobacco (25). Aerial photography using both conventional and infrared films was used during the second world war to detect camouflage, i.e. plant material that had been cut and showed symptoms of water stress and wilting. This experience was put to good use after the war in studies of vegetation stresses and diseases in agriculture and forestry. Colwell (71) demonstrated the potential of aerial photography using panchromatic and infrared films to detect and quantify crop diseases such as cereal rusts and virus diseases of citrus. Other important pioneer work in aerial photography is reported by (28, 30, 31, 44–47, 132a, 141, 143, 153–158, 215, 229, 300–301, 316, 344, 358, 359).

The success of the southern corn blight watch project in the USA (*Helminthosporium maydis*) demonstrated the efficacy of large-scale application of aerial IR-photography to crop disease surveillance, and provided support for future remote sensing in agriculture (23, 240).

Aerial infrared photography has been widely used in surveys of infectious and noninfectious crop diseases in many countries. Clark et al (68) used aerial IR-photography to estimate damage by diseases such as spot blotch of barley, crown rust and barley yellow dwarf virus of oats (BYDV), and powdery mildew of wheat in field plot experiments. Blazquez & Edwards (35) used IR-color photography and spectral reflectance for studies of tomato and potato diseases. Blazquez & Edwards (36) made densitometric studies of color and color infrared (CIR) photographs of diseased cucumber leaves, and Blazquez (33) made similar studies of CIR photographs of a diseased citrus grove. Blazquez et al (41) and Greaves et al (127) studied BYDV and cereal aphid infestation in winter wheat by aerial photography. Kinder (181) used 70 and 35 mm aerial CIR photography of commercial cranberry (*Vaccinium* spp) cultivations affected by various stresses, diseases, and pests. Aerial photography has been applied to a greater extent in forestry than in agriculture, perhaps because there are fewer alternative assessment methods available in forestry.

Aerial photography is a valuable tool in selecting the most suitable area for field plot experiments. Stereo-photography has a potential for photogrammetric measurements of the size and architecture of healthy and stressed plants. Also, the use of retro-reflection photography for nondestructive assessment of dew on the leaf surface (217, 218; H. E. Nilsson, unpublished) has an apparent application in phytopathology.

Early films were mainly analyzed using densitometry, but more advanced image processing and spectral analysis, including enhancements to facilitate evaluation, have become popular.

In the mid-1950s I embarked on studies of take-all and herbicide damage in cereals using 35 mm B&W (black and white) infrared films, and from the 1970s used Kodak Ektachrome infrared film 2443 in Sweden (e.g. 253; unpublished data) and in Tanzania for studies of copper deficiency, water stress, and various infectious diseases of wheat. Together with a yellow filter I used a polarizing filter that improved the stress discrimination substantially. The B&W "Kodak High Speed Infrared Film 2481" (and Kodak Wratten filters #87c or #89b) were used for discrimination of vegetation and bare soil.

Recent experiments have used small radio-piloted aircraft as platforms for low-altitude aerial photography to survey crop diseases or monitor environmental problems (102–106, 123, 147, 174, 221, 224, 365). The system developed by Fouché in South Africa has three videocameras on board the aircraft, one wide-angle camera for navigation, another for taking measurements, and a separate videocamera reading the aircraft instruments. An image of the altitude, compass course, flight angle, flight position, etc, is shown on the ground monitor display. Similar experiments have recently been started in Uppsala using a model aircraft with a 3.5-m wingspan that flies very slowly only a few meters above the field plots. There is a wide angle videocamera for navigation and an IR-camera for taking measurements. The video images are transmitted by radio to a ground-based TV monitor (8).

Radio-piloted aircraft with high-capacity videocameras and equipped with Global Positioning System (GPS) for recording accurate positions are likely in the future. A Swedish laser optic navigation system, estimated to have a GPS accuracy as close as 10 cm, is currently being tested in the European EUREKA project "Autofarming/Elmar" (R Holmqvist, personal information). Such a GPS and radio-piloted remote sensing system might allow surveys of agricultural fields for stressed and diseased areas, in order to apply control measures only where needed. An alternative platform for remote sensing of field plot experiments is a light-weight ground vehicle, with a robot run by solar-powered elements and controlled by the navigation system.

SATELLITE-BORNE AND AERIAL RADIOMETRY

The American Landsat 1 was launched on 23 July, 1972, and was succeeded by Landsat 2, 3, 4, 5, and 6 (No. 6 has been lost) and the European SPOT satellite (Système Probataire pour l'Observation de la Terre). Landsat 1, 2, and 3 had sensors in four spectral bands (500–600, 600–700, 700–800 and 800–1100 nm), and Landsat Thematic Mapper 4 and 5 have two additional MIR-bands and one in the thermal region (450–520, 520–600, 630–690, 760–900, 1550–1750, 2080–2350 nm and 10.4–12.5 μm), in addition to visible and NIR-bands. The Landsat satellites also have other sensors (11). The SPOT-satellite has three broad-band color sensors (500–590, 620–660, 770–870 nm) with 20 m ground resolution and one panchromatic sensor (500–900 nm) with 10 m resolution.

These satellites have collected extensive valuable remote sensing data from agriculture and forestry, including data on plant stresses, diseases, and pests. SPOT gives the best geometric resolution, whereas the spectral resolution of the more recent Landsats is better adapted for vegetation studies (there are more spectral bands suitable for plant pigments as well as for water absorption bands in the MIR region and a band in the thermal region). These high-altitude systems are of value for surveillance over large areas but their geometric resolution is insufficient for application as an operational tool in phytopathometry. The measurements are often impeded by cloud cover and the period from measurement until the data are available for evaluation in phytopathometry is long. The SPOT can be tilted up to 27°, but it is risky to compare data from different measurement angles.

The satellites pass over Sweden too early in the morning for phytopathometry. Hence, measurements are made over canopy that sometimes still has dew on the leaves and other times after the dewfall has dried. Dewfall has a profound influence on spectral reflectance (302, 303; personal observations); it is therefore difficult to compare data accurately. There are also wide differences in growing conditions between agricultural regions. In the south, the growing period is about 240 days, around Uppsala-Stockholm it is about 180–200 days, and in the north about 100–150 days. In southern Sweden, spring cereals are sown in the second half of March or first week of April, around Uppsala in late April or early May, and in the north in late May or the first week of June. There are also regional differences in soil, climate, day length, and agricultural practices and use of cultivars. The development of a cereal crop is delayed by about three to four days for each 100 m increase in altitude above sea level (283). It should also be emphasized that 3-4 days or a week is a long period where substantial changes can happen in development and spectral properties of an agricultural and horticultural crop, which influences the accuracy in evaluation of the spectral data for assessment of growth and predicting of diseases and stresses.

Airborne radiometry is widely used in agriculture and forestry for surveillance of pests, diseases, and environmental stress (162). An advantage over satellite measurements is that the most suitable time and height can be selected and coordinated with conventional ground assessments. In recent years, airborne imaging radiometers such as Airborne Visible to Infrared Imaging System (AVIRIS) have been used both in the USA (76, 120, 362, 363) and in Europe (14, 16). Airborne hyperspectral imaging radiometer systems have been reported for application in vegetation studies (285, 340): these are also of interest in phytopathometry.

GROUNDBASED AND HAND-HELD REMOTE SENSING

Some radiometers can be hand-held or otherwise groundbased, e.g. Exotech 100AX multispectral radiometer (MSR), which has four spectral bands similar to Landsat 1–3 (500–600, 600–700, 700–800 and 800–1100 nm), or Barnes Modular Multiband Radiometer, which has eight spectral bands similar to Landsat Thematic Mapper (450–520, 520–600, 620–700, 730–930, 1150–1300, 1500–1800, 2000–2400 nm and 10.4–12.5 μm).

Using an Exotech 100AX MSR mounted on an expandable hand-held pole supplied with a water level for control of vertical measurements, Nilsson (252–254) obtained good correlation between spectral reflectance data and various diseases of cereals and oilseed rape. Similar agreement exists between Exotech 100A RMS data and stripe rust, stem rusts and barley yellow dwarf virus of wheat (293, 319).

There are numerous reports of ground-based radiometers being used in field experiments on various agricultural crop species in studies of plant growth, amount of biomass, the effects of various growing conditions, and damage by nutrient and pollution stress, diseases and pests. However, insufficient information on growth stage, soil coverage, disease identification and assessment, growing conditions, or measurement conditions often makes reported results difficult to compare and evaluate. The radiometers were apparently truck-mounted in most of these studies (See Reference 11).

Li-Cor Li-1800 portable radiometer has a spectral range 300–1100 nm, while the Spectron Engineering SE590 has a range of 400–1100 nm and the version SE/CE393 has a range of 1100–2500 nm. Finally, the Analytical Spectral Devices' ASP Personal spectrometer II and ASP FieldSpec-FR are very fast instruments with fibre optic probes for the range of 400–1050 nm (512 steps of 1.4 nm) and 350-2500 nm, respectively. VIRAF-II is a new high-resolution multi-purpose radiometer for fast measurements in the range of 400-910 nm (52). A Swedish fibre optic system being developed (so far for the laboratory) has a very fast operation with a resolution of up to 30,000 steps in the range 375-1055 nm ("MES 3000", Now Optics AB, Kista, Sweden; personal information). These high resolution instruments are interesting for studies of biochemical leaf components and spectral details of disease symptoms. A review of earlier radiometers and radiometry was given by Milton (233). Some of the above radiometers may be supplemented with GPS.

Some instruments only measure spectral data from one direction at a time. To estimate incoming sunlight (irradiance), frequent radiance measurements are required over a calibration plate (e.g. $BaSO_4$ coated standard white plate) with known spectral reflectance properties: calibrations must be made correctly (91, 167). Measurements of the canopy must be alternated with those of the calibration plate, followed by calculation of sunlight intensity for the same moment as each canopy measurement. Although laborious, this approach can give good results on clear sunny days, but not with changing cloud cover.

An excellent solution for agricultural field plot experiments is the *Cropscan Radiometer System,* developed by VD Pederson (Cropscan Inc., USA). The light-weight radiometer measures irradiance and radiance simultaneously and is thus less dependant of varying cloudiness. The sensors record 30 nm wide bands centred around 450, 500, 550, 600, 650, 700, 750 and 800 nm. Detailed descriptions of the Cropscan system are given in (255–257, 287, 288, 290, 291).

A newer version of the Cropscan system uses an improved datalogger ("DLC" using PC-cards for program and data) and a 16-channel radiometer. I have supplemented my Cropscan system with extra sensors for simultaneous recordings of air temperature, relative air humidity and wind speed as well as a sensor for IR-thermometry of the canopy temperature (257). The IR-thermometer faces 45° to the canopy to minimize soil interaction.

In making measurements over vegetation with good soil coverage, e.g. cereals, clover, alfalfa, grasses, oilseed rape or peas, the radiometer is usually kept ~ 1.5 m above the canopy (252–257, 261), whereas over potato and sugar beet canopies, the radiometer is kept at a height of 30–60 cm above individual plants to avoid soil reflectance. Very good correlations have been demonstrated between reflectance of individual plants and visually assessed percentage of leaf area infected by late blight (*Phytophthora infestans*) in potatoes (58, 259, 329, 330; HE Nilsson, unpublished), mild yellow virus, and powdery mildew (*Erysiphe polygoni* DC. (syn. *E. betae* Vanha) Weltzien) in sugar beet (260). Cropscan radiometry has also been useful in studies of the accuracy of a new fungicide sprayer and in dose-response studies of fungicides and herbicides in potatoes, oilseed rape, and wheat (259), on seed-treatment dosages for controlled infection levels of barley stripe disease (*Pyrenophora graminea* S. Ito

& Kuribay) in six-row barley (HE Nilsson & L Johnsson; unpublished), and in seed treatment field plot experiments for control of aphids (*Rhopalosiphum padi* L.) and infection by barley yellow dwarf virus in oats (58). Several of my field experiments in cereals demonstrate a significant correlation between grain yield or diseases and the differences in Cropscan-data recorded on various days (254–256, 262; and unpublished). Other reports on use of the Cropscan radiometer include the following references (134, 135, 255–261, 266–268, 270–275, 277, 287–289, 292, 339). In Australia it has been used with good results for estimation of cotton boll yield, when the plants were wilted and brown (P Cull, personal information).

The Cropscan radiometer system is designed for use in field plot experiments; it has great potential to monitor plant growth and development, to detect need of supplementary assessments, to rationalize and facilitate assessments, and to increase the precision and objectiveness of early detection of stress and need-adapted control measures. Interactive radiometric measurements and visual observations and notations are also possible. The Cropscan system is easy to use and takes < 2 s to record spectral data from a test area. In over 20,000 Cropscan measurements since 1982, my average measuring rate is ~ 240 test areas per hour in ordinary field plot experiments.

A helicopter as platform for the Cropscan radiometer was tested for surveying of crop growth and various stress areas in large agricultural fields in western Sweden in 1994. These data were compared with both conventional ground data and hand-held Cropscan data recorded along the flight lines over the fields. Effects such as shading by the outer ends of the helicopter rotor blade were also taken into account (A Carlsson, personal information).

Laboratory measurements of the diffuse optical reflectance of potato tubers in the visible and near-infrared bands (500–2000 nm) to detect *Polyscytalum pustulans*, *Phoma exigua* var. *foveata*, and *Fusarium solani* var. *coeruleum* have been reported (238, 307), and in the range 400 to 2600 nm on watermelon leaves (*Citrullus vulgaris* Schrad.) infected by downy mildew (*Pseudoperonospora cubensis* (Berk. & Curt.) Rostow), and *Fusarium* wilt (*Fusarium oxysporum* f. *niveum* (E.F.Sm.) Snyder & Hansen) (37; see also 137 for review of remote sensing for crop protection).

VIDEOGRAPHY

Although there has been interest in using video cameras for remote sensing for more than a decade (89, 90, 92, 96, 112, 216, 219, 366, 369), the use of aerial videography has expanded during the past five to six years. Important developments in technique and methods have been made by the research group (e.g. Drs DE Escobar, JH Everitt, HW Gausman, AJ Richardson, and CL Wiegand) at USDA-ARS at Weslaco, Texas, as well as by Dr C King and

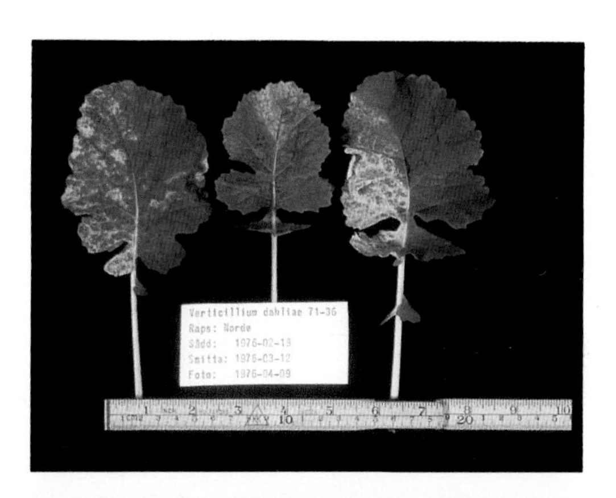

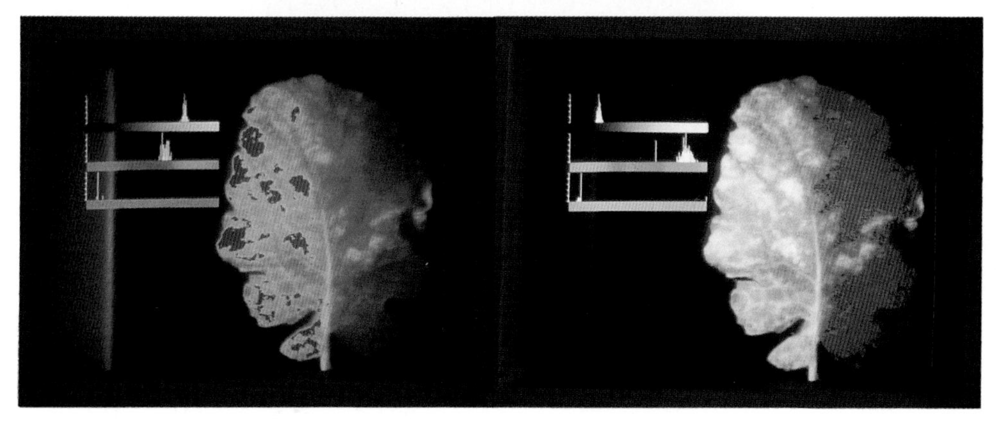

Figure 1 Three leaves of oilseed rape (*Brassica napus* var *oleifera* Metzger) infected by *Verticillium dahliae* Kleb., and studied using Leica CBA8000 image analyzer.

The *left* leaf has an area of 6755 mm^2. Spectral signatures of specific spots were identified and areas with identical color were displayed by the image analyzer and marked as red in Figures 2 and 3. The red areas are 13.4 and 42.0%, respectively. The gray scale, spectral signature and color saturation of the marked areas are indicated to the *left* in Figures 2 and 3.

collaborators at Carleton University, Ottawa, Canada, and Dr CMV Neale at Utah State University, Logan, Utah.

The Biovision Colour IR System (226; Coyote Enterprises, Inc., Texas) is a single lens camera. The light is split into in three beams that are recorded by separate sensors, each with 100-nm spectral band width. The images can be examined on a monitor separately or in combination, where the image will be similar to that recorded by Kodak Ektachrome-Infrared film 2443. The Xybion System is a single lens solid-state camera with a fast rotating wheel with six band-pass filters (219, 338). Other video systems are combinations of two or more separate videocameras with band-pass filters in the visible and NIR ranges (93, 95–97, 241, 243, 244, 265, 357).

The image resolution of digital cameras is seldom more than 512 × 512 pixels (picture elements) (93, 219), and is not comparable with a good film such as Ektachrome Infrared film 2443, although the image can be improved by incorporating image texture into the data analysis process (377). Hart et al (136) and Wiegand et al (371) found both CIR-video and CIR-photography good for assessments of stresses, diseases, and pests of agricultural crops. Kettler et al (178) found differences in interpretation between CIR photography and CIR videography; CIR videography had the best spectral sensitivity but spatial superiority was evident upon visual interpretation of the CIR-film.

Good correlation between IR/R ratio and various vegetation indices from aerial videography and hand-held multispectral reflectance data of field plots has been demonstrated for alfalfa (10), grasses (94, 95), corn (366), and wheat leaf rust (366). Wiegand et al (366, 369) indicated that automatic gain control in video cameras may complicate multitemporal analyses of crops and cultivars. Computer software can enhance and process automation routines for digital multispectral video imagery (34, 244).

New systems with greater resolution have recently been developed: the B&W camera Kodak Videk Megaplus (183); the Kodak CDS200-IR; the Kodak Professional DCS420-IR and Kodak DCS460; and the DALSA IA-D9–2048 Megasensor camera. A recent and excellent technical review of various digital frame video camera systems (DFC) is given by King (182).

The principal limitation of DFCs in low-cost remote sensing is the large digital image size. An 8-bit 1024 × 1024 DFC produces images of over 1 Mbyte each, and with 16-bit, 4000 × 4000 imagers produce over 32 Mbytes per image. The latter are obviously out of range using ordinary desk top or portable computer data transfer and storage rates. However, enhanced computer and memory developments may overcome these problems. A common 1280 × 1024 sensor produces images that fit on a high-density 3.5" diskette, and modern PC-cards and portable optomagnetic disk systems provide data portability and ease of distribution. Another disadvantage of digital imaging is that high-quality hardcopy production of images is expensive.

The Kodak CDS420-IR camera has a spectral range of 400–1020 nm, whereas the CDS460 model so far operates only in the visible range (resolution = 1012 × 1524, and 2048 × 3072 pixels, respectively). Both cameras can be connected directly to a portable Macintosh computer. A "color wheel accessary" and band-pass filters allow for exposures in several spectral bands. Special software from Kodak allows the user to acquire the image information from the camera's card reader into ADOBE Photoshop software on the computer screen. Exposure quality can be tested and image processing functions allow the images, such as those of disease symptoms, to be enhanced, selected, and examined in the field. Images of different spectral ranges can be analyzed, as discussed above on radiometric data (such as IR/R). The cameras have a built-in microphone that allows sound notations with each individual image. Newer versions of portable computers with improved screen, memory, battery capacities and image analysis software should enhance application of remote sensing as much as IR-photography.

Videoimages, unlike IR-films, are instantly accessible for evaluation. Stutte et al (331) report that videography allowed rapid screening of nutrient stress (Ca^{2+}, Fe^{2+} and N^{-}) of various horticultural crops. A narrow-band 680 nm filter generally produced greatest sensitivity for early stress detection, and pseudocolor look-up tables permitted real-time screening of the reflectance. Stutte & Stutte (332) developed an "Image Capture and Analysis System" for near-real time quantification of field crop stress areas using videography. Wiegand et al (367) reported on the success of airborne multispectral videography in monitoring crop growth and assessment of salt stress in sugarcane.

Another interesting video system for close-range studies is the Japanese Keyence Portable Fiber Optic Microscope VH-6100 (Keyence Corporation, Osaka). The fiber optic probe can be fitted with various microscope and macroscope lenses. The system is fully portable, and the image can be displayed on a color screen or stored on floppy disks. Another version is the Finlay Portable TV-microscope (Finlay Microvision Co. Ltd, Southam, Warwickshire, UK).

IR-THERMOGRAPHY

Canopy and plant leaf temperatures have been studied for about 150 years, but it was not until the 1920s that researchers correlated leaf temperature to evapotranspiration (159). Monteith & Szeicz (234), Tanner (334) and Wiegand & Namken (368) first used IR-thermography (recording of thermal emittance in the spectral range of 8–14 μm, sometimes 3–5 μm) to study plant leaf temperature. The advent of lightweight portable instruments in the 1960s and 1970s allowed use of IR-thermography to study relationships between water stress, leaf temperature, canopy-air temperature, and soil water status in various crop species (159). Millard et al (231, 232) used airborne thermal imagery to demonstrate variations in canopy temperature in a barley field. Jackson et al (168) and Idso et al (149–151) developed the term water stress degree day (SDD) as compared with growing degree day (GDD). Jackson et al (163, 167) developed the term crop water stress index (CWSI). O'Toole & Hatfield (284) studied the effect of wind and demonstrated that CWSI values measured at low windspeed overestimated the water stress, but underestimated it when measured at high windspeed. However, numerous researchers have used this concept (19, 149, 151, 159, 162, 232, 235), which is also important in pathology as water stress is involved in many infectious plant diseases. Another crop water stress index is reviewed by Campbell & Norman (55).

Pinter et al (306) found that infections of sugar beet roots by *Pythium aphanidermatum* (Edson) Fitzp. and of cotton roots by *Phymatotrichum omnivorum* increased leaf temperatures by 3–4°C above that of healthy plants. Similar findings have been observed on bean plants infected by root rots caused by *Fusarium solani* (Mart.) Appel & Wr., *Pythium ultimum* Trow, and *Rhizoctonia solani* Kühn in the greenhouse by Tu & Tan (346), and by Mengistu et al (227) with brown stem rot of soybean caused by *Phialophora gregata* (Allington & Chaberl.) W.Gams. Tu & Tan (346) suggested screening for root rot resistance by IR-thermography, and in Sweden, I proposed IR-thermography and IR-videography for screening of the most vigorous tree seedlings before their distribution for reforestation. Fouché (104) and Fouché & Booysen (106) demonstrated that use of computer processed images from low-altitude airborne IR-thermography and IR-videography is a quick method of assessing diseases and moisture stress in soybean and wheat crops as well as avocado trees, and also in demonstrating the spatial variation in moisture stress and in estimating stress conditions for irrigation scheduling. Duczek (85) examined the infection by common root rot, *Fusarium culmorum* (W.G.Sm. Sacc.) of two cultivars of spring wheat and barley periodically from 1983 to 1985; he found significant differences in disease ratings between the cultivars, but no significant differences in leaf temperature recorded using a hand-held IR-thermometer (at Feekes' growth stages 10, 11.1 and 11.2).

I started to use IR-thermography by using a hand-held in 1980 in studies of various plant diseases (250). The technique has subsequently been used in field experiments and some greenhouse experiments mainly with two IR-thermometers, Telatemp Ag42s (251–257, 262, 263) and Everest 4003 (257). The latter paper reviews results of my experiments on IR-thermography during 1980–1988 comprising nine crop species and various biotic and abiotic stresses. As an example, in one field experiment, comprising 30 oat cultivars with differences in root growth, root capacity, and drought tolerance, and grown in a sandy soil at controlled soil moisture levels, a difference of 10°C in flag leaf temperatures was recorded between cultivars with the best and the least tolerance to drought. Even greater differences in leaf temperature were recorded in a greenhouse experiment with cucumber plants inoculated or not inoculated with bacterial infections in their vessels. The leaf temperature of the healthy plants was 16–17°C and of the infected plants 23–24°. When a fan temporarily circulated the air (temp. = 20°), the leaves of the infected plants remained at about 23° while the leaf temperature of the healthy plants fell to 8–9° but later increased to their previous level.

In field experiments with IR-thermometry at Uppsala in spring wheat and barley comprising cultivars with morphological differences (foliar development and root growth), disparities in intensity of irrigation and nitrogen fertilization resulted in substantial variation in leaf and canopy temperatures. Sparse N-level and no irrigation caused weak root growth, particularly in one barley cultivar, and consequently increased leaf temperature. After heavy rain this one cultivar produced new green tillers with lower leaf temperature than the previous leaves. There were also differences among the barley cultivars in amount of awns on their spikes and in ripening. On some cultivars the awns remained green longer, which also influenced the observed canopy temperature (262, 263).

Awns and leaf pubescence as well as leaf color can affect energy balance and canopy temperature, and therefore the results of IR-thermometry (27, 99, 138, 237, 375).

In a preliminary experiment, I observed that the flag leaves of oats infected by barley yellow dwarf virus were 3–4° warmer than visually healthy leaves. However, it can be difficult to evaluate how much of the temperature increase was due to the virus infection or to effects by the virus vectors, e.g. various insects. Water stress in plants influences various physiological parameters, photosynthetic efficiency, and levels of osmotic compounds, etc (286), which, in a recent review by Cabrera et al (53), is suggested to increase susceptibility to aphids. It is also mentioned that aphid infestation causes drought-stress symptoms in leaves of barley, even in the presence of sufficient root moisture. Cabrera et al (53) also report that increasing aphid infestations of barley can substantially influence amounts of soluble sugars, proline, glycine-betaine, chlorophyll, and carbon assimilation and photosynthetic rate as well as the leaf water potential in barley. The water potentials of barley leaves infested with only 18 aphids (*S. gramineum*) per plant were significantly different from the noninfested control, and 104 aphids per plant decreased leaf water potential from –0.25 to –0.86 MPa in comparison with noninfested. The mechanism by which the greenbugs cause such a drop in leaf water potential is unknown, but the insects damage the cuticle and epidermis, the cells lose water, and it is concluded that severe aphid attacks can cause water stress. When the aphids suck leaf cell sap they utilize free amino acids (S Brishammar & J. Weibull, personal information) and eject a lot of soluble sugars, which, as a layer of honeydew on the leaf surface, influence the light reflectance and evapotranspiration.

Differences of 3, 4, or 5°C in flag leaf temperature of cereals due to root and vascular diseases, such as barley infected by *Pyrenophora graminea* and wheat infected by *Cephalosporium gramineum* Nis. & Ika., have often been recorded in my field experiments (251, 252, 256). Similar effects have been recorded in oilseed rape infected by *Verticillium dahliae* Kleb. and *Sclerotinia sclerotiorum* (Lib.) de Bary (253). A slight to moderate infection by take-all disease (*Gaeumannomyces graminis* (Sacc.) Arx & Olivier var. *tritici* Walker) in wheat once caused an increase in flag leaf temperature of 7°C over healthy plants (257).

Significant correlations have been demonstrated in several of my field experiments between cumulative "Canopy-Air temperature" (x) and disease severity (y) or grain yield (z) (252, 254, 262, 263; HE Nilsson, unpublished), similar to those of SDD discussed above.

On numerous occasions I have observed that slight wind gusts of 2–5 m/s or more cause a momentary decrease in leaf temperature, and that the temperature drop as well as the time needed to return to the previous value is correlated with the severity of root and vascular diseases or, in other words, to tolerance to water stress. Although temperature drops and intervals are most interesting to study, wind gusts may also complicate routine measurements of canopy temperatures and mask detection of minor disease infections, and make evaluations difficult.

Smith et al (325) reported that stripe rust on wheat initially reduced stomatal closure and disrupted the cuticle. Evapotranspiration from the leaves increased and the infected leaves were 0.2–1.0°C cooler than the controls during early disease development. As the disease progressed, however, infected leaves senesced more rapidly, evapotranspiration decreased, and leaf temperature increased. We can also assume that dense superficial hyphae and hyphal mats on the leaf surface can influence leaf temperature recorded by IR-thermography.

The literature contains many reports on differences between races and strains of various pathogens, fungi and viruses, on their temperature or infection optima. They may give so-called typical or specific symptoms at one air temperature but react differently at only a few degrees higher or lower air temperature. Uneven soil moisture in pot experiments may result in variations of leaf temperature, and consequently in the development of spores, in spore germination, and in disease symptoms. Research into discrimination between races and strains of leaf pathogens as well as cultivars should, with reference to the above findings, be supplemented with IR-thermography and microclimate studies, including adequate control of wind movements and relative air humidity (258).

IR-thermography should be considered a valuable tool in plant pathology, phytopathometry, and epidemiology. IR-thermography should be considered mainly as a supplement to other remote sensing methods rather than a stand-alone method to evaluate factors such as cultivar differences in susceptibility or resistance to stress.

ACTIVE AND PASSIVE REMOTE SENSING

The discussion thus far has been on passive remote sensing, i.e. the use of films or electronic instruments to measure the electromagnetic solar energy reflected from the vegetation. A prerequisite for detection and assessment of stress and disease is the homogenicity of the canopy and certain solar light conditions (see above). Active remote sensing methods, developed over the past 15 years, direct intensive energy pulses of specific wavelengths against the vegetation; the resulting interaction is monitored and the physiological status of the vegetation is analyzed. Among such active remote sensing methods are Light Detection and Ranging (LIDAR), which uses laser light, analogous to Radio Detection and Ranging (RADAR) remote sensing (130).

REMOTE SENSING USING LASER-INDUCED FLUORESCENCE

The fluorescence of certain plant constituents has long been used in the study and elucidation of photosynthesis and related physiological mechanisms and reactions. Early fluorescence measurements used plant extracts, homogenates, or characterized elements of the plant cell. The development of high-intensity light sources such as the laser has made it possible to excite measurable fluorescence in the intact plant, also by remote sensing.

In contrast to reflectance spectra where only spectral changes resulting from the absorbance of light by chlorophyll and water are apparent, fluorescence spectra have maxima that relate to specific plant constituents and their electronic states. As the concentration and oxidation states of many plant constituents are functions of the vigor of the plant, fluorescence measurements can often be related to biochemical and physiological changes induced by stress or pathogen (61, 63, 64, 192, 195, 199).

In November 1989, the "EUREKA project LASFLEUR" started as a European cooperative research effort to investigate the future application of far-field laser-induced plant fluorescence for synoptic, airborne environmental monitoring of vegetation and the state of health of plants in agriculture and forestry (59, 60, 131, 132, 201, 327, 350).

Analysis of the fluorescence ratio of R/NIR (such as F690/F730 or

F685/F740 nm) allows measurement of the amount of leaf chlorophyll and thus stress effects on the leaf (51, 52, 87, 88, 193, 198–200, 314).

UV-laser-illuminated leaves emit not only red chlorophyll fluorescence, they also fluoresce blue and green (63, 64, 200). The red chlorophyll fluorescence spectra of green leaves, long used in photosynthesis research (176), possesses two fluorescence maxima near 690 nm (F690) and 735 nm (F735) (199). The ratio F690/F735 can be used as a stress indicator of the photosynthetic apparatus and applied as a nondestructive indicator of the in vivo chlorophyll content (192, 193).

The blue fluorescence of green vegetation, detected as a genuine plant signature almost 60 years ago (176), possesses a maximum near 450 nm (F450) and a shoulder in the green region near 530 nm (F530) (64). The ratio F450/F690, which can differ between plant types because of growth conditions, appears to be a suitable stress indicator of plants accessible for remote sensing (64, 65, 131, 132, 197, 198, 201). The nature of the blue fluorescing plant substance(s) is not clear, but there is agreement that various plant phenolics are the major compounds responsible (327). Phenolics are of special importance in plant pathology.

Remote measurements of fluorescence have been widely used in vegetation studies: heat damage (317); freezing or chilling injury (6, 184, 324); high and low thermal stress and water stress of oak (*Quercus ilex* L.) and soybean (*Glycine max* (L.) Merri.) (228). Kharuk et al used remote measurements of chlorophyll fluorescence in studies of the aging process and ozone damage (179). Krajicek & Vrbova (188) and Valentini et al (350) made detailed studies of laser-induced fluorescence spectra of various plant species with environmental stress. McFarlane et al (223), with a Fraunhofer line discriminator, used sunlight-induced fluorescence in a 656.3 nm Fraunhofer line to study water stress of lemon trees.

Using a pulsed 337 nm nitrogen laser, McMurtrey et al (225) discriminated between levels of nitrogen fertilization and N-deficiency of field corn (*Zea mays* L.) by analyzing fluorescence at 440, 525, 685 and 740 nm. Similar techniques have been used to detect other nutrient deficiencies in corn and soybean (62, 64, 65) and rice plants (333). EW Chappelle & JE McMurtrey (personal communication) demonstrated differences in fluorescence spectra due to N, Mg, and K deficiencies and powdery mildew of cereal leaves.

Significant differences in fluorescence in the range 720–750 nm were observed in studies of pea and bean leaves infected by viruses using a very high resolution "MES 5000" radiometer (HE Nilsson, unpublished).

Daley et al (82) and Osmond et al (282) used video imaging in studies of variation in photosynthetic activity and the effects of virus infection on leaves by close range, nondestructive measurements of chlorophyll fluorescence. The great potential of high-speed video imaging of the fluorescence influenced by the effects of virus infection on photosynthetic activity was demonstrated by PF Daley during my visit in 1992 (80, 81).

A Swedish system for multicolor imaging of vegetation fluorescence following laser excitation was developed for remote measurements at distances of up to 50 m (87, 171). Fluorescence spectra at selected points within the detection area are measured with an image-intensified diode array system. Image processing allows extraction of information related to the physiological status of vegetation. A similar experiment comprising monitoring of healthy and chlorotic leaves of *Fagus silvatica* and *Picea abies* is reported by Edner et al (88), and of green and slightly senescent leaves of *Acer platanoides* (88, 171). The advanced instrument system has apparent potential for studies of plant stress and diseases.

RADAR AND MICROWAVE REMOTE SENSING

The use of radar and microwave radiometry in remote sensing of soils and vegetation has aroused considerable interest in recent years; this technique is less influenced by cloud conditions and has potential in studies of moisture conditions of soils and plant canopies, including water stress, and should be of interest to plant pathologists (43, 142, 209).

DIGITAL IMAGE ANALYSIS

Digital image analysis has become an important tool in biological research. Examples of its application include analysis of satellite images, aerial photographs and videographs, macroscopic and microscopic images, reconstruction of multispectral 3D-images in confocal microscopy and fluorescence life studies, acoustic images, nucleic magnetic resonance images (NMRI), images in electron microscopy, and animation of time-lapse images (recorded in very slight illumination to avoid the problems caused by too intensive light). Inoué et al (152) reviewed video technology and its application in video microscopy, including ultramicroscopy and confocal microscopy.

Image processing reduces the total information to a manageable amount, enhances edges and other details, makes geometric corrections, etc, prior to analysis of measurements or identification of size, area, shape, pattern or counting specific details. Digital images can be added or subtracted, and differences between two images displayed. One image can be divided by another, pixel for pixel, in order to make ratios such as between a red and an infrared image—or other calculations as discussed above for radiometry.

One great advantage is that we can display particular areas in specific or contrast colors and thus facilitate visual interpretation and interactive analysis by the operator and the computer. Such contrast-enhanced images are also an excellent tool in lectures and demonstrations in plant pathology. The enhanced and analyzed images can be printed out or stored in a mass-memory (e.g. optical disk). If the computer is connected to a network it is easy to transmit images between distant researchers.

Today there is good software and hardware available on the market for digital image analysis, both for IBM/PC-compatibles and Macintosh computers. Various software versions can be readily applied to phytopathology, and in many cases it is fairly easy to modify them for particular purposes. The investment may appear costly, but in relation to what we can do via this technology in phytopathology and phytopathometry the investment is profitable both scientifically and economically, not forgetting increased personal creativity.

Numerous papers have been published on image analysis in airborne remote sensing, most of them concern forestry and fewer agricultural crops. Some of the reports concern phytopathology. Jackson et al (158) gave an early demonstration of how aerial IR-color photographs can be enhanced and color-coded to facilitate visual interpretation of the distribution and severity of root rot in pea fields. Changes of winter-killing in alfalfa from one year to another are shown by Basu et al (20) using similar methods. Among more recent reports are on aerial IR-photography and computer-aided image analysis to study environmental stress in sunflower and Dutch elm disease (202, 203), tree defects (29), barley yellow dwarf virus and cereal aphid infestations in winter wheat (127), citrus stress (191, 310), diseases of tomato, potato, cucumber and watermelon plants and citrus trees (35, 36, 38, 41), damage by nematodes to cotton, and root rots of cotton, alfalfa and pecans (191). Blazquez & Hedley (39) and Blazquez et al (40) made computer-aided spectrophotometric measurements of 35 mm color IR-films of tomato leaves and found good correlation between readings of the ratio of 480/610 nm and leaf infection by bacteria and by late blight. Similar studies of citrus tree health are reported by Blazquez (32) using ratios of spectral peaks at 510–520 and 600–620 nm.

Digital image analysis at macro- and microscopic levels in plant pathology is attracting increasing interest and has been used in studies of various infectious diseases and stress of many plant species (13, 15, 38, 49, 115, 116, 187, 190, 204–7, 230, 239, 248–250, 268, 272, 321, 326, 328, 342, 349, 356), but there is no room in this chapter for further details.

By using digital image analysis in connection with confocal microscopy it is also possible to nondestructively analyze living plant cells and tissues, e.g. events of cytoplasmic actions of Ca^{2+} and H^{+} in milliseconds (21, 341; MA Browne, personal information at the Royal Microscopical Society meeting Micro-94 in London 1994). In modern instruments we can get near-simultaneous excitations of several laser lights and record fluorescence from spots in the living cell not only in its *x*, *y*, and *z* coordinates but also analyze spectral signatures and functions in time-lapse, and study the physiology of disease phenomena nondestructively, also at very high speed. It is a version of analysis that in some aspects is much different from, but has also much in common with, remote sensing.

NMR IMAGING

NMRI (Nuclear Magnetic Resonance Imaging) analysis is a rather new but very advanced nondestructive image analysis technology with great potential in phytopathology. Pfeffer & Gerasimowicz (299) reviewed NMR-spectroscopy and its application in agricultural research and gave examples of NMRI studies of living roots and fruits of bean plants, while Gross et al (128) discussed NMRI applications to studies of roots and stems, growth and transport in the vessels, and effect of chemical agents. Brown et al (48) used NMRI in studies of changing water content in vivo in *Pelargonium,* Morrison et al (236) studied the distribution of sugar in unripe and partially ripe grapes by NMR micro-imaging, and Johnson et al (172) studied tomato root galls caused by *Meloidogyne incognita* using NMRI. Bottomley et al (42) used NMRI in studies of water transport in root systems of *Vicia faba* L. with light-stressed foliage. Another example of noninvasive histochemistry of plant material by NMRI is reported by Sarafis et al (315). Various versions of NMRI microscopy for anatomical studies of ripening fruits of raspberry and gooseberry fruits (*Ribes* spp.) are reported by Williamson et al (372–374). Goodman et al (121, 122) report applications of NMRI for noninvasive anatomical microscopy and studies of infection by *Botrytis cinerea* Pers. & Fr. in fruits of raspberry (*Rubus idaeus* L.). Halloin et al (133) used NMRI in studies of disease development in sugarbeets caused by *Rhizoctonia solani.*

SUMMARY

Remote sensing comprises several nondestructive methods to acquire and analyze spectral properties of vegetation from various distances, ranging from satellites to ground-based platforms. Multispectral imaging allows nondestructive analyses of plants, pathogens, and plant diseases to be made down to microscopic and ultramicroscopic levels. Techniques and methods of multispectral radiometry, photography, videography, IR-thermography, active and passive remote sensing, radar and microwave remote sensing, laser-induced fluorescence, multispectral image analysis, and NMR all have a great potential in phytopathology and phytopathometry.

Although we cannot yet identify a specific disease or stress by current versions of remote sensing, we can often detect (even previsually) and quantify its intensity and also facilitate, rationalize and increase accuracy of research in phytopathology. Early detection is important for need-adapted control measures. Measurements using, for example, Cropscan radiometry in the summer can provide valuable data nondestructively, not only for monitoring plant growth and indications of supplementary studies, but also be of great value if the field plots cannot be harvested later on due to bad weather conditions or other reasons. Aerial photography is a valuable tool in selecting the most suitable area for field plot experiments.

More research is needed into how various synergistic and competitive (masking) diseases and environmental effects influence spectral properties of crops and cultivars. Access to cheaper and more accurate instruments for routine applications, personnel training and good international collaboration to facilitate the dissemination of information are all vital. The accuracy of conventional disease assessment methods must be enhanced for use in comparisons with remote sensing data. A disease or stress measurement using remote sensing in the field is still primarily a comparison of data from stress areas with those from the surrounding healthy canopy. In the future, more advanced instruments and greater experience should facilitate and increase accuracy in plant disease research.

ACKNOWLEDGMENTS

I thank F Nutter and my colleagues, B Gerhardson and S Brishammar, for their critical reading of the manuscript and for their valuable comments. Special thanks to VD Pederson for numerous discussions, advice, and support on the Cropscan Radiometer System. Sincere thanks are also due to N Rollison and H Gould for linguistic assistance.

Literature Cited

1. Anonymous. 1976. *Manual of Plant Growth Stages and Disease Assessment Keys*. Middlesex, UK: Minist. Agric. Fish. Food
2. Anonymous. 1978. *Crop spectra from LACIE field measurements*. NASA-Johnson Space Cent., Houston, TX. Large Area Crop Inventory Experiment. *LACIE Rep. 00469,JSC-13734*. 196 pp.
3. Deleted in proof
4. Deleted in proof
5. Aase JK, Siddoway FH. 1981. Assessing winter wheat dry matter production via spectral reflectance measurements. *Remote Sens. Environ.* 11:267–77
6. Abbott JA, Campbell TA, Massie DR. 1994. Delayed light emission and fluorescence responses of plants to chilling. *Remote Sens. Environ.* 47:87–97
7. Adcock TE, Nutter FW Jr, Banks PA. 1989. Measuring herbicide injury to soybeans using a radiometer. *Weed Sci.* 38:625–27
8. Alness K. 1992. Modellflygplan mäter odlingar. "*SLU-just nu*". In-house *J. Swed. Univ. Agric. Sci.*, Uppsala, p. 2
9. Analytis S, Kranz J. 1972. Bestimmung

des optimalen Stichprobenumfanges für phytopathologische Untersuchungen. *Phytopathol. Z.* 74:349–57
10. Artan GA, Neale CMU. 1992. An assessment of video-based vegetation indices for the retrieval of biophysical properties of alfalfa. *Proc. Bienn. Workshop Color Aerial Photogr. Videogr. Plant Sci., 13th,* pp. 135–46. Falls Church, VA: Am. Soc. Photogramm. Remote Sens.
11. Asrar G. 1989. *Theory and Applications of Optical Remote Sensing.* New York: Wiley. 734 pp.
12. Asrar G, Kanemasu ET, Yoshida M. 1985. Estimates of leaf area index from spectral reflectance of wheat under different cultural practices and solar angle. *Remote Sens. Environ.* 17:1–11
13. Bacchi LM, Berger RD, Davoli TA. 1992. Color digitization of video images of bean leaves to determine the intensity of rust caused by *Uromyces appendiculatus. Phytopathology* 82:1162
14. Bach H, Demicran A, Mauser W. 1994. The use of AVIRIS data for the determination of agricultural plant development and water content. *Eur. Space Agency (ESA), Earth Obs. Q.* 46:9–10
15. Ball T, Gardner JS, Johnson DA, Hess WM. 1992. Image analysis of uredin-iospores which infect mentha. *Phytopathology* 82:1168
16. Banninger C. 1990. Nature of the reflectance red edge in laboratory and FLI airborne imaging spectrometer data from stressed coniferous forest. In *Image Processing '89. Proc. Workshop, Nevada, 1989,* ed. C Elvidge, pp. 23–28. Falls Church, VA: Am. Soc. Photogramm. Remote Sens.
17. Baret F, Guyot G. 1991. Potentials and limits of vegetation indices for LAI and APAR assessment. *Remote Sens. Environ.* 35:161–73
18. Baret F, Guyot G, Begue A, Maurel P, Podaire A. 1988. Complementary of mid-infrared with visible and near-infrared for monitoring wheat canopies. *Remote Sens. Environ.* 26:213–25
19. Bartholic JF, Namken LN, Wiegand CL. 1972. Aerial thermal scanner to determine temperatures of soils and crop canopies differing in water stress. *Agron. J.* 64:603–8
20. Basu PK, Jackson HR, Wallen VR. 1978. Alfalfa decline and its cause in mixed hay fields determined by aerial photography and ground survey. *Can. J. Plant Sci.* 58:1041–48
21. Batten BE. 1994. Super video rate confocal microscopy and flash photolysis: tools for physiological studies. *Proc. R. Microsc. Soc.* 29:248 (Abstr.)
22. Bauer ME. 1985. Spectral inputs to crop identification and condition assessment. *Proc. IEEE* 73:1071–85
23. Bauer ME, Mroczynski RP, MacDonald RB, Hoffer RM. 1971. Detection of southern corn blight using color infrared aerial photography. *Proc. Bienn. Workshop Color Aerial Photogr. Plant Sci., 3rd, Gainesville, FL,* pp. 114–26. Falls Church, VA: Am. Soc. Photogramm. Remote Sens.
24. Bausch WC. 1993. Soil background effects on reflectance-based crop coefficients for corn. *Remote Sens. Environ.* 46:213–22
25. Bawden FC. 1933. Infra-red photography and plant virus diseases. *Nature* 132:168
26. Bégué A. 1994. Leaf area index, intercepted photosynthetically active radiation, and spectral vegetation indices: A sensitivity analysis for regular-clumped canopies. *Remote Sens. Environ.* 46:45–59
27. Benci JF, Aase JK, Ferguson AH. 1973. Aerodynamic and energy balance comparisons between awned and nonawned barley. *Agron. J.* 65:373–77
28. Blakeman RH. 1990. The identification of crop disease and stress by aerial photography. See Ref. 326a, pp. 229–54
29. Blanchette RA. 1982. New technique to measure tree defect using an image analyzer. *Plant Dis.* 66:394–97
30. Blazquez CH. 1972. Remote sensing of foliar diseases of vegetable crops with infrared color photography. *Fla. State Hortic. Soc.* 85:123–26
31. Blazquez CH. 1976. The role of remote sensing in phytopathology and crop production. *Proc. Am. Phytopathol. Soc.* 3:245 (Abstr.)
32. Blazquez CH. 1991. Measurements of citrus tree health with a scanning densitometer from aerial color infrared photographs. *Plant Dis.* 75:370–72
33. Blazquez CH. 1993. Correlation of densitometric measurements of aerial color infrared photography with visualgrades of citrus groves. *Plant Dis.* 77:477–79
34. Blazquez CH. 1994. Detection of problems in high power voltage transmission and distribution lines with an infrared scanner/video system. *Proc. Bienn. Workshop Color Aerial Photogr. Videogr. Resour. Monit.,* 14th, pp. 103–14. Falls Church, VA: Am. Soc. Photogramm. Remote Sens.
35. Blazquez CH, Edwards GJ. 1983. Infrared color photography and spectral re-

flectance of tomato and potato diseases. *J. Appl. Photogr. Eng.* 9:33–37
36. Blazquez CH, Edwards GJ. 1984. Densitometric studies of color and color infrared photographs of diseased cucumber leaves and plants. *J. Imaging Technol.* 10:79–84
37. Blazquez CH, Edwards GJ. 1986. Spectral reflectance of healthy and diseased watermelon leaves. *Ann. Appl. Biol.* 108:243–49
38. Blazquez CH, Edwards GJ, Nemec S. 1986. Image analysis of xylem alterations in scaffold citrus roots infected with *Fusarium solani*. *Phytopathology* 76:1092–93
39. Blazquez CH, Hedley LE. 1986. Late blight detection in tomato fields with 35 mm color infrared aerial photography. *Phytopathology* 76:1093
40. Blazquez CH, Hedley LE, Benary AJ. 1986. Computer aided spectrophotometric measurements of color and color infrared transparencies of tomato bacterial leaf spot. *Phytopathology* 76:1077
41. Blazquez CH, Richardson AJ, Nixon PR, Escobar D. 1988. Densitometric measurements and image analysis of aerial color infrared photographs. *J. Imaging Technol.* 14:37–42
42. Bottomley PA, Rogers HH, Foster TH. 1986. NMR imaging shows water distribution and transport in plant root systems *in situ*. *Proc. Natl. Acad. Sci. USA* 83:87–89
43. Bouman BAM. 1991. *Linking X-band radar backscattering and optical reflectance with crop growth models*. PhD thesis. Wageningen Agric. Univ., The Netherlands. 169 pp.
44. Brenchley GH. 1964. Aerial photography for the study of potato blight epidemics. *World Rev. Pest Control* 3: 68–84
45. Brenchley GH. 1966. The aerial photography of potato blight epidemics. *J. R. Aeronaut. Soc.* 70:1082–86
46. Brenchley GH. 1968. Aerial photography for the study of plant diseases. *Annu. Rev. Phytopathol.* 6:1–22
47. Brenchley GH, Dodd CV. 1962. Potato blight recording by aerial photography. *Nat. Agric. Advis. Serv., Q. Rev.* 57:21–25
48. Brown JM, Johnson GA, Kramer PJ. 1986. In vivo magnetic resonance microscopy of changing water content in *Pelargonium hortorum* roots. *Plant Physiol.* 82:1158–60
49. Burke MK, LeBlanc DC. 1988. Rapid measurement of fine root length using photoelectronic image analysis. *Ecology* 6:1286–89
50. Buschmann C, Lichtenthaler HK. 1988. Reflectance and chlorophyll fluorescence signatures of leaves. See Ref. 192, pp. 325–32
51. Buschmann C, Lichtenthaler HK. 1988. Complete fluorescence spectra determined during the induction kinetic using a diode-array detector. See Ref. 192, pp. 77–91
52. Buschmann CE, Nagel KS, Kocsanyi L. 1994. Spectrometer for fast measurement of in vivo reflectance, absorptance, and fluorescence in the visible and near-infrared. *Remote Sens. Environ.* 48:18–24
53. Cabrera HM, Argandona VH, Corcuera LJ. 1993. Metabolic changes in barley seedlings at different aphid infestation levels. *Phytochemistry* 35:317–19
54. Campbell CL, Madden LV. 1990. *Introduction to Plant Disease Epidemiology*. New York: Wiley. 532 pp.
55. Campbell GS, Norman JM. 1990. Estimation of plant water status from canopy temperature: an analysis of the inverse problem. See Ref. 326a, pp. 255–71
56. Card DH, Peterson DL, Matson PQ. 1988. Prediction of leaf chemistry by the use of visible and near infrared reflectance spectroscopy. *Remote Sens. Environ.* 26:123–47
57. Carlson RE, Yarger DN, Shaw RH. 1971. Factors affecting spectral properties of leaves with special emphasis on leaf water status. *Agron. J.* 63:486–89
58. Carlsson A, Nilsson HE. 1993. Remote sensing of agricultural field plot experiments. *Abstr. 6th ISPP Congr., Montreal, Canada*, p. 115
59. Cecchi GP, Mazzinghi P, Pantani L, Valentini R, Tirelli D, Angelis P. 1994. Remote sensing of chlorophyll α fluorescence of vegetation canopies. 1. Near and far field measurement techniques. *Remote Sens. Environ.* 47:18–28
60. Cecchi GP, Pantani L, Bazzani M, Raimondi V. 1994. The high spectral resolution fluorescence LIDAR FLIDAR-3 and the monitoring of the environment. *Proc. Int. Airborne Remote Sens. Conf. Exhib., 1st, Strasbourg, France*, III: 689–96
61. Chappelle EW, Lichtenthaler HK. 1994. Fluorescence measurements of vegetation. *Remote Sens. Environ.* 47:1
62. Chappelle EW, McMurtrey JE III, Kim MS. 1991. Identification of the pigment responsible for the blue fluorescence band in the laser unduced fluorescence (LIF) spectra of green plants, and the potential use of this band in remotely estimating rates of photosynthesis. *Remote Sens. Environ.* 36:213–18

63. Chappelle EW, McMurtrey JE III, Wood FM Jr, Newcomb WW. 1984. Laser-induced fluorescence of green plants. 2. LIF caused by nutrient deficiencies in corn. *Appl. Opt.* 23:139–42
64. Chappelle EW, Wood FM Jr, McMurtrey JE III, Newcomb W. 1984. Laser-induced fluorescence of green plants. 1. A technique for the remote detection of plant stress and species differentiation. *Appl. Opt.* 23:134–38
65. Chappelle EW, Wood FM Jr, Newcomb WW, McMurtrey JE III. 1985. Laser-induced fluorescence of green plants. 3. LIF spectral signatures of five major plant types. *Appl. Opt.* 24:74–80
66. Chehbouni JQA, Huete AR, Kerr YH, Sorooshian S. 1994. A modified soil adjusted vegetation index. *Remote Sens. Environ.* 48:119–26
67. Cihlar J, St-Laurent L, Dyer JA. 1991. Relation between the normalized difference vegetation index and ecological variables. *Remote Sens. Environ.* 35: 279–98
68. Clark RV, Galway DA, Paliwal YC. 1981. Aerial infrared photography for disease detection in field plots of barley, oats and wheat. *Phytopathology* 71: 867
69. Cochran WG. 1980. *Sampling Techniques.*. New York: Wiley. 611 pp. 3rd ed.
70. Cohen WB. 1991. Response of vegetation indices to changes in three measures of leaf water stress. *Remote Sens. Environ.* 57:195–202
71. Colwell RN. 1956. Determining the prevalence of certain cereal crop diseases by means of aerial photography. *Hilgardia* 26:223–86
72. Curran PJ. 1980. Multispectral remote sensing of vegetation amount. *Prog. Phys. Geogr.* 4:315–41
73. Curran PJ. 1983. Estimating green LAI from multispectral aerial photography. A hand-held 35-mm camera provided acceptable results for small (36 m^2) study sites. *Remote Sens. Environ.* 49: 1709–20
74. Curran PJ. 1985. *Principles of Remote Sensing*. UK: Longman Sci. Tech. 282 pp.
75. Curran PJ. 1989. Remote sensing of foliar chemistry. *Remote Sens. Environ.* 30:271–78
76. Curran PJ, Dungan JL. 1990. An image recorded by the airborne visible/infrared imaging spectrometer (AVIRIS). *Int. J. Remote Sens.* 2:929–31
77. Curran PJ, Dungan JL, Macler BA, Plummer SE. 1991. The effect of a red leaf pigment on the relationship between red edge and chlorophyll concentration. *Remote Sens. Environ.* 35:69–76
78. Curran PJ, Dungan JL, Macler BA, Plummer SE, Peterson DL. 1992. Reflectance spectroscopy of fresh whole leaves for the estimation of chemical concentration. *Remote Sens. Environ.* 39:153–66
79. Curran PJ, Williamson HD. 1987. GLAI estimation using measurements of red, near infrared, and middle infrared radiance. *Remote Sens. Environ.* 53:181–86
80. Daley PF. 1993. Chlorophyll fluorescence analysis and imaging in plant stress and disease. *Abstr. 6th ISPP Congr., Montreal, Canada,* p. 11
81. Daley PF. 1995. Chlorophyll fluorescence analysis and imaging in plant stress and disease. *Can. J. Plant Pathol.* 16:In press
82. Daley PF, Raschke K, Ball JT, Berry JA. 1988. Topography of photosynthetic activity of leaves obtained from video images of chlorophyll fluorescence. *Plant Physiol.* 90:1233–38
83. Demetriades-Shah TH, Kanemasu ET. 1989. Remote sensing of crop condition. *Proc. Symp. Climate Agric. Systems Approach to Decision Making, Charleston, SC,* ed. A Weiss, pp. 34–49
84. Downing HG, Carter GA, Holladay KW, Cibula WG. 1993. The radiative equivalent water thickness of leaves. *Remote Sens. Environ.* 46:103–7
85. Duczek LJ. 1987. Infrared thermometry of leaf temperature measurements and common root rot of spring wheat and spring barley. *Plant Soil* 101:287–90
86. Dusek DA, Jackson RD, Musick JJ. 1985. Winter wheat vegetation indices calculated from combinations of seven spectral bands. *Remote Sens. Environ.* 18:255–67
87. Edner H, Johansson J, Svanberg S, Wallinder E. 1994. Fluorescence lidar multi-color imaging of vegetation. *Appl. Opt.* 33:2471–79
88. Edner H, Johansson J, Svanberg S, Wallinder E, Bazzani M, et al. 1992. Laser-induced fluorescence monitoring of vegetation in Tuscany. *EARSeL Adv. Remote Sens.* 1:119–30
89. Edwards GJ. 1982. Near infrared aerial video evaluation for freeze damage. *Proc. Fla. State Hortic. Soc.* 95:1–3
90. Edwards GJ, Blazquez CH. 1986. Preliminary experiments with remote sensing to detect citrus canker. *Proc. Fla. State Hortic. Soc.* 98:16–18
91. Epema GF. 1991. Studies of errors in field measurements of the bidirectional reflectance factor. *Remote Sens. Environ.* 35:37–49

92. Escobar DE, Bowen HW, Gausman HW, Cooper GR. 1983. Use of a near-infrared video recording system for the detection of freeze-damaged citrus leaves. *J. Rio Grande Valley Hortic. Soc.* 36:61–66
93. Everitt JH, Escobar DE. 1990. The status of video systems for remote sensing applications. *Proc. Bienn. Workshop Color Aerial Photogr. Videogr. Plant Sci. Related Fields, 12th, Nevada, 1989,* pp. 6–29. Falls Church, VA: Am. Soc. Photogramm. Remote Sens.
94. Everitt JH, Escobar DE, Alaniz MA, Davis MR. 1992. Comparison of ground reflectance and aerial video data for estimating range phytomass and cover. See Ref. 10, pp. 127–34
95. Everitt JH, Escobar DE, Blazquez CH, Hussey MA, Nixon PR. 1986. Evaluation of the mid-infrared (1.45 to 2.0 μm) with a black-and-white infrared video camera. *Photogramm. Eng. Remote Sens.* 52:1655–60
96. Everitt JH, Escobar DE, Villarreal R, Noriega JR, Davis MR. 1991. Airborne video systems for agricultural assessment. *Remote Sens. Environ.* 35:231–42
97. Everitt JH, Nixon PR. 1985. False color video imagery: A potential remote sensing tool for range management. *Photogramm. Eng. Remote Sens.* 51:675–79
98. Everitt JH, Richardson AJ. 1987. Canopy reflectance of seven rangeland plant species with variable leaf pubescence. *Photogramm. Eng. Remote Sens.* 53: 1571–75
99. Ferguson H, Eslik RF, Aase JK. 1973. Canopy temperatures of barley as influenced by morphological characteristics. *Agron. J.* 65:425–28
100. Filajdic N, Sutton TB. 1994. Optimum sampling size for determining different aspects of Alternaria blotch of apple caused by *Alternaria mali. Plant Dis.* 78:719–24
101. Forbes GA, Jeger MJ. 1987. Factors affecting the estimation of disease intensity in simulated plant structures. *J. Plant Dis. Prot.* 94:113–20
102. Fouché PS. 1986. Agricultural research with R/C aircraft. *Landbou Weekblad* 415:30–32
103. Fouché PS. 1992. Assessment of crop stress conditions using low altitude aerial color infra-red photography and computer processing. See Ref. 10, pp. 18–25
104. Fouché PS. 1994. Estimation of crop water stress in soybean and wheat using low altitude aerial infrared themometry and color infrared photography. See Ref. 34, pp. 174–84
105. Fouché PS, Booysen NW. 1990. Remotely piloted aircraft for low altitude aerial surveillance in agriculture. See Ref. 93, pp. 277–85
106. Fouché PS, Booysen NW. 1994. Low altitude surveillance of agricultural crops using inexpensive remotely piloted aircraft. See Ref. 60, pp. 315–26
107. Fredericksen TS, Skelly JM. 1994. Relation of visible and physiological foliar injury from ozone exposure in hardwood tree species. *Phytopathology* 84:11: 1371
108. Gamon JA, Penuelas J, Field CB. 1992. A narrow-waveband spectral index that tracks diurnal changes in photosynthetic efficiency. *Remote Sens. Environ.* 41: 35–44
109. Gao BC, Goetz AFH. 1994. Extraction of dry leaf spectral features from reflectance spectra of green vegetation. *Remote Sens. Environ.* 47:369–74
110. Gausman HW, Cardenas R. 1969. Effect of leaf pubescence of *Gynura aurantiaca* on light reflectance. *Bot. Gaz.* 130:158–62
111. Gausman HW, Cardenas R. 1973. Light reflectance by leaflets of pubescent, normal, and glabrous soybean lines. *Agron. J.* 65:837–38
112. Gausman HW, Escobar DE, Bowen RL. 1983. A video system to demonstrate interactions of near-infrared radiation with plant leaves. *Remote Sens. Environ.* 13:363–66
113. Gausman HW, Escobar DE, Rodriguez RR. 1973. Discrimination among plant nutrient deficiencies with reflectance measurements. Reprinted in *Color Aerial Photography in the Plant Sciences and Related Fields: A Compendium 1967–1983,* ed. GJ Edwards, 1988, pp. 112–26. Falls Church, VA: Am. Soc. Photogramm. Remote Sens. 306 pp.
114. Gausman HW, Menges RM, Escobar DE, Everitt JH, Bowen RL. 1977. Pubescence affects spectra and imagery of silverleaf sunflower (*Helianthus argophyllus*). *Weed Sci.* 256:437–40
115. Gerten DM, Wiese MV. 1984. Video image analysis of lodging and yield loss in winter wheat relative to foot rot. *Phytopathology* 74:872
116. Gerten DM, Wiese MV. 1987. Microcomputer-assisted video image analysis of lodging in winter wheat. *Photogramm. Eng. Remote Sens.* 53:83–88
117. Gilligan CA. 1980. Size and shape of sampling units for estimating incidence of sharp eyespot (*Rhizoctonia cerealis*) in plots of wheat. *J. Agric. Sci.* 99:461–64
118. Gilligan CA. 1980. Size and shape of sampling units for estimating incidence

of stem canker on oil seed rape stubble in field plots after swathing. *J. Agric. Sci.* 94:493–96

119. Gilligan CA. 1986. Use and misuse of the analysis of variance in plant pathology. *Adv. Plant Pathol.* 5:225–61
120. Goetz AFH, Vane G, Solomon JE, Rock BN. 1985. Imaging spectrometry for earth remote sensing. *Science* 228: 1147–53
121. Goodman BA, Williamson B, Chudek JA. 1992. Non-invasive observation of the development of fungal infection in fruit. *Protoplasma* 166:107–9
122. Goodman BA, Williamson B, Chudek JA. 1992. Nuclear magnetic resonance (NMR) microimaging of raspberry fruit: further studies on the origin of the image. *New Phytol.* 122:529–35
123. Gottwald R, Teddes WL. 1986. MADDSAP-1, A versatile remotely piloted vehicle for agricultural research. *J. Econ. Entomol.* 79:857–63
124. Goward S, Markham B, Dye DG, Dulaney W, Yang J. 1991. Normalized difference vegetation index measurements from the advanced very high resolution radiometer. *Remote Sens. Environ.* 35:257–77
125. Grant L, Daughtry CST, Vanderbilt VC. 1987. Variations in the polarized leaf reflectance of *Sorgum bicolor*. *Remote Sens. Environ.* 21:333–39
126. Grant L, Daughtry CST, Vanderbilt VC. 1987. Polarized and non-polarized leaf reflectances of *Coleus blumei*. *Environ. Exp. Bot.* 27:139–45
127. Greaves DA, Hooper AJ, Walpole BJ. 1983. Identification of barley yellow dwarf virus and cereal aphid infestation in winter wheat by aerial photography. *Plant Pathol.* 32:159–72
128. Gross D, Lehman V, Mattingly MM, Rohr G. 1991. *NMR Microscopy, Applications in Biology, Biomedicine and Material Sciences*. Karlsruhe, Germany: Bruker Analytische Messtechnik GMBH. 41 pp.
129. Gu X-F, Guyot G. 1993. Effect of diffuse irradiance on the reflectance factor of reference panels under field conditions. *Remote Sens. Environ.* 45:249–60
130. Günther KP. 1990. Vegetation stress monitoring by fluorescence. *ESA SP-301*, pp. 135–42
131. Günther KP, Dahn HG, Lüdeker W. 1994. Remote sensing vegetation status by laser-induced fluorescence. *Remote Sens. Environ.* 47:10–17
132. Günther KP, Lüdeker W, Dahn HG. 1991. Design and testing of a spectral resolving fluorescence LIDAR system for remote sensing of vegetation. *Proc. 5th Int. Colloq. Phys. Meas. Signat. Remote Sens., Courchevel*, pp. 723–26. Noordwijk, The Netherlands: ESA Publ.

132a. Hagler TB, Downs SW. 1979. Detection of plant stress through multispectral photography. *Ala. Coop. Ext. Serv.*, IM:2:79, pp. 1–12

133. Halloin JM, Cooper TG, Potchen EJ. 1992. A study of disease development in *Rhizoctonia solani* infected sugarbeets using magnetic resonance imaging. *Phytopathology* 82:1160
134. Hansen JG. 1991. Use of multispectral radiometry in wheat yellow rust experiments. *EPPO Bull.* 21:651–58
135. Hansen JG, Jorgensen LN, Simonsen J. 1992. Multispectral radiometry, a source of additional data in field fungicide trials. *Tidsskr. Planteavls Specialserie, Beretning S-2207–1992*, pp. 39–37
136. Hart WG, Everitt JH, Escobar DE, Davis MR, Garza MG. 1988. Comparing imaging systems for assessment of diverse conditions of agricultural resources. *Proc. 1st Workshop Videogr., 1st*, pp. 160–65. Terra Haute, IN: Am. Soc. Photogramm. Remote Sens.
137. Hatfield JL, Pinter PJ Jr. 1993. Remote sensing for crop protection. *Crop Prot.* 12:403–13
138. Hatfield JL, Pinter PJ Jr, Chasseray E, Ezra CE, Reginato RJ, et al. 1984. Effects of panicles on infrared thermometer measurements of canopy temperature in wheat. *Agric. For. Meteorol.* 32:97–105
139. Herbert TJ. 1995. Stereophotogrammetry of plant leaf angles. *Photogramm. Eng. Remote Sens.* 61:89–92
140. Herbert TT. 1982. The rationale for the Horsfall-Barratt Plant disease assessment scale. *Phytopathology* 72:1269
141. Hoffer RM. 1989. Color and infrared photography for vegetation assessment. See Ref. 93, pp. 1–5
142. Holmes MG. 1990. Applications of radar in agriculture. See Ref. 326a, pp. 307–31
143. Hooper A. 1980. Aerial photography as an aid to pest forecasting. *J. R. Aeronaut. Soc.* 84:136–49
144. Horler DNH, Dockray M, Barber J. 1983. The red edge of plant leaf reflectance. *Int. J. Remote Sens.* 4:273–88
145. Horsfall JG, Barratt RW. 1945. An improved grading system for measuring plant diseases. *Phytopathology* 35:655
146. Horsfall JG, Cowling EB. 1978. Pathometry: The measurement of plant disease. In *Plant Disease, An Advanced Treatise*, ed. JG Horsfall, EB Cowling, 2:119–36. New York: Academic
147. Huning JR, Cameron AO. 1994. The

NASA airborne science program in support of Mission to Planet Earth. See Ref. 60, I:448–58
148. Hunt ER Jr. 1991. Airborne remote sensing of canopy water thickness scaled from leaf spectrometer data. *Int. J. Remote Sens.* 12:643–49
149. Idso SB, Jackson RD, Pinter PJ Jr, Reginato RJ, Hatfield JL. 1981. Normalizing the stress-degree-day for environmental variability. *Agric. Meteorol.* 24:45–55
150. Idso SB, Jackson RD, Reginato RJ. 1977. Remote sensing of crop yields. *Science* 196:19–25
151. Idso SB, Reginato RJ, Hatfield JL, Walker GK, Jackson RD, Pinter PJ Jr. 1980. A generalization of the stress-degree-day concept of yield prediction to accommodate a diversity of crops.*Agric. Meteorol.* 21:205–11
152. Inoué S, Walter RJ Jr, Berns MW, Ellis GW, Hansen E. 1987. *Video Microscopy.* New York: Plenum. 584 pp.
153. Jackson HR. 1964. Detection of plant disease symptoms. *J. Biol. Photogr. Assoc.* 32:45–58
154. Jackson HR, Hodgson WA, Wallen VR, Philpotts LE, Hunter J. 1971. Potato late blight intensity levels as determined by microdensitometer studies of false-color-aerial photographs. *J. Biol. Photogr. Assoc.* 39:101–6
155. Jackson HR, Wallen VR. 1970. Comparison of optical density differences in aerial photographs between plant canopy and soils with varying surface moisture. *J. Biol. Photogr. Assoc.* 47:43–47
156. Jackson HR, Wallen VR. 1975. Microdensitometer measurements of sequential aerial photographs of field beans infected with bacterial blight. *Phytopathology* 65:961–68
157. Jackson HR, Wallen VR. 1976. The epiphytology of late blight of potato monitored by sequential color infrared aerial photography. *J. Appl. Photogr. Eng.* 2:207–12
158. Jackson HR, Wallen VR, Downer JF. 1978. Analysis and electronic area measurement of complex aerial photographic images. *J. Appl. Photogr. Eng.* 4:101–6
159. Jackson RD. 1982. Canopy temperature and crop water stress. *Adv. Irrig.* 1:43–85
160. Jackson RD. 1983. Spectral indices in n-space. *Remote Sens. Environ.* 13:409–21
161. Jackson RD. 1983. Assessing moisture stress in wheat with hand-held radiometers. *SPIE* 356:138–42
162. Jackson RD. 1986. Remote sensing of biotic and abiotic plant stress. *Annu. Rev. Phytopathol.* 24:265–87
163. Jackson RD, Idso SB, Reginato RJ, Pinter PJ Jr. 1981. Canopy temperature as a crop water stress indicator. *Water Resour. Res.* 17:1133–38
164. Jackson RD, Jones CA, Uehara G, Santo LT. 1981. Remote detection of nutrient and water deficiencies in sugarcane under variable cloudiness. *Remote Sens. Environ.* 11:327–31
165. Jackson RD, Pinter PJ Jr. 1986. Spectral response of architecturally different wheat canopies. *Remote Sens. Environ.* 20:43–56
166. Jackson RD, Pinter PJ Jr, Idso SB, Reginato RJ. 1979. Wheat spectral reflectance, interaction between crop configuration, sun elevation and azimuth angle. *Appl. Opt.* 18:3730–32
167. Jackson RD, Pinter PJ Jr, Reginato RJ, Idso SB. 1980. Hand-held radiometry. *Agric. Rev. Man. West. Ser. USDA Sci. Educ. Admin.* 19:1–66
168. Jackson RD, Reginato RJ, Idso SB. 1977. Wheat canopy temperature: A practical tool for evaluating water requirements. *Water Resour. Res.* 13:651–56
169. Jackson RD, Slater PN, Pinter PJ Jr. 1983. Discrimination of growth and water stress in wheat by various vegetation indices through clear and turbid atmospheres. *Remote Sens. Environ.* 13: 187–208
170. James WC. 1971. A manual of assessment keys for plant diseases. *Can. Dep. Agric. Publ.* 1458:197
171. Johansson J. 1993. *Fluorescence spectroscopy for medical and environmental diagnostics,* pp. 1–106. Diss. Dep. Phys., Inst. Tech., Lund Univ., Sweden
172. Johnson GW, Bailey JE, Bruck RI, Matyac CA. 1986. Studies of diseased root tissue using nuclear magnetic resonance imaging. *Phytopathology* 76:1067
173. Karlsson E. 1968. *Mätteknik,* p. 113. Univ. fvrl. Uppsala
174. Kasile JD. 1994. Mini-platform/mini-format aerial photography. See Ref. 60, I:363–71
175. Kauth RJ, Thomas GS. 1976. The tasseled cap - A graphic description of the spectral-temporal development of agricultural crops as seen by Landsat. *Proc. Symp. Machine Processing of Remotely Sensed Data,* pp. 41–51. LARS, Purdue Univ., West Lafayette, IN
176. Kautsky H, Hirsch A. 1934. Chlorophyllfluoreszenz und Kohlensäureassimilation. I. Mitteilung: das Fluoreszenzverhalten grüner Pflanzen. *Biochem. Z.* 247:423–34

177. Kennedy BW. 1986. Alteration of leaf angles and circadian leaf movements in seedling sunflowers infected by *Pseudomonas tagetes*. *Phytopathology* 76: 1085–86
178. Kettler DJ, Escobar DE, Everitt JH. 1992. Insights into interpretability differences between color infrared photography and color infrared video data. See Ref. 10, pp. 199–206
179. Kharuk VI, Morgun VN, Rock BN, Williams DL. 1994. Chlorophyll fluorescence and delayed fluorescence as potential tools in remote sensing: A reflection of some aspects of problems in comparative analysis. *Remote Sens. Environ.* 47:98–105
180. Kimes DS, Markham BL, Tucker CJ, McMurtrey JE. 1981. Temporal relations between spectral response and agronomic variables of a corn canopy. *Remote Sens. Environ.* 11:401–11
181. Kinder CR. 1992. Remote sensing as a cranberry marsh photography and videography in plant sciences. See Ref. 10, pp. 68–78
182. King D. 1994. Digital frame cameras: the next generation of low cost remote sensors. See Ref. 34, pp. 19–28
183. King D, Walsh P, Ciuffreda F. 1992. Elevation determination using airborne digital frame camera imagery. See Ref. 10, pp. 147–57
184. Klosson RJ, Krause GH. 1981. Freezing injury in cold-acclimated and unhardened spinnach leaves. II. Effects of freezing on chlorophyll fluorescence and light scattering reactions. *Planta* 151:347–52
185. Knipling EB. 1967. Physical and physiological basis for differences in reflectance of healthy and diseased plants. See Ref. 113, pp. 8–32
186. Knipling EB. 1970. Physical and physiological basis for the reflectance of visible and near-infrared radiation from vegetation. *Remote Sens. Environ.* 1: 155–59
187. Kokko EG, Conner RL, Kozub GC, Lee B. 1993. Quantification by image analysis of subcrown internode discoloration in wheat caused by common root rot. *Phytopathology* 83:976–81
188. Krajicek V, Vrbova M. 1994. Laser-induced fluorescence spectra of plants. *Remote Sens. Environ.* 47:51–54
189. Kranz J. 1988. Measuring plant disease. In *Experimental Techniques in Plant Disease Epidemiology*, ed. J Kranz, J Rotem, pp. 35–50. New York: Springer
190. Lamari L. 1994. Quantification of foliar diseases using true colour computer image analysis. *Phytopathology* 84:1068
191. Lee YJ. 1989. Aerial photography for the detection of soilborne disease. *Can. J. Plant Pathol.* 11:173–76
192. Lichtenthaler HK, ed. 1988. *Applications of Chlorophyll Fluorescence in Photosynthesis Research, Stress Physiology, Hydrology and Remote Sensing.* London: Kluwer. 366 pp.
193. Lichtenthaler HK. 1988. In vivo chlorophyll fluorescence as a tool for stress detection in plants. See Ref. 192, pp. 129–42
194. Lichtenthaler HK. 1988. Remote sensing of chlorophyll fluorescence in oceanography and in terrestrial vegetation: An introduction. See Ref. 192, pp. 287–97
195. Lichtenthaler HK, Bauschmmann C, Rinderle U, Schmuck G. 1984. Application of chlorophyll fluorescence in ecophysiology. *Radiat. Environ. Biophys.* 25:297–308
196. Lichtenthaler HK, Hak R, Rinderle U. 1990. The chlorophyll fluorescence ratio F690/F730 in leaves of different chlorophyll content. *Photosynth. Res.* 25: 295–98
197. Lichtenthaler HK, Lang M, Stober F. 1991. Nature and variation of blue fluorescence spectra of terrestrial plants. *Proc. Int. Geosci. Remote Sens. Symp. IGARSS-91*, pp. 2283–86. Helsinki, Espoo, Finland
198. Lichtenthaler HK, Lang M, Stober F. 1991. Laser-induced blue fluorescence and red chlorophyll fluorescence signatures of differently pigmented leaves. See Ref. 132, pp. 727–30
199. Lichtenthaler HK, Rinderle U. 1988. The role of chlorophyll fluorescence in the detection of stress conditions in plants. *CRC Crit. Rev. Anal. Chem.* 19: 29–85
200. Lichtenthaler HK, Stober F. 1990. Laser-induced chlorophyll fluorescence and blue fluorescence of green vegetation. *Proc. 10th EARSeL Symp., Toulouse*, pp. 234–41. EARSeL, Boulogne-Billancourt
201. Lichtenthaler HK, Stober F, Lang M. 1992. The nature of the different laser-induced fluorescence signature of plants. *EARSeL Adv. Remote Sens.* 1: 20–32
202. Lillesand TM, Meisner DM, French DW, Johnsson JL. 1981. Evaluation of digital photographic enhancement for Dutch elm disease detection. *Photogramm. Eng. Remote Sens.* 48:1581–92
203. Lillesand TM, Seely ML, Lindstrom OM, Goldblatt M, Johnson WL, Meisner DE. 1983. Environmental and spectral data relating to sunflower crop

condition assessment. *RSL Rep.* 83:2:1–47. Remote Sens. Lab., Univ. Minn., St. Paul
204. Lindow SE. 1983. Estimating disease severity of single plants. *Phytopathology* 73:1576–81
205. Lindow SE, Anderson GL. 1986. Microcomputer measurements of pathogen injury to weeds. *Weed Sci.* 34(Suppl. 1):38–42
206. Lindow SE, Webb RR. 1981. Use of digital video image analysis in plant disease assessment. *Phytopathology* 71: 890
207. Lindow SE, Webb RR. 1983. Quantification of foliar plant disease symptoms by microcomputer-digitized video image analysis. *Phytopathology* 73: 520–24
208. Lord D, Desjardins RL, Dubé PA. 1988. Sun-angle effects on the red and near-infrared reflectances of five different crop canopies. *Can. J. Remote Sens.* 14:46–55
209. Luzi G, Paloscia S, Pampaloni P. 1990. Microwave radiometry for monitoring agricultural crops. See Ref. 326a, pp. 308–53
210. Madden LV. 1980. Quantification of disease progression. *Prot. Ecol.* 2:159–76
211. Madden LV. 1986. Statistical analysis and comparison of disease progress curves. In *Plant Disease Epidemiology*, ed. KJ Leonard, WE Fry, pp. 55–84. New York: Macmillan
212. Madden LV. 1995. Modelling yield losses at the field scale. *Proc. 6th ISPP Congr., Montreal, 1993. Can. J. Plant Pathol.* 16:In press
213. Madden LV, Knoke JK, Louie R. 1984. Experimental design for determination of yield losses due to maize dwarf mosaic virus. *Phytopathology* 74:809
214. Malthus TJ, Andrieu B, Danson FM, Jaggard KW, Steven MD. 1993. Candidate high spectral resolution infrared indices for crop cover. *Remote Sens. Environ.* 46:204–12
215. Manzer FE, Cooper GR. 1967. Aerial photographic methods of potato disease detection. *Maine Agr. Exp. Stn. Bull. 646*, pp. 1–14
216. Manzer FE, Cooper GR. 1982. Use of portable video-taping for aerial infrared detection of potato diseases. *Plant Dis.* 66:665–67
217. Mattsson JO. 1974. *Climatic information in night-recorded aerial photography with special regard to registration made in retroreflected light.* Lunds Univ. Naturgeografiska Inst., Rapp. Not. No. 23:1–56
218. Mattsson JO, Cavallin Ch. 1972. Retroreflection of light from drop-covered surfaces and image-producing device for registration of this light. *Lund Stud. Geogr. Ser. A Phys. Geogr.* 53: 285–94
219. Mausel PW, Everitt JH, Escobar DE, King DJ. 1992. Airborne videography: current status and future perspectives. *Photogramm. Eng. Remote Sens.* 58: 1189–95
220. Mauser W, Bach H. 1994. Imaging spectroscopy in hydrology and agriculture - determination of model parameters. In *Imaging Spectrometry - a Tool for Environmental Observations*, ed. J Hill, J Mégier, pp. 261–83. Brussels/Luxembourg: ECSC, EEc, EAEC
221. McCreight RW, Waring RH, Chen CF. 1994. Airborne environmental analysis using ultralight aircraft system. See Ref. 60, I:384–92
222. McDaniel KC, Haas RH. 1982. Assessing mesquite-grass vegetation condition from Landsat. *Photogramm. Eng. Remote Sens.* 48:441–50
223. McFarlane JC, Watson RD, Theisen AF, Jackson RD, Ehrler WL, et al. 1980. Plant stress detection by remote measurement of fluorescence. *Appl. Opt.* 19: 3287–89
224. McGeer T. 1994. Very small autonomous aircraft for economical long-range deployment of lightweight instruments. See Ref. 60, I:329–39
225. McMurtrey JE III, Chappelle EW, Kim MS, Meisinger JJ, Corp LA. 1994. Distinguishing nitrogen fertilization levels in field corn (*Zea mays* L.) with actively induced fluorescence and passive reflectance measurements. *Remote Sens. Environ.* 47:36–44
226. Meisner DE, Lindstrom OM. 1985. Remote sensing research in agriculture and forestry: Design and operation of a color infrared aerial video system. *Res. Rep. 84:1.* Remote Sensing Lab., Univ. Minn., St. Paul. 9 pp.
227. Mengistu A, Tachibana H, Epstein AH, Bidne KG, Hatfield JD. 1987. Use of leaf temperature to measure the effect of brown stem rot and soil moisture stress and its relation to yields of soybeans. *Plant Dis.* 71:632–34
228. Methy M, Olioso A, Trabaud L. 1994. Chlorophyll fluorescence as a tool for management of plant resources. *Remote Sens. Environ.* 47:2–9
229. Meyer MP, Calpouzos L. 1968. Detection of crop diseases. *Photogramm. Eng. Rem. Sens.* 34:554–56
230. Michaels TE. 1986. Measuring ozone sensitivity of white bean using digitized

video analysis. *Ann. Rep. Bean Improv. Coop.* 29:20–21

231. Millard JP, Goettelman RC, LeRoy MJ. 1980. Infrared-temperature variability in a large agricultural field. *NASA Tech. Memo. 81222*:1–23
232. Millard JP, Jackson RD, Goettelman RC, Reginato RJ, Idso SB. 1978. Crop water-stress assessment using an airborne thermal scanner. *Photogramm. Eng. Remote Sens.* 44:77–85
233. Milton EJ. 1987. Principles of radiometry. *Int. J. Remote Sens.* 8:1807–27
234. Monteith JL, Szeicz G. 1962. Radiative temperature in the heat balance of natural surfaces. *Q. J. R. Meteorol. Soc.* 88:496–507
235. Moran MS, Clarke TR, Inoue Y, Vidal A. 1992. Estimating crop water deficit using the relation between surface-air temperature and spectral vegetation index. *Remote Sens. Environ.* 49:246–64
236. Morrison IM, Christie WW, Goodman BA. 1992. Chemistry. *Scott. Crop Res. Inst., Annu. Rep. 1992*, pp. 54–58
237. Morrison M, Voldeng H. 1994. Soybean pubescence - does it make a difference? In *Resource Capture by Crops*, ed. JL Monteith, RK Scott, MH Unsworth, pp. 428–30. UK: Nottingham Univ. Press. 469 pp. *Poster-report 48th Easter School, Nottingham*, p. 1 (Abstr.)
238. Muir AY, Porteus RL, Wastie RL. 1982. Experiments in the detection of incipient diseases in potato tubers by optical methods. *J. Agric. Eng. Res.* 27:131–38
239. Mulesky M, Chism B, Piermarini M, Hagedorn C. 1992. A computerized image analysis program for agricultural applications. *Phytopathology* 82: 1155
240. Myers VI. 1983. Remote sensing applications in agriculture. In *Manual of Remote Sensing*,, ed. RN Colwell, 2:2111–28. Falls Church, VA: Am. Soc. Photogramm. Remote Sens. 2nd ed.
241. Myhre RJ. 1994. A color airborne video system developed within the US Forest Service. See Ref. 34, pp. 9–18
242. Myneni RB, Asrar G. 1994. Atmospheric effects and spectral vegetation indices. *Remote Sens. Environ.* 47:390–402
243. Neale CMU. 1992. An airborne multispectral video/radiometer remote sensing system for natural resource monitoring. See Ref. 10, pp. 119–26
244. Neale CMU, Kuiper J, Tarbet KL, Qiu X. 1994. Image enhancement and processing automation routines for digital multispectral video imagery. See Ref. 34, pp. 29–36
245. Neblette CB. 1927. Aerial photography for the study of plant diseases. *Photo-Era Mag.* 58:346
246. Neblette CB. 1928. Airplane photography for plant disease surveys. *Photo-Era Mag.* 59:175
247. Nilsson HE. 1969. Studies of root and foot rot diseases of cereals and grasses. I. On resistance to *Ophiobolus graminis* Sacc. *Lantbrhvgsk. Annlr.* 35:275–807
248. Nilsson HE. 1980. Remote sensing and image processing for disease assessment. *Prot. Ecol.* 2:271–74
249. Nilsson HE. 1980. Application of remote sensing and image analysis at macroscopic and microscopic levels in plant pathology. See Ref. 337, pp. 76–84
250. Nilsson HE. 1983. Remote sensing of vegetation damage - Plant pathology approaches and crop loss assessment. *Proc. RNRD Symp. Appl. Remote Sens. Resour. Manage., Seattle*, pp. 381–402. Falls Church, VA: Am. Soc. Photogramm. Remote Sens.
251. Nilsson HE. 1983. Temperaturstegring hos plantbestaånd av korn angripet av strimsjuka. *Växtskyddsnotiser* 47:46–48
252. Nilsson HE. 1984. Remote sensing of 6-row barley infected by barley stripe disease. *Växtskyddsrapp. Jordbruk* 36: 1–49
253. Nilsson HE. 1985. Remote sensing of oil seed rape infected by *Sclerotinia*-stem rot and *Verticillium*-wilt. *Växtskyddsrapp. Jordbruk* 33:1–33
254. Nilsson HE. 1985. Remote sensing of 2-row barley infected by net blotch disease. *Växtskyddsrapp. Jordbruk* 34:1–101
255. Nilsson HE. 1986. Remote sensing in plant pathology: information on methods development. *Växtskyddsrapp. Jordbruk* 39:95–104. In Swedish
256. Nilsson HE. 1987. Experiments on hand-held radiometry and IR-thermography of winter wheat in field plot experiments. *Växtskyddsrapp. Jordbruk* 44:1–48
257. Nilsson HE. 1991. Hand-held radiometry and IR-thermography of plant diseases in field plot experiments. *Int. J. Remote Sens.* 12:545–57
258. Nilsson HE. 1995. Remote sensing and image analysis methods in phytopathometry. See Ref. 212
259. Nilsson HE, Alness K. 1995. Hand-held radiometry of field experiments in potatoes damaged by a herbicide and by late blight. *Plant Pathol.* Submitted
260. Nilsson HE, Bramstorp A. 1994. Remote sensing of field experiments in sugar beets. *Betodlaren Organ Sver. Be-*

todlares Centralförening 57:131–35. In Swedish
261. Nilsson HE, Johnsson L. 1988. Remote sensing of plant diseases in field plot experiments. *Växtskyddsrapp. Jordbruk* 49:111–22
262. Nilsson HE, Linnér H. 1987. Remote sensing of wheat and barley in a field plot experiment with different levels of nitrogen fertilization and irrigation. *Växtskyddsrapp. Jordbruk* 45:1–246
263. Nilsson HE, Linnér H. 1987. IR-thermography of canopy temperatures of wheat and barley at different nitrogen fertilization and irrigation. *Växtskyddsrapp. Jordbruk* 43:1–49
264. Noodén LD. 1988. The phenomena of senescence and aging. In *Senescence in Plants*, ed. LD Noodén, AC Leopold, pp. 1–50. New York: Academic. 526 pp.
265. Nowling S, Tueller PT. 1994. A low-cost multispectral airborne video image system for vegetation monitoring on range and forest lands. See Ref. 34, pp. 1–8
266. Nutter FW Jr. 1987. Detection of plant disease gradients using a hand-held, multispectral radiometer. *Phytopathology* 77:643
267. Nutter FW Jr. 1989. Detection and measurement of plant disease gradients in peanut with multispectral radiometer. *Phytopathology* 79:958–63
268. Nutter FW Jr. 1990. Remote sensing and image analysis for crop loss assessment. In *Crop Loss Assessment in Rice*, pp. 93–105. Manila, The Philippines: Int. Rice Res. Inst.
269. Nutter FW Jr. 1993. *DISEASE PRO* - A computer program for training of visual disease assessment. Demonstration at 6th ISPP Congr., Montreal, Canada
270. Nutter FW Jr, Alderman SC. 1987. Use of late leafspot disease gradients to evaluate disease assessment schemes for accuracy, precision, resolution and speed. *Phytopathology* 77:1700
271. Nutter FW Jr, Cunfer BM. 1988. Quantification of barley yield losses caused by *Rhynchosporium secalis* using visual versus remote sensing assessment methods. *Phytopathology* 78:1530
272. Nutter FW Jr, Gleason ML, Jenco JH, Christians NC. 1991. Effect of visual, remote sensing, and image analysis assessment methods on intra-rater and inter-rater reliability estimates in the dollar spot - bentgrass pathosystem. *Phytopathology* 81:1182
273. Nutter FW Jr, Gleason ML, Jenco JH, Christians NC. 1993. Assessing the accuracy, intra-rater repeatability, and inter-rater reliability of disease assessment systems. *Phytopathology* 83:806–12
274. Nutter FW Jr, Littrell RH, Brenneman TB. 1990. Utilization of a multispectral radiometer to evaluate fungicide efficacy to control late leaf spot in peanut. *Phytopathology* 80:102–8
275. Nutter FW Jr, Littrell RH, Pederson VD. 1985. Use of a low-cost, multispectral radiometer to estimate yield loss in peanuts caused by late leafspot (*Cercosporium personatum*). *Phytopathology* 75:502
276. Nutter FW Jr, Littwiller D. 1993. *ALFALFA.PRO* - A computerized disease assessment training and evaluation program for foliar disease of alfalfa. Demonstration at 6th ISPP Congr., Montreal, Canada
277. Nutter FW Jr, Pederson VD. 1984. Yield loss modeling: Tolerance to pathogens. *Phytopathology* 74:872–73
278. Nutter FW Jr, Schultz PM. 1995. Improving the accuracy and precision of disease assessments: selection of methods and use of computer-aided training programs. *Can. J. Plant Pathol.* 16:In press
279. Nutter FW Jr, Worawitlikit O. 1989. *DISEASE.PRO* - A computer program for evaluation and improving a person's ability to assess disease proportion. *Phytopathology* 79:1135
280. O'Brien RD, van Bruggen AHC. 1992. Accuracy, precision and correlation to yield loss of disease severity scales for corky root of lettuce. *Phytopathology* 82:91–96
281. Olson CE Jr. 1969. Early remote detection of physiologic stress in forest stands. See Ref. 113, pp. 173–88
282. Osmond CB, Berry JA, Balachandran S, Büchen-Osmond C, Daley PF, Hodgson RAJ. 1990. Potential consequences of virus infection for shade-sun acclimation in leaves. *Bot. Acta* 103:226–29
283. Osvald H. 1959. *Åkerns nyttoväxter.* Svensk Litteratur AB, Stockholm. 596 pp.
284. O'Toole JC, Hatfield JL. 1983. Effect of wind on the crop water stress index derived by infrared thermometry. *Agron. J.* 75:811–17
285. Otten LJ III, Rafert TB, Sellar RG, Holbert E. 1994. Visible hyper-spectral imaging for airborne environmental sensing. See Ref. 60, II:284–95
286. Paleg LG, Aspinal D, eds. 1981. *The Physiology and Biochemistry of Drought Resistance in Plants.* Sydney: Academic. 492 pp.
287. Pederson VD. 1984. Multispectral radi-

ometry using a 12-bit analog-to-digital converter interfaced with a portable microcomputer. *Phytopathology* 74:872

288. Pederson VD. 1986. Estimation of efficacy of foliar fungicides on barley by multispectral radiometry. *Phytopathology* 76:957
289. Pederson VD. 1986. Spectroradiometric readings of barley canopies grown from fungicide treated seed. *Phytopathology* 76:1147
290. Pederson VD, Fiechtner G. 1980. A low-cost compact data acquisition system for recording visible and infrared reflection from barley crop canopies. See Ref. 337, pp. 71–75
291. Pederson VD, Gudmestad N. 1977. Evaluation of foliar diseases of barley with multispectral sensors. *Proc. Am. Phytopathol. Soc.* 4:149
292. Pederson VD, Nutter FW Jr. 1982. Low-cost, portable multispectral radiometer for assessment of onset and severity of foliar disease of barley. *Proc. SPIE Int. Soc. Opt. Eng.* 356:126–30
293. Pennypacker SP, Scharen AL, Sharp EL, Sands DC. 1982. Spectral classification of wheat infected with barley yellow dwarf and stripe rust. *Phytopathology* 72:1006
294. Penuelas J, Gamon JA, Fredeen AL, Merino J, Field CB. 1994. Reflectance indices associated with physiological changes in nitrogen- and water-limited sunflower leaves. *Remote Sens. Environ.* 48:135–46
295. Perry CR, Lautenschlager LF. 1984. Functional equivalence of spectral vegetation indices. *Remote Sens. Environ.* 14:169–82
296. Deleted in proof
297. Peterson DL, Aber JD, Matson PA, Card DH, Swanberg N, et al. 1988. Remote sensing of forest canopy and leaf biochemical contents. *Remote Sens. Environ.* 24:85–108
298. Peterson DL, Running SW. 1989. Applications in forest science and management. In *Theory and Applications of Optical Remote Sensing*, ed. G Asrar, pp. 429–73. New York: Wiley
299. Pfeffer PE, Gerasimowicz WV. 1989. *Nuclear Magnetic Resonance in Agriculture*. Boca Raton, FL: CRC. 441 pp.
300. Philpotts LE, Wallen VR. 1969. IR-color for crop disease identification. *Photogr. Eng.* 55:1116–25
301. Philpotts LE, Wallen VR. 1970. The use of color infrared photography in estimating loss in white bean production in Huron County, Ontario, 1968. *Can. Farm Econ.* 5:4
302. Pinter PJ Jr. 1986. Effect of dew on canopy reflectance and temperature. *Remote Sens. Environ.* 19:187–205
303. Pinter PJ Jr, Jackson RD. 1981. Dew and vapor pressure as complicating factors in the interpretation of spectral radiance from crops. *Proc. 15th Int. Symp. Remote Sens. Environ.* 15:547–54
304. Pinter PJ Jr, Jackson RD, Ezra CE. 1985. Sun-angle and canopy-architecture effects on the spectral reflectance of six wheat cultivars. *Int. J. Remote Sens.*, 6:1813–25
305. Pinter PJ Jr, Jackson RD, Idso SB, Reginato RJ. 1983. Diurnal patterns of wheat spectral reflectances. *IEEE Trans. Geosci. Remote Sens.* 21:156–63
306. Pinter PJ Jr, Stanghellini ME, Reginato RJ, Idso SB, Jenkins AD, Jackson RD. 1979. Remote detection of biological stress in plants with infrared thermography. *Science* 205:585–87
307. Porteus RL, Muir AY, Wastie RL. 1981. The identification of diseases and defects in potato tubers from measurements of optical spectral reflectance. *J. Agric. Eng. Res.* 26:151–60
308. Qi J, Chehbouni A, Huete AR, Kerr YH, Sorooshian S. 1994. A modified soil adjusted vegetation index. *Remote Sens. Environ.* 48:119–26
309. Railyan V Ya, Korobov RM. 1993. Red edge structure of canopy reflectance spectra of *Triticale*. *Remote Sens. Environ.* 46:173–82
310. Richardson AJ, Blazquez CH. 1989. Estimating citrus grove production using image processing techniques. *J. Imaging Technol.* 15:272–76
311. Richardson AJ, Wiegand CL. 1977. Distinguishing vegetation from soil background. *Photogramm. Eng. Remote Sens.* 43:1541–52
312. Richardson AJ, Wiegand CL, Wanjura DF, Dusek D, Steiner JL. 1992. Multisite analysis of spectral-biophysical data for sorghum. *Remote Sens. Environ.* 41:71–82
313. Richter H. 1993. *Supraliminal contrast functions and volontary negative modulation of accomodation in the visual system*. PhD thesis. Dep. Psychol., Uppsala Univ., Sweden. 115 pp.
314. Rinderle U, Lichtenthaler HK. 1988. The chlorophyll fluorescence ratio F690/F735 as a possible stress indicator. See Ref. 192, pp. 189–96
315. Sarafis V, Rumpel H, Pope J, Kuhn W. 1990. Noninvasive histochemistry of plant materials by magnetic resonance microscopy. *Protoplasma* 159:70–73
316. Schneider CL, Safir GR. 1975. Infrared aerial photography estimation of yield potential in sugar beets exposed to

blackroot diseases. *Plant Dis. Rep.* 59: 627–31
317. Schreiber U, Berry JA. 1977. Heat-induced changes of chlorophyll fluorescence in intact leaves correlated with damage of the photosynthetic apparatus. *Planta* 136:233–38
318. Seck M, Teng PS, Roelfs AP. 1985. The role of wheat leaves in grain yield and leaf rust losses. *Phytopathology* 75: 1299
319. Sharp EL, Perry CR, Scharen AL, Boatwright GO, Sands DC, et al. 1985. Monitoring cereal rust development with a spectral radiometer. *Phytopathology* 75:936–69
320. Sherwood RT, Berg CC, Hoover MR, Zeiders KE. 1983. Illusions in visual assessment of *Stagonospora* leaf spot of orchardgrass. *Phytopathology* 73:173–77
321. Shine JM Jr, Comstock JC. 1993. Digital image analysis system for determining tissue-blot immunoassay results for ratoon stunting disease of sugarcane. *Plant Dis.* 77:511–13
322. Shokes FM, Berger RD, Smith DH, Rasp JM. 1987. Reliability of disease assessment procedures: A case study with late leafspot of peanut. *Oléagineux* 42:245–51
323. Slater PN, Jackson RD. 1982. Atmospheric effects on radiation reflected from soil and vegetation as measured by orbital sensors using various scanning directions. *Appl. Opt.* 21:3923–31
324. Smillie RM, Hetherington SE. 1983. Stress tolerance and stress-induced injury in crop plants measured by chlorophyll fluorescence in vivo. *Plant Physiol.* 72:1043–50
325. Smith RCG, Heritage AD, Stapper M, Barrs HD. 1986. Effect of stripe rust (*Puccinia striiformis* West.) and irrigation on the yield and foliage temperature of wheat. *Field Crops Res.* 14:39–51
326. Spomer IA, Smith MAL. 1989. Image analysis morphometric measurements for tissue water status and other determinations. *Agron. J.* 81:906–10
326a. Steven MD, Clark JA, eds. 1990. *Applications of Remote Sensing in Agriculture.* London: Butterworths. 427 pp.
327. Stober F, Lang M, Lichtenthaler HK. 1994. Blue, green, and red fluorescence emission signatures of green, etiolated, and white leaves. *Remote Sens. Environ.* 47:65–71
328. Stonehouse J. 1994. Assessment of Andean bean diseases using visual keys. *Plant Pathol.* 43:519–27
329. Strömberg A, Brishammar S, Nilsson HE. 1994. *Induced systemic resistance to potato late blight.*. PhD thesis. Swed. Univ. Agric. Sci., Uppsala. *Eur. J. Phytopathol.* Submitted
330. Strömberg A, Nilsson HE, Brishammar S. 1993. Induced systemic resistance to potato late blight in a field experiment studied with visual assessments and remote sensing. *Abstr. 6th ISPP Congr., Montreal,* p. 115
331. Stutte GW, Bors R, Stutte CA. 1990. ICAS quantification of nutrient stress in horticultural crops. See Ref. 93, pp. 109–17
332. Stutte GW, Stutte CA. 1990. Quantification of field crop stress areas using videography. See Ref. 93, pp. 88–98
333. Subhash N, Mohanan CN. 1994. Laser-induced red chlorophyll fluorescence signatures as nutrient stress indicator in rice plants. *Remote Sens. Environ.* 47: 45–50
334. Tanner CB. 1963. Plant temperatures. *Agron. J.* 55:210–11
335. Taubenhaus JJ, Ezekiel WN, Neblette CB. 1929. Airplane photography in the study of cotton root rot. *Phytopathology* 19:1025–29
336. Teng PS, ed. 1987. *Crop Loss Assessment and Pest Management.* St. Paul, MN: Am. Phytopathol. Soc. 270 pp.
337. Teng PS, Krupa SV, eds. 1980. Assessment of losses which constrain production and crop improvement in agriculture and forestry. *Proc. EC Stakman Commemorative Symp. Crop Loss Assessment. Misc. Publ. 7.* St. Paul: Agric. Exp. Stn., Univ. Minn. 327 pp.
338. Thomasson JA, Bennett CW, Jackson BD, Mailander MP. 1994. Differentiating bottomland tree species with multispectral videography. *Photogramm. Eng. Remote Sens.* 60:55–59
339. Thomsen A. 1992. Estimation of leaf-area-index (LAI) from radiation measurements. *Tidsskr. Planteav. Specialserie Beret. S-2207*:29–37
340. Thunen JG, Woody LM. 1994. New sensor technology for acquiring hyperspectral imagery. See Ref. 60, II:322–31
341. Tlalka M, White N, Errington R, Fricker M. 1994. Confocal excitation ratio imaging of Ca^{2+} and H^{+} during blue-light responses in plants. *Proc. R. Microsc. Soc.* 29:236
342. Todd LR, Kommedahl T. 1994. Image analysis and visual estimates for evaluating disease reactions of corn to *Fusarium* stalk rot. *Plant Dis.* 78:876–78
343. Todd SW, Hoffer RM, Milchunas DG. 1993. Comparison of four vegetation indices for estimating aboveground biomass on grazed rangeland. *Proc. ASPRS-ACM Annu. Conv., New Or-*

leans. Falls Church, VA: Am. Soc. Photogramm. Remote Sens. (Mimo, 10 pp.)
344. Toler RW, Smith DB, Harlan JC. 1981. Use of aerial infrared photography to evaluate crop disease. *Plant Dis.* 65:24–31
345. Tomerlin JR, Howell TA. 1988. *DISTRAIN*: A computer program for training people to estimate disease severity on cereal leaves. *Plant Dis.* 72:455–59
346. Tu JC, Tan CS. 1985. Infrared thermometry for determination of root rot severity on beans. *Phytopathology* 75: 840–44, 1281
347. Tucker CJ. 1979. Red and photographic infrared linear combinations for monitoring vegetation. *Remote Sens. Environ.* 8:127–50
348. Tucker CJ. 1980. Remote sensing of leaf water content in the near infrared. *Remote Sens. Environ.* 10:23–40
349. Twidwell EK, Johnson KD, Bracker CE, Patterson JA, Cherney JH. 1989. Plant tissue degradation measurement using image analysis. *Agron. J.* 81:837–40
350. Valentini R, Cecchi G, Mazzinghi P, Mugnozza GS, Agati G, et al. 1994. Remote sensing of chlorophyll *a* fluorescence of vegetation canopies. 2. Physiological significance of fluorescence signal in response to environmental stresses. *Remote Sens. Environ.* 47:29–35
351. Vanderbilt VC. 1985. Measuring plant canopy structure. *Remote Sens. Environ.* 18:281–94
352. Vanderbilt VC, DeVenecia KJ. 1988. Specular, diffuse, and polarized imagery of an oat canopy. *IEEE Trans. Geosci. Remote Sens.* 26:451–62
353. Vanderbilt VC, Grant L, Biehl LL, Robinson BF. 1985. Specular, diffuse, and polarized light scattered by two wheat-canopies. *Appl. Opt.* 24:2408–18
354. Vanderbilt VC, Grant L, Daughtry CST. 1985. Polarization of light scattered by vegetation. *Proc. IEEE* 73:1012–24
355. Vanderbilt VC, Silva LF, Bauer ME. 1990. Canopy architecture measured with a laser. *Appl. Opt.* 29:99–106
356. Venette JR, Venette RC. 1991. Image analysis for evaluation of bean rust severity. *Phytopathology* 81:1213
357. Vlcek WA, King D. 1985. Development and use of a 4-camera video system. *Proc. 19th Int. Symp. Remote Sens. Environ., Ann Arbor, Mich.*, pp. 483–89
358. Wallen VR, Jackson HR. 1971. Aerial photography as a survey technique for the assessment of bacterial blight of field beans. *Can. Plant Dis. Surv.* 51: 163–69
359. Wallen VR, Jackson HR, Basu PK, Baenziger H, Dixon RG. 1977. An electronically scanned aerial photographic technique to measure winter injury in alflafa. *Can. J. Plant Sci.* 57:647–51
360. Wayne MT, Campbell CL. 1987. Sampling procedures for determining severity of alfalfa leaf spot diseases. *Phytopathology* 77:157–62
361. Weber GE, Jörg E. 1991. Errors in disease assessment—A survey. *Phytopathology* 81:1238
362. Wessman CA. 1990. Evaluation of canopy biochemistry. In *Remote Sensing of Biosphere Functioning*, ed. RJ Hobbs, HA Mooney, pp. 135–56. New York: Springer
363. Wessman CA, Aber JD, Peterson DL, Melillo JM. 1988. Remote sensing of canopy chemistry and nitrogen cycling in temperate forest ecosystems. *Nature* 335:154–56
364. Wessman CA, Aber JD, Peterson DL, Melillo JM. 1988. Foliar analysis using near infrared reflectance spectroscopy. *Can. J. For. Res.* 18:6–11
365. Whitmyre G, Murphy FJ. 1973. Ultra low aerial application from a radio controlled model airplane. *Proc. NJ Mosq. Exterm. Assoc.* 16:44–51
366. Wiegand CL, Escobar DE, Everitt JH. 1992. Comparison of vegetation indices from aerial video and hand-held radiometer observations for wheat and corn. See Ref. 10, pp. 98–109
367. Wiegand CL, Escobar DE, Lingle SE. 1994. Detecting growth variation and salt stress in sugarcane using videography. See Ref. 34, pp. 9–18
368. Wiegand CL, Namken LN. 1966. Influences of plant moisture stress, solar radiation, and air temperature on cotton leaf temperature. *Agron. J.* 58:582–86
369. Wiegand CL, Rhoades JD, Escobar DE, Everitt JH. 1992. Photographic and videographic observations for determining and mapping the response of cotton to soil salinity. *Remote Sens. Environ.* 49:212–24
370. Wiegand CL, Richardson AJ, Escobar DE, Gerbermann AH. 1991. Vegetation indices in crop assessments. *Remote Sens. Environ.* 35:105–19
371. Wiegand CL, Scott AW Jr, Escobar DE. 1988. Comparison of multispectral videography and color infrared photography versus crop yield. See Ref. 136, pp. 235–47
372. Williamson B, Goodman BA, Chudek JA. 1992. Nuclear magnetic resonance (NMR) microimaging of ripening red raspberry fruits. *New Phytol.* 129:21–28
373. Williamson B, Goodman BA, Chudek JA. 1993. The structure of mature goose-

berry (*Ribes grossularia*) fruits revealed non-invasively by NMR microscopy. *Micron* 24:377–83

374. Williamson B, Goodman BA, Chudek JA, Johnston DJ. 1992. Nuclear magnetic resonance (NMR) microimaging of soft fruits infected by *Botrytis cinerea*. In *Advances in Botrytis Research*, ed. K Verhoeff, NE Malathrakis, B Williamson, pp. 140–44. *Proc. 10th Botrytis Symp., Crete, Pudoc*. The Netherlands: Wageningen
375. Woolley JT. 1964. Water relations of soybean leaf hairs. *Agron. J.* 56:569–71
376. Woolley JT. 1971. Reflectance and transmittance of light by leaves. *Plant Physiol.* 47:656–62
377. Yuan X. 1988. Spectral and textural characteristics of video imaging. See Ref. 136, pp. 13–16

Annu. Rev. Phytopathol. 1995. 33:529–64

PLANT DISEASE INCIDENCE: DISTRIBUTIONS, HETEROGENEITY, AND TEMPORAL ANALYSIS

L.V. Madden
Department of Plant Pathology, The Ohio State University, The Ohio Agricultural Research and Development Center, Wooster, Ohio 44691

G. Hughes
Institute of Ecology and Resource Management, University of Edinburgh, West Mains Road, Edinburgh EH9 3JG, Scotland

KEY WORDS: dispersion, quantitative epidemiology, spatial patterns, statistical modeling, time series

---

ABSTRACT

The statistical properties of disease incidence are reviewed and used to characterize spatial patterns of diseased entities (e.g. plants), satisfy assumptions of statistical analyses, and quantify change in incidence over time. Frequency of diseased plants can be represented by the binomial or, more commonly, the beta-binomial distribution. Spatial patterns of disease can be described by the aggregation parameter of the beta-binomial distribution, index of dispersion and related statistics, parameters of the binary form of the power law, as well as measures of spatial autocorrelation between sampling units. Disease incidence over time can be represented by continuous- and discrete-time nonlinear disease progress models, such as the logistic, and by autoregressive, integrated, moving average models. Ultimately, simultaneous spatio-temporal analyses can be performed to understand the dynamics of disease incidence in populations.

0066-4286/95/0901-0529$05.00

INTRODUCTION

The assessment of disease intensity is required in virtually every epidemiological study. As stated by Kranz (61), "Without quantification of disease no studies in epidemiology, no assessment of crop losses and no plant disease surveys and their applications would be possible." Measurements are also needed to characterize temporal and spatial dynamics of disease and to evaluate various management tactics such as the use of fungicides and resistant cultivars.

Much has been written about the approaches for assessing diseases (10), determining the most precise and accurate measure of disease (54, 89), and properly analyzing and interpreting disease data (62). Most of the discussion has logically focused on disease severity, that is, the area (in either absolute units or as a percentage) of host tissue that is diseased (usually assessed by visual symptoms). Disease incidence, that is, the number or proportion of diseased plants or plant parts, has received far less attention. For fungi that cause lesions or leaf spots, it is often assumed that severity, compared to incidence, better characterizes the population dynamic process that governs disease increase over time and spread in space (10). There is often a strong relationship between severity and incidence, however, especially at low mean severity (17, 107, 108), implying that incidence can provide meaningful and relatively easy-to-obtain information on epidemics.

Incidence is the primary measure for systemic as well as many root diseases. For instance, the intensity of viral diseases is routinely quantified as incidence (67). Determination of disease severity of virus-infected plants is problematical, and most rating systems remain controversial (10). Disease incidence is also much easier to determine than is severity. It is, therefore, more conveniently used in survey situations, when many observations are needed, or when nonexperts are used to collect data.

Despite its importance in plant pathology, the statistical properties of disease incidence have not been thoroughly explored until recently (44–49). Knowledge of the distribution of incidence (or any binary variable) is needed to assess the effects of treatments on disease properly (5, 22, 47, 125), to describe and quantify spatial patterns (46), to determine crop losses (44, 72), and to determine efficient sampling plans (19, 48). The purpose, therefore, of this article is to review some of the statistical properties of disease incidence. Distributions for representing incidence are reviewed, and methods for quantifying spatial heterogeneity are discussed. It is shown how distributional properties are used when analyzing the temporal component of epidemic data. Special attention is given to time-series analysis because of its potential value in epidemiology.

DEFINITION AND CONCEPTS

Campbell & Madden (10), paraphrasing from several sources, define disease incidence as "the number of plant units that are visibly diseased . . . , usually relative to the total number of units assessed." Seem (107) defines incidence "as the proportion (0 to 1) or percentage (0 to 100) of diseased entities within a sampling unit." The key to these and other related definitions (16, 61, 89) is that incidence data are *binary*. That is, each observation (entity in Seem's terminology: e.g. plant) can take on one of two possible forms: A plant is diseased or it is not. An older term for binary is quantal. Assessments for incidence generally have been based on visual symptoms, but increasing use is being made of biochemical and molecular techniques for assessment (e.g. 38). Although it is sometimes important to relate individual observations to experimental treatments or use the observations to determine spatial patterns (e.g. 10, 76, 86), interest often centers on a group of units or entities. For instance, one can determine the disease status of each of the 10 seedlings in a pot or 100 plants in a quadrat, and then determine the proportion diseased. The individual binary observations are thus grouped within the sampling units (pots and quadrats), and the data can be referred to as grouped binary or proportion data (20). We use the term incidence whether or not the number of diseased observations is divided by the total number of observations.

In epidemiology, the observations often are plants. However, incidence can be determined for leaves, tillers, fruit, flowers, or seeds. Whenever one can determine and record the disease status of individual observations, incidence can be calculated. It is important, however, to make clear what the observations are. If one assessed individual grape leaves for downy mildew symptoms, leaf disease incidence—not plant disease incidence—would be determined.

STATISTICAL DISTRIBUTIONS FOR DISEASE INCIDENCE

Binomial Distribution

Disease incidence is a measure of the probability (π) of a plant or other plant entity being diseased. This probability is obviously a function of the pathogen, host, and environment. With an unrestricted random sample (19, 52), where each of n plants (or plant entities = observations) is observed for disease, the estimate of this probability is simply the total number diseased (X) divided by n ($\hat{\pi} = X/n$). For convenience from this point on, we typically assume that the observations or entities are plants. The probability of a plant not being diseased is $1 - \pi$. These probabilities constitute the two states of the Bernoulli distribution for describing the probabilities of individual observations taking on one of two classes (e.g. diseased or healthy).

In most situations, investigators are more interested in properties of groups of plants than in properties of single plants. Thus, investigators must determine the probability of 0, 1, 2, . . ., and so on, diseased plants in a sample of n. When each observation can be assumed to be independent, the probability distribution is the binomial, a distribution originally derived more than 280 years ago by Bernoulli (see 20). Independent observations in this context is another way of saying that there is a random distribution with a constant (fixed) π. The binomial distribution is given in Table 1. Lower-case x is a specific value, such as 5 diseased plants, and $P(X = x)$ is the probability of exactly x diseased plants in a sample with n plants. Distributions such as the binomial are used in statistical hypothesis testing, calculating confidence intervals, and sampling. Cochran (18), analyzing data later presented in detail by Bald (4), appears to have been the first to use the binomial distribution to characterize the spatial pattern of diseased plants.

The expected value (mean, μ) and variance [$v(X)$] of X for the binomial distribution are given in Table 2. The variance of $\hat{\pi}$ is given by $v(\hat{\pi}) = \pi(1 - \pi)/n$, which is $v(X)/n^2$. Standard error of $\hat{\pi}$ [s.e.($\hat{\pi}$)] is $v(\hat{\pi})^{1/2}$. The variance of X is a direct function of π. The variance is low at π close to 0 and 1, and is highest at $\pi = 0.5$ (or a mean of $n/2$). Coefficient of skewness, a measure of distribution asymmetry, depends on π and n (Table 2; Figure 1). A symmetrical curve has a skewness coefficient of 0; positive skewness means a long tail to the right. Skewness is very low except for $n\pi$ close to 0 and n. This has great bearing on statistical comparison of treatments.

Frequencies

One cannot determine if the binomial distribution is appropriate if the data are collected as an unrestricted random sample of individual plants, because there is insufficient information on the observed distribution. To evaluate goodness of fit, one needs data from N separate sampling units. These can be quadrats in a field, but the sampling units need not be associated with specific spatial location. The n seeds of N separate plants could be assessed for seed infection, a form of cluster sampling (19). For each of the $i = 1, \ldots N$ sampling units, one must determine the numbers of diseased entities (X_i) and total entities (n_i). The total number of entities for all sampling units is nN if n is fixed, and the total number diseased entities is ΣX_i. Mean disease incidence (an estimate of the probability of a plant being diseased, p) is given as $\hat{p} = \Sigma X_i/nN$. The symbol p is used here to distinguish cluster sampling from unrestricted random sampling of individual plants. With cluster sampling, one can visualize a separate value of π (π_i) for each sampling unit ($\hat{\pi}_i = X_i/n_i$). When one can assume that

Table 1 Some discrete distributions for binary and count data (one or more versions given)

Distribution	$P(X=x)=$	Parameters
Binomial	$\binom{n}{x}\pi^x(1-\pi)^{n-x}$	π, n
Beta-binomial	$\binom{n}{x}\dfrac{\prod_{i=0}^{x-1}(p+i\theta)\prod_{i=0}^{n-x-1}(1-p+i\theta)}{\prod_{i=0}^{x-1}(1+i\theta)}$	$p\ [=\alpha/(\alpha+\beta)]$ (analogous to π), $\Theta\ [=1/(\alpha+\beta)]$
	$\binom{n}{x}\dfrac{\Gamma(\alpha+\beta)\Gamma(\alpha+x)\Gamma(\beta+n-x)}{\Gamma(\alpha)\Gamma(\beta)\Gamma(\alpha+\beta+n)}$	α, β, n
	$\dfrac{\binom{-\alpha}{x}\binom{-\beta}{n-x}}{\binom{-\alpha-\beta}{n}}$	α, β, n

Poisson	$e^{-\mu}\left(\frac{\mu^x}{x!}\right)$	μ [equivalent to $n\pi$ (or np) at low π (p) and high n]
Negative binomial	$\binom{k+x-1}{x}\left(\frac{\mu}{k}\right)^x\left(1+\frac{\mu}{k}\right)^{-(k+x)}$	μ, k (equivalent to α at high n and low p)
	$\binom{k+x-1}{x}(p')^k(1-p')^x$	k, p' [$=k/(\mu+k)$]

Table 2 Summary of some properties of four discrete distributions[a]

Distribution	Mean (M)	Variance $[V = v(X)]$	Coefficient of skewness
Binomial	$n\pi$	$n\pi(1 - \pi)$	$(1 - 2\pi)/(n\pi(1 - \pi))^{1/2}$
Beta-binomial	np	$np(1 - p)[1 + \rho(n - 1)]$	$\frac{V^{-1/2}(\alpha + \beta + 2n)(\beta - \alpha)}{(\alpha + \beta + 2)(\alpha + \beta)}$
	$(n\alpha)/(\alpha + \beta)$	$\frac{np(1 - p)(1 + \theta n)}{1 + \theta}$	
		$\frac{\alpha\beta n(\alpha + \beta + n)}{(\alpha + \beta)^2(\alpha + \beta + 1)}$	
Poisson	μ	μ	$\mu^{-1/2}$
Negative binomial	μ	$\mu[1 + (\mu/k)]$	$\frac{1 + 2(\mu/k)}{[\mu(1 + (\mu/k))]^{1/2}}$
	$k(1 - p')/p'$	$k(1 - p')/p'^2$	

[a] See Table 1 for formulations of the distributions. One or more versions of equations are given.

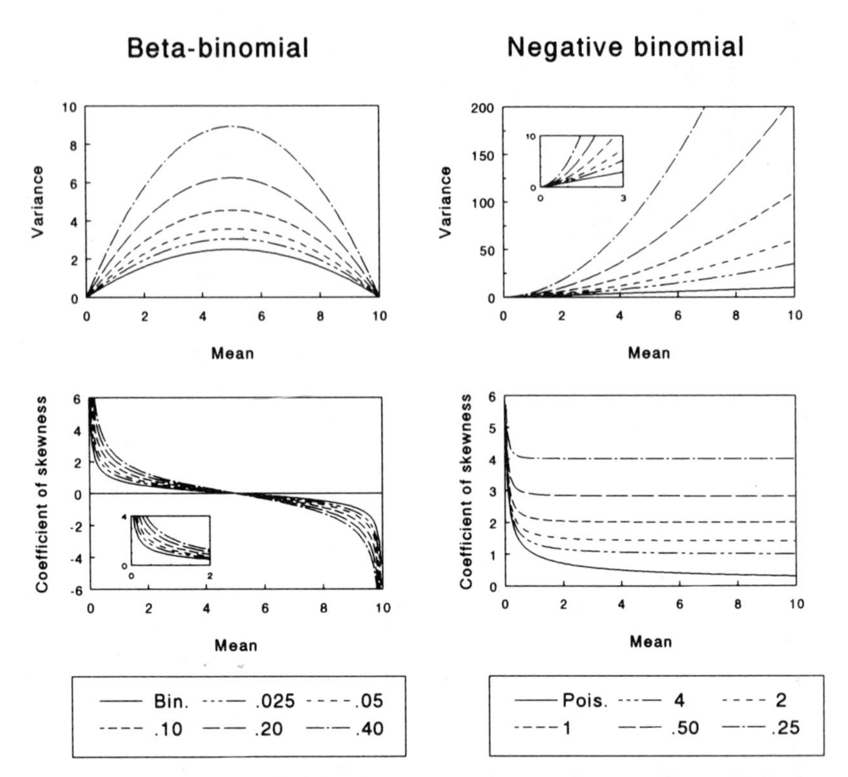

Figure 1 Variances and coefficients of skewness vs the mean for some discrete distributions. *Left-hand side:* beta-binomial distribution with values of θ (see Tables 1 and 2) of 0 (= binomial) to 0.4; $n = 10$ in all cases. *Right-hand side*: negative binomial distribution with values of k of 0.25 to ∞ (= Poisson). Inserts are used to better separate curves at low means.

these π_is are all equal, as one does for the binomial, then p and π are equivalent. The variance of $X[v(X)]$ is $np(1-p)$, and s.e.$(\hat{p})$ is $[p(1-p)/nN]^{1/2}$, because there are now nN observations. If n_i varies, then one can use the mean of the *ns* $(\bar{n})$ in the formulas for variances as an approximation.

A frequency distribution is determined by calculating the number of sampling units with $X = 0, 1, \ldots, n$ diseased entities. Examples are plotted in Figure 2 for citrus decline (27), which does not have a known pathogen, and for root rot of bean (caused by *Rhizoctonia solani*) (12). The expected frequencies are determined by $N \times P(X = x)$. As seen in the figure, the observed and expected frequencies are very similar. Formal testing is done with a χ^2 goodness-of-fit test (124).

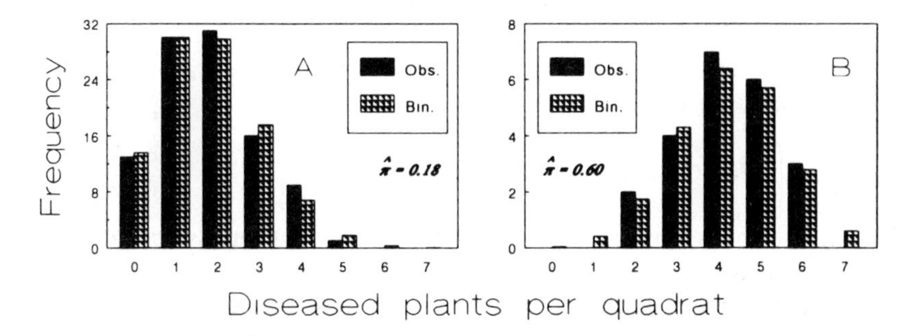

Figure 2 Example frequency distributions for (*A*) *Rhizoctonia* root rot of beans (field CF2-78-12 in Reference 12; $n = 10$ plants; $N = 100$ quadrats) and (*B*) citrus decline in Reference 27; $n = 7$; $N = 22$ quadrats (of arbitrary dimension). "Obs." is the observed frequency, and "Bin." is the expected frequency for a binomial distribution (Table 1) with the shown estimated π parameter.

Beta-Binomial Distribution

If the probability of a plant (or other entity) being diseased is not constant, then the binomial is not appropriate for representing the distribution of diseased entities per sampling unit. Reasons for nonconstancy of π include variation in host, pathogen, or environmental conditions throughout a field; and the multiplication of the pathogen in sampling units being proportional to the (possibly random) initial disease or inoculum in the field. The mechanism need not be in terms of spatial processes, however. If the n seeds from N different plants were each assessed for seed infection, variation among plants could result in π (measured as the proportion infected) varying from plant to plant.

The most common way of dealing with a nonconstant π is to assume that π is a random variable (limited by 0 and 1) with a beta probability density function $[f(\pi)]$ (111). This can be written as:

$$f(\pi) = [\pi^{\alpha-1}(1-\pi)^{\beta-1}]/Be(\alpha,\beta)$$

in which α and β are positive parameters (constants) and $Be(\alpha,\beta)$ is the beta function (8). $Be(\alpha,\beta)$ can also be written as $\Gamma(\alpha)\Gamma(\beta)/\Gamma(\alpha+\beta)$, with $\Gamma(\cdot)$ being the gamma function. This density is very flexible in that many different shapes may be represented, depending on α and β. In the context of disease incidence, the beta density is also a consequence of the power-law relationship between the variance and p (46; see below).

The distribution of diseased plants can be determined by the compounding or mixing of the binomial distribution with the beta density function. If $b(\pi,n)$ represents the binomial distribution, then the mixture distribution is derived as:

$$P(X = x) = \int b(\pi,n)f(\pi)d\pi,$$

which results in the beta-binomial distribution. This distribution can be written in various ways, the simplest being:

$$P\,(X=x) = \binom{n}{x} Be\,(\alpha + x,\, \beta + n - x)/Be\,(\alpha,\beta). \qquad 1.$$

Other expressions are given in Table 1. The beta-binomial can also be generated as a contagious distribution (30, 100).

It is generally more useful to reparameterize the beta-binomial with new parameters $p = \alpha/(\alpha + \beta)$ and $\theta = 1/(\alpha + \beta)$. Here, p is the expected or average value of π, and has the moment estimate described above ($\Sigma X_i/nN$). The θ parameter is an index of aggregation; it ranges from 0 to ∞, although typically it is much less than 1. Another useful parameterization is to replace θ with $\rho = \theta/(1 + \theta) = 1/(\alpha + \beta + 1)$; ρ has an upper limit of 1. Variability or heterogeneity, i.e. overdispersion or extrabinomial variation, increases as θ (or ρ) increases. The variance for the beta-binomial is larger than for the binomial at any value of p (with $\theta > 0$) (Table 2). As with the binomial, the variance is symmetric around $np = n/2$ (at a fixed θ), is highest at $n/2$, and is 0 at $X = 0$ and $X = n$ (Figure 1). As θ goes to 0, $v(X)$ goes to $np(1 - p)$, and the beta-binomial reduces, in the limit, to the binomial distribution (Figure 3). The moment estimate of θ is:

$$\hat{\theta} = [s^2 - np(1 - p)]/[n^2p(1 - p) - s^2], \qquad 2.$$

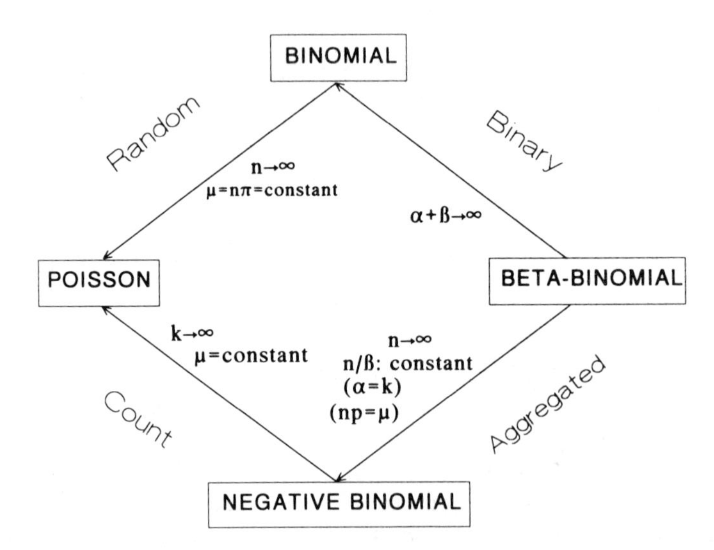

Figure 3 Relationships among the four discrete distributions in Table 1.

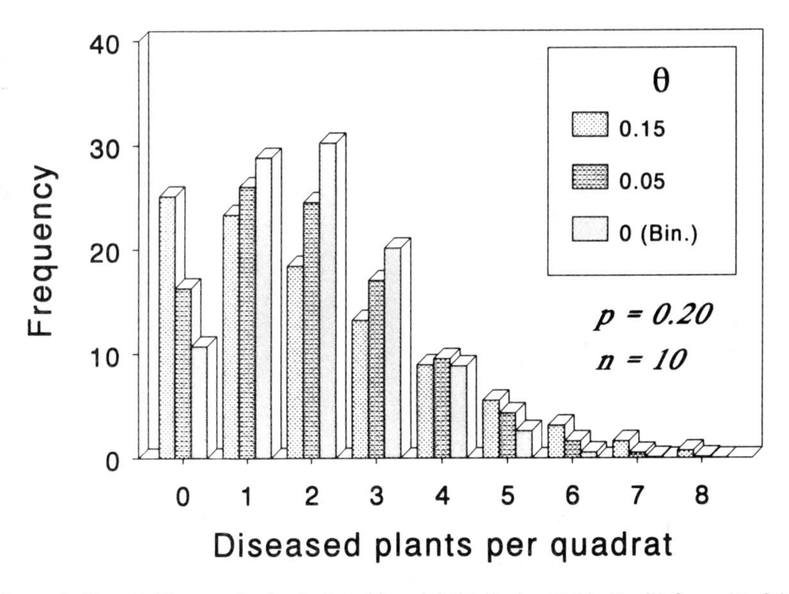

Figure 4 Expected frequencies for the beta-binomial distribution (Table 1) with θ equal to 0.15, 0.05, and 0 (= binomial distribution). In all cases, there were $N = 100$ sampling units (e.g. quadrats), $n = 10$ entities (e.g. plants) per sampling unit, and p (or π) equaled 0.20.

in which s^2 is the observed (empirical) variance of X. Maximum likelihood estimates of p and θ, as well as their standard errors, can be obtained with an iterative Newton-Raphson procedure (112).

Compared to the binomial, the beta-binomial distribution has higher probabilities (and hence frequencies) at Xs near 0 and n, and lower probabilities in the mid-range of X, say, near the mean. An example of this for generated data is in Figure 4 for π or p equal to 0.2, and $\theta = 0$ (binomial), 0.05, and 0.15. Skewness of the beta-binomial increases with θ and can be considerably greater than for the binomial, especially at np near 0 and n (Figure 1; Table 1). At np below $n/2$ ($p > 0.5$), skewness is positive, that is, there is a long tail to the right; at np above $n/2$, skewness is negative.

HISTORICAL DEVELOPMENT The theoretical genesis of the beta-binomial distribution in the form used here is attributed to Skellam (111). It turns out that the beta-binomial is mathematically identical to distributions known as the negative hypergeometric, inverse hypergeometric, and Polya-Eggenberger, although these distributions were developed in other contexts (30, 42, 55), generally for sampling purposes. Kemp & Kemp (60) showed how the binomial and beta-binomial are part of a family known as generalized hypergeometric distributions.

Irwin (51) and Griffiths (41) used the beta-binomial to describe the distribution of infectious disease of humans. Here, each family was a sampling unit. Kemp & Kemp (59) were the first to apply the beta-binomial to ecological data by using the model to describe the proportion of an area covered by a plant species. The data consisted of the number of pins contacted (out of n = 10 pins) by a particular plant species in sampling frames that were randomly

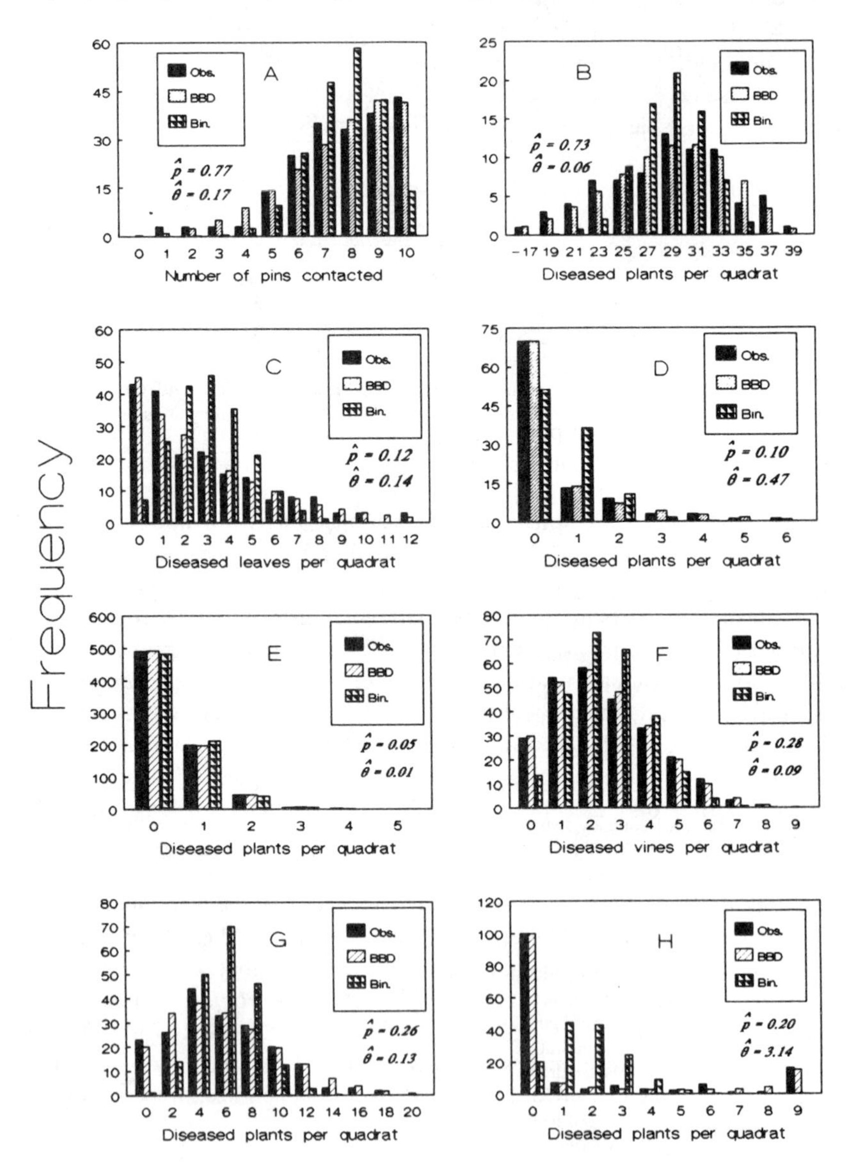

placed in an area. Contact with a pin is analogous to an individual being diseased. The beta-binomial provided a much better fit to the data than did the binomial (59); an example is given in Figure 5*A*. Note that the binomial predicts much higher probabilities, and hence frequencies, around 4–7 pins compared to the observed data or the beta-binomial predictions. Moreover, the frequencies for the binomial are too low at the extremes.

Crowder (22) described seed germination with the beta-binomial, in which the sampling units were separated plants producing the seed. Williams (125) used the same distribution to describe mortality of rat pups in litters; here, each litter was a sampling unit. These papers are important because they relate the distribution of a response variable (seed germination, mortality) to imposed treatments. This work has served as the basis for considerable research, especially in toxicology, for statistically determining the effects of experimental treatments on overdispersed binary data (20, 47, 82, 126).

PLANT PATHOLOGY EXAMPLES Shiyomi & Takai (110) and Takai & Shiyomi (117) used the beta-binomial distribution to represent virus-diseased rice plants. Their articles were published in Japanese, and apparently had no influence outside of Japan on plant disease epidemiology studies. Likewise, Qu et al (100) used some of Bald's data (4) to demonstrate the use of the beta-binomial. Hughes & Madden (46) applied the beta-binomial distribution to the published incidence of virus-diseased plants from four pathosystems. Their derivation of the beta-binomial was based on the power-law relationship between the observed variance of disease incidence and p (see below).

Some examples of the beta-binomial fitted to disease incidence data are shown in Figure 5. The beta-binomial provided a good fit to the incidence of diseases caused by viruses, fungi, and bacteria. The beta-binomial also fitted the output of a computer simulator, which consisted of a bimodal frequency distribution, that is, quadrats that were mostly disease-free or that had 100%

←

Figure 5 Example frequency distributions demonstrating the fit of the beta-binomial ("BBD") and binomial ("Bin.") to observed ("Obs.") data. (*A*) Pins in an ecological "sampling frame" contacted by the plant species *Carex hebes*; n = 10, N = 200 frame samplings (see Table 1 in Reference 59). (*B*) Tobacco plants infected by potyviruses; n = 40 plants, N = 75 quadrats (field B-I-84 in References 46, 73). Note that frequencies are pooled in groups of two ("19" for "19–20," "21" for "21–22," . . .) and that frequencies of 0–17 are pooled into the "–17" category. (*C*) Strawberry leaves infected by *Phomopsis obscurans*; n = 25 randomly selected leaves per quadrat, N = 192 quadrats (Figure 2 in Reference 69). (*D*) Soybean plants infected by *Pseudomonas syringae* pv. *glycines*; n = 6 plants, N = 100 quadrats (compiled from Figure 2*A* in Reference 99). (*E*) Lettuce plants infected by *Sclerotinia minor*; n = 9 plants, N = 740 quadrats (Figure 2 in Reference 77). (*F*) Grape vines infected by *Eutypa lata*; n = 9 vines, N = 304 quadrats (compiled from Figure 3*A* in Reference 84). (*G*) Rice plants with yellow dwarf symptoms; n = 25 plants, N = 200 quadrats (compiled from Figure 4 in Reference 109; original data in Figure 2 of Ref. 116). (*H*) Output of a computer simulator (90) in which most quadrats consisted of no diseased plants or all diseased plants; n = 9, N = 144.

incidence (Figure 5*H*). Moreover, this distribution also described several data sets of incidence of grape leaves infected by downy mildew (70); because n was variable, it was not possible to calculate expected frequencies in this case.

Poisson and Negative Binomial Distributions

Virtually all research in statistical ecology assumes that the Poisson distribution is the appropriate model for a random pattern (11, 58, 91, 124). It is easily shown that the Poisson distribution is obtained, in the limit, from the binomial when n goes to infinity and $n\pi$ (mean: μ) is constant and small (97). The Poisson has several desirable properties, including a simple mathematical form (Tables 1 and 2), and equality of the mean and variance [$\mu = n\pi = v(X)$] (Table 2). To further see the relation between the binomial and Poisson distributions, note that the variance of the binomial is approximately $n\pi$ if π is small (because $1 - \pi \approx 1$).

The Poisson could be an approximate model for binary data when n is much larger than the mean. Generally, this means that the distribution is appropriate for count data that have no (effective) upper bound, or one much larger than the observed data. Classic examples are numbers of plants, insects, nematodes, and fungal propagules in sampling units (11, 124).

If the mean is not constant, one can derive several other distributions in a fashion analogous to how the beta-binomial was derived from the binomial. The negative binomial is the most common distribution for overdispersion of count data (5, 7, 8, 10, 11, 58, 65; Tables 1 and 2). The variance of the negative binomial is greater than that of the Poisson and is a function of the parameter k (Figure 1; Table 1). The variance increases monotonically with the mean, and the separation from the Poisson variance increases as the mean increases. The coefficient of skewness of the negative binomial is greater than for the Poisson but, unlike the variances, skewness eventually stabilizes (Figure 1; Table 2).

Binary vs Count Distributions

Plant pathologists have often used the Poisson, negative binomial, and similar count distributions to characterize the heterogeneity of disease incidence (35, 77, 105, 114), even though incidence is a binary variable. The cited examples are used to demonstrate the situation, not to criticize the authors. The distinctions between count and binary data, or between random and overdispersed binary data, have not been fuly explored in most fields.

Differences between the variances and coefficients of skewness of the beta-binomial and negative binomial distributions might imply that serious errors in interpretation would occur when fitting the negative binomial to overdispersed binary data. Consider the overdispersed "pin"-contact data in Figure 5*A*. These data have negative skewness, which can never be depicted by the negative binomial or the Poisson. Goodness-of-fit statistics for the count distributions would indicate significant lack of fit to the data. The same situation would occur when attempting to fit these count distributions to disease incidence data when p is large. For instance, Rouse et al (105) could only fit the negative binomial to powdery mildew incidence during the early part of an epidemic, presumably because none of the count distributions could describe data for higher means.

At low p there are strong similarities between binary and count distributions. Just as the binomial approaches the Poisson in the limit, the beta-binomial approaches the negative binomial when n goes to infinity and n/β [$= \mu/k$; with $\beta = (1-p)/\theta$] is held constant and nonzero, with $\alpha = k$ (Figure 3). This situation can occur when mean incidence (np or $n\pi$) is low relative to n, and n is large. In terms of variances or coefficients of skewness, the beta-binomial and negative binomial are very similar at low means (Figure 1), as are the binomial and Poisson. Therefore, use of the count distributions for describing disease incidence may be an acceptable approximation when incidence is low.

Effects of the use of incorrect distributions can be seen with a simple simulation study. Assuming that the beta-binomial distribution is appropriate, pseudo-random data were generated with a Monte Carlo simulation for $n = 10$, $N = 64$, $\theta = 0.15$, and $p = 0.05, \ldots, 0.95$ (LV Madden, unpublished). The beta-binomial provided an acceptable fit to all data sets ($P > 0.05$) (Table 3). Estimates of p and θ were close but not identical to the theoretical values; larger N would result in more accurate estimates of p and θ.

At low p (≤ 0.10), the Poisson provided a poor fit to the data, but the negative binomial did provide a good fit for $p \leq 0.40$ (Table 3), in qualitative agreement with the beta-binomial distribution results. The estimate of the negative bino-

Table 3 Fit of the binomial (Bin.), beta-binomial (BBD), Poisson (Pois.) and negative binomial (NBD) distributions to simulated data from the BBD ($\theta = 0.15$; $n = 10$; $N = 64$), together with estimated parameters of the BBD and k parameter of the NBD

	Goodness of fit probability (P)[a]				BBD[b]				NBD
p	*Bin.*	*BBD*	*Pois.*	*NBD*	p	θ	α	β	k
0.05	0.00	0.23	0.00	0.30	0.073	0.180	0.41	5.15	0.52
0.10	0.00	0.56	0.00	0.93	0.076	0.233	0.33	3.97	0.42
0.20	0.00	0.74	0.09	0.79	0.212	0.154	1.38	5.12	2.56
0.40	0.03	0.99	0.68	0.97	0.408	0.147	2.78	4.03	13.10
0.80	0.00	0.53	0.23	—[c]	0.815	0.248	3.28	0.75	—
0.95	0.03	0.12	0.00	—	0.965	0.135	7.15	0.26	—

[a] Pois. and NBD were fitted using DISCRETE (34); Bin and BBD were fitted using BBD (69).
[b] Parameters p and θ were estimated directly; $\alpha(=p/\theta)$ and $\beta(=[1 - p]/\theta)$ were calculated from estimates of p and θ.
[c] Not possible to fit the NBD to the data because the mean was greater than the variance.

mial k parameter was similar to the estimate of α of the beta-binomial (Table 2) at $p \leq 0.1$, but then k increased as p increased above 0.1, incorrectly suggesting declining aggregation as mean increased. The Poisson appeared to provide a good fit for data sets with $0.20 \leq p \leq 0.80$, even though the date were highly overdispersed. This might inadvertently suggest a random pattern. Also, the negative binomial could not be successfully fitted to the data for $p \geq 0.80$ in this example, even though the data were overdispersed. The combination of results for $p = 0.80$ (estimated as 0.815) with the Poisson and negative binomial would incorrectly imply that the data were generated by a random process.

Theoretical and simulation results suggest that use of count distributions for incidence data should be limited to cases when mean incidence is 0.20 or less. In general, one should use the proper discrete distributions whenever possible, especially now that personal computer–based software for model fitting is available (34, 69).

OTHER DISCRETE DISTRIBUTIONS For count data, many alternatives to the negative binomial distribution for overdispersed data are well known (124). There has been much less research on distributions for overdispersed binary data and, consequently, fewer alternatives to the beta-binomial have been developed. Some possibilities have been published (1, 55, 101). More work is needed to determine the fit of the beta-binomial to other plant disease data sets.

There also are distributions for underdispersed data. Athough the binomial is the appropriate model for random binary data, it may be an appropriate model for underdispersed count data as well. For counts with no upper bound, if the probability of an individual entering a quadrat (from movement, birth, etc) decreases as density of individuals increases, then one can represent these data with the binomial distribution (91, 92). Here, n is not the number of plants in a quadrat, but an arbitrary constant based on the probability of entering a quadrat. Such a derivation does not apply to binary data. Models have been proposed for count and binary data exhibiting underdispersion—specifically, a regular spatial pattern (109). These models have not been used much, possibly due to their complexity.

CHARACTERIZING HETEROGENEITY OF DISEASE INCIDENCE

The magnitude of overdispersion may be of interest in epidemiology and other specialities in plant pathology. When characterizing spatial patterns, the degree of aggregation is often quantified by one or more indices of aggregation (10, 11, 29, 65, 92). Even in more controlled studies, unrelated to a spatial analysis, the extent of heterogeneity among sampling units (e.g. infected seeds from separate plants) relative to that expected for a binomial distribution can be determined and used to analyze treatment effects properly (22). There are many ways of characterizing heterogeneity of count data, but no general agreement on the best method to use (10, 29, 43, 50, 65, 85, 92). There are analogous versions for some of these techniques for binary data, as discussed here.

Beta-Binomial Parameter (θ)

When the beta-binomial fits disease incidence data, one can use the estimate of θ as an index of aggregation. In this sense, θ is used in an analogous fashion to the reciprocal of the k parameter of the negative binomial. As noted in the previous section, at low p and high n, $\alpha(= p/\theta)$ of the beta-binomial is similar to k of the negative binomial. Shiyomi & Takai (110) used α as an index of virus-infected rice plants. If maximum likelihood is used to estimate θ (112), its standard error can also be obtained from the inverse of the information matrix. Equality of θ to 0 [i.e. binomial (random) distribution] can then be done with a t-test, $t = \theta/\text{s.e.}(\theta)$. If N is not large, s.e.(θ) may be fairly large, even when heterogeneity is high. With small N, it is probably desirable to characterize heterogeneity with one of the following indices.

Index of Dispersion

For count data, the variance-to-mean (VM) ratio is routinely calculated to measure aggregation and test for overdispersion. Proper interpretation of this ratio requires understanding that the mean is the expected (theoretical) variance for the Poisson distribution. Thus, VM is the observed variance divided by the theoretical variance for a random distribution of count data. More generally, D is defined as the ratio of the observed to the theoretical variance and is called the index of dispersion (32). $D = 1$ for random data, $D > 1$ for overdispersed data, and $D < 1$ for underdispersed data. For count data, $D = VM$. For binary data,

$$D = s^2/[np(1 - p)] \qquad 3.$$

because $np(1 - p)$ is the variance of the binomial distribution (with π estimated as p). Use of D to assess heterogeneity goes back to at least 1925 (32). When mean incidence is low, $1 - p \approx 1$, so the theoretical variance is approximately np, which is the mean. Thus, for low incidence, VM is a reasonable approximation to D. At moderate to high incidence, $D > 1$ (overdispersion) even when $VM < 1$. Thus, VM should not be used at moderate to high p because the mean has no direct relevance to D for binary data.

If there is a random distribution (i.e. the null hypothesis), such that p (or the mean) is constant, $(N - 1)D$ has a chi-square distribution with $N - 1$ degrees of freedom (97). A large value of $(N - 1)D$ indicates that one can accept the alternative hypothesis of overdispersion. For instance, the *Phomopsis* data in Figure 5*C* have a D of 3.96 and a chi-square statistic of 757, highly significant ($P < 0.001$) with 191 degrees of freedom.

C(α) TEST STATISTIC A statistic related to the chi-square test of D is based on a so-called $C(\alpha)$ test of Neyman (88). It is derived from the likelihood function for a particular distribution (93, 118). For disease incidence, the distribution is, as described above, the beta-binomial. A standard normal test statistic (z) is calculated as:

$$z = [n(N - 1)D - nN]/[2N(n^2 - n)]^{1/2}$$

when n is constant. Unlike the chi-square test of D for overdispersion or heterogeneity, the $C(\alpha)$ test has a more specific alternative hypothesis, namely that the overdispersion is described by the beta-binomial. The test is one-sided and potentially more powerful because of the specific alternative (70).

Morisita's Index

Several other statistics have been developed for count data, such as Morisita's index (I_δ) (83). Interpretation for incidence data remains uncertain, except for low p. Morisita modified I_δ for the situation in which there can be no more than n possible individuals in a quadrat (83). The new index is given as $I_B = I_\delta/[1 - (1/n)]$. Shiyomi & Takai (110) stated that $I_B \approx 1 + (1/\alpha)$ for low p. This is equivalent to the known relationship between I_δ and negative binomial k, $I_\delta \approx 1 + (1/k)$ (122). The correction for finite n does not take into account how close p is to 1 (or np is to n). To handle this factor, θ (or ρ) should be used.

Power Law

For count data, it is well established that there is a straight-line relationship between the logarithm of the observed variance ($v = s^2$) and the observed mean (m) (119). We use m and v here to designate the observed (calculated) statistics, in contrast to parameters of specific distributions (e.g. μ for the mean in the Poisson distribution). To see this relationship, one needs a collection of data sets, not just a single mean and variance. This empirical Taylor power law has been found to describe numerous data sets representing many different taxa (96, 119–122) and is supported by some theoretical work (94, 113). Attempts to use Taylor's power law with disease incidence or severity data have given mixed results. In some cases, a significant straight-line relationship was found (108, 123). Sometimes, however, no significant relationship between v and m (on a log scale) occurred or the relationship was nonlinear (65, 71, 73), presumably because the original power law was developed for unbounded counts.

Recently, Hughes & Madden (45) modified the power law for disease incidence (or any binary variable). They showed that the power law can be written generally as

$$\log(v) = \log(A) + b\log(v_r), \quad 4.$$

in which v_r is the theoretical variance for a random distribution (see Table 2 for the Poisson and binomial) and A and b are parameters. A random distribution is indicated by $b = 1$ and $A = 1$ [$\log(A) = 0$], that is, $\log(v) = \log(v_r)$, or $v = v_r$. This is identical to $D = 1$ for all data sets in a collection. If $b = 1$ and $A > 1$, then D is fixed, and equal to A, for all data sets. If $b > 1$, then $\log(v)$ increases with $\log(v_r)$ at a rate greater than v_r. In this context, b is considered an index of aggregation (119, 120), although interpretation of b should not be divorced from the value of A. Some workers have debated the mechanisms that would produce different b values (e.g. 26)

For count data, v_r equals the mean because one assumes that the Poisson is the appropriate random distribution (Table 2). Then, equation 4 is the classic power law with m used as the estimate of v_r:

$$\log(v) = \log(A) + b\log(m). \quad 5a.$$

For disease incidence data as proportions, $v_r = p(1 - p)/n$, so that

$$\log(v) = \log(A) + b\log[p(1 - p)/n]. \quad 5b.$$

For incidence data represented as numbers of diseased entities, $p(1 - p)/n$ is multiplied by n^2, which gives $np(1 - p)$. Equation 5b can then be written as

$$\log(v) = \log(A) + b\log[np(1 - p)]. \quad 5c.$$

When p is low, $np(1 - p) \approx np$, which is the mean incidence. Thus, at low incidence, the original power law would be an appropriate approximation for disease incidence. The use of equation 5a could be very misleading at high p. When n is not fixed within a data set, one can use $\bar{n}$ in the equations.

With constant n, equation 5b can be written as

$$\log(v) = \log(a) + b\log[p(1 - p)], \quad 6a.$$

in which $a = An^{-b}$. Asymmetric forms can then be incorporated with an expanded model:

$$\log(v) = \log(a) + b_1\log(p) + b_2\log(1 - p), \quad 6b.$$

in which b_1 and b_2 are parameters. The estimate of p is used in these equations.

Regression analysis is used to estimate $\log(A)$ and b (or b_1 and b_2). Most authors have used ordinary least-squares regression, but geometric-mean regression has been advocated because neither $\log(v)$ nor $\log(v_r)$ is obviously the dependent variable (70, 121).

PLANT PATHOLOGY EXAMPLES Hughes & Madden (45) found that the binary form of the power law (equation 5b or 6a) fitted data from several virus disease epidemics of tobacco and maize (e.g. Figure 6*B,E*). For these data, the parameter b was significantly greater than one and precisely estimated. Figure 6*A* shows a data set where the best-fitting line is very similar to the binomial (random) line [i.e. $b = 1$ and $\log(A) = 0$]. Many of the other data sets we have analyzed indicate aggregation ($A > 1$ and/or $b > 1$). In addition to the work on virus diseases, the incidence of downy mildew of grape leaves (Figure 6*C*; 70) as well as the incidence of leather rot of strawberry fruit (Figure 6*D*; 103) have been fitted. In the latter case, the estimate of b is much greater than 1, but the data exhibit some curvature at the higher v_r values. This would be a case where the more general equation 6b could be appropriate. A final example is incidence of strawberry leaves infected by *Phomopsis obscurans*, for which variation around the line was higher than for the other examples, probably because each variance was calculated from only $N = 4$ sampling units. Generally, N should be 15 or greater for precisely calculating variances (79, 120). Yang & TeBeest (128) also demonstrated examples such that the variance increased with p to a maximum and then declined; they did not quantify the relationship with regression analysis. Further studies are needed to determine how well the binary form of the power law describes data from other pathosystems.

DISTRIBUTION AND INDICES If the binary form of the power law is appropriate, then the mean and variance of the beta density function for variable π are given by functions of A (or a) and b (46). Therefore, one would predict that the distribution of diseased plants would be described by the beta-binomial. This prediction is supported by the good fit of the beta-binomial to a wide range of incidence data sets (46; see above). The link between the distribution approach and the approach based just on p and v supports the broad applicability of these methods for characterizing heterogeneity of disease incidence. It is interesting to note that for count data, linkage between the power law and probability distributions has proved to be more problematical (7, 56, 57, 94, 95).

The θ parameter of the beta-binomial can be written as a function of the power law parameters $a(= An^{-b})$ and b:

$$\theta = [a - f(p)/n]/[f(p) - a], \qquad 7.$$

in which $f(p) = [p(1-p)]^{1-b}$. This is a more accurate version of equation 6 in Hughes & Madden (46). At large n, equation 7 is approximated well by $a/[f(p) - a]$. When $b = 1$ and $A = 1$ ($a = n^{-1}$), θ is 0 at all p, as one would expect. When $b = 1$ and $A > 1$ ($a = An^{-1}$), then θ is constant at all p. At $b > 1$, θ increases to a maximum at $p = 0.5$, and then declines. The situation is more complex if the more general equation 6b is appropriate. The relationship

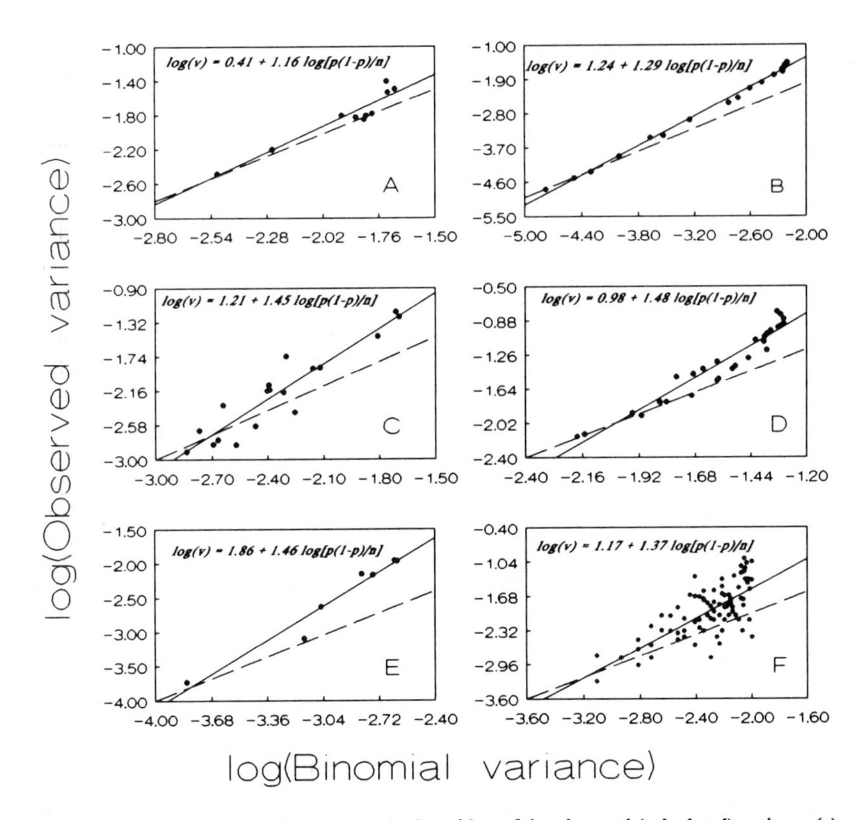

Figure 6 Example relationships between the logarithm of the observed (calculated) variance (v) and the logarithm of the theoretical variance for a binomial (i.e. random) distribution of disease incidence (v_r). Solid line: ordinary least-squares regression fit to data; broken line: $v = v_r$. (*A*) Bean plants infected by *Rhizoctonia solani* (cause of root rot); points are for different fields compiled from data in Reference 12 and from CL Campbell (personal communication). (*B*) Tobacco plants infected by potyviruses; points are different times in field A-N-85 in (46, 73). (*C*) Grape leaves per shoot infected by *Plasmopara viticola*; points are for different plots at one assessment time in 1990 (70). (*D*) Strawberry fruit infected by *Phytophthora cactorum* (cause of leather rot); points are for different times in three different commercial fields in Reference 103. (*E*) Maize plants infected by maize dwarf mosaic virus; points are different times in field M-5 in Reference 71. (*F*) Strawberry leaves infected by *Phomopsis obscurans*; points are different plots in a commercial field (at the same time), each based on $N = 4$ sampling units of $n = 25$ leaves (LV Madden, unpublished).

between the power-law and other measures of aggregation can be obtained by using the variances predicted from p, a, and b in the formulas for D and other indices.

Intracluster Correlation

In the approaches discussed so far, aggregation has been characterized by the variance of diseased plants per sampling unit being inflated above the variance of binomial distribution with the same mean. Such variance inflation arises when there is an excess of observed over expected (binomial) frequencies in the upper and lower tails of the frequency distribution of diseased plants per sampling unit, and a corresponding deficit around the mean. Thus, aggregation can be seen as the tendency of plants in a sampling unit to have similar disease status. This tendency can be measured directly, without reference to the binomial distribution, by calculating the intracluster correlation coefficient ρ (e.g. 20). Like the regular (product moment) correlation coefficient, ρ has a maximum value of 1 (although it is seldom near 1). The minimum value of ρ, however, is $-1/(n-1)$. Values of $\rho > 0$ represent aggregation.

Calculating ρ directly as a correlation coefficient is rather tedious (20), and a MINITAB (81) macro (command file) was written to make the calculation easier (49). This parameter, however, can be estimated from $D[= 1 + \rho(n - 1)]$ if n is constant, or as the ρ parameter $[= \theta/(1 + \theta)]$ of the beta-binomial distribution, if this distribution provides a good fit to the data. It could also be predicted from the power-law relationship between the observed and theoretical variances (through equation 7). Information relating to the spatial locations of the sampling units is not required in order to characterize aggregation in this way, although the size of sampling units can affect the intracluster correlation.

Spatial Autocorrelation

Information on the spatial locations of sampling units can be used to assess similarity of disease incidence between sampling units (65). If the presence of diseased plants in a sampling unit means that it is more (or, indeed, less) likely that plants in a neighboring sampling unit will be diseased, disease incidence is exhibiting spatial autocorrelation (10, 37, 38, 40, 65, 102). The goal of this analysis is to determine the statistical dependence of a mean disease incidence (suitably transformed; see below) at location i on disease incidence at locations s units away (e.g. $i-2$ if $s=2$) (102). Whereas ρ (and similar statistics) measure intracluster correlation, spatial autocorrelation discussed here measure intercluster correlations. The related semi-variance from geostatistics (15) provides a mirror image of the autocorrelation (9). It is not possible in the available space to explain the methodology for these analyses more fully; several articles provide a thorough review of the subject (9, 10, 37, 40, 102, 103).

A comprehensive assessment of the spatial pattern of disease incidence can be made by simultaneously assessing aggregation on the basis of similarity of disease status both within and between sampling units. First steps towards this were outlined by Madden (65). Since then, the availability of computer software for fitting the beta-binomial distribution to disease incidence data (69) and for carrying out spatial autocorrelation analysis (40, 102) has made such a study a more practical proposition.

If the positions of plants (or plant entities, such as stems), and not just the sampling units, are known in one- or two-dimensional space, then several other methods can be used to ascertain spatial patterns. Sequences of diseases and healthy plants can be determined and tested, as well as distances between diseased plants (10, 19, 67, 73, 76, 86, 99).

TEMPORAL ANALYSES

Disease incidence data often are collected over time. One can often expect large changes in disease during single growing seasons or from one season to the next. A thorough evaluation of the population dynamics of disease requires knowledge and use of information on distributional properties of the chosen measure of disease intensity. Just as there often is intra- and intercluster spatial correlation of disease incidence, one expects to find a temporal autocorrelation of incidence. How one analyzes and interprets this autocorrelation depends on the experimental situation and disease under investigation. Relevant aspects of the appropriate analyses are presented here.

Disease Progress Curves

Characterization of disease incidence (or severity) over time is of fundamental importance in epidemiology for describing and understanding disease dynamics as well as for evaluating treatments (10, 68). Quantification typically is done by analyzing a disease progress curve, that is, an ordered sequence of disease values over time (t). Disease incidence can be represented as X_t (for counts of diseased plants or plant units) or p_t [for the (mean) proportion diseased] at time t. X_t or p_t is represented as a function of t to determine population-defining parameters such as initial disease (X_0; p_0), asymptotic disease level [n or Kn, with K being a fraction of n ($0 < K \leq 1$)], and the relative rate of disease increase (r_*). Thorough reviews of the analysis of disease progress curves (including many example curves) have been published (10, 36, 66, 68); only the aspects of this field relevant to disease incidence are discussed here.

MODELS The expected probability of a plant being diseased at time t (p_t) can be represented as $p_t = f(t)$, in which $f(t)$ is some function of time. This function can be the integrated form of nonlinear models such as the exponential, logistic, monomolecular, Gompertz, Richards, or Weibull (68), depending on the determined or postulated mechanism of disease increase. Sometimes, one can transform p_t[$= g(p_t)$] to obtain a model that is linear. For simplicity, we write Z_t for the transformation of p_t[i.e. $Z_t = g(p_t)$].

As an example, for logistic disease development, p_t can be written as

$$p_t = K/[1 + A \exp(-r_L t)], \qquad \text{8a.}$$

in which $A = (K - p_0)/p_0$. The logistic transformation can be used to derive a linear form for this model:

$$\ln[p_t/(K - p_t)] = -\ln(A) + r_L t. \qquad \text{8b.}$$

Here, $Z_t = \ln[p_t/(K - p_t)]$. The linear, nonlinear, and differential equation forms for the common models have been published (10, 68).

STATISTICAL ASSUMPTIONS AND ANALYSES Ordinary least-squares regression analysis (nonlinear or linear), with an assumed normal distribution for p_t or Z_t, usually is used to estimate parameters, determine predicted values for disease incidence, and compare epidemics (68). Such a distributional assumption may not be justified in some circumstances. For a large number of observations (n) at each time and p not too small, the normal distribution is a good approximation to the binomial (19, 106). Large sample theory would also predict that the beta-binomial could be approximated by the normal, but this prediction has not been fully studied.

There are several rules postulated for determining whether a normal approximation is tenable (106). One popular but conservative rule is that the distribution of X is approximated by the normal, with mean np and variance $np(1 - p)$, if $np(1 - p) > 9$. The normal distribution is thus tenable at $p = 0.5$ if $n > 36$. Other cutoffs include $p = 0.25$, $n = 48$; $p = 0.10$, $n = 100$; and $p = 0.01$, $n = 909$. The increasing skewness of the binomial as p (or π) goes to 0 (Table 2; Figure 1) is a reason for the larger n at low incidence. It is expected that n would have to be larger for the beta-binomial to be approximated by the normal distribution. Because p_t may go from $p_t < 0.001$ to ~1.0 during an epidemic, one would need a large n to satisfy the normality assumption. Fortunately, regression analysis is very robust to some departure from normality, especially if other assumptions are met (87). Because disease progress curves often are generated based on large n, normality may be a reasonable assumption. However, when n is small, other methods should be used (see below).

Weighted regression analysis should be used for disease progress curve modeling because the variance is proportional to $p(1 - p)$ (33, 36). The weights are proportional to the inverse of the variance. For analysis of incidence (p) with nonlinear models, it is, therefore, reasonable to use $[p(1 - p)]^{-b}$ as weights, assuming that the modified power law (equation 5) is appropriate. When $b = 1$, this reduces to the weight for the binomial distribution, $[p(1 - p)]^{-1}$.

If linear regression is used, determining the weight is a little more complicated. Large-sample statistical theory shows that the variance of a transformation of p $[g(p)]$ equals the product of two terms: variance of p; and the square of the first derivative of $g(p)$ with respect to p $[= \{g'(p)\}^2]$. For the logistic transformation and an assumed modified power law, the variance is proportional to $[p(1 - p)]^{b-2}$; the weight, then, is $[p(1 - p)]^{2-b}$.

A final but important assumption is that the incidence values at different times are independent, which is not possible with disease progress curves (66). In reality, p_t is highly correlated with the previous value (p_{t-1}), and less correlated with the value two times earlier (p_{t-2}), and so on. This temporal autocorrelation is often "removed," or corrected for, by fitting the proper disease progress model to the data. (In the time-series literature, this is called removing the trend.) Autocorrelation of the residuals should always be evaluated after fitting a model to data that varies over time, however. This is done by calculating the Durbin-Watson statistic or autocorrelation coefficient (87). With significant autocorrelation of the residuals standard errors of estimated parameters are too low, which can lead to false conclusions about differences between epidemics. Madden (66) thoroughly reviews this topic and shows how to correct for autocorrelation of the residuals. Marcus (75) gives one additional approach for adjusting for autocorrelation with a logistic disease progress model in its nonlinear form.

As an alternative to ordinary least-squares regression, one can use maximum likelihood estimation with the assumed error distribution (e.g. binomial) for nonlinear models, such as equation 8a (104). More specialized software is needed for these calculations, but the method produces parameter estimates with many desirable statistical properties, if the proper distribution is selected. When the disease progress model is used in its linearized form (e.g. equation 8b), and a distribution other than the normal is assumed (within a class of distributions), model fitting can be done using the methodology of Generalized Linear Models (GLMs; 20, 25). This is still relatively rare in plant pathology (but see 115).

Transformations

When analyzing disease progress curves, transformation of p_t is based on the linearization of the appropriate disease progress model (68). In other, more empirical analyses, no specific disease progress model is assumed, and an investigator merely wishes to stabilize variances prior to analysis of variance (ANOVA) or regression analysis.

When the variance of p is proportional to $p(1-p)$, the angular transformation [$Z = \arcsin(p^{1/2})$] is appropriate (13). This transformation can be used if X has a binomial or a beta-binomial distribution with constant θ (or ρ) at all treatment levels (22). If the modified power law is appropriate, then the choice of transformation is more difficult. At $b = 1$, the angular transformation is appropriate; at $b = 2$, the logit transformation stabilizes variances. At $1 < b < 2$, no exact transformation is obtainable (45). A numerical result is to use $Z = b(1.62 + 4.71p - 4.32p^2 + 2.55p^4) - 1.62$. A preliminary numerical evaluation of the effect of b using a generalized transformation (3) indicates that when $b \leq 1.2$, the angular transformation is satisfactory; at $b > 1.6$, the logit is a reasonable approximation (LV Madden, unpublished). As an alternative, GLMs can be used instead of least-squares regression analysis and ANOVA (20, 82, 126).

Time-Series Analysis

Disease incidence (or severity) data can be collected over time in situations in which classical disease progress models do not apply. For instance, the proportion of maize plants with northern leaf blight lesions may be determined at the flowering stage in a given location every year for 20 years. Incidence could range from 0 to 1, and there would be no expected increase over time (years in this case) from population dynamic processes. Nevertheless, it is highly likely that there would be a temporal dependence of the disease data. High disease incidence in a given year would indicate a greater inoculum level to overwinter and infect plants the following year. Figures 2 and 3 in Ref. 4 and Figure 8.2 in Reference 10 demonstrate this situation. Yang & Zeng (129) present a detailed time-series analysis of wheat stripe rust ratings over 41 years at five locations in China. A well-studied example from medical epidemiology is the monthly incidence of children with measles in a country over several years (2).

There are several techniques for analyzing serially dependent data when deterministic models (such as equation 8) are not appropriate. The most common method is known as Box-Jenkins time-series analysis (23, 24, 28). This approach is discussed at some length here because of its power and flexibility.

BOX-JENKINS ANALYSIS The form of the data for analysis is relatively simple: observations of some variable (such as disease incidence) over a discrete set of time points. One assumes that the variable is continuous, which is only an approximation for disease incidence. The approximation is good at large n at each time. Normality is not strictly needed for calculating autocorrelations, but estimates of parameters and their standard errors in models are affected by the distribution (23). Special techniques are available for binary or other non-normal variables (21, 53, 64), some following the approach of GLMs. Unlike for the methods discussed here, commercial computer software is not available for these non-normal methods, and they are not discussed further. Time points for the data are assumed to be equally spaced, although the spacing need not be exact. The major restriction for this type of analysis is that a large number of time points, generally at least 30, is needed. Some aspects of the analysis may require at least three times this number.

An example data set from plant pathology is given in Figure 7. Fargette et al (31) recorded the incidence of African cassava mosaic virus (ACMV) in cassava plants 2 months after planting in plots planted monthly in the Ivory Coast. Data were recorded over 65 months, but some periods in the first 2 years had missing values. The most complete section of the data set is shown,

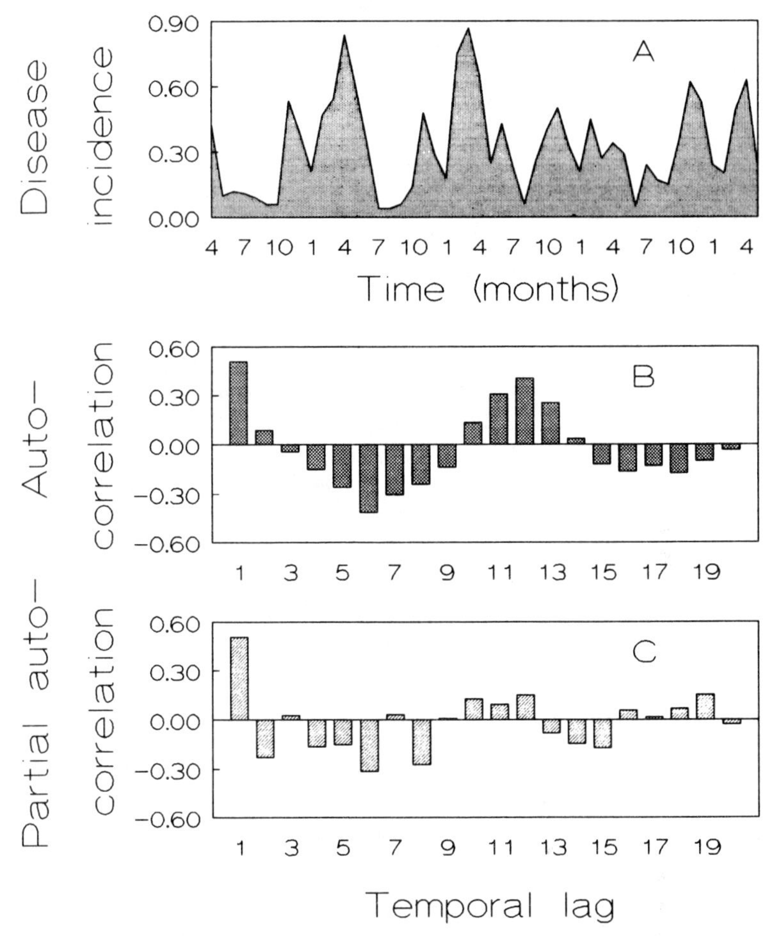

Figure 7 Proportion of cassava plants infected by African cassava mosaic virus (*A*) 2 months after planting, in plots planted monthly, together with (*B*) the temporal autocorrelation and (*C*) partial autocorrelation of logit-transformed disease incidence. Data from Figure 2 in Reference 31 and D Fargette (personal communication).

with a few missing values replaced by the average for that month over the other years. Visual appraisal of the time series shows that extreme (high or low) values are often found in short sequences (Figure 7*A*). A strong periodicity is also seen: Values separated by about 12 months often are very similar, while values separated by 6 months often are highly dissimilar.

The dependency of data over time is assessed with autocorrelation functions (*acf*) and partial autocorrelation functions (*pacf*). The *acf* at lag 1 [$r(1)$] represents the correlation between adjacent values of some variable (e.g. at times

1 and 2, 2 and 3, etc). This is directly analogous to the spatial autocorrelations (10, 40, 102) discussed above. For ACMV, this value is relatively high (~0.5) and positive, reflecting the similarity of adjacent values (Figure 7*B*). The *acf* at lag j [$r(j)$] represents the correlation of values j apart. For ACMV, the next highest correlation is at $j = 6$. The negative value for $r(6)$ and positive value for $r(12)$ reflect the periodicity in the data already noted.

The *pacf* at lag $j[\psi(j)]$ represents the correlation between values j apart, which is not due to the intermediate values (24, 28). This function "removes" the effect of intermediate values so that a dependency (statistical, not necessarily biological) between two nonadjacent times can be assessed. As noted above, $r(1)$ and $r(6)$ were significant; $\psi(1)$ and $\psi(6)$ also were moderate to large (Figure 7*B*,*C*). It can be seen that $r(7)$ was relatively large, implying a dependency of incidence values seven times apart. However, $\psi(7)$ was virtually 0 (Figure 7*C*). The value of $r(7)$, therefore, was high because the disease values at lags 6 and 7 were correlated with each other because they are one lag apart [see $r(1)$], not because of the dependency of disease incidence with incidence seven time periods earlier. This is a type of "carryover effect."

BOX-JENKINS MODELING It is often possible to describe the temporal results with an autoregressive, integrated, moving-average model (ARIMA). To perform the analysis, the variable of interest must be "stationary." Stationarity has a strict mathematical definition, but for practical purposes, a stationary series is one in which the mean and variance do not change systematically with time (23, 24). Constant variance can generally be approximately achieved with a proper transformation, as discussed above. The logit transformation was used in the two examples here because it stabilizes the variance when the power-law b is large (equation 5), and it also produces an additive scale. Weighting is not simple with this analysis and is usually not done. Constant mean indicates that there is no trend in the data. [With classical disease progress curves, the trend is removed by fitting one of the disease progress models (e.g. equation 8) to the data.] Differencing is often used here to achieve stationarity. If Z_t is the (possibly) transformed value of disease incidence at time t, then $Y_t = Z_t - Z_{t-1}$ is the first difference. (By definition, the zeroth difference is simply Z_t.) This new variable is then analyzed. This approach was taken by Yang & Zeng (129) with the wheat rust data in China.

The modeling approach entails defining Y_t (or Z_t) as a function of u prior Y_ts (e.g. $Y_{t-1}, \ldots, Y_{t-u}$), v prior error terms or random variations ($\varepsilon_{t-1}, \ldots \varepsilon_{t-v}$), error at time t (ε_t), and a constant (δ). This can be written as

$$Y_t = \phi_1 Y_{t-1} + \ldots + \phi_u Y_{t-u} + \varepsilon_t + \Theta_1 \varepsilon_{t-1} + \ldots + \Theta_v \varepsilon_{t-v} + \delta, \qquad 9.$$

in which the ϕs and Θs are parameters. The model is conveniently referred to as an ARIMA (u,d,v). The d is the order of the difference ($d = 0,1,2$) and equals 0 when no differencing is done. For example, an ARIMA(1,0,0) (i.e. autoregressive) model can be written as $Z_t = \phi_1 Z_{t-1} + \varepsilon_t + \delta$, and an ARIMA(0,0,2) (moving-average) model as $Z_t = \Theta_1\varepsilon_{t-1} + \Theta_2\varepsilon_{t-2} + \varepsilon_t + \delta$.

Identification of the model process is heavily based on the *acf* and *pacf*. For instance, an ARIMA(1,0,0) model has an *acf* that starts high at lag 1 (positive or negative), and then declines exponentially to 0; the *pacf* has only one large value, at lag 1 (28). In general, autoregressive models have declining *acf*s, but *pacf*s with large values only at the significant lag terms. The opposite is true for moving-average models: The *pacf*s have ψs that start high and decline to 0, whereas the *acf*s have large *r*s only at the significant temporal lag terms. Mixtures of autoregressive and moving-average terms give mixtures of *r*s and ψs, making identification more difficult. When model identification is not obvious, several possible models are fitted to the data, using maximum likelihood, and the residuals are evaluated to determine the most appropriate model. For an appropriate model, the *acf* and *pacf* of the residuals reveal no obvious correlations.

MODELING EXAMPLE The *acf* and *pacf* did not indicate an obvious process for ACMV (31) (Figure 7). The large $\psi(1)$ followed by very low ψs suggests an autoregressive model of order 1 ($u = 1$). However, the *r*s did not decline to 0 as lags increased. Rather, a first-order moving average was indicated ($v = 1$). Both ARIMA(1, 0, 0) and ARIMA(0, 0, 1) were adequate. Fitting an ARIMA(1, 0, 1) resulted in a nonsignificant autoregressive term, however.

The periodicity of the series (Figure 7*A*) could be seen in the *acf* at lags greater than 1. The large negative $r(6)$ was followed by a somewhat smaller positive $r(12)$, and a smaller $r(18)$ (Figure 7*B*). This is representative of an autoregressive model with a periodic component: $Z_r = \phi_1 Z_{t-6} + \varepsilon_t + \delta$. This can be visualized in the *acf* by eliminating the five lags between every sixth lag, that is by acting as if the data consisted of observations every 6 months rather than every month. The moderately large negative value for $\psi(6)$ and the low ψs at lags of 12 and 18 confirm that an autoregressive rather than a moving-average model is appropriate for the periodic component. The most appropriate model, with estimated parameters replacing symbols, was found to be:

$$Z_t = 0.44Z_{t-6} - 0.62\varepsilon_{t-1} - 1.31 + \varepsilon_t, \qquad 10.$$

which had a mean square error of 0.85 (47 degrees of freedom) (LV Madden, unpublished).

The moving-average term in the model (ε_{t-1}) implies that random occurrences or disturbances at $t - 1$ (prior month) influenced (transformed) disease incidence at t (the current month). These disturbances would be directly related to disease incidence at $t - 1$. Examples would include environmental fluctuations and vector flights in the field plots. Moving-average terms are used to account for "local" or short-term influences on the times series. They imply not a lack of autocorrelation between disease incidence values, rather that the autocorrelation does not extend beyond the lag order of the moving-average term. For an ARIMA(0, 0, 1) model, as an example, $r(1)$ equals $\Theta_1/(1 + \Theta_1^2)$, and all other rs are 0 (23).

OTHER APPROACHES When there is a cyclical or periodic nature to a time series, spectral analysis may be appropriate for characterizing the data (98). This method partitions the series into components of different frequencies by modeling the data as a summation of sine and cosine waves. To use this approach properly, a series with at least 100 times should be used, and more than 200 times is preferable.

Another approach is to describe incidence with a discrete version of a population-dynamic model. This does not require cyclical series. One discrete time version of the logistic model can be written as:

$$Z_t = Z_{t-1}\exp[r_L(1 - Z_{t-1}/\kappa)], \qquad 11.$$

in which Z_t is a (possible) transformation of p_t, r_L is the logistic rate parameter, and κ is the carrying capacity (14, 78). Different transformations of p_t, such as the logit and angular, could be used to satisfy statistical requirements for analysis (see above). When disease incidence is less than κ, disease increases; when greater than κ, it decreases. Therefore, κ is not simply the maximum incidence (asymptote) that is used in continuous time models (e.g. equation 8).

Quite complex time series can be generated by equation 11, including cycles, and chaotic behavior at large r_L. Small values of r_L, on the other hand, generate very stable series with little fluctuation. Cheke & Holt (14) and Bigger (6) should be read for an insightful presentation of several time-series approaches for modeling of insect dynamics.

Spatio-Temporal Analyses

With many diseases, one can expect great changes in incidence (or severity) over the course of a season. If the power law (equations 4–6) is appropriate for characterizing heterogeneity, then one would expect a high dependence of aggregation on time. Characterizing this dynamic aggregation can lead to a better understanding of disease progress in fields and permit evaluation of how aggregation affects rate of disease development in time and space and how temporal dynamics affect disease aggregation (46, 63, 65, 71, 73, 127). More theoretical and empirical work still is needed here.

If the location of sampling units is known and disease incidence is observed over time, the temporal and spatial autocorrelation approaches already described can be expanded into a combined analysis (102, 103). In this spatio-temporal analysis, (transformed) disease incidence at location *i* and time $t(Z_{i,t})$ is correlated with (*a*) Z at the previous times at location $i(Z_{i,t}-1, Z_{i,t-2}, \ldots)$ and (*b*) Z at the neighboring locations (e.g. $i - 1$) at the previous times ($Z_{i-1,t-1}$, $Z_{i-2,t-1}, \ldots, Z_{i-1,t-2}, Z_{i-2,t-2}, \ldots$). Both autocorrelations and partial autocorrelations can be calculated to determine statistical dependencies and identify the form of the spatio-temporal ARIMA (STARIMA) model that best describes the data. Unlike the purely temporal case, specialized software is needed for the correlations (102), and parameter estimation is more difficult. So far, only a few investigators (39, 74, 80, 103) have used this approach for characterizing plant diseases or inoculum. It has the potential, however, for providing a framework for conceptualizing and understanding some of the dynamic processes that occur during epidemics when there is not a clearly defined original source (focus) of disease.

CONCLUDING COMMENTS

Incidence data are frequently collected in epidemiological studies of plant disease because they provide a convenient and useful assessment of disease intensity. However, the characteristics of disease incidence data should be taken into account for analysis. These characteristics originate from the statistical properties of binary data, and from the fact that cropping practices and the dispersal mechanisms of plant pathogens result in diseased plants, or plant units, that are often clustered. In this article, a number of statistical methods are gathered together to illustrate the coherent framework of a logically consistent set of analyses that is now available to plant disease epidemiologists for quantifying disease incidence data.

Analysis of epidemiological data is often concerned with characterizing the way in which the level of disease at one time, or in one place, influences the level of disease at a subsequent time, or in another place. Perhaps the most frequently carried out analysis is the fitting of disease progress curves to data from sequences of disease assessments over time. When the data take the form of disease incidence, the method of parameter estimation should take into account the dependence of the variance on the mean implied by the binomial distribution. If comparisons are to be made between disease progress curves based on incidence data, heterogeneity will also have to be taken into account (115). Because some methods of doing this involve the incorporation be the beta-binomial distribution (22, 125) or a variance-mean relationship (82, 126) into the analysis, a description of spatial pattern of disease incidence can be seen to underlie a full analysis of disease progress over time.

The idea of autocorrelation has been seen to be common both to Box-Jenkins analysis of time-series data and to analyses of spatial patterns. With time-series data, the dependency can only be such that the current level of disease depends on the level(s) at prior times. With spatial data, however, the dependency of the level of disease in one place might be multidirectional (37, 38, 103). Incorporation of the statistical properties of incidence is usually achieved by the use of sample sizes that are sufficiently large for incidence to be able to be regarded as a continuous variable. For spatial analyses of disease incidence, the simultaneous study of correlations both within and between sampling units presents interesting new possibilities for epidemic quantification.

Literature Cited

1. Altham PME. 1978. Two generalizations of the binomial distribution. *Appl. Stat.* 27:162–67
2. Anderson RM, Grenfell BT, May RM. 1984. Oscillatory fluctuations in the incidence of infectious disease and the impact of vaccination: time series analysis. *J. Hyg. Comb.* 93:587–608
3. Aranda-Ordaz FJ. 1981. On two families of transformations to additivity for binary response data. *Biometrika* 68:357–63
4. Bald JG. 1937. Investigations on spotted wilt of tomatoes. III. Infection in field plots. *Commonw. Aust. Counc. Sci. Ind. Res. Bull.* 6
5. Barclay HJ. 1992. Modelling the effects of population aggregation on the efficiency of insect pest control. *Res. Popul. Ecol.* 34:131–41
6. Bigger M. 1993. Time series analysis of variation in abundance of selected cocoa insects and fitting of simple linear predictive models. *Bull. Entomol. Res.* 83:153–69
7. Binns MR. 1986. Behavioural dynamics and the negative binomial distribution. *Oikos* 47:315–18
8. Boswell MT, Ord JK, Patil GP. 1979. Chance mechanisms underlying univariate distributions. In *Statistical Ecology: Statistical Distributions in Ecological Work,* ed. JK Ord, GP Patil, C Taillie, 4:1–156. Fairland, MD: Int. Co-op. Publ.
9. Burrough PA. 1987. Spatial aspects of ecological data. In *Data Analysis in Community and Landscape Ecology,* ed. RHG Jongman, CJF ter Braak, OFR van Tongeren, pp. 213–51. Wageningen, The Netherlands: PUDOC
10. Campbell CL, Madden LV. 1990. *Introduction to Plant Disease Epidemiology.* New York: Wiley Intersci. 532 pp.
11. Campbell CL, Noe JP. 1985. The spatial analysis of soilborne pathogens and root diseases. *Annu. Rev. Phytopathol.* 23: 129–48
12. Campbell CL, Pennypacker SP. 1980. Distribution of hypocotyl rot caused in snapbean by *Rhizoctonia solani. Phytopathology* 70:521–25
13. Chanter DO. 1975. Modifications of the angular transformation. *Appl. Stat.* 24: 354–59
14. Cheke RA, Holt J. 1993. Complex dynamics of desert locust plagues. *Ecol. Entomol.* 18:109–15
15. Chellemi DO, Rohrbach KG, York PL, Sonodor RM. 1988. Analysis of the spatial pattern of plant pathogens and diseased plants using geostatistics. *Phytopathology* 78:221–26
16. Chiarappa L, ed. 1981. *Crop Loss Assessment Methods,* Suppl. 3. CAB, UK: FAO
17. Chuang TY, Jeger MJ. 1987. Relationship between incidence and severity of banana leaf spot in Taiwan. *Phytopathology* 77:1537–41
18. Cochran WG. 1936. The statistical analysis of field counts of diseased plants. *J. R. Stat. Soc.* 3:49–67

19. Cochran WG. 1977. *Sampling Techniques*. New York: Wiley. 3rd ed.
20. Collett D. 1991. *Modelling Binary Data*. London: Chapman & Hall
21. Cox DR, Snell EJ. 1989. *Analysis of Binary Data*. London: Chapman & Hall. 2nd ed.
22. Crowder MJ. 1978. Beta-binomial anova for proportions. *Appl. Stat.* 27:34–37
23. Cryer JD. 1986. *Time Series Analysis*. Boston: PWS-Kent
24. Diggle PJ. 1990. *Time Series: A Biostatistical Introduction*. Oxford: Oxford Univ. Press
25. Dobson AJ. 1990. *An Introduction to Generalized Linear Models*. London: Chapman & Hall
26. Downing JA. 1986. Spatial heterogeneity: evolved behavior or mathematical artefact? *Nature* 323:255–57
27. DuCharme EP. 1971. Tree loss in relation to young tree decline and sand hill decline of citrus in Florida. *Proc. Fla. State Hortic. Soc.* 84:48–52
28. Dunstan FDJ. 1993. Time series analysis. In *Biological Data Analysis: A Practical Approach*, ed. JC Fry, pp. 243–310. Oxford: IRL Oxford Univ. Press
29. Dutilleul P, Legendre P. 1993. Spatial heterogeneity against heteroscedasticity: an ecological paradigm versus a statistical concept. *Oikos* 66:152–71
30. Eggenberger F, Pólya G. 1923. Über die Statistick verketteter Vorgänge. *Z. Angew. Math. Mech.* 3:279–89
31. Fargette D, Jeger M, Fauquet C, Fishpool LDC. 1994. Analysis of temporal progress of African cassava mosaic virus. *Phytopathology* 84:91–98
32. Fisher RA. 1925. *Statistical Methods for Research Workers*. New York: Hafner. 362 pp.
33. Fulton WC. 1979. On comparing values of Vanderplank's r. *Phytopathology* 69: 1162–64
34. Gates CE. 1988. Discrete, a computer program for fitting discrete frequency distributions. In *Lecture Notes in Statistics*, ed. L McDonald, B Manly, J Lockwood, J Logan, pp. 458–66. Berlin: Springer-Verlag
35. Gaunt RE, Cole MJ. 1992. Spatial analysis of wheat stripe rust epidemics. *Crop Prot.* 11:131–37
36. Gilligan CA. 1990. Comparison of disease progress curves. *New Phytol.* 115: 223–42
38. Gottwald TR, Miller C, Brlansky RH, Gabriel DW, Civerolo EL. 1989. Analysis of the spatial distribution of citrus bacterial spot in a Florida citrus nursery. *Plant Dis.* 73:297–303
37. Gottwald TR, Graham JH. 1990. Spatial pattern analysis of epidemics of citrus bacterial spot in Florida citrus nurseries. *Phytopathology* 80:181–90
39. Gottwald TR, Reynolds KM, Campbell CL, Timmer LW. 1992. Spatial and spatiotemporal autocorrelation analysis of citrus canker epidemics in citrus nurseries and groves in Argentina. *Phytopathology* 82:843–51
40. Gottwald TR, Richie SM, Campbell CL. 1992. LCOR2—Spatial correlation analysis software for the personal computer. *Plant Dis.* 76:213–15
41. Griffiths DA. 1973. Maximum likelihood estimation for the beta-binomial distribution and an application to the household distribution of the total number of cases of a disease. *Biometrics* 29:637–48
42. Guenther WC. 1975. The inverse hypergeometric—a useful model. *Stat. Neerl.* 29:129–44
43. Heltshe JF, Ritchey TA. 1984. Spatial pattern detection using quadrat samples. *Biometrics* 40:877–85
44. Hughes G. 1990. Characterizing crop responses to patchy pathogen attack. *Plant Pathol.* 39:2–4
45. Hughes G, Madden LV. 1992. Aggregation and incidence of disease. *Plant Pathol.* 41:657–60
46. Hughes G, Madden LV. 1993. Using the beta-binomial distribution to describe aggregated patterns of disease incidence. *Phytopathology* 83:759–63
47. Hughes G, Madden LV. 1993. Maximum likelihood methods for the analysis of aggregated disease incidence data. *Phytopathology* 83:1365 (Abstr.)
48. Hughes G, Madden LV. 1994. Aggregation and incidence of disease: some implications for sampling. *Aspects Appl. Biol.* 37:25–31
49. Hughes G, Madden LV. 1994. Intracluster correlation and the problem of sample size determination. *Phytopathology* 84:1080 (Abstr.)
50. Hurlbert SH. 1990. Spatial distribution of the montane unicorn. *Oikos* 58:257–71
51. Irwin JO. 1954. A distribution arising in the study of infectious diseases. *Biometrika* 41:266–68
52. Ives PM, Moon RD. 1987. Sampling theory and protocol for insects. In *Crop Loss Assessment and Pest Management*, ed. PS Teng, pp. 49–75. St. Paul, MN: APS Press
53. Jacobs PA, Lewis PAW. 1983. Stationary discrete autoregressive moving average times series generated by mixtures. *J. Time Ser. Anal.* 4:19–36

54. James WC, Teng PS. 1979. The quantification of production constraints associated with plant diseases. In *Applied Biology*, ed. TH Coaker, 4:201–67. New York: Academic
55. Johnson NL, Kotz S, Kemp AW. 1992. *Univariate Discrete Distributions*. New York: Wiley Intersci. 565 pp. 2nd ed.
56. Kemp AW. 1987. Families of discrete distributions satisfying Taylor's power law. *Biometrics* 43:693–99
57. Kemp AW. 1988. Families of distributions for repeated samples of animal counts: response. *Biometrics* 44:888–90
58. Kemp CD. 1971. Properties of some discrete ecological distributions. In *Statistical Ecology: Spatial Patterns and Statistical Distributions*, ed. GP Patil, EC Pielou, WE Waters, 1:1–22. University Park, PA: Penn. State Univ. Press
59. Kemp CD, Kemp AW. 1956. The analysis of point quadrat data. *Aust. J. Bot.* 4:167–74
60. Kemp CD, Kemp AW. 1956. Generalized hypergeometric distributions. *J. R. Stat. Soc. B* 18:202–11
61. Kranz J. 1988. Measuring plant disease. In *Experimental Techniques in Plant Disease Epidemiology*, ed. J Kranz, J Rotem, pp. 35–50. Berlin: Springer-Verlag
62. Kranz J, ed. 1990. *Epidemics of Plant Diseases: Mathematical Analysis and Modeling*. Berlin: Springer-Verlag. 2nd ed.
63. Kuno E. 1988. Aggregation pattern of individuals and the outcomes of competition within and between species: differential equation models. *Res. Popul. Ecol.* 30:69–82
64. Li WK. 1994. Time series models based on generalized linear models: some further results. *Biometrics* 50:506–11
65. Madden LV. 1989. Dynamic nature of within-field disease and pathogen distributions. In *Spatial Components of Plant Disease Epidemics*, ed. MJ Jeger, pp. 96–126. Englewood Cliffs, NJ: Prentice Hall
66. Madden LV. 1986. Statistical analysis and comparison of disease progress curves. In *Plant Disease Epidemiology*, ed. KJ Leonard, WE Fry, pp. 55–84. New York: Macmillan
67. Madden LV, Campbell CL. 1986. Descriptions of virus disease epidemics in time and space. In *Plant Virus Epidemics*, ed. GD McLean, RG Garrett, WG Ruesink, pp. 273–93. Sydney: Academic
68. Madden LV, Campbell CL. 1990. Nonlinear disease progress curves. In *Epidemics of Plant Diseases: Mathematical Analysis and Modeling*, ed. J Kranz, pp. 181–229. Berlin: Springer-Verlag. 2nd ed.
69. Madden LV, Hughes G. 1994. BBD—computer software for fitting the beta-binomial distribution to disease incidence data. *Plant Dis.* 78:536–40
70. Madden LV, Hughes G, Ellis MA. 1995. Spatial heterogeneity of the incidence of grape downy mildew. *Phytopathology* 85:269–75
71. Madden LV, Louie R, Knoke JK. 1987. Temporal and spatial analysis of maize dwarf mosaic epidemics. *Phytopathology* 77:148–56
72. Madden LV, Nutter FW Jr. 1995. Modeling crop losses at the field scale. *Can. J. Plant Pathol.* 17: In press
73. Madden LV, Pirone TP, Raccah B. 1987. Analysis of spatial patterns of virus diseased tobacco plants. *Phytopathology* 77:1413–17
74. Madden LV, Reynolds KM, Pirone TP, Raccah B. 1988. Modeling of tobacco virus epidemics as spatio-temporal autoregressive integrated moving-average processes. *Phytopathology* 78:1361–66
75. Marcus R. 1991. Deterministic and stochastic logistic models for describing increase of plant diseases. *Crop. Prot.* 10:155–59
76. Marcus R, Fishman S, Talpaz H, Salomon R, Bar-Joseph M. 1984. On the spatial distribution of citrus tristeza virus disease. *Phytoparasitica* 12:45–52
77. Marois JJ, Adams PB. 1985. Frequency distribution analysis of lettuce drop caused by *Sclerotinia minor*. *Phytopathology* 75:957–61
78. May RM. 1974. Biological populations with nonoverlapping generations: stable points, stable cycles, and chaos. *Science* 186:645–47
79. McArdle BH, Gaston KJ, Lawton JH. 1990. Variation in the size of animal populations: patterns, problems and artefacts. *J. Anim. Ecol.* 59:439–54
80. Mihail J. 1989. *Macrophomina phaseolina:* spatio-temporal dynamics of inoculum and of disease in a highly susceptible crop. *Phytopathology* 79: 848–55
81. Minitab, Inc. 1991. *MINITAB Reference Manual*, Release 8. State College, PA: Minitab, Inc.
82. Moore DF. 1987. Modelling the extraneous variance in the presence of extrabinomial variation. *Appl. Stat.* 36:8–14
83. Morisita M. 1962. I_δ-index, a measure of dispersion of individuals. *Res. Popul. Ecol.* 4:1–7
84. Munkvold GP, Duthie JA, Marois JJ.

1993. Spatial patterns of grapevines with *Eutypa* dieback in vineyards with or without perithecia. *Phytopathology* 83:1440–53
85. Myers JH. 1978. Selecting a measure of dispersion. *Environ. Entomol.* 7:619–21
86. Nelson SC, Marsh PL, Campbell CL. 1992. 2DCLASS, a two-dimensional distance class analysis software for the personal computer. *Plant Dis.* 76:427–32
87. Neter J, Wasserman W, Kutner MH. 1983. *Applied Linear Regression Models.* Homewood, IL: Irwin
88. Neyman J. 1959. Optimal asymptotic tests of composite hypotheses. In *Probability and Statistics*, ed. U Granander, pp. 213–34. New York: Wiley
89. Nutter FW Jr, Teng PS, Royer MH. 1993. Terms and concepts for yield, crop loss and disease thresholds. *Plant Dis.* 77:211–15
90. Partner PLR. 1994. *Modelling mechanisms of change in crop populations.* PhD thesis. Univ. Edinburgh
91. Patil GP, Joshi SW, eds. 1968. *A Dictionary and Bibliography of Discrete Distributions.* Edinburgh: Oliver & Boyd. 268 pp.
92. Patil GP, Stiteler WM. 1974. Concepts of aggregation and their quantification: a critical review with some new results and applications. *Res. Popul. Ecol.* 15: 238–54
93. Paul SR, Liang KY, Self SG. 1989. On testing departure from binomial and multinomial assumptions. *Biometrics* 45: 231–36
94. Perry JN. 1988. Some models for spatial variability of animal species. *Oikos* 51: 124–30
95. Perry JN, Taylor LR. 1985. Adès: new ecological families of species-specific frequency distributions that describe repeated spatial samples with an intrinsic power-law variance-mean property. *J. Anim. Ecol.* 54:931–53
96. Perry JN, Taylor LR. 1986. Stability of real interacting populations in space and time: implications, alternatives and the negative binomial k_c. *J. Anim. Ecol.* 55:1053–68
97. Pielou EC. 1977. *Mathematical Ecology.* New York: Wiley-Intersci.
98. Platt T, Denman KL. 1975. Spectral analysis in ecology. *Annu. Rev. Ecol. Syst.* 6:189–210
99. Poushinsky G, Basu PK. 1984. A study of distribution and sampling of soybean plants naturally infected with *Pseudomonas syringae* pv. *glycinea. Phytopathology* 74:319–26
100. Qu Y, Beck GJ, Williams GW. 1990. Polya-Eggenberger distributions: parameter estimation and hypothesis tests. *Biom. J.* 32:229–42
101. Qu Y, Green T, Piedmonte R. 1993. Symmetric Bernoulli distributions and generalized binomial distributions. *Biom. J.* 35:523–533
102. Reynolds KM, Madden LV. 1988. Analysis of epidemics using spatio-temporal autocorrelation. *Phytopathology* 78: 240–46
103. Reynolds KM, Madden LV, Ellis MA. 1988. Spatio-temporal analysis of epidemic development of leather rot of strawberry. *Phytopathology* 78:246–52
104. Ross GJS. 1990. *Nonlinear Estimation.* Berlin: Springer-Verlag
105. Rouse DI, MacKenzie DR, Nelson RR, Elliot VJ. 1981. Distribution of wheat powdery mildew incidence in field plots and relationship to disease severity. *Phytopathology* 71:1015–20
106. Schader M, Schmid F. 1989. Two rules of thumb for the approximation of the binomial distribution by the normal distribution. *Am. Stat.* 43:23–24
107. Seem RC. 1984. Disease incidence and severity relationships. *Annu. Rev. Phytopathol.* 22:137–50
108. Seem RC, Gilpatrick JD. 1980. Incidence and severity relationships of secondary infections of powdery mildew on apple. *Phytopathology* 70:951–54
109. Shiyomi M. 1981. Mathematical ecology of spatial pattern of biological populations. *Bull. Natl. Inst. Agric. Sci. A* 27:1–29
110. Shiyomi M, Takai A. 1979. The spatial pattern of infected or infested plants and negative hypergeometric series. *Jpn. J. Appl. Entomol. Zool.* 23:224–29 (In Japanese)
111. Skellam JG. 1948. A probability distribution derived from the binomial distribution by regarding the probability of success as variable between sets of trials. *J. R. Stat. Soc. B* 10:257–61
112. Smith DM. 1983. Maximum likelihood estimation of the parameters of the beta binomial distribution. *Appl. Stat.* 32: 192–204
113. Soberón MJ, Loevinsohn M. 1987. Patterns of variations in the numbers of animal populations and the biological foundations of Taylor's law of the mean. *Oikos* 48:249–52
114. Strandberg J. 1973. Spatial distribution of cabbage black rot and the estimation of diseased plant populations. *Phtopathology* 63:998–1003
115. Sweetmore A, Simons SA, Kenward M. 1994. Comparison of disease progress

curves for yam anthracnose (*Colletotrichum gloeosporioides*). *Plant Pathol.* 43:206–15

116. Takai A. 1963. On the distribution pattern of the yellow dwarf disease of the rice plant. *Jpn. J. Ecol.* 13:151–56 (In Japanese)
117. Takai A, Shiyomi M. 1980. Model and analysis of spatial pattern of infected or infested plants. *Jpn. J. Appl. Entomol. Zool.* 24:234–40 (In Japanese)
118. Tarone RE. 1979. Testing the goodness of fit of the binomial distribution. *Biometrika* 66:585–90
119. Taylor LR. 1961. Aggregation, variance and the mean. *Nature* 189:732–35
120. Taylor LR, Perry JN, Woiwod IP, Taylor RAJ. 1988. Specificity of the spatial power-law exponent in ecology and agriculture. *Nature* 332:721–22
121. Taylor LR, Woiwod IP. 1982. Comparative synoptic dynamics I. Relationships between inter- and intra-specific spatial and temporal variance/mean population parameters. *J. Anim. Ecol.* 51:879–906
122. Taylor LR, Woiwod IP, Perry JN. 1978. The density-dependence of spatial behaviour and rarity of randomness. *J. Anim. Ecol.* 47:383–406
123. Thal WM, Campbell CL. 1986. Spatial patterns analysis of disease severity data for alfalfa leaf spot caused primarily by *Leptosphaerulina briosiana*. *Phytopathology* 76:190–94
124. Upton G, Fingleton B. 1985. *Spatial Data Analysis, By Example.* Vol. 1: *Point Pattern and Quantitative Data.* Chichester, UK: Wiley
125. Williams DA. 1975. The analysis of binary responses from toxicological experiments involving reproduction and teratogenicity. *Biometrics* 31:949–52
126. Williams DA. 1982. Extra binomial variation in logistic linear models. *Appl. Stat.* 31:144–48
127. Yang XB, TeBeest DO. 1992. Dynamic pathogen distribution and logistic increase of plant disease. *Phytopathology* 82:380–83
128. Yang XB, TeBeest DO. 1992. Taylor's power law and the variance-mean relationship of plant diseases. *Phytopathology* 82:1162 (Abstr.)
129. Yang XB, Zeng SM. 1992. Detecting patterns of wheat stripe rust pandemics in time and space. *Phytopathology* 82: 571–76

SUBJECT INDEX

D

CUMULATIVE INDEXES

CONTRIBUTING AUTHORS, VOLUMES 24–33

CHAPTER TITLES, VOLUMES 24–33

BIOLOGICAL AND CULTURAL CONTROL

SPECIAL TOPICS

ANNUAL REVIEWS

a nonprofit scientific publisher
4139 El Camino Way
P.O. Box 10139
Palo Alto, CA 94303-0139 • USA

ORDER FORM

ORDER TOLL FREE
1.800.523.8635
from USA and Canada

Fax: 1.415.855.9815

Annual Reviews publications may be ordered directly from our office; through stockists, booksellers and subscription agents, worldwide; and through participating professional societies. **Prices are subject to change without notice. We do not ship on approval.**

- **Individuals:** Prepayment required on new accounts. in US dollars, checks drawn on a US bank.
- **Institutional Buyers:** Include purchase order. Calif. Corp. #161041 • ARI Fed. I.D. #94-1156476
- **Students / Recent Graduates: $10.00 discount from retail price, per volume.** ***Requirements:*** **1.** be a degree candidate at, or a graduate within the past three years from, an accredited institution; **2.** present proof of status (photocopy of your student I.D. or proof of date of graduation); **3.** Order direct from Annual Reviews; **4.** prepay. This discount **does not** apply to standing orders, *Index on Diskette,* Special Publications, ARPR, or institutional buyers.
- **Professional Society Members:** Many Societies offer *Annual Reviews* to members at reduced rates. Check with your society or contact our office for a list of participating societies.
- **California orders** add applicable sales tax. • **Canadian orders** add 7% GST. Registration #R 121 449-029.
- **Postage paid** by Annual Reviews (4th class bookrate/surface mail). UPS ground service is available at $2.00 extra per book within the contiguous 48 states only. UPS air service or US airmail is available to any location at actual cost. UPS requires a street address. P.O. Box, APO, FPO, not acceptable.
- **Standing Orders:** Set up a standing order and the new volume in series is sent automatically each year upon publication. Each year you can save 10% by prepayment of prerelease invoices sent 90 days prior to the publication date. Cancellation may be made at any time.
- **Prepublication Orders:** Advance orders may be placed for any volume and will be charged to your account upon receipt. Volumes not yet published will be shipped during month of publication indicated.

NOTE: For copies of individual articles from any *Annual Review,* or copies of any article cited in an *Annual Review,* call **Annual Reviews Preprints and Reprints (ARPR)** toll free 1-800-347-8007 (fax toll free 1-800-347-8008) from the USA or Canada. From elsewhere call 1-415-259-5017.

ANNUAL REVIEWS SERIES *Volumes not listed are no longer in print*		**Prices, postpaid, per volume. USA/other countries**	Regular Order Please send Volume(s):	Standing Order Begin with Volume:
❑ *Annual Review of*	**ANTHROPOLOGY**			
Vols. 1-20	(1972-91)	$41 / $46		
Vols. 21-22	(1992-93)	$44 / $49		
Vol. 23-24	(1994 and Oct. 1995)	$47 / $52	Vol(s). ______	Vol. ______
❑ *Annual Review of*	**ASTRONOMY AND ASTROPHYSICS**			
Vols. 1, 5-14, 16-29	(1963, 67-76, 78-91)	$53 / $58		
Vols. 30-31	(1992-93)	$57 / $62		
Vol. 32-33	(1994 and Sept. 1995)	$60 / $65	Vol(s). ______	Vol. ______
❑ *Annual Review of*	**BIOCHEMISTRY**			
Vols. 31-34, 36-60	(1962-65,67-91)	$41 / $47		
Vols. 61-62	(1992-93)	$46 / $52		
Vol. 63-64	(1994 and July 1995)	$49 / $55	Vol(s). ______	Vol. ______
❑ *Annual Review of*	**BIOPHYSICS AND BIOMOLECULAR STRUCTURE**			
Vols. 1-20	(1972-91)	$55 / $60		
Vols. 21-22	(1992-93)	$59 / $64		
Vol. 23-24	(1994 and June 1995)	$62 / $67	Vol(s). ______	Vol. ______